INTERNATIONAL SERIES OF MONOGRAPHS ON PHYSICS

SERIES EDITORS

J. BIRMAN	CITY UNIVERSITY OF NEW YORK
S. F. EDWARDS	UNIVERSITY OF CAMBRIDGE
R. FRIEND	UNIVERSITY OF CAMBRIDGE
M. REES	UNIVERSITY OF CAMBRIDGE
D. SHERRINGTON	UNIVERSITY OF OXFORD
G. VENEZIANO	CERN, GENEVA

D1794261

International Series of Monographs on Physics

143. W. Götze: *Complex dynamics of glass-forming liquids—a mode-coupling theory*
142. V.M. Agranovich: *Excitations in organic solids*
141. W.T. Grandy: *Entropy and the time evolution of macroscopic systems*
140. M. Alcubierre: *Introduction to 3 + 1 numerical relativity*
139. A. L. Ivanov, S. G. Tikhodeev: *Problems of condensed matter physics—quantum coherence phenomena in electron-hole and coupled matter-light systems*
138. I. M. Vardavas, F. W. Taylor: *Radiation and climate*
137. A. F. Borghesani: *Ions and electrons in liquid helium*
136. C. Kiefer: *Quantum gravity, Second edition*
135. V. Fortov, I. Iakubov, A. Khrapak: *Physics of strongly coupled plasma*
134. G. Fredrickson: *The equilibrium theory of inhomogeneous polymers*
133. H. Suhl: *Relaxation processes in micromagnetics*
132. J. Terning: *Modern supersymmetry*
131. M. Mariño: *Chern-Simons theory, matrix models, and topological strings*
130. V. Gantmakher: *Electrons and disorder in solids*
129. W. Barford: *Electronic and optical properties of conjugated polymers*
128. R. E. Raab, O. L. de Lange: *Multipole theory in electromagnetism*
127. A. Larkin, A. Varlamov: *Theory of fluctuations in superconductors*
126. P. Goldbart, N. Goldenfeld, D. Sherrington: *Stealing the gold*
125. S. Atzeni, J. Meyer-ter-Vehn: *The physics of inertial fusion*
123. T. Fujimoto: *Plasma spectroscopy*
122. K. Fujikawa, H. Suzuki: *Path integrals and quantum anomalies*
121. T. Giamarchi: *Quantum physics in one dimension*
120. M. Warner, E. Terentjev: *Liquid crystal elastomers*
119. L. Jacak, P. Sitko, K. Wieczorek, A. Wojs: *Quantum Hall systems*
118. J. Wesson: Tokamaks, *Third edition*
117. G. Volovik: *The Universe in a helium droplet*
116. L. Pitaevskii, S. Stringari: *Bose-Einstein condensation*
115. G. Dissertori, I.G. Knowles, M. Schmelling: *Quantum chromodynamics*
114. B. DeWitt: *The global approach to quantum field theory*
113. J. Zinn-Justin: *Quantum field theory and critical phenomena, Fourth edition*
112. R.M. Mazo: Brownian *motion—fluctuations, dynamics, and applications*
111. H. Nishimori: *Statistical physics of spin glasses and information processing—an introduction*
110. N.B. Kopnin: *Theory of nonequilibrium superconductivity*
109. A. Aharoni: *Introduction to the theory of ferromagnetism, Second edition*
108. R. Dobbs: *Helium three*
107. R. Wigmans: *Calorimetry*
106. J. Kbler: *Theory of itinerant electron magnetism*
105. Y. Kuramoto, Y. Kitaoka: *Dynamics of heavy electrons*
104. D. Bardin, G. Passarino: *The Standard Model in the making*
103. G.C. Branco, L. Lavoura, J.P. Silva: *CP Violation*
102. T.C. Choy: *Effective medium theory*
101. H. Araki: *Mathematical theory of quantum fields*
100. L. M. Pismen: *Vortices in nonlinear fields*
99. L. Mestel: *Stellar magnetism*
98. K. H. Bennemann: *Nonlinear optics in metals*
96. M. Brambilla: *Kinetic theory of plasma waves*
94. S. Chikazumi: *Physics of ferromagnetism*
91. R. A. Bertlmann: *Anomalies in quantum field theory*
90. P. K. Gosh: *Ion traps*
88. S. L. Adler: *Quaternionic quantum mechanics and quantum fields*
87. P. S. Joshi: *Global aspects in gravitation and cosmology*
86. E. R. Pike, S. Sarkar: *The quantum theory of radiation*
83. P. G. de Gennes, J. Prost: *The physics of liquid crystals*
73. M. Doi, S. F. Edwards: *The theory of polymer dynamics*
69. S. Chandrasekhar: *The mathematical theory of black holes*
51. C. Møller: *The theory of relativity*
46. H. E. Stanley: *Introduction to phase transitions and critical phenomena*
32. A. Abragam: *Principles of nuclear magnetism*
27. P. A. M. Dirac: *Principles of quantum mechanics*
23. R. E. Peierls: *Quantum theory of solids*

Complex Dynamics of Glass-Forming Liquids

A mode-coupling theory

WOLFGANG GÖTZE

Fakultät für Physik,
Technische Universität München

OXFORD
UNIVERSITY PRESS

OXFORD
UNIVERSITY PRESS

Great Clarendon Street, Oxford OX2 6DP,
United Kingdom

Oxford University Press is a department of the University of Oxford.
It furthers the University's objective of excellence in research, scholarship, and education by publishing worldwide. Oxford is a registered trade mark of Oxford University Press in the UK and in certain other countries

First published 2009
First published in paperback 2012

Impression: 1

British Library Cataloguing in Publication Data
Data available

Library of Congress Cataloging in Publication Data
Data available

ISBN 978–0–19–923534–6 (hbk.)
978–0–19–965614–1 (pbk.)

Printed in Great Britain by
CPI Group (UK) Ltd, Croydon, CR0 4YY

PREFACE

During the past 20 years, a number of novel spectroscopic techniques have been introduced for the study of liquid motion within a time interval, which extends by three or more orders of magnitude the one characteristic for normal condensed-matter dynamics. Similar progress has been made in the research on colloid dynamics. Furthermore, the rapid development of computers and the invention of new codes made it possible to apply the versatile method of molecular-dynamics simulations to the study of the motion of systems of several thousand particles for times extending the natural scale by six and more orders of magnitude. All these studies have established a new field of statistical physics, which can be referred to as glassy dynamics of liquids: upon cooling or densifying the many-particle system, there evolve complex control-parameter-sensitive spectra and correlation-decay patterns as precursors of the liquid–glass transformation. This holds, provided the system is complicated enough so that phase transitions to an ordered state can be avoided.

In parallel to the indicated activities, a theory for the evolution of glassy dynamics in liquids has been developed, which is known as mode-coupling theory (MCT). It is based on equations of motion for correlation functions, whose subtleties are due to the interplay of nonlinearities with diverging retardation effects. There appear bifurcation singularities, which cause fascinating scenarios for a slow dynamics. It has been demonstrated that these MCT scenarios reproduce some of the features of the glassy dynamics of liquids. The microscopic version of the MCT is definitive in the sense that the long-time parts of correlation functions can be evaluated provided the equilibrium structure functions are available as input information. There is a set of examples showing that, occasionally, MCT can describe data for glassy dynamics on a 10%-accuracy level. The theme of this book is a comprehensive discussion of that theory.

Chapter 1 presents a series of observations, which demonstrate various facets of glassy dynamics. The examples are selected so that one can understand the challenge of this field of condensed-matter physics without consideration of a non-elementary data analysis.

The correlation-function formalism is the mathematical tool used throughout this book for a description of the dynamics. Chapter 2 explains this method, in particular the representation of response functions in terms of relaxation kernels. In Chapter 3, the most important correlation functions for simple liquids are explained, their general properties are derived, and their relevance for the description of the hydrodynamic-limit behaviour is demonstrated. The material of Chapters 2 and 3 is discussed in several books already. The subject is taken up here again not only because a self-contained presentation is intended; it is also the purpose to list the conventions, definitions, and the familiar results, which are used repeatedly in the following text. Thereby, a separation is made of the well established body of knowledge on the microscopic theory of liquids from the results on the special topic of this book.

The foundation of the mode-coupling theory is presented in Chapter 4 giving equal weight to two routes of reasoning. On one route, the MCT is discussed as a well defined mathematical model for a statistical description of a dynamics. The model deals with a set of coupled equations of motion, which determine a set of correlation functions. These equations are regular; they are specified by a number of characteristic frequencies for the short-time dynamics and by coupling constants quantifying the nonlinearities. The equations exhibit bifurcations, which can be classified completely. The models appear worth studying because they describe subtle scenarios for the bifurcation dynamics, which are not known from other mathematical-physics studies. The other route deals with the motivation of the models within the microscopic theory of simple liquids. These models are obtained from the request to formulate equations of motion for the density-fluctuation correlators for such low temperatures or such high densities that there is no separation of the time scales for the motion of the fluctuations of the densities from those of the forces. Chapter 5 extends the microscopic version of the MCT to one dealing with mixtures and with molecular liquids. Furthermore, it is shown how quantities like shear moduli, tagged-particle correlators, and mean-squared displacements can be calculated.

The basic bifurcation of MCT describes a transition from a liquid to an amorphous solid. The complex slow MCT-bifurcation dynamics evolves if control parameters are shifted towards and through critical values. The essence of the transition scenario can be understood by asymptotic solution of the equations of motion for frequencies below the band of normal-system excitations and for control parameters approaching the critical values. In Chapter 6, this asymptotic-solution theory is developed. A pair of power-law decay processes, a pair of divergent time scales and a pair of interconnected scaling laws are shared by all models. These general laws, which are obtained as leading asymptotic formulas, are proposed by MCT as the paradigm for a discussion of the evolution of the glassy dynamics in liquids. The amplitudes are calculated, which discriminate the various correlation functions and which determine the range of validity of the leading-order asymptotic results. Modifications of the asymptotic-solution theory for an explanation of the dynamics for states near more complicated bifurcation points are discussed also.

The exposition of the theory in Chapters 4–6 is interrupted occasionally by discussions of some diagrams, which exhibit quantitative comparisons of data with results of the MCT. One purpose for these interruptions is to corroborate the claim that the MCT bifurcation scenario shares a number of facets with those observed for glassy liquids. Another intention is to identify features of the data, which become obvious only if the experimental facts are confronted with non-elementary results of a theory. Finally, it is the purpose to motivate the citation of some references as an entry to the literature, which deals with tests of the relevance of MCT for an understanding of the complex dynamics in glass-forming liquids.

The mode-coupling theory has been used to explain the complex dynamics of cooled van der Waals systems, ionic melts, alkali-silicate melts and water for times extending from 1 picosecond to 20 nanoseconds and beyond. Also the sluggish dynamics of colloidal suspensions and of dense polymer melts was analysed within this approach. Hence, this book appeals to students and researchers interested in the physics of liquids, chemical physics and biophysics.

It is assumed that the reader has some basic knowledge of classical mechanics and statistical physics, of calculus and linear algebra. MCT studies the implications for the dynamics of bifurcation scenarios of certain nonlinear integro-differential equations, which so far have not been analysed in the mathematical literature. Therefore, it is the intention to present in detail the deduction, which leads from the basic equations to results relevant for the description of measurable properties of liquids. Needed but less familiar mathematical theorems are formulated explicitly; repeatedly used formulas and some very technical derivations are listed in some appendices.

My work on the dynamics of glass-forming liquids started in collaboration with Alf Sjölander, Lennart Sjögren, and Ulf Bengtzelius when I spent a sabbatical term 1983 at Chalmers University in Göteborg. I remember with gratitude the many discussions with these colleagues and their friendly hospitality during that and later stays at their institute.

The discoveries on glassy relaxation by Herman Z. Cummins, Walter Kob, Bill van Megen, Francesco Sciortino, and Piero Tartaglia have been crucial for my studies on MCT. Discussions with them and the results of their research provided a big stimulus for the work done by our group in Garching. I thank them and their collaborators cordially for their support.

It is my pleasure to thank all my former coworkers who participated in the development of the MCT. In particular, I thank S.-H. Chong, M. Sperl and Th. Voigtmann, who constructed for me a large number of the figures shown in this book. I am also grateful to J. Baschnagel, G. Hinze, W. Kob and J. Wiedersich for sending me the figures and the data I asked for. In this thanks, I include Frau H. Sprzagala, who organized with efficiency and friendliness the work of our group in Garching and who also transformed the manuscript for this book into the typescript.

During the preparation of this book, Rolf Schilling has read chapter after chapter and helped me with critique and suggestions. This help as well as the many discussions with him during the past years have been a great encouragement for me. I am very indebted to him.

I dedicate this book to Jana.

Garching, Spring 2008 W.G.

CONTENTS

1

GLASSY DYNAMICS OF LIQUIDS–FACETS OF THE PHENOMENON

Most liquids transform in a solid if they are cooled or compressed. Often, one observes a first-order phase transition in a crystalline state, which is connected with discontinuities of certain thermodynamic functions like the density. The origin of such transition is a discontinuous change of the equilibrium structure, namely the abrupt appearance of long-range order of the particle positions. A different possibility is the transformation of the liquid in an amorphous solid, a state called a glass. This liquid–glass transformation is not connected with discontinuities or other singularities of the conventional thermodynamic functions. Nor has it ever been observed that the glass formation is accompanied by a qualitative change of the equilibrium structure. However, the cooling or compression of the liquid towards the glass state is connected with the appearance of peculiar dynamical processes. They manifest themselves in low-frequency spectra or in long-time-decay functions of the perturbed system, which exhibit features unknown from other areas studied in physics. These processes shall be referred to as complex dynamics of glass-forming liquids or as glassy dynamics. In this chapter, a series of figures shall be discussed, which illustrate this concept. The figures are selected so that they demonstrate such facets of the phenomenon, which can be appreciated without appeal to a non-elementary data analysis.

1.1 Stretching of the dynamics

The most general feature of glassy dynamics is the stretching of excitation spectra or response functions over intervals of the frequency ω or the time t, respectively, which extend over several orders of magnitude. The stretching phenomenon shall be demonstrated in this section by two examples.

The natural time scale t_{mic} for the intermolecular dynamics of conventional condensed matter is a picosecond (ps). This means that the natural scale for excitation frequencies is a terahertz (THz). In condensed matter, particles are packed so closely that the average distance between the constituents exceeds their diameter by only, say, 30% or less. Therefore, in all motions, complexes of several particles are involved. As a result, typical excitation frequencies for intermolecular vibrations are smaller than the ones for intramolecular motions, say, by about an order of magnitude. Figure 1.1 exhibits a representative example. The upper panel shows the spectra of liquid toluene measured at four temperatures by inelastic light scattering. One notices the lowest intramolecular excitation line at a frequency $\Delta\nu$ between 6 and 7 THz. The intermolecular vibrational spectra are barely visible near the origin of the frequency axis. The toluene molecules

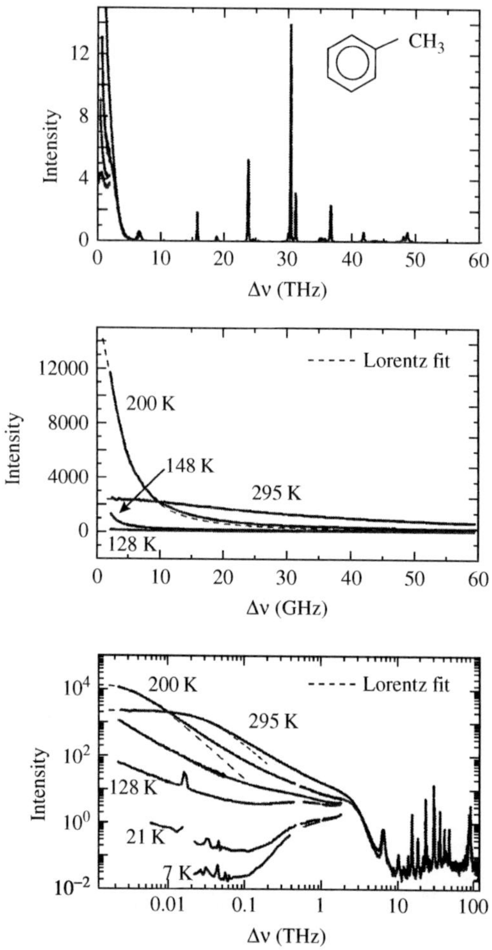

FIG. 1.1. Depolarized-light-scattering intensity as a function of the frequency-change $\Delta\nu = \omega/(2\pi)$ for liquid toluene (melting temperature $T_m = 178$ K, glass-transformation temperature $T_g = 117$ K) at temperatures $T/\mathrm{K} = 295$, 200, 148, 128, and toluene glass at $T/\mathrm{K} = 21$ and 7. The diagrams are constructed (Wiedersich 2003) from data measured by Wiedersich *et al.* (2000). The dashed lines are fits to the spectra for 200 K and 295 K by Eq. (1.1a).

attract each other by weak van der Waals forces. For short distances, there is a strong repulsion. The band for vibrations and librations for such van der Waals systems – be it a liquid, a crystal, or a glass – extends between about 0.1 THz up to some THz. One might expect to observe a nearly frequency-independent weakly temperature-dependent background spectrum for frequencies below 0.1 THz. However, this is not the case as is demonstrated in the other two panels of the figure.

The frequency scale chosen for the middle panel is 10 gigahertz (GHz), i.e., it exhibits dynamics on a time scale of 0.1 nanosecond (ns). The vibrational spectrum is located above the upper end of the $\Delta\nu$-axis. Instead of the expected weak $\Delta\nu$-independent background spectrum, the liquid for $T = 200\,\mathrm{K}$ exhibits a huge spectrum, which increases strongly with decreasing $\Delta\nu$. Heating to 295 K, the spectrum broadens but still describes a large intensity. Cooling to 148 K, the major part of the spectrum is shifted below 5 GHz. Within the frequency interval discussed in the middle panel, the spectral functions of angular frequency $\omega = 2\pi\Delta\nu$, which shall be abbreviated as $\phi''(\omega)$, can be fitted well by Lorentzians:

$$\phi''(\omega) = f \cdot \tau/[1 + (\omega\tau)^2]. \tag{1.1a}$$

Here $f = (2/\pi)\int_0^\infty \phi''(\omega)d\omega$ is a measure for the total spectral intensity, and τ is a time scale for the process. This scale is three orders of magnitude larger than the natural time scale for the dynamics. If the 200 K spectrum is viewed on the natural scale for condensed-matter dynamics, i.e., if it is discussed with a resolution of, say, 0.05 THz, it cannot be distinguished from a spectral peak of width zero. An idealized zero-width peak would describe an elastic light-scattering process. Therefore, spectra as shown in the middle panel are often referred to as quasi-elastic ones. Glassy dynamics deals with the phenomena leading to quasi-elastic spectra, which depend sensitively on control parameters like the temperature or the density. Spectra, which can be described by Eq. (1.1a), are the Fourier transforms of an exponential,

$$\phi(t) = f\exp[-(t/\tau)], \tag{1.1b}$$

$\phi''(\omega) = \int_0^\infty \cos(\omega t)\phi(t)dt$. Such a decay function $\phi(t)$ does not deal with oscillations. A process, which can be described by Eqs. (1.1a,b), shall be referred to as a simple relaxation.

In order to exhibit adequately the intermolecular vibrational spectrum of toluene, the scale of the upper panel should not have been expanded by a factor 1000 but by a factor 10. In order to present the major part of the $T = 295\,\mathrm{K}$ spectrum, the scale should have been expanded by a factor 100. It is impossible to exhibit properly the spectra for the three mentioned temperatures on a common linear axis for the frequency. An adequate way to view the evolution of glassy dynamics upon cooling or compression is the choice of a logarithmic axis for the frequency or for the time. This is done in the lower panel of Fig. 1.1. In addition to the liquid spectra for the above mentioned temperatures, also spectra for the glass at $T = 21\,\mathrm{K}$ and 7 K are presented.

The lower panel exhibits intramolecular excitation lines of toluene within the interval 5 THz $< \Delta\nu <$ 100 THz. The 7 K spectrum shows the intermolecular-vibration spectrum of the amorphous solid for 0.1 THz $< \Delta\nu <$ 5 THz. The glass spectrum for $T = 21\,\mathrm{K}$ shows that there is some background spectrum for $\Delta\nu <$ 0.1 THz, which increases with decreasing frequency. These spectra for $\Delta\nu <$ 0.1 THz are related to the low-temperature anomalies of glasses; their study requires the application of quantum mechanics. The research on low-temperature anomalies of glasses will not be discussed in this book. The glassy dynamics to be considered in the following deals with phenomena yielding the toluene spectra shown for $T = 128\,\mathrm{K}$ or larger temperatures. In this regime, the equations of motion of classical mechanics are the basis for all discussions.

The mentioned fit of the 200 K spectrum by a Lorentzian suggests that the dynamics deals with a simple relaxation process specified by a decay rate $1/\tau$ of some GHz. Equation (1.1a) implies $\phi''(\omega\tau \gg 1) = (f/\tau)/\omega^2$. Thus, a $1/\Delta\nu^2$ variation of the spectrum is suggested for, say, $\Delta\nu > 10$ GHz. Such power law appears as a straight line of slope (-2) in the double-logarithmic presentation used in the lower panel. The suggestion leads to contradictions to the experimental findings. The measured spectrum for $\Delta\nu > 20$ GHz is much larger than the high-frequency tail of the Lorentzian. The extrapolation of the tail to $\Delta\nu = 2$ THz leads to an underestimation of the spectrum by almost a factor 100. The success of the Lorentzian fits shown in the middle panel is a delusion, which is caused by choosing a too small frequency interval. For $\Delta\nu = 2$ THz, the 200 K spectrum exceeds the intermolecular-vibration spectrum by more than a factor 2. Decreasing $\Delta\nu$, the 200 K spectrum increases further. At $\Delta\nu = 20$ GHz, it exceeds the vibrational spectrum by about three orders of magnitude. Only if even lower frequencies are considered, does there occur the crossover to the Lorentzian. Comparison of the representation of the data in the middle panel with that in the lower one demonstrates that the spectrum for 200 K cannot be exhibited adequately on a linear frequency axis. The dynamics is not characterized by some typical frequency. This feature is called stretching of the dynamics. Choosing a logarithmic frequency axis is the conventional procedure to present stretched spectra.

The measured spectra exhibit a crossover frequency ω_0 so that $\phi''(\omega) \approx \phi''(\omega = 0)$ for $\omega \leq \omega_0$. For $\omega \leq \omega_0$, the frequency variation can be neglected, i.e., there is a white-noise spectrum there. For the 295 K spectrum, $\omega_0/2\pi$ is about 10 GHz. Stretching implies that $\omega_0/2\pi$ is several orders of magnitude smaller than the natural scale $1/t_{\mathrm{mic}}$ for the dynamics. In this sense, stretching implies slowness of the dynamics. One does not observe stretching for normal liquid states. Stretching evolves upon compression or upon cooling. The latter case is demonstrated in the middle panel for a change of the temperature from 295 K to 200 K. Sensitivity of some parts of the spectra with respect to changes of the thermodynamic state is also an outstanding feature of stretching.

Two digressions might be adequate. First, sound waves can propagate in liquids and in glasses. Shear waves can propagate in amorphous solids but not in liquids. The theory of these vibrational excitations will be discussed in detail in Chapter 3. The mentioned waves are characterized by a wave number q and an oscillation frequency Ω_q proportional to q. Since Ω_q can be arbitrarily small for q tending to zero, there is no low-frequency gap for the vibrational spectrum of homogeneous matter. However, the instrument used to measure the spectra in Fig. 1.1 was designed so that single-wave excitation was excluded by selection rules for the light-scattering process. Multi-wave excitations contribute to the scattering intensity for low frequencies, but they produce a small weakly temperature-dependent background only. This suggests that wave excitations are not relevant for the explanation of the quasi-elastic spectra in liquids.

Second, most liquids mentioned in this book have some transition point to a crystalline state. Toluene, for example, has a melting temperature at ambient pressure $T_{\mathrm{m}} = 178\,\mathrm{K}$. Figure 1.1 demonstrates that there is glassy dynamics in liquid toluene for temperatures 117 K above as well as 30 K below T_{m}. Never has there been observed any peculiarity for the evolution of the glassy dynamics related to T_{m}. But, often, glassy dynamics is particularly well developed for temperatures below T_{m}, i.e., for the supercooled state. Therefore, one is interested to carry out experiments for these metastable states. To do so, one has to choose sufficiently complicated systems. Simple liquids like argon, benzene or a sodium-chloride melt cannot be supercooled conveniently. The cited toluene is one of the simplest van der Waals liquids, which can be supercooled. More complicated systems like salol or orthoterphenyl can be supercooled easily. For this reason, the latter liquids are used regularly as model systems for experiments on glassy dynamics.

Figure 1.2 demonstrates the stretching phenomenon in the time domain; it reproduces as full lines a response function $\chi(t)$ for supercooled salol. This function was obtained by optical-Kerr-effect spectroscopy. A subpicosecond pulse of polarized light is used to create an optical anisotropy of the liquid. The response

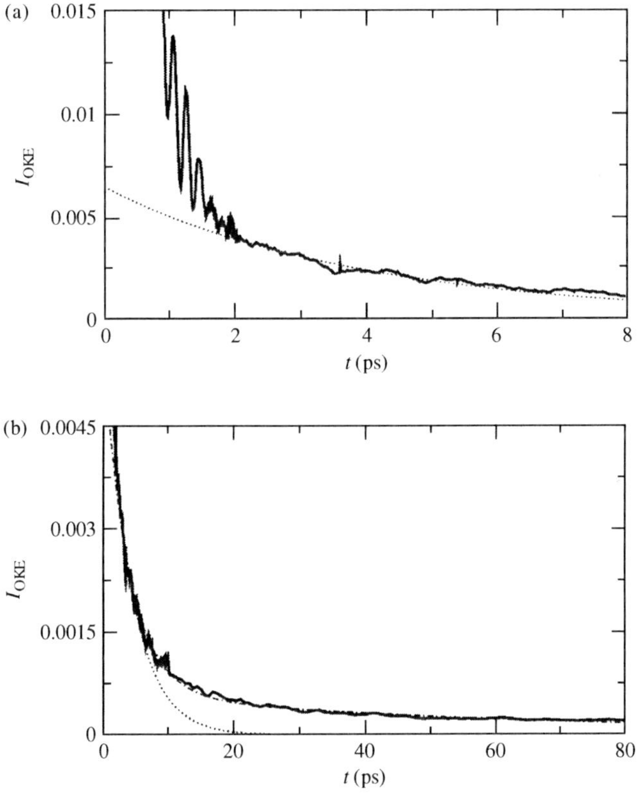

FIG. 1.2. *Continued.*

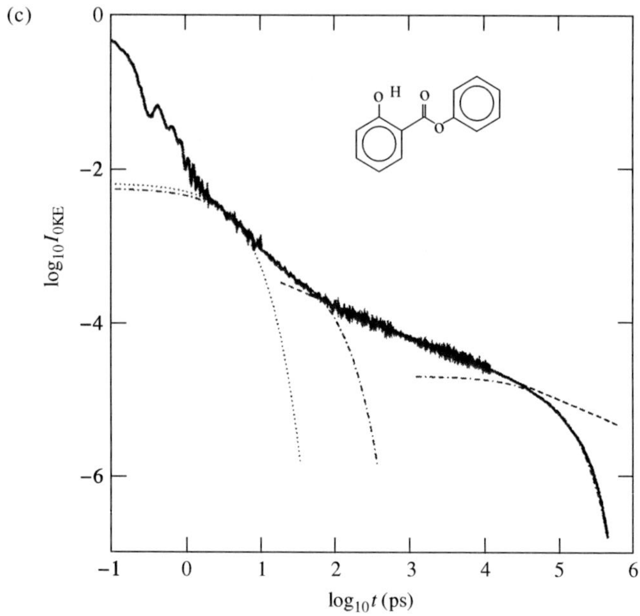

FIG. 1.2. The full lines reproduce an optical-Kerr-effect response function measured by Hinze *et al.* (2000) at $T = 257$ K for salol ($T_{\mathrm{m}} = 318$ K, $T_g = 218$ K), which is normalized to unity for $t = 0.01$ ps. The dotted lines exhibit the response function $6.5 \times 10^{-3} \exp[-(t/(4 \text{ ps})]$. The dashed-dotted lines in panels (b) and (c) show $5.0 \times 10^{-3} \exp[-t/(4 \text{ ps})] + 6.0 \times 10^{-4} \exp[-t/(60 \text{ ps})]$. The dashed-dotted line shown in panel (c) for $\log(t) > 3$ exhibits $2.0 \times 10^{-5} \exp[-t/(9 \times 10^4 \text{ ps})]$.

The dashed straight line in panel (c) has slope $-$ 0.41, and it represents the response function for a von Schweidler-law-decay function, Eq. (1.4a), for an exponent $b = 0.59$ (Sperl 2004a).

by a birefringence is measured as intensity I_{OKE} of a scattered probe beam for various delay times t. Function $\chi(t)$ is proportional to the negative derivative of a decay function, $\chi(t) \propto -d\phi(t)/dt$. It is proportional to the Fourier transform of a susceptibility spectrum $\chi''(\omega)$. The latter is proportional to the fluctuation spectrum $\phi''(\omega)$ multiplied by the frequency: $\chi''(\omega) \propto \omega\phi''(\omega)$. The precise form of the relations between correlators or decay functions $\phi(t)$, fluctuation spectra $\phi''(\omega)$, susceptibilities $\chi(\omega) = \chi'(\omega) + i\chi''(\omega)$, and response functions $\chi(t)$ will be discussed in the following Chapter 2.

Figure 2.1(a) presents the response on a ps scale. The signal exhibits oscillatory decay, and it reaches the 0.5% level at about 2 ps. For larger times, $I_{\mathrm{OKE}}(t)$ does not continue the oscillatory variation. Rather, it exhibits a monotonically decreasing tail. The dotted line shows that the tail can be fitted by an exponential, which is specified by a relaxation time $\tau = 4$ ps. The increase of the time from 1 ps to 8 ps deals with the decay from about 0.005 to about 0.0009. To exhibit the decay of the exponential tail below its 10% level, the scale of the time axis has to be increased. Figure 1.2(b) exhibits the response on a 10 ps scale. The diagram shows that the description of the tail as a simple relaxation process becomes inadequate for times exceeding about 10 ps. There appears a long-time tail. The dashed-dotted line demonstrates that the response between 2 ps and 80 ps can be described well by a sum of two exponentials. They are specified by relaxation times which differ by more than an order of magnitude. The first process is the above mentioned one with $\tau = 4$ ps, and the second process is specified by a relaxation time $\tau = 60$ ps. Within the interval 10 ps $< t <$ 80 ps, the second tail describes a decay from about 0.000 51 to about 0.000 16. The scale of the time axis should be chosen larger, if one wanted to document the decay below the 20% level of the second-tail process. On the other hand, the time scale should be chosen smaller, since Fig. 1.1(b) cannot exhibit the beginning of the monotonic behaviour. There is no adequate scale to exhibit the dynamics on a linear time axis. This is the manifestation of stretching. The presentation of the data on a logarithmic time axis in Fig. 1.2(c) corroborates this conclusion. The data demonstrate glassy dynamics for a time increase by more than five orders magnitude. The stretched decay starts at about 2 picoseconds and ends at about 0.4 microseconds. The former time specifies the scale for intermolecular vibrations and the latter approaches the scale characterizing the functionally-important dynamics of biopolymers.

It is tempting to generalize Eq. (1.1b) to a sum of simple relaxation processes: $\phi(t) = \sum_n f_n \exp[-\gamma_n t], f_n > 0, \gamma_n = 1/\tau_n \geq 0$. The sum can be understood as an approximation to a representation of $\phi(t)$ as Laplace transform of a non-negative weight function $\rho(\gamma)$:

$$\phi(t) = \int_0^\infty \exp[-\gamma t]\rho(\gamma)d\gamma, \quad \rho(\gamma) \geq 0. \tag{1.2a}$$

A process that can be described by such decay function $\phi(t)$ shall be called a general relaxation process. This formula can be interpreted as superposition of simple relaxation processes where the relaxation rate $\gamma = 1/\tau$ occurs with probability density $\rho(\gamma)/\int_0^\infty \rho(\gamma)d\gamma$. Functions, which can be written in form of Eq. (1.2a), are called completely monotonic. They have the property

$$[-\partial/\partial t]^\ell \phi(t) > 0, \quad \ell = 0, 1, \ldots . \tag{1.2b}$$

According to Bernstein's theorem, Eq. (1.2b) implies Eq. (1.2a); the two preceding equations are equivalent (Feller (1971), Sec. XIII. 4). The simple relaxation process is reproduced by a point distribution located at $\gamma = \tau^{-1}$. Stretching of the dynamics means that the distribution of rates γ or times $\tau = 1/\gamma$ is spread over large intervals. A stretched distribution function $\rho(\gamma)$ cannot be described by a peak of some width Δ located at some characteristic decay rate γ_0 so that $\Delta/\gamma_0 \ll 1$.

For times t exceeding 2 ps, the signal $I_{\mathrm{OKE}}(t)$ in Fig. 1.2(c) can be fitted within the accuracy of the data by a sum of six exponentials. The first two contributions of such a representation are shown by the dotted and dashed-dotted lines. The contribution for the largest relaxation time is exhibited as dashed-dotted line for $t \geq 10^3$ ps. It shows that the final part of the measured decay can be considered as a simple relaxation process. The described data interpolation (Sperl 2004a) indicates that glassy dynamics, as opposed to vibrational dynamics, is a general relaxation process.

The evolution of the spectra of glassy dynamics within the full gigahertz band upon cooling a liquid was measured first by H.Z. Cummins and collaborators. Examples of their findings are discussed below in connection with Figs. 1.3 and 4.28. The spectra have been obtained by inelastic light scattering combining a Raman and a tandem-Fabry–Perot spectrometer. Depolarized scattering was detected in order to exclude sound-wave excitations, and the backward-scattering signal was used in order to eliminate shear-wave contributions. The data shown in Fig. 1.1 have been obtained by the same technique using an upgrading of the original instrumentation. Optical-Kerr-effect spectroscopy was introduced for the study of glassy dynamics by Torre *et al.* (1998).

1.2 Power-law relaxation

In this section, some data shall be considered, which indicate that the stretching of the glassy dynamics in liquids is caused by the succession of two power-law-relaxation processes.

The evolution of the glassy-dynamics spectra upon cooling the mixed molten salt $0.4\,Ca(NO_3)_2\,0.6K(NO_3)$ (CKN) is shown in Fig. 1.3. The interaction among the constituents of this mixture of ions consists of a strong short-range repulsion complemented by Coulomb forces. The data are derived from the cross section $I(\omega)$ for depolarized-light scattering. The spikes exhibited for a frequency near 15 GHz and a similar spike of the 128 K spectrum shown in the lower panel of Fig. 1.1 are due to unwanted sound-wave excitations. The spikes appear because of non-perfect polarization filters. The measured spectral intensities $I(\omega)$ for CKN (Li *et al.* 1992) are quite similar to the ones shown in the lower panel of Fig. 1.1. But, Fig. 1.3 does not exhibit the $I(\omega)$, which are proportional to the fluctuation spectra $\phi''(\omega)$; it shows the susceptibility spectra $\chi''(\omega) \propto \omega I(\omega)$. It will be explained in Sec. 2.5 that $\chi''(\omega)$ can be measured, in principle, by absorption spectroscopy. Therefore, susceptibility spectra are also referred to as loss spectra. The band of intermolecular vibrations of CKN manifests itself by a nearly temperature-independent loss peak with a maximum position between 3 and 4 THz. For normal-liquid dynamics, one would expect that the fluctuation spectrum reaches its zero-frequency limit below this band, say $\phi''(\omega < 0.1\text{ THz}) \approx \phi''(\omega = 0)$. Such an ω-independent spectrum manifests itself as a loss spectrum

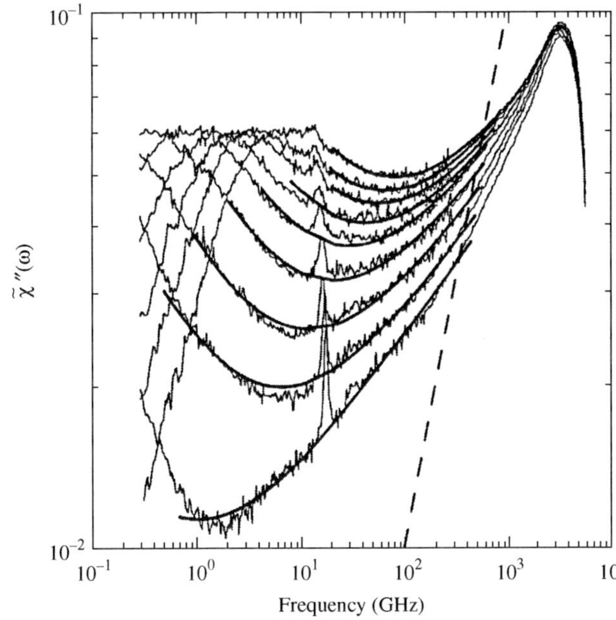

FIG. 1.3. Double-logarithmic presentation of the loss spectra $\tilde{\chi}''(\omega) \propto \chi''(\omega)$ of 0.4 $Ca(NO_3)_2$ 0.6 K (NO_3) (CKN, $T_{\rm m} = 483$ K, $T_g = 333$ K) measured by depolarized-light scattering for temperatures $T/°\mathrm{C} = 195$, 180, 170, 160, 150, 140, 130, 120, and 110 (from top to bottom). The full lines are fits to the minima by Eq. (1.7a) with $a = 0.27$ and $b = 0.46$. Reproduced from Li *et al.* (1992). A dashed straight line of slope unity is added to indicate an upper-limit estimate of a white-noise-induced spectrum for normal-liquid dynamics.

proportional to ω. Thus, a white-noise-induced spectrum appears as a straight line with slope unity in the double-logarithmic plot used. Such a line is added in dashes to the figure in order to indicate an upper-limit estimation for a regular low-frequency spectrum. For frequencies below 0.2 THz, the measured spectra are much larger than such an estimate; and this is one manifestation of the glassy dynamics of CKN.

The loss spectra exhibit maxima at low frequencies. These low-frequency peaks are another manifestation of the glassy dynamics. The positions $\omega_{\rm max}$ of these maxima shift to smaller values if the temperature is lowered. The maximum position can be used to introduce a temperature-dependent time scale $\tau_{\rm max} = 1/\omega_{\rm max}$. These scales $\omega_{\rm max}$ or $\tau_{\rm max}$ quantify the sensitive temperature dependence of the low-frequency end of the loss spectra and the large-time end of the decay functions $\phi(t)$, respectively. Notice that the maximum intensity $\chi''(\omega_{\rm max})$ does not vary with changes of T. This means that height of the fluctuation spectrum $\phi''(\omega_{\rm max}) \propto \chi''(\omega_{\rm max})/\omega_{\rm max}$ increases with decreasing T

proportional to $1/\omega_{\text{max}}$. In the original literature, the dynamics leading to the control-parameter-sensitive low-frequency loss peak is usually referred to as the α-process of the glassy system.

Between the low-frequency-loss peak and the loss peak of intermolecular vibrations at high frequencies, the spectra have a minimum at some frequency ω_{min}. With decreasing temperature, ω_{min} decreases and so does the minimal spectral intensity $\chi_{\text{min}} = \chi''(\omega_{\text{min}})$. But, ω_{min} depends less sensitively on T than does ω_{max}. Decreasing T from 195 °C to 150 °C, ω_{max} decreases by about a factor 10, while ω_{min} decreases by about a factor 3.

The dynamics leading to the low-frequency loss peak was discussed first by R. Kohlrausch about 150 years ago. He measured decay functions $\phi(t)$ of the dielectric polarization of a glass and demonstrated that the strong deviations of his data from the simple relaxation law could be described well – though not perfectly – by the function

$$\phi(t) = f \exp[-(t/\tau)^{\beta}]. \tag{1.3}$$

This Kohlrausch function, which is also called the stretched exponential, quantifies stretching by an exponent β, where $0 < \beta < 1$. The parameter β is referred to as stretching exponent or Kohlrausch exponent. For $t/\tau \gg 1$, the Kohlrausch function decays so strongly to zero that its Fourier transform varies smoothly for small frequencies: $\phi''(\omega\tau \ll 1) = \phi''(\omega = 0) + O(\omega^2)$. Therefore, the susceptibility spectrum $\chi''(\omega)$ increases linearly with frequency on the low-frequency wing of the loss peak. This feature of the dynamics is exhibited also by the spectrum of the simple-relaxation process. Such white-noise-induced susceptibility spectrum is exhibited in Fig. 1.3, for example, by the 195 °C spectrum for frequencies below 1 GHz. The stretching of the loss-peak spectrum is a property of its high-frequency wing. It is due to the decay for times, which are small compared to $\tau_{\text{max}} \approx \tau$. For $\omega\tau \gg 1$, the Kohlrausch spectrum is due to the algebraic singularity exhibited by Eq. (1.3) for small times: $\phi(t) = f[1-(t/\tau)^{\beta}+O(t^{2\beta})]$. The Fourier transform of such power-law variation, say $f(t) = -(t/\tau)^b$, yields a corresponding power-law decrease for the susceptibility spectrum: $\omega f''(\omega) = \omega \int_0^\infty \cos(\omega t) f(t) dt = -[\int_0^\infty \cos(\xi)\xi^b d\xi]/(\omega\tau)^b$. Factors as given by the bracket are not relevant for the considerations of this chapter. They are derived in Appendix A.7 and they will be discussed in Chapter 6. But, let us note already here that a decay function

$$\phi(t) = f - h(t/\tau)^b, \quad 0 < b < 1, \quad f, h > 0, \tag{1.4a}$$

is equivalent to the power-law susceptibility spectrum

$$\chi''(\omega) = h \sin(\pi b/2)\Gamma(1+b)/(\omega\tau)^b. \tag{1.4b}$$

$\Gamma(x)$ denotes the gamma function.

It was discussed by E. von Schweidler about a century ago that the high-frequency wings of dielectric-loss peaks can often be fitted well by Eq. (1.4b). Since the $t/\tau \to 0$ asymptote of Eq. (1.3) reproduces von Schweidler's law, the Kohlrausch spectrum exhibits a high-frequency tail according to Eq. (1.4b) with with β being the von Schweidler exponent b. With increasing ω, the function $1/\omega^b$ decreases more slowly than $1/\omega$. The latter decay is obtained from the Lorentzian for $\omega\tau \gg 1$. Hence, Kohlrausch's loss spectrum is asymmetric. The stretching is due to the von Schweidler-law tail, Eq. (1.4b). This feature is in agreement with the results shown in Fig. 1.3. By extensive studies of dielectric-loss spectra for a variety of glass-forming liquids, G. Williams and collaborators have established the Kohlrausch function as a reasonable description of the low-frequency-loss-peak dynamics.

Another fit function for the low-frequency dielectric susceptibilities of glassy liquids was introduced by R.H. Cole and D.W. Davidson about half a century ago:

$$\chi(\omega)/\chi_s = f/[1 - i(\omega\tau)]^{\beta_{\mathrm{CD}}}. \tag{1.5}$$

Here, χ_s is the static susceptibility, which determines the dielectric constant by $\epsilon = 1 + 4\pi\chi_s$. It is given by the total integral of the fluctuation spectrum, $\chi_s = \int_0^\infty \phi''(\omega)d\omega/\pi$, as will be shown in Chapter 2. The other three parameters of this function have a similar meaning as the ones noted for the Kohlrausch function. Parameter $f, 0 < f < 1$, is the percentage of the integrated fluctuation spectrum contributed by the loss-peak dynamics. Parameter τ is a characteristic time scale for the process causing the loss peak, and the Cole–Davidson exponent $\beta_{\mathrm{CD}}, 0 < \beta_{\mathrm{CD}} < 1$, quantifies stretching. Similar to what was discussed above, the Cole–Davidson function describes a crossover from a white-noise-induced loss at small frequencies, $\chi''(\omega\tau \ll 1) = f\chi_s\beta_{\mathrm{CD}}(\omega\tau)$, to von Schweidler decay for large frequencies, $\chi''(\omega\tau \gg 1) = f\chi_s \sin(\pi\beta_{\mathrm{CD}}/2)/(\omega\tau)^{\beta_{\mathrm{CD}}}$. It depends on the system, which of the two fit functions (1.3) or (1.5) describes the largest part of the loss-peak dynamics.

The indicated work leading to Eqs. (1.3)–(1.5) deals with the dynamics on time scales which are six and more orders of magnitude larger than the natural scale for condensed matter. Therefore, the evolution of the glassy dynamics out of the normal-condensed-matter dynamics could not be examined. However, Fig. 1.3 indicates that the mentioned classical observations on the low-frequency-loss-peak dynamics remain valid also for the dynamics within the GHz band of the frequencies.

von Schweidler's law is equivalent to a power-law response-function: $\chi(t) \propto -d\phi(t)dt = [hb/\tau](\tau/t)^{1-b}$. The dashed straight line in Fig. 1.2 demonstrates that the optical-Kerr-effect response of salol follows this law for a time increase of about a factor 300. The data for $t \geq 60$ ps exhibit the low-frequency loss-peak dynamics. The response demonstrates a rather abrupt crossover for t around 20 ns from von Schweidler decay to exponential decay. The stretching of the salol loss-peak dynamics is caused predominantly by the t^b-law process.

Within the frequency interval 2 GHz $< \omega <$ 300 GHz, the CKN loss spectrum for $T = 110\,^\circ$C increases with increasing ω. As emphasized above, this spectrum is considerably larger than any reasonable estimation of a white-noise-induced one. Nor can it be fitted by the superposition of the von Schweidler-law tail of the low-frequency loss peak, $\chi''(\omega) \propto 1/\omega^b$, and a white-noise-induced spectrum $\chi''(\omega) \propto \omega^1$. Figure 1.3 shows that the specified part of the spectrum exhibits nearly power-law increase: $\chi''(\omega) \propto \omega^a$, $a \approx 0.2$. Such power-law part of the spectra within the GHz band was detected first by Knaak *et al.* (1988) by neutron-scattering spectroscopy, and this for CKN. The formulas describing an ω^a behaviour are obtained from Eqs. (1.4a,b) by changing exponent b to exponent $-a$. The decay law

$$\phi(t) = f + h(t_0/t)^a, \quad 0 < a < 1, \quad f, h > 0, \tag{1.6a}$$

is equivalent to a susceptibility spectrum

$$\chi''(\omega) = h \sin(\pi a/2)\Gamma(1-a)(\omega t_0)^a. \tag{1.6b}$$

The preceding discussions suggest to describe the loss minimum by the interpolation $\chi''(\omega) = A\omega^a + B\omega^{-b}$. This four-parameter-fit formula can be written more transparently as

$$\chi''(\omega) = \chi_{\min}[b(\omega/\omega_{\min})^a + a(\omega_{\min}/\omega)^b]/(a+b). \tag{1.7a}$$

Such an expression was used first by Sjögren (1990) in order to describe dielectric loss minima for polymer melts in the kHz region. According to Eqs. (1.4a,b) and (1.6a,b), this susceptibility spectrum is equivalent to a decay function $\phi(t) - f = A't^{-a} - B't^b$. Written more transparently, one gets

$$\phi(t) = f + H[(t_{\text{cr}}/t)^a - (t/t_{\text{cr}})^b]. \tag{1.7b}$$

There holds $\chi_{\min} = c_1 H$ and $\omega_{\min} = c_2/t_{\text{cr}}$ with constants c_1 and c_2, which are determined by the exponents a and b.

The full lines in Fig. 1.3 demonstrate that Eq. (1.7a) accounts for the CKN spectra for frequency intervals up to 2.5 decades in width, and this for temperature-independent exponents a and b. The success of the fits must not be overinterpreted as an independent measurement of the two exponents a and b. The fit interval is too small to allow for an identification of the asymptotes $\chi''(\omega \gg \omega_{\min}) \propto \omega^a$ and $\chi''(\omega \ll \omega_{\min}) \propto \omega^{-b}$. The curvature of the $\log \chi''$ versus $\log \omega$ curve in the minimum is given by the product $a \cdot b$. Therefore, one can vary exponent a to some extend without changing the fit quality of the minimum. This holds provided one varies exponent b proportional to $1/a$. Stretching of the dynamics seems to be caused by the crossover between two power laws. Decreasing ω below the band of normal-liquid excitations, the loss spectrum decreases according to the ω^a law. Near the spectral minimum, there is a crossover to the ω^{-b} law. For frequencies near the spectral maximum, there is another crossover to a white-noise-induced ω^1-variation for frequencies tending to zero. These considerations suggest that it is the dynamics causing the loss minimum, which is the essential process leading to relaxation stretching.

The data in Fig. 1.2 support the preceding conclusion. At the end of the vibrational decay, there starts a straight part of the $\log(\chi)$ versus $\log(t)$ curve. It extends over about a one-decade time interval. Near $t = 20$ ps, there starts the crossover to the von Schweidler decay discussed above. The first straight-line piece has a slope close to -1. A fit to Eq. (1.6a) would lead to an exponent a close to zero.

Many properties of liquids have been discovered with the aid of molecular-dynamics-simulation studies (Boon and Yip 1980; Hansen and McDonald 1986; Balucani and Zoppi 1994). This technique is based on the numerical integration of the equations of motion of systems of up to some ten thousand degrees of freedom. At present, typically, the time intervals accessible can extend up to five or six orders of magnitude beyond the natural scale t_{mic} for condensed matter dynamics. The protocols of the orbits for the system in phase space can be used to extract reproducible information on, for example, decay functions and their Fourier transforms. J.-P. Hansen and his collaborators made the first extensive studies showing that this technique can be applied also to analyse the glassy dynamics of cooled equilibrium liquids (Bernu *et al.* 1987; Roux *et al.* 1989).

Starting with a paper by Kob and Andersen (1995a), many simulation studies of the glassy dynamics have been performed for a binary Lennard-Jones mixture (LJM). This mixture is a model of a van der Waals system consisting of 80% big particles and 20% small ones, which are referred to as A and B particles, respectively. The forces between the three possible kinds of interacting pairs, A-A, A-B, and B-B, are derived from Lennard-Jones potentials. The latter depend on the interparticle distance r as

$$V(r) = 4\epsilon[(\sigma/r)^{12} - (\sigma/r)^6]. \tag{1.8a}$$

For technical reasons, a long-distance cutoff is introduced for the potential. Such a potential has been used regularly to describe simple liquids like argon. A possibility to introduce natural scales for the length, the energy, the temperature and the time are $\sigma, \epsilon, \epsilon/k_{\text{B}}$ and $\sqrt{m\sigma^2/48\epsilon} = \tau_0$, respectively. Here, m denotes the particle mass and k_{B} abbreviates Boltzmann's constant. The simulation results are usually cited in these units so that

$$\sigma = 1, \epsilon = 1, \tau_0 = 1, k_{\text{B}} = 1. \tag{1.8b}$$

In the case of liquid argon, these natural units have the following values, called argon units:

$$\sigma = 0.340\,\text{nm}, \epsilon/k_{\text{B}} = 120\text{K}, \tau_0 = 0.311\,\text{ps}. \tag{1.8c}$$

The glassy-dynamics studies for the LJM are done with $m = m_A = m_B$ and the units are formed with the A-A-potential parameters. The parameters $\sigma_{\alpha,\beta}$ and $\epsilon_{\alpha,\beta}, \alpha, \beta = A, B$, are chosen carefully so that crystallization is avoided. The data to be cited for this system have been obtained for $N = 1000$ particles, which are inclosed in a cubic box of side length $L = 9.4\sigma_{AA}$. The particle number density is about that of liquid argon near its triple point.

Figure 1.4 presents results for the LJM at $T = 0.446$. The corresponding temperature in argon units, $T = 53.5\,\text{K}$, is considerably below the melting temperature $T_{\text{m}} = 83.8\,\text{K}$ of argon at ambient pressure. The results are for a density-fluctuation-decay function $\phi(t)$ of wave number $q = 7.2/\sigma_{AA}$ for a tagged large particle. Hence, the function probes dynamics on the length scale of the interparticle spacing. The precise definition of such correlation function and its properties are discussed in Sec. 3.2. The function $\phi_{\text{ND}}(t)$, which is shown as dashed line, is obtained from Newton's equations of motion for the particles. The time scale τ chosen in the diagram is defined by the request $\phi_{ND}(t = \tau) = 1/e$. One recognizes an oscillatory short-time motion indicating $t_{\text{mic}} \approx 10^{-4}\tau$ as natural time scale for the liquid dynamics. The oscillation is damped out within the crossover interval $10^{-4} \leq t/\tau \leq 10^{-3}$. For $10^{-3} \leq t/\tau$, a stretched decay is exhibited, which extends over 4 orders-of-magnitude time increase. Another

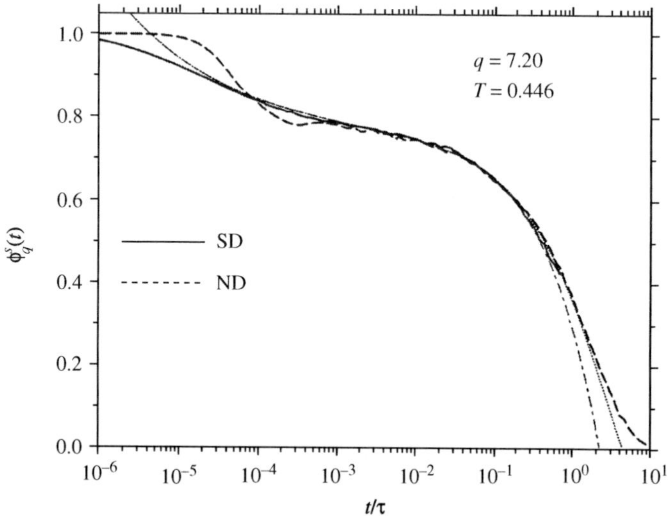

FIG. 1.4. Molecular-dynamics simulation results for the density-fluctuation-decay function $\phi(t) = \phi_q^s(t)$ for wave number $q = 7.20/\sigma_{AA}$ for the tagged big particles of a binary Lennard-Jones mixture (LJM). The model is explained in connection with Eqs. (1.8a,b). The full and dashed lines refer to the same states with temperature $T = 0.446\epsilon_{AA}/k_B$ but to stochastic and Newtonian equations of motion, respectively. The time scale τ is arbitrary but adjusted so that the two results coincide for $t/\tau \approx 0.1$.

The dashed-dotted line is a fit of the data by a function discussed in Sec. 6.1.3 (iii), which is well approximated by the one in Eq. (1.7b) for exponents $a = 0.32$ and $b = 0.62$. The dotted line is obtained by adding a correction term $H'(t/\tau)^{2b}$ to the fit. Reproduced from Gleim *et al.* (1998).

simulation was done for the same thermodynamic state but for a different equation of motion. The result is shown by the full line; the scale τ was adjusted so that the full and dashed lines coincide for $t/\tau \approx 10^{-1}$. To get the full line, Newton's equations of motion have been complemented by a stochastic force with a nearly white-noise spectrum. The strength of this force was increased such that the result for $\phi_{\mathrm{SD}}(t)$ became strength independent except for a change of the over-all time scale. Thereby, a model for a colloid was constructed (Gleim *et al.* 1998). The stochastic force describes the fluctuating forces exerted by the liquid on the solute particle. For an isolated colloid particle, these forces cause Brownian motion. The full line decays from its initial value 1.0 to 0.95, if t/τ increases up to about $4 \cdot 10^{-6}$. This initial behaviour can be fitted well by $\exp[-(t/\tau_{\mathrm{SD}}], \tau_{\mathrm{SD}}/\tau = 8 \cdot 10^{-5}$. Such decay describes the normal-state dynamics of a colloid. For $t/\tau > 10^{-5}$, the observed correlator is much larger than the specified exponential. Thus, one estimates that the correlator $\phi_{\mathrm{SD}}(t)$ for the stochastic dynamics exhibits glassy behaviour for a time exceeding about $10^{-5}\tau$. For the stochastic dynamics, the simulation algorithm is so involved that data beyond $t/\tau \approx 1$ could not be obtained.

The dashed-dotted line in Fig. 1.4 exhibits a formula for the interpolation between the t^{-a} and $-t^{b}$ laws, which is discussed in chapter 6. However, within the time interval of interest, this line is close to the one calculated with Eq. (1.7b) for the exponents $a = 0.32$ and $b = 0.62$. The other parameters have to be chosen as $f = 0.75, H = 0.01$, and $t_{\mathrm{cr}}/\tau = 0.023$. The correlator $\phi_{\mathrm{SD}}(t)$ is described well by Eq. (1.7b) for a time increase exceeding 4 orders of magnitude. Within the one-decade time interval, $3 \cdot 10^{-5} \leq t/\tau \leq 3 \cdot 10^{-4}$, Eq. (1.7b) reduces to Eq. (1.6a) within the accuracy of the drawing. In this sense, the t^{-a}-power law in its pure form is exhibited. On the other hand, Eq. (1.7b) reduces to Eq. (1.4a) only for $t/\tau > 3 \cdot 10^{-1}$. For the data in Fig. 1.4, von Schweidler's law describes the data only for less than half-a-decade time increase. For the description of the stretched relaxation within the large interval $3 \cdot 10^{-4} \leq t/\tau \leq 3 \cdot 10^{-1}$, it is essential that there is a combination of two power laws.

In addition to providing strong evidence for the existence of the t^{-a} process and some support for the two-power-law scenario, the figure demonstrates a remarkable property. For the whole time interval, where both functions $\phi_{\mathrm{ND}}(t)$ and $\phi_{\mathrm{SD}}(t)$ exhibit glassy dynamics, both correlators agree. Thus, the details of the microscopic equations of motion appear irrelevant for the glassy dynamics, except for fixing a time scale. Let us also note the following feature. Within the above specified one-decade time interval, where $\phi_{\mathrm{SD}}(t)$ exhibits the t^{-a} decay, the function $\phi_{\mathrm{ND}}(t)$ is dominated by oscillations.

1.3 Superposition principles

In this section, it shall be discussed that parts of the low-frequency-loss-peak dynamics exhibit a scaling law, which is referred to as superposition principle.

The upper panel of Fig. 1.5 demonstrates the evolution of glassy dynamics upon cooling as it manifests itself in a density-fluctuation-decay function

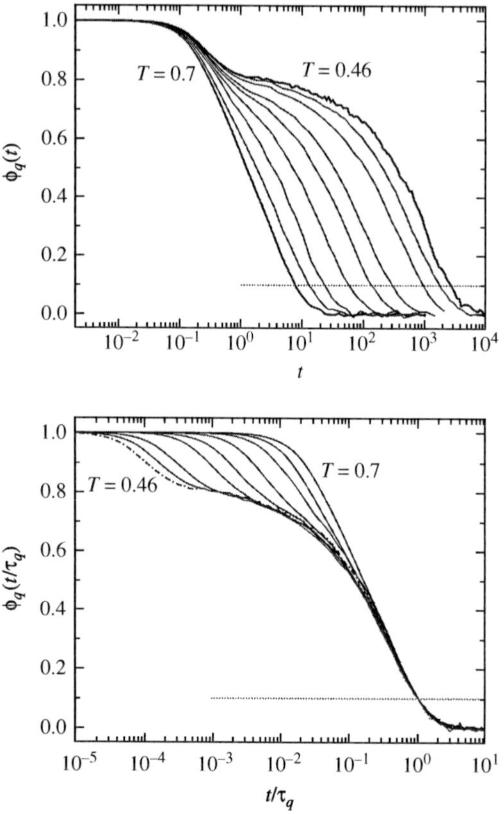

FIG. 1.5. Monomer-density-fluctuation-decay functions $\phi_q(t)$ for wave number $q = 6.9$ obtained by molecular-dynamics simulations for a dense-polymer-liquid model. The model and the units used for the representation of the results are explained in the text. The curves in the upper panel refer to temperatures $T = 0.70$, 0.65, 0.60, 0.55, 0.52, 0.50, 0.48, 0.47, and 0.46 (from left to right). The lower panel shows the same sequence of curves (from right to left) as functions of the rescaled time $\tilde{t} = t/\tau_q$, where the T-dependent scale τ_q is defined by the request $\phi_q(t = \tau_q) = 0.1$. The lower panel is reproduced from Aichele and Baschnagel (2001a) and the upper one shows the original simulation results (Baschnagel 2003).

$\phi_q(t)$. The curves are results of a molecular-dynamics-simulation study for a model of a dense polymer melt. Polymer molecules have many internal degrees of freedom. These lead to a large number of intramolecular vibrations of the isolated polymer chain whose frequencies are smaller or equal to the ones for the intermolecular vibrations of condensed-polymer matter. The model describes the polymer molecules as chains with 10 interaction sites, which represent the monomers. There is an intramolecular monomer-monomer interaction derived from an anharmonic spring potential. In addition, there are van der Waals interactions between all pairs of interaction sites, intra- as well as intermolecular ones, modelled by Lennard-Jones potentials. The units used for the presentation of the results are σ, ϵ, and ϵ/k_B for the length, energy, and temperature, respectively. These choices agree with the ones explained in Eq. (1.8b). Different from the convention used for the presentation of the data for the LJM, the results for the polymer model are cited with $\sqrt{m\sigma^2/\epsilon}$ as unit of the time. The curve for temperature $T = 0.7$ decays from 0.9 to 0.1 within a time increase of 1.6 decades. It is stretched by about a factor 1.8 relative to the one described by the elementary relaxation law. From this curve, the natural time scale for the short-time dynamics of the system is estimated to about $t_{\text{mic}} = 0.5$. Upon cooling to $T = 0.46$, the dynamics slows down by more than a factor 300.

The decay curves shown in the upper panel for $\phi_q(t) \leq 0.70$ and for low temperatures are almost parallel. They are related by a horizontal translation. Since a logarithmic abscissa is used, this translation is equivalent to a rescaling of the time. Such property implies the following. The functions $\phi(t)$, which depend on time t and on temperature, can be represented in terms of a temperature-independent shape function $\tilde{\phi}(\tilde{t})$—also called a master function—and a scale τ:

$$\phi(t) = \tilde{\phi}(t/\tau). \tag{1.9a}$$

Such relation is called a scaling law. The strong dependence of $\phi(t)$ on T is reduced to that of the time scale τ. If the correlation function can be described by a Kohlrausch function, validity of scaling is equivalent to the statement that the strength factor f and the exponent β in Eq. (1.3) are temperature independent. If the shape function $\tilde{\phi}(\tilde{t})$ would be completely monotonic, one could write it in terms of a temperature-independent shape function $\tilde{\rho}(\tilde{\gamma})$ as noted in Eq. (1.2a): $\tilde{\phi}(\tilde{t}) = \int_0^\infty \tilde{\rho}(\tilde{\gamma}) \exp[-\tilde{t}\tilde{\gamma}] d\tilde{\gamma}$. Thus, the above formulated law would be equivalent to the scaling law for the rate distribution $\rho(\gamma)$:

$$\rho(\gamma) = \tau\tilde{\rho}(\gamma\tau). \tag{1.9b}$$

As discussed in the preceding section, there is a one-to-one correspondence between the shape function $\tilde{\phi}(\tilde{t})$ and its spectrum $\tilde{\phi}''(\tilde{\omega})$. The latter is related to the loss spectrum by $\tilde{\chi}''(\tilde{\omega}) \propto \tilde{\omega}\tilde{\phi}''(\tilde{\omega})$. Hence, the scaling law for $\phi(t)$ is equivalent to one for the loss spectrum. Function $\chi''(\omega)$ is determined by a temperature-independent shape function $\tilde{\chi}$ so that the dependence on the control parameters is merely due to that of the scale τ:

$$\chi''(\omega) = \tilde{\chi}(\omega\tau). \tag{1.9c}$$

If the susceptibility can be described by the Cole–Davidson function (1.5), the scaling law holds if and only if $f \cdot \chi_s$ and β_{CD} are temperature independent.

Validity of Eq. (1.9a) means the following. A representation of the decay functions for various temperatures as function of the rescaled time $\tilde{t} = t/\tau$ leads to a temperature-independent result: $\phi(\tilde{t}\tau) = \tilde{\phi}(\tilde{t})$. All ϕ versus $\tilde{t}$ diagrams superimpose on the shape function $\tilde{\phi}$. Therefore, the scaling law is also referred to as the time-temperature-superposition principle. Similarly, presenting the susceptibility spectra as function of the rescaled frequency $\tilde{\omega} = \omega\tau$, all χ'' versus $\tilde{\omega}$ diagrams collapse on a single one: $\chi''(\tilde{\omega}/\tau) = \tilde{\chi}(\tilde{\omega})$. There holds a frequency-temperature-superposition law.

The lower panel of Fig. 1.5 presents a test of Eq. (1.9a). It exhibits the same decay functions as the upper one, but presented as functions of the rescaled time $\tilde{t} = t/\tau_q$. A scale $\tau = \tau_q$ for the long-time decay process is defined as the time required for the decay to 10% of the initial value: $\phi_q(t = \tau_q) = 0.1$. The functions for the temperatures $T = 0.47$ and 0.48 collapse within the accuracy of the data for $\tilde{t} \geq 10^{-3}$. This means that the scaling law holds for an interval of $\tilde{t}$ larger than 3.5 orders of magnitude. In this case, the shape function $\tilde{\phi}_q(\tilde{t})$ describes the decay from 0.8 to zero.

The discussion of Fig. 1.4 suggests that the scaling refers to the low-frequency-loss-peak dynamics. For shorter times, i.e., for the time interval where there occurs the decay from 1.0 to $f_q = 0.8$, the superposition principle does not hold. The result for $T = 0.49$ also scales on the shape function, albeit only for larger rescaled times $t > \tilde{t}_- = 4 \cdot 10^{-3}$. This trend continues. The decay from $0.9 f_q^c$ to $0.1 f_q^c$ is stretched on a time interval of two orders of magnitude. Upon cooling, the scaling-law regime evolves from large to small rescaled times $\tilde{t}$.

The curve for the lowest measured temperature $T = 0.46$ is plotted as dashed-dotted line, because Aichele and Baschnagel (2001a) want to point out that this result does not fit to the trend exhibited by the other temperatures. For t/τ near 0.03, the $\phi(t)$ data are larger than $\tilde{\phi}(t/\tau)$. This indicates that the scaling-law description becomes invalid at some low temperature.

Figure 1.6 exhibits a test of the superposition principle in the frequency domain. The data refer to the loss-peak spectra of CKN shown in Fig. 1.3. Indeed, if presented as function of $\tilde{\omega} = \omega/\omega_{\max}$, the data collapse on a shape function $\tilde{\chi}(\tilde{\omega})$ as formulated by Eq. (1.9c). The time scale is chosen as $\tau = \tau_{\max} = 1/\omega_{\max}$. The spectra near and above the loss minimum do not obey the scaling law. Therefore, Li *et al.* (1992) did not include them in the figure. In analogy to what was discussed above for data in the time domain, the scaling low evolves from low to high rescaled frequencies upon decreasing the temperature. The interval of rescaled frequencies $\tilde{\omega}$ for the validity of the scaling law extends up to a fraction of $\omega_{\min}/\omega_{\max}$. Upon decreasing T, this interval expands to larger values of $\tilde{\omega}$, since $\omega_{\max}$ decreases stronger with T than does $\omega_{\min}$.

The loss spectrum for the simple relaxation process reads $\chi''(\omega) \propto f\omega\tau/[1 + (\omega\tau)^2]$. It is invariant if $\omega\tau$ is changed to $1/\omega\tau$, i.e., if $\log(\omega\tau)$ is changed to $-\log(\omega\tau)$. This susceptibility peak is symmetric if considered on a logarithmic frequency axis. It has a width at half of its maximum of 1.14 decades; one has to increase the frequency by a factor 13.9 if one wants to scan the spectrum in the interval where it exceeds 50% of its maximum value. This elementary spectrum is added as the inner bell-shaped curve to Fig. 1.6 in order to demonstrate two

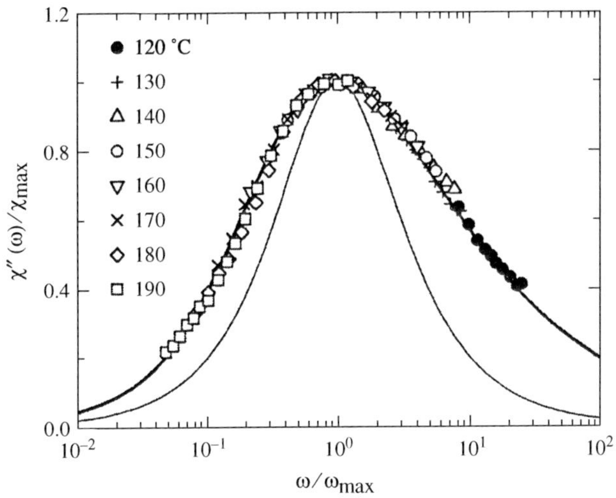

FIG. 1.6. The low-frequency-loss-peak parts of the CKN susceptibility spectra from Fig. 1.3 replotted as $\chi''(\omega)/\chi_{\rm max}$ versus $\omega/\omega_{\rm max}$. Here $\omega_{\rm max}$ denotes the position of the low-frequency maximum of the loss spectrum and $\chi_{\rm max} = \chi''(\omega_{\rm max})$. The solid line through the data shows the spectrum calculated for the Kohlrausch function, Eq. (1.3), with stretching exponent $\beta = 0.55$. The inner bell-shaped curve shows the Lorentzian susceptibility spectrum of a simple relaxation process. Reproduced from Li *et al.* (1992).

important features of glassy dynamics: the measured peak is much broader than suggested by Eq. (1.1a), and it is not symmetric. The loss peak shown for CKN has a width at half maximum of two decades. If one wants to scan the spectrum within the frequency interval where it exceeds 50% of its maximum value, one has to increase the frequency by a factor 100. Even though the loss peak deals only with a part of the glassy-dynamics spectrum, the underlying dynamics is stretched so strongly that the spectrum cannot be viewed adequately on a linear frequency axis. The loss-peak spectrum is asymmetric because the stretching is due to that of its high-frequency wing. The full line through the data is the Kohlrausch spectrum for an exponent $\beta = 0.55$. It demonstrates that Eq. (1.3) accounts for the shape function $\tilde{\chi}(\tilde{\omega})$.

Neutron-scattering spectroscopy plays a distinguished role in the research of condensed matter for two reasons. First, it can probe the dependence of the dynamics on the diameters of the complexes of particles, which are involved in the motion. This is done by measuring the scattering intensity as function of the wave number q of the density fluctuations created by the scattering process. Here, $2\pi/q$ varies on the natural length scale of condensed matter, i.e., it varies, say, between 0.1 and 10 times the particle diameter. Second, the coupling of the neutrons to the particles is very simple and can be quantified by a single number, namely the scattering length. As a result, there is a trivial relation between the scattering intensity and the particle coordinates (Lovesey 1984). The general properties of the measured spectrum $\phi_q''(\omega)$ and of its Fourier transform $\phi_q(t)$, the density-fluctuation-decay function, are discussed in Chapter 3. Neutron-scattering studies of the glassy dynamics of a liquid have first been done by Mezei *et al.* (1987) using a spin-echo spectrometer. An upgrading of this instrument was applied to obtain the data for $\phi_q(t)$ exhibited in Fig. 1.7 for the time interval marked by IN11. The shown decay curves are normalized to $\phi_q(t = 0) = 1$. They are measured for the van der Waals liquid orthoterphenyl (OTP). A time-of-flight spectrometer was used to measure $\phi_q''(\omega)$ on a frequency interval larger than two decades. These data were Fourier-transformed to get the decay curves within the interval marked by IN5.

Figure 1.7 displays dynamics on a time interval, which corresponds closely to the frequency interval displayed in Fig. 1.3. For times smaller than 2 ps, the decay curves depend on temperature only moderately, as expected for a normal liquid. But, contrary to what is expected for normal-liquid behaviour, the $\phi_q(t)$ do not decay to zero for times of order of, say, 10 ps. In agreement with what was discussed in connection with Fig. 1.2, the decay is stretched if t exceeds, say, 2 ps. The figure exhibits glassy dynamics for the three-decade time interval 2 ps $< t <$ 2 ns. Let us consider the data for the two lowest temperatures. These curves can be fitted reasonably with the Kohlrausch function as is shown by the dotted lines. The same $f = 0.65$ and $\beta = 0.57$ can be used, i.e., the curves are

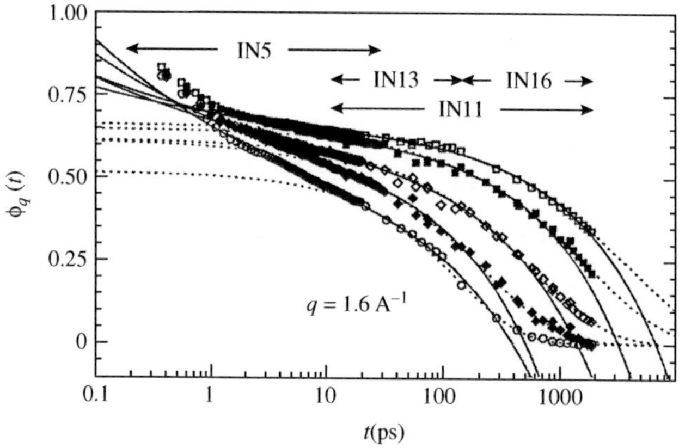

FIG. 1.7. Total-density-fluctuation decay functions $\phi_q(t)$ for wave number $q = 16$ nm^{-1} of deuterated orthoterphenyl (OTP, $T_{\rm m} = 329$ K, $T_g = 243$ K) measured by coherent neutron-scattering spectroscopy and normalized to $\phi_q(t = 0) = 1$. The data refer to temperatures $T/K = 293, 298, 305, 313$, and 320 (from top to bottom). The dotted lines are fits of the long-time decay by Kohlrausch functions. The full lines are fits of the intermediate-time decay by functions discussed in Sec. 6.1.3 (iii). Reproduced from Tölle (2001).

connected by the time-temperature superposition principle. The same holds, if the curves for $T = 305\,\mathrm{K}$ and $T = 313\,\mathrm{K}$ are included; but now one has to choose f somewhat smaller. There is some uncertainty in the choice of f, since the long-time process is not separated well enough from the decay at shorter times.

1.4 Two-step relaxation through a plateau

The results reported above suggest that the essential part of the glassy dynamics deals with a two-step-relaxation process describing the crossing of a plateau of the decay-function versus $\log(t)$ diagrams. The plateau reflects transient arrest of density fluctuations. In this section, it will be shown that the deviations of the fluctuations from the transiently arrested parts occur so that there are no correlations between the variations due to changes of time and the ones due to changes of the positions.

The discussion given above for Figs. 1.4, 1.5, and 1.7 imply that the long-time part of the dynamics, which causes the low-frequency loss peak, deals with the relaxation from some plateau to zero. In Eqs. (1.3), (1.4a), (1.7b), this plateau value is denoted by f. The decay of $\phi(t)/\phi(t = 0)$ below f depends strongly on the temperature as described by the scale τ in the superposition-principle formula (1.9a). This process cannot start for times at the end of the normal-liquid-behaviour interval, since the normal-liquid dynamics depends only weakly on temperature. Evidence was presented that there is a glassy-relaxation process in between the two mentioned processes, namely a decay towards the plateau dominated by a t^{-a} law, Eqs. (1.6a). The stretched part of the glassy dynamics appears as a two-step process. A power-law decay towards the plateau, specified by an exponent a, is the first step. A power-law decay below the plateau, specified by an exponent b, is the second one. A possibility to define a time scale t_{cr} for the merging of the two steps is the one specifying the plateau crossing: $\phi(t_{\mathrm{cr}})/\phi(t = 0) = f$. The discussion of Eqs. (1.7a,b) has shown that the plateau-crossing time t_{cr} is the scale specifying the loss minimum, $t_{\mathrm{cr}} \propto 1/\omega_{\mathrm{min}}$.

The decay of the normalized function $\phi(t)/\phi(t = 0)$ from 0.95 to 0.85, which is exhibited by the dashed curve in Fig. 1.4, requires an increase of t by about a factor 3. A similar statement holds for the elementary relaxation process, Eq. (1.1b). The decay of the dashed curve in Fig. 1.4 from 0.2 to 0.1 requires a time increase of about a factor 2. All three examples demonstrate normal-dynamics behaviour for a decay of 10% of the total one. Quite a different behaviour is shown in Fig. 1.4 for the decay around the plateau $f = 0.75$. The decay from 0.80 to 0.70 requires an increase of t by more than a factor 70. This variation of $\phi(t)$ is stretched so much that it cannot be exhibited adequately on a linear t axis. The relaxation curves for the two lowest temperatures, which are shown in Fig. 1.5 or 1.7, demonstrate that the $\phi(t)$ versus $\log(t)$ curves for $t \sim t_{\mathrm{cr}}$ become flatter if T decreases. Let us define a plateau region $t_- \leq t \leq t_+$ by the request that the decay function deviates from f by less than some positive margin ϵ, $|\phi(t_\pm)/\phi(t = 0) - f| = \epsilon$.

The mentioned figures suggest that $\log(t_+/t_-)$ increases upon lowering T. In this sense, the plateau is defined the better the lower is T.

In order to determine a plateau value f unambiguously, some theory is required. Such theory should justify a formula for the description of the plateau crossing or of the approach towards the plateau. A fit of the data on a large time interval should allow to deduce f. Fits of the second relaxation step by a Kohlrausch function have been used often with this intention. The dotted lines in Fig. 1.7 are examples for such procedure. The fit by the formula (1.7b), which was discussed in connection with Fig. 1.4, is another approach with the same motivation. According to this formula, the essential features of the two-step relaxations around the plateau are the following. For times short on scale $t_{\rm cr}$, the t^{-a} process dominates. If the increase of $t_{\rm cr}$ due to a decrease of T is compensated by a decrease of H in the sense that $A' = Ht_{\rm cr}^a$ varies only weakly with temperature, $\phi(t)$ could be matched with the normal-liquid decay for short times. For times long on scale $t_{\rm cr}$, the $-t^b$ process dominates. The amplitude H and the scale $t_{\rm cr}$ combine to the factor $B' = Ht_{\rm cr}^{-b}$ of the von Schweidler decay. Introducing a time τ by the formula $\tau^{-b} = Ht_{\rm cr}^{-b}$, the large-time asymptote gets the form of Eq. (1.4a). This is the initial part of the second decay process. It is consistent with the superposition principle; and τ denotes the temperature dependent scale. One gets $\tau/t_{\rm cr} \propto 1/H^{1/b}$. Thus, τ increases stronger with decreasing T than does $t_{\rm cr}$. This explains the expansion of the plateau interval upon lowering the temperature. Since $\chi_{\rm min} \propto H$ and $\omega_{\rm max}/\omega_{\rm min} \propto t_{\rm cr}/\tau$, the decrease of both parameters with decreasing T is a necessary implication of the two-power-law scenario. Both features are exhibited in Fig. 1.3.

Let us come back to the decay function for a tagged-particle density fluctuation $\phi_q(t)$, which is normalized to $\phi_q(t = 0) = 1$. Here $q = |\vec{q}|$ is the modulus of the wave vector $\vec{q}$ of the fluctuation. This function is considered for a special values for q in Fig. 1.4. By Fourier transformation from the space of wave vectors $\vec{q}$ in the space of positions $\vec{r}$, one gets a function $\phi(r, t)$. Because of isotropy, it depends on vector $\vec{r}$ merely via its length $r = |\vec{r}|$. A molecular-dynamics simulation yields $\phi(r, t)$ as directly as $\phi_q(t)$. The precise definitions and the properties of the functions $\phi_q(t)$ and $\phi(r, t)$ will be discussed in Sec. 3.2. There, it is shown that $\phi(r, t)$ denotes the probability density for a tagged particle to travel a distance r within the time increase t. The integral $\int_{r_1}^{r_2} 4\pi r^2 \phi(r, t) dr$ is the percentage of the particles to be found at time t within the distances $r_1 \leq r \leq r_2$ from the origin if they started their motion for $t = 0$ at $r = 0$. Hence, this function is useful for getting a picture of the liquid dynamics. In principle, this function can also be measured by incoherent neutron-scattering spectroscopy (Lovesey 1984). The introduction of $\phi(r, t)$ makes it explicit that information on the q-dependence of $\phi_q(t)$ yields information on space-dependent features of the particle motion.

Signorini *et al.* (1990) considered the ratio $R(r, t) = \Delta\phi(r, t)/\Delta\phi(r', t)$ formed with the differences $\Delta\phi(r, t) = \phi(r, t) - \phi(r, t')$. Here, r' is some reference distance; and t' is some reference time chosen from the interval where $\phi_q(t)$ is close to its plateau. These authors showed for their simulation results for a model of the molten salt CKN that $R(r,t)$ is time independent. This holds for times t chosen from that interval where the density-fluctuation-decay functions $\phi_q(t)$ are close to their plateaus. The time-independence of $R(r,t)$ is equivalent to the following form for the probability density

$$\phi(r,t) = F(r) + H(r)G(t). \tag{1.10a}$$

The function $G(t)$ is not unique; it can be modified by adding constants. This can be used to require $G(t_{\mathrm{cr}}) = 0$ for some time t_{cr} chosen within the above mentioned plateau regime. Fourier transformation casts this relation in the equivalent form

$$\phi_q(t) = F_q + H_q G(t). \tag{1.10b}$$

This shows that F_q has the meaning of the plateau for the function $\phi_q(t)$. Here, F_q and H_q are related to $F(r)$ and $H(r)$ by Fourier transformation. Up to some time-independent term $F(r)$, the probability distribution $\phi(r,t)$ factorizes. One factor, $H(r)$, depends on the distance but it is independent of time. The other factor, $G(t)$, depends on the time but it is independent of the distance. The total probability $\phi(r,t)$ consists of an arrested background distribution $F(r)$ and a remainder $\delta\phi(r,t) = \phi(r,t) - F(r)$. The latter does not exhibit any correlation between the variations in time with those in space.

Figure 1.8 demonstrates the factorization property for the A particles of the LJM for an increase of the time by a factor larger than 600. For short times, the particles in a liquid move freely with an averaged velocity v. There is the deterministic correlation between change in time and change in position: $r = vt$. For very long times, the particle motion over long distances is described by the diffusion equation. This yields a stochastic relation between change of time and change of position $r \propto \sqrt{D \cdot t}$, as will be discussed in Chapter 3. Here D denotes the particle diffusivity. In both mentioned limits for t, $\phi(r,t)$ is a Gaussian function where the width of the distribution increases proportional to t or $\sqrt{t}$, respectively. Therefore, within the regime of normal-liquid short-time dynamics and within the long-time regime, $R(r,t)$ depends strongly on t, i.e., the factorization property is not valid. It was shown explicitly for the example discussed in Fig. 1.8 that $R(r,t)$ develops strong time dependencies if t decreases below 3 and if it increases above 1900 (Kob and Andersen 1995a). The factorization property is a characteristic feature of glassy dynamics for the intermediate-time regime. If the temperature decreases, this time interval expands. For the simulation results discussed in connection with Fig. 1.5, the factorization property was demonstrated also in form of Eq. (1.10b) by Aichele and Baschnagel (2001b).

The mentioned diffusion equation is an example for a description of a correlation of density-fluctuation changes due to variations of the time with those due to variations of the position. The sound-wave equation for the fluctuations of the total density is another example for a description of space-time correlations. Beginning with the detection of the roton spectrum of liquid helium, a major part of the neutron-scattering studies of liquids and solids concerns the determination of dispersion laws, more generally, of the changes of the $\phi''_q(\omega)$ versus

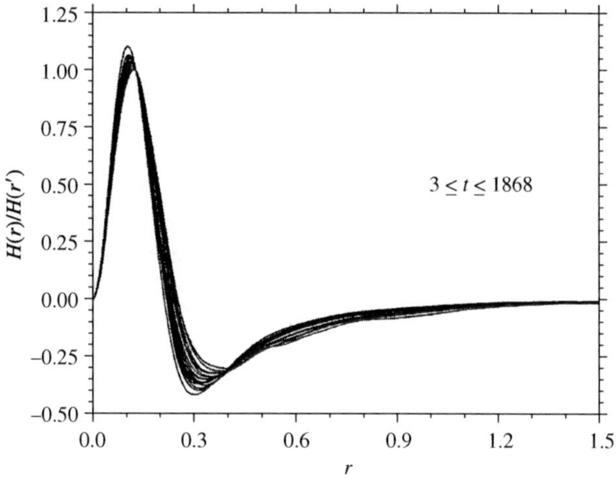

FIG. 1.8. The ratio $\Delta\phi(r,t)/\Delta\phi(r',t) = H(r)/H(r')$ formed with the differences $\Delta\phi(r,t) = \phi(r,t) - \phi(r,t')$ of the probability density $\phi(r,t)$ for the motion of a tagged particle over a distance r within the time increase t. The results are obtained by molecular-dynamics simulation for the A particles of the LJM at temperature $T = 0.466$. The different curves refer to times from the interval $3 \leq t \leq 1868$ chosen roughly with an equal distribution of $\log t$. The reference time is $t' = 3000$ and the reference distance is $r' = 0.13$. The model and the units are defined in connection with Eqs. (1.8a,b). Reproduced from Kob and Andersen (1995a).

ω curves due to changes of q. Dispersion laws formulate stringent correlations between the variations with changes of q and with ω. The factorization property states complete absence of such correlations. Therefore, Eqs. (1.10a,b) formulate a property, which distinguishes glassy dynamics qualitatively from the dynamics of other states of condensed matter. The time interval for this property is the same as that for a stretched two-step decay through a plateau.

1.5 The cage effect

In a liquid, each particle is surrounded by a shell of neighbours, which form a cage. The transient localization of the particles in their cages appears to be the mechanism causing the plateaus of the various decay functions of glassy states. In this section, it shall be demonstrated how the cage effect manifests itself for the mean-squared displacement of a particle.

Let $\vec{r}(t)$ denote the position of a tagged particle in a liquid. This particle moves along a complicated orbit because the neighbouring particles exert complicated forces. The latter fluctuate erratically as function of time. A quantity characterizing the tagged-particle motion statistically is the mean-squared displacement $\delta r^2(t) = \langle[\vec{r}(t) - \vec{r}(t=0)]^2\rangle$. This is the basic concept of A. Einstein's theory of Brownian motion. For the case of a mesoscopic Brownian particle, one can assume that the time scale of interest for the description of $\delta r^2(t)$ is much larger than that for the force fluctuations. Under this time-scale-separation condition, the forces can be treated as stochastic. There holds Einstein's law $\delta r^2(t) = 6D \cdot t$, where D denotes the tagged particle's diffusivity. For the particle motion in a liquid, the time dependence of $\delta r^2(t)$ is more complicated. A detailed discussion of $\delta r^2(t)$ is given in Sec. 3.2. There, it is shown that the cited diffusion law is valid only in the limit of long times. For very short times, on the other hand, one can write $\vec{r}(t) = \vec{r}(t = 0) + \vec{v}t$. Here terms of order t^2 are neglected, and $\vec{v}$ denotes the initial velocity of the particle. Hence, one gets the ballistic-motion law for the short-time asymptote, $\delta r^2(t) = 3v_{th}^2 t^2$, with $v_{th} = \sqrt{k_B T/m}$ abbreviating the thermal velocity of the particle with mass m.

Figure 1.9 exhibits molecular-dynamics-simulation results for $\delta r^2(t)$ for the A-particles of the Lennard-Jones mixture (LJM). A double-logarithmic presentation is chosen so that power-law variation $\delta r^2(t) \propto t^x$ manifests itself by a straight line of slope x. The ballistic law is observed if $\delta r^2(t)/\sigma_{AA}^2 < 0.01$. This means that interaction effects do not influence the averaged displacement seriously if the latter remains shorter than about 10% of the effective particle diameter σ_{AA}. The straight $\log(\delta r^2(t))$ versus $\log(t)$ lines of slope 2 shift down with decreasing temperature since $v_{th}^2 \propto T$. The straight lines of slope 1 for the diffusion asymptote are exhibited for $\delta r^2(t) > \sigma_{AA}^2$. This means that the assumption of stochastic motion becomes valid if the particle has moved a distance longer than its diameter. The diffusion asymptotes shift down with decreasing temperature. The shift becomes very large if T decreases below 1. This reflects the drastic slowing down of the long-time motion upon cooling within the regime of glassy dynamics. The subtleties of the dynamics show up in the regime where $\delta r^2(t)/\sigma_{AA}^2$ increases from about 0.01 to about 1.

The two curves shown in Fig. 1.9 for $T = 4$ and 5 exhibit a dynamics similar to that for normal liquids. There is a crossover from ballistic to diffusive power-law variation for $\delta r^2(t)$ near 0.05 σ_{AA}^2. The $\log(\delta r^2(t))$ versus $\log(t)$ curves are downward bent. In particular, the diffusion asymptote for large times is approached from below. A F(x) versus x curve is called downward bent or upward bent if the second derivative F"(x) is negative or positive, respectively. The decrease of $\delta r^2(t)$ with decreasing temperature in the diffusion regime is only a bit stronger than that in the ballistic one. The major part of the temperature dependence of D is caused by the trivial change of the natural scale v_{th} for the velocities.

A more subtle crossover in form of a change from downward bent to upward bent behaviour of the $\log(\delta r^2(t))$ versus $\log(t)$ curve is shown for $T = 1$. This curve exhibits an inflection point. In particular, all curves for $T \leq 1$ approach the diffusion asymptote for large times from above. These features are precursors of the development of a two-step scenario for the crossing of a plateau of value δr_p^2 near $0.03\sigma_{AA}^2$. This scenario is exhibited fully for temperatures $T = 0.6$ or lower. For $T = 0.466$, $\delta r^2(t)$ stays close to the plateau for the two-orders-of-magnitude time interval $3 \leq t \leq 300$. This feature of the glassy dynamics is an extreme manifestation of the cage-effect. The neighbors of the tagged particle form a

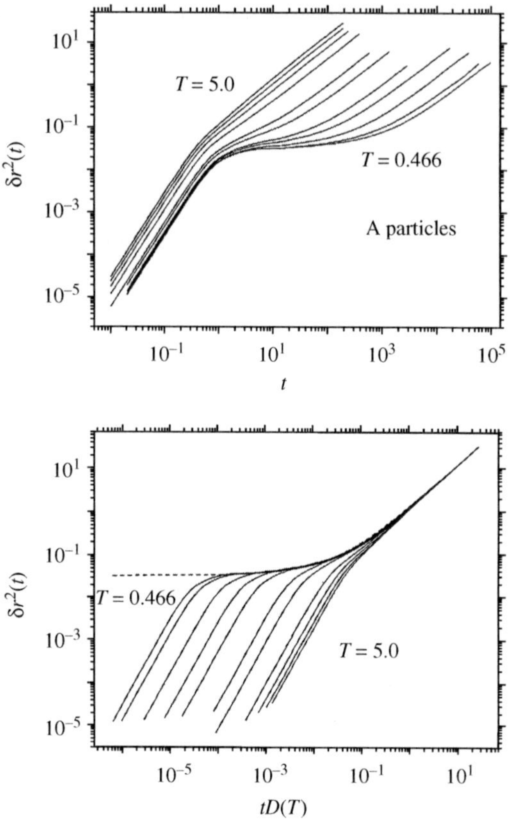

FIG. 1.9. The upper panel shows mean-squared displacements $\delta r^2(t)$ obtained by molecular-dynamics simulations for the A particles of the LJM for temperatures $T = 5.0$, 4.0, 3.0, 2.0, 1.0, 0.8, 0.6, 0.55, 0.50, 0.475, and 0.466 (from left to right). The definitions of the model and of the units are discussed in connection with Eqs. (1.8a–c). The lower panel shows the same data as the upper one (from right to left), but for each temperature the time t is rescaled to $\tilde{t} = t \cdot D(T)$ with $D(T)$ denoting the A-particle diffusivity at temperature T. The dashed line shows a fit of the long-time parts by $r_p^2 + A\tilde{t}^b + 6\tilde{t}, b = 0.48$. Reproduced from Kob and Andersen (1995a).

cage, which localizes the particle for long periods of time. The time required for $\delta r^2(t)$ to leave the transient cage increases strongly upon cooling.

The lower panel of Fig. 1.9 exhibits the same data as the upper one, but presented as function of the rescaled time $\tilde{t} = t \cdot D(T)$. Here, $D(T)$ denotes the diffusivity for the respective temperature T. It will be explained in Sec. 3.2 that the strong T-dependence of $D(T)$ is caused by that of a characteristic time scale τ_D for the long-time dynamics of the tagged-particle-density-decay function, $D \propto 1/\tau_D$. Hence, the collapse of the data for $T = 0.466$ and $T = 0.475$ for $\tilde{t} \geq 2 \times 10^{-4}$ expresses the time-temperature superposition principle for the mean-squared displacement. The figure demonstrates the evolution of this scaling law from large to small values of the rescaled time $\tilde{t}$ upon decreasing the temperature in complete analogy to what is demonstrated in Fig. 1.5 for a different function. The dashed line exhibits the function $\delta\tilde{r}^2(\tilde{t}) = r_p^2 + A\tilde{t}^b + 6\tilde{t}$, $b = 0.48$. Within the accuracy of the data, this function is the T-independent shape function entering the scaling law $\delta r^2(t) = \delta\tilde{r}^2(\tilde{t})$. The first two contributions in this three-term fit formula express von Schweidler's law for the mean-squared displacement. The figure suggests the following: plateau formation as result of the cage effect, the cage leaving according to von Schweidler's law, and the crossover to the stochastic dynamics for the longest times are the phenomena dealt with by the second relaxation step of the glassy dynamics of the LJM.

The importance of the cage effect can be appreciated already by a look at a typical configuration of the particles in a liquid. For the sake of simplicity, a two-dimensional analogue of a hard-sphere system is chosen to produce Fig. 1.10. It deals with a liquid of hard disks moving in a plane with a density close to that where nucleation phenomena would alter the equilibrium drastically. The distribution shows a representative part of one generated by a Monte Carlo

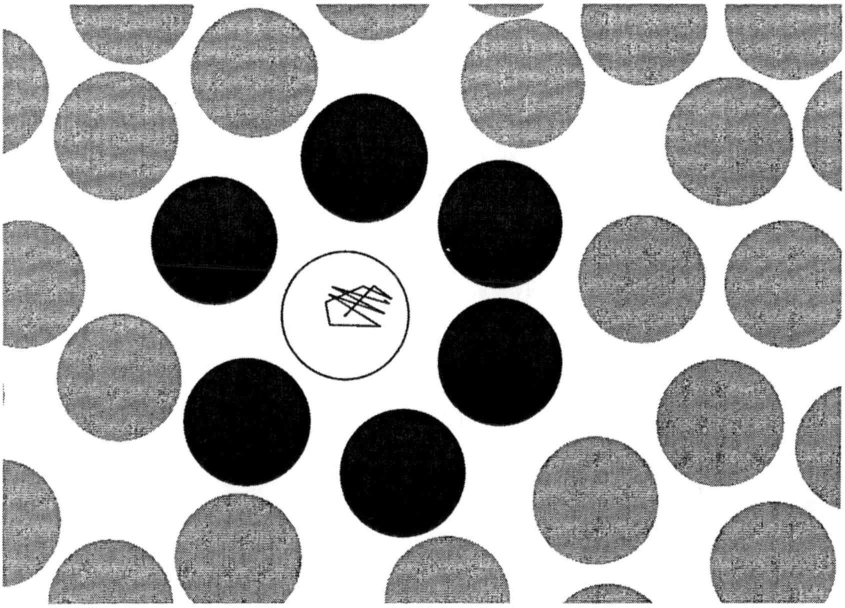

FIG. 1.10. A typical part of a configuration of hard disks filling the plane with a packing fraction φ slightly below 0.69. The open circle marks a tagged particle and the six black ones are its nearest neighbours. The polygon is an orbit of the tagged-particle's centre up to the tenth collision event. It starts with some typical direction and is calculated under the assumption that the neighbours are fixed in the plane (Voigtmann 2003a).

algorithm. This is a numerical procedure to produce configurations so that the averages approach the ones prescribed by Boltzmann's statistics. A tagged particle is shown as open circle. It is surrounded by its six nearest neighbours, which are marked in black. If one assumes these neighbours to be fixed, the tagged particle would be localized. It could merely rattle in a cage. A representative orbit for the centre of such a caged particle is shown in the figure. It extends over 10 collision events and demonstrates that $\mid \vec{r}(t) - \vec{r}(t=0) \mid$ is bounded by about 30% of the disk diameter. This number is of the order as $\sqrt{\delta r_p^2}$ deduced from the plateau in Fig. 1.9, if one identifies the disk diameter with the Lennard-Jones potential's length parameter σ_{AA}. For a simple liquid, the motion of a particle over distances as large as its diameter requires that one of its neighbours moves over a distance of similar size. This implies that the neighbours of the neighbour move. The long-distance motion is a cooperative phenomenon with several particles involved. Function $\delta r^2(t)$ increases with t much more slowly than expected for ballistic motion. Figure 1.9 demonstrates that the slowing-down of the mean-squared displacement becomes dramatic if the temperature decreases. In a LJM, this temperature decrease is to a large extent equivalent to an increase of the effective particle diameter for the collision. For $T = 0.466$, hundreds of collisions occur, before cooperative motions destroy the transient localization.

The above cited molecular-dynamics-simulation studies have demonstrated that the glassy dynamics of the LJM and of the polymer-melt model share two facets. The first one concerns the evolution of the second relaxation step for the density-fluctuation-decay functions and the form of the shape function in the time-temperature superposition principle. The second feature concerns a factorization property for the near-plateau dynamics. Let us compare now the mean-squared displacements for the two systems with the intention to document a pronounced feature of the second relaxation step of the decamer dynamics, which has no analogue for the LJM. Mean-squared displacements for a tagged particle in the centre of the decamer chains, $\delta r^2(t) = g_1(t)$, are shown as full lines in Fig. 1.11. They are exhibited as function of the rescaled time $\tilde{t} = D \cdot t$. As before, D denotes the diffusivity of the tagged particle. The inset shows from right to left a sequence of results calculated for falling temperature. It exhibits the evolution of the scaling-law description, $g_1(t) = \delta\tilde{r}^2(\tilde{t})$, of the second relaxation process of a two-step scenario as explained in the lower panel of Fig. 1.9. The full line in the main frame shows the result for the lowest temperature studied. A dashed line $6r_{\text{sc}}^2 + A(Dt)^b$, $b = 0.75$ is added in order to demonstrate that $g_1(t)$ increases above the plateau $\delta r_p^2 = 6r_{\text{sc}}^2 \approx 0.054$ according to a von Schweidler law. The result for $\delta r^2(t)$ from Fig. 1.9 for the lowest temperature for the LJM is added to the figure as dashed-dotted line. Contrary to the $\delta r^2(t)$ for the mixture, the mean-squared displacement for the decamer does not show a direct crossover from von Schweidler-law behaviour to diffusion behaviour. Rather, the molecular liquid exhibits a sublinear power-law increase, $g_1(t) \propto t^x$, specified by an exponent $x = 0.63$. This power-law interval begins at the end of the von

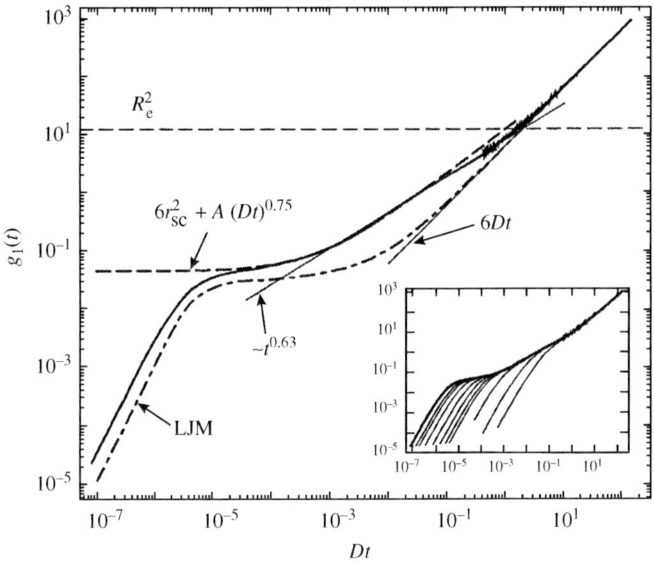

FIG. 1.11. Molecular-dynamics-simulation results for the mean-squared displacement $\delta r^2(t) = g_1(t)$ for an inner monomer of the $T = 0.480$-dense-polymer-melt model defined in connection with Fig. 1.5 (full line) and for an A-particle of the $T = 0.466$-LJM defined in connection with Fig. 1.4 (dashed-dotted line). The inset exhibits results for the polymer model for the temperatures $T = 0.48$, 0.49, 0.50, 0.52, 0.55, 0.60, 0.65, 0.70, 1.0, 2.0, and 4.0 (from left to right). The horizontal axis present the rescaled times $\tilde{t} = D \cdot t$, where D denotes the particle diffusivity at the respective temperature. The heavy dashed line exhibits a fit of the above-plateau increase by the von Schweidler-law part of Eq. (6.123): $\delta r_s^2(t) = 6[r_{sc}^2 + h_{\text{MSD}}\tilde{t}^b]$. The quantity R_e^2 is the averaged end-to-end distance squared of the decamer chains. Reproduced from Baschnagel *et al.* (2000).

Schweidler-law regime and it ends at the beginning of the diffusion regime. The power-law variation is part of the scaling-law regime. This implies that the exponent x does not depend on the control parameter T for the thermodynamic state. The power law-part $\tilde{t}^x$ starts at $\tilde{t} = 2 \cdot 10^{-3}$; and it deals with an increase of $\tilde{t}$ by nearly three orders of magnitude.

The Rouse theory of polymer dynamics describes the motion of a single long chain of monomers in a viscous liquid. The Newtonian friction experienced by the monomers is assumed to be so large that all intramolecular vibrations are overdamped. Hence, one considers a system of coupled simple relaxators, which are placed on a one-dimensional array. The resulting cooperative motions of the molecule yields a mean-squared displacement exhibiting a sublinear long-time asymptote $g_1(t) \propto t^{1/2}$ (Doi and Edwards (1986), Sec. 6.4.4). The new facet of glassy dynamics demonstrated by Fig. 1.11 is the following. Within the dense decamer melt, the mentioned slow intramolecular motions of the large molecules become a part of the second step of the two-step scenario for the glassy dynamics. Their time scale τ_D is that of the diffusivity; and τ_D increases strongly upon cooling because of the increasing importance of the cage effect for the monomer motion. The sublinear power-law variation of the mean-squared displacement becomes a feature of the temperature-independent shape function entering the scaling law for the motion at long times. The power-law exponent 1/2 of the Rouse theory is changed to some larger number x.

There is the possibility of a coherent motion of a monomer along the decamer chain together with its two neighbours. Therefore, in a dense liquid, transient localization of a particle in a cage is less effective than it would be in the absence of the intramolecular binding. As a result, the localization length r_p in the transient cage is larger and the time interval for the transient trapping is shorter for the polymer model than for the LJM. The crossover to the stochastic diffusion dynamics sets in for the LJM as soon as $\delta r^2(t)$ reaches the particle diameter. This is not the case for the decamer because of the binding-induced coherence effects. There enters a second length scale R_e besides the localization length in the cage. The quantity R_e^2 is the average of the square of the vector connecting the two ends of the chain molecule. The mean-squared displacement $\delta r^2(t)$ has to exceed R_e^2 before the coherence is destroyed. Since R_e^2 exceeds r_p^2 considerably, there is such huge $\tilde{t}$ interval between the end of the von Schweidler-law process and the beginning of the diffusion.

1.6 Crossover phenomena

In this section, data shall be discussed, which show that the evolution of the glassy dynamics upon cooling or compression exhibits a crossover phenomenon near some characteristic temperature T_c or density n_c, respectively. For $T > T_c$ or $n < n_c$, respectively, there holds the superposition principle in a good approximation. Thus, the plateau height and the stretching of the low-frequency loss-peak dynamics change only weakly with changes of the control parameter T and n, if at all. However, for T decreasing below T_c or n increasing above n_c, the superposition principle is violated considerably. The plateau height may rise steeply or the stretching may increase noteworthily with increasing $\epsilon = (T_c - T)/T_c$ or $\epsilon = (n - n_c)/n_c$.

Let us relate the plateau height f of a normalized decay function $\phi(t), \phi(t = 0) = 1$, to the spectra for the process. To proceed, the description of the second process shall be extrapolated as $\phi^{ex}(t)$ to times smaller than $\tau' = 1/\omega_{min}$. This can be done, e.g., with Eq. (1.3) or (1.4a) so that $\phi^{ex}(t = 0) = f$. Using the representation of $\phi(t)$ and $\phi^{ex}(t)$ as Fourier transforms of their spectra $\phi''(\omega)$ and $\phi^{ex\prime\prime}(\omega)$, respectively, one gets $1 = (2/\pi) \int_0^\infty \phi''(\omega) d\omega$ and $f = (2/\pi) \int_0^\infty \phi^{ex\prime\prime}(\omega) d\omega$. Since $\chi''(\omega) = \omega\phi''(\omega)$, one gets equivalently $1 = (2/\pi) \int_{-\infty}^{\infty} \chi''(\omega) dx$, $f = (2/\pi) \int_{-\infty}^{\infty} \chi^{ex\prime\prime}(\omega) dx$ with $x = \ln \omega$. For $\omega \ll \omega_{min}$, there holds $\chi^{ex\prime\prime}(\omega) = \chi''(\omega)$. For $\omega \gg \omega_{max}$, $\chi^{ex\prime\prime}(\omega)$ decays exponentially as function of x : $\chi^{ex\prime\prime}(\omega) \propto$

$1/\omega^\alpha = \exp[-\alpha x]$. Here $\alpha = \beta$ or $\alpha = b$, depending on the formula used for the extrapolation. Therefore, the contribution to f given by $\int_{\ln\omega_{\rm min}}^{\infty} \chi^{\rm ex\prime\prime}(\omega)dx = \int_{\omega_{\rm min}}^{\infty} \phi^{\rm ex\prime\prime}(\omega)d\omega$ can be neglected. Hence, one can write

$$f = \int_0^{\omega_{\rm min}} \phi''(\omega)d\omega \Big/ \int_0^\infty \phi''(\omega)d\omega, \tag{1.11a}$$

$$= \int_{-\infty}^{\ln\omega_{\rm min}} \chi''(\omega)d\ln\omega \Big/ \int_{-\infty}^{\infty} \chi''(\omega)d\ln\omega. \tag{1.11b}$$

The plateau value f is the weight of the loss-peak spectrum, considered as function of $x = \ln\omega$, relative to the weight of the total spectrum. It is the weight of the fluctuation spectrum integrated over the frequencies in the loss-peak region relative to the total integral of that spectrum. There is an inherent uncertainty in the definition of f. The extrapolation procedure, which lead to the cutoff $\omega_{\rm min}$, is to some extend arbitrary. But, the uncertainty is the smaller the better the separation of the two relaxation steps. One can characterize f also by the spectra outside the loss-peak region:

$$1 - f = \int_{\omega_{\rm min}}^{\infty} \phi''(\omega)d\omega \Big/ \int_0^\infty \phi''(\omega)d\omega, \tag{1.11c}$$

$$= \int_{\ln\omega_{\rm min}}^{\infty} \chi''(\omega)d\ln\omega \Big/ \int_{-\infty}^{\infty} \chi''(\omega)d\ln\omega. \tag{1.11d}$$

Figure 1.12 exhibits plateau values measured for the above discussed molten salt CKN. The wave numbers q are many orders of magnitude smaller than the natural wave-vector scale for the microscopic dynamics. Therefore, no dependence on q is detected and the results represent the limit $f_{q=0}$. The data are obtained by stimulated-light-scattering spectroscopy. A pair of intense picosecond laser pulses with wave vectors $\vec{q}_1$ and $\vec{q}_2$ are crossed in the probe. By electrostriction and heating, they produce a density fluctuation with wave vector $\vec{q} = \vec{q}_1 - \vec{q}_2$. A probe beam measures the intensity $I(t)$ of this density fluctuation of wave vector $\vec{q}$ as function of the delay time t. The decay function $\phi_q(t) = I(t)/I(t=0)$ depends on the wave number $q = |\vec{q}|$ only because of isotropy of the system. The result is analysed in terms of the elastic modulus, which is proportional to a stress susceptibility. This concept is explained in Sec. 3.3 within the generalized-hydrodynamics theory. The plateau of the stress-decay function, which is easily extracted from the modulus, determines $f_{q=0}$. The $f_{q=0}$ versus T plot exhibits a crossover within the interval $375 < T/\mathrm{K} < 385$. For $T > 385\,\mathrm{K}$, $f_{q=0}$ is temperature independent as expected from the validity of the superposition principle. However, for $T < 375\,\mathrm{K}$, the superposition principle is violated.

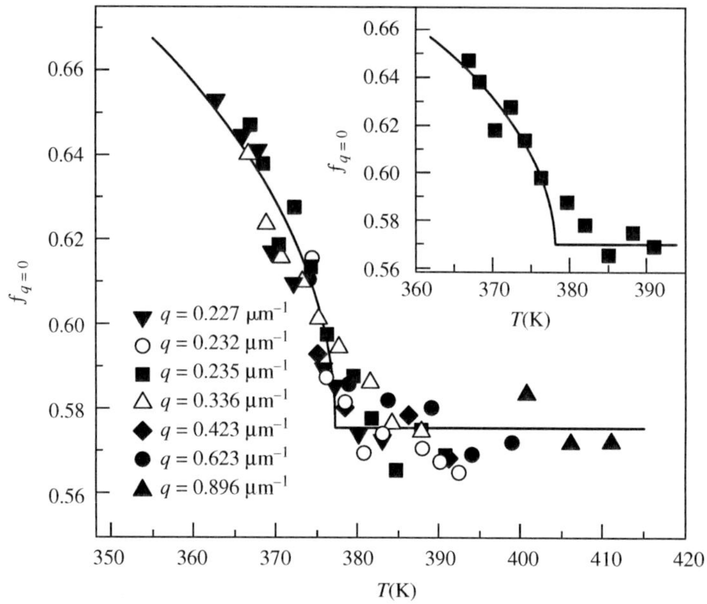

FIG. 1.12. Plateau value $f_{q=0}$ of CKN as function of temperature T measured by impulsive-stimulated-light-scattering spectroscopy for the small wave numbers q indicated. The full lines exhibit the functions: $f_{q=0}-$ const. for $T > T_c$, $f_{q=0}$ – const. $\propto \sqrt{T_c - T}$ for $T < T_c$, where $T_c \approx 378\,\mathrm{K}$. The inset shows the results for the representative wave number $q = 0.235\,\mu\mathrm{m}^{-1}$. Reproduced from Yang and Nelson (1996).

The plateau value increases by about 15% upon decreasing T. This increase can be described by a square-root law: $f(T) - f(T = T_c) \propto \sqrt{T_c - T}, T < T_c$, as is shown by the lines. Such fit suggests the value for the crossover temperature $T_c = (378\pm2)\,\mathrm{K}$ (Yang and Nelson 1996). Measurements of the dielectric modulus of CKN demonstrate a similar cusp anomaly located at a crossover temperature T_c, which is near the cited value (Pimenov *et al.* 1996). These observations suggest that the characteristic temperature T_c is a property of the thermodynamic state of the liquid rather than one of the probing variable. This suggestion is supported by extensive neutron-scattering studies of OTP determining f_q via Eq. (1.11a). Data for this system have been discussed above in connection with Fig. 1.7. The measurement of f_q was done for incoherent scattering (Petry *et al.* 1991) and coherent one (Tölle *et al.* 1997). In both experiments, a considerable range of wave numbers q was examined probing the dynamics on the length scale of the interparticle distances. The data are consistent with the mentioned square-root anomaly for a common $T_c \approx 290\,\mathrm{K}$. Detection of the square-root cusp was reported first by Frick *et al.* (1988) for incoherent neutron-scattering data obtained for a dense polymer melt.

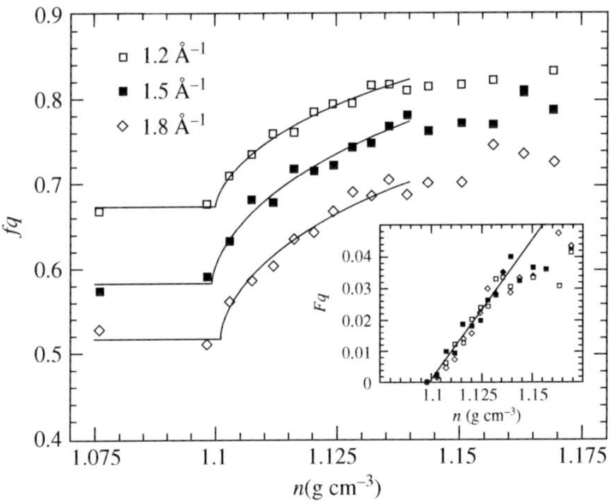

FIG. 1.13. Plateau values f_q for three wave numbers q as function of density n measured by incoherent-neutron-scattering spectroscopy for OTP at temperature $T = 301$ K. The lines for $n \geq n_c = 1.100$ g/cm^3 exhibit fits according to $f_q(n) - f_q(n_c) = \bar{h}_q\sqrt{(n - n_c)}$. The inset shows $F_q = [(f_q(n) - f_q(n_c))/\bar{h}_q]^2$ versus n. Reproduced from Tölle (2001).

The evolution of the glassy dynamics within the GHz band upon compression of the system for a fixed temperature was studied for the two van der Waals liquids isopropylbenzene (Li *et al.* 1995) and orthoterphenyl (OTP) (Tölle *et al.* 1998) using depolarized light scattering and incoherent-neutron-scattering spectroscopy, respectively. Qualitatively, the scenario observed for increasing the density n is the same as described above for decreasing the temperature T. Figure 1.13 shows the plateau values f_q of OTP measured for three wave numbers for T about 10 K larger than the above mentioned crossover value T_c at ambient pressure. Increasing the pressure to about 40 MPa, the density increases by about 2%. The slowest relaxation process obeys the superposition principle, i.e., f_q is n-independent. Increasing n above a crossover density $n_c = 1.100\,\mathrm{g/cm^3}$, f_q increases. An increase of n by 2% leads to an increase of f_q by 20% or more, depending on q. The value n_c does not depend on the wave number. The increase can be described by the square-root law

$$f_q - f_q^c = \tilde{h}_q\sqrt{\epsilon}, \tag{1.12}$$

where $\epsilon = (n - n_c)/n_c$. This is shown by the full lines in Fig. 1.13, in particular by the diagram in the inset.

It is difficult to measure the cusp anomaly. It may deal with a change of only some percent of f. This parameter f has to be extracted from data for spectra, which vary over several orders of magnitude within the temperature range of interest. Usually, one does not measure the normalized quantities. For example, the cross-section for the above-mentioned coherent neutron scattering is given by the dynamical structure factor $S_q(\omega) = S_q\phi_q''(\omega)$. The normalization S_q is the static structure factor, a quantity discussed in Sec. 3.3. The normalization constant may vary with T or n, and this variation can overwhelm that of f.

Applying quasi-optical techniques, A. Loidl and collaborators have measured the dielectric 'constant', $\epsilon(\nu) = \epsilon'(\nu) + i\epsilon''(\nu)$, of several liquids for frequencies ν varying within the complete GHz band. Their data exhibit the evolution of glassy dynamics upon cooling as it is probed by the response of the system's dipole moment. Combining such results with those obtained with the classical instrumentation, they documented the dynamics for frequency variations over more than 17 orders of magnitude. Figure 1.14 exhibits results for glycerol. This is a liquid where hydrogen-bond formation is a crucial part of the interaction in addition to the short-ranged repulsion between the molecules. The static dielectric constant, $\epsilon_0 = \epsilon(\nu \to 0) = \epsilon'(\nu \to 0)$, of glycerol is very large and it increases upon cooling by a factor more than 3, as is shown in the lower panel. Let us express the dielectric constant in terms of a dynamic polarizability $\chi(\omega) : \epsilon(\nu) = \epsilon_\infty + 4\pi\chi(\omega), \omega = 2\pi\nu$. Here, ϵ_∞ combines the vacuum value unity with the small contributions due to electronic degrees of freedom. The polarizability is a dynamical susceptibility, $\chi(\omega) = \chi'(\omega) + i\chi''(\omega)$, formed with the dipole moment. Its real part $\chi'(\omega)$, which describes reactive response, is connected with the loss spectrum $\chi''(\omega)$ by a spectral representation. These general properties of susceptibilities are discussed in Chapter 2. The mentioned increase of ϵ_0 is due to that of the static susceptibility $\chi_s = \chi(\omega = 0) = \chi'(\omega = 0)$. The latter varies according to a Curie–Weiss law, $\chi_s \propto 1/[T - T_0], T_0 \approx 100\,\mathrm{K}$, as is shown in the right inset of the lower panel.

The full lines in Fig. 1.14 show simultaneous fits of $\epsilon''(\nu)$ and $\epsilon'(\nu)$ within the low-frequency-loss-peak region by the Cole–Davidson expression for $\chi(\omega)$. The fit describes $\epsilon'(\nu)$ very well. It describes the loss spectra for $\epsilon''(\nu) > 1$, i.e., the weight factor $\chi_s f$ accounts for not less than 95% of the spectral weight. The shown increase of the peak weight with decreasing T, and the corresponding increase of the flat low-frequency parts of the ϵ' versus $\log \nu$ curves, is due to the increase of the static susceptibility χ_s. The variation of the peak weight of the loss spectrum disappears, if one considers a normalized spectrum $\epsilon''(\nu)/\chi_s$. A possible anomaly of the temperature variation of the large f disappears in the noise of the data. This was shown explicitly by Adichtchev *et al.* (2003), who analysed also the loss spectrum for frequencies above the loss minimum. Using Eq. (1.11d), they determined $(1 - f)$. Thereby, the cusp anomaly has been detected and the crossover temperature for glycerol has been identified: $T_c \approx 290\,\mathrm{K}$.

The data analysis of Schneider *et al.* (1998) unfolds another facet of the crossover at T_c. It is exhibited by the left inset of the lower panel. For $T > T_c$, the stretching exponent β_{CD} has a nearly temperature independent value of about 0.7. This implies the validity of the scaling law for the normalized low-frequency loss-peak dynamics. For $T < T_c, \beta_{\mathrm{CD}}$ decreases almost linearly with T, i.e., the scaling law is invalid. Indeed, the upper panel shows that the widths

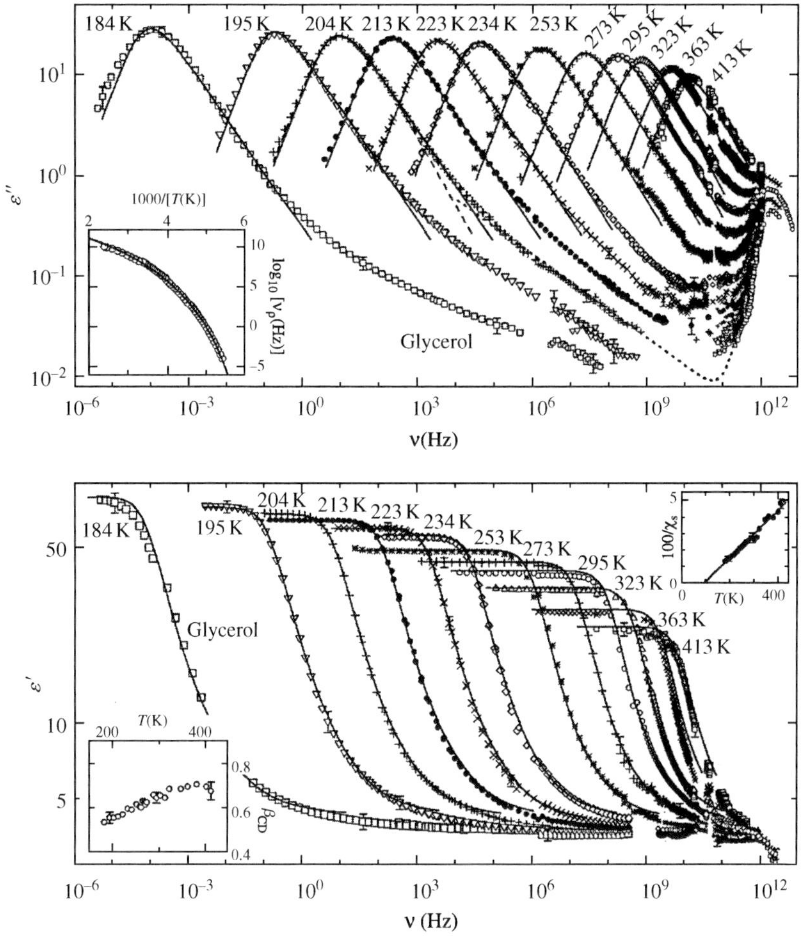

FIG. 1.14. Absorptive part ϵ'' and reactive part ϵ' of the dielectric constant as function of frequency $\nu = \omega/(2\pi)$ for various temperatures T measured for glycerol ($T_{\rm m} = 291$ K, $T_g = 185$ K). The full lines are fits by the Cole–Davidson function for the susceptibility, specified by the stretching exponent β_{CD}, Eq. (1.5). The dashed line in the upper panel exhibits a fit of the 204 K spectrum by the one of the Kohlrausch process, Eq. (1.3), with the stretching exponent $\beta = 0.68$. The inset in the upper panel shows the loss peak position ν_p as a function of the reciprocal temperature and its fit by $\nu_p = 5.6 \times 10^{13} \times \exp[-2160\ \text{K}/(T - 131\ \text{K})]$ Hz. The right inset in the lower panel shows the inverse static susceptibility together with its fit by a Curie–Weiss law: $1/\chi_s \propto (T - T_0)$. The left inset shows the Cole–Davidson exponent $\beta_{\rm CD}$ as function of temperature. Reproduced from Schneider *et al.* (1998).

of the loss peaks at half-the-maximum values increases if T decreases below 300 K. For example: the von Schweidler decay describes the 204 K spectrum for $20 < \nu/\text{Hz} < 2000$ with an exponent $b \sim 0.5$. This b is considerably smaller than the exponent 0.7 describing the analogous data for $T \geq 295$ K. Extrapolating the β_{CD} versus T behaviour from the $T < T_c$ regime to, e.g., $T = 413$ K, one would underestimate the stretching of the spectrum at this temperature. Similarly, extrapolating the $\beta_{\text{CD}} \approx 0.7$ behaviour from the $T > T_c$ regime to $T = 204$ K, one underestimates the loss-peak stretching at this temperature. Shifting the control parameter T through the value T_c, there occurs a qualitative change of the glassy dynamics.

The dielectric-loss spectra of glycerol for T considerably below T_c show another fascinating feature. There appears a second power-law part on the high-frequency wing of the loss peak. For example: the 204 K data can be described by $\chi'' \propto 1/\omega^{b'}$ for a three-decade frequency interval $\nu > 10^6$ Hz. Here, the exponent $b' \approx 0.2$ is smaller than the von Schweidler exponent $b \approx 0.5$ for that temperature. The T-dependence of the stretching for temperatures near but below T_c and the excess wings for frequencies around 1 GHz have first been detected for glycerol and for a number of van der Waals systems by Dixon *et al.* (1990).

Some digression shall be added concerning the dynamics on macroscopic time scales as exhibited in Fig. 1.14 for $T \leq 204$ K. The scale for the long-time relaxation can be quantified as $\tau = 1/\nu_p$, where ν_p denotes the loss-peak position. This scale increases by almost a factor 100 if the temperature decreases from 204 K to 195 K. A further decrease of the temperature T to 184 K implies a further increase of τ by about a factor 1000 so that the relaxation time τ exceeds two hours. The inset in the upper panel shows that this accelerated increase of τ can be fitted by $\tau \propto \exp[E_0 K/(T - 131\,\text{K})]$. Extrapolating the fit function for τ to temperatures below, say, 165 K, the relaxation time reaches geological scales. In this case, the dielectric relaxation can be ignored in all laboratory experiments. A measurement yields $\epsilon(\omega) = \epsilon_\infty$ as if the dipole dynamics was frozen. Such scenario is observed for other probing variables as well, in particular for the shear dynamics. For a liquid, the shear dynamics is quantified by the viscosity η. This is a transport coefficient whose strong increase with decreasing temperature is due to the increase of a shear-stress relaxation time τ. Details are discussed in Sec. 3.3. Probed on a time scale larger than τ, shear deformations of the liquid relax to zero. But, probed on a scale smaller than τ, relaxation of shear cannot be detected and the system behaves like an isotropic solid, i.e., it is a glass. Let $T(t)$ denote the system's temperature as function of the time t in a cooling experiment. Cooling with a rate $\gamma = [\partial T(t)/\partial t]/T$, a crossover from a liquid-like behaviour to an amorphous-solid-like behaviour occurs at a temperature T^*, obeying $\gamma = 1/\tau(T^*)$. This T^* depends only weakly on γ since τ depends so strongly on T. The value T^* depends on the probing variable and on the details of the cooling procedure. For a typical cooling experiment with γ of a fraction of a degree per second, $\tau(T^*)$ is of the order of 10^3 to 10^5 seconds. Cooling to below T^*, one produces a glass. This is an amorphous solid whose non-equilibrium state depends on the history of its preparation. It is the convention to define the glass-transformation temperature T_g as that where the viscosity reaches the value 10^{12} Pas. For glycerol, $T_g = 185\,\text{K}$.

The data for glycerol are typical for conventional liquids in the sense that T_g is smaller than T_c. For T near T_c, the time scale for the slowest relaxation process is about four to five orders of magnitude larger than the time scale for intermolecular vibrations. In this book, the evolution of glassy dynamics is of interest, which is demonstrated in the diagrams preceding Fig. 1.14. Changes of the time scale exceeding t_{mic} by, say, eight or more orders of magnitude shall not be discussed in the following text. Since the time scale for T_g phenomena exceeds the natural time scale for liquid dynamics by about 15 orders of magnitude, the discussions in this book are not relevant for a description of the conventional liquid–glass transformation. Experiments on the liquid–glass transformation for colloidal suspensions and their theoretical explanations will be discussed, however.

1.7 Hard-sphere systems: the paradigms

Systems of hard spheres can be prepared as colloidal suspensions. Photon-correlation spectroscopy can be used to measure the decay function for solute-density fluctuations $\phi_q(t)$. In this section, data shall be considered, which exhibit the evolution of the glassy dynamics of hard-sphere colloids upon increasing the particle-number density ρ. The results are similar to what was discussed above for molecular liquids. The crossover at a critical density deals with a transition from the dynamics of a liquid for $\rho < \rho_c$ to the dynamics of an amorphous solid for $\rho > \rho_c$.

Colloids are systems of mesoscopic particles suspended in some liquid. The number density ρ of the solute particles can be varied in wide limits. If ρ is of the order $1/d^3$, where d characterizes the particle diameter, the colloidal suspension is an example for condensed matter. If d is chosen of the order of the wavelength of light, photon-correlation spectroscopy can be used to measure the above mentioned decay functions, where q is of the order $2\pi/d$ or smaller. A single colloid particle performs Brownian motion quantified by its diffusivity D_0. The natural time scale is the Brownian time $\tau_b = d^2/(24D_0)$. It is the time, for which Einstein's law, $\delta r^2(t) = 6D_0 t$, predicts an increase of the mean-squared displacement to $(d/2)^2$. For $d \sim 500$ nm, τ_b is of the order of 30 milliseconds. Photon correlations can be measured for delay times t up to about one hour. Consequently, one can explore slow dynamics for times exceeding the natural scale by up to five orders of magnitude. There is a huge separation of the time scale and length scale relevant for the description of the solute particles from those for the description of the solvent molecules. Therefore, the solvent degrees of freedom can be eliminated and the liquid can be treated as an inert back ground for the discussion of the solute. The solvent determines the constant D_0 and influences the effective interaction between the solute particles. There appear velocity-dependent forces between the particles moving in a liquid, which are called hydrodynamic interactions.These do not influence the equilibrium structure, but they enter the equations of motion for the solute.

Conceptually, the simplest colloidal suspension consists of particles whose interaction potential is close to that of hard spheres. This means that there is no interaction potential between particles if their distance exceeds the sum of their radii; and if the distance is shorter, the potential is practically infinite. If all particles have the same diameter d, there is only one control parameter for the thermodynamics, namely the particle-number density ρ. Usually, one characterizes the state by the packing fraction $\varphi = \pi\rho d^3/6$, i.e., by the volume filled by the spheres relative to the available volume for the system. The temperature is irrelevant for the description, it enters merely via the solvent parameter D_0. The upper limit for φ is the value 0.740... for close packing. The system has a freezing point $\varphi_F \approx 0.494$ and a melting point $\varphi_M \approx 0.545$. If φ increases through φ_F, a first order phase transition occurs to a face-centred cubic crystal. For $\varphi_F < \varphi < \varphi_M$, the stable state is a mixture of liquid and crystal, with the volume of the latter increasing with increasing φ. For $\varphi_M < \varphi$, the crystalline solid is the stable phase. A strictly one-component hard-sphere colloid cannot be prepared. The best one can achieve is the preparation of a sample of spheres with a mean value d for the diameters whose relative diameter-distribution width p is small. The parameter p is called polydispersity. If p for such quasi-monodisperse system is smaller than some critical value p_0, the phase behaviour is very close to that for the $p = 0$ system. But, the nucleation rate decreases strongly with p increasing towards p_0. If p is close to p_0, one can compress the system long enough to measure $\phi_q(t)$ for the metastable state. For $p > p_0$, no crystallization is observed. In this case, the system is amorphous for all densities below the close-packing value. The critical value p_0 is about 7%. For more details, the reader can consult a review by Pusey (1991).

van Megen *et al.* (1991) have measured the density-fluctuation-decay functions $\phi_q(t)$, normalized to $\phi_q(t = 0) = 1$, for quasi-monodisperse hard-sphere colloidal suspensions. They observed that there are two amorphous states. For $p > p_0$ and a packing fraction φ smaller than some value φ_g, the functions decay to zero for long times. Such behaviour is expected for a liquid state. However, for $\varphi > \varphi_g$, they measured a non-vanishing long-time limit $f_q = \phi_q(t \to \infty)$, $0 < f_q < 1$. Increasing φ through $\varphi_g, \phi_q(t \to \infty)$ increases abruptly from zero to some $f_q^c > 0$. Increasing φ further, f_q increases as well. A non-vanishing long-time limit implies that the fluctuation spectrum $\phi_q''(\omega)$ exhibits a strictly elastic peak. Such phenomenon is well known for the density-fluctuation functions $\phi_{\vec{q}}(t)$ for crystalline solids. In the latter case, the long-time limit is called Debye–Waller factor $f_{\vec{q}}$. The long-range order in crystals implies that $f_{\vec{q}}$ is anisotropic; and $f_{\vec{q}}$ is non-zero only for wave vectors $\vec{q}$ at the points on the reciprocal lattice. For the amorphous colloids, $f_{\vec{q}}$ is isotropic and non-zero for all q. In this sense, the state for $\varphi > \varphi_g$ is an amorphous solid, i.e., a glass. The value $\varphi = \varphi_g$ marks a transition of the hard-sphere colloidal system from a liquid to a glass. The data suggest that φ_g is close to 0.58. If $p < p_0$, the sample crystallizes for

$\varphi_F < \varphi < \varphi_g$, provided one waits long enough. However, no crystallization can be observed for $\varphi > \varphi_g$. Observation of whether there occurs crystallization for $\varphi > \varphi_F$ or not is a means to decide whether the system is in the state of a metastable liquid or of a glass, respectively.

Aging effects can be observed for the glass, which are connected with particle rearrangements on length scales of order d. But, long-distance motion, which would be necessary for establishing crystalline order, is absent. This observation can be expected by assuming that φ_g is the point of vanishing diffusivity. Such an expectation was confirmed by measuring the diffusivity D of a tagged particle by incoherent-light-scattering experiments (van Megen *et al.* 1998). Long-time measurements of the mean-squared displacements $\delta r^2(t)$ have been possible by tagging the particles with a fluorescent dye. Thereby, the arrest at some plateau r_p^2 was demonstrated for time increases up to $3 \cdot 10^7 t_{\rm mic}$ (Simeonova and Kegel 2004). For times, which exceed the natural scale by more than seven orders of magnitude, the colloid glass exhibits an amorphous arrested structure.

Figure 1.15 exhibits examples for the mentioned decay functions $\phi_q(t)$ of quasi-monodisperse hard-sphere colloidal suspensions. The results for the two samples H6 and H7 demonstrate the liquid–glass transition occurring due to an increase of the packing fraction smaller than 1.5%. The Debye–Waller factor can be estimated as $f_q = \phi_q(t \approx 10^7 \mu\text{s})$, assuming that the decay exhibited for $t > 10^7\,\mu\text{s}$ is due to aging. Figure 1.16 shows the strong increase of f_q with increasing $\epsilon = (\varphi - \varphi_g)/\varphi_g$, and Fig. 1.17 demonstrates that this increase is consistent with the square-root law, Eq. (1.12). The f_q versus q diagram exhibits a pronounced maximum for $qd = 6.8$ indicating that density fluctuations arrest most easily for wavelengths close to the particle diameter.

The Brownian motion of a single colloid particle is described by the simple relaxation law $\phi_q(t) = \exp(-t/\tau_q)$. Here $\tau_q = 1/(D_0 q^2)$ with D_0 denoting the diffusivity. This result is derived in Chapter 3. The formula for $\phi_q(t)$ remains valid for arbitrary φ, albeit only for the short-time regime and after replacement of D_0 by a renormalized q-dependent diffusivity $D(q)$. Such law is shown as dashed line in the upper panel of Fig. 1.15. For packing fractions φ larger than about 0.49, this normal-liquid result describes only 5% of the decay of $\phi_q(t)$. The curves measured for sample F2 are slowed down by about a factor 5 relative to the expectation based on normal-liquid dynamics. Increasing φ further, the long-time decay slows down dramatically. For samples H5 and H6, the decay of $\phi_q(t)$ to 40% of its initial value requires a time, which is at least three orders of magnitude larger than the natural time scale for the dynamics. The decay curves exhibit a two-step-relaxation scenario as discussed above in connection with Fig. 1.4: there is a stretched upward-bent decay towards some plateau followed by a stretched downward-bent decay below the plateau. The second process exhibits the superposition principle, and the plateau value is the Debye–Waller factor f_q^c extrapolated for φ decreasing to φ_g (van Megen and Underwood 1993). The cited findings show that the hard-sphere colloids exhibit an evolution

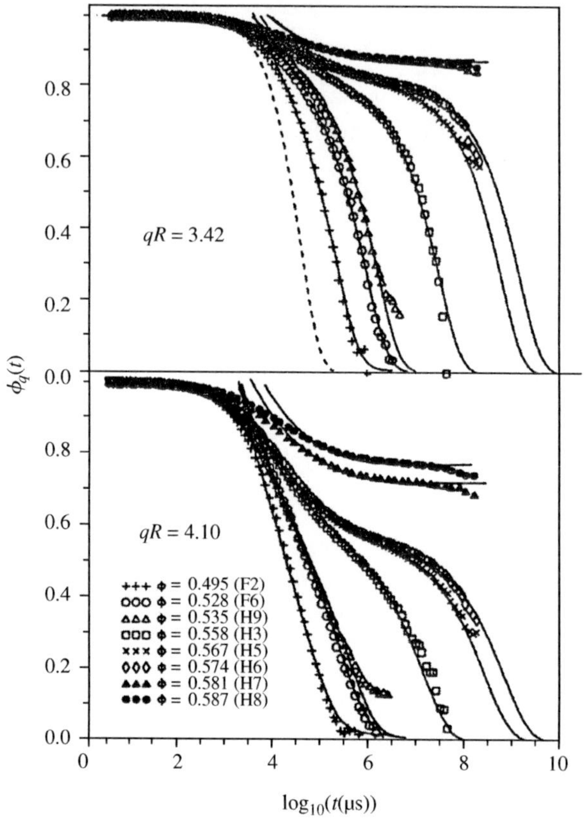

FIG. 1.15. Normalized density-fluctuation-decay functions $\phi_q(t)$ measured for quasi-monodisperse hard-sphere colloids by photon-correlation spectroscopy for eight packing fractions φ and two wave numbers q. The particle radius $R = d/2$ is 205 nm. The dashed line in the upper panel shows the simple relaxation law, $\phi_q(t) = \exp[-(t/\tau)]$, with τ fitted to the data for $t < 1$ ms.
The full lines are discussed in Sec. 6.1.3 (iii) and 6.2.2 (iv). Reproduced from van Megen and Underwood (1993).

of a glassy dynamics upon increasing the density similarly to what was discussed above for molecular liquids. The liquid–glass transition point φ_g for the colloids is the analogue to the crossover density n_c identified for OTP in Fig. 1.13, or, for the inverse of the crossover temperature T_c discussed above for CKN, OTP and glycerol.

The cited data suggest that it is not the complexity of the building blocks of liquids which causes the complexity of the glassy dynamics. The dense hard-sphere systems are the simplest models for condensed matter. They show an evolution of glassy dynamics and, closely tight to it, a glass transition. The

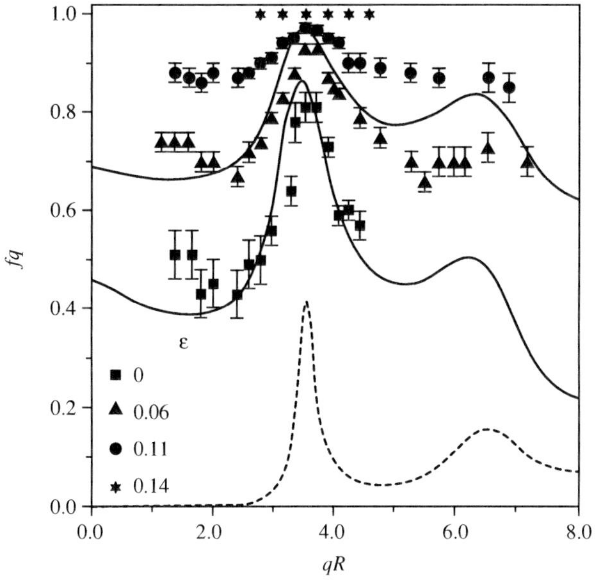

FIG. 1.16. The Debye–Waller factors f_q of a quasi-monodisperse hard-sphere colloidal glass as a function of the wave number q measured by photon-correlation spectroscopy. Here, $R = d/2$ denotes the particle radius and $\epsilon = (\varphi - \varphi_g)/\varphi_g$ is the relative distance of the packing fraction φ from the glass-transition point $\varphi_g = 0.578 \pm 0.004$.

The full lines are discussed in Sec. 4.3.5. The dashed line shows $S_q/10$ with S_q denoting the Percus–Yevick approximation for the structure factor for $\varphi = \varphi_g$. Reproduced from van Megen (1995).

hard-sphere systems with packing fractions below φ_g are paradigms for glass-forming liquids.

Figure 1.18 shows further results, which demonstrate the potential of colloid studies for the unfolding of subtleties of the glassy dynamics. The data demonstrate mixing effects for the complex relaxation of hard-sphere systems. In order to ease an appreciation of the shown relaxation curves, a digression shall be made on the basic formula underlying the data analysis (Pusey 1991). Let us suppose that the N-particle system is a homogeneous isotropic mixture consisting of $N_\alpha = x_\alpha N$ identical spherical particles of species $\alpha, \alpha = 1, \ldots, n, \sum_\alpha x_\alpha = 1$. The interaction of a light wave with a single colloid particle of species α is quantified by a scattering amplitude $b_\alpha(q)$. Here, q denotes the modulus of the wave vector $\vec{q}$ describing the scattering event. The real numbers $b_\alpha(q)$ are determined by the difference of the refraction-index profile of the solute particle and the refraction index of the solvent. The variable, which describes the coupling of the light to the mixture reads $A_{\vec{q}} = \sum_\alpha b_\alpha(q)\rho^\alpha_{\vec{q}}$, with $\rho^\alpha_{\vec{q}}$ denoting the density

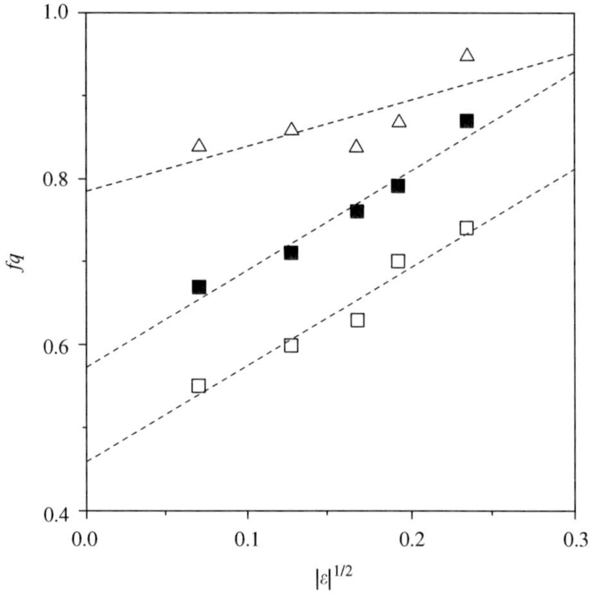

FIG. 1.17. The Debye–Waller factors f_q of a quasi-monodisperse hard-sphere colloidal glass as function of $\sqrt{\epsilon}$. Here, ϵ is the relative distance of the packing fraction from the glass-transition value φ_{g}. From top to bottom, the data refer to $qR = 3.42$, 3.05, 2.68 with $R = d/2$ denoting the particle radius. The dashed lines are linear interpolations of the data. Reproduced from van Megen (1995).

fluctuation with wave vector $\vec{q}$ for species α. The photon-correlation measurement provides the function $\langle A_{\vec{q}}(t)^* A_{\vec{q}}\rangle$, where the brackets denote canonical averaging. Hence, the experiment yields the function

$$\phi_q^m(t) = \sum_{\alpha\beta} b_\alpha(q)\phi_q^{\alpha\beta}(t)b_\beta(q)/N_q. \tag{1.13}$$

The constant N_q ensures $\phi_q^m(t=0) = 1$. The functions $\phi_q^{\alpha\beta}(t) = \langle \rho_{\vec{q}}^\alpha(t)^* \rho_{\vec{q}}^\beta\rangle/N$ characterize the dynamics of the colloidal suspension. They are elements of a real symmetric n-by-n matrix; their general properties are discussed in Sec. 5.2.

The data under discussion are obtained for a binary mixture of hard spheres. The two nearly monodisperse subsystems are characterized by their packing fractions $\varphi_\alpha = (\pi/6)(x_\alpha \rho)d_\alpha^3$, $\alpha = A, B$. Here ρ denotes the total number density; x_α and d_α abbreviate the number concentrations and diameters, respectively, of the two species. The 20 equilibrium states for the relaxation curves presented in the figure are specified by the five total packing fractions $\varphi = \varphi_A + \varphi_B$ analysed and by the four relative packing fraction $\hat{x}_B = \varphi_B/\varphi$ studied. The particles are fabricated so that their optical properties depend on the temperature. If the latter is increased from 6 °C to 26 °C, $b_A(q)$ increases and $b_B(q)$ decreases

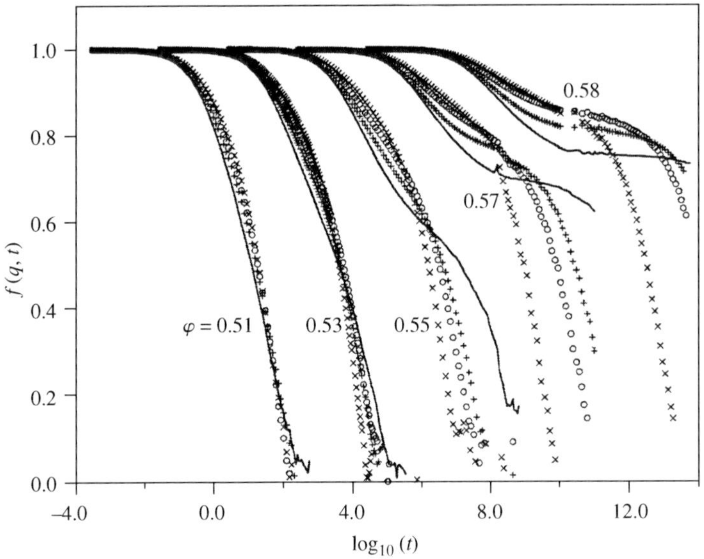

FIG. 1.18. Relaxation functions $f(q,t) = \phi_q^{AA}(t)/\phi_q^{AA}(t=0)$ of the density fluctuations for the big particles measured by photon-correlation spectroscopy for a binary colloidal mixture of hard spheres for a wave number $q = 6.0/d_A$. The particles have diameters d_A and $d_B = 0.6d_A$. The time t is measured in units of $t_{\mathrm{mic}} = d_A^2/(24D_0)$, where D_0 is the diffusivity of a single big particle. The packing fractions φ increase from 0.51 to 0.58 as indicated; successive data sets are shifted by two units along the horizontal axis in order to avoid overcrowding. The results refer to the relative packing fractions for the small particles of $\hat{x}_B = 0.00$ (full lines), 0.05 (pluses), 0.10 (circles), and 0.20 (crosses). Reproduced from Williams and van Megen (2001).

by factors, which can exceed 10. Measurements for the dilute system yields the three quantities $b_\alpha(q)b_\beta(q), \alpha \leq \beta$, as function of temperature. Measuring $\phi_q^m(t)$ for three temperatures, Eq. (1.13) is inverted to calculate the three functions $\phi_q^{\alpha\beta}(t), \alpha \leq \beta$.

The full lines in Fig. 1.18 refer to $\hat{x}_B = 0$. The functions $f(q,t)$ for these quasi-monodisperse systems are in qualitative agreement with the ones shown in Fig. 1.15. They exhibit the evolution of a stretched slow decay upon increasing φ. There is arrest at some critical packing fraction φ_{g}, which is located between $\varphi = 0.57$ and $\varphi = 0.58$. For $\varphi = 0.51$, an increase of $\hat{x}_B$ causes small but systematic changes of the decay curves. These changes increase if φ increases to 0.53 and then to 0.55. A first mixing effect is obvious: the scale for the long-time decay decreases with increasing $\hat{x}_B$. Increasing the fraction x_B of the small particles so that the total packing fraction φ is kept fixed implies an acceleration of the long-time part of the relaxation processes. This indicates an increase of the critical packing fraction upon mixing: $\varphi_g(\hat{x}_B = 0) < \varphi_g(\hat{x}_B = 0.05) < \varphi_g(\hat{x}_B = 0.10) < \varphi_g(\hat{x}_B = 0.20)$. The data for $\varphi = 0.58$ demonstrate explicitly the melting of the hard-sphere glass if a percentage of the big spheres is replaced by small ones so that φ is kept fixed. A second mixing effect concerns the initial part of the glassy relaxation. There is a slowing down of the decay due to mixing in the sense that the $f(q,t)$ versus $\log t$ curves become flatter if $\hat{x}_B$ increases for fixed φ. The combination of the two effects causes the crossing of the relaxation curves referring to the same φ but different $\hat{x}_B$.

For states with $\varphi = 0.57$ and $\hat{x}_B = 0.00$ or 0.05, the $f(q,t)$ versus $\log t$ curves exhibit an inflection point. The value of $f(q,t)$ at the inflection point is expected to approach the height of the plateau of the curves if $\epsilon = [\varphi - \varphi_g(\hat{x}_B)]/\varphi_g(\hat{x}_B)$ decreases further. For $\hat{x}_B = 0.10$ and $\hat{x}_B = 0.20$, ϵ is so large that an inflection point cannot be identified. For $\varphi = 0.58$, all curves for the mixture exhibit an inflection point; and the ones for $\hat{x}_B = 0.05$ and $\hat{x}_B = 0.10$ demonstrate a plateau. The data exhibit a third mixing effect: the height of the plateau increases with $\hat{x}_B$.

1.8 Hard-sphere systems with short-range attraction

Several features of glassy dynamics shall be considered in this section, which are observed in systems with hard-sphere repulsion complemented by a short-range attraction. There are high-density liquid states, which can be driven into a glass by increasing as well as by decreasing the attraction strength. Within the liquid, an increase of the attraction can cause a strong acceleration of the glassy relaxation. The density-fluctuation-decay functions of the liquid can exhibit an enormous stretching without any indication of a two-step-relaxation scenario, and, for specific wave numbers, these functions decay logarithmically. The mean-squared displacement exhibits a power-law variation, which is specified by a control-parameter-dependent exponent.

The interactions among colloidal particles can be changed to some extend by modifying the solvent. This opens the fascinating perspective to study the dependence of the glassy dynamics on tunable details of the interaction potential. For example, adding to the solute of a hard-sphere colloidal suspension a non adsorbing polymer, the hard-core potential is complemented by the Asakura–Oosawa depletion interaction. This is an attraction potential whose strength increases with the polymer density c_p. The range of the attraction Δ is of the order of the diameter of the coils formed by the polymer. If two colloid particles approach each other closely, polymer coils are excluded from the region between them. There results an imbalance of the osmotic pressure, which is equivalent to an attraction force. Figure 1.19 summarizes extensive studies of the states of such systems. In this case, the attraction range Δ is about 9% of the hard-core diameter d. The polydispersity is chosen large enough to keep the liquid in metastable states for long times, but small enough for crystallization of the liquid to be observable. The horizontal axis of the figure represents the colloid packing

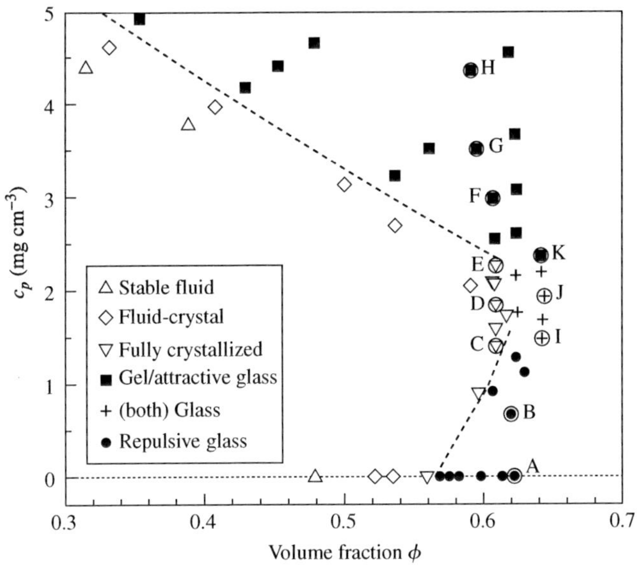

FIG. 1.19. Diagram of the states for a colloid–polymer mixture. The horizontal axis shows the packing fraction φ of the hard-sphere colloids, and the vertical one the polymer density c_p. States marked by filled circles, filled squares and crosses are amorphous solids. States marked by open symbols are liquids, crystals or liquid-crystal mixtures as noted in the inset. The dashed lines are guides to the eye indicating the lines of liquid-to-glass transitions, which one would observe if one would increase properly the polydispersity of the colloidal system. Reproduced from Pham *et al.* (2004).

fraction φ. The vertical axis shows c_p as measure of the attraction strength. The dashed lines are an estimation of the boundary between liquid and glass.

Let us consider the states for the hard-sphere system indicated by the symbols on the horizontal axis. The open symbols document liquid, liquid–crystal mixtures, and fully crystallized states. If the polydispersity would be chosen somewhat larger, all these states would be liquids. The states marked by full circles do not crystallize on time scales of days, sometimes of weeks. As discussed in the preceding section, this observation is the criterion for considering the state as glass. The liquid glass transition occurs at some critical value $\varphi^c_{\rm HSS}$ near 0.564. Other points for the dashed transition lines are identified similarly.

For $c_p > 2.5$ mg cm^{-3}, one finds a transition line, which decreases with increasing φ. This behaviour is similar to the one discussed for the van der Waals systems OTP in connection with Fig. 1.13, anticipating that the effect of increasing the attraction strength is essentially equivalent to the one of decreasing the temperature. However, the transition line does not decrease monotonically to the hard-sphere transition point at $c_p = 0, \varphi = \varphi^c_{\rm HSS}$. There are liquid states

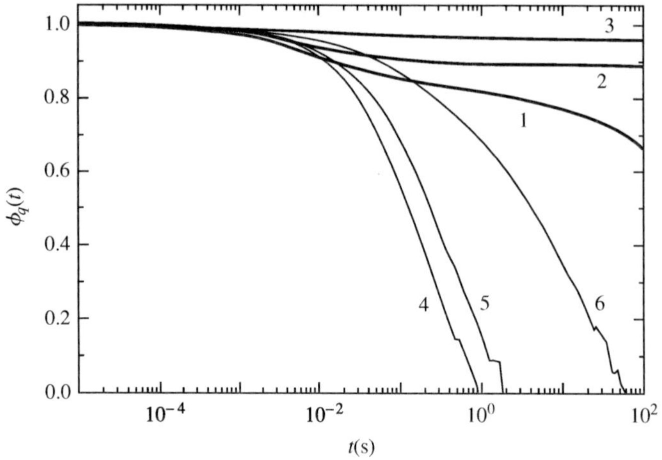

FIG. 1.20. Normalized density-fluctuation-decay function $\phi_q(t)$ measured for the delay time t by photon-correlation spectroscopy for a binary mixture of hard-sphere colloidal suspensions. The wave number q is close to the position of the first structure-factor peak. The curves 1, 2 and 3 refer to packing fractions $\varphi = 0.592$, 0.62 and 0.67, respectively. The curves 4, 5 and 6 are measured for the same sequence of packing fractions, but the solvent contains a certain fixed percentage of a non-adsorbing polymer. Reproduced from Eckert and Bartsch (2002).

for $c_p > 0$ and $\varphi > \varphi^c_{\mathrm{HSS}}$, e.g., the states marked by C, D and E. Keeping the density of states fixed at that of state D, the system experiences a transition from the liquid to a glass if c_p increases up to, say, the polymer concentration of state G. Decreasing c_p to zero, one arrives at a hard-sphere glass. Hence, there must be a second liquid-glass transition occurring upon decreasing c_p. Indeed, there is a second transition line starting at $c_p = 0$ and $\varphi = \varphi^c_{\mathrm{HSS}}$, which increases with increasing φ. Decreasing c_p for $\varphi \approx 0.6$ from $c_p \approx 3.5\,\mathrm{mg\,cm^{-3}}$ to zero there occurs a reentry phenomenon. Near state E, there is a transition from a glass to a liquid and then, near state C, there is a transition from the liquid to a glass. For the high-density-liquid states, the attraction causes two mechanisms with opposite effects; one stabilizes the glass and the other the liquid. Near state E, the former dominates. But, near state C, the latter mechanism is more important.

Figure 1.20 exhibits an example showing the drastic change of the long-time dynamics of a dense hard-sphere suspension due to switching on the short-range attraction. In this case, the ratio of the polymer-coil diameters to the averaged hard-sphere diameters is about 0.05. The density-fluctuation decay curves 1, 2 and 3 refer to three binary mixtures of quasi-monodisperse hard spheres for increasing packing fraction φ. A mixture is chosen in order to avoid crystallization. This system exhibits a liquid–glass transition at a packing fraction

$\varphi_g = 0.595$ (Eckert and Bartsch 2002). Curve 1 shows the two-step relaxation process as discussed above in connection with Figs. 1.4, 1.7 and 1.15. The natural time scale is $t_{mic} \approx 0.01\,s$ and the plateau is about 0.8. The decay towards the plateau is stretched to the two-decade interval $0.01\,s < t < 1\,s$. The decay below the plateau to the 5% level requires a time increase by an estimated factor 10^4. Curves 2 and 3 refer to the glass state. They exhibit a stretched decay to the Debye–Waller factors 0.87 and 0.95, respectively. Adding the attraction potential, curve 1 is changed to curve 4. The latter exhibits a stretched decay to zero. However, the time for the decay from 1.0 to 0.7 is shortened by about a factor 1000 relative to that for the system with the simple hard-core interaction. Adding the Asakura–Oosawa attraction to the two systems with $\varphi > \varphi^c$, the hard-sphere glass melts. This is shown by the change of curves 2 and 3 to curves 5 and 6, respectively.

Curves 4, 5 and 6 exhibit the evolution of the glassy dynamics upon increasing the packing fraction φ for a colloid with short-range attraction. Upon increasing φ from 0.592 to 0.67, the time for the decay of $\phi_q(t)$ to 10% of its initial value increases by almost a factor 100. The curve 6 requires a time increase of nearly three orders of magnitude for the decay from 0.9 to 0.1. Despite this slow decay and pronounced stretching, curve 6 does not exhibit a plateau nor any indication of a two-step process. The $\phi_q(t)$ versus $\log t$ curve for state 6 is more stretched than that for state 5. The long-time decay of the correlators for these two states are not connected by the superposition principle. Measurements by Pham *et al.* (2004) of the decay functions $\phi_q(t)$ for three wave numbers $qd = 3.0, 7.7$ and 8.6 for the liquid states marked by C, D and E in Fig. 1.18 demonstrate the same features.

Let us consider molecular-dynamics simulation results for a liquid of particles interacting via a hard sphere repulsion complemented by a shell of a constant attraction potential of short range (Zaccarelli *et al.* 2002). In order to bypass crystallization, a mixture of particles with size ratio 1.2 is chosen. The thermodynamic state is specified by three control parameters: the packing fraction φ of the spheres, the ratio Γ of the attraction-potential strength and the thermal energy ($k_B T$), and the ratio δ of the attraction-shell width Δ and the hard-core diameter d. The described square-well potentials model the interactions discussed above for colloidal suspensions. The parameter Γ is the analogue of c_p used in the diagram of states in Fig. 1.19. The dynamics for the simulations is Newtonian like in a conventional liquid rather than a stochastic one as would be appropriate for a colloid. The results for the diffusivity for the system with $\delta \approx 0.03$ yield a liquid–glass boundary in qualitative agreement with Fig. 1.19. In particular, the liquid regime for $\varphi > \varphi^c_{HSS}$ has been identified.

Figure 1.21 exhibits the density-fluctuation-decay functions $\phi_q(t)$ for a representative set of wave numbers q for two states of the square-well system denoted S_1 and S_2. These states refer to liquids with the same Γ and the same $\varphi > \varphi^c_{HSS}$. But, they have a different attraction range: S_1 refers to $\delta \approx 0.03$ and S_2 to $\delta \approx 0.04$. Estimating the time scale for normal-liquid dynamics by $t_{mic} = 0.2$,

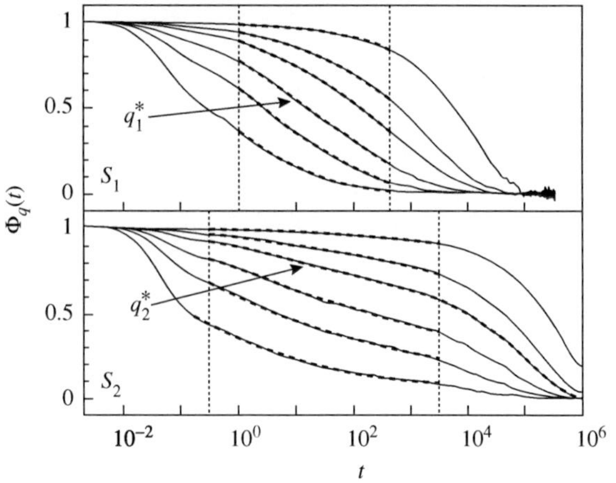

FIG. 1.21. Molecular-dynamics simulation results for the density-fluctuation-decay functions of the large species of a binary mixture of particles of equal mass interacting via a square-well potential. The hard-sphere diameter of the particles is $1.2d$ and d, respectively. The two states S_1 and S_2 considered agree in density, attraction potential strength, and thermal energy. They differ in the ratio δ of attraction-well width Δ and hard-core diameter: $\delta \approx 0.03$ and 0.04, respectively. The units of length and time are chosen so that diameter d and thermal velocity are unity. The wave numbers are $q = 6.7$, 11.7, 16.8, 23.5, 33.5, and 50.3 (from top to bottom). The vertical lines mark the time interval where the curves for wave numbers q_1^* and q_2^*, respectively, exhibit logarithmic decay. The dashed lines within the same time intervals exhibit fits of the data by Eq. (1.14). Reproduced from Sciortino *et al.* (2003).

the $\phi_q(t)$ versus $\log(t)$ curves exhibit glassy dynamics for increases of time by about six and seven orders of magnitude for states S_1 and S_2, respectively. The curves differ from the simulation results shown in Figs. 1.4 and 1.5 and from the measured results for hard-sphere colloids shown in Figs. 1.15 and 1.20 in the sense that there is no two-step-relaxation scenario. The simulation result shown for $qd = 11.7$ for state S_1 is downward bent within the three-decade time interval dealing with the decay from 0.9 to 0.1. This is very similar to the behaviour exhibited by curve number 6 in Fig. 1.20. The decay curves $\phi_q(t)$ referring to states S_1 and S_2 are not related by a superposition principle. The cited simulation studies show that the above specified new facets of glassy dynamics are not colloid-specific effects. Rather, they result from the short-range attraction. They occur for systems with a short-time dynamics, which is Newtonian as well as stochastic.

A new facet exhibited by the simulation data is the logarithmic decay, $\phi_{q^*}(t) \propto -\ln(t/\tau)$. It occurs for wave numbers $q_1^* = 23.5/d$ and $q_2^* = 16.8/d$ for states S_1 and S_2, respectively. It is impossible to describe the stretched decay curves shown for $q \geq q^*$ by a Kohlrausch function. But, the dashed lines in Fig. 1.21 demonstrate that a major part of the data can be fitted by a quadratic polynomial of $\ln(t)$:

$$\phi_q(t) = F_q - H_q^{(1)} \ln(t/\tau) - H_q^{(2)} \ln^2(t/\tau). \tag{1.14}$$

Here F_q and $H_q^{(1)}$ are positive. Coefficient $H_q^{(2)}$ vanishes for $q = q^*$; and it is positive and negative for $q < q^*$ and $q > q^*$, respectively. For the two states S_1 and S_2, the amplitudes F_q are the same and the ratio of the amplitudes $H_q^{(1)}$ is independent of the wave number (Sciortino *et al.* 2003).

Logarithmic decay in glassy systems has been reported first for photon-correlation measurements of $\phi_q(t)$ for micellar solutions (Mallamace *et al.* 2000). It is exhibited also by simulation data obtained for the tagged-particle-density-decay functions for a system with a strong repulsion complemented by a short-range attraction (Puertas *et al.* 2002).

Figure 1.22 demonstrates that the mean-squared displacements of the square-well system differ qualitatively from those shown in Fig. 1.9 for a van der Waals liquid. Instead of intervals with near-plateau variations of $\delta r^2(t)$, there are intervals of sublinear power-law increase:

$$\delta r^2(t) \propto t^\alpha. \tag{1.15}$$

For states S_1 and S_2, the exponents α are about 0.44 and 0.28, respectively.

There are three relevant length scales for the description of the $\log \delta r^2(t)$ versus $\log t$ curves. Two of them are the particle diameter d and the characteristic length r_p for the displacements of the tagged particle in the cage provided by its

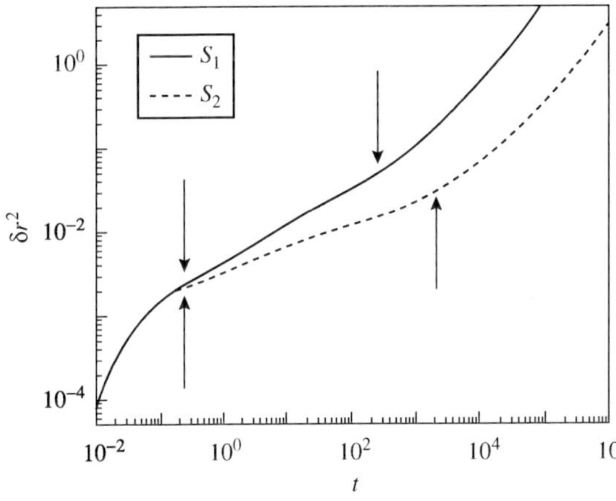

FIG. 1.22. Mean squared displacements δr^2 for the large particles for the states S_1 and S_2 of the system defined in connection with Fig. 1.21. The arrows indicate the time intervals, where the data exhibit the power-law increase according to Eq. (1.15). Reproduced from Sciortino *et al.* (2003).

transiently-arrested neighbours. The corresponding values for the mean-squared displacements have the same meaning and nearly the same size as discussed in connection with Fig. 1.9. If cooperative motion permits delocalization of the tagged particle, $\delta r^2(t)$ starts a stretched increase above $r_p^2 \approx 3 \times 10^{-2}\, d^2$. If the mean-squared displacement exceeds d^2, there holds Einstein's law for stochastic motion: $\delta r^2(t) \propto t$. Small changes of the equilibrium distribution due to small changes of control parameters like δ merely cause changes of the time scale for the $\delta r^2(t)$ versus t curves. The specified parts of the curves are connected by the superposition principle. For a van der Waals system, the control-parameter-insensitive transient dynamics describes the increase of $\sqrt{\delta r^2(t)}$ up to about 70% of r_p. A long time interval is required till $\delta r^2(t)$ reaches r_p^2; and even more time has to pass till $\delta r^2(t)$ reflects the possibility for delocalization. Within this long time interval, the mean-squared displacement increases from about $0.5\, r_p^2$ to about $2r_p^2$. The plateau of the curves in Fig. 1.9 demonstrate the described role of r_p^2. For the square-well system, the well width $\Delta = \delta d$ defines a third length scale. Figure 1.22 demonstrates that the transient motion terminates if $\delta r^2(t)$ reaches about $\Delta^2 \approx 2 \times 10^{-3} d^2 \approx r_p^2/20$. The power law (1.15) describes the increase of $\delta r^2(t)$ from Δ^2 to about $2r_p^2$. Within this interval, correlation decay is dominated by sticking and unsticking processes experienced by the tagged particle due to collisions with its cage-forming neighbours.

The results shown in Figs. 1.19–1.22 establish the existence of a new type of condensed amorphous matter. This matter exhibits facets of a glassy dynamics, which differ qualitatively from those demonstrated in the other diagrams discussed in this chapter.

2

CORRELATION FUNCTIONS

The subject of this book is the dynamics of amorphous condensed matter for densities, temperatures, and excitation energies chosen such that quantum-mechanical effects are not important. Therefore, Hamilton's equations of motion for particles moving in some box are the basis of all the following discussions.

Hamilton's mechanics is formulated with the goal to discuss orbits, i.e., to determine the position P_t of the system in phase space at time t if the position at time t_0 is some point P_0. Assuming certain regularity properties of Hamilton's function, the solutions of Hamilton's equations of motion yield the coordinates of P_t as differentiable functions of those of P_0, and vice versa. This general theorem of classical mechanics implies that variations of the coordinates of P_t are smaller than some given arbitrarily small value ϵ, provided the corresponding variations of those of P_0 are kept below some margin ϵ_0. In other words, variations of the coordinates of P_0 within a margin ϵ_0 yield corresponding variations of those of P_t that are smaller than a finite margin ϵ. This ϵ depends on t. If, for fixed t, ϵ_0 decreases to zero, also ϵ decreases to zero. Let Δt denote a natural time scale for the dynamics of the system, for example, a picosecond for a liquid. Restricting t to the order of Δt, reproducible numerical solutions of Hamilton's equations for typical interactions between molecules in liquids provide orbits for up to several thousand particles.

Our understanding of classical mechanics and our intuition on the properties of orbits is built mainly on understanding the solutions of Hamilton's equations for a variety of integrable systems. Typically for these systems, the mentioned ϵ increases linearly with t for large times. If one wants to extend the time interval for studies of orbits within a given error margin ϵ by, say, an order of magnitude, one has to decrease the error margin ϵ_0 of the coordinates of the point P_0 by a factor 10. However, integrability of the equations of motion is an exception. In particular, liquids are described by non-integrable equations of motion. Such systems may have unstable orbits. An orbit is called unstable if the above mentioned ϵ increases, loosely formulated, exponentially with time: $\epsilon = \epsilon_0 \exp[\gamma(t/\Delta t)]$, $\gamma > 0$. In order to extend the time interval for calculations for such orbit for a fixed ϵ by an order of magnitude, $\ln(\epsilon_0)/\gamma$ has to be decreased by about a factor 10. But discriminating coordinates of P_0 on exponentially small scales ϵ_0 is difficult. The data shown in Chapter 1 demonstrate that a study of the glassy dynamics of liquids not only requires a change of $(t/\Delta t)$ by one or two orders of magnitude, but by four or more. The corresponding request to discriminate orbits by their initial coordinates on scales where $\ln(\epsilon_0)$ varies by several orders of magnitude cannot be fulfilled. Consequently, it cannot be the goal of a discussion to describe the coordinates of unstable orbits for time increases of interest for a description of glassy liquids.

To explain the relevance of the preceding conclusion, some qualitative consideration shall be presented for a system of identical particles, which interact via a short-ranged centrosymmetric repulsion potential. The orbits describe a succession of binary collision events. The average time between two successive collisions of some tagged particle defines the natural scale Δt for the motion. The mentioned binary collision process deals with a known phenomenon for an integrable system. All features relevant for the present purpose can be obtained by considering the collision of a point particle from a fixed hard sphere of radius d, where $d/2$ is a measure of the radius of the repulsion potential. The particle moves along a ray with constant speed, and the ray is reflected by the sphere. The ray is specified by an angle of direction, say α, and by the coordinates of some initial point, which has some distance $R > d$ from the centre of the sphere. The reflection changes α to some other angle β. The cartesian coordinate differences for the particle moving on two orbits through the same initial point vary linearly with t, independent of whether there is a collision with the sphere or not. An elementary calculation shows that $|d\beta/d\alpha| \geq K$, where $K > 1$. A small change ϵ_0 of the initial coordinate α is magnified by the collision to a change of β by ϵ_1, where $\epsilon_1 \geq K\epsilon_0$. For the many particle system under consideration, there occur about $n = t/\Delta t$ collisions within the time interval t. Thereby, the angle change ϵ_0 is magnified to $\epsilon_n \geq K^n\epsilon_0 = \exp[\gamma(t/\Delta t)]\epsilon_0$, $\gamma = \ln(K) > 0$. The variation ϵ of the coordinate α increases exponentially. This reasoning due to E. Borel makes it plausible that practically all orbits of a system with hard-sphere-like interactions are unstable.

Let us reconsider Fig. 1.10 as quantitative demonstration of the instability phenomenon. The figure shows a representative part of a configuration of a two-dimensional caricature of a hard-sphere system. Let us consider the orbit of the tagged particle, shown as open circle. Let us assume also the six nearest neighbours as immobile. An orbit is a succession of straight-line pieces connecting positions of collision points. The orbit is enclosed in a cage. The six wall pieces are circles of radius d whose centres agree with those of the neighbour particles. These cages are shown in three replicas in Fig. 2.1. The first replica shows the orbit up to the first 10 collisions. The orbit starts at the point marked by a dot; it is shown also in Fig. 1.10. It illustrates rattling of the localized particle within a time interval of length of about 10 Δt. Actually, the figure exhibits two orbits, one as full line and the other as a dotted one. These orbits start at the same point, but they differ in their initial direction by about 10^{-7}. During the time $t \leq 10\Delta t$, the orbits are so close to each other that they cannot be distinguished within the line width used in the figure. The middle replica exhibits the continuation of the orbit from collision 10 to 17. A tiny divergence of the orbits becomes visible after collision number 13, which is magnified after collision number 14. The third replica continues the orbits from the collisions number 17, which are marked by dots, to collision number 27. In collision number 19, the two orbits hit different

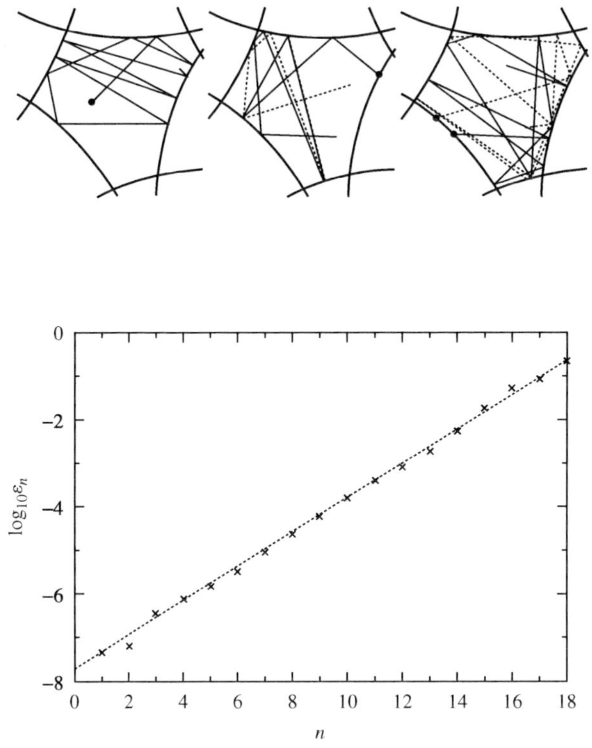

FIG. 2.1. Three replicas of pieces of circles of radius d around the centres of the immobile six nearest neighbours of the tagged particle for the configuration of hard disks of diameter d shown in Fig. 1.10. The full and dotted lines show two orbits of the centre of the moving tagged particle starting at the same point, which is marked by a dot in the first replica. The initial direction of the two orbits differ by about 10^{-7}. The left, middle and right replica show the orbits from collision number 1 till number 10, from number 10 till number 17, and from number 17 till number 27, respectively. The collision points number 10 and 17 are marked by dots in the middle and right replica, respectively. The diagram shows ϵ_n as function of n, where ϵ_n is the angle formed by the two collision points number n for the two orbits and the centre of the collision partner in the shell of the six cage-forming neighbours. The straight dotted line shows the function $\epsilon_n = \epsilon_1 \exp[\gamma(n-1)]$ with γ adjusted in order to demonstrate an interpolation for the sequence of crosses (Voigtmann 2003a).

neighbour particles. As long as the two orbits hit the same neighbour for collision number n, one can specify the difference of the orbits by the angle ϵ_n formed by the collision points and the centre of the neighbour. The diagram in Fig. 2.1 demonstrates that the increase of ϵ_n follows an exponential very closely. This illustration of E. Borel's argument shows in particular that the instability of orbits is not related to the fact that condensed-matter systems deal with many particles. Rather, the instability results from the repulsive interactions, which are typical for particles in liquids and glasses. A theorem proven by Ya. G. Sinai implies that the orbits of a particle moving in a cage as discussed in Fig. 2.1 are unstable for almost all initial conditions. Thus, there is no reason to assume that the long-time dynamics of amorphous matter can be explained directly by the properties of some individual orbit.

The discussion above makes it evident that it is a basic problem for every theory of amorphous matter to define meaningful concepts for a quantitative description of the long-time dynamics. J.C. Maxwell and L. Boltzmann suggested to define such quantities within a statistical description of the solutions of Hamilton's equations. Boltzmann's kinetic theory for gases is an example for a microscopic approach towards amorphous-matter dynamics allowing to identify reproducible long-time phenomena and to quantify them by parameters, which can be calculated from the interaction potentials. A. Einstein's theory of Brownian motion is the first example for a statistical description of long-time phenomena in condensed matter. Very loosely formulated, the goal of a statistical description of Hamiltonian systems is the calculation of the percentage of orbits that pass at time t a neighbourhood of some point P in phase space under the condition that the orbits have passed at time t_0 a neighbourhood of some point P_0.

A well established mathematical framework for a statistical description of a dynamics is provided by the theory of correlation functions. All discussions in this book shall be done within this frame. The formalism of correlation functions shall be explained in this chapter. In particular, it is the aim to introduce various notations and conventions to be used for fluctuation spectra, susceptibilities, loss spectra, and relaxation kernels.

2.1 The evolution of dynamical variables

In this section, the vector space of dynamical variables will be introduced. The method of canonical averaging is used to define a metric for this space. Hamilton's equations of motion shall be reformulated in form of a one-parameter transformation group of unitary operators. The generator of this group is the hermitian Liouville operator.

Within the framework of classical mechanics for systems with f degrees of freedom, the states are described by points $P = (\boldsymbol{p}, \boldsymbol{q})$ in phase space. Here, $\boldsymbol{p} = (p_1, \ldots, p_f)$ denotes the set of generalized momenta and $\boldsymbol{q} = (q_1, \ldots, q_f)$ is the set of coordinates. Complex-valued functions of the states shall be denoted by capital letters like, e.g., $A = A(P) = A(\boldsymbol{p}, \boldsymbol{q})$. Complex rather than real functions are considered for the sake of mathematical convenience. For the beginning, it shall be assumed that all functions considered are continuous and have continuous partial derivatives of any order. The special functions, which vanish outside a bounded region in phase space, shall be referred to as restricted dynamical variables. Within the set of restricted dynamical variables, linear combinations of two variables, say A and B, formed with complex coefficients, say a and b, are defined as usual: $(aA+bB)(P)=aA(P)+bB(P)$. Thereby, the set gets the structure of a linear function space. The restricted dynamical variables are the vectors of this space. The standard concepts for vector spaces shall be used like linear dependence, subspace, and so on.

The definition of a Poison bracket will be helpful. It is a prescription to construct from two functions, say A and B, a third one according to $\{A,B\}(P)=\sum_\alpha[(\partial A(P)/\partial q_\alpha)(\partial B(P)/\partial p_\alpha)-(\partial A(P)/\partial p_\alpha)(\partial B(P)/\partial q_\alpha)]$. In order to derive a useful formula, let us consider three functions A, B, and C. One of them is assumed to be a restricted dynamical variable. Partial integration yields

$$\iint d\boldsymbol{p}d\boldsymbol{q}\left[C\left(\frac{\partial A}{\partial q_\alpha}\right)\left(\frac{\partial B}{\partial p_\alpha}\right)\right]=-\iint d\boldsymbol{p}d\boldsymbol{q}\left[\left(\frac{\partial C}{\partial q_\alpha}\right)\left(\frac{\partial B}{\partial p_\alpha}\right)+C\left(\frac{\partial^2 B}{\partial p_\alpha \partial q_\alpha}\right)\right]A.$$

Subtracting the analogous identity obtained by interchanging the role of $\partial/\partial q_\alpha$ and $\partial/\partial p_\alpha$, and summing over α, one arrives at

$$\iint d\boldsymbol{p}d\boldsymbol{q}\{A,B\}(P)C(P)=\iint d\boldsymbol{p}d\boldsymbol{q}A(P)\{B,C\}(P). \tag{2.1}$$

The original view on the dynamics is built on the concept of an orbit in phase space. This is a mapping of the axis of times t on the states: $t\to P_t=(\boldsymbol{p}(t),\boldsymbol{q}(t))$. An orbit describes the motion of the system from P_{t_1} to P_{t_2} if the time changes from t_1 to t_2. The properties of the orbits are governed by Hamilton's function H. The $2f$ coordinates of P_t obey Hamilton's equations of motion: $\partial_t q_\alpha(t)=\partial H(P_t)/\partial p_\alpha$, $\partial_t p_\alpha(t)=-\partial H(P_t)/\partial q_\alpha$, $\alpha=1,\ldots,f$. Ordinary derivatives $dF(t)/dt$ of functions $F(t)$ shall also be denoted by $\partial_t F(t)$. Throughout this book, it will be assumed that H is time independent. This means that the dynamics is autonomous; the system is isolated from the environment. As a result, Hamilton's function is a constant of motion, $H(P_{t_1})=H(P_{t_2})$. The known theorems for ordinary differential equations ensure the following. For every time t_0 and every point P_0, there is some $\tau>0$ so that Hamilton's equations define a unique solution P_t for $|\,t-t_0\,|\le\tau$, obeying $P_{t_0}=P_0$. Let us restrict all discussions to such Hamilton's functions, for which the same τ can be chosen for all states P_0. In this case, for $|\,t-t_0\,|\le\tau$, the solutions define a mapping of the phase space in itself. The mapping can be iterated N times to get an extension on the interval $|\,t-t_0\,|\le N\cdot\tau$. Since N is arbitrary, one concludes that the solutions P_t of Hamilton's equations, solved with the condition $P_{t=0}=P$, define a mapping of the phase space in itself. One may think of P_t as points of a $2f$-dimensional fluid and refer to the mapping $t\to P_t$ as a flow of the fluid in phase space. The original formulation of the task of the dynamics is: characterize the flow in phase space!

The concepts reviewed above can be used to define the time evolution of the restricted dynamical variables. This is a mapping of every vector, say A, on another one denoted by $A(t)$, which is defined by $A(t)(P) = A(P_t)$. Written with phase-space coordinates, this means: $A(t)(\boldsymbol{p}, \boldsymbol{q}) = A(\boldsymbol{p}(t), \boldsymbol{q}(t))$. A so called explicit time dependence of dynamical variables shall not be considered, unless emphasized by a change of the notation. Obviously, this mapping is a linear transformation of the vector space of restricted variables in itself, i.e., it is a linear operator: $(aA + bB)(P_t) = aA(P_t) + bB(P_t)$. Linear operators shall be indicated by script letters. The operator of the mapping $A = A(t = 0) \to A(t)$ shall be denoted by $\mathcal{U}(t)$. It is defined by

$$A(t) = \mathcal{U}(t)A. \tag{2.2}$$

$\mathcal{U}(t)$ is called the time-translation or time-evolution operator. By construction, $\mathcal{U}(t = 0)$ is the unit operator. Standard considerations for differential equations lead to the result

$$\mathcal{U}(t_1)\mathcal{U}(t_2) = \mathcal{U}(t_1 + t_2). \tag{2.3a}$$

In particular, the operator maps the space on itself; it has an inverse given by

$$\mathcal{U}(t)^{-1} = \mathcal{U}(-t). \tag{2.3b}$$

These equations mean that the set of all operators $\mathcal{U}(t)$ presents a one-parameter transformation group.

One gets $\partial_t A(P_t) = \Sigma_\alpha[(\partial A(P_t)/\partial q_\alpha)\partial_t q_\alpha + (\partial A(P_t)/\partial p_\alpha)\partial_t p_\alpha]$. Substituting Hamilton's equation of motion, one obtains the equation of motion for the restricted dynamical variables: $\partial_t A(t) = \{A(t), H\}$. The right-hand side is a linear function of the vector A. Defining a linear operator $\mathcal{L}$ in the space of restricted dynamical variables by

$$\mathcal{L}A = i\{H, A\}, \tag{2.4a}$$

Hamilton's equations of motion can be written more concisely as a linear differential equation for the vectors:

$$-i\partial_t A(t) = \mathcal{L}A(t). \tag{2.4b}$$

Because of Eq. (2.2) and the arbitrariness of A, Eq. (2.4b) is equivalent to the equation of motion for the time-evolution operator:

$$-i\partial_t \mathcal{U}(t) = \mathcal{L}\mathcal{U}(t). \tag{2.4c}$$

The original aim of the dynamics can also be formulated as this: solve Eq. (2.4c) with the initial condition $\mathcal{U}(t = 0) = 1$! This is equivalent to: solve Eq. (2.4b) for all initial conditions $A(t = 0) = A$! The operator $\mathcal{L}$ is called the Liouville operator, the Liouvillian, or the generator of the time-translation group.

A statistical description of the system can be based on Boltzmann's probability density $\rho(P)$. This is a function defined on the phase space by

$$\rho(P) = \exp[-H(P)/(k_\mathrm{B}T)]/Z. \tag{2.5a}$$

Here T denotes the temperature of the system and k_B abbreviates Boltzmann's constant. It will be assumed that all functions $\rho(P) \cdot H(P)^n$, $n = 0, 1, \ldots$, are absolutely integrable over the whole phase space. The normalization constant is the positive finite number $Z = \int d\boldsymbol{p}d\boldsymbol{q} \exp[-H(P)/(k_\mathrm{B}T)]$. Averages or expectation values of some function, say X, are defined by

$$\langle X \rangle = \int d\boldsymbol{p}d\boldsymbol{q} X(\boldsymbol{p},\boldsymbol{q})\rho(\boldsymbol{p},\boldsymbol{q}). \tag{2.5b}$$

The Hamilton function governs the formation of averages and it determines the equations of motion. This leads to some important formulas. Let us apply Eq. (2.1) with $A = \rho$, $B = Y(t)$, and $C = X$. Here, Y is some restricted dynamical variable. The right-hand side reads $\langle\{Y(t), X\}\rangle$. The left-hand side can be reformulated by using $\{\rho, Y(t)\} = [-\rho/(k_\mathrm{B}T)]\{H, Y(t)\}$. As a result, one gets Yvon's theorem:

$$\big\langle X\mathcal{L}Y(t)\big\rangle = i(k_\mathrm{B}T)\big\langle\{X, Y(t)\}\big\rangle. \tag{2.6}$$

Specializing to $X = 1$, one obtains

$$\langle\mathcal{L}Y(t)\rangle = 0, \tag{2.7a}$$

or $\partial_t\langle Y(t)\rangle = 0$. This means that averages of dynamical variables are time independent:

$$\langle Y(t_0)\rangle = \langle Y(t_0 = 0)\rangle. \tag{2.7b}$$

Specializing to $Y = A^*(t_1)B$, where A and B denote two restricted dynamical variables, one gets

$$\langle A^*(t_1 + t_0)B(t_0)\rangle = \langle A^*(t_1)B\rangle. \tag{2.7c}$$

Time-dependent functions of the kind used in this equation are of utmost importance in the following text. Therefore, it might be worthwhile to recall the meaning of the condensed notations: $\langle A^*(t)B\rangle = \int d\boldsymbol{p}d\boldsymbol{q}\, A(\boldsymbol{p}(t),\boldsymbol{q}(t))^*\, B(\boldsymbol{p},\boldsymbol{q})\rho(\boldsymbol{p},\boldsymbol{q})$. Here, $P_t = (\boldsymbol{p}(t), \boldsymbol{q}(t))$ is the state for time t on that orbit, which passes at time $t = 0$ the state $P_{t=0} = (\boldsymbol{p}(t = 0), \boldsymbol{q}(t = 0)) = (\boldsymbol{p}, \boldsymbol{q}) = P$.

The concept of averaging shall be used to define a complex-valued function of two vectors, say A and B:

$$(A \mid B) = \langle A^*B\rangle. \tag{2.8}$$

The number $(A \mid B)$ is called the overlap of the functions A and B. This function is linear in the second variable: $(A \mid b_1B_1 + b_2B_2) = b_1(A \mid B_1) + b_2(A \mid B_2)$. It is hermitian: $(A \mid B)^* = (B \mid A)$. One gets $(A \mid A) = \int dp dq \rho(\boldsymbol{p}, \boldsymbol{q}) \mid A(\boldsymbol{p}, \boldsymbol{q}) \mid^2 \geq 0$. Since the integrand is continuous and since $\rho(\boldsymbol{p}, \boldsymbol{q}) > 0$, $(A \mid A) = 0$ can hold only for the trivial case that $A(P) = 0$ for all P. Consequently, $(A \mid B)$ defines a scalar product. Thereby, the vector space of restricted dynamical variables gets the structure of a unitary space. The scalar product leads to the usual definition of a metric: the norm of a vector A is given by $||A|| = \sqrt{(A \mid A)}$. The distance δ between two vectors, say X and Y, is defined as $\delta = ||X - Y||$. Consequently, one has a definition of the concept: variable X approximates variable Y within an error δ. Writing explicitly

$$||X - Y|| = \left[\int d\boldsymbol{p} d\boldsymbol{q} \rho(\boldsymbol{p}, \boldsymbol{q}) \mid X(\boldsymbol{p}, \boldsymbol{q}) - Y(\boldsymbol{p}, \boldsymbol{q}) \mid^2 \right]^{1/2}, \tag{2.9}$$

one recognizes $||X - Y||$ as the familiar mean-squared deviation between the two functions $X(\boldsymbol{p}, \boldsymbol{q})$ and $Y(\boldsymbol{p}, \boldsymbol{q})$. But, the differences squared are weighted with the probability density $\rho(\boldsymbol{p}, \boldsymbol{q})$ from Eq. (2.5a). In this sense, δ is a statistical information on the distance between X and Y. Notice that δ may be very small even if $\mid X(P) - Y(P) \mid$ is huge for P near some point P_0.

A mapping can be defined in the vector space, called conjugation, $A \to A_c$:

$$A_c(\boldsymbol{p}, \boldsymbol{q}) = A(\boldsymbol{p}, \boldsymbol{q})^*. \tag{2.10a}$$

This mapping leaves fixed the norm, but changes the scalar product to its complex conjugate value

$$(A_c \mid B_c) = (A \mid B)^*. \tag{2.10b}$$

For later reference, some general concepts shall be recollected in this paragraph. Let $g : g_{m,n}$, $m, n = 1, 2, \ldots$ denote a finite or infinite matrix. Let $a_1, a_2, \ldots$ denote a sequence of complex numbers. The a_n can be arbitrary, except that only a finite number of them should be non zero. If for all such sequences

$$\sum_{m,n} a_m^* g_{m,n} a_n \geq 0, \tag{2.11}$$

matrix g is called positive semidefinite. The matrix is called positive definite if the equality sign can hold only in such case that $a_n = 0$ for all n. Choosing all but two of the elements of the sequence as zero, the inequality formulates conditions for a 2-by-2 matrix. One derives as one condition $g_{m,n}^* = g_{n,m}$, i.e., matrix g is hermitian. Furthermore, the determinant must not be negative, i.e., $g_{m,n} g_{n,m} \leq g_{m,m} g_{n,n}$. Let $A_1, A_2, \ldots$ be a finite or infinite set of vectors. The matrix g defined by

$$g_{m,n} = (A_m \mid A_n), \tag{2.12}$$

is hermitian. The finite linear combination $A = \Sigma_n a_n A_n$ yields $(A \mid A) \geq 0$ and the equality holds only for $A = 0$. One arrives at Eq. (2.11), i.e., matrix g

is positive semidefinite. The equality holds in formula (2.11) if and only if the vectors are linearly dependent. Applying the result for $A_1 = X$ and $A_2 = Y$, the mentioned determinant condition yield the Schwarz inequality

$$(X \mid Y)(Y \mid X) \leq (X \mid X)(Y \mid Y). \tag{2.13}$$

The equality holds if and only if one of the vectors is a multiple of the other.

Let us understand some most important features of the time evolution, which can be formulated with the aid of the metric. Using Eq. (2.7c) with $t_0 = t$ and $t_1 = 0$, one finds for every pair of restricted dynamical variables A and B:

$$(A(t) \mid B(t)) = (A \mid B). \tag{2.14a}$$

The time-translation operator $\mathcal{U}(t)$ is isometric. Since $\mathcal{U}^{-1}$ is defined on the whole space of restricted dynamical variables, it is unitary. In particular, the distance between two vectors does not change in time, $||X(t) - Y(t)|| = ||X - Y||$. If X approximates Y at time zero with an error δ, the variable $X(t)$ approximates $Y(t)$ with the same error. This holds for every time t, despite the possible instability behaviour of the system's orbits. Equation (2.14a) is equivalent to $0 = \partial_t(A(t) \mid B(t)) = (\partial_t A(t) \mid B(t)) + (A(t) \mid \partial_t B(t))$. Because of Eq. (2.4b), Eq. (2.14a) holds if and only if

$$(\mathcal{L}A(t) \mid B(t) = (A(t) \mid \mathcal{L}B(t)). \tag{2.14b}$$

The Liouville operator $\mathcal{L}$ is hermitian.

This paragraph shall be used for a digression with the aim to illustrate a facet of the preceding formulas. Let us assume that the space of dynamical variables has a finite dimensionality d. In such space, there is an isomorphismus between the algebra of all operators $\mathcal{O}$ and the one of all d-by-d matrices $\hat{\boldsymbol{O}}$. The one-to-one mapping $\mathcal{O} \leftrightarrow \hat{\boldsymbol{O}}$ can be achieved by introducing an orthonormalized basis, i.e., a set of d vectors $E_1, \ldots, E_d$, obeying $(E_m \mid E_n) = \delta_{m,n}$. The matrix $\hat{\boldsymbol{O}}$ is given by its elements $O_{m,n} = (E_m \mid \mathcal{O}E_n)$. These matrix elements specify the operator by $\mathcal{O} = \Sigma_{m,n} \mid E_m)O_{m,n}(E_n \mid$. Here and in the following, the convenient notation of bras and kets is used: the operator $\mathcal{P} = \mid X)(Y \mid$ is defined as $\mathcal{P}A = X(Y \mid A)$. Let matrix $\hat{\boldsymbol{L}}$ be given by $L_{m,n} = (E_m \mid \mathcal{L}E_n)$, and let matrix $\hat{\boldsymbol{U}}(t)$ be given by $U_{m,n}(t) = (E_m \mid \mathcal{U}(t)E_n)$. These matrices correspond to the Liouville operator $\mathcal{L}$ and the time translation operator $\mathcal{U}(t)$, respectively. Denoting the adjoint of matrix $\hat{\boldsymbol{O}}$ by $\hat{\boldsymbol{O}}^\dagger, O^\dagger_{m,n} = O^*_{n,m}$, one notices that Eq. (2.14a) is equivalent to

$$\hat{\boldsymbol{U}}(t)^\dagger = \hat{\boldsymbol{U}}(t)^{-1}, \tag{2.15a}$$

i.e., $\hat{\boldsymbol{U}}(t)$ is a unitary matrix. Equation (2.14b) is equivalent to

$$\hat{\boldsymbol{L}}^\dagger = \hat{\boldsymbol{L}}, \tag{2.15b}$$

i.e., $\hat{\boldsymbol{L}}$ is a hermitian matrix. Equation (2.4c) is equivalent to

$$\partial_t \hat{\boldsymbol{U}}(t) = i\hat{\boldsymbol{L}}\hat{\boldsymbol{U}}(t). \tag{2.16a}$$

These are d^2 linear differential equations with constant coefficients. They are to be solved for the d^2 functions $U_{m,n}(t)$ with the initial conditions $U_{m,n}(t=0) = \delta_{m,n}$. The solution is

$$\hat{\boldsymbol{U}}(t) = \exp[i\hat{\boldsymbol{L}}t]. \tag{2.16b}$$

The matrix exponential is defined by its series expansion $\hat{\boldsymbol{U}}(t) = \sum_{\ell=0}^{\infty}(i\hat{\boldsymbol{L}}t)^{\ell}/\ell!$. The series converges for all t. The Eq. (2.16a) and also the theorem $\hat{\boldsymbol{U}}(t_1)\hat{\boldsymbol{U}}(t_2) = \hat{\boldsymbol{U}}(t_1+t_2)$ are verified by manipulations with the power series. Equation (2.16b) provides the solution of the dynamical problem.

Since $\hat{\boldsymbol{L}}$ is hermitian, there exists an orthonormalized basis of eigenvectors, $E'_1, \ldots, E'_d$, corresponding to a set of real eigenvalues $\Omega_1, \ldots, \Omega_d$. One gets $(E'_m \mid \hat{\boldsymbol{L}}E'_n) = \Omega_m \delta_{m,n}$, and the Liouville operator reads

$$\mathcal{L} = \sum_m \Omega_m \mid E'_m)(E'_m \mid. \tag{2.17a}$$

The solution of Eq. (2.16a) with the initial condition $\hat{U}(t=0) = 1$ is given by

$$\mathcal{U}(t) = \sum_m \exp[i\Omega_m t] \mid E'_m)(E'_m \mid. \tag{2.17b}$$

Knowledge of a matrix is equivalent to knowing all the matric elements that can be formed with pairs of vectors, say A and B. Equation (2.17b) provides these functions

$$(A \mid \mathcal{U}(-t)B) = \sum_m \exp[-i\Omega_m t](A \mid E'_m)(E'_m \mid B). \tag{2.17c}$$

Equations (2.17a,b) are the spectral representations of the Liouville operator $\mathcal{L}$ and of the time-translation group $\mathcal{U}(t)$, respectively. If the space of dynamical variables would be of finite dimensionality d, the problem of the dynamics could be reduced to that of the principal-axis transformation of the Liouvillian. The matrix elements $(A(t) \mid B)$ would be almost periodic functions of the time, where the underlying frequencies are the eigenvalues of $\hat{L}$. There are systems whose dynamics can be characterized by such discrete spectrum of an operator $\mathcal{L}$, e.g., a system of harmonic oscillators. But, such systems are of no relevance for the following discussions in this book.

The unitary vector space of dynamical variables of a liquid does not have a finite dimensionality. The generalization of Eqs. (2.17a–c) to this case is not trivial since new limit concepts have to be introduced and new limit theorems have to be derived. The general mathematical problems appearing here are the same as the ones appearing for the formulation of quantum mechanics. The basic mathematical problems have been solved within the spectral theory of operators in Hilbert spaces. This theory is available since long in excellent books, like the one by Akhiezer and Glazman (1993). To proceed, one uses a routine method to extend the unitary function space to a Hilbert space. The vectors in this extended space are the general dynamical variables. The set of restricted dynamical variables is dense in the Hilbert space. Every general dynamical variable X can be approximated arbitrarily well by a restricted one Y. One finds that the $X(\boldsymbol{p}, \boldsymbol{q})$ are Lebesque-measurable functions, which are square integrable with the weight function $\rho(\boldsymbol{p}, \boldsymbol{q})$. More precisely, the vectors X are the equivalence classes of such functions that differ only for points P from sets of measure zero.

Linear operators $\mathcal{O}$ in a unitary space are called bounded, if there is some number c so that $||\mathcal{O}Y|| \leq c||Y||$ for all vectors Y. Such operators can be extended to the whole Hilbert space. For a general vector X, one has to find a sequence Y_ℓ, $\ell = 1, 2, \dots$ of restricted dynamical variables with $Y_\ell \to X$. Then, one can define $\mathcal{O}X$ as the limit of $\mathcal{O}Y_\ell$. Because of Eq. (2.14a), the evolution operator $\mathcal{U}(t)$ is bounded, and thus it can be extended. The extension is the unitary time translation operator. Equation (2.14a) remains valid, the matrix elements $(A \mid \mathcal{U}(-t)B)$ remain continuous in t, and Eqs. (2.3a,b) remain valid. Thus, the dynamics is specified by a one-parameter group of unitary operators.

The Liouville operator $\mathcal{L}$ cannot be extended on the whole Hilbert space, since it is unbounded. It can be extended somewhat, e.g., on vectors where the right-hand side of Eq. (2.4a) can be defined as a square integrable function. But, it is difficult to characterize its maximal domain of definition explicitly. One route to its definition proceeds via analyzing the matrix elements $\phi_{A,B}(t) = (A \mid \mathcal{U}(-t)B)$ with the goal to find the generalization of Eq. (2.17c). These functions $\phi_{A,B}(t)$ are called correlation functions in the physicists' literature.

It is not the subject of this book to substantiate the remarks in the preceding three paragraphs. The following discussions will be restricted to the special cases of pure continuous spectra, where the sum in Eq. (2.17c) can be replaced by an integral. For this special case, the desired results shall be derived within the framework of classical analysis.

2.2 Correlation-function description of the dynamics

Correlation functions are the basic quantities to be used in his book for the description of the dynamics. These functions will be defined in this section, and their elementary general properties will be compiled. Using Laplace transforms of the correlation functions, the resolvent operator for the Liouville operator will be introduced.

The correlation function of two dynamical variables, say A and B, also called the $A - B$ correlator, is the scalar product of vector $A(t)$ with vector $B = B(t = 0)$:

$$\phi_{A,B}(t) = (A(t) \mid B). \tag{2.18a}$$

If adequate, the function $\phi_{A,B}(t)$ will also be referred to as the correlator in the time domain. Because of Eq. (2.2), this function can be expressed by the $A-B$ matrix element of the time-evolution operator

$$\phi_{A,B}(t) = (A \mid \mathcal{U}(-t)B). \tag{2.18b}$$

In order to write $(\mathcal{U}(t)A \mid B)$ in this form, the group property (2.3a) and the unitary property (2.14a) were used: $\phi_{A,B}(t) = (\mathcal{U}(t+t_1)A \mid \mathcal{U}(t_1)B)$. This identity will be applied routinely in the following without explicit reference. A distinguished role is played by correlators with $B = A$. These are called the autocorrelation functions of the variables A or the autocorrelators:

$$\phi_A(t) = \phi_{A,A}(t). \tag{2.19}$$

Applying the Schwarz inequality (2.13) with $X = A(t)$ and $Y = B$, one gets

$$\mid \phi_{A,B}(t) \mid^2 \leq ||A||^2 ||B||^2. \tag{2.20a}$$

Thus, the correlators in the time domain are bounded continuous functions of t. In particular, the autocorrelators are bounded by their initial values $\phi_A(t = 0) = ||A||^2$,

$$\mid \phi_A(t) \mid \leq \phi_A(t=0). \tag{2.20b}$$

Equation (2.18b) implies that $\phi_{A,B}(t)$ is a linear function of the dynamical variable B and an antilinear one of the variable A:

$$\phi_{A,b_1B_1+b_2B_2}(t) = \phi_{A,B_1}(t) \cdot b_1 + \phi_{A,B_2}(t) \cdot b_2, \tag{2.21a}$$

$$\phi_{a_1A_1+a_2A_2,B}(t) = a_1^* \phi_{A_1,B}(t) + a_2^* \phi_{A_2,B}(t). \tag{2.21b}$$

For every fixed t, the correlator $\phi_{A,B}(t)$ is a bilinear form defined on the whole Hilbert space. As usual, one checks that this form is determined by its diagonal elements. Introducing the four variables $X_\pm = (A \pm B)/2$, $Y_\pm = (A \pm iB)/2$, one finds

$$\phi_{A,B}(t) = \big[\phi_{X_+}(t) - \phi_{X_-}(t)\big] - i\big[\phi_{Y_+}(t) - \phi_{Y_-}(t)\big]. \tag{2.22}$$

General correlators are linear combinations of four autocorrelators. Therefore, many properties of the general correlators are simple implications of those of the autocorrelators.

Knowing all matrix elements of a bounded operator is equivalent to knowing this operator. Thus, one concludes from Eq. (2.18b) that knowing all correlators is equivalent to knowing the time-translation operators $\mathcal{U}(t)$. The set of all correlation functions $\phi_{A,B}(t)$ provides a complete statistical description of the dynamics. To formulate this conclusion more explicitly, let us introduce an orthonormalized basis for the Hilbert space, $E_1, E_2, \ldots$ Every dynamical variable, say Y, can be written as $Y = \sum_m E_m(E_m, Y)$. This means that $Y_{m_0} = \sum_{m=1}^{m_0} E_m(E_m \mid Y)$ converges towards Y in the limit of m_0 tending to infinity. Applying this result to some time dependent dynamical variable, say $X(t)$, one gets $X(t) = \sum_m E_m(E_m(-t) \mid X)$. Expanding $E_m(-t)$, one obtains $(E_m(-t) \mid X) = \sum_n (E_n \mid X)(E_m(-t) \mid E_n)$. This is an absolutely convergent series of complex numbers. Introducing the $E_m - E_n$ correlator, $Y_{m,n}(t) = (E_m(t) \mid E_n)$, one can combine the equations to

$$X(t) = \sum_m E_m \left[\sum_n Y_{m,n}(-t)(E_n \mid X) \right]. \tag{2.23a}$$

Let us note this limit relation more pedantically. For an arbitrary small $\epsilon > 0$, one can find a pair of numbers n_0, m_0 so that the dynamical variable

$$X_{m_0 n_0}(t) = \sum_{m=1}^{m_0} \sum_{n=1}^{n_0} E_m Y_{m,n}(-t)(E_n \mid X) \tag{2.23b}$$

approximates $X(t)$ within an error not larger than ϵ:

$$||X(t) - X_{m_0,n_0}(t)|| \leq \epsilon. \tag{2.23c}$$

Since the set of restricted dynamical variables is dense in the Hilbert space, one can choose a basis from that set. In this case, the E_n are in the domain of any power of the Liouville operator $\mathcal{L}$. The set of correlators $Y_{m,n}(t)$, $m, n, = 1, 2, \ldots$, provides a complete statistical description of the dynamics of the Hamiltonian system under discussion.

To continue the list of simple properties of the correlators in the time domain, one can use the hermiticity of the scalar product, $(A(t) \mid B)^* = (B \mid A(t))$ in order to get

$$\phi_{A,B}(t)^* = \phi_{B,A}(-t). \tag{2.24}$$

This shows that one could restrict the discussions of correlators to the time interval $t \geq 0$. Equation (2.10b) implies $(A(t) \mid B)^* = (A_c(t) \mid B_c)$, i.e.,

$$\phi_{A,B}(t)^* = \phi_{A_c,B_c}(t). \tag{2.25}$$

If the variable A belongs to the domain of the Liouville operator, Eq. (2.4b) yields an equation of motion for the correlator:

$$i\partial_t \phi_{A,B}(t) = \phi_{\mathcal{L}A,B}(t). \tag{2.26a}$$

Similarly, if B belongs to the domain of $\mathcal{L}$, there holds

$$i\partial_t \phi_{A,B}(t) = \phi_{A,\mathcal{L}B}(t). \tag{2.26b}$$

One can iterate these results in order to derive

$$(i\partial_t)^\ell \phi_{A,B}(t) = \phi_{\mathcal{L}^m A,\mathcal{L}^n B}(t)\,, \qquad \ell = n+m, \tag{2.26c}$$

provided A belongs the the domain of $\mathcal{L}^m$ and B of that of $\mathcal{L}^n$. This condition insures that $\phi_{A,B}(t)$ has bounded continuous derivatives of order $1, 2, \ldots, \ell$.

There is the concept of a positive-definite function, which will be needed in the following discussions. A complex valued function, say $\phi(t)$, which is defined for all real t, is called positive definite if

$$\sum_{j,k=1}^{N} \xi_j^* \phi(t_j - t_k)\xi_k \geq 0 \tag{2.27a}$$

holds for every choice of a finite number of values $t_1 < t_2 < \cdots < t_N$ and every set of N complex number $\xi_1, \xi_2, \ldots, \xi_N$. If one introduces the matrix $g_{j,k} = \phi(t_j - t_k)$ and remembers the discussion of Eq. (2.11), one notices that the definition is equivalent to requiring the N-by-N matrix with elements $g_{j,k}$ to be positive semidefinite. Applying the relation (2.27a) for $N = 2$ with $t_1 = t$ and $t_2 = 0$, one finds the 2-by-2 matrix with elements $g_{1,1} = g_{2,2} = \phi(0)\,, g_{1,2} = \phi(t)$, and $g_{2,1} = \phi(-t)$ to be positive semidefinite. This finding is equivalent to:

$$0 \leq \phi(0), \quad \phi(t)^* = \phi(-t), \quad |\phi(t)| \leq \phi(0). \tag{2.27b}$$

Let us consider some dynamical variable A in order to construct $X = \sum_{j=1}^{N} \xi_j A(t_j)$. On the one hand, one gets $0 \leq (X \mid X) = \sum_{j,k=1}^{N} \xi_j^* (A(t_j) \mid A(t_k))\xi_k$. On the other hand, one can write $(A(t_j) \mid A(t_k)) = (A(t_j - t_k) \mid A) = \phi_A(t_j - t_k)$. One concludes that the autocorrelation functions in the time domain are continuous positive-definite functions.

Laplace transforms of complex valued functions of the time t, say $F(t)$, to functions of the complex frequency z are a tool to be used routinely in this book. The frequencies z are chosen from the whole complex plane, except for the values on the real axis. The set of admissible functions for the Laplace transforms has to be defined by some restriction. For the discussions in this chapter, it is adequate to require that $F(t)$ is continuous and bounded for all t. The Laplace transform of $F(t)$ shall be defined for Im $z \gtrless 0$ by

$$LT[F(t)](z) = \pm i \int \theta(\pm t) \exp[izt] F(t) dt. \tag{2.28}$$

Here and in the following, an integral with unspecified time interval, $\int \cdots dt$, is meant to be extended over the full axis of time, $\int_{-\infty}^{\infty} \cdots dt$. Furthermore, $\theta(t)$ denotes Heaviside's function, which is 0, 1/2, and 1 for $t < 0$, $t = 0$ and $t > 0$, respectively. If no misunderstanding is to be expected, the simplified notation $F(z) = LT[F(t)](z)$ shall be used. The simplest properties of the Laplace transform are discussed in Appendix A.1. Here, it shall be recalled that $F(z)$ is a holomorphic function for all z off the real axis. The Laplace transform is a linear invertible mapping of the linear space of admissible functions in the linear space of functions that are holomorphic for Im $z \neq 0$.

The correlators $\phi_{A,B}(t)$ are continuous and bounded functions of the time t. Therefore, the functions of frequency z

$$\phi_{A,B}(z) = LT[\phi_{A,B}(t)](z) \tag{2.29}$$

are well defined for Im $z \neq 0$ for all pairs of dynamical variables A and B. The function $\phi_{A,B}(z)$ is holomorphic . It shall be called the $A - B$ correlator in the frequency domain. Similarly, the function $\phi_A(z) = LT[\phi_A(t)](z) = \phi_{A,A}(z)$ is called the A autocorrelator in the frequency domain.Writing $z = \omega + i\eta$ with $\omega = \text{Re}\, z$ and $\eta = \text{Im}\, z$, one can estimate $\mid \theta(\pm t)\exp[izt]F(t) \mid \leq \theta(\pm t)\exp[-\eta t] \mid F(t) \mid$. The inequality (2.20a) for the correlators in the time domain has a counter part for the correlators in the frequency domain, which reads

$$\mid \phi_{A,B}(z) \mid \leq ||A||\ ||B||\ /\mid \text{Im}\, z \mid. \tag{2.30}$$

The symmetries of the correlators in the time domain can be translated to symmetries of the correlators in the frequency domain. Combining Eq. (2.24) with Eqs. (A.5b,c), one gets

$$\phi_{A,B}(z)^* = \phi_{B,A}(z^*). \tag{2.31}$$

Thus, the correlators for Im $z < 0$ can be expressed by those for Im $z > 0$. One could restrict the discussions of the correlators to the upper half of the z-plane. Equation (2.25) can be combined with Eq. (A.5c) to

$$\phi_{A_c,B_c}(z) = -\phi_{A,B}(-z^*)^*. \tag{2.32}$$

Because of the linearity of the Laplace transform, Eqs. (2.21a,b) hold also if one replaces each of the correlators $\phi_{X,Y}(t)$ by its Laplace transform $\phi_{X,Y}(z) = LT[\phi_{X,Y}(t)](z)$. The resulting two equations imply that $\phi_{A,B}(z)$ is a bilinear form of the pairs of vectors A and B for every frequency z. Equation (2.30) means that this function is bounded. This in turn is equivalent to the statement that the function is continuous: if a sequence of vectors A_ℓ, $\ell = 1, 2, \ldots$, converges to the vector A in the sense of the Hilbert space metric, also $\phi_{A_\ell,B}(z)$ converges to $\phi_{A,B}(z)$. The corresponding statement holds also with respect to the second variable: $B_\ell \to B$ implies $\phi_{A,B_\ell}(z) \to \phi_{A,B}(z)$. According to a theorem of Hilbert-space analysis (Akhiezer and Glazman (1993), Sec. 21), there is a one-to-one correspondence between bounded bilinear forms and bounded operators. This operator, which shall be denoted by $\mathcal{R}(z)$, is defined by

$$\phi_{A,B}(z) = (A \mid \mathcal{R}(z)B). \tag{2.33}$$

Let us recall that the equations of motion for the restricted dynamical variables, Eq. (2.4b), are determined by the Liouville operator $\mathcal{L}$. These equations define the group of unitary time translations $\mathcal{U}(t)$. The latter determine the functions $\phi_{A,B}(t)$ via Eq. (2.18b), and these determine $\phi_{A,B}(z)$ via a Laplace transform. Hence, the operator $\mathcal{R}(z)$ is determined by $\mathcal{L}$. It is called the resolvent of $\mathcal{L}$. While the Liouville operator cannot be defined on the whole Hilbert space, operator $\mathcal{R}(z)$ is bounded and defined everywhere. The rules for calculations with bounded operators follow the ones for the algebra of complex matrices. For example, $(A \mid \mathcal{R}(z)B)^* = (B \mid \mathcal{R}(z)^\dagger A)$, where $\mathcal{O}^\dagger$ denotes the adjoint of operator $\mathcal{O}$. Equation (2.31) yields $(A \mid \mathcal{R}(z)B)^* = (B \mid \mathcal{R}(z^*)A)$. Since A and B are arbitrary, one gets $\mathcal{R}(z)^\dagger = \mathcal{R}(z^*)$. This result reflects the fact that the time translation group is built with unitary operators.

Let us assume that A and B belong to the subspace of restricted dynamical variables. This implies that A and B are in the domain of $\mathcal{L}$. Because of Eqs. (2.26a,b), $\phi_{A,B}(t)$ has a continuous first derivative. Therefore, the Laplace transform defines a one-to-one mapping of $\phi_{A,B}(t)$ to $\phi_{A,B}(z)$, and Eq. (A.4) provides an explicit expression for the inverse transformation: $LT^{-1}[\phi_{A,B}(z)](t) = \phi_{A,B}(t)$. Since $\phi_{A,B}(t)$ and $\phi_{A,B}(z)$ are continuous bilinear forms defined on the whole Hilbert space, and since the set of restricted dynamical variables is dense in this space, the mapping $\phi_{A,B}(t) \to LT[\phi_{A,B}(t)](z)$ is one-to-one for all vectors A and B from the space of general dynamical variables. This observation provides an a posteriori justification for referring to $\phi_{A,B}(t)$ and $\phi_{A,B}(z)$ as the $A - B$ correlators in the time domain and frequency domain, respectively.

Using Eq. (A.7a) for the Laplace transform of a derivative, Eq. (2.26b) leads to $z\phi_{A,B}(z) + (A \mid B) = \phi_{A,\mathcal{L}B}(z)$. Equivalently, one gets $(A \mid \mathcal{R}(z)[\mathcal{L} - z]B) = (A \mid B)$. All vectors B from the domain of $\mathcal{L}$ are also in the domain of the operator $\mathcal{O} = \mathcal{R}(z)[\mathcal{L} - z]$. On this dense domain, $\mathcal{O}$ is bounded, since $|(A|\mathcal{O}B)| \leq ||A||\ ||B||$. Hence, $\mathcal{O}$ can be continued on the whole Hilbert space, and there it is the unit operator. One arrives at the first resolvent identity:

$$\mathcal{R}(z)[\mathcal{L} - z] = 1. \tag{2.34a}$$

Interchanging the vectors A and B in $(A|\mathcal{R}(z)[\mathcal{L} - z]|B)^* = (B|A)$ and replacing z by z^*, one gets $(A \mid [\mathcal{L} - z]\mathcal{R}(z)B) = (A \mid B)$. One carries out the analogous reasoning as above in order to arrive at the second resolvent identity:

$$[\mathcal{L} - z]\mathcal{R}(z) = 1. \tag{2.34b}$$

2.3 Spectral representations

The concept of a spectrum is of utmost importance in all discussions of a dynamics. In this section, it will be shown how to express continuous spectra in terms of the correlators and vice versa. The necessary mathematical methods are provided by the theory of Fourier transformations. The formulas needed from this theory and the conventions used in this book are explained in Appendix A.2. It will be indicated also how the results have to be formulated for the general case with spectra referring to an arbitrary Hamiltonian dynamics.

To derive the desired formulas within the frame of classical calculus, two restrictive assumptions shall be made. These will be used throughout this section, unless stated otherwise. Firstly, it will be assumed that the correlators in the time domain, say $\phi_{A,B}(t)$, are absolutely integrable. Hence, their Fourier transforms $\hat{\phi}_{A,B}(\omega) = FT[\phi_{A,B}(t)](\omega)$ can be defined for all real frequencies ω by Eq. (A.9). The $\hat{\phi}_{A,B}(\omega)$ are bounded and continuous functions. Secondly, it will be assumed that $\hat{\phi}_{A,B}(\omega)$ is absolutely integrable. As a result, its Fourier transform can be defined, and, up to a change of signs in the argument and a factor 2π, it yields the inversion of the first transformation, Eq. (A.15). Consequently, the transformation of $\phi_{A,B}(t)$ to $FT[\phi_{A,B}(t)](\omega)$ is a one-to-one mapping.

The spectrum or the spectral function of the $A-B$ correlator shall be denoted by $\phi''_{A,B}(\omega)$. This function is defined as half of the correlator's Fourier transform: $\hat{\phi}_{A,B}(\omega) = 2 \cdot \phi''_{A,B}(\omega)$. Explicitly, Eq. (A.9) expresses the spectrum as linear integral transformation of the correlator

$$\phi''_{A,B}(\omega) = \frac{1}{2} \int \exp[i\omega t] \phi_{A,B}(t) dt. \tag{2.35a}$$

Equation (A.15) provides the inversion of this formula, and this is the spectral representation of the correlator in the time domain

$$\phi_{A,B}(t) = \frac{1}{\pi} \int \exp[-i\omega t] \phi''_{A,B}(\omega) d\omega. \tag{2.35b}$$

Here and in the following, integrals over the frequency with unspecified limits, $\int \cdots d\omega$, are meant to extend over the whole frequency axis, $\int_{-\infty}^{\infty} \cdots d\omega$. The spectrum of the autocorrelator of some variable A is called the A-fluctuation spectrum, and it shall be denoted by $\phi''_A(\omega) = \phi''_{A,A}(\omega)$. Because of Eq. (2.22), a general $A - B$ spectrum can be written as linear combination of four fluctuation spectra.

Let us note some elementary implications of the preceding definitions. The hermiticity relation (2.24), is equivalent to

$$\phi''_{A,B}(\omega)^* = \phi''_{B,A}(\omega). \tag{2.36}$$

This implies that the fluctuation spectra are real functions: $\phi''_A(\omega) = \phi''_A(\omega)^*$. The symmetry equation (2.25) is equivalent to

$$\phi''_{A,B}(\omega)^* = \phi''_{A_c,B_c}(-\omega). \tag{2.37}$$

If $\omega^L \phi''_{A,B}(\omega)$ is absolutely integrable for some $L = 0, 1, 2, \ldots$, spectral moments $C^{(\ell)}_{A,B}$ can be defined for all $\ell = 0, 1, \ldots, L$:

$$C^{(\ell)}_{A,B} = \frac{1}{\pi} \int \omega^{\ell} \phi''_{A,B}(\omega) d\omega. \tag{2.38a}$$

From Eq. (2.35b), one infers that $\phi_{A,B}(t)$ has continuous derivatives up to order L. In particular, there holds

$$C^{(\ell)}_{A,B} = (i\partial_t)^\ell \phi_{A,B}(t = 0). \tag{2.38b}$$

Using Eq. (2.26c), one can write

$$C^{(\ell)}_{A,B} = (\mathcal{L}^m A|\mathcal{L}^n B), \qquad m + n = \ell, \tag{2.38c}$$

provided A is the domain of $\mathcal{L}^m$ and B in that of $\mathcal{L}^n$.

The distinguished role of autocorrelation functions becomes obvious, if one acknowledges Bochner's theorem. It is derived in Appendix A.3 in connection with Eqs. (A.22a,b): the Fourier transform $\hat{\phi}(\omega)$ of the function $\phi(t)$ is non negative if and only if $\phi(t)$ is positive definite. It is shown above in connection with Eq. (2.27b) that autocorrelators $\phi_A(t)$ are positive definite functions. Consequently, fluctuation spectra are non-negative:

$$\phi''_A(\omega) \geq 0. \tag{2.39}$$

The spectra can be zero for some intervals of frequencies. However, if $\phi''_A(\omega)$ is zero for all ω, Eq. (2.35b) yields $||A||^2 = \phi_A(t = 0) = 0$. A fluctuation spectrum can be identical to zero only for the trivial case that the underlying variable A is the null vector in Hilbert space.

The spectrum $\phi''_{A,B}(\omega)$ of a correlator with $A \neq B$ need not be positive; not even need it be real. However, there are restrictions generalizing Eq. (2.39). To identify these inequalities, let us consider a set of dynamical variables $A_1, A_2, \ldots$ These variables shall be used to define a matrix of correlators

$$\phi_{m,n}(t) = (A_m(t)|A_n), \qquad m, n = 1, 2, \ldots . \tag{2.40a}$$

From that matrix, one can derive the matrix of correlators in the frequency domain

$$\phi_{m,n}(z) = LT[\phi_{m,n}(t)](z), \tag{2.40b}$$

and also the matrix of the spectra

$$\phi''_{m,n}(\omega) = \tfrac{1}{2} FT[\phi_{m,n}(t)](\omega). \tag{2.40c}$$

The later matrix is hermitian because of Eq. (2.36): $\phi''_{m,n}(\omega)^* = \phi''_{n,m}(\omega)$. Let us note also that $\phi_{m,n}(t = 0) = g_{m,n}$ is an element of the positive semidefinite matrix discussed above in connection with Eq. (2.12). As done there, a set of complex numbers $a_1, a_2, \ldots$ shall be chosen to construct $A = \sum_n a_n A_n$. Using Eq. (2.39), one gets

$$\sum_{m,n} a^*_m \phi''_{m,n}(\omega) a_n \geq 0. \tag{2.41}$$

For every ω, the matrix of the spectra is positive semidefinite. Applying this finding for two variables $A_1 = A$ and $A_2 = B$, the determinant condition for positive semidefinite matrices yields Bogoliubov's inequality for spectra:

$$|\phi''_{A,B}(\omega)|^2 \leq \phi''_A(\omega)\phi''_B(\omega). \tag{2.42}$$

To find a relation between a correlator in the frequency domain and its spectrum, one can use Eq. (A.18) for the boundary values of $\phi_{A,B}(\omega \pm i\epsilon)$ for ϵ tending to zero with $\epsilon > 0$. The equation shows that $\lim_{\epsilon\to 0} \phi_{A,B}(\omega \pm i\epsilon) = \phi_{A,B}(\omega \pm i0) = \pm iFT[\theta(\pm t)\phi_{A,B}(t)](\omega)$ exists as continuous function of ω. It is more convenient to express the two limits in terms of their average, $[\phi_{A,B}(\omega + i0) + \phi_{A,B}(\omega - i0)]/2$, to be denoted by $\phi'_{A,B}(\omega)$, and of their difference. The latter is proportional to the spectrum because of Eq. (A.19). Hence,

$$\phi_{A,B}(\omega \pm i0) = \phi'_{A,B}(\omega) \pm i\phi''_{A,B}(\omega). \tag{2.43}$$

Equations (2.36) and (2.37) have their counterparts for the average function $\phi'_{A,B}(\omega)$, which follow from Eqs. (2.31) and (2.32):

$$\phi'_{A,B}(\omega)^* = \phi'_{B,A}(\omega), \tag{2.44}$$

$$\phi'_{A_c,B_c}(\omega)^* = -\phi'_{A,B}(-\omega). \tag{2.45}$$

For autocorrelators, not only is the spectrum real but also the average function $\phi'_{A,A}(\omega)$. For this case, Eq. (2.43) presents the limits in terms of their real and imaginary parts:

$$\phi'_A(\omega) = \mathrm{Re}\phi_A(\omega \pm i0), \tag{2.46a}$$

$$\phi''_A(\omega) = \pm \mathrm{Im}\, \phi_A(\omega \pm i0). \tag{2.46b}$$

If the frequency z is shifted across the real axis, the real part of $\phi_A(z)$ varies continuously while the imaginary part changes its sign. In order to reverse Eq. (2.43), i.e., in order to express the correlator in the frequency domain in terms of its spectrum, one can substitute Eq. (2.35b) into Eq. (2.28). As discussed in connection with Eq. (A.20), this yields the spectral representation of the correlator in the frequency domain

$$\phi_{A,B}(z) = \frac{1}{\pi} \int [\omega - z]^{-1} \phi''_{A,B}(\omega) d\omega. \tag{2.47}$$

The holomorphic function $\phi_{A,B}(z)$ is expressed as a Cauchy integral.

There is the concept of a positive-analytic function that will be of use below. A function of the complex frequency z, say $\phi(z)$, is called positive analytic, if three conditions are fulfilled: $\phi(z)$ is holomorphic for all z off the real axis, $\phi(z)^* = \phi(z^*)$, and

$$\mathrm{Im}\, \phi(z) \gtrless 0 \quad \text{for } \mathrm{Im}\,(z) \gtrless 0. \tag{2.48}$$

Because of Eq. (2.31), the autocorrelator of some variable A in the frequency domain obeys $\phi_A(z)^* = \phi_A(z^*)$. It is holomorphic for $\mathrm{Im}\, z \neq 0$. The spectral representation yields for $z = x \pm iy$: $\pm \mathrm{Im}\, \phi_A(z)/y = \frac{1}{\pi} \int [(\omega - x)^2 + y^2]^{-1} \phi''_A(\omega) d\omega$. Because of Eq. (2.39), the integral is non negative. It can vanish only, if the continuous integrand vanishes for all ω. This means that $\phi''_A(\omega)$ is zero for all ω. According to the discussion of Eq. (2.39), this can happen only for the trivial case of $||A|| = 0$. One concludes that the auto-correlation functions $\phi_A(z)$ for non vanishing dynamical variables A are positive-analytic functions. Let us note for later reference a special implication of this finding:

$$\phi_A(z) \neq 0\,, \quad \text{if } ||A|| \neq 0 \quad \text{and} \quad \mathrm{Im}\, z \neq 0. \tag{2.49}$$

The same reasoning, which was explained in connection with Eq. (2.41), can be used to generalize the preceding discussion to matrices of correlators. Using the notations from Eq. (2.40b), one gets for Im $z \gtrless 0$

$$\pm \mathrm{Im} \sum_{m,n} a_m^* \phi_{m,n}(z) a_n \geq 0. \tag{2.50}$$

If the variables $A_1, A_2, \ldots$ are linearly dependent, there exists a non-trivial set of coefficients $a_1, a_2, \ldots$ so that $A = \sum a_n A_n = 0$. With this set of coefficients, there holds the equality signs in the formulas (2.41) and (2.50) for all ω and all z, respectively. If there holds the equality sign in formula (2.50) for a non-trivial set of coefficients and for some non-real $z = z_0$, one concludes from Eq. (2.49) that the variables are linearly dependent.

Let us come back to the case where the Liouville operator can be reduced to a matrix describing transformations in a d-dimensional subspace of dynamical variables. According to Eq. (2.17c), the autocorrelator of some variable A reads $\phi_A(t) = \sum_{m=1}^{d} \exp[-i\Omega_m t]\Delta_A^{(m)}$. Here, the Ω_m are real eigenfrequencies and $\Delta_A^{(m)} = |(A|E'_m)|^2$ are the squares of the overlaps of variable A with the eigenvector E'_m. A Fourier transform does not exist. The Laplace transform is the elementary function $\phi_A(z) = \sum_{m=1}^{d} [\Omega_m - z]^{-1} \Delta_A^{(m)}$. It is holomorphic everywhere except for simple poles on the real frequency axis. The limits $\phi_A(\omega \pm i0)$ do not exist for $\omega = \Omega_m$, $m = 1, \ldots, d$. It is trivial to modify the preceding theory so that the spectrum is a discrete one – as formulated above – plus a continuous one. But this would still not be the general case. To treat the latter, one has to make use of the concept of a Stieltjes integral as explained in Appendix A.3. Let us conclude this section by citing the relevant formulas.

As shown in Sec. 2.2, the autocorrelator of some dynamical variable, say X, is a continuous positive definite function $\phi_X(t)$. According to Bochner's theorem, there is a monotonically increasing function of frequency $\sigma_X(\omega)$, which is bounded between $\sigma_X(\omega \to -\infty) = 0$ and $\sigma_X(\omega \to \infty) = \phi_X(t = 0) = ||X||^2$ so that the correlator is a Fourier–Stieltjes transform:

$$\phi_X(t) = \int \exp[-i\omega t] d\sigma_X(\omega). \tag{2.51a}$$

The mapping $\phi_X(t) \leftrightarrow \sigma_X(\omega)$ is one-to-one provided one requires the convention for the behaviour at possible discontinuities $\omega = \Omega_\ell$: $\sigma_X(\Omega_\ell) = [\sigma_X(\Omega_\ell + 0) + \sigma_X(\Omega_\ell - 0)]/2$. According to Eq. (2.22), a general $A-B$ correlator can be written as superposition of four autocorrelators. Since the Stieltjes integral is linear with respect to the weight function, one arrives at the spectral representation

$$\phi_{A,B}(t) = \int \exp[-i\omega t] d\sigma_{A,B}(\omega). \tag{2.51b}$$

Here, the weight function $\sigma_{A,B}(\omega)$ reads

$$\sigma_{A,B}(\omega) = [\sigma_{X_+}(\omega) - \sigma_{X_-}(\omega)] + i[\sigma_{Y_-}(\omega) - \sigma_{Y_+}(\omega)]. \tag{2.51c}$$

Equation (2.51b) is the substitute for Eq. (2.35b) for the general case. It reduces to the case of a continuous spectrum if and only if $\sigma_{A,B}(\omega)$ has a continuous derivative $d\sigma_{A,B}(\omega)/d\omega = \phi''_{A,B}(\omega)/\pi$. The case of a mere discrete spectrum, which was discussed in the preceding paragraph, is reproduced by a step function $\sigma_{A,B}(\omega) = \sum_{m=1}^{d} \Delta^{m}_{A,B}\theta(\omega - \Omega_m)$, where $\Delta^{(m)}_{A,B} = (A|E'_m)(E'_m|B)$. The introduction of Stieltjes integrals is the classical method to cope also with the possibility that $\sigma(\omega)$ is continuous but not differentiable. Laplace transformation of Eq. (2.51b) yields the spectral representation of the correlator in the frequency domain for the general case:

$$\phi_{A,B}(z) = \int [\omega - z]^{-1} d\sigma_{A,B}(\omega). \tag{2.51d}$$

As explained in Appendix A.3, this formula can be derived directly within the theory of positive analytic functions. As before, Eq. (2.51d) implies that the auto correlator $\phi_A(z)$ is positive analytic and obeys Eq. (2.30).

Equation (2.51b) is the key result of the spectral theory for the Liouville operator. If the vectors A and B are restricted dynamical variables, one can derive from Eqs. (2.4c, 2.18b) the spectral representation for matrix elements of $\mathcal{L}$

$$(A|\mathcal{L}B) = \int \omega d\sigma_{A,B}(\omega). \tag{2.52}$$

This is the proper generalization of Eq. (2.17a) for an infinite dimensional space of vectors. This formula can be used to proof that $\mathcal{L}$ can be extended to a self-adjoint operator in the Hilbert space so that the general theory of functions of self-adjoint operators can be applied. Formula (2.51b) leads to Stone's theorem, which says that Eq. (2.16b) is generalized to

$$\mathcal{U}(t) = \exp[i\mathcal{L}t]. \tag{2.53}$$

Equation (2.51d) also leads to the resolvent identities formulated in Eqs. (2.34a,b). The proofs of the cited general theorems can be found in the book of Akhiezer and Glazman (1993).

2.4 Memory-kernel descriptions of correlators

2.4.1 *Zwanzig–Mori equations*

Often, one is interested in a situation where there is a distinguished set of variables, say $A_1, A_2, \ldots$, whose dynamics should be treated differently from the one for the remaining variables. In this section, the formalism shall be developed for achieving the corresponding asymmetric splitting of the equations of motions. It shall be requested that the variables are linearly independent and that they are in the domain of the Liouville operator $\mathcal{L}$. In most applications, the set of variables is finite, but the following considerations hold for infinite sets as well.

Let us remember from the discussion of Eq. (2.12) that the matrix $\boldsymbol{g}$ formed with the elements $g_{m,n} = (A_m|A_n)$, $m, n = 1, 2, \ldots$, is hermitian and positive definite. It is called the metric matrix for the set of variables $A_1, A_2 \ldots$. There is the well known algorithm of orthonormalizing the distinguished set of variables. In the first step one introduces $E_1 = A_1/C_1$. With $C_1 = [g_{1,1}]^{1/2}$, vector E_1 has unit length. One can invert: $A_1 = C_1 E_1$. In the second step, one chooses $E_2 = [A_2 - E_1(E_1|A_2)]/C_2$. One gets $(E_1|E_2) = 0$. Taking $C_2 = [(g_{2,2}g_{1,1} - g_{1,2}g_{2,1})/g_{1,1}]^{1/2}$, E_2 is normalized. One can invert: $A_2 = C_2E_2 + E_1 g_{1,2}/C_1$. In the third step, one chooses $E_3 = [A_3 - E_2(E_2|A_3) - E_1(E_1|A_3)]/C_3$, etc. One constructs recursively a matrix $\boldsymbol{T}$ and its inverse $\boldsymbol{T}'$ so that

$$E_m = \sum_n A_n \, T_{n,m}, \tag{2.54a}$$

$$A_n = \sum_\ell E_\ell \, T'_{\ell,n}. \tag{2.54b}$$

Here $T_{n,m} = 0$ and $T'_{n,m} = 0$ for $n > m$. Since $(E_m|E_n) = \delta_{m,n}$, one gets matrix $\boldsymbol{g}$ factorized

$$g_{m,n} = \sum_\ell T'^{*}_{\ell,m} \, T'_{\ell,n}. \tag{2.54c}$$

This corresponds to the matrix identity $\boldsymbol{g} = \boldsymbol{T}'^{\dagger}\boldsymbol{T}'$, with $\boldsymbol{T}'^{\dagger}$ denoting the adjoint matrix of $\boldsymbol{T}'$. Taking the inverse, one gets $\boldsymbol{g}^{-1} = \boldsymbol{T}\boldsymbol{T}^{\dagger}$, i.e.,

$$(\boldsymbol{g}^{-1})_{m,n} = \sum_\ell T_{m\ell} \, T^{*}_{n\ell}. \tag{2.54d}$$

Using the notations of bras and kets, the operator $\mathcal{P} = \sum_m |E_m)(E_m|$ is the projector on the subspace spanned by the distinguished vectors. Substituting Eq. (2.54a) and using Eq. (2.54d), one can write

$$\mathcal{P} = \sum_{mn} |A_m)(\boldsymbol{g}^{-1})_{m,n}(A_n|. \tag{2.55}$$

The following manipulations are of interest only if $\mathcal{P}$ is not the unity operator, i.e., if the set of distinguished variables does not span the complete Hilbert space. In this case, one can choose another set of orthonormalized vectors, say $F_1, F_2, \ldots$ such that the set $\{E_1, E_2, \ldots, F_1, F_2, \ldots\}$ is an orthonormalized basis. This implies $(E_m, F_n) = 0$ and $1 = \sum_m |E_m)(E_m| + \sum_n |F_n)(F_n|$. Let $\mathcal{Q} = \sum_n |F_n)(F_n|$ be the projector on the space spanned by the $F_1, F_2, \ldots$. There hold the general relations for projectors: $\mathcal{P} = \mathcal{P}^\dagger$ and $\mathcal{Q} = \mathcal{Q}^\dagger$, $\mathcal{P}\mathcal{P} = \mathcal{P}$ and $\mathcal{Q}\mathcal{Q} = \mathcal{Q}$. The two operators $\mathcal{P}$ and $\mathcal{Q}$ project on orthogonal subspaces,

$$\mathcal{P}\mathcal{Q} = \mathcal{Q}\mathcal{P} = 0. \tag{2.56a}$$

They provide a resolution of unity

$$\mathcal{P} + \mathcal{Q} = 1. \tag{2.56b}$$

Each of the two projectors characterizes uniquely the subspace spanned by the distinguished variables.

The projector $\mathcal{Q}$ shall be used to define a reduced Liouville operator by

$$\mathcal{L}_\mathcal{Q} = \mathcal{Q}\mathcal{L}\mathcal{Q}. \tag{2.57a}$$

Since $\mathcal{L}_\mathcal{Q} = \mathcal{L} - \mathcal{P}\mathcal{L} - \mathcal{L}\mathcal{P} + \mathcal{P}\mathcal{L}\mathcal{P}$, the domain of definition of $\mathcal{L}$ is contained in that of $\mathcal{L}_\mathcal{Q}$. Operator $\mathcal{L}_\mathcal{Q}$ is self-adjoint: ${\mathcal{L}_\mathcal{Q}}^\dagger = \mathcal{L}_\mathcal{Q}$. Therefore, the spectral theory of self-adjoint operators in Hilbert space can be used, as indicated in the preceding section. For every pair of vectors from the domain of $\mathcal{L}$, say A and B, a weight function $\sigma^\mathcal{Q}_{A,B}(\omega)$ can be constructed so that

$$(A|\mathcal{L}_\mathcal{Q}B) = \int \omega d\sigma^\mathcal{Q}_{A,B}(\omega). \tag{2.57b}$$

Operator functions can be used to define a one-parameter group of time translations

$$\mathcal{U}_\mathcal{Q}(t) = \exp[i\mathcal{L}_\mathcal{Q}t]. \tag{2.57c}$$

Similarly, the reduced resolvent operator

$$\mathcal{R}_\mathcal{Q}(z) = [\mathcal{L}_\mathcal{Q} - z]^{-1} \tag{2.57d}$$

can be defined for complex frequencies z obeying Im $z \neq 0$. These expressions are equivalent to the spectral representations for

$$\phi^\mathcal{Q}_{A,B}(t) = (A|\mathcal{U}_\mathcal{Q}(-t)B), \tag{2.57e}$$

$$\phi^\mathcal{Q}_{A,B}(z) = (A|\mathcal{R}_\mathcal{Q}(z)B). \tag{2.57f}$$

They are the analogues to Eqs. (2.51b,d). All the theorems studied in the preceding section remain valid. In particular, $\phi^\mathcal{Q}_{A,B}(z)$ is the Laplace transform of $\phi^\mathcal{Q}_{A,B}(t)$. The functions $\phi^\mathcal{Q}_A(t) = \phi^\mathcal{Q}_{A,A}(t)$ and $\phi^\mathcal{Q}_A(z) = \phi^\mathcal{Q}_{A,A}(z)$, are positive definite and positive analytic, respectively.

Also $\phi^{\mathcal{Q}}_{A,B}(t)$ and $\phi^{\mathcal{Q}}_{A,B}(z)$ will be called correlation functions. The dynamics described by $\mathcal{L}_Q$ is referred to as the reduced one. It is a dynamics that moves vectors in the subspace spanned by $F_1, F_2, \ldots$ into this subspace. If $A = \mathcal{P}A$, one can use Eq. (2.56a) to get $-i\partial_t \mathcal{U}_Q(t)A = \mathcal{U}_Q(t)\mathcal{L}_Q A = \mathcal{U}_Q(t)\mathcal{QLQP}A = 0$. This means that all vectors from the subspace spanned by the $A_1, A_2, \ldots$ are constants of motion for the reduced dynamics. The introduced terminology generalizes the concept of a dynamics. The generalization is not defined by Hamilton's equations of motion. Rather, it is defined via a unitary one-parameter group of time translation or, equivalently, by the generator $\mathcal{L}_Q$ of this group.

Inserting Eq. (2.56b) into Eq. (2.34b), one gets the second resolvent identity in the form $(\mathcal{L} - z)\mathcal{PR}(z) + (\mathcal{L} - z)\mathcal{QR}(z) = 1$. Multiplying this equation from the left and from the right by $\mathcal{P}$, one obtains

$$(\mathcal{PLP} - z)\mathcal{PR}(z)\mathcal{P} + \mathcal{PLQR}(z)\mathcal{P} = \mathcal{P}. \tag{2.58a}$$

Multiplying the second resolvent identity from the left by $\mathcal{Q}$ and from the right by $\mathcal{P}$, one obtains $(\mathcal{L}_Q - z)\mathcal{QR}(z)\mathcal{P} = -\mathcal{QLPR}(z)\mathcal{P}$. Multiplying this result from the left by $\mathcal{QR}_Q(z)$, one finds

$$\mathcal{QR}(z)\mathcal{P} = -\mathcal{QR}_Q(z)\mathcal{QLPR}(z)\mathcal{P}. \tag{2.58b}$$

Substituting Eq. (2.56b) into Eq. (2.34a), one obtains for the first resolvent identity: $\mathcal{R}(z)\mathcal{P}(\mathcal{L} - z) + \mathcal{R}(z)\mathcal{Q}(\mathcal{L} - z) = 1$. Multiplying this result from the left and from the right by $\mathcal{Q}$, one gets a counterpart of Eq. (2.58a):

$$\mathcal{QR}(z)\mathcal{PLQ} + \mathcal{QR}(z)\mathcal{Q}(\mathcal{QLQ} - z) = \mathcal{Q}. \tag{2.58c}$$

Multiplying the first resolvent identity from the left by $\mathcal{P}$ and from the right by $\mathcal{Q}$ leads to $\mathcal{PR}(z)\mathcal{PLQ} + \mathcal{PR}(z)\mathcal{Q}(\mathcal{L}_Q - z) = 0$. This can be multiplied from the right by $\mathcal{R}_Q(z)\mathcal{Q}$ in order to get a counterpart of Eq. (2.58b):

$$\mathcal{PR}(z)\mathcal{Q} = -\mathcal{PR}(z)\mathcal{PLQR}_Q(z)\mathcal{Q}. \tag{2.58d}$$

The above derived four implications of the resolvent identities shall be combined to derive two new ones. The first identity is obtained by using Eq. (2.58b) for the elimination of $\mathcal{QR}(z)\mathcal{P}$ in Eq. (2.58a):

$$\left[z - \mathcal{PLP} + \mathcal{PLQR}_Q(z)\mathcal{QLP}\right]\mathcal{PR}(z)\mathcal{P} = -\mathcal{P}. \tag{2.59a}$$

The bracket is an operator that maps vectors from the space spanned by the distinguished variables into this space. The operator $\mathcal{PR}(z)\mathcal{P}$ has the same property. Therefore, the identity can be considered as one for operators defined in the subspace, which is spanned by $A_1, A_2, \ldots$ The identity means that $\mathcal{PR}(z)\mathcal{P} = \mathcal{R}'(z)\mathcal{P}$, where $\mathcal{R}'(z)$ is the resolvent of $\mathcal{PLP} - \mathcal{PLQR}_Q(z)\mathcal{QLP}$. The second identity is an expression for the complete resolvent. Using Eq. (2.56b), it can be written as sum of four terms: $\mathcal{R}(z) = \mathcal{PR}(z)\mathcal{P} + \mathcal{PR}(z)\mathcal{Q} + \mathcal{QR}(z)\mathcal{P} + \mathcal{QR}(z)\mathcal{Q}$.

The second term on the right hand side shall be taken from Eq. (2.58d) and the third one from Eq. (2.58b). To get an expression for the fourth term, one multiplies Eq. (2.58c) from the right by $\mathcal{R}_{\mathcal{Q}}(z)\mathcal{Q}$: $\mathcal{QR}(z)\mathcal{Q} = \mathcal{QR}_{\mathcal{Q}}(z)\mathcal{Q} - \mathcal{QR}(z)\mathcal{PLQR}_{\mathcal{Q}}(z)\mathcal{Q}$. The factor $\mathcal{QR}(z)\mathcal{P}$ shall be eliminated with the aid of Eq. (2.58b). Altogether, there are five contributions:

$$\mathcal{R}(z) = \mathcal{QR}_{\mathcal{Q}}(z)\mathcal{Q} + \big[\mathcal{P} - \mathcal{QR}_{\mathcal{Q}}(z)\mathcal{QLP}\big]\mathcal{PR}(z)\mathcal{P}\big[\mathcal{P} - \mathcal{PLQR}_{\mathcal{Q}}(z)\mathcal{Q}\big]. \quad (2.59b)$$

Equation (2.59a) demonstrates that $\mathcal{PR}(z)\mathcal{P}$ can be expressed in terms of the reduced resolvent $\mathcal{R}_{\mathcal{Q}}(z)$. As a result, Eq. (2.59b) provides a formula expressing the complete resolvent $\mathcal{R}(z)$ in terms of the reduced one. Equations (2.59a,b) are a rewriting of Frobenius' formulas for the inversion of a block matrix $\mathcal{M} = \mathcal{L} - z$.

Because of Eq. (2.55), all operators $\mathcal{O}$ in Eq. (2.59a) can be expressed in terms of matrix elements formed with the distinguished variables $(A_m|\mathcal{O}A_n)$. Let us specify these matrix elements. Since $\mathcal{P}A_n = A_n$, one gets $(A_m|\mathcal{P}A_n) = g_{m,n}$. Remembering Eqs. (2.33), (2.40a,b), the matrix element of the resolvent leads to the matrix correlator in the frequency domain: $(A_m|\mathcal{PR}(z)\mathcal{P}A_n) = \phi_{m,n}(z)$. Using $\mathcal{P}^2 = \mathcal{P}$, one gets $(A_m|\mathcal{PLPPR}(z)\mathcal{P}A_n) = (A_m|\mathcal{LPR}(z)A_n) = \sum_\ell \Omega_{m,\ell}\phi_{\ell,n}(z)$. Here, a matrix $\boldsymbol{\Omega}$ of frequencies $\Omega_{m,\ell}$ is defined by $\Omega_{m,\ell} = \sum_k (A_m|\mathcal{L}A_k)(\boldsymbol{g}^{-1})_{k,\ell}$. Introducing a matrix $\hat{\boldsymbol{\Omega}}$ by the formula for its elements

$$\hat{\Omega}_{m,k} = (A_m|\mathcal{L}A_k), \quad (2.60a)$$

the result for $\boldsymbol{\Omega}$ can be written as

$$\boldsymbol{\Omega} = \hat{\boldsymbol{\Omega}}\boldsymbol{g}^{-1}. \quad (2.60b)$$

Operator $\mathcal{L}$ is hermitian and this implies $\hat{\boldsymbol{\Omega}}^\dagger = \hat{\boldsymbol{\Omega}}$. Thus, there holds the symmetry relation for matrix $\boldsymbol{\Omega}$:

$$\sum_\ell \Omega_{m,\ell}g_{\ell,n} = \sum_\ell \Omega^*_{n,\ell}g_{m,\ell}. \quad (2.60c)$$

The third term in the bracket of Eq. (2.59a) leads to a frequency dependent matrix, $\boldsymbol{M}(z)$, given by the reduced resolvent $\mathcal{R}_{\mathcal{Q}}(z)$. Defining the fluctuating forces

$$\dot{A}_k^{\mathcal{Q}} = \mathcal{Q}\partial_t A_k(t=0) = i\mathcal{QL}A_\ell, \quad (2.61a)$$

one gets $M_{m,\ell}(z) = \sum_k(\dot{A}_m^{\mathcal{Q}}|\mathcal{R}_{\mathcal{Q}}(z)\dot{A}_k^{\mathcal{Q}})(\boldsymbol{g}^{-1})_{k,\ell}$. Let us use the frequency-dependent coefficients in order to define a z-dependent matrix $\hat{\boldsymbol{M}}(z)$ by

$$\hat{M}_{m,k}(z) = (\dot{A}_m^{\mathcal{Q}}|\mathcal{R}_{\mathcal{Q}}(z)\dot{A}_k^{\mathcal{Q}}). \quad (2.61b)$$

These functions can be represented as Laplace transforms of the corresponding correlators in the time domain as explained in connection with Eq. (2.57e), i.e., $\hat{M}_{m,k}(z) = LT[\hat{M}_{m,k}(t)](z)$, where

$$\hat{M}_{m,k}(t) = (\dot{A}_m^{\mathcal{Q}} | \mathcal{U}_{\mathcal{Q}}(-t) \dot{A}_k^{\mathcal{Q}}). \tag{2.61c}$$

One can write

$$\boldsymbol{M}(z) = \hat{\boldsymbol{M}}(z)\boldsymbol{g}^{-1}, \qquad \boldsymbol{M}(t) = \hat{\boldsymbol{M}}(t)\boldsymbol{g}^{-1}. \tag{2.61d}$$

Since $\mathcal{U}_{\mathcal{Q}}(t)$ is unitary and $\mathcal{U}_{\mathcal{Q}}(-t)^{-1} = \mathcal{U}_{\mathcal{Q}}(t)$, one gets $(A_m^{\mathcal{Q}} | \mathcal{U}_{\mathcal{Q}}(-t) A_k^{\mathcal{Q}})^* = (A_k^{\mathcal{Q}} | \mathcal{U}_{\mathcal{Q}}(t) A_m^{\mathcal{Q}})$. Thus, there hold the symmetry relations

$$\sum_\ell M_{m,\ell}(t) g_{\ell,n} = \sum_\ell M_{n,\ell}(-t)^* g_{m,\ell}. \tag{2.61e}$$

Laplace transformation and using Eqs. (A.5b,c) yields the equivalent symmetries in the frequency domain

$$\sum_\ell M_{m,\ell}(z) g_{\ell,n} = \sum_\ell M_{n,\ell}(z^*)^* g_{m,\ell}. \tag{2.61f}$$

These identities are generalized Onsager relations.

Multiplying Eq. (2.59a) from the left by $(A_m|$ and from the right by $|A_n)$ and using Eqs. (2.60b), (2.61b), one arrives at

$$\sum_\ell \left[z\delta_{m,\ell} - \Omega_{m,\ell} + M_{m,\ell}(z) \right] \phi_{\ell,n}(z) = -g_{m,n}. \tag{2.62a}$$

Using formulas (A.6a,b) and (A.7a), one gets the equations

$$i\partial_t \phi_{m,n}(t) - \sum_\ell \Omega_{m,\ell} \phi_{\ell,n}(t) + i \sum_\ell \int_0^t M_{m,\ell}(t - t') \phi_{\ell,n}(t') dt' = 0, \tag{2.62b}$$

which have to be solved with the initial conditions $\phi_{m,n}(t = 0) = g_{m,n}$. The preceding formulas are the Zwanzig–Mori equations of motion in the frequency domain for the functions $\phi_{m,n}(z)$ and in the time domain for the correlators $\phi_{m,n}(t)$, respectively, (Zwanzig 1961a,b; Mori 1965a,b).

The Zwanzig–Mori equations (2.62a) express the special correlation functions $\phi_{m,n}(z) = (A_m \mid \mathcal{R}(z) A_n)$ in terms of overlaps of dynamical variables, namely $\Omega_{m,n}$, and of matrix elements of the reduced resolvent, namely $M_{m,n}(z)$. The analogous result can be obtained for general correlators in the frequency domain, say $\phi_{X,Y}(z)$. Because of Eq. (2.33), one gets the result by taking the $X - Y$ matrix element of Eq. (2.59b). The projector $\mathcal{P}$ is substituted from Eq. (2.55). Overlaps of dynamical variables are due to the first terms in the brackets of Eq. (2.59b). The contribution from the left bracket shall be denoted by

$$\Omega_m^X = \sum_k (X \mid A_k)(\boldsymbol{g}^{-1})_{k,m}. \tag{2.63a}$$

The hermiticity of matrix $\boldsymbol{g}^{-1}$ and of the scalar product imply $\Omega_m^{X*} = \sum_k (\boldsymbol{g}^{-1})_{m,k}(A_k \mid X)$. Therefore, the z-independent contribution from the right bracket in Eq. (2.59b) is given by

$$\Omega_n^{Y*} = \sum_\ell (\boldsymbol{g}^{-1})_{n,\ell}(A_\ell|Y). \tag{2.63b}$$

The second term in the left bracket of Eq. (2.59b) yields a reduced-resolvent matrix element, which shall be denoted by

$$M_m^X(z) = i\sum_k (\mathcal{Q}X \mid \mathcal{R}_\mathcal{Q}(z)\dot{A}_k^\mathcal{Q})(\boldsymbol{g}^{-1})_{k,m}. \tag{2.63c}$$

The fluctuating forces $\dot{A}_k^\mathcal{Q}$ appear as result of $\mathcal{L}A_k$, Eq. (2.61a). The hermiticity of the reduced Liouvillian implies $\mathcal{R}_\mathcal{Q}(z^*)^\dagger = \mathcal{R}_\mathcal{Q}(z)$ and thus $M_m^X(z^*)^* = -i\sum_k(\boldsymbol{g}^{-1})_{mk}(\dot{A}_k^\mathcal{Q}|\mathcal{R}_\mathcal{Q}(z)\mathcal{Q}X)$. Hence, the z-dependent contribution from the right bracket in Eq. (2.59b) is given by

$$M_n^Y(z^*)^* = -i\sum_\ell (\boldsymbol{g}^{-1})_{n\ell}(A_\ell^\mathcal{Q}|\mathcal{R}_\mathcal{Q}(z)\mathcal{Q}Y). \tag{2.63d}$$

The contribution due to the first term in Eq. (2.59b) shall be abbreviated by

$$M_{X,Y}(z) = (\mathcal{Q}X|\mathcal{R}_\mathcal{Q}(z)|\mathcal{Q}Y). \tag{2.63e}$$

The desired result reads

$$\phi_{X,Y}(z) = M_{X,Y}(z) + \sum_{m,n}\left[\Omega_m^X + M_m^X(z)\right]\phi_{m,n}(z)\left[\Omega_n^Y + M_n^Y(z^*)\right]^*. \tag{2.64}$$

The Zwanzig–Mori equations of motion formalize the intention to describe the dynamics of a system as that of two coupled subsystems. The first subsystem deals with the set of distinguished variables. The second one deals with the variables perpendicular to those of the first one. If the changes of the distinguished variables $\lim_{\tau\to 0}[A_\ell(\tau) - A_\ell(\tau = 0)]/\tau$ would be a vector in the first subsystem, $\dot{A}_\ell^\mathcal{Q}$ in Eq. (2.61a) would be zero. The dynamics for the first subsystem would be independent of that for the second one. In this case, the functions $M_{m,\ell}(t)$ would be zero. Assuming that there is a finite number d of distinguished variables, the equation of motion (2.62b) would reduce to that discussed above in connection with Eqs. (2.15a)–(2.17c). For the special case that the two subsystems are related to different types of particles, the $\dot{A}_\ell^\mathcal{Q}(t) = \mathcal{U}_\mathcal{Q}(t)\mathcal{Q}\partial_t A_\ell$ would be combination of forces, which the second subsystem exerts on the first one. But these are not the complete forces. The part of the forces, which can be written as a combination of the distinguished variables, is subtracted. The subtracted part is accounted for by the matrix of frequencies $\Omega_{n,m}$. Moreover, the fluctuating forces do not evolve as described by $\mathcal{U}(t)$. Their time evolution is generated by the reduced Liouvillian $\mathcal{L}_Q$. If the fluctuating forces are non-zero, the time variation of the first subsystem creates a disturbance of the second subsystem. The latter varies in time as described by the evolution operator $\mathcal{U}_Q(t)$. The change of the second system causes a change of the forces on the first system. This force, which acts at time t on the variable A_m, depends on the perturbation created by the first system at time t', where $t' \leq t$. The specified coupling effect of $A_\ell(t')$ with $A_m(t)$ is described by the functions $M_{m,\ell}(t-t')$. Therefore, the fluctuating-force correlators $M_{m,\ell}(t)$ are also called memory kernels. For $M_{m,\ell}(t) = 0$, the correlators $\phi_{m,\ell}(t)$ are almost periodic functions, Eq. (2.17c). The functions $M_{m,\ell}(t)$ or $M_{m,\ell}(z)$ are also called relaxation kernels.

The Zwanzig–Mori formalism has reduced the problem of calculating the correlators of the distinguished variables to two problems. Firstly, one has to determine the memory kernels and, secondly, one has to solve the equations of motion (2.62b). For given kernels, the latter equations are linear integro-differential equations of the Volterra type. Via Laplace transformation, the problem of solving these equations can be reduced to the problem (2.62a) of a matrix inversion.

There are many possibilities to select sets of distinguished variables, i.e., to formulate Zwanzig–Mori equations. The mere formulation of these equations does not necessarily imply a simplification of the problem to study the dynamics of a given system. But the formalism supplies a frame for studies provided one has a motivation for the selection of a distinguished set of variables and for a treatment of the memory kernels. The original applications of the Zwanzig–Mori formalism are built on the understanding that the memory kernels are simpler than the correlators of the distinguished variables. Examples from the theory of liquids demonstrating this situation are discussed in Chapter 3. The application to be used in this book will not be based on the understanding of an asymmetry between the spectra of the correlators and those of the kernels. On the contrary, the theory will be used with the understanding that the glassy-dynamics correlators $\phi_{m,n}(t)$ and their kernels $M_{m,n}(t)$ exhibit the same subtleties. This is discussed in Chapter 4.

2.4.2 *Models for correlation functions*

In the preceding section, the general properties of correlation function have been used in order to derive the equations of motion (2.62a,b). These equations are specified by matrices of frequencies $\Omega_{m,n}$ and of kernels $M_{m,n}(z)$. Some general properties of these matrices have been derived. In this section, the reversed view shall be adopted. The mentioned matrices $\boldsymbol{\Omega}$ and $\boldsymbol{M}(z)$ are considered as given quantities. They specify equations of the type derived by R. Zwanzig and H. Mori for a matrix of functions $\phi_{m,n}(z)$. Properties of $\boldsymbol{\Omega}$ and $\boldsymbol{M}(z)$ shall be formulated, which guarantee the existence of a unique solution $\boldsymbol{\phi}(z)$. It will be shown that the $\phi_{m,n}(t)$ and their Laplace transforms $\phi_{m,n}(z)$ exhibit the general properties of correlation-function matrices in the time domain and frequency domain, respectively. These considerations, which are due to Kadanoff and Martin (1963), are the mathematical basis of motivations of models for a statistical description of a dynamics.

The essence of the approach shall be explained first by considering a single distinguished variable only. Let $\phi(t)$ denote a function, which is defined for all times t. It shall be continuous, bounded, and have some determining function as first derivative. For convenience, it shall be normalized: $\phi(t=0)=1$. It shall be requested that the function solves the following equation of motion for $t \gtrless 0$:

$$\partial_t\phi(t) + i\Omega\phi(t) \pm \nu\phi(t) + \int_0^t M(t-t')\phi(t')dt' = 0. \tag{2.65a}$$

Here, Ω and ν denote two real numbers and $\nu \geq 0$. Function $M(t)$ shall be continuous and positive definite. Because of Eq. (2.27b), there holds $|M(t)| \leq C$ with $C = M(0)$. According to Appendix A.3, the requested properties for $M(t)$ are equivalent to requirement that there is a positive-analytic Laplace transform $M(z) = LT[M(t)](z)$, which obeys the inequality: $|M(z)| \leq C/|\text{Im}\, z|$. From Eqs. (A.6a,b), (A.7a) one concludes that there is a unique solution $\phi(t)$ if and only if its Laplace transform $\phi(z)$ is uniquely determined by the equation

$$\big[z - \Omega \pm i\nu + M(z)\big]\phi(z) = -1, \qquad \text{Im}\, z \gtrless 0. \tag{2.65b}$$

For non-real values of z, $M(z)$ is holomorphic. Hence, function $G(z) = [z - \Omega \pm i\nu + M(z)]$ is holomorphic, and Eq. (2.65b) defines uniquely a meromorphic function $\phi(z)$. The symmetry of positive-analytic functions, $M(z)^* = M(z^*)$, implies the symmetry $\phi(z)^* = \phi(z^*)$. For Im $z \gtrless 0$, there holds Im $M(z) \gtrless 0$. Consequently, $\pm\text{Im}\, G(z) = \pm[\text{Im}\, z \pm \nu \pm \text{Im}\, M(z)] \geq \pm\text{Im}\, z > 0$. The function $G(z)$ does not have a zero, i.e., $\phi(z)$ is holomorphic. Furthermore, Im $\phi(z) = \text{Im}\, G(z)/|G(z)|^2 \gtrless 0$ for Im $z \gtrless 0$. The solution $\phi(z)$ is positive analytic. Since $|\text{Im}\, G(z)| \geq |\text{Im}\, z|$, there holds the fundamental bound condition $|\phi(z)| \leq 1/|\text{Im}\, z|$. According to Appendix A.3, the unique solution $\phi(t)$ of Eq. (2.65a) is continuous and positive definite. According to Bochner's theorem, it exhibits a spectral representation (2.51a). In this sense, it can be considered as a model of a correlation function of some dynamical variable A: $\phi(t) = (A(t)|A)$. It is normalized: $\phi(t=0) = (A|A) = 1$. The equation of motion (2.65a) for $\phi(t)$ together with the initial condition is equivalent to the fraction representation for the Laplace transform: $\phi(z) = -1/G(z)$.

The fraction representation implies

$$\phi(z) = (-1/z)\Big\{1 + \big[(\Omega \mp i\nu)/z\big] + O(1/(z\text{Im}\, z))\Big\}. \tag{2.66a}$$

If the kernel has continuous boundary values $M(\omega \pm i0)$ and if $\nu + M''(\omega) \neq 0$, also the functions $\phi(\omega \pm i0) = \phi'(\omega) \pm i\phi''(\omega)$ are continuous. The high-frequency tail of the spectral function reads

$$\phi''(\omega) = (\nu/\omega^2)\big[1 + O(1/\omega)\big]. \tag{2.66b}$$

Specializing the spectral representation (2.35b) to $t = 0$, one gets the sum rule

$$\frac{1}{\pi}\int \phi''(\omega)d\omega = 1. \tag{2.66c}$$

Because of the $1/\omega^2$-tail of the $\phi''(\omega)$ versus ω function, one cannot interchange the time derivative with the integration in Eq. (2.35b). It is left to the reader to show that the inversion formula (A.4) can be combined with Eq. (2.66b) in order to get

$$P\frac{1}{\pi}\int \omega\phi''(\omega)d\omega = \Omega. \tag{2.66d}$$

Here, the principle-value integration is defined by $P\int \ldots d\omega = \lim_{R\to\infty} \int_{-R}^{R} \ldots d\omega$. Changing the function $M(z)$, one can change the non-negative spectrum $\phi''(\omega)$ drastically. However, the two sum rules (2.66c,d) do not change at all.

Equations (2.65a,b) are generalizations of the Zwanzig–Mori equations in the sense that the positive analytic function $M(z)$ is complemented by a function $M^{(0)}(z) = \pm i\nu$, Im $z \gtrless 0$. Both functions $M(z)$ and $M^{(0)}(z)$ are positive analytic. Contrary to $M(z)$, function $M^{(0)}(z)$ does not vanish for Im $z \to \pm\infty$. Function $M^{(0)}(z)$ can neither be written as Laplace transform of a continuous function, nor can it be written as a spectral integral. There exists a spectrum in the sense that $\nu = [M^{(0)}(\omega + i0) - M^{(0)}(\omega - i0)]/(2i)$. Such a frequency-independent limit is called a white-noise spectrum.

In the following generalizations of the preceding considerations, notations and results for matrix functions are used. These are explained in Appendix A.5. It is the aim to discuss d-by-d matrix functions $\boldsymbol{\phi}(t)$, which combine d^2 functions of the time $\phi_{m,n}(t)$; $m, n = 1, \ldots, d$. These functions are defined for all times t, they are continuous, bounded, and have determining functions $\partial_t\phi_{m,n}(t)$ as first derivatives. The initial value $\boldsymbol{\phi}(t = 0) = \boldsymbol{g}$ is requested to be a hermitian matrix, which is positive definite: $\boldsymbol{g} > \boldsymbol{0}$. This implies that $\boldsymbol{g}^{-1}$ exists and is positive definite as well. Let $\hat{\boldsymbol{\Omega}}$ and $\hat{\boldsymbol{\nu}}$ denote two hermitian matrices; the latter shall be positive semidefinite: $\hat{\boldsymbol{\nu}} \geq \boldsymbol{0}$. Furthermore, $\hat{\boldsymbol{M}}(t)$ shall abbreviate a continuous positive-definite matrix function. According to Appendix A.5, these restrictions are equivalent to the request that the Laplace transform $\hat{\boldsymbol{M}}(z) = LT[\hat{\boldsymbol{M}}(t)](z)$ is a positive-analytic matrix function obeying the condition: $|M_{k,\ell}(z)| \leq C/|\text{Im}\, z|$; $k, \ell = 1, \ldots, d$. Here, C denotes a finite positive number. Two further frequency matrices are defined by $\boldsymbol{\Omega} = \hat{\boldsymbol{\Omega}}\boldsymbol{g}^{-1}$ and $\boldsymbol{\nu} = \hat{\boldsymbol{\nu}}\boldsymbol{g}^{-1}$. A frequency-dependent kernel shall be introduced by $\boldsymbol{M}(z) = \hat{\boldsymbol{M}}(z)\boldsymbol{g}^{-1}$; it is the Laplace transform of a matrix function $\boldsymbol{M}(t) = \hat{\boldsymbol{M}}(t)\boldsymbol{g}^{-1}$. By construction, there holds the symmetry relation (2.60c) and a corresponding one with $\boldsymbol{\Omega}$ replaced by $\boldsymbol{\nu}$. Similarly, the relations

(2.61e,f) are valid. It is required that $\boldsymbol{\phi}(t)$ solves the set of integrodifferential equations of the Volterra type for $t \gtrless 0$:

$$\partial_t \boldsymbol{\phi}(t) + i\boldsymbol{\Omega}\boldsymbol{\phi}(t) \pm \boldsymbol{\nu}\boldsymbol{\phi}(t) + \int_0^t \boldsymbol{M}(t-t')\boldsymbol{\phi}(t')dt' = 0. \tag{2.67a}$$

This equation together with the initial condition is equivalent to the matrix equation, which expresses the $\boldsymbol{\phi}(z)$ in terms of the kernel $\boldsymbol{M}(z)$:

$$\big[z\mathbf{1} - \boldsymbol{\Omega} \pm i\boldsymbol{\nu} + \boldsymbol{M}(z)\big]\boldsymbol{\phi}(z) = -\boldsymbol{g}, \qquad \mathrm{Im}\, z \gtrless 0. \tag{2.67b}$$

As discussed above, these equations generalize the Zwanzig–Mori equations for correlation functions by the addition of a white-noise kernel $\boldsymbol{\nu}$.

The symmetries of the preceding equations are more transparent if one maps the function $\boldsymbol{\phi}(t)$ invertibly on another function $\hat{\boldsymbol{\phi}}(t)$ according to

$$\hat{\boldsymbol{\phi}}(t) = \boldsymbol{g}^{-1}\boldsymbol{\phi}(t)\boldsymbol{g}^{-1}. \tag{2.68}$$

Since $\boldsymbol{P} = \boldsymbol{g}^{-1}$ is hermitian, one concludes from the discussion of Eq. (A.45c) that $\hat{\boldsymbol{\phi}}(t)$ is a positive-semidefinite matrix if and only if this is true for $\boldsymbol{\phi}(t)$. Hence, $\hat{\boldsymbol{\phi}}(t)$ is a positive-definite continuous bounded matrix function if and only if the function $\boldsymbol{\phi}(t)$ has this property. Similarly, one concludes from Appendix A.5 that $\hat{\boldsymbol{\phi}}(z) = LT[\hat{\boldsymbol{\phi}}(t)](z)$ is positive analytic and obeys the bound condition (A.52) if and only if $\boldsymbol{\phi}(z)$ exhibits these properties. Multiplication of Eq. (2.67a) by $\boldsymbol{g}^{-1}$ from the right casts the equation of motion in the equivalent equation:

$$\boldsymbol{g}\partial_t \hat{\boldsymbol{\phi}}(t) + i\hat{\boldsymbol{\Omega}}\hat{\boldsymbol{\phi}}(t) \pm \hat{\boldsymbol{\nu}}\hat{\boldsymbol{\phi}}(t) + \int_0^t \hat{\boldsymbol{M}}(t-t')\hat{\boldsymbol{\phi}}(t')dt' = 0. \tag{2.69a}$$

It has to be solved with the initial condition $\hat{\boldsymbol{\phi}}(t=0) = \boldsymbol{g}^{-1}$. Equivalently, there holds

$$\big[z\boldsymbol{g} - \hat{\boldsymbol{\Omega}} \pm i\hat{\boldsymbol{\nu}} + \hat{\boldsymbol{M}}(z)\big]\hat{\boldsymbol{\phi}}(z) = -\mathbf{1}. \tag{2.69b}$$

Since $\hat{\boldsymbol{M}}(z)$ is holomorphic for Im $z \neq 0$, the same is true for the matrix function $\boldsymbol{G}(z) = [z\boldsymbol{g} - \hat{\boldsymbol{\Omega}} \pm i\hat{\boldsymbol{\nu}} + \hat{\boldsymbol{M}}(z)]$. Using the definition (A.51) and the hermiticity of $\boldsymbol{g}, \hat{\boldsymbol{\Omega}}$, and $\hat{\boldsymbol{\nu}}$, one gets: $\pm\mathrm{Im}\,\boldsymbol{G}(z) = \pm\mathrm{Im}\, z\boldsymbol{g} + \boldsymbol{\nu} \pm \mathrm{Im}\,\hat{\boldsymbol{M}}(z)$. Matrix $\hat{\boldsymbol{M}}(z)$ is positive analytic and this implies $\pm\mathrm{Im}\,\boldsymbol{M}(z) \geq 0$. Since $\boldsymbol{g} > 0$ and $\hat{\boldsymbol{\nu}} \geq 0$, one concludes that $\pm\mathrm{Im}\,\boldsymbol{G}(z) \geq |\mathrm{Im}\, z|\boldsymbol{g} > 0$ for Im $z \gtrless 0$. Consequently, matrix $\boldsymbol{G}(z)$ is invertible, i.e., $\hat{\boldsymbol{\phi}}(z) = -\boldsymbol{G}(z)^{-1}$ is holomorphic for all non-real frequencies z. The symmetry $\hat{\boldsymbol{M}}(z)^\dagger = \hat{\boldsymbol{M}}(z^*)$ implies the corresponding one for $\hat{\boldsymbol{\phi}}(z)$. Since $\mathrm{Im}\,\hat{\boldsymbol{\phi}}(z) = [\boldsymbol{G}(z)^{-1\dagger} - \boldsymbol{G}(z)^{-1}]/(2i) = \boldsymbol{G}(z)^{-1\dagger}\,[\mathrm{Im}\,\boldsymbol{G}(z)]\boldsymbol{G}(z)^{-1}$, one can use Eqs. (A.45b,c) with $\boldsymbol{M} = \mathrm{Im}\,\boldsymbol{G}(z)$ and $\boldsymbol{P} = \boldsymbol{G}(z)^{-1}$ in order to conclude: Im $\hat{\boldsymbol{\phi}}(z) \gtrless 0$ for Im $z \gtrless 0$. In summary, Eq. (2.69b) defines a positive-analytic matrix function $\hat{\boldsymbol{\phi}}(z)$. The mentioned bound condition $\boldsymbol{M}(z) \leq \boldsymbol{M}(t=0)/|\mathrm{Im}\, z|$ leads to the formula

$$\hat{\boldsymbol{\phi}}(z) = -[z\boldsymbol{g}]^{-1}\Big\{\mathbf{1} + \big[(\boldsymbol{\Omega} \mp i\boldsymbol{\nu})/z\big] + \boldsymbol{O}(1/|z\,\mathrm{Im}\, z|)\Big\}. \tag{2.70}$$

Matrix $\hat{\boldsymbol{\phi}}(z)$ exhibits the bound condition (A.52). Therefore, it is the Laplace transform of a continuous positive-definite matrix function $\hat{\boldsymbol{\phi}}(t)$. The elements of $\boldsymbol{\phi}(t)$ provide models of correlation functions formed for a set of independent variables $A_1, A_2, \ldots : \phi_{m,n}(t) = (A_m(t)|A_n)$.

2.5 Linear-response theory

The scalar product of two dynamical variables, $(A|B)$, the $A - B$ correlator in the time domain, $\phi_{A,B}(t)$, and the boundary values of the correlator in the frequency domain, $\phi'_{A,B}(\omega) + i\phi''_{A,B}(\omega)$, determine the response of the system to external perturbations. Therefore, these quantities are measurable in principle. It is crucial to appreciate these features of the correlation-function description of the dynamics in order to establish a physical understanding of the outcome of theories formulated within this frame. It is equally important to know these features in order to correlate the theory with the results of experiments. This section is devoted to a derivation of the results of the response theory, which were obtained in full generality by Kubo (1957).

Let us start with a problem, which is – from a mathematical point of view – elementary. The Hamilton function $H(\boldsymbol{p}, \boldsymbol{q})$ shall be modified by some addition $\Delta H_b(\boldsymbol{p}, \boldsymbol{q})$, which will be referred to as a perturbation. For the sake of simplicity, it shall be assumed to be a linear combination of dynamical variables $B_n(\boldsymbol{p}, \boldsymbol{q})$, $n = 1, 2, \ldots$:

$$\Delta H_b(\boldsymbol{p}, \boldsymbol{q}) = -\sum_n b_n B_n(\boldsymbol{p}, \boldsymbol{q}). \tag{2.71}$$

In other words, a system shall be considered, that is specified by the Hamiltonian $H_b(\boldsymbol{p}, \boldsymbol{q}) = H(\boldsymbol{p}, \boldsymbol{q}) + \Delta H_b(\boldsymbol{p}, \boldsymbol{q})$. Here and in the following, an index b is attached whenever a dependence on the parameters $b_1, b_2, \ldots$ is to be emphasized. The function ΔH_b has to be real. However, it is often convenient to write the perturbation as superposition of complex quantities. Therefore, the b_n as well as the $B_n(\boldsymbol{p}, \boldsymbol{q})$ are permitted to be complex. The coefficient b_n is referred to as external field conjugate to the variable B_n, or, coupling to the variable B_n. Boltzmann's probability density of the perturbed system reads $\rho_b(\boldsymbol{p}, \boldsymbol{q}) = \exp[-H_b(\boldsymbol{p}, \boldsymbol{q})/(k_\mathrm{B}T)]/Z_b$, where $Z_b = \iint d\boldsymbol{p}d\boldsymbol{q}\exp[-H_b(\boldsymbol{p}, \boldsymbol{q})/(k_\mathrm{B}T)]$. One can expand $\rho_b(P) = \rho(P)[1 - (\Delta H_b(P) - \langle \Delta H_b \rangle)/(k_\mathrm{B}T) + O(b^2)]$. Thus, the expectation value of some variable X for the perturbed system reads: $\langle X \rangle_b = \int d\boldsymbol{p}, d\boldsymbol{q}\rho_b(\boldsymbol{p}, \boldsymbol{q})X(\boldsymbol{p}, \boldsymbol{q}) = \langle X \rangle - \langle X(\Delta H_b - \langle \Delta H_b \rangle)\rangle/(k_\mathrm{B}T) + O(b^2)$. It is convenient to introduce some new notation. The deviation of some variable X from its expectation value is called the fluctuation of X; it shall be abbreviated by

$$\delta X = X - \langle X \rangle. \tag{2.72a}$$

If X_1 denotes the trivial variable given by $X_1(\boldsymbol{p}, \boldsymbol{q}) = 1$, one gets $\langle X_1 \rangle = (X_1|X_1) = 1$. Thus, δX is the projection of X perpendicular to $X_1 : \delta X = X - X_1(X_1|X)$. The scalar product of the fluctuations of two variables, say A and B, is used to define a quantity called thermodynamic $A - B$ susceptibility:

$$\chi^T_{A,B} = (\delta A|\delta B)/(k_\mathrm{B}T). \tag{2.72b}$$

Using $(A|\delta B) = (\delta A|B) = (\delta A|\delta B) = (A|B) - \langle A^* \rangle \langle B \rangle$, and replacing X by the conjugate A^* of some variable A, one obtains the desired expansion formula

$$\langle A^* \rangle_b - \langle A^* \rangle = \sum_n \chi^T_{A,B_n} b_n + O(b^2). \tag{2.73}$$

Perturbing the system by an external field of strength b that couples to the dynamical variable B, $\Delta H_b = -bB$, and measuring the expectation value of variable $\delta A^* = A^* - \langle A^* \rangle$, one can determine $\chi^T_{A,B}$. This is done by extrapolating the ratio $\langle \delta A^* \rangle_b / b$ to the limit of vanishing b. The non-trivial request for the experiment is the guarantee that the system is connected to some thermostat so that expectation values are determined by the distribution function $\rho_b(P)$ for temperatures T.

The mathematical properties of the susceptibility follow from those discussed in Sec. 2.3 for the scalar product. Let us assume, for example, that the variables $A_n = \delta B_n$, $n = 1, 2, \ldots$, are linearly independent. The susceptibility matrix formed with these variables shall be denoted by

$$\chi^T_{m,n} = (A_m|A_n)/(k_\mathrm{B}T). \tag{2.74}$$

Since this matrix is proportional to the one discussed in Eq. (2.12), $\chi^T_{m,n} = g_{m,n}/(k_\mathrm{B}T)$, it is positive definite. In particular, $\chi^T_{B,B} = \langle |\delta B|^2 \rangle / (k_\mathrm{B}T) > 0$ if B is not a multiple of X_1. Let us consider a perturbation $\Delta H_b = -b \cdot B$ for fixed b, which is so small that the $O(b^2)$ terms in Eq. (2.73) can be neglected. If $\chi^T_{B,B}$ is large, a large change of $\langle \delta B \rangle$ is caused by the perturbation. In this sense, the system exhibits a soft response. If $\chi^T_{B,B}$ is small, the response is small, i.e., the system reacts stiffly.

In order to extend the discussion to more general deviations of the system from equilibrium, one needs a theory for the calculation of the probability density for general situations. The statistical description shall be achieved by the formation of averages,where, however, the weight function may depend on the time as well. It shall be denoted by $\rho_t(\boldsymbol{p}, \boldsymbol{q})$. To motivate an equation for $\rho_t(P)$, two results from classical mechanics have to be remembered. First, Hamilton's equations remain valid also for non-autonomous systems. To deal with such situation, the Hamilton function has to be generalized to one depending on time: $H_t(\boldsymbol{p},\boldsymbol{q})$. The orbits are solutions of the set of first-order differential equation $\partial_t q_\alpha(t) = \partial H_t(P_t)/\partial p_\alpha$, $\partial_t p_\alpha(t) = -\partial H_t(P_t)/\partial q_\alpha$, $\alpha = 1,\ldots,f$. Solved for the initial condition $P_{t_0} = P_0$, they describe the flow in phase space $P_0 \to P_t$ in the same way as explained in Sec. 2.1 for isolated systems. Second, there holds Liouville's theorem: the Jacobian determinant of the mapping $P_0 \to P_t$ is unity: $\partial(\boldsymbol{p}(t),\boldsymbol{q}(t))/\partial(\boldsymbol{p}(t_0),\boldsymbol{q}(t_0)) = 1$. Equivalently, there holds: if a $(2f)$-dimensional domain G_0 of the phase space is mapped on a domain $G(t)$, its volume remains unchanged: $\iint_{G(t)} d\boldsymbol{p}d\boldsymbol{q} = \iint_{G_0} d\boldsymbol{p}d\boldsymbol{q}$. For non-autonomous systems as well as for autonomous ones, the flow in phase space is that of an incompressible fluid.

It is plausible to request the conservation of probability, i.e., it is natural to postulate that the probability to find the system in $G(t)$ at time t is the same as that to find it in G_0 at time t_0. This means $\iint_{G(t)} \rho_t(\boldsymbol{p},\boldsymbol{q})d\boldsymbol{p}d\boldsymbol{q} = \iint_{G_0} \rho_{t_0}(\boldsymbol{p},\boldsymbol{q})d\boldsymbol{p}d\boldsymbol{q}$. Postulating this equation is equivalent to postulating that the time derivative of the integral is zero, i.e., $(d/dt)\iint_{G_0} \rho_t(\boldsymbol{p}(t),\boldsymbol{q}(t))d\boldsymbol{p}(t_0)$ $d\boldsymbol{q}(t_0) = 0$. Since G_0 is arbitrary, the time derivative of the integrand has to vanish: $\partial_t\rho_t(\boldsymbol{p}(t),\boldsymbol{q}(t)) + \sum_\alpha(\partial_t p_\alpha(t))\partial\rho_t(P_t)/\partial p_\alpha + (\partial_t q_\alpha(t))\partial\rho_t(P_t)/\partial q_\alpha(t) = 0$. Substituting Hamilton's equations, one arrives at Liouville's equation of motion for the probability density

$$\partial_t\rho_t(P) = \{H_t(P),\, \rho_t(P)\}. \tag{2.75}$$

If $H_t(P) = H(P)$ is time independent, any function F of the Hamilton function, $\rho(P) = F(H(P))$, is a time independent solution of Eq. (2.75). The Liouville equation has to be complemented with the postulate that only Boltzmann's functions $F(x) \propto \exp[-x/(k_{\mathrm{B}}T)]$ shall be considered as equilibrium solution. Notice that Eq. (2.75) agrees with the equation for the change of a dynamical variable $\partial_t A(P_t)$, except for a change of the sign on the right-hand side.

As a first application of Liouville's equation, let us consider a system whose probability density at time $t = 0$ is the Boltzmann distribution $\rho_{\mathrm{b}}(\boldsymbol{p},\boldsymbol{q})$ for the modified Hamiltonian $H_{\mathrm{b}}(\boldsymbol{p},\boldsymbol{q})$. For time $t > 0$, there should be no perturbation, i.e., the system should exhibit the dynamics as governed by the Hamilton function $H(\boldsymbol{p},\boldsymbol{q})$. Hence, one has to solve Eq. (2.75) for the isolated system $\partial_t\rho(P) = \{H(P),\, \rho_t(P)\}$. According to Eqs. (2.4a,c), the solution reads $\rho_t(P) = \mathcal{U}(-t)\rho_{t=0}(P)$. In the paragraph following Eq. (2.71), the initial value is identified as $\rho_{t=0}(\boldsymbol{p},\boldsymbol{q}) = \rho(\boldsymbol{p},\boldsymbol{q})[1 + \sum_n b_n\delta B_n/(k_{\mathrm{B}}T) + O(b^2)]$. Hence, the expectation value of some variable A^*, $\langle A^*\rangle_t = \iint \rho_t(\boldsymbol{p},\boldsymbol{q})A(\boldsymbol{p},\boldsymbol{q})^* d\boldsymbol{p}d\boldsymbol{q}$, can be expressed in terms of $(A|\mathcal{U}(-t)\delta B_n) = (\delta A|\mathcal{U}(-t)\delta B_n)$. The latter matrix elements are correlators as defined in Eq. (2.18b). One arrives at the desired result:

$$\langle A^*\rangle_{b,t} - \langle A^*\rangle = \sum_n \phi_{\delta A,\delta B_n}(t)\cdot b_n/(k_{\mathrm{B}}T) + O(b^2). \tag{2.76a}$$

The correlators $\phi_{\delta A,\delta B_n}(t) = \phi_{A,B_n}(t) - \langle A^*\rangle\langle B_n\rangle$ determine the time evolution of the expectation values $\langle\delta A^*\rangle_{b,t}$ of the isolated system out of a specific initial state. The latter is the equilibrium state corresponding to the modified Hamilton function $H - \sum_n b_n B_n$. The correlators describe $\langle \delta A^* \rangle_{b,t}$ in leading order in the external fields b_n. Measuring $\langle \delta A^* \rangle_{b,t}$ for an initial state corresponding to the perturbation $\Delta H_b = -bB$, one can determine the correlator $\phi_{\delta A, \delta B}(t)$. This is done by extrapolation of the ratio $\langle \delta A^* \rangle_{b,t}/b$ for b tending to zero.

If one uses Eqs. (2.73), (2.74) and introduces the normalization matrix $g_{m,n}$ from Eq. (2.12), one can write for the initial condition $\langle A_m^* \rangle_{b,t=0} = \langle A_m^* \rangle_b = \langle A_m^* \rangle + \sum_n g_{m,n} b_n/(k_\text{B}T) + O(b^2)$. This formula can be used in order to express the perturbation amplitudes $b_1, b_2, \ldots$ in terms of the initial values $\langle \delta A_1^* \rangle_b$, $\langle \delta A_2^* \rangle_b, \ldots$. One gets

$$\langle \delta A^* \rangle_{b,t} = \sum_{m,n} \phi_{\delta A, A_m}(t) (\boldsymbol{g}^{-1})_{m,n} \langle \delta A_n^* \rangle_b + O(b^2). \tag{2.76b}$$

Let us consider the more complicated situation where there is a time dependent perturbation $\Delta H_t(\boldsymbol{p}, \boldsymbol{q})$. For the beginning of the considerations, it shall be assumed that there is some time t_0 with the following property. For $t \leq t_0$, the perturbation vanishes. For these early times, the system is described statistically by Boltzmann's distribution $\rho(P)$ for the Hamilton function $H(\boldsymbol{p}, \boldsymbol{q})$. For $t > t_0$, the motion is governed by the time dependent Hamilton function $H_t(\boldsymbol{p}, \boldsymbol{q}) = H(\boldsymbol{p}, \boldsymbol{q}) + \Delta H_t(\boldsymbol{p}, \boldsymbol{q})$. The task is the evaluation of $\rho_t(\boldsymbol{p}, \boldsymbol{q})$ for $t > t_0$ so that expectation values of dynamical variables, say $\langle A^* \rangle_t$, can be calculated. These will, in general, depend on the time. Notice that it is the formulation of the problem, which introduces a distinction between the dynamics in the past and that in the future.

The Liouville equation can be written as $\partial_t \rho_t(P) - \{H(P), \rho_t(P)\} = I_t(P)$, where $I_t(P) = \{\Delta H_t(P), \rho_t(P)\}$. With respect to the time dependence, this expression is an inhomogeneous linear first-order differential equation. Variable $I_t(P)$ is the inhomogeneity. A solution of the homogeneous equation reads $\rho_t^H(P) = \mathcal{U}(t_0 - t)\rho_{t_0}(P)$. The standard procedure of solving the inhomogeneous equation is the method of variation of constants: one introduces a function $\tilde{\rho}_t(P) = \mathcal{U}(t - t_0)\rho_t(P)$, i.e., one represents the solution of the Liouville equation in the form $\rho_t(P) = \mathcal{U}(t_0 - t)\tilde{\rho}_t(P)$. The initial condition for the new function reads $\tilde{\rho}_{t_0}(P) = \rho(P) = \rho_{t_0}$. The Liouville equation is equivalent to the equation for $\tilde{\rho}_t(P) : \partial_t \tilde{\rho}_t(P) = \mathcal{U}(t - t_0) I_t(P)$. Integrating this equation over t, one finds $\tilde{\rho}_t(P) = \tilde{\rho}_{t_0}(P) + \int_{t_0}^t \mathcal{U}(t' - t_0) I_{t'}(P) dt'$. Multiplying this equation with $\mathcal{U}(t_0 - t)$, one arrives at

$$\rho_t(P) = \rho(P) + \int_{t_0}^t \mathcal{U}(t' - t) \big\{ \Delta H_{t'}(P), \, \rho_{t'}(P) \big\} dt'. \tag{2.77a}$$

One checks that a solution of this equation is a solution of Eq. (2.75) as well. Hence, the linear integral equation (2.77a) is equivalent to the Liouville equation for the specified Hamilton function together with the initial condition $\rho_{t_0} = \rho$.

Iteration casts the integral equation in the equivalent form:

$$\rho_t(P) = \rho(P) + \int_{t_0}^{t} \mathcal{U}(t'-t)\{\Delta H_{t'}(P),\, \rho(P)\}dt'$$
$$+ \int_{t_0}^{t} \mathcal{U}(t'-t) \int_{t_0}^{t'} \Big\{\Delta H_{t'}(P), \mathcal{U}(t''-t')\{\Delta H_{t''}(P),\, \rho_{t''}(P)\}\Big\}dt''dt'. \tag{2.77b}$$

The preceding formulas shall be used to study a perturbation of the form

$$\Delta H_t(\boldsymbol{p},\boldsymbol{q}) = -\sum_n b_n(t)B_n(\boldsymbol{p},\boldsymbol{q}). \tag{2.78a}$$

It generalizes the problem discussed in connection with Eq. (2.71) in the sense that the external fields depend on time t. Let us request that the functions $b_n(t)$ are piecewise continuous and absolutely integrable. Under this condition, the limit $t_0 \to -\infty$ can be performed in Eqs. (2.77a,b). Since $\{B_n(P),\, \rho(P)\} = -\{B_n(P),\, H(P)\}\, \rho(P)/(k_\mathrm{B}T) = -\dot{B}_n(P)\rho(P)/(k_\mathrm{B}T)$, and since $\mathcal{U}(\tau)|X\rho) = |X(\tau)\rho)$, one gets

$$\mathcal{U}(t'-t)\,\{\Delta H_{t'}(P),\, \rho(P)\} = \sum_n b_n(t')\dot{B}_n(P_{t'-t})\rho(P)/(k_\mathrm{B}T). \tag{2.78b}$$

Above and in some of the following derivations, the notation $\dot{B}(P) = \{B(P), H(P)\} = \partial_t B(P)$ is used. Substituting Eq. (2.77a) into $\langle A^*\rangle_t = \iint \rho_t(P)A^*(P)\, d\boldsymbol{p}d\boldsymbol{q}$, one finds for the expectation value of some variable A^*:

$$\langle A^*\rangle_t - \langle A^*\rangle = \sum_n \int \chi_{A,B_n}(t-t')b_n(t')dt' + O(b^2). \tag{2.79}$$

The preceding result generalizes Eq. (2.73). The coefficient $\chi_{A,B_n}(t-t')$ connects the fluctuation of variable A^* at time t, $\langle \delta A^*\rangle_t = \langle A^*\rangle_t - \langle A^*\rangle$, with the perturbing field b_n at time t'. The coefficient is called the $A - B_n$ response function. It is the product of $\theta(t-t')/(k_\mathrm{B}T)$ and $\langle A^*\dot{B}_n(t'-t)\rangle = -\partial_t\langle A^*(t-t')B_n\rangle$. Because of Eq. (2.18a), the response function is given by the derivative of the correlator:

$$\chi_{A,B}(t) = -\theta(t)\partial_t\phi_{A,B}(t)/(k_\mathrm{B}T). \tag{2.80a}$$

Using Yvon's theorem (2.6) and Eq. (2.26b), one can write equivalently

$$\chi_{A,B}(t) = -\theta(t)\langle\{A^*(t), B\}\rangle. \tag{2.80b}$$

There holds $\phi_{A,B}(t) = \phi_{\delta A,\delta B} + \langle A^*\rangle\langle B\rangle$ and $\{A^*, B\} = \{\delta A^*, \delta B\}$. Hence, on the right-hand sides of Eqs. (2.80a,b), one can replace the variables A and B by their fluctuations δA and δB, respectively. The Heaviside function $\theta(t)$ expresses causality: the response at time t is caused by the perturbing fields at times t', which precede t.

In principle, a measurement of $\chi_{A,B}(t)$ can be performed as follows. One applies a perturbation field $b(t)$ coupling to the variable B, $\Delta H_{t'} = -b(t')B$, and measures the induced change of the expectation value of variable A^*, $\langle \delta A^* \rangle_t$. The field is written as a strength factor b times a unit impulse of duration $\epsilon > 0$, denoted as $\varphi_\epsilon(t) : b(t) = b\varphi_\epsilon(t)$. The impulse is constructed as $\varphi_\epsilon(t) = \varphi(t/\epsilon)/\epsilon$, where the shape function $\varphi(t)$ is normalized as $\int \varphi(t)dt = 1$. For ϵ tending to zero, the φ_ϵ versus t diagram exhibits a spike near $t = 0$. Its height increases proportional to $1/\epsilon$ and its width decreases proportional to ϵ so that $\int \varphi_\epsilon(t)dt = 1$. One gets $I_\epsilon(t) = \int \chi_{A,B}(t - t')b(t')dt' = b \int \chi_{A,B}(t - \epsilon\tau)\varphi(\tau)d\tau$. Since $\chi(t)$ is bounded and continuous for $t > 0$, the integrand has an ϵ-independent majorant proportional to $\varphi(\tau)$. Therefore, the limit $\epsilon \to 0$ can be performed under the integral: $\lim_{\epsilon\to 0} I_\epsilon(t) = b\chi_{A,B}(t)$. Consequently, one gets $\chi_{A,B}(t)$ as extrapolation of $\langle \delta A^* \rangle_t / b$ for $b \to 0$ and $\epsilon \to 0$.

Let us express the probing fields $b(t)$ as superpositions of their Fourier components $\hat{b}(\omega)$,

$$b(t) = \frac{1}{2\pi} \int \exp[-i\omega t]\hat{b}(\omega)d\omega, \tag{2.81a}$$

and let us calculate the Fourier components of the response $\hat{A}(\omega)$:

$$\hat{A}(\omega) = \int \exp[i\omega t]\langle \delta A^* \rangle_t dt. \tag{2.81b}$$

The inversion formulas (A.9), (A.15) for Fourier transforms allow to express $\hat{b}(\omega)$ and $\langle \delta A^* \rangle_t$ as Fourier integrals of $b(t)$ and $\hat{A}(\omega)$, respectively. The convolution theorem (A.12) for Fourier transforms can be used to cast Eq. (2.79) in the equivalent form

$$\hat{A}(\omega) = \sum_n \chi_{A,B_n}(\omega)\hat{b}_n(\omega) + O(b^2). \tag{2.81c}$$

Here, $\chi_{A,B}(\omega) = FT[\chi_{A,B}(t)](\omega)$ denotes the Fourier transform of the response function. Equation (2.81c) is analogous to Eq. (2.73). In all terms of this formula, however, the frequency ω appears as a parameter. The dependence on this parameter reflects the time dependence of all quantities entering Eq. (2.79). Therefore, $\chi_{A,B}(\omega)$ is referred to as dynamical susceptibility. From Eqs. (2.80a), (A.18) one infers that $\chi_{A,B}(\omega)$ is proportional to the boundary value of the Laplace transform of $\partial_t\phi_{A,B}(t)$. Equation (A.7a) motivates the definition of the $A - B$ response function in the frequency domain:

$$\chi_{A,B}(z) = [z\phi_{\delta A,\delta B}(z) + (\delta A|\delta B)]/(k_B T). \tag{2.82a}$$

Using this abbreviation, one can write the dynamical susceptibility in the form $\chi_{A,B}(\omega) = \chi_{A,B}(\omega + i0)$, i.e.,

$$\chi_{A,B}(\omega) = \chi'_{A,B}(\omega) + i\chi''_{A,B}(\omega). \tag{2.82b}$$

In particular, there is a simple relation between the $A-B$ susceptibility spectrum $\chi''_{A,B}(\omega)$ and the $\delta A - \delta B$ fluctuation spectrum:

$$\chi''_{A,B}(\omega) = \omega\phi''_{\delta A,\delta B}(\omega)/(k_{\mathrm{B}}T). \tag{2.82c}$$

According to Eq. (2.81c), the dynamical susceptibility $\chi_{A,B}(\omega)$ can be measured, in principle, as follows. One applies a perturbation $\Delta H_t = -b(t)B$ and calculates the Fourier components $\hat{b}(\omega)$ of the perturbing field. One measures $\hat{A}(\omega)$ or one calculates this quantity from the measurement of $\langle \delta A^* \rangle_t$. The dynamical susceptibility $\chi_{A,B}(\omega)$is obtained by extrapolating the ratio $\hat{A}(\omega)/\hat{b}(\omega)$ for vanishing perturbation strength $b(t)$.

Let us consider a dynamical variable A that is conserved for the isolated system. Since $A(t) = A$, there holds $\partial_t\phi_{A,B}(t) = \partial_t(A(t)|B) = 0$, i.e., there is no linear response: $\chi_{A,B}(t) = 0$. Substituting Eq. (2.77a) into Eq. (2.77b), one notices that the response in quadratic order is given by the last line with $\rho_{t''}(P)$ replaced by the conserved variable $\rho(P)$. Using Eq. (2.78a), one gets $\mathcal{U}(t''-t')\left\{\Delta H_{t''},\rho\right\}(P) = +\sum_n b_n(t'')\dot{B}_n(P_{t''-t'})\rho(P)/(k_{\mathrm{B}}T)$. Substituting the real function $\Delta H_{t'}(P) = -\sum_m b^*_m(t')B_m(P)^*$ and performing the limit $t_0 \to -\infty$, one gets for the second-order contribution to $\langle A^* \rangle_t - \langle A^* \rangle$:

$$-\iint \left[\sum_{m,n}(b^*_m(t')b_n(t'')/(k_{\mathrm{B}}T))\theta(t-t')\theta(t'-t'')\left[\iint A^*(P)\right.\right.$$
$$\left.\left.\times \left\{B^*_m(P_{t'-t}),\dot{B}_n(P_{t''-t})\rho(P)\right\} d\mathbf{p}d\mathbf{q}\right]\right] dt''dt'.$$

Partial integration yields

$$\iint A^*(P)\left\{B^*(P),C(P)\right\} d\mathbf{p}d\mathbf{q} = \iint \left\{A^*(P),B^*(P)\right\} C(P)d\mathbf{p}d\mathbf{q},$$

and this allows the rewriting of the phase-space integral as

$$\langle\{A^*(P),B^*_m(P_{t'-t})\}\,\dot{B}_n(P_{t''-t})\rangle = \langle\{A^*(P_{t-t''}),B^*_m(P_{t'-t''})\}\,\dot{B}_n(P)\rangle.$$

Since $A^*(t'-t'') = A^*$, and since the amplitudes $b_n(t)$ are assumed as absolutely integrable, one can perform the limit $t \to \infty$. Thereby, one arrives at a formula for the total change of variable A, $\langle \Delta A^* \rangle_{\mathrm{tot}} = \langle A^* \rangle_{t\to+\infty} - \langle A^* \rangle_{t\to-\infty}$:

$$\langle \Delta A^* \rangle_{\mathrm{tot}} = -\sum_{m,n}\iint \left[(b^*_m(t')b_n(t)/(k_{\mathrm{B}}T))\theta(t'-t'')\phi_{Y_m,X_n}(t'-t'')\right]dt'dt''$$
$$+\,O(b^3). \tag{2.83}$$

Here, the correlators are formed with the variables $X_n = \partial_t B_n$ and $Y_m = \{A, B_m\}$.

The preceding formula shall be applied to calculate the total change of the energy $\langle \Delta H \rangle_{\text{tot}}$. It is the change of the expectation value of the Hamilton function as produced by switching on the external fields in the past and then switching them off in the future, $b_n(t \to \pm\infty) = 0$. Because of Eq. (2.4a,b), one gets $Y_m = -X_m$. One can eliminate the function $\theta(t' - t'')$ in the integrand of Eq. (2.83) if one writes $\langle \Delta H \rangle_{\text{tot}} = (\langle \Delta H \rangle_{\text{tot}} + \langle \Delta H \rangle_{\text{tot}}^*)/2$, interchanges in one of the terms n and m as well as t' and t'', and uses Eq. (2.24):

$$\langle \Delta H \rangle_{\text{tot}} = [1/(2k_{\text{B}}T)] \sum_{m,n} \iint b_m(t')^* \phi_{X_m, X_n}(t' - t'') b_n(t'') dt' dt'' + O(b^3). \tag{2.84a}$$

Using Parseval's theorem and the convolution theorem for Fourier transforms, the double integral over the times can be transformed in one over a frequency: $\int \hat{b}_m(\omega)^* \hat{\phi}_{X_m, X_n}(\omega) \hat{b}_n(\omega) d\omega/(2\pi)$. This is discussed also in connection with the derivation of Eq. (A.17). One can write $\phi_{X_m, X_n}(t) = -\partial_t^2 \phi_{m,n}(t)$. Introducing $A_m = \delta B_m$, $m = 1, 2, \ldots$, $\phi_{m,n}(t) = \phi_{A_m, A_n}(t)$ denotes an element of the matrix correlator studied above in connection with Eq. (2.40a). Using Eq. (A.14a) and the connection between spectrum and Fourier transform, one gets

$$\langle \Delta H \rangle_{\text{tot}} = [1/(2\pi k_{\text{B}}T)] \int d\omega \sum_{m,n} \hat{b}_m(\omega)^* \omega^2 \phi''_{m,n} \hat{b}_n(\omega) d\omega + O(b^3). \tag{2.84b}$$

This formula explains the fact that the spectral matrix is positive definite, Eq. (2.41). No matter how the perturbing fields $b_n(t)$ are chosen, $\langle \Delta H \rangle_{\text{tot}}$ cannot be negative in the leading quadratic order in the field strength. The system can only absorb energy, it cannot decrease its energy. The formula underlies all absorption-spectroscopic techniques for measuring spectra. One disturbs the system by external fields. The fields shall be characterized by an overall strength b, by some polarization factors β_n, $n = 1, 2, \ldots$, and by a shape function $\xi_{\Omega,\epsilon}(t)$: $b_n(t) = b\beta_n \xi_{\Omega,\epsilon}(t)$. The shape function is the Fourier-back transform of $\xi_{\Omega,\epsilon}(\omega)$ considered in Eq. (A.21a). It identifies a frequency band that is centered at frequency Ω and that has a width tending to zero for vanishing ϵ. The proof of Eq. (A.21c) shows that the extrapolation of $\langle \Delta H \rangle_{\text{tot}}[2\pi k_{\text{B}}T/(b^2\Omega^2)]$ to the limit of $b \to 0$ and $\epsilon \to 0$, yields $\sum_{m,n} \beta_m \phi''_{m,n}(\Omega)\beta_n$. From the data measured for various choices of β_n, one can derive all the numbers $\phi''_{m,n}(\Omega)$. Since the functions $\chi''_{A,B}(\omega) = \omega\phi''_{\delta A,\delta B}(\omega)/(k_{\text{B}}T)$ determine the energy absorption of the system, they are referred to as the dissipative parts of the dynamic susceptibility or as the loss spectra. The functions $\chi'_{A,B}(\omega)$ are called the reactive parts. The identity (2.82c) is called the fluctuation–dissipation theorem.

2.6 The arrested parts of correlation functions

For the models of dynamical systems to be studied in the following chapters, it will be of importance to quantify the property that a perturbed system does not approach its canonical equilibrium state for large times. This quantification shall be done by parameters, which are called the arrested parts of the correlators. In this section, these parameters will be defined and their general properties shall be discussed.

The considerations of this section are based on the following theorem: for every pair of dynamical variables, say A and B, the following limit for the $A-B$ correlator in the frequency domain exists as a finite number

$$\lim_{\epsilon\to 0}(-i\epsilon)\phi_{A,B}(i\epsilon) = F_{A,B}. \tag{2.85}$$

The number $F_{A,B}$ is called the arrested part of the $A-B$ correlator. For autocorrelators, the limit obeys the inequalities

$$(A|A) \geq F_{A,A} \geq 0. \tag{2.86}$$

The proof of these results is based on the general form of the spectral representation for $\phi_{A,B}(z)$. Autocorrelators are positive analytic functions; and for such functions, the proof is given in Appendix A.3 in connection with Eq. (A.36). Because of Eq. (2.22), $\phi_{A,B}(z)$ is a linear combination of auto correlators so that Eq. (2.85) appears as a corollary.

Let us note a speciality. If A is in the domain of the Liouville operator $\mathcal{L}$, one concludes form Eqs. (2.26a), (A.7a): $\phi_{\mathcal{L}A,B}(z) = [z\phi_{A,B}(z) + (A|B)]$. Hence,

$$\lim_{\epsilon\to 0}\phi_{\mathcal{L}A,B}(i\epsilon) = [(A|B) - F_{A,B}]. \tag{2.87a}$$

This formula implies

$$F_{\mathcal{L}A,B} = 0. \tag{2.87b}$$

If a variable X is proportional to the time derivative of some other variable, say $X \propto \partial_t A \propto \mathcal{L}A$, there holds $F_{X,B} = 0$.

Up to a factor $(k_{\mathrm{B}}T)$, the function $\phi_{\mathcal{L}A,B}(z)$ is the dynamical $A-B$ susceptibility, introduced in Eq. (2.82a). Therefore, Eq. (2.87a) guaranties the existence of the ϵ-to-zero limit of $\chi_{A,B}(i\epsilon)$. This limit is called the static $A-B$ susceptibility:

$$\chi^0_{A,B} = \lim_{\epsilon\to 0}\chi_{A,B}(i\epsilon). \tag{2.88a}$$

Explicitly, one can write

$$\chi^0_{A,B} = [(\delta A|\delta B) - F_{\delta A,\delta B}]/(k_{\mathrm{B}}T). \tag{2.88b}$$

Because of Eq. (2.72b), the first term in this bracket determines the thermodynamic susceptibility $\chi^T_{A,B}$. Consequently, the quantity $F_{\delta A,\delta B}$ is proportional to the difference of thermodynamic and static susceptibility: $F_{\delta A,\delta B} = [\chi^T_{A,B} - \chi^0_{A,B}](k_{\mathrm{B}}T)$. The inequalities (2.86) are equivalent to

$$\chi^T_{A,A} \geq \chi^0_{A,A} \geq 0. \tag{2.89}$$

Equations (2.21a,b) hold also for correlators in the frequency domain, since the Laplace transform is a linear mapping. Hence, Eq. (2.85) implies the linearity of $F_{A,B}$ with respect to the variable B,

$$F_{A,b_1B_1+b_2B_2} = F_{A,B_1}b_1 + F_{A,B_2}b_2, \tag{2.90a}$$

and the antilinearity with respect to the variable A,

$$F_{a_1A_1+a_2A_2,B} = a_1^* F_{A_1,B} + a_2^* F_{A_2,B}. \tag{2.90b}$$

The functions $F_{A,B}$ are a bilinear form defined on the Hilbert space of dynamical variables. Because of Eqs. (2.31, 2.85), this form is hermitian

$$F_{A,B}^* = F_{B,A}. \tag{2.90c}$$

Equation (2.86) shows that the form is positive semidefinite.

For a set of variables $A_1, A_2, \ldots$, correlators $\phi_{m,n}(z)$ can be defined as is explained in connection with Eqs. (2.40, 2.41). These correlators can be considered as elements of a matrix correlator $\boldsymbol{\phi}(z)$. Notations and ordering properties for matrix correlators shall be used as explained in Appendix A.5. The matrix $\boldsymbol{F}$ of arrested parts is defined in analogy to Eq. (2.85)

$$\lim_{\epsilon\to 0}(-i\epsilon)\boldsymbol{\phi}(i\epsilon) = \boldsymbol{F}. \tag{2.91a}$$

For every array of complex numbers $a_1, a_2, \ldots, \phi_a(z) = \sum_{mn} a_m^* \phi_{m,n}(z) a_n$ is an autocorrelator. This result can be used to generalize Eq. (2.86):

$$\boldsymbol{g} \geq \boldsymbol{F} \geq \boldsymbol{0}. \tag{2.91b}$$

If $\langle A_n \rangle = 0$, $n = 1, 2, \ldots$, Eq. (2.82a) determines the matrix $\boldsymbol{\chi}(z)$ of dynamical-susceptibility elements $\chi_{m,n}(z) = \chi_{A_n,A_m}(z)$:

$$\boldsymbol{\chi}(z) = [z\boldsymbol{\phi}(z) + \boldsymbol{g}]/(k_{\rm B}T). \tag{2.92a}$$

The static limit $\boldsymbol{\chi}^0 = \lim_{\epsilon\to 0} \boldsymbol{\chi}(i\epsilon)$ exists; it is given by

$$\boldsymbol{\chi}^0 = [\boldsymbol{g} - \boldsymbol{F}]/(k_{\rm B}T). \tag{2.92b}$$

This formula implies the inequalities

$$\boldsymbol{\chi}^T \geq \boldsymbol{\chi}^0 \geq 0. \tag{2.92c}$$

This paragraph shall be used for the definition of an averaged long-time limits of a function $\phi(t)$. It will be sufficient to restrict the discussions to functions $\phi(t)$ that are continuous and bounded, say, $|\phi(t)| \leq c$. Let us introduce a shape function $w(\tau)$, which shall be defined for $\tau \geq 0$. It shall be bounded, piecewise continuous, absolutely integrable, and normalized by $\int_0^\infty w(\tau)d\tau = 1$. For every $\epsilon > 0$, a weight function shall be defined by $w_\epsilon(t) = \epsilon w(\epsilon t)$. It is obtained by rescaling of the shape function, and it has all the properties mentioned for $w(\tau) = w_{\epsilon=1}(\tau)$. The integral $I(\epsilon) = \int_0^\infty w_\epsilon(t)\phi(t)dt$ is well defined; it is the average of $\phi(t)$ formed with the weight $w_\epsilon(t)$ for the interval $0 \le t < \infty$. For every finite positive t_+, there holds $\lim_{\epsilon\to 0} \int_0^{t_+} w_\epsilon(t)\phi(t)dt = 0$. Consequently, if there exists $F = \lim_{\epsilon\to 0} I(\epsilon)$, there holds $F = \lim_{\epsilon\to 0} \int_{t_+}^\infty w_\epsilon(t)\phi(t)dt$. The value F does not depend on the behaviour of $\phi(t)$ on finite time intervals. The number F quantifies the large-time behaviour of the function $\phi(t)$. It shall be called an averaged long-time limit of $\phi(t)$: $F = \text{limav}_{t\to\infty}\, \phi(t)$. An averaged long-time limit for t tending to large negative values is defined by $\text{limav}_{t\to -\infty}\, \phi(t) = \text{limav}_{t\to\infty}\, \phi(-t)$. The integrand in $I(\epsilon) = \int_0^\infty w(\tau)\phi(\tau/\epsilon)d\tau$ has the ϵ-independent integrable majorant $c|w(\tau)|$. Therefore, the limit $\epsilon \to 0$ can be performed under the integral, provided the conventional long-time limit $\phi(\infty) = \lim_{t\to\infty} \phi(t)$ exists. In this case, $F = \phi(\infty)$. The concept of $\text{limav}_{t\to\infty}$ extends the concept $\lim_{t\to\infty}$ to certain functions, which do not have a conventional long-time limit. The limit formed with the shape function $w(\tau) = \exp[-\tau]$ is called Abel's limit of $\phi(t)$.

Let $\phi(t)$ denote a continuous positive-definite function. The spectral theorem discussed in Appendix A.3 leads to Eq. (A.35). This means that the averaged long-time limit exists for the shape function $w(\tau) = \theta(1-\tau)$:

$$\underset{t\to\infty}{\text{limav}}\, \phi(t) = \lim_{\epsilon\to 0} \epsilon \int_{t_+}^{1/\epsilon} \phi(t)dt. \tag{2.93a}$$

Expressing $\phi(i\epsilon)$ explicitly as Laplace integral (2.28), one can rewrite Eq. (A.36) as

$$\underset{t\to\infty}{\text{limav}}\, \phi(t) = \lim_{\epsilon\to\infty} \epsilon \int_{t_+}^{\infty} \exp[-\epsilon t]\phi(t)dt. \tag{2.93b}$$

The cited formulas imply also $\text{limav}_{t\to\infty}\, \phi(t) = F = \text{limav}_{t\to -\infty}\, \phi(t)$. Because of Eq. (2.22), every correlator is a sum of continuous positive-definite functions. Therefore, the preceding conclusions hold for all correlation functions:

$$\underset{t\to\pm\infty}{\text{limav}}\, \phi_{A,B}(t) = F_{A,B}. \tag{2.94}$$

The $F_{A,B}$ are Abel's limits of the correlators, and they agree with the averaged long-time limit defined in Eq. (2.93a). The limits for times tending to $+\infty$ agree with the limits for times tending to $-\infty$. There are models for a dynamics, where the correlators contain almost periodic parts, as mentioned in connection with Eq. (2.17a–c). In this case, the conventional long-time limits of the correlators do not exist. The above specified averaged long-time limits exist for all dynamical systems. It is left to the reader to extend the proof of Eq. (A.35) so that $w(\tau) = \theta(1-\tau)$ is replaced by an arbitrary shape function $w(\tau)$.

The number $F_{A,B}$ is the rest of the $A - B$ correlations if the time tends to very large values. This holds, provided $\lim_{t\to\infty} \phi_{A,B}(t)$ exists. If the limit does not exist, $F_{A,B}$ specifies a remainder obtained by an averaging procedure of the long-time tail. Equation (2.85) shows that a simple pole, $-F_{A,B}/z$, is the strongest singularity of a correlator in the origin of the frequency plane.

In the remainder of this section, the preceding results shall be combined with the findings of Sec. 2.5. It is the goal to relate the arrested parts $F_{A,B}$ to information that is, in principle, accessible to measurements. Most directly, one can combine Eq. (2.94) with Eq. (2.76b) in order to get

$$\operatorname*{limav}_{t\to\infty}\langle\delta A^*\rangle_{b,t} = \sum_{m,n} F_{\delta A,A_m}(g^{-1})_{m,n}\langle\delta A_n^*\rangle_b + O(b^2). \tag{2.95}$$

Here $\langle\delta A^*\rangle_{b,t}$ is the expectation value of variable δA^* for the isolated system. The initial condition is the canonical equilibrium for a Hamilton function, which is modified by a perturbation ΔH_b. The A_m denote the fluctuations of the perturbing variables $\delta B_m = B_m - \langle B_m\rangle$ so that $A_m = \delta A_m$. If all the arrested parts $F_{\delta A,A_m}$ $m = 1, 2, \ldots$, vanish, the averaged long-time limit of the perturbation vanishes in leading order in the perturbing fields b_n. In this sense, the system approaches its original equilibrium state. If one of the arrested parts is non-zero, the perturbation $\langle\delta A^*\rangle_{b,t}$ need not vanish, no matter how large is t. In this case, the system does not return to its original equilibrium state. Measuring $\langle\delta A^*\rangle_{b,t}$ for $t_+ \le t \le \tau$, one can calculate $\int_{t_+}^{\tau}\langle\delta A^*\rangle_{b,t}dt/\tau$. Extrapolating to $\tau\to\infty$, one gets the right hand side of Eq. (2.95). It relates the averaged long-time limit of $\langle\delta A^*\rangle_{b,t}$ with the initial conditions $\langle\delta A_n^*\rangle_b = \langle\delta A_n^*\rangle_{b,t=0}$. Extrapolating to $b_n \to 0$ for different choices of $\langle\delta A_n^*\rangle_b$, one can derive the $F_{\delta A,A_m}$.

Let us consider a time dependent perturbation of the equilibrium described by Eq. (2.78a). For $t \le 0$, the coupling amplitudes shall have the special form

$$b_n(t) = b_n \exp[\epsilon t], \tag{2.96a}$$

$\epsilon > 0$. For $t = 0$, the Hamiltonian function is modified by the ϵ-independent ΔH_b considered in Eq. (2.71). The relative speed of the switching-on process reads $\partial_t b_n(t)/b_n(t) = \epsilon$; it tends to zero for ϵ tending to zero. This limit is called an adiabatic switching-on of the perturbation ΔH_b. A first question of interest concerns the state of the system at $t = 0$. It is characterized by the averages of variables A^* at $t = 0$. The value $\langle\delta A^*\rangle_{b,t=0}^{\epsilon}$, is obtained by substituting Eq. (2.96a) into Eq. (2.79). If the B_n are in the domain of the Liouville operator $\mathcal{L}$, Eq. (2.80a) relates the result to the Laplace transform of $\partial_t\phi_{A,B_n}(t)$ for $z = i\epsilon$. The result is $[z\phi_{A,B_n}(z) + (A|B_n)]$. Using the Eq. (2.82a) for the dynamical susceptibility, one arrives at $\langle\delta A^*\rangle_{b,t=0}^{\epsilon} = \sum_n \chi_{A,B_n}(i\epsilon)b_n + O(b^2)$. Here, B_n can be replaced by $\delta B_n = A_n$. According to the discussion of Eq. (2.88a), the adiabatic limit can be performed with the result:

$$\lim_{\epsilon\to 0}\langle\delta A^*\rangle_{b,t=0}^{\epsilon} = \sum_n \chi_{A,B_n}^0 b_n + O(b^2). \tag{2.96b}$$

This formula is analogous to Eq. (2.73). But, the thermodynamic susceptibilities χ^T_{A,B_n} are replaced by the static susceptibilities χ^0_{A,B_n}. A measurement of the χ^0_{A,B_n} can be done, as explained in Sec. 2.5 for the χ^T_{A,B_n}. But now, there are two extrapolations involved: the limit $\epsilon \to 0$ and the limit $b_n \to 0$ have to be performed.

If the arrested part $F_{\delta A,B}$ vanishes, Eq. (2.88b) yields $\chi^0_{A,B} = \chi^T_{A,B}$. In this case, the thermodynamic susceptibility can be measured by the adiabatic switching-on process of the otherwise isolated system; one does not have to worry about keeping the temperature constant. If $F_{\delta A,\delta A} \neq 0$, one gets from Eq. (2.88b) $\chi^0_{A,A} < \chi^T_{A,A}$. If the δA autocorrelator has a non-vanishing arrested part, the system responds more stiffly to an adiabatic than to an isothermal perturbation.

Because of the great relevance of Eq. (2.96b), an additional comment might be of interest. Let $\eta_n(\tau)$ be a set of functions defined for $\tau \leq 0$, which are continuous and have continuous time derivatives $\dot{\eta}_n(\tau)$. The $\eta_n(\tau)$ shall be absolutely integrable, which implies $\eta_n(\tau \to -\infty) = 0$. The normalization convention $\eta_n(t = 0) = 1$ shall be imposed. The coupling amplitudes in Eq. (2.78a) shall be defined for every $\epsilon > 0$ as

$$b_n(t) = b_n \eta_n(t \cdot \epsilon). \tag{2.96c}$$

Again, the perturbation ΔH_b from Eq. (2.71) is switched on adiabatically for $\epsilon \to 0$. But now, the detailed form $\eta_n(\tau)$ is rather general; in particular, it can be different for different n. One can show that Eq. (2.96b) holds also for this case. The proof is built on the above mentioned generalization of Eq. (A.35).

A second question appearing in connection with Eq. (2.96a) is the following. What is the long-time variation of $\langle \delta A^* \rangle_{b,t}$ if the system evolves freely for $t \geq 0$ after an initial condition had been prepared by an adiabatic switching-on procedure? This questions concerns the analogue of Eq. (2.95). The answer is implied as a special case of the following theorem: for all perturbations defined by Eq. (2.78a), there holds

$$\operatorname*{limav}_{t \to \infty} \langle \delta A^* \rangle_{b,t} = O(b^2). \tag{2.97}$$

The anticipated absolute integrability of the perturbing fields $b_n(t)$ implies $\lim_{t \to -\infty} b_n(t) = 0$. Hence, the perturbation ΔH_t describes a general switching-on process for a modification of the Hamilton function. The absolute integrability implies also $\lim_{t \to \infty} b_n(t) = 0$. Hence, the perturbation deals with a general switching-off process of the modification. Thus, during the course of time, the Hamilton function $H_t(\boldsymbol{p}, \boldsymbol{q})$ changes from the equilibrium function $H(\boldsymbol{p}, \boldsymbol{q})$ to the modified function $H(\boldsymbol{p}, \boldsymbol{q}) + \Delta H(\boldsymbol{p}, \boldsymbol{q})$ for $t = 0$. Here $\Delta H(\boldsymbol{p}, \boldsymbol{q})$ is given by Eq. (2.71) with $b_n = b_n(t = 0)$. If the time increases further, $H_t(\boldsymbol{p}, \boldsymbol{q})$ changes back to $H(\boldsymbol{p}, \boldsymbol{q})$. As a result of the perturbations, expectation values of dynamical variables vary with time in a complicated manner. But, in the long-time limit, the averaged expectation values of all variables return to their equilibrium values in linear order in the fields b_n: $\operatorname{limav}_{t \to \infty} \langle A^* \rangle_{b,t} = \langle A^* \rangle + O(b^2)$. In this sense, one can summarize: a weakly disturbed system returns to its equilibrium for times tending to infinity. Quite different from the results described by Eq. (2.95), no linear perturbation remains for long times. Let us emphasize that no assumption on adiabaticity for the switching-on nor for the switching-off procedure is made. It is not even assumed that the amplitudes $b_n(t)$ can be differentiated. The results hold in particular for perturbations that are switched off suddenly at $t = 0$ after they had been switched on adiabatically for $t \leq 0$. The theorem does not hold beyond terms of linear order. For example, in quadratic order, the energy may increase, as was discussed in connection with Eq. (2.84b).

The proof of the theorem starts with splitting the perturbing fields in two parts $b(t) = b^-(t) + b^+(t)$. Here, $b_n^-(t) = \theta(-t)b_n(t)$ describes the switching-on process for $t \leq 0$, and $b_n^+(t) = \theta(t)b_n(t)$ describes the switching-off of the perturbation for $t \geq 0$. According to Appendix A.2, the Fourier transforms $\hat{b}^{\pm}(\omega) = \int \exp[i\omega t]b_n^{\pm}(t)dt$ are continuous functions of the real frequency ω that are bounded, say $|\hat{b}^{\pm}(\omega)| \leq c$. Substitution of this splitting into Eq. (2.79) and application of Eq. (2.80a) yields $\langle \delta A^* \rangle_{b,t} = \sum_n [F_n^-(t) + F_n^+(t)]/(k_{\mathrm{B}}T) + O(b^2)$. Here, the abbreviations have been used: $F_n^{\pm}(t) = -\int \theta(t-t')\partial_t \phi_{A,B_n}(t-t')b_n^{\pm}(t')dt'$. The proof will be done by showing that Abel's long-time limits of $F_n^{\pm}(t)$ vanish, i.e., it will be shown that $\lim_{\epsilon \to 0} I_n^{\pm}(\epsilon) = 0$, where $I_n^{\pm}(\epsilon) = -i\epsilon LT[F_n^{\pm}(t)](i\epsilon)$ for $\epsilon > 0$. To ease the notation, the index n shall be dropped from now on. Using Eq. (2.26a), one gets $F^+(t) = i\int_0^t \phi_{\mathcal{L}A,B}(t-t')b^+(t')dt'$. The Laplace transform can be done with Eqs. (A.6a,b): $I^+(\epsilon) = -i\epsilon\phi_{\mathcal{L}A,B}(i\epsilon)b^+(i\epsilon)$, where $b^+(z)$ denotes the Laplace transform of $b^+(t)$. According to Eq. (A.18), $\lim_{\epsilon \to 0} b^+(i\epsilon)$ is the finite number $i\hat{b}^+(\omega = 0)$. According to Eq. (2.85), there holds $\lim_{\epsilon \to 0} -i\epsilon\phi_{\mathcal{L}A,B}(i\epsilon) = F_{\mathcal{L}A,B}$. Because of Eq. (2.87b), one arrives at the desired result $\lim_{\epsilon \to 0} I^+(\epsilon) = 0$. Since B is in the domain of the Liouville operator $\mathcal{L}$, the first spectral moment for the $A - B$ correlator exists. One can use the spectral representation, Eq. (2.51b) to write $\partial_t \phi_{A,B}(t-t') = \int \exp[-i\omega(t-t')](-i\omega)d\sigma_{A,B}(\omega)$. Therefore, $F^-(t) = i\int \exp[-i\omega t]\hat{b}^-(\omega) \cdot \omega d\sigma_{A,B}(\omega)$. Hence, $I^-(\epsilon)/\epsilon = i\int [\omega/(i\omega+\epsilon)]\hat{b}^-(\omega)d\sigma_{A,B}(\omega)$. The integrand has c as an ϵ-independent integrable majorant. Therefore, the limit $\epsilon \to 0$ can be carried out under the integral yielding a finite number. One gets the desired result $\lim_{\epsilon \to 0} I^-(\epsilon) = 0$ and finishes the proof.

3

ELEMENTS OF LIQUID DYNAMICS

It seems desirable to have some knowledge of normal-liquid behaviour as the basis of an adequate appreciation of the discoveries on glassy dynamics. Therefore, it is the first goal of this chapter to discuss the most simple results of liquid dynamics. Concepts like mean-squared displacements and diffusion of a tagged particle, transverse-current diffusion and sound waves shall be explained. This will be done within a microscopic theory using the statistical description of the dynamics by means of correlation functions. Thereby, a second goal shall be achieved, namely the definition of the correlators for densities, currents and forces, the representation of these correlators in terms of relaxation kernels, and the derivation of some general relations for these functions. These results will be needed in the following chapters of this book. The diffusion equation and the linearized Navier–Stokes equations will be derived pedantically as examples for hydrodynamic-limit descriptions of homogeneous matter. This will be done in order to demonstrate that the Zwanzig–Mori theory of correlation functions offers an efficient formalism for deriving useful specific results from general regularity assumptions.

It is a further aim of this chapter to explain how Langevin's theory can be incorporated in the correlation-function formalism in order to derive improvements of the hydrodynamic-limit results. This theory is built on the assumption of the separation of the time scales for the dynamics of, for example, stresses from that of the generalized fluctuating forces causing this dynamics. As a result, Maxwell's theory of visco-elastic behaviour will be obtained. Some molecular-dynamics-simulation results shall be considered in order to demonstrate that the onset of glassy dynamics is connected with the time-scale-separation assumption to become illegitimate.

3.1 Preliminaries

3.1.1 *Homogeneous isotropic systems without chirality*

The considerations of this book are restricted to the bulk properties of homogeneous isotropic matter, which does not exhibit chirality. These restrictions are equivalent to the validity of certain symmetry properties of the correlation functions, which will be formulated in this section. Implications of the time-reversal symmetry of Hamilton's equations of motion shall also be derived.

The systems to be discussed consist of N particles interacting via centrosymmetric pair potentials. In addition, there is some wall potential V_L ensuring the confinement of the system in some box. The latter shall be chosen as a cube with length L of the edges. The particles shall be labelled by Greek letters $\kappa, \sigma, \ldots$. Vectors in three-dimensional space shall be indicated by letters with arrows. The three vector components in some Cartesian reference system shall be denoted by letters like $m, n, \ldots$. For example, the particle centres are abbreviated by $\vec{r}_\kappa$, $\kappa = 1, 2, \ldots N$, and the three coordinates of $\vec{r}_\kappa$ are denoted by r_κ^n, $n = 1, 2, 3$. If μ_κ denotes the mass of particle κ and if $\vec{v}_\kappa$ abbreviates its velocity, its momentum is $\vec{p}_\kappa = \mu_\kappa \vec{v}_\kappa$. The state of the system is a point $P = (\boldsymbol{p}, \boldsymbol{q})$ of the $6N$-dimensional phase space, which is specified by the $3N$ momentum coordinates $\boldsymbol{p} = (\vec{p}_1, \ldots, \vec{p}_N)$ and by the $3N$ configuration coordinates $\boldsymbol{q} = (\vec{r}_1, \ldots, \vec{r}_N)$. Hamilton's function is the sum of kinetic energy, interaction potential and wall potential: $H(\boldsymbol{p}, \boldsymbol{q}) = \frac{1}{2}\sum_\kappa [\vec{p}_\kappa^{\,2}/\mu_\kappa] + \frac{1}{2}\sum_{\kappa \neq \sigma} V_{\kappa\sigma}(|\vec{r}_\kappa - \vec{r}_\sigma|) + V_L$. Here, $V_{\kappa\sigma}(r)$ is the interaction potential between particles κ and σ as function of their distance $r = |\vec{r}|$. Boltzmann's probability density (2.5a) factorizes in one part $\rho_c(\boldsymbol{q})$ describing weights of configurations and another part $\rho_0(\boldsymbol{p})$ specifying the weights for the momentum averages: $\rho(\boldsymbol{p}, \boldsymbol{q}) = \rho_c(\boldsymbol{q})\rho_0(\boldsymbol{p})$. The latter part reads: $\rho_0(\boldsymbol{p}) = Z_0^{-1} \exp[-\sum_\kappa \vec{p}_\kappa^{\,2}/(2\mu_\kappa k_{\mathrm{B}} T)]$, $Z_0 = \Pi_\kappa (2\pi\mu_\kappa k_{\mathrm{B}} T)^{3/2}$. Averages of momenta can be done independently from those of configurations. Averages over monomials of momenta can be calculated elementarily. In particular, averages over monomials of odd degree vanish. Averages of quadratic monomials are given by

$$\langle v_\kappa^m v_\sigma^n \rangle = \delta^{m,n} \delta_{\kappa,\sigma} (v_{th}^\kappa)^2, \tag{3.1a}$$

with

$$v_{th}^\kappa = \sqrt{k_{\mathrm{B}} T/\mu_\kappa} \tag{3.1b}$$

denoting the thermal velocity of particle κ at temperature T. The preceding sentences define the space of dynamical variables, its metric, and the time evolution in the sense of Sec. 2.1.

Time-reversal invariance is a symmetry built on the following property of Hamilton's equations of motion. If $P_t = (\boldsymbol{p}(t), \boldsymbol{q}(t))$ is an orbit for the initial condition $P_0 = (\boldsymbol{p}(t=0), \boldsymbol{q}(t=0)) = (\boldsymbol{p}^0, \boldsymbol{q}^0)$, $Q_t = (\boldsymbol{p}(-t)_{\mathrm{in}}, \boldsymbol{q}(-t))$ is an orbit for the initial condition $Q_0 = (\boldsymbol{p}_{\mathrm{in}}^0, \boldsymbol{q}^0)$. Here, the inversion of the momentum $\boldsymbol{p}$ is defined by $\boldsymbol{p}_{\mathrm{in}} = (-\vec{p}_1, \ldots, -\vec{p}_N)$. Boltzmann's weight function is invariant under the transformation $\boldsymbol{p} \to \boldsymbol{p}_{\mathrm{in}}$. Consequently, defining the time-reversal operator $\mathcal{S}_{\mathrm{re}}$ for every dynamical variable X by

$$[\mathcal{S}_{\mathrm{re}} X](\boldsymbol{p}, \boldsymbol{q}) = X(\boldsymbol{p}_{\mathrm{in}}, \boldsymbol{q}), \tag{3.2a}$$

one finds the symmetry property

$$(\mathcal{S}_{\mathrm{re}} X | \mathcal{S}_{\mathrm{re}} Y) = (X, Y). \tag{3.2b}$$

The scalar product (2.8) of every pair of dynamical variables X and Y is invariant under operation of $\mathcal{S}_{\mathrm{re}}$. The operator $\mathcal{S}_{\mathrm{re}}$ maps the space of variables on itself since $\mathcal{S}_{\rm re}^{-1} = \mathcal{S}_{\rm re}$; therefore, it is unitary. The cited property of the equations of motion implies the symmetry property for the $A - B$ correlators:

$$([\mathcal{S}_{\rm re}A](t)|\mathcal{S}_{\rm re}B) = (A(-t)|B). \tag{3.2c}$$

Expressing the time evolution in terms of the unitary operator $\mathcal{U}(t)$ and exploiting Eq. (3.2b), one can reformulate the preceding equations as follows: $(\mathcal{U}(t)\mathcal{S}_{\rm re} A|\mathcal{S}_{\rm re}B) = (\mathcal{S}_{\rm re}^{-1}\mathcal{U}(t)\mathcal{S}_{\rm re}A|B) = (\mathcal{U}(-t)A|B)$. Since A and B are arbitrary dynamical variables, one gets the identity: $\mathcal{S}_{\rm re}^{-1}\mathcal{U}(t)\mathcal{S}_{\rm re} = \mathcal{U}(-t)$. Equivalently, there holds

$$\mathcal{U}(t)\mathcal{S}_{\rm re} = \mathcal{S}_{\rm re}\mathcal{U}(-t). \tag{3.2d}$$

A dynamical variable X is said to be of definite time-inversion parity if it is an eigenvector of $\mathcal{S}_{\rm re}$:

$$\mathcal{S}_{\rm re}X = \epsilon_T^X X. \tag{3.3a}$$

Since $\mathcal{S}_{\rm re}\mathcal{S}_{\rm re}X = X$, one gets $\epsilon_T^X = 1$ or $\epsilon_T^X = -1$. In the former case, one calls the variable X of even time-inversion parity, and in the latter of an odd one. If X and Y have a definite parity, Eq. (3.2b) implies $(X|Y) = \epsilon_T^X \epsilon_T^Y (X|Y)$. Hence, variables of opposite parity are perpendicular:

$$(X|Y) = 0, \quad \text{if } \epsilon_T^X \epsilon_T^Y = -1. \tag{3.3b}$$

Equation (3.2c) implies that the correlator of two variables with definite time inversion parity is an even or odd function of time depending on whether the parity is equal or opposite:

$$(A(t)|B) = \epsilon_T^A \epsilon_T^B (A(-t)|B). \tag{3.3c}$$

Let us formulate the transformation properties of the correlation functions, which reflect the covariance of Hamilton's equations of motion for the symmetry group of the Euklidian geometry. First, there are the translation operators $\mathcal{S}_{\rm tr}(\vec{d})$ for a vector $\vec{d}$. They are defined for every dynamical variable X by

$$[\mathcal{S}_{\rm tr}(\vec{d})X](\boldsymbol{p}, \boldsymbol{q}) = X(\boldsymbol{p}, \boldsymbol{q}_{\vec{d}}). \tag{3.4a}$$

Here, the translated configuration coordinate is given by $\boldsymbol{q}_{\vec{d}} = (\vec{r}_1 - \vec{d}, \ldots, \vec{r}_N - \vec{d})$. Second, there are the rotation operators $\mathcal{S}_{\rm ro}(D)$ which are specified by orthogonal 3-by-3 matrices D of determinant unity. The matrix elements shall be denoted by $D^{m,n}$ so that the coordinates of the rotation $D\vec{r}$ of some vector $\vec{r}$ are given by $(D\vec{r})^m = \sum_n D^{mn} r^n$. The definition of the rotation operator in the space of dynamical variables X reads

$$[\mathcal{S}_{\rm ro}(D)X](\boldsymbol{p}, \boldsymbol{q}) = X(\boldsymbol{p}_D, \boldsymbol{q}_D). \tag{3.4b}$$

The rotated phase-space points are given by $\boldsymbol{p}_D = (D^{-1}\vec{p}_1, \ldots, D^{-1}\vec{p}_N)$ and $\boldsymbol{q}_D = (D^{-1}\vec{r}_1, \ldots, D^{-1}\vec{r}_N)$. Finally, the space-inversion operator $\mathcal{S}_{\text{in}}$ is defined by

$$[\mathcal{S}_{\text{in}}X](\boldsymbol{p}, \boldsymbol{q}) = X(\boldsymbol{p}_{\text{in}}, \boldsymbol{q}_{\text{in}}). \tag{3.4c}$$

The inversion for the configuration coordinates is determined as above for the momentum coordinates: $\boldsymbol{q}_{\text{in}} = (-\vec{r}_1, \ldots, -\vec{r}_N)$. The Euclidean measure $d\boldsymbol{p}d\boldsymbol{q}$ in the definition (2.5b) of averages is invariant under all transformations $\mathcal{S}$ defined above. Hence, one gets

$$\langle \mathcal{S}X \rangle = \langle X \rangle_{\mathcal{S}}. \tag{3.5a}$$

The index $\mathcal{S}$ indicates that the average is calculated with H replaced by $\mathcal{S}^{-1}H$ in Boltzmann's weight function (2.5a). This implies the formula for the scalar product:

$$(\mathcal{S}X|\mathcal{S}Y) = (X|Y)_{\mathcal{S}}. \tag{3.5b}$$

Hamilton's equations of motion exhibit covariance with respect to translations: if $(\boldsymbol{p}(t), \boldsymbol{q}(t))$ is an orbit with initial condition $(\boldsymbol{p}^0, \boldsymbol{q}^0)$, then $(\boldsymbol{p}(t), \boldsymbol{q}(t)_{\vec{d}})$ is an orbit with initial condition $(\boldsymbol{p}^0, \boldsymbol{q}^0_{\vec{d}})$ for a system whose Hamiltonian is $\mathcal{S}_{\text{tr}}(-\vec{d})H$. A corresponding statement holds for the other symmetry operators. Hence, the correlators exhibit the property

$$([\mathcal{S}A](t)|\mathcal{S}B) = (A(t)|B)_{\mathcal{S}}. \tag{3.5c}$$

Here, the index means that averages as well as time translations refer to the transformed Hamiltonian $\mathcal{S}^{-1}H$. The right-hand sides of Eqs. (3.5a–c) deal with averages, overlaps and correlations of dynamical variables for a system obtained from the original one by translation, rotation, or space inversion. These quantities are the same as calculated for the system specified by Hamiltonian H, with, however, the dynamical variables translated by $-\vec{d}$, rotated in the inverse directions, or with variables obtained by space inversion.

The discussions in this book shall be restricted to bulk properties of amorphous matter. Variables dealing with surface properties shall not be considered. In order to formulate this restriction precisely, appeal to the thermodynamic limit shall be made. For the sake of simplicity, let us consider a system of identical particles. It is assumed that the wall potential V_L does not influence the properties of the system outside a surface layer of an L-independent thickness δL. Only such dynamical variables are considered, which do not depend on the particle coordinates within this surface layer. The limit $L \to \infty$ and $N \to \infty$ is considered so that the averaged density N/L^3 is kept fixed. The contribution of the particles within the surface layer vanishes proportional to $\delta L/L$ relative to that outside the layer. This means that there is no influence of the wall potential in the specified limit. Since the kinetic energy and the interaction potential are invariant under the three types of transformations $\mathcal{S}$, the Hamiltonians H and

$\mathcal{S}^{-1}H$ differ merely by the wall potential. In the thermodynamic limit, this difference does not enter the formulas (3.5a–c). Hence, the index $\mathcal{S}$ can be dropped in the cited equations. One arrives at the symmetry relations

$$\langle \mathcal{S}X\rangle = \langle X\rangle, \quad (\mathcal{S}X|\mathcal{S}Y) = (X|Y), \quad ([\mathcal{S}A](t)|\mathcal{S}B) = (A(t)|B). \tag{3.6a}$$

Here, $\mathcal{S}$ can be any of the symmetry operations defined in Eqs. (3.4a–c). Since the operators $\mathcal{S}^{-1}$ are defined on the whole space of dynamical variables, the second identity means that the operators $\mathcal{S}$ are unitary. Expressing the time evolution in terms of the operator $\mathcal{U}(t)$, one gets: $(\mathcal{U}(t)\mathcal{S}A|\mathcal{S}B) = (\mathcal{S}^{-1}\mathcal{U}(t)\mathcal{S}A|B) = (\mathcal{U}(t)A|B)$. Because A and B are arbitrary, there holds $\mathcal{S}^{-1}\mathcal{U}(t)\mathcal{S} = \mathcal{U}(t)$, or

$$\mathcal{U}(t)\mathcal{S} = \mathcal{S}\mathcal{U}(t). \tag{3.6b}$$

The unitary operators $\mathcal{U}(t)$ and $\mathcal{S}$ commute.

It is very difficult to answer the question, whether the thermodynamic limit exists and whether or not the invariance properties are valid for given density and temperature. A system for which the thermodynamic-limit assumptions are valid and for which Eqs. (3.6a) hold for all translations $\mathcal{S} = \mathcal{S}_{\text{tr}}(\vec{d})$ is called homogeneous. If the equations hold for all rotations $\mathcal{S} = \mathcal{S}_{\text{ro}}(D)$, the system is called isotropic. If the equations are violated for the space inversion, the system is called chiral. The discussions of this book are restricted to systems, for which the relations (3.6a) are valid for all three groups of transformations $\mathcal{S}$.

3.1.2 *Densities and density fluctuations*

The dynamics of homogeneous systems is described most conveniently by fluctuations. These are obtained by Fourier transformation of densities from the position space to the space of wave vectors. In this section, the conventions for these transformations will be explained. The symmetries will be derived, which are exhibited by the correlators formed with the fluctuations of particle densities and particle-current densities.

Basic quantities for the description of amorphous-matter dynamics are the densities $\rho^\kappa(\vec{r})$ and the current-densities $\vec{j}^\kappa(\vec{r})$ of some particle with number κ:

$$\rho^\kappa(\vec{r}) = \delta(\vec{r} - \vec{r}_\kappa), \qquad \vec{j}^\kappa(\vec{r}) = \vec{v}_\kappa \delta(\vec{r} - \vec{r}_\kappa). \tag{3.7}$$

These quantities are not dynamical variables as defined in Chapter 2. Rather, they are generalized dynamical variables defined for the underlying three-dimensional space of position vectors $\vec{r}$. The concept of the three-dimensional distribution function $\delta(\vec{r})$ is used. This means that $\rho^\kappa(\vec{r})$ and $\vec{j}^\kappa(\vec{r})$ specify a linear mapping from the space of properly defined test functions $\xi(\vec{r})$ in the space of dynamical variables. The mapping is defined by the formulas $A = \xi(\vec{r}_\kappa)$ and $V^n = v^n_\kappa \xi(\vec{r}_\kappa)$, $n = 1, 2, 3$. The mapping is conventionally written in the form of $A = \int \rho^\kappa(\vec{r})\xi(\vec{r})d^3\vec{r}$ and $V^n = \int j^{\kappa n}(\vec{r})\xi(\vec{r})d^3\vec{r}$. Let us consider two dynamical variables A and B formed with two test functions $\xi(\vec{r})$ and $\eta(\vec{r})$ from two densities

$\rho^\kappa(\vec{r})$ and $\rho^\sigma(\vec{r})$: $A = \xi(\vec{r}_\kappa)$ and $B = \eta(\vec{r}_\sigma)$. The correlator $\phi_{\xi,\eta}(t) = (A(t)|B)$ is well defined by the formulas of Sec. 2.2. Because of Eqs. (2.21a,b), it is a bilinear hermitian form defined on the space of test functions. Hence, the formula $\phi_{\xi,\eta}(t) = \iint \xi^*(\vec{r}_1)\phi(\vec{r}_1,\vec{r}_2;t)\eta(\vec{r}_2)d^3\vec{r}_1 d^3\vec{r}_2$ defines a generalized function $\phi(\vec{r}_1,\vec{r}_2;t)$ for every time t. This generalized function is a distribution defined on the space of test functions $\zeta(\vec{r}_1,\vec{r}_2)$, which are explained on the six-dimensional space spanned by the pairs of vectors $(\vec{r}_1,\vec{r}_2)$. This generalized function is called the correlator of the underlying generalized dynamical variables, and it is denoted by $\phi(\vec{r}_1,\vec{r}_2,t) = (\rho^\kappa(\vec{r}_1,t)|\rho^\sigma(\vec{r}_2))$. In a similar manner, correlators are formed with two current densities or with one density and one current density. It is assumed that the reader has some knowledge of generalized functions in order to accept the preceding sentences and a few similar ones to follow. Manipulations as above will be done without further comments. Generalized variables like $\rho^\kappa(\vec{r})$ and $\vec{j}^\kappa(\vec{r})$ will be treated as dynamical variables, which depend on $\vec{r}$ as parameter.

Densities as defined above can be eliminated in favour of fluctuations using Fourier transformations from space of position vectors $\vec{r}$ to the reciprocal space of wave vectors $\vec{k}$. The procedure is known from continuum mechanics or field theory. This and the following three paragraphs formulate some reminders. The functions $e_{\vec{k}}(\vec{r}) = \exp[-i\vec{k}\vec{r}]$ form a complete orthogonal set on the box of position vectors $\vec{r} = (r^1, r^2, r^3)$, $-L/2 \leq r^\ell \leq L/2$, $\ell = 1,2,3$. The vectors $\vec{k}$ are points on the discrete set

$$\vec{k} = 2\pi(n^1,n^2,n^3)/L, \quad n^\ell = 0, \pm 1, \pm 2, \ldots, \quad \ell = 1,2,3. \tag{3.8a}$$

Orthogonality means that $\int_L e^*_{\vec{k}}(\vec{r})e_{\vec{p}}(\vec{r})d^3\vec{r} = L^3\delta_{\vec{k},\vec{p}}$. Here, $\delta_{\vec{k},\vec{p}}$ is unity for $\vec{k} = \vec{p}$ and zero otherwise. The L at the integral indicates integration over the mentioned box. If some function $F(\vec{r})$ can be represented as uniformly and absolutely convergent Fourier series,

$$F(\vec{r}) = L^{-3}\sum_{\vec{k}} \exp[-i\vec{k}\vec{r}]F_{\vec{k}}, \tag{3.8b}$$

the orthogonality property yields the formula for the Fourier coefficients

$$F_{\vec{k}} = \int_L \exp[i\vec{k}\vec{r}]F(\vec{r})d^3\vec{r}. \tag{3.8c}$$

Completeness means that sufficiently regular function $F(\vec{r})$ can be represented as uniformly and absolutely convergent Fourier series. A sufficient regularity condition is that $F(\vec{r})$ and its partial derivatives are continuous and vanish on the box boundary. If $F(\vec{r})$ and $G(\vec{r})$ obey this condition, one derives Parseval's theorem

$$\int_L F^*(\vec{r})G(\vec{r})d^3\vec{r} = L^{-3}\sum_k F^*_{\vec{k}}G_{\vec{k}}. \tag{3.8d}$$

Let $F(\vec{r})$ be a piecewise continuous function defined on the three-dimensional position-vector space. If $F(\vec{r})$ is absolutely integrable, one can define the Fourier transform for every $\vec{q}$ from the three-dimensional space of wave vectors:

$$\hat{F}(\vec{q}) = \int \exp[i\vec{q}\vec{r}]F(\vec{r})d^3\vec{r}. \tag{3.9a}$$

This concept is a generalization of the one-dimensional Fourier transform defined in Eq. (A.9) of Appendix A.2. As explained there, one shows that $\hat{F}(\vec{q})$ is continuous in $\vec{q}$. Similarly, if $\hat{F}(\vec{q})$ is absolutely integrable, there holds the inversion formula in analogy to Eq. (A.15):

$$F(\vec{r}) = \frac{1}{(2\pi)^3}\int \exp[-i\vec{q}\vec{r}]\hat{F}(\vec{q})d^3\vec{q}. \tag{3.9b}$$

It is assumed that the reader is familiar with the elementary properties of three-dimensional Fourier transformations as they can be obtained by generalizations of Eqs. (A.9)–(A.17). For later reference, let us consider the special case that function $\hat{F}(\vec{q}) = \hat{f}_q$ depends on the wave vector merely through its modulus $q = |\vec{q}|$. The angular integrals in Eq. (3.9b) can be performed to show that $F(\vec{r}) = f_r$ depends on the position vector only through its modulus $r = |\vec{r}|$:

$$f_r = (1/2\pi^2)\int_0^\infty \left[\sin(qr)/(qr)\right]q^2\hat{f}_q dq. \tag{3.9c}$$

If $F(\vec{r}) = f_r$ depends on the position vector $\vec{r}$ only via its modulus $r = |\vec{r}|$, Eq. (3.9a) yields $\hat{F}(\vec{q}) = \hat{f}_q$ with $\hat{f}_q$ given by the inversion formula

$$\hat{f}_q = (4\pi)\int_0^\infty \left[\sin(qr)/(qr)\right]r^2 f_r dr. \tag{3.9d}$$

Let $F_L(\vec{r})$ denote the restriction of the absolutely integrable function $F(\vec{r})$ to the box. This means $F_L(\vec{r}) = F(\vec{r})$ for $\vec{r}$ within the box and $F_L(\vec{r}) = 0$ for $\vec{r}$ outside. If $F^L_{\vec{k}}$ denotes the Fourier transform of $F_L(\vec{r})$ calculated according to Eq. (3.8c) and if $\hat{F}^L(\vec{q})$ denotes the one calculated according to Eq. (3.9a), one gets $F^L_{\vec{k}} = \hat{F}^L(\vec{k})$. In this sense, function $\hat{F}^L(\vec{q})$ is a continuous interpolation of $F^L_{\vec{k}}$ from the discrete set of wave vectors $\vec{k}$ onto the whole space of wave vectors $\vec{q}$. Function $|F(\vec{r})|$ is an L-independent absolutely-integrable majorant of $G(\vec{r}) = \exp[i\vec{q}\vec{r}]F_L(\vec{r})$. Hence, for the integral of $G(\vec{r})$, one can interchange the $L \to \infty$ limit with the integration in order to get $\lim_{L\to\infty}\hat{F}^L(\vec{q}) = \hat{F}(\vec{q})$. Consequently, for fixed $\vec{k}$ but L increasing to infinity, the right-hand side of Eq. (3.8c) converges towards $\hat{F}(k)$. Similarly, the right-hand side of Eq. (3.8b) is a Riemann-sum approximation for the integral in Eq. (3.9b). Therefore, the following two 'equations' hold as thermodynamic-limit formulas $L^{-3}\sum_{\vec{k}}\ldots = (2\pi)^{-3}\int\ldots d^3\vec{k}$, $L^3\delta_{\vec{k},\vec{p}} = (2\pi)^3\delta(\vec{k}-\vec{p})$. Switching from one side to the other in these 'formulas' will be done in the following without further mention of the underlying thermodynamic-limit concepts.

Fourier transformation of the densities in Eqs. (3.7) yields dynamical variables in the sense of Sec. 2.1:

$$\rho_{\vec{q}}^{\kappa} = \exp[i\vec{q}\vec{r}_{\kappa}], \qquad \vec{j}_{\vec{q}}^{\kappa} = \vec{v}_{\kappa} \exp[i\vec{q}\vec{r}_{\kappa}]. \tag{3.10}$$

These quantities are called density fluctuations and current-density fluctuations, respectively, for wave vector $\vec{q}$. Let us come back to the variable $A = \xi(\vec{r}_{\kappa})$ considered above for the test functions $\xi(\vec{r})$. Expressing the test function in terms of its Fourier components $\xi_{\vec{k}}$ according to Eq. (3.8b), one can write $A = L^{-3}\sum_k \rho_{\vec{k}}^{\kappa *}\xi_{\vec{k}}$. Since the right-hand side is a linear functional of $\xi(\vec{r})$, it defines uniquely the generalized function $\rho^{\kappa}(\vec{r})$ via $A = \int_L \rho^{\kappa *}(\vec{r})\xi(\vec{r})$. Appeal to Eq. (3.8d) is the motivation to call the generalized function $\rho^{\kappa}(\vec{r})$ the Fourier back transform of $\rho_{\vec{k}}^{\kappa} : \rho^{\kappa}(\vec{r}) = L^{-3}\sum_{\vec{k}} \exp[-i\vec{k}(\vec{r} - \vec{r}_{\kappa})]$. Similar formulas hold for the three components of $\vec{j}_{\vec{q}}^{\kappa}$.

The various symmetry relations, which are formulated in Sec. 3.1.1, imply symmetries for the correlation functions formed with density and current-density fluctuations. These shall be specified in the remainder of this section. Some details of the proofs are transferred to Appendix B. Let $Z_{\vec{q}}$ denote a set of dynamical variables, which are explained for all wave vectors $\vec{q}$. Such a dynamical-variable-valued function of $\vec{q}$ is called a field fluctuation if there holds the transformation law for all translations $\mathcal{S}_{\text{tr}}(\vec{d})$:

$$\mathcal{S}_{\text{tr}}(\vec{d})Z_{\vec{q}} = \exp[-i\vec{q}\vec{d}]Z_{\vec{q}}. \tag{3.11a}$$

The variable $Z_{\vec{q}}$ is a common eigenvector of all the unitary translations operators $\mathcal{S}_{\text{tr}}(\vec{d})$. The density fluctuations $\rho_{\vec{q}}^{\kappa}$ and the three current-fluctuation components $j_{\vec{q}}^{\kappa n}$, $n = 1, 2, 3$, are examples for field fluctuations. Because of Eq. (3.6b), the same is true for the time-translated variables $Z_{\vec{q}}(t) : \mathcal{S}_{\text{tr}}(\vec{d})Z_{\vec{q}}(t) = \exp[-i\vec{q}\vec{d}]Z_{\vec{q}}(t)$. Let us consider the correlations of two field fluctuations, say $A = X_{\vec{k}}$ and $B = Y_{\vec{p}}$. The condition of homogeneity is formulated by Eq. (3.6a) with $\mathcal{S} = \mathcal{S}_{\text{tr}}(\vec{d})$. According to Eq. (B.2b), this condition is equivalent to the absence of correlations between fluctuations referring to different wave vectors:

$$(X_{\vec{k}}(t)|Y_{\vec{p}}) = \delta_{\vec{k},\vec{p}}(X_{\vec{k}}(t)|Y_{\vec{k}}). \tag{3.11b}$$

Two other versions of the preceding result are worth mentioning. First, let us assume that the two fields can be transformed to variables $X(\vec{r})$ and $Y(\vec{r})$, respectively, according to Eq. (3.8b). These are dynamical variables defined for all positions $\vec{r}$. The double transform $L^{-6}\sum_{\vec{k}} \exp[i\vec{k}\vec{r}_1]\sum_{\vec{p}}\exp[-i\vec{p}\vec{r}_2]$ of Eq. (3.11b) yields

$$(X(\vec{r}_1, t)|Y(\vec{r}_2)) = F_{X,Y}(\vec{r}_2 - \vec{r}_1, t)/L^3, \tag{3.12a}$$

where the function $F_{X,Y}(\vec{r},t)$ is defined by

$$F_{X,Y}(\vec{r},t) = L^{-3}\sum_{\vec{q}} \exp[-i\vec{q}\vec{r}](X_{\vec{q}}(t)|Y_{\vec{q}}). \tag{3.12b}$$

Homogeneity means that there holds $(X(\vec{r}_1+\vec{d},t)|Y(\vec{r}_2+\vec{d})) = (X(\vec{r}_1,t)|Y(\vec{r}_2))$ for all translation vectors $\vec{d}$.

Second, let $X^{(i)}_{\vec{k}_i}$, $i = 1,2,\ldots,I$, and $Y^{(j)}_{\vec{p}_j}$, $j = 1,2,\ldots,J$, denote sets of I and J field fluctuations, respectively. The products $X_{\vec{k}} = X^{(1)}_{\vec{k}_1}\cdot X^{(2)}_{\vec{k}_2}\cdots X^{(I)}_{\vec{k}_I}$ and $Y_{\vec{p}} = Y^{(1)}_{\vec{p}_1}\cdot Y^{(2)}_{\vec{p}_2}\cdots Y^{(J)}_{\vec{p}_J}$ are field fluctuations for wave vectors $\vec{k} = \vec{k}_1+\vec{k}_2+\cdots+\vec{k}_I$ and $\vec{p} = \vec{p}_1+\vec{p}_2+\cdots+\vec{p}_J$, respectively. Hence, Eq. (3.11b) implies

$$\left\langle \left[X^{(1)}_{\vec{k}_1}(t)\cdots X^{(I)}_{\vec{k}_I}(t)\right]^* \left[Y^{(1)}_{\vec{p}_1}\cdots Y^{(J)}_{\vec{p}_J}\right]\right\rangle = 0 \quad \text{if } \vec{k} \neq \vec{p}. \tag{3.13}$$

A field fluctuation $X_{\vec{q}}$ is said to have even or odd conjugation parity, if there holds

$$X_{\vec{q}c} = \epsilon^X_c X_{-\vec{q}} \tag{3.14a}$$

for $\epsilon^X_c = 1$ or -1, respectively. The conjugation is defined in Eq. (2.10a). If $X_{\vec{q}}$ has even parity, the variable $iX_{\vec{q}}$ has an odd one. Let the density $X(\vec{r})$ be related to its Fourier transform $X_{\vec{q}}$ as explained with Eqs. (3.8b,c). One concludes that

$$X(\vec{r})_c = \epsilon^X_c X(\vec{r}). \tag{3.14b}$$

Hence, $\epsilon^X_c = 1$ or -1 means that $X(\vec{r})$ is real or imaginary, respectively. Both the density fluctuations $\rho^\kappa_{\vec{q}}$ and the current fluctuations $\vec{j}^\kappa_{\vec{q}}$ have even conjugation parity, i.e., $\rho^\kappa(\vec{r})$ and $\vec{j}^\kappa(\vec{r})$ are real. If $X_{\vec{q}}$ and $Y_{\vec{q}}$ are two field fluctuations with definite conjugation parity ϵ^X_c and ϵ^Y_c, respectively, one derives from Eq. (2.10b) the symmetry relation for the scalar product

$$(X_{\vec{q}}|Y_{\vec{q}})^* = \epsilon^X_c\epsilon^Y_c(X_{-\vec{q}}\,|\,Y_{-\vec{q}}). \tag{3.14c}$$

Similarly, Eq. (2.25) yields for the correlator of the two fields:

$$(X_{\vec{q}}(t)\,|\,Y_{\vec{q}})^* = \epsilon^X_c\epsilon^Y_c(X_{-\vec{q}}(t)\,|\,Y_{-\vec{q}}). \tag{3.14d}$$

A field fluctuation $X_{\vec{q}}$ is said to have a definite space-inversion parity if it transforms according to

$$\mathcal{S}_{\text{in}}X_{\vec{q}} = \epsilon^X_P X_{-\vec{q}}. \tag{3.15a}$$

Since the space inversion operators have the property $\mathcal{S}_{\text{in}}\mathcal{S}_{\text{in}} = 1$, there holds $\epsilon^X_P = 1$ or $\epsilon^X_P = -1$. In the former case, the parity is called even; in the later one, it is called odd. The density fluctuations $\rho^\kappa_{\vec{q}}$ have even parity and the current fluctuations $j_{\vec{q}}^{\kappa m}$ have an odd one. Let $X_{\vec{q}}$ and $Y_{\vec{q}}$ be a pair of fields with definite space-inversion parity ϵ_P^X and ϵ_P^Y, respectively. The absence-of-chirality condition is valid for the correlation function of the two fields if and only if

$$(X_{\vec{q}}(t)|Y_{\vec{q}}) = \epsilon_P^X \epsilon_P^Y (X_{-\vec{q}}(t)|Y_{-\vec{q}}). \tag{3.15b}$$

A field fluctuation, say $A_{\vec{q}}$, is called a scalar field if it transforms under all rotations D according to the equation

$$\mathcal{S}_{\rm ro}(D) A_{\vec{q}} = A_{D\vec{q}}. \tag{3.16a}$$

The density fluctuations $\rho_{\vec{q}}^{\kappa}$ are examples for a scalar-field fluctuations. If a triple of field fluctuations, say $V_{\vec{q}}^m$, $m = 1, 2, 3$, transforms according to

$$\mathcal{S}_{\rm ro}(D) V_{\vec{q}}^m = \sum_k V_{D\vec{q}}^k D^{km}, \tag{3.16b}$$

it is called a vector field $\vec{V}_{\vec{q}}$ with coordinates $V_{\vec{q}}^m$. The triple of the current-fluctuation coordinates $j_{\vec{q}}^{\kappa m}$ is an example. If a matrix of fields $T_{\vec{q}}^{mn}$, $m, n = 1, 2, 3$, transforms according to the rule

$$\mathcal{S}_{\rm ro}(D) T_{\vec{q}}^{mn} = \sum_{k\ell} T_{D\vec{q}}^{k\ell} D^{km} D^{\ell n}, \tag{3.16c}$$

it is called a tensor field of rank two. The isotropy conditions imply restrictions for the averages, scalar products, and correlations formed with fields. They shall be formulated here only for the cases, which are needed in this book.

The density fluctuations are scalar fields. According to Eq. (B.4b), their correlations are independent of the wave-vector direction. One can write

$$(\rho_{\vec{q}}^{\kappa}(t)|\rho_{\vec{q}}^{\sigma}) = F_q^{(0)\kappa\sigma}(t). \tag{3.17a}$$

Here, the function on the right-hand side depends on the wave vector $\vec{q}$ merely via its modulus $q = |\vec{q}|$. Similarly, one can express the correlations between density and the vector field of current fluctuations in terms of functions independent of the direction of $\vec{q}$. Equations (B.6a,c) yield

$$(\rho_{\vec{q}}^{\kappa}(t)|j_{\vec{q}}^{\sigma m}) = G_q^{(1)\kappa\sigma}(t)(q^m/q), \tag{3.17b}$$

$$(j_{\vec{q}}^{\kappa m}(t)|\rho_{\vec{q}}^{\sigma}) = G_q^{(2)\kappa\sigma}(t)(q^m/q). \tag{3.17c}$$

The absence-of-chirality condition (3.15b) is obeyed automatically by the formulas above, because $\rho_{\vec{q}}^{\kappa}$ has even space-inversion parity and $j_{\vec{q}}^{\kappa m}$ has an odd one. But, this condition simplifies the representation (B.7b) of the current correlators. Because of Eq. (B.11), one can write

$$(j_{\vec{q}}^{\kappa m}(t)|j_{\vec{q}}^{\sigma n}) = F_q^{(1)\kappa\sigma}(t)\big[(q^m/q)(q^n/q)\big] + F_q^{(2)\kappa\sigma}(t)\big[\delta^{mm} - (q^m/q)(q^n/q)\big]. \tag{3.17d}$$

The functions $F_q^{(1,2)\kappa\sigma}(t)$ depend on the wave vector via the modulus only.

The densities and the current densities are real fields. Hence, Eq. (3.14c) implies that the density–density and current–current correlators are real,

$$F_q^{(a)\kappa\sigma}(t) = F_q^{(a)\kappa\sigma}(t)^*, \tag{3.18a}$$

$a = 0, 1, 2$, while the mixed-type correlators are imaginary,

$$G_q^{(b)\kappa\sigma}(t) = -G_q^{(b)\kappa\sigma}(t)^*, \tag{3.18b}$$

$b = 1, 2$. The densities have even time-inversion parity and the currents have an odd one. Hence, Eq. (3.3c) implies that the correlators of fields of the same type are even functions of the time,

$$F_q^{(a)\kappa\sigma}(t) = F_q^{(a)\kappa\sigma}(-t), \tag{3.18c}$$

while the mixed ones are odd functions,

$$G_q^{(b)\kappa\sigma}(t) = -G_q^{(b)\kappa\sigma}(-t). \tag{3.18d}$$

Using the preceding equations, the hermiticity condition (2.24) implies

$$F_q^{(a)\kappa\sigma}(t) = F_q^{(a)\sigma\kappa}(t), \tag{3.18e}$$

$$G_q^{(1)\kappa\sigma}(t) = G_q^{(2)\sigma\kappa}(t). \tag{3.18f}$$

If a correlator $F(t)$ is an even function of time, the formula (2.35a) for its spectrum can be simplified to

$$F''(\omega) = \int_0^\infty \cos(\omega t) F(t) dt. \tag{3.19a}$$

Thus, the spectrum $F''(\omega)$ is an even function of frequency ω, and the spectral representation (2.35b) specializes to

$$F(t) = \frac{2}{\pi} \int_0^\infty \cos(\omega t) F''(\omega) d\omega. \tag{3.19b}$$

The reverse statement holds also: if $F''(\omega)$ is an even function of frequency, the spectral representation shows that $F(t)$ is an even function of the time. Equations (3.19a,b) imply that $F(t)$ is real if and only if $F''(\omega)$ is real. The spectral representation for the correlator in the frequency domain, Eq. (2.47), shows that $F''(\omega) = F''(-\omega)$ implies

$$F(z) = -F(-z), \tag{3.20a}$$

while $F''(\omega)^* = F''(\omega)$ leads to

$$F(z) = F(z^*)^*. \tag{3.20b}$$

Because of Eq. (2.43), the reverse statements hold also: Eq. (3.20a) implies that the spectrum $F''(\omega)$ is an even function of ω, and Eq. (3.20b) implies that the spectrum is real. Combining the preceding equations, one gets a symmetry with respect to reflections of $z = x + iy$ at the imaginary axis

$$F(x + iy) = -F(-x + iy)^*. \tag{3.20c}$$

The equations of this paragraph apply for the three correlators $F_q^{(a)\kappa\sigma}(t)$, their spectra, and their Laplace transforms, respectively.

If a correlator $G(t)$ is an odd function of t, Eq. (2.35a) leads to

$$G''(\omega) = i \int_0^\infty \sin(\omega t) G(t) dt. \tag{3.21a}$$

The spectrum obeys $G''(\omega) = -G''(-\omega)$ and the spectral representation gets the form

$$G(t) = (-i)\frac{2}{\pi} \int_0^\infty \sin(\omega t) G''(\omega) d\omega. \tag{3.21b}$$

Hence, if $G''(\omega)$ is an odd function of ω, $G(t)$ is an odd one of t. Equations (3.21a,b) show that the spectrum is real if and only if $G(t)$ is imaginary. If an odd spectrum is used in the spectral representation for the correlator $G(z)$, one can derive the symmetry relation

$$G(z) = G(-z). \tag{3.22a}$$

If the spectrum is real, the analogue of Eq. (3.20b) holds: $G(z) = G(z^*)^*$. Combination of both results modifies Eq. (3.20c) to

$$G(x + iy) = G(-x + iy)^*. \tag{3.22b}$$

The equations of this paragraph apply for $G_q^{(b)\kappa\sigma}(t)$, $G_q^{(b)\kappa\sigma''}(\omega)$ and $G_q^{(b)\kappa\sigma}(z)$, respectively, $b = 1, 2$.

3.2 Tagged-particle dynamics

3.2.1 *Basic concepts and general equations*

The simplest functions describing statistically some features of the dynamics are obtained by correlating the phase-space coordinates of a special particle with the phase-space coordinates of the same particle. The special particle is referred to as a tagged one and, usually, the indicated correlation functions are called self-correlations. In this section, the simplest self-correlators are defined, namely the tagged-particle-density-fluctuation correlator and the velocity correlator. Also, the mean-squared displacement will be introduced.

Let us denote all quantities referring to the tagged particle under consideration by an index s. For example, $\vec{r}_s = (r_s^1, r_s^2, r_s^3)$ and $\vec{v}_s = (v_s^1, v_s^2, v_s^3)$ denote its position and its velocity, respectively. The most important dynamical variables to be considered are the tagged-particle-density fluctuations for wave vectors $\vec{q} : \rho^s_{\vec{q}} = \exp[i\vec{q}\vec{r}_s]$. They are the spatial Fourier transforms of the tagged-particle density $\rho^s(\vec{r}) = \delta(\vec{r}-\vec{r}_s)$. These variables have been discussed above in Eqs. (3.7), (3.10); merely the general index κ is specialized to s. There holds the equation of motion

$$\partial_t \rho^s_{\vec{q}} = i\vec{q}\vec{j}^s_{\vec{q}}. \tag{3.23a}$$

Here, $\vec{j}^s_{\vec{q}} = \vec{v}_s \exp[i\vec{q}\vec{r}_s]$ is the tagged-particle-current-density fluctuation. It is the spatial Fourier transform for wave vector $\vec{q}$ of the tagged-particle-current density $\vec{j}^s(\vec{r}) = \vec{v}_s\delta(\vec{r}-\vec{r}_s)$ as introduced in connection with Eqs. (3.7), (3.10). The equation of motion is the Fourier transform of the differential equation

$$\partial_t \rho^s(\vec{r},t) + \vec{\partial}\vec{j}^s(\vec{r},t) = 0. \tag{3.23b}$$

This is a continuity equation as is familiar from other areas of physics. It presents a field-theoretical formulation of a conservation law, namely the conservation of the probability to find the tagged particle somewhere in space. Let us introduce the projection $j^s_{\vec{q}}$ of the current fluctuation on the wave-vector direction $\vec{q}/q$, $q = |\vec{q}|$:

$$j^s_{\vec{q}} = (\vec{q}/q)\vec{j}^s_{\vec{q}}. \tag{3.23c}$$

The continuity equation formulates a proportionality of $j^s_{\vec{q}}$ and the time derivative of $\rho^s_{\vec{q}}$:

$$-i\partial_t \rho^s_{\vec{q}} = \mathcal{L}\rho^s_{\vec{q}} = qj^s_{\vec{q}}. \tag{3.23d}$$

Here, $\mathcal{L}$ denotes the Liouville operator as introduced in Eq. (2.4a).

Three types of scalar products can be formed with the above defined dynamical variables. Since $\rho^{s*}_{\vec{q}}\rho^s_{\vec{q}} = 1$, the density fluctuations are normalized

$$(\rho^s_{\vec{q}}|\rho^s_{\vec{q}}) = 1. \tag{3.24a}$$

Since $\rho^{s*}_{\vec{q}}\vec{j}^s_{\vec{q}} = \vec{v}_s$, the average yields the orthogonality relation

$$(\rho^s_{\vec{q}}|j^{sm}_{\vec{q}}) = 0, \qquad m = 1,2,3. \tag{3.24b}$$

Since $j^{sm*}_{\vec{q}} j^{sn}_{\vec{q}} = v^m_s v^n_s$, one gets from Eqs. (3.1a,b)

$$(j^{sm}_{\vec{q}}|j^{sn}_{\vec{q}}) = \delta^{m,n}(v^s_{th})^2, \qquad m,n = 1,2,3. \tag{3.24c}$$

Here, v^s_{th} denotes the thermal velocity of the tagged particle with mass μ_s.

The tagged-particle-density correlator is the autocorrelation function defined via Eq. (2.18a) for $A = B = \rho^s_{\vec{q}}$. It is the function from Eq. (3.17a) formed with

$\kappa = \sigma = s$. From the discussion in Sec. 3.1.2, one infers that this correlator depends on the wave vector via the modulus q only,

$$\phi_q^s(t) = (\rho_{\vec{q}}^s(t)|\rho_{\vec{q}}^s). \tag{3.25a}$$

It is a real even function of the time t. From the discussion of Eqs. (3.19a,b) one concludes, that the spectrum $\phi_q^{s\,\prime\prime}(\omega)$ is a real and even function of the frequency ω. According to Eq. (3.23d), the variable $\rho_{\vec{q}}^s$ is in the domain of $\mathcal{L}$. Hence, one can use Eqs. (2.26a,b) in order to express the $\rho_{\vec{q}}^s - j_{\vec{q}}^s$-correlators as time derivatives of the density correlator

$$i\partial_t \phi_q^s(t)/q = (\rho_{\vec{q}}^s(t)|j_{\vec{q}}^s) = (j_{\vec{q}}^s(t)|\rho_{\vec{q}}^s). \tag{3.25b}$$

Similarly, one can use Eq. (2.26c) with $m = n = 1$ in order to express the autocorrelator of the current projections,

$$\psi_q^s(t) = (j_{\vec{q}}^s(t)|j_{\vec{q}}^s), \tag{3.25c}$$

as second time derivative of the density correlator:

$$-\partial_t^2 \phi_q^s(t)/q^2 = \psi_q^s(t). \tag{3.25d}$$

From Eqs. (3.24a), (3.25a) one gets the correlator's initial value: $\phi_q^s(t = 0) = 1$. Similarly, one gets from (3.24b), (3.25b) a vanishing first Taylor coefficient: $\partial_t \phi_q^s(t = 0) = 0$. Hence, one can use Eq. (A.7b) in order to express the Laplace transform of the current correlator, $\psi_q^s(z)$, in terms of the density correlator in the frequency domain, $\phi_q(z)$:

$$\psi_q^s(z) = (z/q^2)[z\phi_q^s(z) + 1]. \tag{3.25e}$$

Equations (3.25c,d) yield the second Taylor coefficient of the density correlator: $\partial_t^2 \phi_q^s(t = 0) = -q^2(j_{\vec{q}}^s|j_{\vec{q}}^s) = -q^2(v_{th}^s)^2$. The latter equality follows from Eqs. (3.23c), (3.24c). Anticipating that the third time derivative of $\phi_q^s(t)$ exists as continuous bounded function for $t \geq 0$, one arrives at the formula:

$$\phi_q^s(t) = 1 - \tfrac{1}{2}(\Omega_{sq} t)^2 + O(|t|^3). \tag{3.26a}$$

The frequency

$$\Omega_{sq} = q v_{th}^s \tag{3.26b}$$

determines the leading-order decay of the correlations for short times. This frequency is independent of interaction effects; it merely reflects the width of Maxwell's velocity distribution.

Expressing the scalar product in Eq. (3.25a) as average $\langle \rho_{\vec{q}}^s(t)^* \rho_{\vec{q}}^s \rangle$, one can write the correlator of the density fluctuations as

$$\phi_q^s(t) = \Big\langle \exp\big[i\vec{q}[\vec{r}_s(t) - \vec{r}_s(t = 0)]\big]\Big\rangle. \tag{3.27a}$$

It is the Fourier transform from position space to the reciprocal space of the average of a generalized dynamical variable as discussed above in connection with Eq. (3.7):

$$\phi^s(\vec{r},t) = \Big\langle \delta\big(\vec{r} - [\vec{r}_s(t) - \vec{r}_s(t=0)]\big) \Big\rangle. \tag{3.27b}$$

There holds the normalization condition

$$\int \phi^s(\vec{r},t)d^3\vec{r} = \phi^s_{q=0}(t) = 1. \tag{3.27c}$$

The generalized function $\phi^s(\vec{r},t)$ is the probability density for a motion of the particle by a displacement $\vec{r}$ if the time changes from zero to t. Notice, that this probability depends on the vector $\vec{r}$ merely via its modulus $r = |\vec{r}|$, as was explained in connection with Eqs. (3.9c,d). For $t = 0$, the probability density is a point distribution, $\phi^s(\vec{r}, t = 0) = \delta(\vec{r})$. If the time increases, the particles move away from their initial positions. Thereby, the probability distribution spreads out. A parameter, which quantifies this spread, is the averaged quadratic radius

$$\delta r_s^2(t) = \int r^2 \phi^s(\vec{r},t)d^3\vec{r}. \tag{3.28a}$$

This quantity is called the mean-squared displacement; and Eq. (3.27b) yields the formula

$$\delta r_s^2(t) = \Big\langle \big[\vec{r}_s(t) - \vec{r}_s(t=0)\big]^2 \Big\rangle. \tag{3.28b}$$

The functions $\phi^s_q(t)$, $\phi^s(\vec{r},t)$ and $\delta r_s^2(t)$ are paradigms for quantities characterizing the dynamics statistically. Molecular-dynamics-simulation results for these functions have been discussed above in connection with Figs. 1.4, 1.5, 1.8, 1.9, 1.11, and 1.22, respectively. In principle, $\phi^s_q(t)$ and $\phi^{s\,\prime\prime}_q(\omega)$ can be measured by incoherent-neutron-scattering experiments (Lovesey 1984).

Let us abbreviate the tagged-particle's coordinate differences by $d^m(t) = r_s^m(t) - r_s^m(t = 0)$. There is a rotation $\mathcal{S}_{\text{ro}}(D)$, which transforms the direction m in direction n. The equations for isotropy (3.6a,b) yield $\langle d^m(t)d^m(t)\rangle = \langle d^n(t)d^n(t)\rangle$. Similarly, there is a rotation around the axis m which inverts the direction of axis n. Isotropy implies, therefore, that $\langle d^m(t)d^n(t)\rangle = 0$ for $m \neq n$. Thus, one can write

$$\Big\langle [r_s^m(t) - r_s^m(t=0)][r_s^n(t) - r_s^n(t=0)] \Big\rangle = 2\delta^{m,n}\Delta_s(t). \tag{3.29a}$$

There holds

$$\Delta_s(t) = \frac{1}{2}\Big\langle \big[\vec{e}(\vec{r}_s(t) - \vec{r}_s(t=0))\big]^2 \Big\rangle \tag{3.29b}$$

for any unit vector $\vec{e}$. Up to a factor 6, the function $\Delta_s(t)$ is the mean-squared displacement

$$\delta r_s^2(t) = 6\Delta_s(t). \tag{3.29c}$$

The expansion of the exponential in Eq. (3.27a) yields an expansion of the correlator in powers of q. There are no contributions of odd powers because of absence of chirality. Using Eq. (3.29b) with $\vec{e} = \vec{q}/q$, one gets

$$\phi_q^s(t) = 1 - q^2\Delta_s(t) + O(q^4). \tag{3.30}$$

For small q, the density-fluctuation correlations decrease with increase of the wave number. In leading order, this decrease is determined by the mean-squared displacement. Since $\phi_q^s(t)$ is an even function of time, there holds also $\Delta_s(t) = \Delta_s(-t)$. Comparing Eqs. (3.26a), (3.30), one obtains the ballistic law as the short-time asymptote

$$\Delta_s(t) = \tfrac{1}{2}(v_{th}^s t)^2 + O(|t|^3). \tag{3.31}$$

Differentiating Eq. (3.29a), one gets $\partial_t \Delta_s(t)\delta^{m,n} = \langle v_s^m(t)[r_s^n(t) - r_s^n(t = 0)]\rangle$. There are the initial conditions $\Delta_s(t = 0) = \partial_t \Delta_s(t = 0) = 0$. Using the stationarity condition (2.7c), one gets $\partial_t \Delta_s(t) = \langle v_s^m(t = 0)[r_s^m(t = 0) - r_s^m(-t)]\rangle$. Differentiating once more and using Eq. (2.7c) again leads to

$$\partial_t^2 \Delta_s(t) = K_s(t). \tag{3.32a}$$

Here, the velocity correlator $K_s(t)$ is introduced by

$$(v_s^m(t)|v_s^n) = \delta^{m,n} K_s(t). \tag{3.32b}$$

The initial value of this function is given by the thermal velocity:

$$K_s(t = 0) = (v_{th}^s)^2. \tag{3.32c}$$

Since $\Delta_s(t) = \Delta_s(-t)$, also $K_s(t)$ is an even function of the time. Equation (3.32a) together with the initial conditions mentioned for $\Delta_s(t)$ is equivalent to a representation of $\Delta_s(t)$ in terms of the velocity correlations:

$$\Delta_s(t) = \int_0^t (t - t')K_s(t')dt'. \tag{3.32d}$$

Since $LT[t](z) = -i/z^2$, one can use Eqs. (A.6a,b) in order to rewrite the preceding formula as one connecting $\Delta_s(z) = LT[\Delta_s(t)](z)$ with the velocity correlator in the frequency domain:

$$\Delta_s(z) = -K_s(z)/z^2. \tag{3.32e}$$

3.2.2 *Tagged-particle diffusion*

This section shall be used to demonstrate a procedure for deriving the diffusion law for the tagged-particle density from the Zwanzig–Mori equations of motion. This law is obtained from postulates of some rather general regularity properties concerning the time and wave-number dependencies of certain relaxation kernels. The diffusion law is the simplest example establishing a connection between a microscopic description of the dynamics by correlation functions and a well-known phenomenological description of a liquid-dynamics phenomenon.

Let us postulate that the velocity correlator is absolutely integrable, i.e., $\int_0^\infty |K_s(t)|dt = C < \infty$. The autocorrelator $K_s(t)$ is continuous and bounded by its initial value $(v_{th}^s)^2$. Hence, the request $C < \infty$ is equivalent to one for the large-time behaviour of $K_s(t)$. It is discussed in Appendix A.2 that the Fourier transform $FT[K_s(t)](\omega) = \hat{K}(\omega)$ exists as a continuous function of the frequency ω. It determines the spectrum $K_s''(\omega) = \hat{K}(\omega)/2$, which is non-negative and can be written as $K_s''(\omega) = \int_0^\infty \cos(\omega t) K_s(t)dt$. It is an even function of ω, and there holds $0 \leq K_s''(\omega) \leq C$. The zero-frequency velocity spectrum shall be denoted by D_s:

$$D_s = K_s''(\omega = 0). \tag{3.33a}$$

It provides an integral information on the dynamics, namely

$$D_s = \int_0^\infty K_s(t)dt. \tag{3.33b}$$

According to Eq. (3.32d), $\Delta_s(t)/t$ is given as the integral over $\theta(t - t')(1 - t'/t)K_s(t')$. This integrand has $|K_s(t')|$ as integrable t-independent majorant. Hence, one can perform the long-time limit under the integral. Thereby, one finds:

$$\lim_{t\to\infty} \Delta_s(t)/t = D_s. \tag{3.33c}$$

A further implication of the postulated integrability is that the correlator Laplace transform $K_s(z)$ has continuous boundary values $K_s(\omega \pm i0) = K_s'(\omega) \pm iK_s''(\omega)$; this is discussed in connection with Eq. (2.43). Equations (2.28), (3.33b) identify D_s as the zero-frequency limit of the velocity correlator in the frequency domain

$$\lim_{z\to 0} K_s(z) = \pm iD_s, \qquad \text{Im z} \gtrless 0. \tag{3.33d}$$

It appears as an exceptional case that the non-negative function $K_s''(\omega)$ vanishes for $\omega = 0$. It can be expected only for special idealized descriptions of matter. Examples for such idealizations are the model of a crystalline solid without defect motion and the idealized glass states to be discussed in detail in this book. Let us postulate here that a liquid state deals with a non-vanishing value for D_s. Equation (3.33c) shows that the number D_s quantifies the leading asymptotic increase of the mean-squared displacement for large times:

$$\Delta_s(t) \sim D_s t, \quad t \to \infty. \tag{3.33e}$$

The general concept of an asymptotic equality is explained in Appendix A.7. One concludes from Eqs. (3.32e), (3.33d) that the large-time divergency of $\Delta_s(t)$ is equivalent to a second-order zero-frequency pole of the mean-squared displacement's Laplace transform:

$$\Delta_s(z) \sim \mp i D_s/z^2, \quad z \to 0, \quad \text{Im z} \gtrless 0. \tag{3.33f}$$

To derive a further implication of the $D_s > 0$ assumption, the current correlator from Eq. (3.25c) shall be considered. Choosing $\vec{q} = (0, 0, q)$, one can write $\psi_q^s(t) = \langle v_s^3(t) v_s^3(t=0) \exp[-iq[r_s^3(t) - r_s^3(t=0)]]\rangle$. Regularity of $\psi_q^s(t)$ with respect to the q-dependence shall be postulated in the sense that the limit of a vanishing wave number can be interchanged with the averaging procedure and with the Laplace transformation. This leads to $\lim_{q \to 0} \psi_q^s(t) = K_s(t)$ and

$$\lim_{q \to 0} \psi_q^s(z) = K_s(z). \tag{3.34a}$$

Combination of this result with Eq. (3.33d) yields

$$\lim_{z \to 0} \lim_{q \to 0} \psi_q^s(z) = \pm i D_s, \quad \text{Im z} \gtrless 0. \tag{3.34b}$$

In Sec. 2.6 it was shown that $[z\phi_q(z) + 1]$ has a finite limit for z approaching 0. Therefore, Eq. (3.25e) yields the formula

$$\lim_{q \to 0} \lim_{z \to 0} \psi_q^s(z) = 0. \tag{3.34c}$$

Considered as a function of the variable pair (q, z), Im $z > 0$, the current correlator $\psi_q^s(z)$ is discontinuous at $q = 0$, $z = 0$. The parameter D_s quantifies the discontinuity. This finding can also be reformulated as a singular behaviour of the density-fluctuation spectrum. According to Eq. (2.46b), the spectra of the autocorrelators $\psi_q^s(z)$ and $\phi_q^s(z)$ can be obtained as imaginary parts for $z = \omega + i0$. Since Eq. (3.25e) implies $\psi_q^{s\prime\prime}(\omega) = [\omega^2/q^2]\phi_q^{s\prime\prime}(\omega)$, one gets from Eq. (3.34b)

$$\lim_{\omega \to 0} \lim_{q \to 0} [\omega^2/q^2] \phi_q^{s\prime\prime}(\omega) = D_s, \tag{3.34d}$$

while Eq. (3.34c) yields

$$\lim_{q \to 0} \lim_{\omega \to 0} [\omega^2/q^2] \phi_q^{s\prime\prime}(\omega) = 0. \tag{3.34e}$$

The mathematical origin of the above specified singularity for low frequencies and small wave numbers is the factor $[1/q^2]$ in Eq. (3.25d). This factor is caused by the factor q in front of $j^s_{\vec{q}}$ in Eq. (3.23d). Hence, the identified singular behaviours are implications of the conservation law for the density.

The formalism, which is explained in Sec. 2.4, is well suited to get control over the conservation-law-induced singularities identified above. The procedure shall be explained in three steps. First, the Zwanzig–Mori equation of motion in the frequency domain shall be formulated for the single distinguished variable $A = \rho^s_{\vec{q}}$. The normalization constant $g = (A|A)$ is unity in this case. Variable A has a definite time inversion parity and, thus, $\mathcal{L}A$ has an opposite one. Hence, there holds $(A|\mathcal{L}A) = 0$. As a result, the frequency Ω in Eq. (2.60a) vanishes. For the same reason, the projector $\mathcal{Q}$ can be dropped in Eq. (2.61a), and Eq. (3.23d) yields for the fluctuating force: $\dot{A}^{\mathcal{Q}} = iqj^s_{\vec{q}}$. The equation of motion (2.62a) can be written as fraction representation of the correlator:

$$\phi^s_q(z) = -1/[z + q^2 K^s_q(z)]. \tag{3.35a}$$

Equation (2.61b) yields the expression of the kernel $K^s_q(z)$ as matrix element of a resolvent

$$K^s_q(z) = (j^s_{\vec{q}}|\mathcal{R}_{\mathcal{Q}}(z) j^s_{\vec{q}}). \tag{3.35b}$$

The resolvent $\mathcal{R}_{\mathcal{Q}}(z) = [\mathcal{L}_{\mathcal{Q}} - z]^{-1}$ refers to a time evolution generated by the reduced Liouvillian $\mathcal{L}_{\mathcal{Q}} = \mathcal{Q}\mathcal{L}\mathcal{Q}$. The operator $\mathcal{Q} = 1 - |A)(A|$ projects perpendicular to the density fluctuations. The kernel $K^s_q(z)$ is the Laplace transform of $K^s_q(t) = (j^s_{\vec{q}}|\mathcal{U}_{\mathcal{Q}}(-t) j^s_{\vec{q}})$, Eq. (2.61c). From Eqs. (3.23c, 3.24c), one gets

$$K^s_q(t = 0) = (v^s_{th})^2. \tag{3.35c}$$

The continuity equation leads to Eq. (3.25e) for the current correlator. Therefore, one can express also this correlation function in terms of the kernel $K^s_q(z)$:

$$\psi^s_q(z) = zK^s_q(z)/[z + q^2 K^s_q(z)]. \tag{3.36a}$$

One can use this result in order to express the kernel in terms of the correlator: $K^s_q(z) = \psi^s_q(z)/[1 - q^2\psi^s_q(z)/z]$. Hence, Eq. (3.34a) leads to the formula

$$\lim_{q\to 0} K^s_q(z) = K_s(z). \tag{3.36b}$$

The kernel $K^s_q(z)$ is a correlation function referring to the reduced dynamics. However, its zero-wave-number limit is a correlator of the dynamics defined by the Liouvillian $\mathcal{L}$, i.e., by Hamilton's equations of motion. The limit is the velocity correlator introduced in the preceding section.

The time translation operator $\mathcal{U}_{\mathcal{Q}}(t)$ treats $\rho^s_{\vec{q}}$ as a constant of motion; it does not lead to a continuity equation for $\rho^s_{\vec{q}}$ anymore. The reason, which causes the low-frequency-small-wave-number discontinuity of the correlators formed with $\mathcal{U}(t)$, is absent for the dynamics generated by the reduced Liouvillian $\mathcal{L}_{\mathcal{Q}}$. Therefore, it seems justified to assume that function $K^s_q(z)$ depends continuously on the pair of variables (q, z) for Im $z \gtrless 0$. Hence, as the second step of the Zwanzig–Mori analysis, the existence of $\lim_{(q,z)\to(0,0)} K_q^s(z) = d_\pm$, Im $z \gtrless 0$, shall be postulated. Equations (3.33d), (3.36b) imply $d_\pm = \pm i D_s$, and one gets

$$\lim_{(q,z)\to(0,0)} K_q^s(z) = \pm i D_s, \quad \mathrm{Im}\ \mathrm{z} \gtrless 0. \tag{3.37}$$

It might be worthwhile to rewrite this equation more explicitly as follows. For every margin $\epsilon > 0$, there exist a length scale r^* and a time scale t^*, so that

$$|K_q^s(z) \mp i D_s| \leq \epsilon D_s, \quad \mathrm{Im}\ \mathrm{z} \gtrless 0, \tag{3.38a}$$

whenever

$$qr^* \leq 1, \quad |zt^*| \leq 1. \tag{3.38b}$$

Naturally, r^* and t^* are the larger the smaller is ϵ. These scales are of the order of magnitude of the natural ones for condensed matter: $t^* \approx t_{\mathrm{mic}}$ and $r^* \approx d$.

One checks that Eqs. (3.36a, 3.37) imply Eqs. (3.34b,c). Similarly, Eqs. (3.35a), (3.37) imply Eqs. (3.34d,e). Formulas (3.35a), (3.36a) provide a representation of the correlators for density and current fluctuations in terms of a continuous function $K_q^s(z)$ so that the conservation-law-induced discontinuities of the correlators and their spectra are reproduced.

A digression will be helpful below. It is discussed in connection with Eqs. (2.57e,f) that kernels like $K_q^s(z)$ are positive-analytic functions, i.e., they are holomorphic off the real frequency axis and obey

$$K_q^s(z)^* = K_q^s(z^*), \tag{3.39a}$$

$$\mathrm{Im}\ \mathrm{K_q^s(z)} \gtrless 0, \quad \mathrm{Im,}\ \mathrm{z} \gtrless 0. \tag{3.39b}$$

Since Eq. (3.35a) implies $K_q^s(z) = -[z + \phi_q^s(z)^{-1}]/q^2$, the kernel $K_q^s(z)$ depends on the wave vector through its modulus only and the symmetry (3.20a) of the correlator implies the analogous one of the kernel

$$K_q^s(-z) = -K_q^s(z). \tag{3.39c}$$

The reverse statements are also true. If a positive analytic the function $K_q^s(z)$ is substituted in Eq. (3.35a), the function $\phi_q^s(z)$ has the same property. This was discussed in connection with Eq. (2.65b). Formula (3.39a) implies the corresponding symmetry for $\phi_q^s(z)$. This means that $\phi_q^s(t)$ is an even real function of the time. If the kernel does not increase too strongly for large frequencies so that

$$\lim_{y\to\infty} K_q^s(iy)/y = 0, \tag{3.39d}$$

Eq. (3.35a) reproduces the correct initial value for the density correlator: $\phi_q(t = 0) = 1$. This is demonstrated in connection with Eq. (2.65b). These considerations show that the Zwanzig–Mori representation (3.35a) of the correlator suggests the construction of models for the density-fluctuation dynamics in the sense discussed in Sec. 2.4.2. Every function $K_q^s(z)$ obeying Eqs. (3.39a–d) can be substituted into Eq. (3.35a) so that $\phi_q^s(t)$ is a correlation function with all general symmetries. And it has the correct value for $t = 0$. The limit for the model kernel in Eq. (3.37) defines the spectral discontinuity according to Eqs. (3.34d,e). Formula (3.36b) implies a model for the velocity correlator obeying Eq. (3.33a).

As the third and final step of the Zwanzig–Mori procedure, a model for the density correlator shall be constructed. The kernel $K_q^s(z)$ shall be replaced by its limit for $q = 0$ and $z = 0$. Let us indicate here and in the following text such model by two bars, one indicating the zero-frequency limit and the other the zero-wave-number limit: $\overline{\overline{K}}{}_q^s(z) = \pm iD_s$, Im $z \gtrless 0$. This leads to the simplest model in the sense discussed in the preceding paragraph:

$$\overline{\overline{\phi}}{}_q^s(z) = -1/[z \pm iD_s q^2], \quad \text{Im z} \gtrless 0. \tag{3.40a}$$

The spectrum is a Lorentzian:

$$\overline{\overline{\phi}}{}_q^{s\,\prime\prime}(\omega) = D_s q^2/[\omega^2 + (D_s q^2)^2]. \tag{3.40b}$$

Fourier transformation yields the model for the correlator in the time domain

$$\overline{\overline{\phi}}{}_q^s(t) = \exp[-|t| D_s q^2]. \tag{3.40c}$$

The model describes an elementary relaxation process (1.1b) specified by the relaxation time

$$\tau_q^s = 1/[D_s q^2]. \tag{3.40d}$$

Expanding Eq. (3.40c) for small q and comparing the result with Eq. (3.30), one finds that the specified model for the density-fluctuation dynamics implies a linear law for the mean-squared displacement

$$\overline{\Delta}_s(t) = D_s |t|. \tag{3.40e}$$

The model describes asymptotically the density correlator for frequencies and wave number approaching zero. According to Eq. (3.38a), it describes the inverse of the correlator with a relative error ϵ, $[\phi_q^s(z)^{-1} - \overline{\overline{\phi}}{}_q^s(z)^{-1}]\tau_q^s < \epsilon$, provided (q, z) are restricted by the inequalities (3.38b). If wave numbers are so small that $q \leq 1/r^*$ and if frequencies are so low that $|z| \leq 1/t^*$, the correlator $\phi_q^s(z)$ can be replaced by the model $\overline{\overline{\phi}}{}_q^s(z)$ within an error specified by

$$|\phi_q^s(z) - \overline{\overline{\phi}}{}_q^s(z)| \leq \epsilon(D_s q^2)|\phi_q^s(z)\overline{\overline{\phi}}{}_q^s(z)|. \tag{3.41}$$

The Fourier transform of the Gaussian $\overline{\overline{\phi}}^s_q(t)$ is the Gaussian

$$\overline{\overline{\phi}}_s(\vec{r},t) = (2\rho\sqrt{\pi})^{-3}\exp[-r^2/(4\rho^2)], \qquad \rho = \sqrt{D_s|t|}. \tag{3.42a}$$

This function is a regularization of the δ-distribution. For $t \to 0$, it approaches the point distribution $\delta(\vec{r})$ in the sense of a distribution limit

$$\lim_{t\to 0}\overline{\overline{\phi}}_s(\vec{r},t) = \delta(\vec{r}). \tag{3.42b}$$

The Function $\overline{\overline{\phi}}^s_q(t)$ is the unique solution of the differential equation $\partial_t\overline{\overline{\phi}}^s_q(t) \pm D_s q^2\overline{\overline{\phi}}^s_q(t) = 0$, $t \gtrless 0$, obeying the initial condition $\overline{\overline{\phi}}^s_q(t=0) = 1$. Fourier transformation casts this equation in a field equation for the probability density

$$\partial_t\overline{\overline{\phi}}_s(\vec{r},t) \mp D_s\vec{\partial}\vec{\partial}\,\overline{\overline{\phi}}_s(\vec{r},t) = 0, \quad t \gtrless 0, \tag{3.42c}$$

to be solved with Eq. (3.42b) as the initial condition. This differential equation for the model function $\overline{\overline{\phi}}_s(\vec{r},t)$ is the diffusion equation, which is familiar from other areas of physics. It specifies a function of space and time in terms of a single parameter D_s, which is called diffusivity or diffusion constant. In the present case, D_s is called the self-diffusion constant or tagged-particle diffusivity. The function $\overline{\overline{\phi}}_s(\vec{r},t)$ is the well-known fundamental solution or Green's function of the diffusion equation. Parameters like D_s are called transport coefficients; they specify the equations, which describe the propagation of fields.

One concludes that the correlations of density fluctuations for small frequencies and small wave numbers vary as described by a diffusion equation. The transport coefficient D_s, which characterizes the interaction of the tagged particle with the surroundings, is given by an integral over the velocity correlator. The relation (3.33b) is called a Green–Kubo equation. The diffusivity specifies the mean-squared displacement by Eq. (3.40e), which is Einstein's law for Brownian motion.

The essential point underlying the preceding derivations concerns the separation of the two relevant time scales governing the dynamics. The first scale $t^* \approx t_{\text{mic}}$ is the natural one for particle motion, say a picosecond for a conventional liquid. The second scale τ^s_q describes the relaxation of density fluctuations on a length scale $2\pi/q$. Because of the conservation law for the tagged-particle density, the probability in a volume can change only by flow through its surface. Since the ratio of a surface layer to the volume decreases to zero for increasing volume, one can assume $1/\tau^s_q$ to vanish for q tending to zero. Thus, τ^s_q is arbitrarily well separated from t^* if the wave numbers are arbitrarily small:

$$\lim_{q\to 0} t^*/\tau^s_q = 0. \tag{3.43}$$

Thereby, t^*/τ^s_q is identified as a small parameter of the problem. The Zwanzig–Mori formula (3.35a) expresses the density correlator in terms of the function

$K_q^s(z)$, whose frequency dependence occurs on scale $1/t^*$. Discussing only long-wavelength fluctuations, one can replace $K_q^s(z)$ by the velocity correlator $K_s(z)$. Discussing the dynamics on scale τ_q^s, one can replace the velocity correlator by its zero-frequency limit, i.e., one can replace it by the positive analytic function

$$\overline{K}_s(z) = \pm i D_s, \qquad \operatorname{Im} z \gtrless 0. \tag{3.44a}$$

This means that the velocity spectrum is replaced by a white noise one.

$$\overline{K}_s''(\omega) = D_s. \tag{3.44b}$$

The white-noise assumption leads to the results formulated in Eqs. (3.40a–e), in particular to the quantitative result for the scale τ_q^s. It is emphasized in Sec. 2.4 that there is a number of possibilities to formulate Zwanzig–Mori formulas. The special possibility chosen in Eq. (3.35a) is motivated by the insight explained above. It is a virtue of the formalism that it establishes a frame to analyse the consequences of the understanding of the time-scale-separation feature. The formalism provides the tools to detach unproven regularity assumptions, like the one of the integrability of $|K_s(t)|$, from proven identities, like Eq. (3.33c) for the mean-squared-displacement asymptote.

The values of $\overline{\overline{\phi}}_q^s(z)$ for large frequencies z enter Eq. (A.4) for the determination of $\overline{\overline{\phi}}_q^s(t)$ as Laplace back transformation. Similarly, the spectrum $\overline{\overline{\phi}}_q^{s\,\prime\prime}(\omega)$ for large frequencies ω enters the representation of $\overline{\overline{\phi}}_q^s(t)$ via Fourier transformation. However, the approximation of $\phi_q^s(z)$ by $\overline{\overline{\phi}}_q^s(z)$ or the one of $\phi_q^{s\,\prime\prime}(\omega)$ by $\overline{\overline{\phi}}_q^{s\,\prime\prime}(\omega)$ is justified only for small values of $|z|$ or $|\omega|$, respectively. For large values of $|z|$ or $|\omega|$, the models for the correlators in the frequency domain or for the spectra may differ strongly from the functions $\phi_q^s(z)$ or $\phi_q^{s\,\prime\prime}(\omega)$, respectively. These differences may cause features of the model functions $\overline{\overline{\phi}}_q^s(t)$, which are qualitatively different from the corresponding ones of $\phi_q^s(t)$. The differentiability properties of $\overline{\overline{\phi}}_q^s(t)$ provide examples for such artifacts. It was shown in connection with Eq. (3.25d), that $\phi_q^s(t)$ has continuous time derivatives up to order 2. This implies a decrease of the density-fluctuation spectrum for large frequencies stronger than ω^{-2}, since Eq. (A.14b) yields $\lim_{|\omega|\to\infty} \phi_q^{s\,\prime\prime}(\omega)\omega^2 = 0$. But, the model spectrum yields $\lim_{|\omega|\to\infty} \overline{\overline{\phi}}_q^{s\,\prime\prime}(\omega)\omega^2 = D_s q^2$. For large frequencies, the relative error of the model spectrum diverges: $\lim_{|\omega|\to\infty} \overline{\overline{\phi}}_q^{s\,\prime\prime}(\omega)/\phi_q^{s\,\prime\prime}(\omega) = \infty$. This mistreatment of the large-frequency behaviour ruins the possibility to perform a second time derivative of $\overline{\overline{\phi}}_q^s(t)$. According to Eq. (3.40c), the model correlator exhibits a cusp for $t = 0$, $\overline{\overline{\phi}}_q^s(t) - 1 = -|t| D_s q^2 + O(t^2)$, instead of the quadratic small-time variation of the correlator $\phi_q^s(t)$, which is quantified by Eq. (3.26a). Let us emphasize, that the identified huge relative error occurs for frequencies ω, where the spectra are negligible compared to the ones for low-frequency: $\lim_{|\omega|\to\infty} \overline{\overline{\phi}}_q^{s\prime\prime}(\omega) D_s q^2 = \lim_{|\omega|\to\infty} \phi_q^{s\prime\prime}(\omega) D_s q^2 = 0$. Similarly, Einstein's formula (3.40e) for the mean-squared displacement extrapolates the large time asymptote to all times. Thereby, a cusp of the $\overline{\Delta}_s(t)$ versus t curve is created for $t = 0$ instead of the correct ballistic short-time variation, which is quantified by Eq. (3.31). The origin of the cusp artifact is the white-noise model for the velocity spectrum. This implies a function $\overline{K}_s(z)$, which is constant on the frequency half planes. However, Eqs. (A.8a,b) show that there is no determinating function $\overline{K}_s(t)$ whose Laplace transform is given by Eq. (3.44a). Whenever a white-noise model is introduced to describe a correlator, the latter or one of its derivatives exhibits a cusp artifact at $t = 0$.

The diffusion equation had been formulated within the continuum theory of matter long before the atomistic structure of liquids had been established. The derivation of Eq. (3.42c) is a justification of the classical approach. There are, however, some reservations for this statement. Within the phenomenological theory, the field equation is claimed for all times with the upper alternative for the sign. The field equation is claimed for tagged matter densities. These claims lead to the problem of compatibility of time-reversal-invariant fundamental equations of motion and the irreversibility features implied by the phenomenological diffusion equation. Equation (3.42c) does not deal with densities, but with the correlation of the density at points $\vec{r}+\vec{r}_1$ at times $t+t_1$ with that at position $\vec{r}_1$ at time t_1. Time reversal symmetry implies, that this correlation decays with times increasing into the future in the same manner as it does with times decreasing into the past. Furthermore, the diffusion equation holds only for the description of spectra in a regime of small wave numbers and small frequencies. This means that only a coarse-grained version of $\overline{\overline{\phi}}_s(\vec{r}, t)$ is a good approximation of the coarse-grained correlator $\phi_s(r, t)$. The scales for the coarse graining for the variations in space and time are the numbers r^* and t^*, respectively, noted in the relations (3.38b). For the coarse-graining of the time variation, this is explained in Appendix C.

The derivation of the diffusion equation and related results like Eqs. (3.33a–d), (3.40a–e) can be done for every conserved density of definite time-inversion parity. These conditions ensure Eqs. (3.23d), (3.24b). The right-hand sides of Eqs. (3.24a,c) could be different constants or even q-dependent functions, but this would merely yield trivial factors in the final results. An example demonstrating this understanding will be discussed in Sec. 3.3.2 for the transverse-current dynamics.

Finally, two notorious situations shall be mentioned for which the above made regularity assumptions are not justified. In these cases, the diffusion-law description of the dynamics of the fluctuations for small wave numbers and small frequencies is not valid. First, there may exist a further conservation law in addition to the one considered above. The Eqs. (3.35a,b), (3.36a) remain valid. But, since the projector $\mathcal{Q}$ eliminates only one of the conserved densities, $K_q^{s\prime\prime}(\omega)$ may exhibit a similar singularity as discussed for $\phi_q^{s\prime\prime}(\omega)$ in Eqs. (3.34d,e). This would render Eq. (3.38a) invalid. Second, the state of the system may be a critical point for a phase transition. In this case, there may exist an order-parameter fluctuation, say $A_{\vec{q}}$, for which the static susceptibility $(A_{\vec{q}}|A_{\vec{q}})$ diverges for q tending to zero. As a result, a length scale r^* with the properties anticipated in Eqs. (3.38a,b) does not exist. Both situations can be handled by expanding the formalism. The equation (2.62a) has to be used in its general matrix form. In the first case, all coupled conserved fluctuations have to be included in the set of distinguished variables. After this modification of the projector $\mathcal{Q}$, the regularity assumption can be made for the matrix of relaxation kernels. This extension of the theory will be demonstrated for two and three coupled conserved fluctuations in Sec. 3.3.5 in connection with the discussion of the density fluctuations in simple liquids. In the second case, the set of distinguished variables has to be enlarged even more by including order-parameter fluctuations; examples are explained in the book by Forster (1975) in connection with the phenomenon of spontaneously broken symmetries.

3.2.3 *The friction coefficient*

The description of the tagged-particle density correlations by a diffusion-law Green's function does not imply a meaningful result for the velocity correlator $K_s(t)$. To eliminate this shortcoming, an improvement of the theory was developed by P. Langevin, which shall be discussed in this section.

The force $\vec{f}_s$ acting on the tagged particle can be evaluated from the potential $V_s(\vec{r}) = \sum_{\sigma \neq s} V_{s\sigma}(|\vec{r} - \vec{r}_\sigma|)$:

$$\vec{f}_s = -(\vec{\partial} V_s)(\vec{r}_s). \tag{3.45a}$$

It shall be assumed that $V_s(\vec{r})$ is sufficiently regular so that $\partial^m V_s(\vec{r}_s)$ as well as $\partial^n \partial^m V_s(\vec{r}_s)$, $m, n = 1, 2, 3$, are dynamical variables in the sense defined in Sec. 2.1. There is the equation of motion

$$-i\partial_t \mu_s \vec{v}_s = \mu_s \mathcal{L} \vec{v}_s = -i\vec{f}_s. \tag{3.45b}$$

Hence, the regularity assumption ensures that the velocity $\vec{v}_s$ is in the domain of the Liouville operator. Iteration of the argumentation shows that also $\vec{f}_s$ is in the domain of $\mathcal{L}$. From Eq. (2.26c) one concludes that the velocity correlator $K_s(t) = (v_s^1(t)|v_s^1)$ has continuous derivatives up to order 4. Since $K_s(t)$ is an even function of t, and since its initial value is given by $(v_{th}^s)^2$, one can quantify a Taylor-expansion formula by a positive frequency to be denoted as Ω_E^s:

$$K_s(t)/(v_{th}^s)^2 = 1 - \tfrac{1}{2}(\Omega_E^s t)^2 + O(t^4). \tag{3.46a}$$

Equation (3.32b) yields: $-\partial_t^2 K_s(t)\delta^{m,n} = (i\partial_t)^2 (v_s^m(t)|v_s^n) = (f_s^m(t)|f_s^n)/(\mu_s)^2$. Specializing to $t = 0$, one arrives at

$$(f_s^m|f_s^n)/[\mu_s v_{th}^s]^2 = \delta^{m,n} \Omega_E^{s^2}. \tag{3.46b}$$

Equation (3.46a) is the analogue of Eq. (3.26a). The frequency Ω_E^s, as opposed to its analogue Ω_{sq}, depends on the particle interactions.

Re-expressing f_s^n as a time derivative of v_s^n and using Yvon's theorem (2.6), one gets $(f_s^m|f_s^n) = i\mu_s\langle f_s^m \mathcal{L} v_s^n\rangle = -(k_{\rm B}T)\mu_s\langle\{f_s^m, v_s^n\}\rangle$. The Poisson bracket yields $\{f_s^m, \mu_s v_s^n\} = \partial f_s^m/\partial r_s^n = -\partial^m\partial^n V_s(\vec{r}_s)$. Consequently,

$$\Omega_E^{s^2} = (1/3\mu_s)\big\langle\vec{\partial}\vec{\partial}V_s(\vec{r}_s)\big\rangle. \tag{3.47}$$

The number $\Omega_E^{s^2}$ is the average of the squares of the frequencies of small oscillations of the tagged particle under the restriction that all other particles are fixed at their respective positions $\vec{r}_\sigma$. Such an oscillator picture for a treatment of the dynamics was suggested by A. Einstein as a basis of his explanation of the low-temperature specific heat of solids. Therefore, Ω_E^s is called the Einstein frequency of the system. It defines the natural scale for the frequencies of the condensed system in the sense discussed in connection with Fig. 1.1. The average in Eq. (3.47) is not very sensitive to details of the particle arrangements. The value for, say, liquid argon near its triple point can be estimated by assuming a face-centred-cubic-crystal arrangement. The Lennard-Jones potential with the parameters from Eqs. (1.8a,c) yields a value for $\Omega_E/2\pi$ of the order of 1 THz. And this number is typical for all van der Waals liquids, glasses and crystals. The mentioned assumptions on small oscillations are unjustified for liquids; in this phase, the displacements of the particles diverge for large times. A large displacement of the tagged particle requires a cooperative motion of the neighbouring particles, as is explained in connection with Fig. 1.10. Handling this cooperative motion for long times is the complicated problem of liquid dynamics. Formulas (3.46a), (3.47) do not reflect this complication.

Let us apply the procedure from Sec. 2.4.1 in order to express the correlator for the single distinguished variable $A = v_s^1/v_{th}^s$, $(A(t)|A) = K_s(t)/(v_{th}^s)^2$, in terms of a relaxation kernel. The variable A is normalized, $(A|A) = 1$, and it has a definite time-inversion parity. The further reasoning follows strictly the one explained for the derivation of Eqs. (3.35a–c). Comparison of the equations of motions (3.23d) and (3.45b) shows that $-if_s^1/\sqrt{k_{\rm B}T}$ is the analogue of $j_{\vec{q}}^s$ and $1/\sqrt{\mu_s}$ is that of q. Writing the kernel in Eq. (2.62a) as $X_s(z)/\mu_s$, the function $X_s(z)$ is the analogue of $K_q^s(z)$. There holds

$$K_s(z)/(v_{th}^s)^2 = -1/[z + X_s(z)/\mu_s], \tag{3.48a}$$

$$X_s(z) = (f_s^1|\mathcal{R}_s(z)f_s^1)/(k_{\rm B}T). \tag{3.48b}$$

The resolvent $\mathcal{R}_s(z) = [\mathcal{L}_s - z]^{-1}$ is formed with the reduced Liouvillian $\mathcal{L}_s = \mathcal{QLQ}$. The projector $\mathcal{Q} = 1 - |v_s^1)(v_{th}^s)^{-2}(v_s^1|$ eliminates the tagged-particle velocity v_s^1 from the dynamics. The kernel $X_s(z)$ is the Laplace transform of

$$X_s(t) = (\mathcal{U}_{\mathcal{Q}}(t)f_s^n|f_s^n)/(k_{\rm B}T). \tag{3.48c}$$

It correlates the forces, which are derived conventionally from the interaction potentials. However, the forces do not evolve in time as described by Hamilton's equations of motion. The time evolution is described by the unitary translation operator $\mathcal{U}_Q(t)$ generated by $\mathcal{L}_s$. It treats the velocity component v_s^1 as constant. Only that part of the forces is considered, which fluctuates around this constant of motion. These facts are the motivation for introducing the terminology of fluctuating force and fluctuating-force correlators for the quantities considered in Eqs. (2.61a) and (2.61b), respectively. The initial value $X_s(t = 0) = (f_s^1|f_s^1)/(k_{\rm B}T)$ is obtained from Eq. (3.46b):

$$X_s(t = 0)/\mu_s = \Omega_E^{s^2}. \tag{3.48d}$$

Let us remember that $K_s(z) = -K_s(-z)$. Equation (3.48a) implies the same symmetry for the kernel $X_s(z)$, i.e., $X_s(t)$ is an even function of time. In the same manner one concludes from the reality of $K_s(t)$ to that of $X_s(t)$. The postulated integrability of $|K_s(t)|$ leads to the conclusion that the velocity correlator in the frequency domain has continuous boundary values $K_s(\omega \pm i0) = K_s'(\omega) \pm iK_s''(\omega)$. The zero-frequency limit of the spectrum D_s is the same as the zero-frequency limit of the correlator, Eqs. (3.33a,d). Since $0 < D_s$ is requested, one derives from Eq. (3.48a) that $X_s(\omega \pm i0) = X_s'(\omega) \pm iX_s''(\omega)$ is a continuous function of ω. The spectrum $X_s''(\omega)$ is non-negative. Its zero-frequency limit is a positive number to be denoted by ξ_s:

$$X_s''(\omega = 0) = \xi_s, \tag{3.49a}$$

$$\lim_{z\to 0} X_s(z) = \pm i\xi_s, \qquad \operatorname{Im} z \gtrless 0. \tag{3.49b}$$

Equation (3.33d) provides an expression of the diffusivity in terms of the fluctuating-force spectrum:

$$D_s = (k_{\rm B}T)/\xi_s. \tag{3.49c}$$

The preceding equations reduce the quantities $K_s(z)$ and D_s to the quantities $X_s(z)$ and ξ_s. Without some additional insight, the formulas (3.48a)–(3.49c) do not provide progress towards understanding the velocity correlator.

Langevin's theory is based on treating the fluctuating-force spectrum as a white-noise one. The results of his approach shall be indicated by an index L, for example

$$X_s^{L''}(\omega) = \xi_s. \tag{3.50a}$$

As discussed in connection with Eqs. (3.44a,b), the kernel is modelled by the positive analytic function

$$X_s^L(z) = \pm i\xi_s, \quad \operatorname{Im} z \gtrless 0. \tag{3.50b}$$

This function is not a correlator; but substituted into Eq. (3.48a), it yields a model for the velocity correlator in the sense discussed in Sec. 2.4.2:

$$K_s^L(z) = -(v_{th}^s)^2/[z \pm i\xi_s/\mu_s], \quad \operatorname{Im} z \gtrless 0. \tag{3.50c}$$

Langevin's model for the force spectrum implies a Lorentzian for the velocity spectrum

$$K_s^{L''}(\omega)/(v_{th}^s)^2 = \tau_s^L/[1 + (\omega\tau_s^L)^2] \tag{3.50d}$$

and an exponential for the velocity correlator in the time domain

$$K_s^L(t) = (v_{th}^s)^2 \exp[-|t|/\tau_s^L]. \tag{3.50e}$$

The time scale for this simple relaxation process is given by

$$\tau_s^L = \mu_s/\xi_s. \tag{3.50f}$$

Equation (3.49b) means the following. For every $\epsilon > 0$, there exists a time t^* so that $|X_s(z) \mp i\xi_s| \leq \epsilon\xi_s$, provided

$$|t^* z| \leq 1. \tag{3.51a}$$

The time t^* specifies the regime of frequencies, where the white noise provides a description of the fluctuating-force spectrum within an relative error ϵ. The formulas (3.48a, 3.50c) lead to

$$|K_s(z) - K_s^L(z)|/D_s \leq \epsilon |K_s(z) K_s^L(z)|/D_s^2. \tag{3.51b}$$

In the limit of small z, this result reproduces Eq. (3.38a) for $q = 0$. Thus, t^* is identical to that introduced in formulas (3.38b). The model provides an asymptotic description of the low-frequency dynamics of the velocity correlations. The error of the description is specified by the preceding inequality.

Substituting Eq. (3.50e) into Eq. (3.32d), one gets Langevin's formula for the mean-squared-displacement function:

$$\Delta_s^L(t) = D_s\{|t| - \tau_s^L + \tau_s^L \exp[-|t|/\tau_s^L]\}. \tag{3.52}$$

This finding is a qualitative improvement of Eq. (3.40e). The $t = 0$ cusp of $\overline{\Delta}_s(t)$ is eliminated, and the correct ballistic short-time variation is reproduced: $\Delta_s^L(t) = (v_{th}^s)^2\{\frac{1}{2}t^2 - \frac{1}{6}|t|^3/\tau_s^L + O(t^4)\}$. The expected cusp artifact due to the white-noise assumption is shifted to the second derivative of $\Delta_s^L(t)$. This remaining artifact is caused by the cusp, which is exhibited by the model for the velocity correlator, $K_s^L(t) - K_s^L(t=0) = -[(v_{th}^s)^2/\tau_s^L]|t| + O(t^2)$. The latter exhibits a qualitative deviation from the regular short-time asymptote of $K_s(t)$, which is noted in Eq. (3.46a).

Langevin's velocity correlator is the solution of the equation of motion

$$\partial_t \mu_s K_s^L(t) \pm \xi_s K_s^L(t) = 0, \qquad t \gtrless 0, \tag{3.53a}$$

for the initial condition $K_s^L(t=0) = K_s(t=0) = (v_{th}^s)^2$. For $t > 0$, this is Newton's equation of motion for a particle of mass μ_s moving in a homogeneous environment. The particle experiences a friction force proportional to the velocity but of opposite sign. The properties of the environment are specified by a single number ξ_s, which is called the friction coefficient. Langevin's theory provides a derivation of Newton's friction law. It establishes a connection of the material's parameter ξ_s with the atomistic theory of matter. Equation (3.49c) is Einstein's relation between diffusivity and friction.

Within the Zwanzig–Mori formalism, Eq. (3.53a) is the analogue of Eq. (3.42c). Therefore, the analogues reservations concerning the preceding conclusions have to be kept in mind, which have been explained in connection with the derivation of the diffusion equation. Equation (3.53a) is not the equation of a phase space coordinate $v_s^n(t)$, but one of the correlation of this coordinate at time $t + t_1$ with that at time t_1. The equation does not describe the full dynamics, but merely the spectra for low frequencies. And the description is not done exactly, but only within some non-zero error margin indicated by ϵ in Eq. (3.51b). The implications of the latter two reservations for the time dependence of $K_s^L(t)$ are discussed in Appendix C. The correlator $K_s^L(t)$ may differ strongly from $K_s(t)$ since there may be strong differences between the spectra $K_s^{L\prime\prime}(\omega)$ and $K_s''(\omega)$ for large ω. A meaningful comparison of the two correlators in the time domain is possible only after suppression of the effects of the high-frequency spectra. This can be done by considering new functions $K_*^L(t)$ and $K_*(t)$, which are derived by smoothening or coarse-graining of $K_s^L(t)$ and $K_s(t)$, respectively, on a time scale of order t^*. Then, $K_*^L(t)$ is an approximation of $K_*(t)$ in the sense of $|K_*^L(t) - K_*(t)| \leq \epsilon \cdot K^L(t = 0) \cdot c$; here c is a number of order unity.

Langevin's dynamics is specified by the time scale τ_s^L. The theory is valid if and only if the white-noise assumption, Eq. (3.49a), is justified for frequencies extending up to some multiple of $1/\tau_s^L$. The theory is based on the existence of the time-scale separation:

$$t^*/\tau_s^L \ll 1. \tag{3.53b}$$

From a mathematical point of view, this condition for the validity of Eqs. (3.50c–e) is the same as the condition from Eq. (3.43) for the validity of Eq. (3.40a–c). Because of the conservation law for the variable $\rho_{\vec{q}}^s$, the condition $t^*/\tau_q^s \ll 1$ can be fulfilled by choosing the wave number q small enough. In general, there is no such parameter like q whose proper adjustment can guarantee the condition (3.53b). Further insight is required in order to decide whether or not the above theory is applicable for a specified liquid at given temperature and density.

There are two important systems where the Langevin theory is valid. The first example deals with the motion of the tagged particle in a gas of scatterers. Here, τ_s^L is the average time between two scattering events and t^* is the duration time for the scattering process. The ratio t^*/τ_s^L is the smaller the more dilute is the gas of scatterers. For the relaxation of the tagged particle's momentum it is not relevant whether the scatterers are moving particles like in a gas or whether they are some frozen arrangement of scattering centers like impurities in an otherwise perfect material. This was anticipated by P. Drude who suggested Eq. (3.50c) for the description of electrical conductivities. This equation or its equivalents are often referred to as Drude's formulas. For this example, Boltzmann's kinetic equation is the general framework to treat the tagged-particle motion in leading order in the small parameter t^*/τ_s^L. The second problem concerns Brownian motion. Here, the diameter d_s of the mesoscopic tagged particle exceeds the diameter d of the liquid molecules by four or more orders of magnitude. The average in $\langle\vec{\partial}\vec{\partial}V_s(\vec{r}_s)\rangle$ increases with increasing d_s like the surface of the Brownian particle, i.e., like $(d_s/d)^2$. The mass μ_s increases like $(d_s/d)^3$. One concludes from Eq. (3.47) that the ratio of the time scale for the Brownian particle, $1/\Omega_E^s \approx \tau_s^L$, and the one for the fluctuating forces, $1/\Omega_E \approx t^*$, increases like $\sqrt{(d_s/d)}$. The time-scale separation parameter t^*/τ_s^L is the smaller the larger is the tagged particle.

3.2.4 *The cage effect and glassy-dynamics precursors of the velocity correlations*

In this section, molecular-dynamics-simulation data shall be discussed, which show that the cage effect in simple liquids causes strong deviations of the mean-squared displacement and of the velocity correlations from the Langevin-theory results. An extension of the theory will be presented. It is built on a white-noise model for a higher-order relaxation kernel, and it copes with some features of the cage effect.

The Langevin formula (3.52) for the mean-squared displacement is equivalent to $\delta r_s^2(t) = A_s\{(t/\tau_s^L) - 1 + \exp[-(t/\tau_s^L)]\}$, $A_s = 6D_s\tau_s^L = 6[v_{th}^s\tau_s^L]^2$. This is a scaling law specified by the time scale τ_s^L and the amplitude scale A_s. Mean-squared displacements of different liquids or of different states of a given liquid merely differ by the values for these scales. The double-logarithmic presentations, i.e., the $\log(\delta r_s^2(t))$ versus $\log(t)$ curves, have the same shape; they differ by translations $\log A_s$ in the vertical direction and $-\log\tau_s^L$ in the horizontal one. The simulation results from Fig. 1.9 do not support such a prediction. For $T \leq 1.0$, the shape of the curves varies sensitively with changes of T.

Figure 3.1 reproduces the mean-squared-displacement results from Fig. 1.9 for the three temperatures $T = 4, 2$ and 1. The dashed lines are the Langevin results with A_s and τ_s^L adjusted for each T so that the small-time ballistic increase and the large-time diffusion behaviour are reproduced. The data shown in Fig. 1.9 for $T = 5$ cannot be distinguished from the Langevin result within the accuracy of the presentation. The theory accounts well for the $T = 4$ data, even though there are systematic errors within the interval $-0.5 < \log_{10}(t) < 0$. For $T = 2$, the errors are serious within the one-decade interval $-0.5 < \log_{10}(t) < 0.5$. The $\log\delta r_s^2(t)$ versus $\log(t)$ curves are downward bent for $T \geq 3$, in qualitative agreement with the Langevin result. However, for $T \leq 2$, the curves exhibit an inflection point. Decreasing T to low values, the inflection-point regime evolves to a plateau, a prominent feature of glassy relaxation discussed in Chapter 1. Therefore, the appearance of the inflection-point behaviour can be considered as the onset of glassy dynamics. The curves for $T \leq 2$ in Fig. 1.9 show a feature of

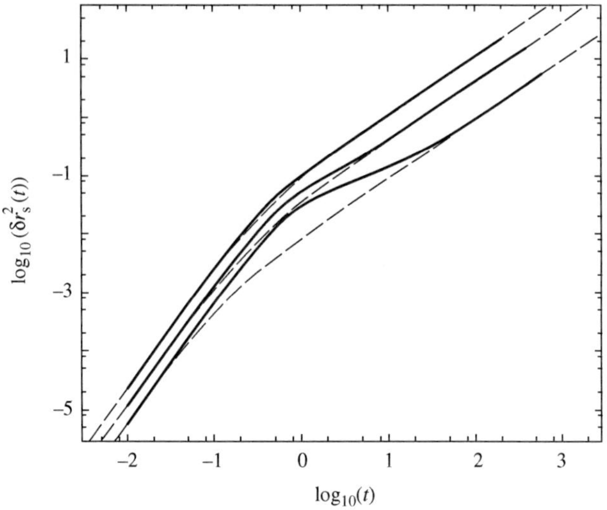

FIG. 3.1. The full lines exhibit the logarithm of the mean-squared displacement $\log_{10}(\delta r_s^2(t))$ versus $\log_{10}(t)$ as obtained by Kob and Andersen (1995a) from molecular-dynamics simulations for the A-particles of a LJM. The results for for temperatures $T = 4.0, 2.0$ and 1.0 (from left to right) are reproduced from Fig. 1.9. The dashed lines exhibit the results $\delta r_s^2(t) = 6\Delta_s^L(t)$ for Eq. (3.52) of Langevin's theory with values for the diffusivity D_s and relaxation time τ_s^L fitted for each temperature.

glassy dynamics, which is not present for larger T. The Langevin-theory results for $T = 1$ cannot deal with the long crossover interval $0.1 < t < 10$.

Let us notice that the temperature $T = 2$ for argon units corresponds to 240 K. This temperature is more than twice the boiling temperature of argon at ambient pressure. Many glass-forming molecular van der Waals liquids, e.g., salol, disintegrate at such high values of T. Benzene consists of robust molecules; and glassy-dynamics precursors have been detected for this system for temperatures up to close to the boiling point (Wiebel and Wuttke 2002).

An addendum to the preceding discussion seems adequate. Decreasing the temperature from $T = 4$ to 2 and 1, the square of the thermal velocity and, hence, the ballistic asymptote decreases by factors 2 and 4, respectively. This trivial slowing down of the dynamics upon cooling explains the downward shifts of the curves in Fig. 3.1 by factors $\log 2$ and $2\ \log 2$, respectively, for $t \leq 0.1$. The corresponding diffusion asymptotes shift down by factors about 3 and 16, respectively. This accelerated decrease of the transport coefficient D_s, i.e., the accelerated slowing down of the long time dynamics is caused by the increasing importance of interaction effects. The effect enhances dramatically for lower temperatures: D_s decreases by more than two orders of magnitude if T decreases from 1 to 0.466. This decrease is directly related to the increase of the length of the plateau upon cooling, which is exhibited in Fig. 1.9. Thus, the accelerated decrease of D_s is a characteristic feature of glassy dynamics reflecting the increasing duration for the transient localization of the particle in the cage formed by its neighbours. The cited data show that the time-scale-separation assumption of the Langevin theory is illegitimate for $T \leq 2$.

The cage effect was documented first in the pioneering work of Rahman (1964). The dots in Fig. 3.2 exhibit his results for the normalized velocity correlator $\psi(t) = K_s(t)/K_s(t=0)$ and for the normalized velocity spectrum $G(\omega) = K_s''(\omega)/K_s''(\omega=0)$. The data refer to liquid argon for a temperature about 11 K above the melting point. They were produced by molecular-dynamics simulations for a system of 864 particles interacting via a Lennard-Jones potential adjusted for this system. The temperature in units from Eq. (1.8c) is about $T = 0.79$, and thus, the dynamics studied is close to the one discussed in Fig. 1.9 for $T = 0.8$. The correlator $\psi(t)$ decreases below the 5% level for times near 0.3 ps, which is of the order of $2\pi/\Omega_E^s$. The spectrum $G(\omega)$ decreases below 0.1 for frequencies $\omega/2\pi$ exceeding about 2 THz.

The long-time asymptote of the cited simulation results for the mean-squared displacement yields the diffusivity $D_s = 2.43 \times 10^{-5}\text{cm}^2\text{s}^{-1}$ and, thereby, the relaxation time $\tau_s^L = D_s/(v_{th}^s)^2 = 0.123$ ps. Thereby, Langevin's formulas for the normalized correlator $\psi^L(t) = \exp[-|t|/\tau_s^L]$ and for the normalized spectrum $G^L(\omega) = 1/[1+(\omega\tau_s^L)^2]$ are determined. The results are shown as dashed lines in Fig. 3.2. The artificial short-time cusp, $\psi^L(t) - 1 = -|t|/\tau_s^L + O(t^2)$, implies a severe underestimation of the velocity correlations for t near τ_s^L. The cusp causes the artificial large high-frequency tail of the Lorentzian spectrum, $G^L(\omega) = 1/(\omega\tau_s^L)^2 + O(1/\omega^4)$. This leads to a large overestimation of $K_s''(\omega)$ for $\omega/2\pi > 2$THz. The most remarkable difference between $K_s^L(t)$ and $K_s(t)$ is noticed for the times exceeding 0.3 ps. Opposed to the monotonically decreasing positive function $K_s^L(t)$, the velocity correlations of liquid argon near its melting temperature become negative and they increase monotonically with time for $t > 0.5$ ps.

For short times, $K_s(t)$ decreases according to Eq. (3.46a). This decay of the correlations yields a positive contribution to the integral of the Green–Kubo equation (3.33b), which is of order $(v_{th}^s)^2/\Omega_E^s$. This contribution varies nearly linearly with T and it explains the order of magnitude of the diffusivity of the liquid at high temperature. If the particles were localized in a cage, the long-time limit of the mean-squared displacement would be finite. Hence, Eq. (3.33c) would yield a vanishing diffusivity D_s. Consequently, there would be negative parts of $K_s(t)$, which cancel the positive initial contribution of the Green-Kubo integral. The temperature-sensitive strong suppression of D_s, which is discussed above as a signature of glassy dynamics, is equivalent to a delicate nearly complete cancellation of the temperature-insensitive short-time contribution to the right-hand side of Eq. (3.33b) and negative temperature-sensitive contributions for longer times. The orbit for the motion of a tagged particle in a cage, which is exhibited in Fig. 2.1, shows indeed that for the majority of collision events at

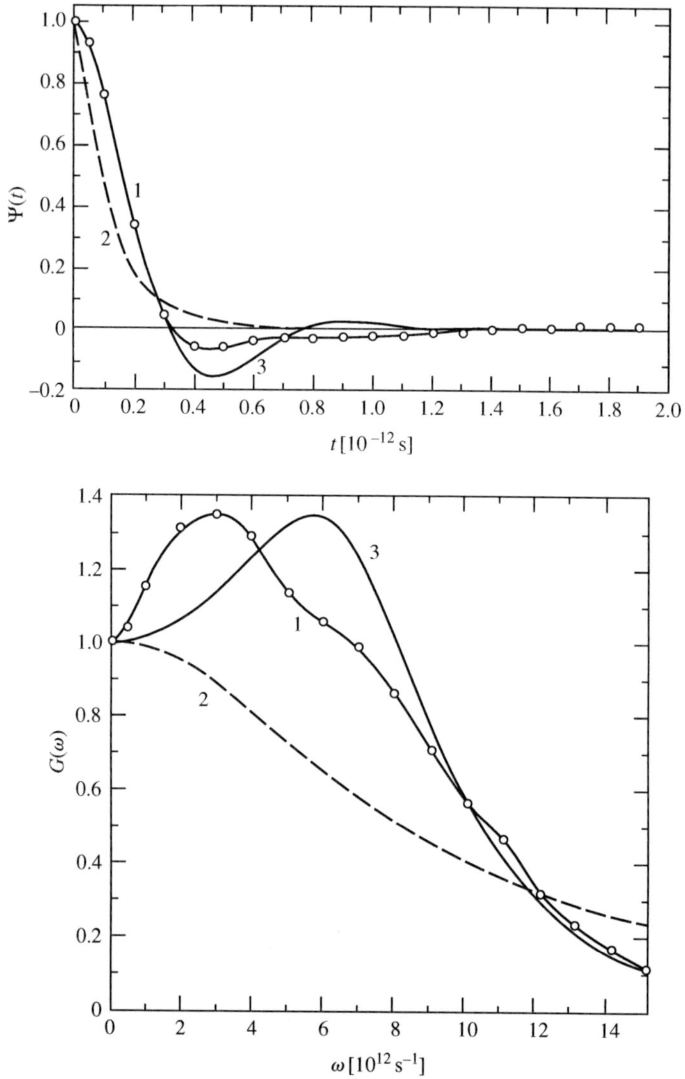

FIG. 3.2. Normalized velocity correlators $\psi(t) = K_s(t)/K_s(t=0)$ and normalized velocity spectra $G(\omega) = K''_s(\omega)/K''_s(\omega = 0)$. The units of the time t are $10^{-12}\,\text{s}$. The unit for the frequency ω is $10^{12}\,\text{s}^{-1}$ so that $\omega = 2\pi$ corresponds to 1 THz. The dots, which are connected by lines with label 1, are molecular-dynamics-simulation results for liquid argon at $T = 94.4\,\text{K}$ (Rahman 1964). The dashed lines with label 2 are the Langevin-theory results, Eqs. (3.50d,e). The full lines with label 3 are the results based on a simple-relaxation model for the fluctuating force dynamics, Eqs. (3.55a,b). Reproduced from Berne *et al.* (1966).

the times t_i, the velocity gets reversed: $\vec{v}_s(t_i - 0)\vec{v}_s(t_i + 0) < 0$. The appearance of negative values of $K_s(t)$ is a manifestation of the cage effect.

To eliminate the cusp artifact of $K_s^L(t)$, the white-noise model for the fluctuating-force spectrum $X_s''(\omega)$ is to be abandoned. Rather, this correlator shall be expressed in terms of a relaxation kernel. The reasoning follows that leading to Eq. (3.48a), and the analogue of this formula is the fraction representation $X_s(z)/[\mu_s \Omega_E^{s^2}] = -1/[z + N(z)]$. The existence of a non-vanishing zero-frequency limit of $X_s(z)$, Eq. (3.49b), guarantees the existence of $\lim_{z\to 0} N(z) = \pm i\nu$, Im $z \gtrless 0$. The frequency ν and the frequency ξ_s/μ_s are related so that frequency Ω_E^s is their geometrical mean:

$$\nu = \Omega_E^{s^2} \mu_s/\xi_s. \tag{3.54a}$$

Berne *et al.* (1966) introduced a model for the velocity dynamics by replacing the relaxation kernel $N(z)$ by the positive-analytic function $\pm i\nu$, Im $z \gtrless 0$. This is equivalent to a white-noise assumption for the higher-order fluctuating-force spectrum, $N''(\omega) = \nu$. The approach is based on the simple-relaxation model for the fluctuating-force correlations:

$$X_s^B(t) = \Omega_E^{s^2} \mu_s \exp[-\nu|t|]. \tag{3.54b}$$

Substituting $X_s^B(z)/\mu_s = -\Omega_E^{s^2}/[z \pm i\nu]$, Im $z \gtrless 0$, into Eq. (3.48a), one arrives at a model for the velocity correlator. For Im $z > 0$, there holds

$$K_s^B(z)/(v_{th}^s)^2 = -1/\{z - \Omega_E^{s^2}/[z + i\nu]\}. \tag{3.55a}$$

The values in the lower half plane for the frequencies can be obtained by observing the general symmetries: $K_s(z)^* = K_s(z^*)$ or $K_s(-z) = -K_s(z)$. The function $K_s^B(z)$ is meromorphic, and it can be continued onto the whole frequency plane. It is holomorphic for Im $z > 0$ and, generically, it has two simple poles $z_\pm$ in the lower half plane. It can be written as sum of partial fractions $K_s^B(z)/(v_{th}^s)^2 = -[A_+/(z - z_+)] - [A_-/(z - z_-)]$. The inverse Laplace transform, Eq. (A.4), yields the correlator in the time domain:

$$K_s^B(t)/(v_{th}^s)^2 = A_+ \exp[-iz_+|t|] + A_- \exp[-iz_-|t|]. \tag{3.55b}$$

Details are discussed in Appendix A.4. This result implies the short-time asymptote (A.38c):

$$K_s^B(t)/(v_{th}^s)^2 = 1 - \tfrac{1}{2}(\Omega_E^s t)^2 \left[1 - \tfrac{1}{3}\nu|t|\right] + O(t^4). \tag{3.55c}$$

The correct leading-order initial decay of $K_s(t)$ is reproduced. From Eq. (3.55a), one obtains the high-frequency asymptote

$$K_s^B(z)/(v_{th}^s)^2 = -(1/z)\{1 + (\Omega_E^s/z)^2[1 - i\nu/z] + O(z^{-4})\}. \tag{3.55d}$$

Introducing the model (3.54b) for the relaxation kernel $X_s(t)$ is equivalent to replacing the simple-relaxation-process correlator (3.50c) of the Langevin theory by the correlator (3.55a) of a harmonic oscillator. The latter is characterized by the oscillator frequency Ω_E^s and the friction constant ν.

The results for $K_s^B(t)$ and $K_s^{B\prime\prime}(\omega)$ are shown as full lines with label 3 in Fig. 3.2 for parameters Ω_E^s and $\nu = \mu_s \Omega_E^{s^2} D_s/(k_B T)$ deduced from the simulation data for Ω_E^2 and D_s. The model describes the initial 90% of the decay, which occurs for $t \leq 0.3$ ps. The high-frequency tail of the spectrum, which follows from Eq. (3.55d) as $G(\omega) = -(\Omega_E^s/\omega)^4 + O(1/\omega^6)$, describes the data for $\omega/2\pi > 1.5$ THz. Let us introduce the ratio r of the two frequencies, which characterize the model dynamics: $r = (\nu/\Omega_E^s) = (\Omega_E^s \tau_s^L)$. For $r < 2$, the frequencies $z_\pm$ have real parts. The correlator $K_s^B(t)$ exhibits oscillations. For the system studied in Fig. 3.2, parameter r is near unity. Therefore, an important implication of the cage effect, namely the appearance of negative velocity correlations for t around 0.5 ps, is reproduced qualitatively by the model. This oscillation explains also the appearance of a maximum for the spectrum $K_s''(\omega)$ at some positive frequency ω_{peak}, which is noted in Eq. (A.39).

Comparison of the full lines with label 3 and the data points in Fig. 3.2 shows that the model $K_s^B(t)$ for the velocity correlations has shortcomings. The negative correlations for 0.4 ps $< t <$ 0.6 ps are overestimated by about a factor 2. The calculated peak frequency of the spectrum is about twice the correct one. A qualitative error is the appearance of a positive oscillation maximum of $K_s^B(t)$ for times near 0.9 ps, while the data exhibit a negative monotonically increasing tail for $t \geq 0.5$ ps. The damped-harmonic-oscillator model for the velocity correlator can describe qualitatively the cage-effect-induced modifications of the dynamics on the natural time scale t_{mic} of the liquid. But, it cannot cope with the onset of glassy dynamics. The latter manifests itself by a non-oscillatory negative long-time tail. This observation is similar to that discussed in the upper two panels of Fig. 1.2 for experimental data. A simple relaxation function can account for the decay function for times up to about $t = 3t_{\text{mic}} \approx 6$ ps. For larger times, there appears an additional long-time tail.

The white-noise assumption for the spectrum $N''(\omega)$ is not legitimate for argon at temperatures near the melting point. The fluctuating-force correlator $X_s(t)$ does not exhibit the time-scale-separation property. To corroborate the preceding conclusion, the normalized correlator $\alpha(t) = X_s(t)/X_s(t=0)$ is shown in Fig. 3.3. It is obtained from the cited simulation data by inverting Eq. (3.48a). The force correlations do not exhibit a simple relaxation process as is anticipated by Eq. (3.54b). The first 90% of the decay occur on the time scale $1/\nu \sim 1/\Omega_E^s$. But, in addition to this decay on scale t_{mic}, there appears a second much slower process causing a positive long-time tail of $\alpha(t)$. The dashed and dotted curves in Fig. 3.3 exhibit two three-parameter fits of $\alpha(t)$ as double relaxation process,

$$X_s^1(t)/(\Omega_E^s)^2 = (1 - A)\exp[-(t/t_1)^2] + A\exp[-(t/t_2)], \tag{3.56a}$$

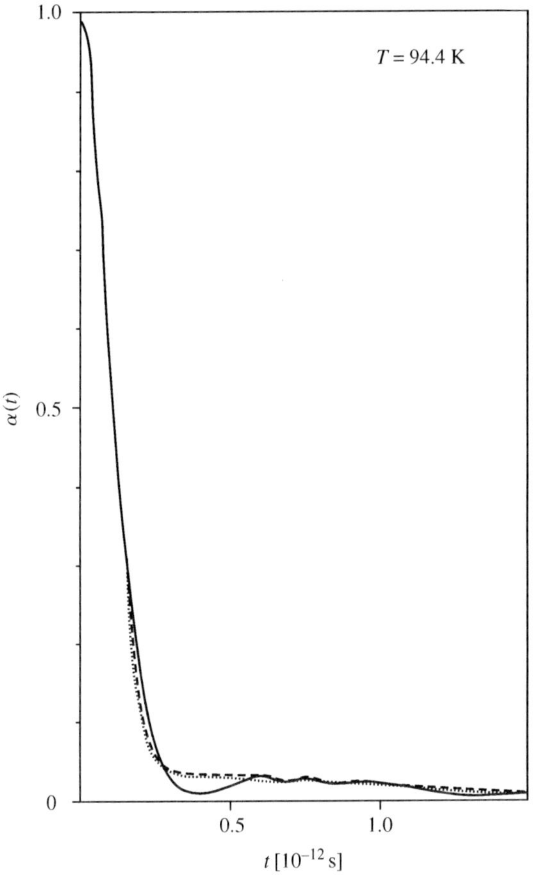

FIG. 3.3. Normalized fluctuating-force correlators of a tagged particle $\alpha(t) = X_s(t)/X_s(t=0)$ as defined in Eqs. (3.48a–c). The full line is determined from the simulation data for liquid argon shown in Fig. 3.2 as dots. The dashed and the dotted lines present three-parameter fits by sums of two relaxation functions according to Eqs. (3.56a) and (3.56b), respectively. Reproduced from Damle and Tillu (1969).

and

$$X_s^2(t)/(\Omega_E^s)^2 = (1-A)\exp[-(t/t_1)^2] + A\exp[-(t/t_2)^2], \tag{3.56b}$$

respectively. Here t_1 is of order $1/\nu$ and t_2 exceeds t_1 by about a factor 5. These models for kernel $X_s(t)$ reproduce the data from Fig. 3.2 (Damle and Tillu 1969). The slow positive monotonically decreasing force correlations for $t > 0.5$ ps are the counterpart of the negative monotonically increasing velocity correlations in the same time interval. They are the analogue of the tail, which is shown for

$t > 8$ ps in panel (b) of Fig. 1.2. These tails remain to be understood if one wants to explain how the cage effect causes the evolution of glassy dynamics.

3.3 Densities and currents in simple liquids

For the remaining part of this chapter, the discussions shall be restricted to systems of identical particles of mass μ. In the following, the theory of hydrodynamic fluctuations and visco-elastic effects shall be discussed.

3.3.1 *Definitions and general equations*

In this section, the conservation laws for the particle number and for the particle momentum shall be formulated using the field fluctuations for the density, the current, and the stress. The correlators for these fluctuations shall be defined. Dispersion laws for three frequencies will be identified, which characterize the short-time decay of density and current correlations.

The thermodynamic state of a system of N identical particles enclosed in a cubic box of edge length L is specified by the averaged number density $\rho = N/L^3$ and by the temperature T. The most important quantities for a description of simple-liquid dynamics are the density $\rho(\vec{r}) = \sum_\kappa \delta(\vec{r} - \vec{r}_\kappa)$ and the current density $\vec{j}(\vec{r}) = \sum_\kappa \vec{v}_\kappa \delta(\vec{r} - \vec{r}_\kappa)$. These are generalized dynamical variables obtained as sums of the quantities considered in Eq. (3.7). Via Fourier transformation from the space of positions $\vec{r}$ to the space of wave vectors $\vec{q}$, one gets field fluctuations, which are dynamical variables in the sense of Sec. 2.1:

$$\rho_{\vec{q}} = \sum_\kappa \exp[i\vec{q}\vec{r}_\kappa], \quad \vec{j}_{\vec{q}} = \sum_\kappa \vec{v}_\kappa \exp[i\vec{q}\vec{r}_\kappa]. \tag{3.57}$$

These variables are sums of the quantities considered in Eq. (3.10). The density fluctuation $\rho_{\vec{q}}$ is a scalar-field fluctuation of even parity for space and time inversion. The current-density fluctuation $\vec{j}_{\vec{q}}$ is a vector-field fluctuation of odd parity for space and time inversion. Both variables have even conjugation parity; $\rho(\vec{r})$ and $\vec{j}(\vec{r})$ are real quantities. There are three types of scalar products to be formed with the specified fluctuations. Since averages over velocities can be done independently from those over the configuration space, one gets from Eq. (3.1a) for the current-current overlaps

$$(j_{\vec{q}}^m | j_{\vec{q}}^n) = N\delta^{m,n} v_{th}^2. \tag{3.58a}$$

Here, $v_{th} = \sqrt{k_B T/\mu}$ abbreviates the thermal velocity. Since $\langle \vec{v}_\kappa \rangle = 0$, the fluctuations of the currents are perpendicular to those of the densities:

$$(j_{\vec{q}}^m | \rho_{\vec{q}}) = 0. \tag{3.58b}$$

These equations are the analogues to Eqs. (3.24b,c). The analogue of Eq. (3.24a) expresses the density-fluctuation normalization. This is a non-trivial quantity.

Because of the isotropy condition, it depends on the wave vector merely through its modulus $q = |\vec{q}|$. One writes

$$(\rho_{\vec{q}}|\rho_{\vec{q}}) = NS_q, \tag{3.58c}$$

where S_q is called the structure factor.

The structure factor is an important quantity characterizing the arrangements of the particles in the amorphous state. Some digressions on properties of S_q shall be made in this and the following four paragraphs . Substituting Eq. (3.57) into Eq. (3.58c), one gets

$$S_q = 1 + \rho \int \exp[i\vec{q}\vec{r}]g(r)d^3\vec{r}, \tag{3.59a}$$

where the function $g(r)$ is defined as the double sum over all particle pairs:

$$\rho g(r) = \frac{1}{N}\sum_{\kappa\neq\sigma}\langle\delta(\vec{r} - \vec{r}_\kappa + \vec{r}_\sigma)\rangle. \tag{3.59b}$$

The discussion of Eqs. (3.9a–d) shows that the average in this equation depends on the position vector merely through its modulus $r = |\vec{r}|$, as is anticipated already by the notation. More generally, one can rewrite double sums formed with pair contributions of some function of position, say $F(\vec{r})$, as integral:

$$\frac{1}{N}\sum_{\kappa\neq\sigma}\langle F(\vec{r}_\kappa - \vec{r}_\sigma)\rangle = \rho\int F(\vec{r})g(r)d^3\vec{r}. \tag{3.59c}$$

The simple sum over all particles σ formed for fixed κ, $\sum_{\sigma\neq\kappa}\langle F(\vec{r}_\kappa - \vec{r}_\sigma)\rangle$, does not depend on κ, since Boltzmann's weight function is invariant under permutation of the N equal particles. Hence, the double sum on the left-hand side can be replaced by a simple one:

$$\frac{1}{N}\sum_{\kappa\neq\sigma}\langle F(\vec{r}_\kappa - \vec{r}_\sigma)\rangle = \sum_{\sigma\neq s}\langle F(\vec{r}_s - \vec{r}_\sigma)\rangle. \tag{3.59d}$$

Here, s can denote any of the system's particles. The preceding equations show that $\rho g(r)$ is the expectation-value density for finding some particle different from particle s in a distance r from that tagged one. Therefore, $g(r)$ is called the pair-distribution function or radial-distribution function.

Equations (3.59a,b) deal with generalized variables and functions. Let us be somewhat more specific in this respect. To proceed, the concept of an amorphous state shall be made more precise by requesting that $g(r)$ is piecewise continuous, that there exists the limit $g(r \to \infty) = c$, and that the function $h(r) = g(r) - c$ is absolutely integrable. The latter condition ensures the existence of the Fourier transform $\hat{h}(\vec{q})$, which depends continuously on $\vec{q}$ and vanishes in the limit of large wave number q. This function is called the pair-correlation function. Because of isotropy, it is independent of the direction of $\vec{q}$. One can write $\hat{h}(\vec{q}) = \rho\tilde{h}_q$. The integral $I = \int_{r_0}^{r_0+R} \rho g(r) d^3\vec{r}$, $R > 0$, is the number of particles expected in the shell $r_0 \leq |\vec{r}_\kappa - \vec{r}_s| \leq r_0 + R$. For the limit $r_0 \to \infty$ one gets $I \to c\rho(4\pi/3)[(r_0+R)^3 - r_0^3]$, and therefore $c = 1$. Hence, there holds

$$\lim_{r\to\infty} g(r) = 1. \tag{3.60a}$$

From Eq. (3.59a) one obtains

$$S_q = 1 + \rho\tilde{h}_q + (2\pi)^3\rho\delta(\vec{q}). \tag{3.60b}$$

Here, $\rho\tilde{h}_q$ is the Fourier transform of the pair-correlation function $h(r) = g(r) - 1$:

$$\rho\tilde{h}_q = \int \exp[i\vec{q}\vec{r}]h(r)d^3\vec{r}. \tag{3.60c}$$

Thus, the structure factor approaches unity for large wave numbers

$$\lim_{q\to\infty} S_q = 1. \tag{3.60d}$$

For non-interacting particles, the average in Eq. (3.58c) is N, i.e., the structure factor is equal to unity. The deviation of $\tilde{h}_q$ from zero is caused by interactions. Approximation theories for the evaluation of S_q and $g(r)$ are discussed in the book by Hansen and McDonald (1986). It is motivated there why these theories are built on a representation of S_q in terms of a quantity c_q by writing

$$S_q = 1/(1 - \rho c_q). \tag{3.61}$$

This representation is called the Ornstein–Zernike relation, and c_q is called the direct correlation function.

Let us consider an external single-particle potentials $u(\vec{r})$. It causes a change of the Hamiltonian

$$\Delta H = -\sum_\kappa u(\vec{r}_\kappa). \tag{3.62a}$$

If the potentials are space independent, there holds $\Delta H = -Nu$. In this case, ΔH describes the change of the chemical potential by u. Hence, $u(\vec{r})$ has the meaning of a space-dependent change of the chemical potential. One can write $\Delta H = -\int u(\vec{r})\rho(\vec{r})d^3\vec{r} = -\sum_{\vec{p}} u_{\vec{p}}^*\rho_{\vec{p}}/L^3$. The latter equality is Parseval's identity (3.8d), where $u_{\vec{p}}$ denotes the Fourier transform of the chemical potential. Ignoring terms of order u^2, Eqs. (2.72a,b), (2.73) determine the expectation value of the density fluctuation created by ΔH for $\vec{q} \neq \vec{0}$: $\langle\rho_{\vec{q}}^*\rangle = \sum_{\vec{p}}(\rho_{\vec{q}}|\rho_{\vec{p}})u_{\vec{p}}^*/[L^3 k_B T]$. Because of the homogeneity condition (3.11b), only the term $\vec{p} = \vec{q}$ contributes. Introducing the function

$$\kappa_q = S_q/(\rho k_B T), \tag{3.62b}$$

one gets

$$\lim_{u_q \to 0} \langle \rho_{\vec{q}}^* \rangle / [u_{\vec{q}}^* \rho^2] = \kappa_q, \quad q \neq 0. \tag{3.62c}$$

Let us recall that the thermodynamic limit is implied in all preceding formulas. In the $q \to 0$ limit, the left-hand side reduces to the thermodynamic derivative $[\partial \langle \rho \rangle / \partial u] / \rho^2$, which is given by the relative change of the density with pressure P, $\kappa_{is} = [\partial \rho / \partial P] / \rho$ (Hansen and McDonald 1986). This second derivative of the thermodynamic potential is the isothermal compressibility κ_{is} of the system:

$$\lim_{q \to 0} \kappa_q = \kappa_{is}. \tag{3.62d}$$

κ_q is the isothermal compressibility for density fluctuations of wave vector $\vec{q}$.

The structure factor S_q can be measured as total cross section for the scattering of neutrons with a momentum transfer $\hbar \vec{q}$ (Lovesey 1984). The upper panel of Fig. 3.4 reproduces results obtained in this manner for liquid argon in a state near its triple point. The lower panel exhibits the pair distribution function evaluated by inverting the Fourier representation (3.59a). Yarnell *et al.* (1973) emphasize that their experimental data agree very well with the ones obtained by molecular dynamics simulations for a Lennard-Jones system with parameters specified in Eq. (1.8c) (Verlet 1968). The strong repulsion causes the excluded-volume effect, i.e., the vanishing of $g(r)$ for distances smaller than about $0.95\ \sigma$. The dense packing requires that a particle is surrounded by 12 nearest neighbours, and this yields a pronounced peak of the $g(r)$ versus r curve. This peak position occurs at a distance slightly below the one for the potential minimum. For r close to 2σ, there is a second maximum for the shell of the six next-to-nearest neighbours. Some further oscillations around unity can be noticed. The described intermediate-range order of the particles in fuzzy shells with radii of about $\sigma, 2\sigma$ and 3σ causes a strong first peak of S_q at a wave number $q_{\max}$ somewhat larger than $2\pi/\sigma$. The sharp increase of the $g(r)$ versus r curve between 3.2 Å and 3.7 Å causes oscillations of the S_q versus q curve around unity. The strong suppression of S_q below the free gas value unity for small wave numbers expresses the fact that the compressibility of a densely packed system of spheres is very low compared to that of a dilute gas : $\kappa_{is} \ll 1/[\rho k_{\mathrm{B}} T]$. The structure factor of a dense Lennard-Jones system is in close agreement with that for a hard-sphere system, where the adopted hard-sphere diameter is slightly decreasing with increasing temperature (Verlet 1968). Hence, the function S_q mainly reflects the solution of the geometrical problem of finding a maximum-entropy packing of spheres with density ρ.

The density-fluctuation correlation function is the most important function to be used in this book for the description of the dynamics. It shall be referred to as density correlator and is defined as

$$\phi_q(t) = (\rho_{\vec{q}}(t) | \rho_{\vec{q}}) / (N S_q). \tag{3.63}$$

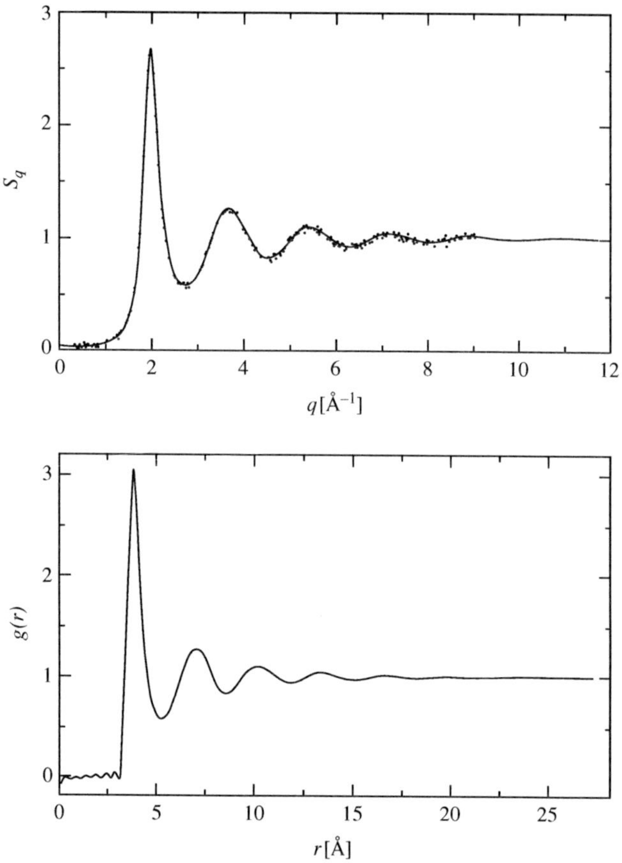

FIG. 3.4. The structure factor S_q as a function of the wave number q of liquid argon at $T = 85\,\mathrm{K}$ obtained by neutron-scattering experiments (upper panel). The line is an interpolation and extrapolation of the data used to evaluate by Fourier transformation the corresponding pair distribution $g(r)$ as a function of the distance r, which is shown in the lower panel. Reproduced from Yarnell *et al.* (1973).

The definition (3.57) implies the representation in terms of the correlation functions of time t and wave number q discussed in Eq. (3.17a): $\phi_q(t) = \sum_{\kappa\sigma} F_q^{(0)\kappa\sigma}(t)/(NS_q)$. Similarly, Eq. (3.17d) yields the expression for the current-density-fluctuation correlations

$$(j_{\vec{q}}^m(t)|j_{\vec{q}}^n)/(Nv_{th}^2) = \phi_{Lq}(t)\big[(q^m/q)(q^n/q)\big] + \phi_{Tq}(t)\big[\delta^{m,n} - (q^m/q)(q^n/q)\big], \tag{3.64a}$$

where $\phi_{L,Tq}(t) = \sum_{\kappa\sigma} F_q^{(1,2)\kappa\sigma}(t)/(Nv_{th}^2)$. According to Eqs. (3.18a, c), the three functions $\phi_q(t)$, $\phi_{Lq}(t)$, and $\phi_{Tq}(t)$ are real and even functions of t. Also the two functions expressing the matrix of the current correlations, are auto-correlators. Using the distinguished wave vector $\vec{q}^* = (0, 0, q)$, one gets

$$\phi_{Lq}(t) = (j_{\vec{q}^*}^3(t)|j_{\vec{q}^*}^3)/(Nv_{th}^2), \tag{3.64b}$$

$$\phi_{Tq}(t) = (j_{\vec{q}^*}^1(t)|j_{\vec{q}^*}^1)/(Nv_{th}^2) = (j_{\vec{q}^*}^2(t)|j_{\vec{q}^*}^2)/(Nv_{th}^2). \tag{3.64c}$$

Therefore, the functions exhibit non-negative spectra, which are even functions of frequency ω. The spectra and the correlators are connected by Eqs. (3.19a,b). The correlators in the frequency domain obey the symmetry relations formulated as Eqs. (3.20a–c). The Eqs. (3.58a,c) imply the normalization $\phi_q(t=0) = \phi_{Lq}(t=0) = \phi_{Tq}(t=0) = 1$. The current field can be split into its longitudinal part, $\vec{j}_{L\vec{q}} = (\vec{q}\vec{j}_{\vec{q}})\vec{q}/q^2$, and its transverse one, $\vec{j}_{T\vec{q}} = (\vec{q} \times \vec{j}_{\vec{q}}) \times \vec{q}/q^2$: $\vec{j}_{\vec{q}} = \vec{j}_{L\vec{q}} + \vec{j}_{T\vec{q}}$. Specializing Eqs. (B.9)–(B.11) one gets

$$(j_{L\vec{q}}^m(t)|j_{L\vec{q}}^n)/(Nv_{th}^2) = \phi_{Lq}(t)\big[(q^m/q)(q^n/q)\big], \tag{3.64d}$$
$$(j_{T\vec{q}}^m(t)|j_{T\vec{q}}^n)/(Nv_{th}^2) = \phi_{Tq}(t)\big[\delta^{m,n} - (q^m/q)(q^n/q)\big]. \tag{3.64e}$$

Therefore, $\phi_{Lq}(t)$ and $\phi_{Tq}(t)$ are called the longitudinal and transverse current correlators, respectively.

From Eq. (3.57), one obtains the equation of motion

$$-i\partial_t \rho_{\vec{q}} = \mathcal{L}\rho_{\vec{q}} = qj_{\vec{q}}. \tag{3.65a}$$

Here, $j_{\vec{q}}$ denotes the current projection on the wave-vector direction

$$j_{\vec{q}} = (\vec{q}/q)\vec{j}_{\vec{q}} = (\vec{q}/q)\vec{j}_{L\vec{q}}, \tag{3.65b}$$

and $\mathcal{L}$ abbreviates the Liouville operator. Fourier transformation casts Eq. (3.65a) in the field equation

$$\partial_t \rho(\vec{r}, t) + \vec{\partial}\vec{j}(\vec{r}, t) = 0. \tag{3.65c}$$

Equations (3.65a–c) are analogues to Eqs. (3.23a–d) for the tagged-particle variables. They express the conservation law for the particle number.

Equation (3.65a) can be combined with Eqs. (2.26a,b) to express density-current correlators in terms of the time derivative of the density correlator:

$$i\partial_t(\rho_{\vec{q}}(t)|\rho_{\vec{q}}) = q(j_{\vec{q}}(t)|\rho_{\vec{q}}) = q(\rho_{\vec{q}}(t)|j_{\vec{q}}). \tag{3.66a}$$

Equations (A.7a), (3.58c), and (3.63) permit us to derive a connection between the density correlator in the frequency domain, $\phi_q(z)$, and the current-density matrix element of the resolvent $\mathcal{R}(z) = 1/[\mathcal{L} - z]$,

$$1 + z\phi_q(z) = [q/(NS_q)](j_{\vec{q}}|\mathcal{R}(z)|\rho_{\vec{q}}). \tag{3.66b}$$

The variables $j_{\vec{q}}$ and $\rho_{\vec{q}}$ can be interchanged on the right-hand side.

Application of Eq. (2.26c) with $m = n = 1$ yields the analogue of Eq. (3.25d): $-\partial_t^2(\rho_{\vec{q}}(t)|\rho_{\vec{q}}) = q^2(j_{\vec{q}}(t)|j_{\vec{q}})$. From Eqs. (3.64d, 3.65b), one finds $(j_{\vec{q}}(t)|j_{\vec{q}}) = Nv_{th}^2\phi_{Lq}(t)$. Introducing a positive frequency Ω_q by

$$\Omega_q^2 = q^2 v_{th}^2/S_q, \tag{3.67}$$

one gets a proportionality of the second time derivative of the density correlator and the longitudinal-current correlator:

$$-\partial_t^2\phi_q(t) = \Omega_q^2\phi_{Lq}(t). \tag{3.68a}$$

In particular, one gets $\partial_t^2\phi_q(t = 0) = -\Omega_q^2$. From Eqs. (3.58b), (3.66a) one obtains $\partial_t\phi_q(t = 0) = 0$. Anticipating that $\phi_{L,q}(t)$ has a continuous bounded time derivative for $t \neq 0$, one can formulate the small-time asymptote of the density correlator

$$\phi_q(t) = 1 - \tfrac{1}{2}(\Omega_q t)^2 + O(|t|^3). \tag{3.68b}$$

The frequency Ω_q quantifies the initial decay of density-fluctuation correlations. The spectral representation, Eq. (3.19b), permits us to express the Taylor coefficients $(\partial/\partial t)^\ell\phi_q(t = 0)$ as an integral of ω^ℓ weighted with the spectral density $\phi_q''(\omega)$. Hence, Eq. (3.68a) yields the sum rule:

$$\int_0^\infty \omega^2\phi_q''(\omega)d\omega \Big/ \int_0^\infty \phi_q''(\omega)d\omega = \Omega_q^2. \tag{3.68c}$$

The frequency Ω_q is the root of the mean-squared spread of the density-fluctuation spectrum. It is an overall measure for the frequency range for the density fluctuations (de Gennes 1959). Equation (A.7b) can be used to write Eq. (3.68a) as an equivalent connection for functions in the frequency domain:

$$\phi_{Lq}(z) = (z/\Omega_q^2)[z\phi_q(z) + 1]. \tag{3.69}$$

The formula (3.68b) is the analogue to Eq. (3.26a). Frequency Ω_q, as opposed to frequency Ω_{sq}, depends on the interaction. The most interesting effect due to the correlations is caused by the first peak for wave numbers near $q_{\max}$. This peak, which is exhibited in Fig. 3.4 for argon for $q_{\max} \approx 2$ Å^{-1}, leads to a strong decrease of Ω_q^2. This is demonstrated by the lower curve in Fig. 3.5. The intermediate-range order in dense liquids causes a shrinking of the spectral width for $q \approx q_{\max}$, a phenomenon called de Gennes narrowing. It is equivalent to a slowing down of the initial decay of the correlations, $1/\Omega_{q_{\max}} > 1/\Omega_{sq_{\max}}$. Upon increasing T or decreasing ρ, the intermediate-range order gets washed out. This

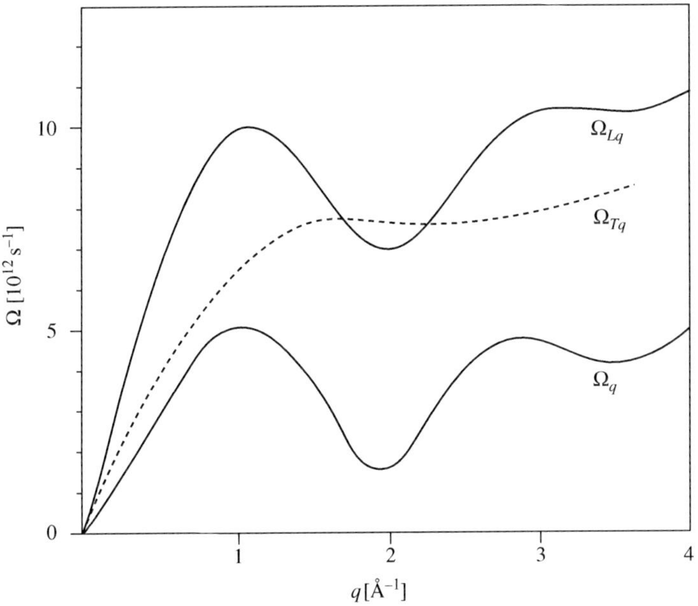

FIG. 3.5. Characteristic frequencies Ω_q, Ω_{Lq}, and Ω_{Tq} of liquid argon calculated according to Eqs (3.67), (3.74a) and (3.74b), respectively. The structure factor S_q and the pair-distribution function $g(r)$ are taken from Fig. 3.4 and the potential $V(r)$ from Eqs. (1.8a,c). Equation (3.74e) yields the Einstein frequency $\Omega_E = 7.4 \times 10^{12}$ s^{-1}. Reproduced from Götze and Lücke (1975).

implies a decrease of the first-peak value $S_{q_{\max}}$ and thereby a widening of the spectral distribution.

A further correlation effect is caused by the suppression of S_q for small wave number, which is demonstrated in Fig. 3.4 for $q < 1$ Å^{-1}. It leads to a speeding up of the correlation decay relative to that for a free gas, or, equivalently, relative to that for the tagged-particle-density fluctuations: $\Omega_q \gg \Omega_{sq}$ for $q \ll q_{\max}$. There is a linear dispersion law for small wave-numbers similar to what is formulated for Ω_{sq} in Eq. (3.26b):

$$\Omega_q = c_{is} q + O(q^3). \tag{3.70a}$$

But the speed c_{is} exceeds v_{th}. According to Eqs. (3.62b,d), it is given by the isothermal compressibility κ_{is} and the mass density $\mu\rho$:

$$c_{is}^2 = 1/(\mu\rho\kappa_{is}). \tag{3.70b}$$

This velocity c_{is} is the sound speed obtained in the continuum theory of matter under the assumption that density variations occur isothermally.

The dynamical variable $\mu\vec{j}_{\vec{q}} = \sum_\kappa \mu\vec{v}_\kappa \exp[i\vec{q}\vec{r}_\kappa]$ is the momentum-density fluctuation. It is the Fourier transform of the momentum density $\mu\vec{j}(\vec{r})$. The time derivative of $\mu\vec{j}_{\vec{q}}$ has the meaning of a force-density fluctuation to be denoted by $\vec{f}_{\vec{q}}$. In analogy to formula (3.45b), the equation of motion reads

$$-i\partial_t\mu\vec{j}_{\vec{q}} = \mu\mathcal{L}\vec{j}_{\vec{q}} = -i\vec{f}_{\vec{q}}. \tag{3.71a}$$

Explicitly, one gets $\vec{f}_{\vec{q}} = \sum_\kappa\{\mu\vec{v}_\kappa(i\vec{q}\vec{v}_\kappa) - \sum_{\sigma\neq\kappa}\vec{\partial}V(|\vec{r}_\kappa - \vec{r}_\sigma|)\}\exp[i\vec{q}\vec{r}_\kappa]$. Symmetrizing the potential-gradient contribution, one obtains

$$\begin{aligned}\vec{f}_{\vec{q}} = i&\sum_\kappa \mu\vec{v}_\kappa(\vec{q}\vec{v}_\kappa)\exp[i\vec{q}\vec{r}_\kappa] \\ &-\frac{1}{2}\sum_{\kappa\neq\sigma}\vec{\partial}V(|\vec{r}_\kappa - \vec{r}_\sigma|)\big\{\exp[i\vec{q}\vec{r}_\kappa] - \exp[i\vec{q}\vec{r}_\sigma]\big\}.\end{aligned} \tag{3.71b}$$

This force vanishes for vanishing wave vector $\vec{q}$. To express this property more explicitly, one can proceed as follows. The distance vectors $\vec{r}_{\kappa\sigma} = \vec{r}_\kappa - \vec{r}_\sigma$ are introduced so that $\vec{r}_{\kappa,\sigma} = (r_\kappa + r_\sigma)/2 \pm \vec{r}_{\kappa\sigma}/2$. Let us abbreviate the derivative of the central symmetric potential by $V'(r) = dV(r)/dr$. There holds

$$f^m_{\vec{q}} = i\sum_n \tau^{mn}_{\vec{q}} q^n, \tag{3.71c}$$

where the tensor-density fluctuation $\tau^{mn}_{\vec{q}}$ can be written as

$$\begin{aligned}\tau^{mn}_{\vec{q}} &= \sum_\kappa \mu v^m_\kappa v^n_\kappa \exp[i\vec{q}\vec{r}_\kappa] \\ &-\sum_{\kappa\neq\sigma}\Big\{\{r^m_{\kappa\sigma}r^n_{\kappa\sigma}V'(|\vec{r}_{\kappa\sigma}|)\}\{\sin[\vec{q}\vec{r}_{\kappa\sigma}/2]/(|\vec{r}_{\kappa\sigma}|\vec{q}\vec{r}_{\kappa\sigma})\} \\ &\times\exp[i\vec{q}(\vec{r}_\kappa + \vec{r}_\sigma)/2]\Big\}.\end{aligned} \tag{3.71d}$$

This tensor is symmetric: $\tau^{mn}_{\vec{q}} = \tau^{nm}_{\vec{q}}$. It shall be assumed that $V(r)$ is sufficiently regular and that it decreases sufficiently strongly for large distances so that $\tau^{mn}_{\vec{q}}$ including its limit $q \to 0$ are dynamical variables in the sense of Sec. 2.1. The components $\tau^{mn}_{\vec{q}}$ can be written as a Fourier transform of a combination $\tau^{mn}(\vec{r})$ of generalized dynamical variables $\delta(\vec{r} - \vec{r}_\kappa)$ and $\delta(\vec{r} - \vec{r}_{\kappa\sigma})$. Fourier-back transformation casts the equation of motion in a field equation

$$\partial_t\mu j^m(\vec{r},t) + \sum_n \partial^n\tau^{mn}(\vec{r},t) = 0. \tag{3.71e}$$

This is the conventional form of a continuity equation for a vector field. All three components of the momentum density $\mu j^m(\vec{r},t)$ obey an equation in analogy to Eq. (3.23b). The three components $\tau^{m1}(\vec{r},t)$, $\tau^{m2}(\vec{r},t)$ and $\tau^{m3}(\vec{r},t)$ are the fluxes of $\mu j^m(\vec{r},t)$. The total-momentum components $\lim_{q\to o} \mu j_{\vec{q}}^m = \int \mu j^m(\vec{r},t)d^3\vec{r}$ are conserved. The momentum-flux tensor $\tau^{mn}(\vec{r},t)$ is also called the stress tensor. The momentum change $\vec{f}_{\vec{q}}$ consists of two contributions. The first term in Eq. (3.71b) is the change because the particles move away and carry along their momentum. The second term is the momentum change due to the forces acting between the pairs κ and σ. Similarly, the stress tensor splits in a kinetic and a potential term.

One can use the equation of motion in order to derive the analogue of Eq. (3.68a). The force–force correlations can be expressed in terms of the second time derivatives of the current–current correlation.

$$-\partial_t^2 (j_{\vec{q}}^m(t)|j_{\vec{q}}^n)/[Nv_{th}^2] = (f_{\vec{q}}^m(t)|f_{\vec{q}}^n)/[N\mu k_{\mathrm{B}}T]. \tag{3.72}$$

In particular, the second Taylor coefficients of the current correlations for $t = 0$ are given by the force-fluctuation overlaps. Let us assume that the force correlations have continuous bounded derivatives for $t \neq 0$. Using Eqs. (3.64b,c), one can formulate the short-time asymptotes in analogy to Eq. (3.68b):

$$\phi_{L,Tq}(t) = 1 - \tfrac{1}{2}(\Omega_{L,Tq}t)^2 + O(|t|^3). \tag{3.73a}$$

The frequencies Ω_{Lq} and Ω_{Tq}, which characterize the initial decay of the longitudinal and the transverse current correlations, respectively, are given by

$$\Omega_{L,Tq}^2 = (f_{\vec{q}^*}^{3,1}|f_{\vec{q}^*}^{3,1})/[N\mu k_{\mathrm{B}}T]. \tag{3.73b}$$

Substituting $f_{\vec{q}^*}^{3,1} = i\mu\mathcal{L}\dot{j}_{\vec{q}^*}^{3,1}$ and using Yvon's theorem, one gets $\Omega_{L,Tq}^2 = -\langle\{f_{\vec{q}^*}^{3,1*}, \dot{j}_{\vec{q}^*}^{3,1}\}\rangle/N$. The Poisson bracket reads: $\langle\{f_{\vec{q}^*}^{m*}, j_{\vec{q}^*}^{m}\}\rangle/N = -v_{th}^2[2q^{*m}q^{*m} + q^2] - [1/(N\mu)]\sum_{\kappa\neq\sigma}\partial^m\partial^m V(|\vec{r}_{\kappa\sigma}|)(1-\exp[i\vec{q}\vec{r}_{\kappa\sigma}])$. Equation (3.59c) yield:

$$\Omega_{Lq}^2 = 3v_{th}^2q^2 + (2\rho/\mu)\int\{\partial^3\partial^3V(|\vec{r}|)g(r)\sin^2[qr^3/2]\}d^3\vec{r}, \tag{3.74a}$$

$$\Omega_{Tq}^2 = v_{th}^2q^2 + (2\rho/\mu)\int\{\partial^1\partial^1V(|\vec{r}|)g(r)\sin^2[qr^3/2]\}d^3\vec{r}, \tag{3.74b}$$

(de Gennes 1959). Figure 3.5 exhibits the $\Omega_{L,Tq}$ versus q curves for liquid argon for a state near the triple point.

In analogy to what is discussed in Eq. (3.70a), there are linear dispersion laws for small wave numbers

$$\Omega_{L,Tq} = c_{L,T}q + O(q^3). \tag{3.74c}$$

The squares of the speeds are inversely proportional to the mass density $\mu\rho$. The remaining factors are structure functions. One writes

$$c_{L,T}^2 = G_{L,T}/(\mu\rho). \tag{3.74d}$$

The G_L and G_T have the dimension of inverse compressibilities. They are called the longitudinal and transverse elastic modulus, respectively.

Using Eqs. (3.59c,d), one can rewrite formula (3.47) for the Einstein frequency as an average of the potential derivatives weighted with the radial distribution function:

$$\Omega_E^2 = (\rho/3\mu)\int \vec{\partial}\vec{\partial}V(|\vec{r}|)g(r)d^3\vec{r}. \tag{3.74e}$$

Application of Lebesque's theorem (A.10) identifies Ω_E^2 as the large-wave-number limit of the interaction contribution to Ω_{Lq}^2 and Ω_{Tq}^2:

$$\lim_{q\to\infty}[\Omega_{Lq}^2 - 3v_{th}^2q^2] = \lim_{q\to\infty}[\Omega_{Tq}^2 - v_{th}^2q^2] = \Omega_E^2. \tag{3.74f}$$

3.3.2 *Transverse-current diffusion*

Like the tagged-particle density, the transverse-current density is a conserved field fluctuation. In this section, the derivations of Sec. 3.2.2 shall be imitated to derive a diffusion equation for the transverse-current correlations. The kinematic viscosity is the diffusion constant and it can be expressed as Green–Kubo integral over stress correlations.

The transverse-current correlator is an autocorrelation function, $\phi_{Tq}(t) = (A(t)|A)$. It is formed with the dynamical variable $A = j_{\vec{q}^*}^1/[v_{th}\sqrt{N}]$, which is normalized, $(A|A) = 1$, and which has a definite time-inversion parity, $\epsilon_T = -1$. Hence, $\mathcal{L}A$ has the opposite time-inversion parity and $(A|\mathcal{L}A) = 0$. The equations of motion (3.71a,c) yield $\mathcal{L}A/q = -if_{\vec{q}^*}^1/[q\mu v_{th}\sqrt{N}] = \tau_{\vec{q}^*}^{13}/[\mu v_{th}\sqrt{N}]$. The application of the Zwanzig–Mori formula (2.62a) for the single distinguished variable A yields a fraction representation for the correlator in the frequency domain. The derivation was explained in Sec. 3.2.2 for $A = \rho_{\vec{q}}^s$. The result is the analogue of Eqs. (3.35a,b):

$$\phi_{Tq}(z) = -1/[z + q^2K_{Tq}(z)], \tag{3.75a}$$

$$K_{Tq}(z) = (\tau_{\vec{q}^*}^{13}|\mathcal{R}_{\mathcal{Q}}(z)\tau_{\vec{q}^*}^{13})/[\mu^2v_{th}^2N]. \tag{3.75b}$$

Since the positive analytic function $K_{Tq}(z)$ can be expressed trivially in terms of $\phi_{Tq}(z)$, one can derive the general symmetries $K_{Tq}(z)^* = K_{Tq}(z^*)$ and $K_{Tq}(-z) = -K_{Tq}(z)$. The spectrum $K_{Tq}''(\omega)$ is a real even non-negative function of frequency. The resolvent $\mathcal{R}_{\mathcal{Q}} = [\mathcal{L}_{\mathcal{Q}} - z]^{-1}$ is formed with the reduced Liouvillian $\mathcal{L}_{\mathcal{Q}} = \mathcal{Q}\mathcal{L}\mathcal{Q}$, where $\mathcal{Q} = 1 - |A)(A|$. The kernel is the Laplace transform of $K_{Tq}(t) = (\exp[i\mathcal{L}_{\mathcal{Q}}t]\tau_{\vec{q}^*}^{13}|\tau_{\vec{q}^*}^{13})/[\mu^2v_{th}^2N]$. The initial value $K_{Tq}(t=0) = (f_{\vec{q}^*}^1|f_{\vec{q}^*}^1)/[q^2\mu^2v_{th}^2N]$ is given by the frequency Ω_{Tq} specified in Eq. (3.73b):

$$K_{Tq}(t=0) = \Omega_{Tq}^2/q^2. \tag{3.75c}$$

To proceed beyond a reformulation of the correlator $\phi_{Tq}(z)$ in terms of a kernel $K_{Tq}(z)$, some regularity assumptions shall be made. These define special properties of those systems to which the discussion shall be restricted. It will be assumed that $K_{Tq}(z)$ is a continuous function of wave number for small values of q. This means that states of the liquid are considered, which do not exhibit long-range-order phenomena. And there is no coupling of $j^1_{\vec{q}^*}$ to other conserved fluctuations. There are the conserved scalar densities for the particle number and energy. They cannot couple to the transverse vector field under discussion because of the isotropy condition (B.10). Because of the isotropy condition (B.9c), there is no coupling to the conserved longitudinal-momentum-density fluctuation. Hence, one can follow the reasoning presented in connection with Eqs. (3.25b–d), (3.34a) in order to arrive at the analogue of Eq. (3.36b). The kernel $K_{Tq}(t)$ describes correlations for a reduced dynamics. Its zero-wave-number limit, however, is a correlator dealing with the full dynamics of the system:

$$\lim_{q\to 0} K_{Tq}(z) = K_T(z), \tag{3.76a}$$

$$K_T(t) = (V_T(t)|V_T), \tag{3.76b}$$

$$V_T = \lim_{q\to 0} \tau^{13}_{\vec{q}^*}/[\mu v_{th}\sqrt{N}]. \tag{3.76c}$$

The explicit form of the shear stress variable V_T can be derived from Eq. (3.71d). Equations (3.74c), (3.75c) determine the normalization:

$$(V_T|V_T) = c_T^2. \tag{3.76d}$$

The transverse-stress correlator or shear correlator $K_T(t)$ plays the same role for the discussion of the dynamics of the conserved density $j^1_{\vec{q}^*}$ as does the velocity correlator $K_s(t)$ for the discussion of the conserved tagged-particle density $\rho^s_{\vec{q}^*}$. However, there is an important difference between the variables defining these correlators: $V_T \propto \lim_{q\to 0} \tau^{13}_{\vec{q}^*}$ has an even time-inversion parity while $v_s^3 \propto \lim_{q\to 0} j^s_{\vec{q}^*}$ has an odd one.

It shall be assumed that $K_T(t)$ is absolutely integrable and that $K_{Tq}(z)$ is a continuous function of the variable pair (q, z), Im $z \gtrless 0$. The former property ensures that the boundary values of $K_T(z)$ for real frequencies ω exist as continuous functions: $K_T(\omega \pm i0) = K'_T(\omega) \pm iK''_T(\omega)$. The spectrum $K''_T(\omega)$ is an even function of frequency ω, while $K'_T(\omega)$ is an odd one. Denoting the zero-frequency spectrum by ν, one gets the limit relations

$$\nu = K''_T(\omega = 0) \tag{3.77a}$$

$$= \lim_{z\to 0}(\mp i)K_T(z), \qquad \text{Im}z \gtrless 0, \tag{3.77b}$$

$$= \lim_{(q,z)\to(0,\pm i0)} (\mp i)K_{Tq}(z). \tag{3.77c}$$

The spectral representation (3.19a) leads to a Green–Kubo equation

$$\nu = \int_0^\infty K_T(t)dt. \tag{3.77d}$$

The case of a vanishing ν shall be excluded by definition of the liquid state, i.e., $\nu > 0$. The quantity ν is the analogue of the tagged-particle diffusivity D_s discussed in Sec. 3.2.2. It is called the kinematic viscosity. It is the custom to also introduce a shear viscosity η by the definition

$$\eta = \nu \cdot (\mu\rho). \tag{3.77e}$$

The continuity property formulated as Eq. (3.77c) implies that the relaxation kernel $K_{Tq}(z)$ can be approximated arbitrarily well by $\pm i\nu$. There holds $|K_{Tq}(z) \mp i\nu|/\nu < \epsilon$, Im $z \gtrless 0$, provided the frequency and the wave number are restricted to sufficiently small values as noted in Eq. (3.38b). This means that the correlations of the transverse currents for small frequencies and small wave numbers can be approximated by the model correlator

$$\overline{\overline{\phi}}_{Tq}(z) = -1/[z \pm i\nu q^2], \quad \mathrm{Im} z \gtrless 0, \tag{3.78a}$$

in the sense explained in connection with Eq. (3.41). This is the correlator of a simple relaxation process as discussed in Eq. (3.40a). Transverse current correlations for small wave numbers relax exponentially with a time scale $\tau_{Tq} = 1/(\nu q^2)$. The low-frequency spectrum is a Lorentzian. The model correlator is obtained by Fourier-transformation of a generalized correlation function $\overline{\overline{\phi}}_T(\vec{r}, t)$. The latter is the solution of the diffusion equation

$$\partial_t \overline{\overline{\phi}}_T(\vec{r}, t) \mp \nu \vec{\partial}\vec{\partial}\, \overline{\overline{\phi}}_T(\vec{r}, t) = 0, \quad t \gtrless 0, \tag{3.78b}$$

for the initial condition $\lim_{t\to 0} \overline{\overline{\phi}}_T(\vec{r}, t) = \delta(\vec{r})$. The proof is the analogue of that for Eq. (3.42c). One concludes that long-wave-length-low-frequency correlations of the transverse currents vary in space and time as diffusion process. The transport coefficient, i.e., the diffusivity for transverse currents, is the kinematic viscosity ν. This is an asymptotic result. It holds in the limit $q \to 0$ because the time scale τ_{Tq} for the current-relaxation is arbitrarily well separated from the time scale t^* characterizing the dynamics, which is generated by $\mathcal{L}_Q$.

One can show that Eq. (3.78b) is equivalent to that part of the linearized Navier–Stokes equations for hydrodynamic flow, which deals with the transverse part of the current field (Boon and Yip 1980; Hansen and McDonald 1986). Thus, the preceding considerations provide a derivation of the mentioned classical hydrodynamic equations within a microscopic theory of liquids. Therefore, the model of the dynamics formulated by the correlation function $\overline{\overline{\phi}}_{Tq}(t)$ is referred to as the hydrodynamic-limit description of the dynamics. This terminology is used also for other functions describing correlations in the limit $q \to 0$, $z \to \pm i0$. The function $\overline{\overline{\phi}}{}^s_q(z)$, Eq. (3.40a), is an example dealing with the hydrodynamic-limit model for the tagged-particle densities. Let us recall that the two bars on the symbols indicate the underlying limit for the two variables q and z being small.

3.3.3 *The generalized-hydrodynamics description of transverse-current correlations*

An essential step in the derivation of some hydrodynamic description is a white-noise assumption for certain relaxation kernels, e.g., $K''_T(\omega) \approx K''_T(\omega = 0) = \nu$. This approximation is justified if the frequencies are restricted to sufficiently small values, $\omega t^* \leq 1$, Eq. (3.38b). Glassy dynamics is characterized by the appearance of structures for the spectra occurring on a frequency scale $1/t^* = 1/\tau$, which is several order of magnitude smaller than the natural THz scale for normal-liquid dynamics. Moreover, $1/\tau$ decreases dramatically with the temperature. These facts are demonstrated in Figs. 1.3 and 1.14. Consequently, for glassy liquids, the regime of validity for the hydrodynamic description may shrink to very small frequency intervals. It is an obvious desire to have a description, which is not limited by the small scale $1/\tau$. One can indeed construct a useful description of the dynamics which is valid for all frequencies z provided that the wave numbers are restricted to small values. Such description is referred to as generalized-hydrodynamics theory. A familiar example for such an approach is Maxwell's theory of electromagnetic fields in continuous media. It is well known that a wealth of phenomena can be described adequately by generalizing the dielectric constant or the conductivity in Maxwell's field equations to frequency-dependent functions. In the same spirit, the generalized-hydrodynamics theories are constructed. In this section, such a theory shall be discussed for the transverse-current dynamics. It will be shown that the appearance of transverse sound waves in liquids is an implication of this theory.

To proceed, it shall be assumed again that the kernel $K_{Tq}(z)$ in Eq. (3.75b) is a uniformly continuous function of q. Hence, for every $\epsilon > 0$, there is a length r^* so that the kernel can be approximated by its $q = 0$ limit within relative error ϵ. There holds

$$|K_{Tq}(z) - K_T(z)| \leq \epsilon K_T(z), \tag{3.79a}$$

provided

$$qr^* \leq 1. \tag{3.79b}$$

Equation (3.75a) motivates the introduction of a model for the transverse-current correlator defined by

$$\overline{\phi}_{Tq}(z) = -1/[z + q^2 K_T(z)]. \tag{3.80}$$

This function is an approximation of the transverse-current correlator for small wave numbers in the sense of $|\phi_{Tq}(z) - \overline{\phi}_{Tq}(z)| \leq \epsilon |K_T(z)| q^2 |\phi_{Tq}(z)\overline{\phi}_{Tq}(z)|$. The function $\overline{\phi}_{Tq}(z)$ shares the following general properties with the correlation function $\phi_{Tq}(z)$. It is the Laplace-transform of a positive-definite function $\overline{\phi}_{Tq}(t)$ obeying the initial condition $\overline{\phi}_{Tq}(t = 0) = 1$. This follows from the discussion in Sec. 2.4.2. The symmetries $K_T(z)^* = K_T(z^*)$ and $K_T(-z) = -K_T(z)$ imply the analogues ones for $\overline{\phi}_{Tq}(z)$. Furthermore, if the frequencies z are reduced to such small values that $K_T(z) \approx K_T(\pm i0)$, Im $z \gtrless 0$, the function $\overline{\phi}_{Tq}(z)$ reduces to $\overline{\overline{\phi}}_{Tq}(z)$ from Eq. (3.78a). This motivates the terminology generalized-hydrodynamics model. It describes the current correlations for all z for q tending to zero. Functions built on such small-q limit shall be indicated by a bar.

Anticipating that $K_T(t)$ has a continuous bounded derivative for $t \neq 0$, one can write $K_T(t) = K_T(t = 0) + O(|t|)$. From Eqs. (A.8d, 3.76d) one obtains the large-frequency asymptote for the kernel:

$$K_T(z) = -c_T^2/z + O(1/z^2). \tag{3.81a}$$

Equation (3.80) can be written in the time domain as Zwanzig–Mori integrodifferential equation in analogy to Eq. (2.62b): $\partial_t \overline{\phi}_{Tq}(t) = -q^2 \int_0^t K_T(t-t')\overline{\phi}_{Tq}(t')dt'$. The asymptotic properties of $K_T(t)$ imply that $\overline{\phi}_{Tq}(t)$ has a bounded continuous third derivative and that $\partial_t^2 \overline{\phi}_{Tq}(t = 0) = -K_T(t = 0)$. Consequently, there holds the short-time expansion

$$\overline{\phi}_{Tq}(t) = 1 - \tfrac{1}{2}(c_T q t)^2 + O(|t|^3). \tag{3.81b}$$

The hydrodynamic-model correlator exhibits a $t = 0$ cusp: $\overline{\overline{\phi}}_{Tq}(t) = \exp[-\nu q^2 |t|]$ $= 1 - [\nu q^2 |t|] + O(t^2)$. This cusp artifact is due to the white noise approximation $\overline{\overline{K}}_{Tq}(z) = \pm i\nu$, Im $t \gtrless 0$. The generalized-hydrodynamics model reproduces correctly the initial decay of the correlations, albeit for the small-q limit only.

In order to identify a qualitatively new feature of the generalized-hydrodynamics description, let us consider the high-frequency behaviour. The leading-order asymptotic result is obtained by replacing the kernel by its $z \to \infty$ limit. This asymptotic treatment uses the first term of Eq. (3.81a). The resulting functions shall be denoted by the symbol ∞. There holds

$$K_T^{\infty}(z) = -c_T^2/z. \tag{3.82a}$$

Substitution in Eq. (3.80) yields a correlator

$$\overline{\phi}_{Tq}^{\infty}(z) = -z/[z^2 - \omega_{Tq}^2]. \tag{3.82b}$$

It has two simple poles on the real frequency axis at $z = \pm\,\omega_{Tq}$, where

$$\omega_{Tq} = c_T \cdot q. \tag{3.82c}$$

The correlator obeys the equation $0 = -\omega_{Tq}^2 \overline{\phi}_{Tq}^{\infty}(z) + z + z^2 \overline{\phi}_{Tq}^{\infty}(z)$. Because of Eq. (A.7b), this result is equivalent to the harmonic-oscillator equation in the time domain

$$\partial_t^2 \overline{\phi}_{Tq}^{\infty}(t) + \omega_{Tq}^2 \overline{\phi}_{Tq}^{\infty}(t) = 0, \tag{3.82d}$$

to be solved for the initial conditions $\overline{\phi}^{\infty}_{Tq}(t=0)=1$, $\partial_t\overline{\phi}^{\infty}_{Tq}(t=0)=0$. The solution of this equation is an undamped monochromatic oscillation $\overline{\phi}^{\infty}_{Tq}(t)=\cos[\omega_{Tq}t]$. The partial fraction representation of the correlator in the frequency domain, $\overline{\phi}^{\infty}_{Tq}(z)=-[A_+/(z-z_+)]-[A_-/(z-z_-)]$, $A_\pm=1/2$, is reproduced by Eq. (2.51d) for a spectrum consisting of two points of equal weight at $\pm\omega_{Tq}$: $\sigma(\omega)=\frac{1}{2}\theta(\omega+\omega_{Tq})+\frac{1}{2}\theta(\omega-\omega_{Tq})$. Equivalently, there holds

$$\overline{\phi}^{\infty\prime\prime}_{Tq}(\omega)=\frac{\pi}{2}\delta(\omega+\omega_{Tq})+\frac{\pi}{2}\delta(\omega-\omega_{Tq}). \tag{3.82e}$$

Fourier transformation from the space of wave vectors to the space of positions $\vec{r}$ casts Eqs. (3.82c,d) into the equivalent wave equation

$$[\partial_t^2-c_T^2\vec{\partial}\vec{\partial}]\overline{\phi}^{\infty}_T(\vec{r},t)=0, \tag{3.83}$$

to be solved for the initial conditions $\overline{\phi}^{\infty}_T(\vec{r},t=0)=\delta(\vec{r})\,,\partial_t\overline{\phi}^{\infty}_T(\vec{r},t=0)=0$.

The preceding Eqs. (3.82b–e), (3.83) are well known formulas describing the propagation of transverse-current fields within the mechanics of ideal isotropic elastic continua. Ideal means that damping phenomena are ignored. Such theory yields the wave equation (3.83), where c_T denotes the transverse-sound speed. Every field fluctuation of given wave vector $\vec{q}$ obeys a harmonic oscillator equation, where there is a linear dispersion law (3.82c) for the frequency. The sound speed is given by Eq. (3.74d), where G_T is the transverse elastic modulus or the shear modulus. Thus, the preceding results describe the propagation of shear waves in the liquid.

The theory of undamped shear waves in elastic media becomes valid in the limit of small q, i.e., for small frequencies. The preceding theory of shear waves in liquids, on the other hand, holds only in the limit of such high frequencies, that $K_T(z)$ can be replaced by its leading term in Eq. (3.81a). On the other hand, the frequency $z=\omega_{Tq}$ has to be so small, that the corresponding wave number $q=|z/c_T|$ obeys the restriction in Eq. (3.79b). Whether or not both specified conditions for the existence of high-frequency transverse sound are approximately fulfilled depends on the detailed form of $K_{Tq}(z)$ for the given state of the liquid. In order to emphasize that Eqs. (3.81, 3.82) deal with high-frequency properties, the transverse modulus G_T is usually denoted by G^∞.

The complete generalized-hydrodynamics approach provides spectra $\overline{\phi}''_{Tq}(\omega)$, which interpolate between the two limiting cases $\overline{\overline{\phi}}''_{Tq}(\omega)$ for small frequencies and $\overline{\phi}^{\infty\prime\prime}_{Tq}(\omega)$ for large ones. Because of the momentum conservation law, the condition of the separation of time scales for current relaxation and for stress fluctuations is fulfilled the better the smaller the wave number q. Hence, for small q, the spectrum exhibits a central diffusion peak as discussed in Eq. (3.40b). In this limit, the function $K_T(\omega+i0)$ is close to its frequency-independent absorptive part $K_T(\omega+i0)\approx i\nu$. Increasing q, the peak widens, and the relevant frequency interval for the spectrum becomes larger. The frequency dependence of $K''_T(\omega)$

and the appearance of a reactive part $K'_T(\omega)$ yield deviations of $\overline{\phi}''_{Tq}(\omega)$ from the Lorentzian-central-peak shape. In particular, the spectrum at large ω will be lower than suggested by the diffusion-peak tail. For large q, the large-frequency parts of $K_T(\omega)$ become crucial. In the extreme case, the absorptive part can be neglected and the kernel is a strongly-ω-dependent purely-reactive function $K_T(\omega + i0) \approx -c_T^2/\omega$. This yields a doublet of sound resonances whose widths are small compared to their resonance positions ω_{Tq}. Decreasing q, the resonance positions shift to smaller q. The crossover phenomena from diffusion behaviour of viscous flow to wave propagation of elastic media are referred to as visco-elastic effects.

In order to demonstrate the similarity of the generalized-hydrodynamics description with the above mentioned theory of electromagnetic fields in continuous media, it might be helpful to reformulate Eq. (3.80). It reads $[z + q^2 K_T(z)]\overline{\phi}_{Tq}(z) = -1$, so that the Fourier transformation leads to the equivalent differential equation

$$(-iz)\overline{\phi}_T(\vec{r}, z) - [-iK_T(z)]\vec{\partial}\vec{\partial}\,\overline{\phi}_T(\vec{r}, z) = i\delta(\vec{r}). \tag{3.84a}$$

The quantity $-iz\overline{\phi}_T(\vec{r}, z) - i\delta(\vec{r})$ is the Laplace transform of $\partial_t\overline{\phi}_T(\vec{r}, t)$, Eq. (A.7a). Hence, the preceding formula is the diffusion equation in the frequency domain. However, the transport coefficient in the diffusion equation is generalized to the frequency-dependent function $[-iK_T(z)]$. The latter is called the frequency-dependent kinematic viscosity; it reduces to ν in the static limit $z \to i0$. Another rewriting is obtained by multiplying Eq. (3.84a) by $(-iz)$:

$$(-z^2)\overline{\phi}_T(\vec{r}, z) - [G_T(z)/(\mu\rho)]\vec{\partial}\vec{\partial}\overline{\phi}_T(\vec{r}, z) = z\delta(\vec{r}). \tag{3.84b}$$

Here, the abbreviation

$$G_T(z) = \mu\rho[-zK_T(z)] \tag{3.84c}$$

is introduced. It is explained in connection with the derivation of Eq. (3.82d) that $-z^2\overline{\phi}_T(\vec{r}, z) - z\delta(\vec{r})$ is the Laplace transform of $\partial_t^2\overline{\phi}_T(\vec{r}, t)$. Therefore, Eq. (3.84c) is the wave equation in the frequency domain. However, the square of the wave speed, c_T^2 in Eq. (3.83), is generalized to a frequency-dependent function $G_T(z)/(\mu\rho)$. This would be the formula for the square of the sound speed of an elastic continuum if $G_T(z)$ would be a shear modulus. Therefore, $G_T(z)$ has the meaning of a frequency- dependent shear modulus. According to Eq. (3.81a), $G_T(z)$ reduces to $G_T = G^\infty$ in the limit z tending to infinity.

Because of Eq. (3.77a), a liquid state is characterized by a finite positive zero-frequency limit ν of the frequency-dependent diffusivity:

$$-iK_T(\omega + i0) = \nu + O(\omega), \quad \text{liquid}. \tag{3.85a}$$

This is equivalent to an imaginary low-frequency modulus, which vanishes linearly with the frequency:

$$G_T(\omega + i0) = -i\omega\eta + O(\omega^2), \quad \text{liquid}. \tag{3.85b}$$

An isotropic elastic continuum is a model for condensed matter, where there are shear waves in the limit of small wave number. The waves are specified by some positive sound speed, say c_T^0. The latter is expressed by a positive static shear modulus G_T^0 with a formula in analogy to Eq. (3.74d): $c_T^0 = \sqrt{G_T^0/(\mu\rho)}$. For this model of a dynamics, the frequency-dependent shear modulus has a positive low-frequency limit:

$$G_T(\omega + i0) = G_T^0 + O(\omega), \quad \text{elastic medium.} \tag{3.86a}$$

Because of Eq. (3.84c), this is equivalent to a $(1/z)$-pole of the correlator for the transverse stresses

$$K_T(\omega + i0) = -[G_T^0/(\mu\rho)]/\omega + O(\omega^0), \quad \text{elastic medium.} \tag{3.86b}$$

The generalized-hydrodynamics approach can describe liquids and elastic continua within the same frame. The latter state is characterized by the fact that the stress correlator $K_T(z)$ exhibits a non-vanishing arrested part, Eq. (2.85). It is given by the square of the transverse-sound speed: $c_T^{02} = \text{limav}_{t\to\infty} K_T(t)$.

3.3.4 *Visco-elastic features and glassy-dynamics precursors of the transverse-current correlators*

In this section, Langevin's theory is applied to motivate a simple relaxation-process model for the shear correlator $K_T(t)$. Using this result, the generalized-hydrodynamics approach leads to Maxwell's theory of visco-elasticity.

According to Eq. (3.76d), the variable $A = V_T/c_T$ is normalized, $(A|A) = 1$. It has a definite time-inversion parity, $\epsilon_T^A = 1$. Application of the Zwanzig–Mori formula (2.62a) yields the fraction representation for the Laplace transform of the normalized stress correlator: $LT[(A(t)|A)](z) = K_T(z)/c_T^2 = -1/[z + M_T(z)]$. Here, $M_T(z)$ is a relaxation kernel constructed with a reduced dynamics for generalized-force fluctuations $\mathcal{L}V_T$. The derivation is the strict analogue to that which leads to Eq. (3.48a) for the velocity correlator $K_s(z)$ with $M_T(z)$ being the analogue of $X_s(z)/m_s$. Let t^* denote the time scale which characterizes the dynamics of the generalized-force fluctuations. Then, a white noise approximation can be applied for frequencies smaller than $1/t^* : M_T''(\omega) = \xi_T$. Here, $\xi_T = M_T''(\omega = 0)$ is the zero-frequency spectrum. Using this expression for all frequencies, one arrives at Langevin's theory for the stress dynamics. It is built on the formula $M_T(z) = \pm i\xi_T$, Im $z \gtrless 0$. As a result, one gets a model for the shear correlator in the frequency domain to be denoted by $K_T^M(z)$:

$$K_T^M(z) = -c_T^2/[z \pm i\xi_T], \quad \text{Im}\, z \gtrless 0. \tag{3.87a}$$

This is the analogue of Eq. (3.50c), and there are the corresponding analogues of Eqs. (3.50d–f). In particular, the stress dynamics is described by a simple relaxation law, $K_T^M(t) = c_T^2 \exp[-t/\tau_T^M]$ - the formula suggested by J. C. Maxwell. His relaxation time for the stress correlation is given by

$$\tau_T^M = 1/\xi_T. \tag{3.87b}$$

Remembering Eqs. (3.77a,e) for the connection of shear viscosity η and the zero-frequency shear spectrum $K_T^{M\prime\prime}(\omega = 0) = c_T^2/\xi_T$ and also the relation (3.74d) between c_T^2 and the high-frequency shear modulus $G^\infty = G_T$, one obtains Maxwell's relation between viscosity and relaxation time:

$$\eta = G^\infty \tau_T^M. \tag{3.87c}$$

The preceding equations formulate a valid extension of the hydrodynamic-limit approach if and only if the time-scale-separation assumption is justified:

$$\tau^M \gg t^*. \tag{3.87d}$$

If one replaces in Eq. (3.80) the kernel $K_T(z)$ by $K_T^M(z)$, one gets the transverse-current-fluctuation correlator in the frequency domain within Maxwell's theory of visco-elasticity:

$$\phi_{Tq}^M(z) = -1/\{z - \omega_{Tq}^2/[z \pm i/\tau_T^M]\}, \quad \mathrm{Im}\, z \gtrless 0. \tag{3.88}$$

The frequency ω_{Tq} denotes the high-frequency-transverse-sound dispersion from Eq. (3.82c). The function $\phi_{Tq}^M(z)$ is the correlator for a harmonic oscillator specified by the frequency $\Omega = \omega_{Tq}$ and the friction constant $\Gamma = 1/\tau_T^M$, which is discussed in Appendix A.4. The essential parameter characterizing the dynamics is $r_q = 1/(\omega_{Tq}\tau_T^M)$. For $r_q > 2$, the correlator $\phi_{Tq}^M(t)$ decays monotonically to zero in qualitative agreement with the viscous relaxation for $q \to 0$. For $r_q < 2$, the correlator exhibits damped oscillations as expected for an elastic medium.

The evolution of the visco-elastic features of $\phi_{Tq}^{M\prime\prime}(\omega)$ with changes of q is demonstrated in Fig. 3.6. For wave numbers so small that $r_q > \sqrt{2}$, the spectrum $\phi_{Tq}^{M\prime\prime}(\omega)$ exhibits a central peak. The peak-width decreases with decreasing wave number, and for $q \to 0$ one gets the diffusion peak. For $r_q < \sqrt{2}$, the spectrum exhibits a doublet, i.e., peaks at positions $\pm\tilde{\omega}_q$, $\tilde{\omega}_q = \sqrt{\omega_{Tq}^2 - (1/\tau_T^M\sqrt{2})^2}$. If q decreases to the critical value $q_c = 1/(\sqrt{2}c_T\tau_T^M)$, $\tilde{\omega}_q$ decreases to zero. If q increases to large values, $\tilde{\omega}_q$ increases to ω_{Tq}, as explained in the preceding section. In this limit, the width of the sound peaks of $\phi_{Tq}^{M\prime\prime}(\omega)$ becomes q-independent.

Figure 3.7 exhibits transverse-current-fluctuation spectra of a Lennard-Jones liquid obtained by molecular-dynamics simulations for an 864-particle system. The state analysed is close to the triple point, i.e., it is close to that discussed above in Fig. 3.2. The data exhibit the evolution of transverse-sound resonances as discussed in the preceding paragraph. The full lines are data fits within Maxwell's theory of visco-elasticity. But Eq. (3.88) is generalized in the sense that ω_{Tq}^2 was replaced by Ω_{Tq}^2 from Eq. (3.74b) and that τ_T^M is fitted for every q separately. The figure demonstrates that the generalized-hydrodynamics approach together with Langevin's theory for the stress dynamics explains qualitatively some essential features of the visco-elastic behaviour of a simple liquid.

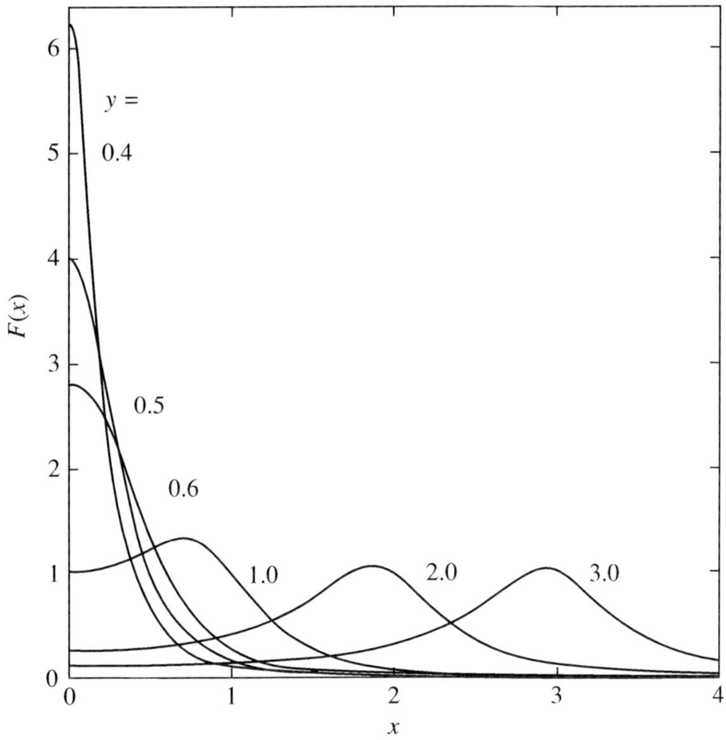

FIG. 3.6. Spectra of the rescaled transverse-current correlations $F = \phi_{Tq}^{M\prime\prime}(\omega)/\tau_T^M$ as a function of the rescaled frequency $x = \omega\tau_T^M$ for some rescaled wave numbers $y = c_T\tau_T^M q$ according to the visco-elasticity formula (3.88). Here, τ_T^M and c_T denote Maxwell's relaxation time for shear and the transverse-high-frequency-sound speed, respectively.

Figure 3.7 demonstrates also that the result $\phi_{Tq}^M(z)$ has shortcomings. Visco-elastic effects are underestimated in the sense that the sound resonances are sharper than described by Eq. (3.88); and, with increasing q, they appear already for a wave number smaller than the calculated value q_c. In order to specify these problems more directly, Fig. 3.8 exhibits as the full line the simulation result for the normalized shear correlator $\beta(t) = K_T(t)/K_T(t=0)$. This function is not caused by a simple-relaxation process as implied by Eq. (3.87a). The initial decay, say the part $1.0 \geq \beta(t) > 0.3$, follows closely an exponential function $\exp[-(t/t_1)]$. But, this decay does not continue to long times. Rather, there is some tail characterized by a time scale t_2, which exceeds t_1 by at least a factor 5. This finding is similar to the one demonstrated in Fig. 3.3 for the normalized force correlations of a tagged particle. The corresponding simulation result $\alpha(t) = X_s(t)/X_s(t=0)$ for the state of the Lennard-Jones liquid under

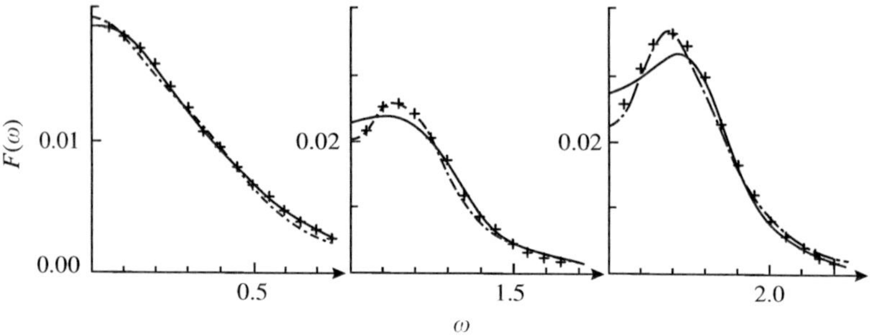

FIG. 3.7. The dashed-dotted lines are molecular-dynamics-simulation data for rescaled transverse-current-fluctuation spectra $F(\omega) = (v_{th}q)^2\phi''_{Tq}(\omega)/2\pi$ of a Lennard–Jones liquid at $T = 0.722$ and $\rho = 0.844$ for wave numbers $q = 0.752, 1.366$ and 1.931 (from left to right). The units are defined in Eqs. (1.8a–c) so that $\omega = 1$ corresponds to 0.51 THz and $q = 1.93$ corresponds to 0.57 Å^{-1} for liquid argon. Notice that the scale of the vertical axis of the first panel differs from that of the others and that the scale of the frequency axis increases from panel to panel.

The full lines are fits by spectra according to Maxwell's theory of visco-elasticity, Eq. (3.88), with ω^2_{Tq} replaced by Ω^2_{Tq}, Eq. (3.74b), and τ^M_T adjusted for every q separately. The crosses exhibit the representation of the correlator by Eq. (3.75a) with a three-parameter fit for the relaxation kernel $K_{Tq}(t)$ according to Eq. (3.89). Reproduced from Levesque *et al.* (1973).

discussion is shown as dashed line in Fig. 3.8. One notices that the time scales t_1 and t_2 for $\beta(t)$ are quite similar to those for $\alpha(t)$. In analogy to what was discussed in connection with Eqs. (3.56a,b), Levesque *et al.* (1973) have analysed their data with Eq. (3.75a) using a two-correlator model for the kernel

$$K_{Tq}(t)/\Omega^2_{Tq} = (1 - A)\exp[-(t/t_1)] + A\exp[-(t/t_2)]. \tag{3.89}$$

Choosing t_1 near $0.6\,\tau_0$, A near 0.05 and allowing both parameters to decrease somewhat with increasing wave number, they could fit perfectly their data for the transverse-current spectra. For the long-time scale, a q-independent value has been chosen, $t_2 = 4.7\,\tau_0$. This fit is shown for three wave numbers by the crosses in Fig. 3.7. One concludes the following. There is a long-time tail for the stress relaxation, which renders the time-scale-separation assumption illegitimate. The appearance of this tail signalises the onset of glassy dynamics. The phenomenon is similar to that discussed in panels (a) and (b) of Fig. 1.2. It is this glassy-dynamics precursor, which causes the visco-elastic effects of the transverse-current spectra to be more pronounced than expected within Maxwell's theory.

Levesque *et al.* (1973) point out the close similarities of their simulation results for the Lennard-Jones model of liquid argon with the results obtained

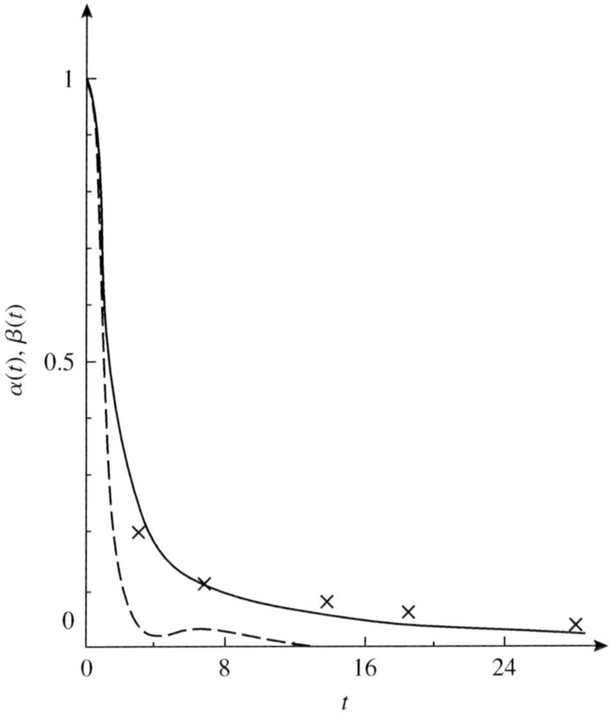

FIG. 3.8. The full and dashed line are the normalized transverse-stress correlator $\beta(t) = K_T(t)/K_T(t \to 0)$ based on definitions (3.76a–c) and the normalized tagged-particle-force kernel $\alpha(t) = X_s(t)/X_s(t \to 0)$ based on definitions (3.48a,b), respectively, obtained by molecular-dynamics simulations for a Lennard-Jones liquid for the same state as discussed in Fig. 3.7. The unit of time is 0.321 τ_0 corresponding to 0.1 ps for argon. The crosses are results for $\beta(t)$ obtained by Alder *et al.* (1970) by molecular-dynamics simulations for a hard-sphere liquid for a packing fraction close to 0.49. Reproduced from Levesque *et al.* (1973).

in the seminal simulation studies of Alder *et al.* (1970) for dense hard-sphere systems. Choosing the hard-sphere diameter d about 2% larger than the length-scale σ of the Lennard-Jones potential, the structure factors S_q of the model for argon is close to that of the hard-sphere system for a packing fraction $\varphi = 0.47$. The tagged-particle diffusivities D_s and the shear viscosities η for both systems do not deviate by more than 10%. The crosses in Fig. 3.8 demonstrate that also the long-time tails for the stress relaxation agree well. The agreement of the long-time decay and of the large values for η and $1/D_s$ of argon near its triple point with the corresponding quantities of the hard-sphere system near its freezing point suggests the following conclusion. The glassy-dynamics precursors of a Lennard-Jones liquid at high density and low temperature is caused by the strong interparticle repulsion.

3.3.5 *Representations of the density correlators in terms of relaxation kernels*

In this section, the correlation functions formed with density fluctuations and longitudinal-current fluctuations shall be expressed in terms of a single relaxation kernel $K_{Lq}(z)$, which describes correlations of fluctuating stresses. This representation reflects the connections between the correlators, which are implied by the continuity equation. In addition, it will be shown how the kernel $K_{Lq}(z)$ can be expressed in terms of three other kernels so that those singularities for small z and q are taken care of, which are implied by the energy-conservation law.

The normalized density correlator, Eq. (3.63), and the normalized longitudinal-current correlator, Eq. (3.64b), are auto-correlation functions formed with the dynamical variables $A_1 = \rho_{\vec{q}^*}/\sqrt{NS_q}$ and $A_2 = j^3_{\vec{q}^*}/\sqrt{Nv^2_{th}}$, respectively. Both variables refer to conserved quantities, namely to the total particle number and to the total particle momentum, respectively. The variables are coupled. Equation (3.66a) expresses $(A_1(t)|A_2) = (A_2(t)|A_1) = (j^3_{\vec{q}^*}(t)|\rho_{\vec{q}^*})/[Nv_{th}\sqrt{S_q}]$ as $i\partial_t(A_1(t)|A_1)/\Omega_q$; and Eq. (3.68a) yields $(A_2(t)|A_2) = -\partial^2_t(A_1(t)|A_1)/\Omega^2_q$. In order to express the expected conservation-law-induced singularities for $z \to 0$ and $q \to 0$, a procedure shall be used in analogy to what was explained in Sec. 3.2.2 for the conserved tagged-particle-density fluctuation $\rho^s_{\vec{q}}$ and in Sec. 3.3.2 for the conserved transverse-momentum-density fluctuation $j^1_{\vec{q}^*}$. Both variables A_1 and A_2 shall be treated simultaneously. The continuity equation (3.65a) formulates the coupling: $\mathcal{L}A_1 = \Omega_q A_2$.

Let us formulate Eq. (2.62a) for the 2-by-2 matrix $\phi_{m,n}(z), m, n = 1$ or 2, which is the Laplace transform of the correlator matrix formed with the pair of distinguished variables A_1 and A_2 : $\phi_{m,n}(t) = (A_m(t)|A_n)$. The normalization matrix is trivial: $g_{m,n} = (A_m|A_n) = \delta_{m,n}$. The projectors $\mathcal{P}_1 = |\rho_{\vec{q}^*})[NS_q]^{-1}(\rho_{\vec{q}^*}|$ and $\mathcal{P}_2 = |j^3_{\vec{q}^*})[Nv^2_{th}]^{-1}(j^3_{\vec{q}^*}|$ on the distinguished variables are orthogonal: $\mathcal{P}_1\mathcal{P}_2 = \mathcal{O}$. If $\mathcal{Q}_1 = 1 - \mathcal{P}_1$ and $\mathcal{Q}_2 = 1 - \mathcal{P}_2$ denote the projectors perpendicular to A_1 and A_2, respectively. $\mathcal{Q}_{12} = \mathcal{Q}_1\mathcal{Q}_2 = 1 - \mathcal{P}_1 - \mathcal{P}_2$ is the projector perpendicular to the two-dimensional space spanned by A_1 and A_2. The reduced Liouvillian reads $\mathcal{L}_{12} = \mathcal{Q}_{12}\mathcal{L}\mathcal{Q}_{12}$, and it defines the resolvent $\mathcal{R}_{12}(z) = [\mathcal{L}_{12} - z]^{-1}$. The variables A_1 and A_2 have time-inversion parity $\epsilon^1_T = 1$ and $\epsilon^2_T = -1$, respectively. Therefore, the variables $\mathcal{L}A_1$ and $\mathcal{L}A_2$ have parities -1 and $+1$, respectively. Variables of opposite time-inversion parity are orthogonal. Therefore, the diagonal elements of the frequency matrix (2.60a) vanish: $\Omega_{1,1} = \Omega_{2,2} = 0$. The off-diagonal elements read $\Omega_{1,2} = (A_1|\mathcal{L}A_2) = \Omega_{2,1} = \Omega_q$. Since $\mathcal{Q}_{12}\mathcal{L}A_1 = 0$, the first fluctuating force from Eq. (2.61a) vanishes: $\dot{A}^{\mathcal{Q}}_1 = 0$. As a result, the 2-by-2 matrix (2.61b) of relaxation kernels has only one non-vanishing element, namely $M_{2,2}(z) = (\dot{A}^{\mathcal{Q}}_2|\mathcal{R}_{12}(z)\dot{A}^{\mathcal{Q}}_2)$ with $\dot{A}^{\mathcal{Q}}_2 = \mathcal{Q}_{12}\partial_t A_2 = \mathcal{Q}_1\partial_t A_2$. Let us denote this kernel by $M_{Lq}(z)$. The Zwanzig–Mori equations yield the four functions $\phi_{m,n}(z)$ in terms of Ω_q and $M_{Lq}(z)$:

$$\phi_q(z) = -1/\Big\{z - \Omega_q^2/[z + M_{Lq}(z)]\Big\}, \tag{3.90a}$$

$$\phi_{Lq}(z) = -z/\Big\{z\big[z + M_{Lq}(z)\big] - \Omega_q^2\Big\}, \tag{3.90b}$$

$$(j_{\vec{q}}|\mathcal{R}(z)\rho_{\vec{q}}) = (\rho_{\vec{q}}|\mathcal{R}(z)j_{\vec{q}}) = -(Nv_{th}^2 q)/\Big\{z\big[z + M_{Lq}(z)\big] - \Omega_q^2\Big\}. \tag{3.90c}$$

The time derivative in the definition of $\dot{A}_2^{\mathcal{Q}}$ can be expressed in terms of forces or stresses using Eqs. (3.71a,c): $\partial_t A_2 = f_{\vec{q}^*}^3/[\mu v_{th}\sqrt{N}] = iq\tau_{\vec{q}^*}^{33}/[\mu v_{th}\sqrt{N}]$. Often, it is more transparent to split off the factor q, which reflects the conservation of momentum. One introduces the kernel $K_{Lq}(z) = M_{22}(z)/q^2$. This kernel is the longitudinal counterpart of the kernel $K_{Tq}(z)$ introduced in Eqs. (3.75a,b) for the transverse-current correlators. One gets

$$M_{Lq}(z) = q^2 K_{Lq}(z), \tag{3.90d}$$

$$K_{Lq}(z) = (\mathcal{Q}_1\tau_{\vec{q}^*}^{33}|\mathcal{R}_{12}(z)\mathcal{Q}_1\tau_{\vec{q}^*}^{33})/[\mu^2 v_{th}^2 N]. \tag{3.90e}$$

The initial value of the relaxation kernel $M_{Lq}(t)$ shall be expressed by a non-negative frequency Δ_q:

$$M_{Lq}(t=0) = \Delta_q^2. \tag{3.91a}$$

One gets $\Delta_q^2[\mu^2 v_{th}^2 N] = (f_{\vec{q}^*}^3|\mathcal{Q}_1 f_{\vec{q}^*}^3) = (f_{\vec{q}^*}^3|f_{\vec{q}^*}^3) - (f_{\vec{q}^*}^3|\rho_{\vec{q}^*})[NS_q]^{-1}(\rho_{\vec{q}^*}|f_{\vec{q}^*}^3)$. The first term yields Ω_{Lq}^2, because of Eq. (3.73b). The second term leads to Ω_q^2, since $-(f_{\vec{q}^*}^3|\rho_{\vec{q}^*}) = i(\mathcal{L}j_{\vec{q}^*}^3|\rho_{\vec{q}^*})\mu = i(j_{\vec{q}^*}^3|\mathcal{L}\rho_{\vec{q}^*})\mu = i(j_{\vec{q}^*}^3|j_{\vec{q}^*}^3)q\mu = iNv_{th}^2 q\mu$. Hence, Δ_q^2 is the difference of the squares of the two characteristic frequencies identified before:

$$\Delta_q^2 = \Omega_{Lq}^2 - \Omega_q^2. \tag{3.91b}$$

From Eqs. (3.70a, 3.74c), one obtains the small-q expansion

$$\Delta_q^2 = [c_L^2 - c_{is}^2]q^2 + O(q^3). \tag{3.91c}$$

Formulas, which have the structure of Eqs. (3.90a–c), will be considered with different intentions in the following. To avoid repetitions of the argumentations, a digression shall be made here. The notation used above for the derivation of the Zwanzig–Mori equations shall be used here as well. Let $\tilde{M}_{Lq}(z)$ denote a positive-analytic function, which obeys the condition:

$$\lim_{y\to\infty} \tilde{M}_{Lq}(iy)/y = 0. \tag{3.92a}$$

Substituting this function in Eqs. (3.90a,b), functions $\tilde{\phi}_q(z)$ and $\tilde{\phi}_{Lq}(z)$ shall be defined. More generally, a 2-by-2 matrix $\tilde{\phi}_{m,n}(z)$ is defined by using Eq. (2.62a) with $\tilde{M}_{1,1}(z) = \tilde{M}_{1,2}(z) = \tilde{M}_{2,1}(z) = 0$ and $\tilde{M}_{2,2}(z) = \tilde{M}_{Lq}(z)$. Furthermore, $\Omega_{1,1} = \Omega_{2,2} = 0$ and $\Omega_{1,2} = \Omega_{2,1} = \Omega_q$ are used. According to the discussion of Eqs. (2.69a–c), one concludes: the $\tilde{\phi}_{m,n}(z)$ are Laplace transforms of continuous functions $\tilde{\phi}_{m,n}(t)$, which have a spectral representation with a positive semi definite matrix of spectra, $\tilde{\phi}''_{m,n}(\omega)$, and which obey the initial condition $\tilde{\phi}_{m,n}(t = 0) = \delta_{m,n}$. According to Eq. (2.69d), there holds: $|\tilde{\phi}_{1,1}(t)| \leq 1, |\tilde{\phi}_{2,2}(t)| \leq 1$ and $|\tilde{\phi}_{1,2}(t)| \leq 1$. Introducing the original notations and noticing the implications of the continuity equation, one gets

$$|\phi_q(t)| \leq 1, \quad |\partial_t^2 \tilde{\phi}_q(t)/\Omega_q^2| \leq 1, \quad |\partial_t \tilde{\phi}_q(t)/\Omega_q| \leq 1. \tag{3.92b}$$

There hold

$$\tilde{\phi}_{Lq}(z) = (z/\Omega_q^2)[z\tilde{\phi}_q(z) + 1], \tag{3.92c}$$

and a relation between $\tilde{\phi}_{1,2}(z) = \tilde{\phi}_{2,1}(z)$ and $\tilde{\phi}_{1,1}(z)$, which corresponds to Eq. (3.66b).

Let us continue the digression with a stronger request for $\tilde{M}_{Lq}(z)$. This kernel shall have the form

$$\tilde{M}_{Lq}(z) = \pm i\nu_q + \hat{M}_{Lq}(z), \quad \text{Im } z \gtrless 0. \tag{3.93a}$$

Here, ν_q is a non-negative number specifying a possible white-noise contribution to $\tilde{M}''_{Lq}(\omega)$. The function $\hat{M}_{Lq}(z)$ shall be positive analytic. There shall be a finite constant C so that there holds for Im $z \neq 0$:

$$|z\hat{M}_{Lq}(z)| \leq C. \tag{3.93b}$$

It is shown in Appendix A.3 that $\hat{M}_{Lq}(z)$ is the Laplace transform of a continuous positive-definite function $\hat{M}_{Lq}(t)$. According to the discussion of Eqs. (2.69), (2.70), the functions $\tilde{\phi}_{m,n}(t)$ have continuous derivatives and there hold Eqs. (2.62b). The equation for $m = n = 1$ reads $i\partial_t\tilde{\phi}_{1,1}(t) = \Omega_q\tilde{\phi}_{2,1}(t)$. Thus, $\tilde{\phi}_{1,1}(t)$ has a continuous second derivative. The equation is equivalent to Eq. (3.66b). The equation for $m = 1$ and $n = 2$ reads $i\partial_t\tilde{\phi}_{1,2}(t) = \Omega_q\tilde{\phi}_{2,2}(t)$. Hence, $\tilde{\phi}_{1,2}(t)$ has a second and $\tilde{\phi}_{1,1}(t)$ has a third derivative, which are continuous. Combined with the first equation, one gets Eq. (3.68a). The correlators in time domain are connected as requested by the continuity equation. The equation for $m = 2$ and $n = 1$ is obeyed because of the symmetry $\tilde{\phi}_{1,2}(t) = \tilde{\phi}_{2,1}(t)$. Using the preceding results, the remaining equation for $m = n = 2$ is equivalent to: $i\partial_t\{-\partial_t^2\tilde{\phi}_{1,1}(t) - \nu_q\partial_t\tilde{\phi}_{1,1}(t) - \Omega_q^2\tilde{\phi}_{1,1}(t) - \int_0^t \hat{M}_{Lq}(t-t')\partial'_t\tilde{\phi}_{1,1}(t')dt'\} = 0$. The curly bracket vanishes for $t = 0$. Hence, there remains a single equation,

$$\partial_t^2\tilde{\phi}_q(t) + \nu_q\partial_t\tilde{\phi}_q(t) + \Omega_q^2\tilde{\phi}_q(t) + \int_0^t \hat{M}_{Lq}(t-t')\partial'_t\tilde{\phi}_q(t')dt' = 0, \tag{3.93c}$$

to be solved for the initial condition

$$\tilde{\phi}_q(t) = 1 - \tfrac{1}{2}(\Omega_q t)^2 + O(t^3). \tag{3.93d}$$

An addendum to the digression concerns the general symmetries. The elementary algebraic relation (3.90a) implies that

$$\tilde{\phi}_q(-z) = -\tilde{\phi}_q(z), \quad \tilde{\phi}_q(z)^* = \tilde{\phi}_q(z^*) \tag{3.94a}$$

hold if and only if there hold the symmetry relations

$$\hat{M}_{Lq}(-z) = -\hat{M}_{Lq}(z), \quad \hat{M}_{Lq}(z)^* = \hat{M}_{Lq}(z^*). \tag{3.94b}$$

Since $\phi_q(t)$ is a real and even function of time, Eqs. (3.94a) hold for the correlator $\phi_q(z)$. Hence, Eqs. (3.94b) hold for the kernel $M_{Lq}(z)$. If one replaces the kernel by $\tilde{M}_{Lq}(z)$ so that the general symmetries (3.94b) are obeyed, the functions $\tilde{\phi}_q(z)$ obey the Eqs. (3.94a). The corresponding symmetries for $\tilde{\phi}_{Lq}(z)$ follow and so do the ones for the density-current correlators.

The double-fraction representation (3.90a) will be of great importance in the following chapter. Therefore, it might be justified to point out that it can be derived alternatively as follows. Function $\phi_q(z)$ is the autocorrelator of the normalized variable $A_1 = \rho_{\vec{q}^*}/\sqrt{NS_q}$, which has a definite time-inversion parity $\epsilon_T^{A_1} = 1$. The continuity equation (3.65a) serves as equation of motion: $-i\partial_t A_1 = qj, j = j^3_{\vec{q}^*}/\sqrt{NS_q}$, where j has time-inversion parity -1. The Zwanzig–Mori formula (2.62a) can be applied for the single distinguished variable A_1. The procedure was explained in Sec. 3.2.2 for the variable $A = \rho^s_{\vec{q}}$. The results are the strict analogue to Eqs. (3.35a–c):

$$\phi_q(z) = -1/[z + q^2 K_q(z)]. \tag{3.95a}$$

The relaxation kernel

$$K_q(z) = (j^3_{\vec{q}^*}|\mathcal{R}_1(z) j^3_{\vec{q}^*})/(NS_q) \tag{3.95b}$$

is the matrix element of j formed with the resolvent $\mathcal{R}_1(z) = [\mathcal{L}_1 - z]^{-1}$. Here, the reduced Liouvillian $\mathcal{L}_1 = \mathcal{Q}_1 \mathcal{L} \mathcal{Q}_1$ treats the density fluctuations as constants of motion. The initial condition for kernel $K_q(t)$ is given by the frequency Ω_q; since $(j^3_{\vec{q}^*}|j^3_{\vec{q}^*})/NS_q = v_{th}^2/S_q$, there holds

$$q^2 K_q(t=0) = \Omega_q^2. \tag{3.95c}$$

The procedure shall be repeated, in order to express $K_q(z)$ in terms of a relaxation kernel, i.e., Eq. (2.62a) is applied for the single distinguished variable $A_2 = j^3_{\vec{q}^*}/\sqrt{Nv_{th}^2}$. The reasoning was explained in Sec. 3.3.2 for the variable $A = j^1_{\vec{q}^*}/\sqrt{Nv_{th}^2}$. The fluctuating force follows from the conservation law for the momentum: $\mathcal{Q}_2 \mathcal{L}_1 j^3_{\vec{q}^*} = \mathcal{Q}_2 \mathcal{Q}_1 \mathcal{L} j^3_{\vec{q}^*} = \mathcal{Q}_1 \tau^{33}_{\vec{q}^*}(q/\mu)$. The new resolvent is formed with the reduced Liouvillian $\mathcal{Q}_2\mathcal{L}_1\mathcal{Q}_2 = \mathcal{Q}_2\mathcal{Q}_1\mathcal{L}\mathcal{Q}_1\mathcal{Q}_2$. It is the quantity $\mathcal{R}_{12}(z)$ introduced above. Remembering Eqs. (3.90d,e), one gets

$$q^2 K_q(z) = -\Omega_q^2/[z + q^2 K_{Lq}(z)]. \tag{3.95d}$$

Combination of this result with Eq. (3.95a) yields the desired formula (3.90a).

The derivation of the hydrodynamic-limit behaviour is based on the assumption of continuity of certain relaxation kernels, where the latter are functions of the pair of variables (q, z), Im $z \gtrless 0$. Let us use the preceding set of formulas for the density correlator in order to emphasize that the mentioned regularity property is an assumption built on physical insight. A necessary bit for the insight is the understanding of the coupling to the conserved fluctuations. For the purpose of simplicity, it shall be assumed in this paragraph that there are no other conserved quantities, which couple to $\rho_{\vec{q}^*}$ and $j^3_{\vec{q}^*}$. Since the reduced Liouvillian $\mathcal{L}_{12}$ treats the mentioned two variables as constants of motion, the kernel $K_{Lq}(z)$ is assumed to be continuous; there exists $\lim_{(q,z)\to(0,\pm i0)} K_{Lq}(z) = \pm i\nu_L$. As before, the non-generic case of a vanishing $\nu_L = K''_{Lq=0}(\omega = 0)$ shall be excluded by definition of the liquid state. From Eqs. (3.70a), (3.90a), one concludes that the behaviour of $K_q(z)$ for small frequencies and small wave numbers can be described by the model $\overline{\overline{K}}_q(z) = -c_{is}^2/[z \pm i\nu_L q^2]$, Im $z \gtrless 0$. The generator $\mathcal{L}_1$ describes a dynamics with a single conserved quantity, namely $j^3_{\vec{q}^*}$. As explained in Sec. 3.3.2 for variable $j^1_{\vec{q}^*}$, $K_q(t)$ describes diffusion. The zero-frequency stress spectrum ν_L is the diffusion constant for the problem. As a result, $\lim_{z\to\pm i0} K_q(z) = \lim_{z\to\pm i0} \overline{\overline{K}}_q(z) = \pm ic_{is}^2/(\nu_L q^2)$ holds for small q. The kernel $K_q(z)$ is singular. Formula (3.95a) is the analogue of Eq. (3.35a). The reduced dynamics, which enters as $\mathcal{R}_1(z)$ in Eq. (3.95b), treats $\rho_{\vec{q}^*}$ as constant of motion. Similarly, $\mathcal{R}_Q(z)$ treats $\rho^s_{\vec{q}}$ as constant of motion. But, contrary to the latter case, in the former case, the reduced Liouvillian $\mathcal{L}_1$ deals with a further conserved variable, namely with $j^3_{\vec{q}^*}$. Therefore, a regularity assumption for the kernel $K_q(z)$ is illegitimate.

Proceeding from Eq. (3.95a) to Eq. (3.95d) was a simple method to include a second conserved quantity. But, the simplicity was based on the property that the second conserved variable is related to the first one by the equation of motion: $\mathcal{L}\rho_{\vec{q}^*} = qj^3_{\vec{q}^*}$. The general procedure for incorporating a further conserved variable is more involved as is demonstrated in the remaining part of this section.

There holds the conservation of energy. Written as field equation, it connects the energy density $E(\vec{r}, t)$ with the energy-current density $\vec{J}(\vec{r}, t)$ by a continuity equation: $\partial_t E(\vec{r}, t) + \vec{\partial}\vec{J}(\vec{r}, t) = 0$. Via Fourier transformation, the fluctuations for the energy density $E_{\vec{q}}$ and for the energy-current density $\vec{J}_{\vec{q}}$ are introduced. The continuity equation gets a form analogues to Eq. (3.65a): $-i\partial_t E_{\vec{q}} = \mathcal{L}E_{\vec{q}} = qJ_{\vec{q}}$, where $J_{\vec{q}} = (\vec{q}\vec{J}_{\vec{q}})/q$ is the projection of $\vec{J}_{\vec{q}}$ on the direction of the wave vector $\vec{q}$. It is more convenient to introduce that part of the dynamical variable $E_{\vec{q}}$, which is perpendicular to the density fluctuations:

$$h_{\vec{q}} = E_{\vec{q}} - \rho_{\vec{q}}(\rho_{\vec{q}}|E_{\vec{q}})/[NS_q]. \tag{3.96a}$$

Combining the continuity equations for $E_{\vec{q}}$ and $\rho_{\vec{q}}$, one can write the third conservation law for the system as

$$-i\partial_t h_{\vec{q}} = \mathcal{L}h_{\vec{q}} = qj^h_{\vec{q}}. \tag{3.96b}$$

Here, $j^h_{\vec{q}} = (\vec{q}\vec{j}^h_{\vec{q}})/q$ with $\vec{j}^h_{\vec{q}} = \vec{J}_{\vec{q}} - \vec{j}_{\vec{q}}(\rho_{\vec{q}}|E_{\vec{q}})/[NS_q]$. The variable $h_{\vec{q}}$ is called the heat-density fluctuation and $j^h_{\vec{q}}$ is the heat-current-density fluctuation. The variable $E(\vec{r})$ is a real scalar field of even parity under time reversal and space inversion. The three components of $\vec{J}(\vec{r})$ form a real vector field of odd parity under time reversal and space inversion. These are the same symmetry properties, which hold for $\rho(\vec{r})$ and $\vec{j}(\vec{r})$, respectively. Consequently, $h_{\vec{q}}$ is a scalar fluctuation of even parity under conjugation as well as under time reversal and space inversion. Similarly, $\vec{j}^h_{\vec{q}}$ is a vector fluctuation of even parity under conjugation and odd parity under time reversal and space inversion. Because of isotropy, the scalar products $(E_{\vec{q}}|\rho_{\vec{q}})$ and $(E_{\vec{q}}|E_{\vec{q}})$ depend on the wave vector merely through the modulus. Because of Eq. (3.14c), the overlap $(E_{\vec{q}}|\rho_{\vec{q}})$ is real. The normalization of the heat fluctuations shall be denoted by

$$g_q = (h_{\vec{q}}|h_{\vec{q}}). \tag{3.96c}$$

Let us complement the set of variables $A_1 = \rho_{\vec{q}^*}/\sqrt{NS_q}$ and $A_2 = j^3_{\vec{q}^*}/\sqrt{Nv^2_{th}}$ by the third variable $A_3 = h_{\vec{q}^*}/\sqrt{g_q}$. The three variables A_1, A_2, A_3 are orthonormalized: $(A_m|A_n) = \delta_{m,n}$, $m, n = 1, 2, 3$. The projector $\mathcal{P}_3 = |A_3)(A_3| = |h_{\vec{q}^*})g_q^{-1}(h_{\vec{q}^*}|$ obeys $\mathcal{P}_3\mathcal{P}_2 = \mathcal{P}_3\mathcal{P}_1 = \mathcal{O}$. If $\mathcal{Q}_3 = 1 - \mathcal{P}_3$, the operator $\mathcal{Q}_{123} = \mathcal{Q}_1\mathcal{Q}_2\mathcal{Q}_3 = \mathcal{Q}_3\mathcal{Q}_{12} = 1 - \mathcal{P}_1 - \mathcal{P}_2 - \mathcal{P}_3$ projects on the space perpendicular to the three conserved variables A_1, A_2, and A_3. The reduced Liouvillian $\mathcal{L}_{123} = \mathcal{Q}_{123}\mathcal{L}\mathcal{Q}_{123} = \mathcal{Q}_3\mathcal{L}_{12}\mathcal{Q}_3$ generates a dynamics, which treats $\rho_{\vec{q}^*}$, $j^3_{\vec{q}^*}$, and $h_{\vec{q}^*}$ as constants of motion. Let us denote the corresponding resolvent by

$$\mathcal{R}_{\mathcal{Q}}(z) = [\mathcal{L}_{123} - z]^{-1}. \tag{3.97}$$

Hydrodynamic-limit results are based on formulas expressing all correlators of interest in terms of matrix elements of the kind $(X|\mathcal{R}_{\mathcal{Q}}(z)Y)$.

As a first step towards an elimination of the conserved variable A_3, the resolvent-matrix element $\phi_{hq}(z) = (h_{\vec{q}^*}|\mathcal{R}_{12}(z)h_{\vec{q}^*}) = g_q(A_3|\mathcal{R}_{12}(z)|A_3)$ shall be expressed in terms of a kernel $M_q(z)$. To proceed, the reasoning leading from Eq. (3.95b) to Eq. (3.95d) shall be copied. The role of $\mathcal{R}_1(z)$ and A_2 in the former equations is taken over by $\mathcal{R}_{12}(z)$ and A_3, respectively. Applying Eq. (2.62a) for the single distinguished variable A_3, one obtains the fraction representation $(A_3|\mathcal{R}_{12}(z)|A_3) = -1/[z + M_q(z)]$. The kernel is formed with the reduced generator $\mathcal{Q}_3\mathcal{L}_{12}\mathcal{Q}_3 = \mathcal{L}_{123}$ so that Eq. (2.61b) leads to $M_q(z) = (\dot{A}^{\mathcal{Q}}_3|\mathcal{R}_{\mathcal{Q}}(z)|\dot{A}^{\mathcal{Q}}_3)$.

According to Eq. (2.61a), the generalized force reads: $\dot{A}_3^{\mathcal{Q}} = i\mathcal{Q}_3\mathcal{L}_{12}A_3$. The orthogonality relation for the three distinguished variables implies $\mathcal{Q}_{12}A_3 = (1 - \mathcal{P}_1 - \mathcal{P}_2)A_3 = A_3$. Hence, $\dot{A}_3^{\mathcal{Q}} = i\mathcal{Q}_3\mathcal{Q}_1\mathcal{Q}_2\mathcal{L}A_3$. Variable $\mathcal{L}A_3$ has odd time-inversion parity. Therefore, it is perpendicular to A_1 and A_3. This implies $\mathcal{Q}_1\mathcal{Q}_3\mathcal{L}A_3 = \mathcal{L}A_3$. From Eq. (3.96b), one gets $\mathcal{L}A_3 = qj^h_{\vec{q}^*}/\sqrt{g_q}$ and the kernel reads $M_q(z) = q^2(\mathcal{Q}_2 j^h_{\vec{q}^*}|\mathcal{R}_{\mathcal{Q}}(z)\mathcal{Q}_2 j^h_{\vec{q}^*})/g_q$. Summarizing, one arrives at the fraction representation

$$\phi_{hq}(z)/g_q = -1/[z + q^2 k_{hq}(z)] \tag{3.98a}$$

with a kernel defined by

$$k_{hq}(z) = (\mathcal{Q}_2 j^h_{\vec{q}^*}|\mathcal{R}_{\mathcal{Q}}(z)\mathcal{Q}_2 j^h_{\vec{q}^*})/g_q. \tag{3.98b}$$

As a second step towards the elimination of the conserved variable $h_{\vec{q}}$, the kernel in Eq. (3.90e) shall be denoted as $K_{Lq}(z) = \psi_{X,X}(z)$. Here, the dynamical variable of concern is $X = \mathcal{Q}_1\tau^{33}_{\vec{q}^*}/[\mu v_{th}\sqrt{N}]$. The dynamics is generated by $\mathcal{L}_{12} : \psi_{X,Y}(z) = (X|[\mathcal{L}_{12} - z]^{-1}|Y)$. This function shall be reformulated according to Eq. (2.64). There is only the single distinguished variable $h_{\vec{q}^*}$. The reduced Liouvillian is $\mathcal{Q}_3\mathcal{L}_{12}\mathcal{Q}_3 = \mathcal{L}_{123}$ and, therefore, the resolvent in Eqs. (2.63c–e) is identical with the one defined in Eq. (3.97). The thermodynamic function from Eq. (2.63a) shall be denoted by

$$b_q = (\mathcal{Q}_1\tau^{33}_{\vec{q}^*}|h_{\vec{q}^*})/[\mu v_{th} g_q\sqrt{N}]. \tag{3.99a}$$

The kernel in Eq. (2.63e) shall be abbreviated as

$$k_{Lq}(z) = (\mathcal{Q}_3\mathcal{Q}_1\tau^{33}_{\vec{q}^*}|\mathcal{R}_{\mathcal{Q}}(z)\mathcal{Q}_3\mathcal{Q}_1\tau^{33}_{\vec{q}^*})/[\mu^2 v^2_{th} N]. \tag{3.99b}$$

The fluctuating force in Eq. (2.61a) reads $\dot{A}^{\mathcal{Q}} = i\mathcal{Q}_3\mathcal{L}_{12}h_{\vec{q}^*}$. This is $\sqrt{g_q}\dot{A}_3^{\mathcal{Q}}$, where $\dot{A}_3^{\mathcal{Q}}$ was determined in the preceding paragraph. Hence, $\dot{A}^{\mathcal{Q}} = iq\mathcal{Q}_2 j^h_{\vec{q}^*}$. The kernel in Eq. (2.63c) shall be denoted by $q \cdot B_q(z)$, i.e.,

$$B_q(z) = -(\mathcal{Q}_3\mathcal{Q}_1\tau^{33}_{\vec{q}^*}|\mathcal{R}_{\mathcal{Q}}(z)\mathcal{Q}_2 j^h_{\vec{q}^*})/[\mu v_{th} g_q\sqrt{N}]. \tag{3.99c}$$

Combining the results, one arrives at the desired expression

$$K_{Lq}(z) = k_{Lq}(z) + [b_q + qB_q(z)]\phi_{hq}(z)[b_q + qB_q(z^*)]^*. \tag{3.99d}$$

Since Laplace transforms vanish for Im $z \to \infty$, one concludes from Eq. (3.98a): $\lim_{\text{Im } z\to\infty}(-z)[b_q + qB_q(z)]\phi_{hq}(z)[b_q + qB_q(z^*)]^* = b_q^2 g_q$. Anticipating that $K_{Lq}(t)$ has a continuous bounded derivative for $|t| \neq 0$, one concludes from Eqs. (A.8c, 3.91a): $\lim_{\text{Im } z\to\infty}(-z)K_{Lq}(z) = \Delta_q^2/q^2$. Therefore

$$\lim_{\text{Im } z\to\infty}[-zk_{Lq}(z)] = (\Delta_q^2/q^2) - b_q^2 g_q. \tag{3.100a}$$

Equivalently, one gets for the non-negative initial value of correlator $k_{Lq}(t)$:

$$k_{Lq}(t = 0) = \Delta_q^2/q^2 - b_q^2 g_q. \tag{3.100b}$$

Proceeding from Eqs. (3.90a,d) to Eq. (3.99d), also the conserved variable $h_{\vec{q}^*}$ has been accounted for.

3.3.6 *Sound waves and heat diffusion*

In this section, the hydrodynamics-limit results will be derived for the correlators, which can be formed with the fluctuations of the density and the longitudinal currents. For small wave numbers and small frequencies, these correlators can be expressed in terms of two transport coefficients, namely the longitudinal viscosity and the heat-diffusion constant. Density fluctuations vary in space and time as sound waves. If there is no coupling between stress and heat, the sound speed is determined by the isothermal compressibility. Coupling to heat fluctuations leads to an increase of the sound speed to the one determined by the adiabatic compressibility. In addition, the density fluctuations have a contribution, which varies in space and time as described by a diffusion equation.

As done before, the absence of long-range-order effects is assumed for the amorphous state. All overlaps formed with two field fluctuations, say $X_{\vec{q}}$ and $Y_{\vec{q}}$, are anticipated to vary smoothly as function of wave vector $\vec{q}$. The reduced dynamics $\mathcal{U}_Q(t) = \exp[i\mathcal{L}_{123}t]$ treats the fluctuations of density, momentum, and energy as constants of motion. Therefore, matrix elements formed with the corresponding resolvent $(X_{\vec{q}}|\mathcal{R}_Q(z)Y_{\vec{q}})$, are assumed to be uniformly continuous in $\vec{q}$ for small q and arbitrary z. They are also assumed to be continuous as function of the variable pair (q, z) for $(q, z) \to (0, \pm i0)$, Im $z \gtrless 0$. Restricting the wave numbers by condition (3.70a), wave-number-dependent functions are replaced by their leading-order Taylor-expansion terms for small q. For example, Eq. (3.70a) is used to replace Ω_q by ω_{isq}. Here,

$$\omega_{isq} = c_{is}q \tag{3.101a}$$

is the linear sound dispersion formed with the isothermal sound speed c_{is}. From Eqs. (3.91b), (3.100a), one infers that the non-negative thermodynamic function $b_q^2 g_q$ has the meaning of a velocity square. Let us express its low-q limit in terms of a parameter called γ:

$$b_q^2 g_q = c_{is}^2(\gamma - 1), \quad \gamma \geq 1. \tag{3.101b}$$

Let us also abbreviate the zero-q limits of the correlators defined in Eqs. (3.98b), (3.99b), and (3.99c) as follows:

$$\lim_{q\to 0} k_{hq}(z) = k_h(z), \quad \lim_{q\to 0} k_{Lq}(z) = k_L(z), \tag{3.102a}$$

and $B(z) = \lim_{q\to 0} B_q(z)$. These functions have the general properties of Laplace transforms of correlation functions. In particular, $k_L(z)$ and $k_h(z)$ are positive analytic and obey the general symmetries formulated in Eqs. (3.20a–c). Combining Eqs. (3.91c), (3.100b), and (3.101b), one finds

$$k_L(t = 0) = c_L^2 - \gamma c_{is}^2. \tag{3.102b}$$

The above formulated assumptions reflect physical insight in the properties of the system under study. Application of these assumptions transforms the general equations of the preceding section into specific results for a description of small-wave-number correlations of liquids. As done in Sec. 3.3.3, the various quantities referring to this small-q description shall be indicated by a bar over the symbols.

For small wave numbers, one can neglect $qB_q(z)$ in comparison to b_q. Therefore, the results of the preceding paragraph motivate the combination of Eqs. (3.98a, 3.99d) to a model for the longitudinal-stress correlations

$$\overline{K}_{Lq}(z) = k_L(z) - c_{is}^2(\gamma - 1)/[z + q^2 k_h(z)]. \tag{3.103a}$$

Substituting the preceding results into Eqs. (3.90a–c), one gets model correlators:

$$\overline{\phi}_q(z) = -\left[z + q^2\overline{K}_{Lq}(z)\right]/\left\{z\left[z + q^2\overline{K}_{Lq}(z)\right] - \omega_{isq}^2\right\}, \tag{3.103b}$$

$$\overline{\phi}_{Lq}(z) = -z/\left\{z\left[z + q^2\overline{K}_{Lq}(z)\right] - \omega_{isq}^2\right\}, \tag{3.103c}$$

$$\overline{\phi}_{j_{\vec q},\rho_{\vec q}}(z) = \overline{\phi}_{\rho_{\vec q},j_{\vec q}}(z) = (Nv_{th}^2 q)/\left\{z\left[z + \overline{K}_{Lq}(z)\right] - \omega_{isq}^2\right\}. \tag{3.103d}$$

From Eqs. (3.102b, 3.103a) one obtains $\lim_{\mathrm{Im}\, z\to\infty}[-z\overline{K}_{Lq}(z)] = \overline{K}_{Lq}(t=0)$ in agreement with the small-q limit (3.91c) of the initial value for the complete kernel:

$$\overline{K}_{Lq}(t=0) = c_L^2 - c_{is}^2. \tag{3.103e}$$

Function $k_h(z)$ is an autocorrelation function in the frequency domain exhibiting the general symmetries of Eqs. (3.20a,b). Therefore, $-1/[z + q^2 k_h(z)]$ is a positive-analytic function, which exhibits the mentioned symmetries, and which can be written as Laplace transform of a positive-definite continuous function. This was explained in connection with an analogous problem for Eq. (3.80). The function approximates the correlator $\phi_{hq}(z)/g_q$ for small wave numbers. Similarly, $k_L(z)$ is an autocorrelator with the proper symmetries, which approximates $k_{Lq}(z)$ for small q. Consequently, $\overline{K}_{Lq}(z)$ is a function with all the properties discussed for $\tilde{K}_{Lq}(z)$ in connection with Eqs. (3.93a)–(3.94b). One concludes that the functions defined in Eqs. (3.103b–d) have the correct analytical properties of correlators, the correct symmetries, and the correct relations implied by the continuity equation. These model functions define a generalized-hydrodynamics description of the density and longitudinal-current dynamics. The terminology and the motivation for the approach are the same as explained in Sec. 3.3.3 for the transverse-current dynamics.

To proceed to some specific result, one needs expression for the two correlators defining $\overline{K}_{Lq}(z)$ via Eq. (3.103a). Such expressions, say $\tilde{k}_L(z)$ and $\tilde{k}_h(z)$, define a model $\tilde{K}_{Lq}(z)$. If $\tilde{k}_L(z)$ and $\tilde{k}_h(z)$ have the same general properties and symmetries as mentioned above, the analogues statement holds for the model $\tilde{K}_{Lq}(z)$ of the kernel. Then, one can appeal to the discussion of Eqs. (3.92a)–(3.93d) and conclude on the properties of the models for $\tilde{\phi}_q(z)$ and the other correlators. Thereby, one obtains special models for the generalized-hydrodynamics behaviour.

The simplest model for the density dynamics is obtained by restricting the frequencies to such small values that the kernels can be replaced by their zero-frequency limits. This leads to the hydrodynamic-limit description as discussed in Sec. 3.2.2 for the tagged-particle density correlators and in Sec. 3.3.2 for the transverse-current dynamics. The zero-frequency spectra shall be denoted as follows: $k''_L(\omega = 0) = \nu_L$ and $k''_h(\omega = 0) = \gamma\chi$. As before, the non-generic possibility $\nu_L = 0$ or $\chi = 0$ shall be excluded by definition of the liquid state. The transport coefficient ν_L is the counterpart of ν from Eq. (3.77a), and it is called the longitudinal kinematic viscosity. The transport coefficient χ is called the thermal diffusivity or thermal-diffusion constant. As done before, the various functions of the hydrodynamic-limit model shall be indicated by a double bar above the symbols. The formulas are defined by:

$$\overline{\overline{k}}_L(z) = \pm i\nu_L, \quad \overline{\overline{k}}_h(z) = \pm i\gamma\chi, \quad \mathrm{Im}\, z \gtrless 0. \tag{3.104a}$$

This leads to:

$$\overline{\overline{K}}_{Lq}(z) = \pm i\nu_L - c_{is}^2(\gamma - 1)/[z \pm i\gamma\chi q^2]. \tag{3.104b}$$

Specializing $\overline{K}_{Lq}(z)$ to $\overline{\overline{K}}_{Lq}(z)$, Eqs. (3.103b–d) formulate the correlators in the hydrodynamic limit. These formulas describe the correlation functions of the fluctuations for frequencies z and wave numbers q restricted to small values as discussed in connection with the conditions (3.38b). Let us remember from the discussion of Secs. (3.2.3) and (3.3.3) that the formulas yield correct results if and only if the time scale for the dynamics of the density and current fluctuations is well separated from the time scale t^* of the dynamics generated by the reduced Liouvillian $\mathcal{L}_{123}$. In the limit of wave number q tending to zero, the separation property is indeed fulfilled. This result will be a byproduct of the following discussion. The remainder of this section is devoted to the analysis of the density spectra $\overline{\overline{\phi}}''_q(\omega)$ resulting from $\overline{\overline{K}}_{Lq}(z)$. The other spectra are obtained as implication of the particle-number conservation law, e.g., $\overline{\overline{\phi}}''_{Lq}(\omega) = [\omega^2/\omega_{isq}^2]\overline{\overline{\phi}}''_q(\omega)$.

Let us start the discussion of the hydrodynamic-limit results with the special case $\gamma = 1$. It deals with correlations for small wave numbers and small frequencies for a state, where couplings of stress fluctuations and heat fluctuations can be ignored; Eqs. (3.99a, 3.101b) imply $(\mathcal{Q}_3\tau_{\vec{q}^*}^{33}|h_{\vec{q}^*}) = 0$. The kernel $\overline{\overline{K}}_{Lq}(z)$ reduces to $\pm i\nu_L$ For Im $z > 0$, the density correlator gets the form:

$$\overline{\overline{\phi}}_q(z) = -1/\Big\{z - \omega_{isq}^2/\big[z + i\nu_L q^2\big]\Big\}, \quad \gamma = 1. \tag{3.105a}$$

This function is the correlator of a harmonic oscillator discussed in Appendix A.4. There, the frequency Ω and the damping constant Γ have to be identified with the q-dependent quantities ω_{isq} and $\nu_L q^2$, respectively. According to Eq. (A.38b), the correlator in the time domain is the solution of the equation

$$\partial_t^2\overline{\overline{\phi}}_q(t) + \nu_L q^2\partial_t\overline{\overline{\phi}}_q(t) + c_{is}^2 q^2\overline{\overline{\phi}}_q(t) = 0, \quad t \geq 0, \tag{3.105b}$$

for the initial conditions $\overline{\overline{\phi}}_q(t = 0) = 1$ and $\partial_t \overline{\overline{\phi}}_q(t = 0) = 0$. Fourier transformation from the wave-vector space to the position space casts the equation in an equivalent field equation for the density-correlation function $\overline{\overline{\phi}}(\vec{r}, t)$,

$$\partial_t^2 \overline{\overline{\phi}}(\vec{r}, t) - \nu_L(\vec{\partial}\vec{\partial})\partial_t \phi(\vec{r}, t) - c_{is}^2(\vec{\partial}\vec{\partial})\phi(\vec{r}, t) = 0, \tag{3.105c}$$

which has to be solved with the initial conditions $\overline{\overline{\phi}}(\vec{r}, t = 0) = \delta(\vec{r})$ and $\partial_t \overline{\overline{\phi}}(\vec{r}, t = 0) = 0$. With increasing time, density correlations propagate in space as described by a wave equation with a damping term. The speed c_{is} of the waves is determined by the isothermal compressibility. The friction is specified by the transport coefficient ν_L. This finding is equivalent to the linearized Navier–Stokes equations for the longitudinal fields of a fluid (Boon and Yip 1980; Hansen and McDonald 1986). The waves described by Eq. (3.105c) are called sound waves or, more explicitly, hydrodynamic sound waves.

The parameter $r_q = \nu_L q^2/\omega_{isq} = (\nu_L/c_{is})q$ characterizes the qualitative behaviour of the density correlator. Let us focus on $r_q < 2$. Introducing the positive functions of wave number

$$\gamma_q = \tfrac{1}{2}\nu_L q^2, \qquad \omega_q = \sqrt{\omega_{isq}^2 - \gamma_q^2}, \tag{3.106a}$$

one gets for the so-called hydrodynamic poles of the density correlator:

$$z_\pm = \pm\omega_q - i\gamma_q. \tag{3.106b}$$

According to Eqs. (A.40, A.41), these numbers specify the analytic continuation of $\overline{\overline{\phi}}_q(z)$ from the upper half plane of complex frequencies onto the complete plane. This continuation is the meromorphic function $\overline{\overline{\phi}}_q(z) = -[A_+/(z - z_+)] - [A_-/(z - z_-)]$, where $A_\pm = [1 \pm i\gamma_q/\omega_q]/2$. The condition $r_q < 2$ is equivalent to the restriction of the wave numbers to such small values that

$$q < 2(c_{is}/\nu_L). \tag{3.106c}$$

The spectrum can be written as sum of two contributions due to the two partial fractions $A_\pm/(z - z_\pm)$:

$$\overline{\overline{\phi}}_q{}''(\omega) = \phi_q^+(\omega) + \phi_q^-(\omega), \tag{3.107a}$$

$$\phi_q^\pm(\omega) = \frac{1}{2}\Big\{a_{1q} + a_{2q}\big[(\omega_q \mp \omega)/\omega_q\big]\Big\}\Big\{\gamma_q/\big[(\omega_q \mp \omega)^2 + \gamma_q^2\big]\Big\}. \tag{3.107b}$$

Here, two coefficients a_{1q} and a_{2q} have been introduced for later convenience. For the present case, there holds $a_{1q} = a_{2q} = 1$.

The existence of sound waves manifests itself by a pair of symmetrically placed peaks of the spectrum $\overline{\overline{\phi}}_q{}''(\omega)$. For $|a_{2q}(\omega_{q+} \mp \omega)/\omega_q| \leq |a_{1q}|\epsilon$, the peaks are Lorentzians within a relative error ϵ. The peak position is $\pm\,\omega_q$ and the half width at half height is γ_q. Considering the deviation of the frequency from ω_q in units of γ_q by writing $(\omega_q \mp \omega)/\gamma_q = x$, one gets $\phi_q^{\pm}(\omega) = \frac{1}{2} a_{1q} \frac{1}{\gamma_q}(1+O(\epsilon))/[1+x^2]$. Here, $O(\epsilon) \leq \epsilon |a_{2q}/a_{1q}|$ if $x \leq \epsilon\omega_q/\gamma_q = \epsilon[2c_{is}/\nu_L]/q$. Hence, the percentage of the peaks, which can be described by a Lorentzian within a relative error ϵ, increases for decreasing q. The described pattern is called the Brillouin doublet.

The critical wave vector q_c, which separates the wave-vector regime with a doublet spectrum, $q < q_c$, from that with a central peak, $q \geq q_c$, is determined from the condition $r_q = \sqrt{2}$, i.e., $q_c = \sqrt{2}c_{is}/\nu_L$. The doublet spectrum of the hydrodynamic sound peaks appears for q tending to zero. Opposed to that, the doublet for the high-frequency transverse sound occurs for q above some critical value. The difference of the two types of spectra is caused by the different q dependence of the damping function. For the transverse current $A_1 \propto \vec{j}_{T\vec{q}}$, the fluctuating force is given by a stress component. This exhibits a q-insensitive spectrum, which can be replaced by its $q = 0$ limit. As a result, there appears a q-independent damping term $\Gamma = 1/\tau_T^M$ in Eq. (3.88). However, for the density fluctuation $A_1 \propto \rho_{\vec{q}}$, the fluctuating force is given by the longitudinal current component $A_2 \propto \vec{j}_{L\vec{q}}$. Since this field fluctuation is conserved, its correlations for q tending to zero depend strongly on wave numbers. There are no spectra for $q = 0$, since homogeneous translations do not require forces. Fluctuations for $q \neq 0$ lead to the damping term $\Gamma = \nu_L q^2$ in Eq. (3.105a), which decreases with q proportional to q^2. Decreasing q, the position of the high-frequency-sound peak for transverse spectra decreases proportional to q, but the width remains fixed. This is shown in Fig. 3.6 for the change of the dimensionless q from 3 to 2. Decreasing q, the position of the hydrodynamic-sound peak for the density spectrum also decreases proportional to q. But, simultaneously, the peak width decreases proportional to q^2. This is shown in Fig. 3.9(a) for a change of the dimensionless q from 2 to 1.

For $\gamma > 1$, the hydrodynamic-limit expression $\overline{\overline{\phi}}_q(z)$ reflects the coupling of heat fluctuations to stress fluctuations. Again, the result for Im $z > 0$ shall be continued as a meromorphic function onto the complete frequency plane. It is convenient to introduce the rescaled frequency

$$\zeta = z/q, \tag{3.108a}$$

and to write the correlator as a fraction of polynomials:

$$N(z) = (\zeta + i\nu_L q)(\zeta + i\gamma\chi q) - c_{is}^2(\gamma - 1), \tag{3.108b}$$

$$D(z) = \zeta(\zeta + i\nu_L q)(\zeta + i\gamma\chi q) - c_{is}^2\gamma(\zeta + i\chi q), \tag{3.108c}$$

$$\overline{\overline{\phi}}_q(z) = (-1/q)N(z)/D(z). \tag{3.108d}$$

There hold the symmetries

$$N(-z^*)^* = N(z), \; -D(-z^*)^* = D(z). \tag{3.108e}$$

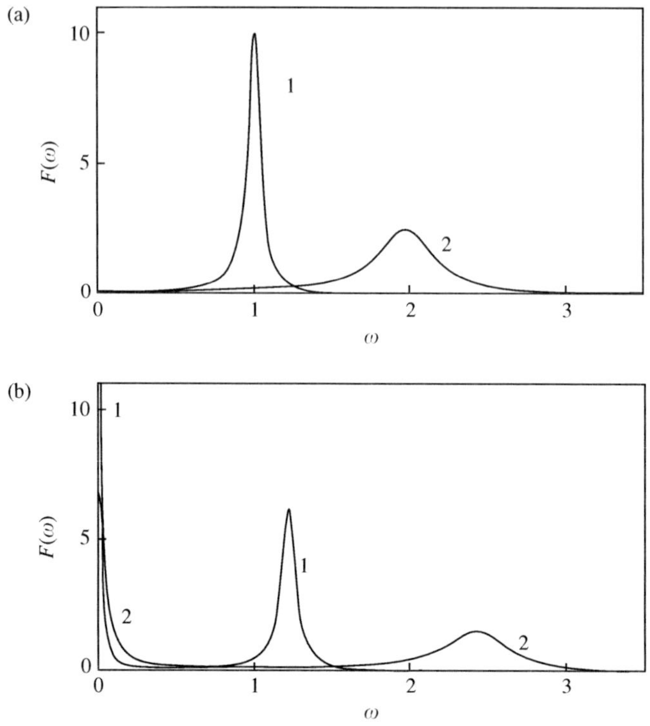

FIG. 3.9. Density-fluctuation spectra $\overline{\overline{\phi}}''_q(\omega) = F(\omega)$ as a function of the frequency ω for the hydrodynamic-limit description explained by Eqs. (3.104a,b). The units of time and length are chosen such that the isothermal sound speed has the value $c_{is} = 1.0$ and that the longitudinal kinematic viscosity has the value $\nu_L = 0.1$. Curves with label 1 and 2 refer to the wave number $q = 1$ and $q = 2$, respectively. Panel (a) exhibits the result (3.105a) for the adiabaticity coefficient $\gamma = 1$. Panel (b) shows the results for $\gamma = 1.5$ and for the thermal diffusivity $\chi = \nu_L/8$.

The analytical continuation exhibits the same symmetry, which is formulated in Eq. (3.20c) for the correlator.

The denominator polynomial can be factorized as $D(z) = [\zeta - \zeta_-(q)][\zeta - \zeta_0(q)][\zeta - \zeta_+(q)]$. Here $\zeta_{-,0,+}(q)$ denote the three roots of $D(z)$ as function of q. One can choose $\zeta_\mp(q = 0) = \mp\sqrt{\gamma}c_{is}, \zeta_0(q = 0) = 0$. The non-degenerate roots of a polynomial can be chosen as analytic functions of the polynomial's coefficients. Hence, there is a critical wave number $q_c > 0$ with the following property. For $q < q_c$, the $\zeta_{-,0,+}(q)$ are pairwise different analytic functions of q. If $D(\tilde{z}) = 0$, one gets $D(-\tilde{z}^*) = 0$ because of Eq. (3.108e). Thus, $\zeta_\pm(q)^* = -\zeta_\mp(q)$ and $\zeta_0(q)^* = -\zeta_0(q)$. One concludes that there are three hydrodynamic poles $z_0 = q\zeta_0(q)$ and $z_\mp = q\zeta_\mp(q)$. The $z_\mp$ can be written as noted in Eq. (3.106b), and

$$z_0 = -i\theta_q. \tag{3.109}$$

The quantities ω_q, γ_q and θ_q are real non-negative analytic functions of q, which exhibit the small-q asymptotes: $\omega_q/q = \sqrt{\gamma}c_{is} + O(q), \gamma_q/q = O(q), \theta_q/q = O(q)$. Hence, for

$$q < q_c, \tag{3.110a}$$

there is the partial-fraction representation $\overline{\overline{\phi}}_q(z) = -[A_0/(z - z_0)] - [A_+/(z - z_+)] - [A_-/(z - z_-)]$. The symmetry equations (3.108e) imply $A_0 = A_0^*$ and $A_+ = A_-^*$. The residues are $A_0 = -q^2N(-i\theta_q)/[\omega_q^2 + (\theta_q - \gamma_q)^2]$ and $A_\pm = \pm q^2N(z_\pm)/[2\omega_q(z_\pm - z_0)]$. Each of the three fractions yields a contribution $\phi_q^\alpha(\omega) = -\text{ Im } [A_\alpha/(z - z_\alpha)]$, $\alpha = -, 0, +$ to the fluctuation spectrum:

$$\overline{\overline{\phi}}_q''(\omega) = \phi_q^-(\omega) + \phi_q^0(\omega) + \phi_q^+(\omega). \tag{3.110b}$$

The second contribution is the spectrum of a simple relaxation process. It contributes the fraction $a_{0q} = A_0$ to the total spectrum:

$$\phi_q^0(\omega) = a_{0q} \cdot \theta_q/[\omega^2 + \theta_q^2]. \tag{3.110c}$$

Writing $A_\pm = [a_{1q} \pm ia_{2q}\gamma_q/\omega_q]/2$, the spectral contributions $\phi_q^\pm(\omega)$ are given by Eq. (3.107b).

One can determine the asymptotic expansion of the zeros $\zeta_{-,0,+}(q)$ in a power series of q, and this yields the corresponding expansions of A_0 and $A_\pm$. There holds:

$$\theta_q = \chi q^2 + O(q^4), \tag{3.111a}$$

$$\omega_q = \sqrt{\gamma}c_{is}q + O(q^3), \tag{3.111b}$$

$$\gamma_q = \tfrac{1}{2}[\nu_L + (\gamma - 1)\chi]q^2 + O(q^4), \tag{3.111c}$$

$$a_{0q} = (\gamma - 1)/\gamma + O(q^2), \tag{3.111d}$$

For sufficiently small q, the coefficients a_{0q}, a_{1q} and a_{2q} are positive. For q tending to zero, function ω_q exhibits a linear dependence on q and γ_q decreases proportional to q^2. From the discussion of Eq. (3.107b) one knows that the contribution $[\phi_q^+(\omega) + \phi_q^-(\omega)]$ describes a Brillouin doublet due to sound excitations. Equations (3.110c), (3.111a) describe a diffusion process as discussed in Sec. 3.2.2. The qualitative change of the density-fluctuation spectrum due to the coupling to heat-fluctuations is the appearance of a central peak due to heat diffusion. This part of the spectrum is called the Rayleigh peak. Figure 3.9(b) exhibits examples for spectra described by Eqs. (3.108a–d).

The part of the total spectrum, which is contributed by the Rayleigh peak, is given by $a_{0q} = \int_{-\infty}^{\infty} \phi_q^0(\omega)d\omega / \int_{-\infty}^{\infty} \overline{\overline{\phi}}_q''(\omega)d\omega$. The remaining part is exhausted by the pair of Brillouin peaks $1 - a_{0q} = \int_{-\infty}^{\infty} [\phi_q^{+\prime\prime}(\omega) + \phi_q^{-\prime\prime}(\omega)]d\omega / \int_{-\infty}^{\infty} \overline{\overline{\phi}}_q''(\omega)d\omega = a_{1q}$. The ratio of the former contribution to the latter is $R_{\text{LP}}(q) = a_{0q}/(1 - a_{0q})$. From Eq. (3.111d), one gets

$$\lim_{q \to 0} R_{\text{LP}}(q) = \gamma - 1. \tag{3.111e}$$

This number is called the Landau–Placzek ratio. Due to the coupling of density-fluctuations to heat-fluctuations, the intensity of the Brillouin peaks is reduced. For $\gamma = 1$, a Brillouin line contributes (1/2) to the total spectral integrals. For $\gamma > 1$, the contribution is $\lim_{q\to 0} a_{1q}/2 = 1/(2\gamma)$.

To understand two further $\gamma > 1$ effects on the sound-wave spectra, the limit $q \to 0$ shall be considered for $\zeta = z/q$ near $c_{is}\sqrt{\gamma}$. This means that $\zeta_0(q)/\zeta = z_0(q)/z$ tends to zero. In this case, the sound-wave peak is arbitrarily well separated from the heat-fluctuation spectrum. Expanding the normalized heat-fluctuation correlator in leading and next-to-leading order in the small parameter $z_0(q)/q$, $-1/[z \pm i\gamma\chi q^2] = -[1/z] \pm i\gamma\chi[q^2/z^2] + \cdots = -[1/z] \pm i\gamma\chi q^2/\omega_q^2 + \cdots$, one gets from Eq. (3.104b): $z\overline{\overline{K}}_{Lq} = -c_{is}^2(\gamma - 1) \pm iz[\nu_L + c_{is}^2(\gamma-1)\gamma\chi q^2/\omega_q^2] + \cdots$ The denominator in Eqs. (3.103b–c) gets the form $\{z^2 + q^2 z\overline{\overline{K}}_{Lq}(z) - \omega_{isq}^2\} = \{[z^2 - q^2 c_{is}^2\gamma] \pm iz[\nu_L + (\gamma - 1)\chi(\gamma c_{is}^2 q^2/\omega_q^2] + \cdots\}$. The roots of the curly bracket are the hydrodynamic poles for the sound propagation. The leading-order term yields the increase of the sound speed from c_{is} to $\sqrt{\gamma}c_{is}$ as noted in Eq. (3.111b). The next-to-leading-order term yields the increase of the sound-peak-width parameter from $\frac{1}{2}\nu_L$ to $\frac{1}{2}[\nu_L + (\gamma - 1)\chi]$ as noted in Eq. (3.111c). In leading order, the heat-fluctuation correlator reads $\phi_{hq}(z)/\phi_{hq}(t = 0) = -1/z$, which is equivalent to $\phi_{hq}(t) = \phi_{hq}(t = 0)$. In this approximation, density fluctuations occur so fast that no heat-diffusion can occur. Such compressions are called adiabatic ones. The adiabatic sound speed c_{ad} is related to the isothermal one by $c_{ad}^2 = \gamma c_{is}^2$. Working out the equilibrium averages in Eq. (3.101b) one finds that $\gamma = C_p/C_v$. This is the adiabaticity coefficient, i.e., the ratio of the specific heat at constant pressure, C_p, and the one at constant volume C_v (Hansen and McDonald 1986). There holds the analogue of Eq. (3.70b), where the isothermal compressibility is replaced by the adiabatic one $\kappa_{ad} = \kappa_{is}/\gamma$. If $\gamma \neq 1$, a compression creates a temperature change. Hence, there is a heat fluctuation, which diffuses in space. Via thermal expansion, this creates a diffusing density fluctuation, which appears as the Rayleigh peak. The diffusion causes a relaxation of the function $\phi_{hq}(t)/\phi_{hq}(t = 0)$. The leading correction to the adiabatic behaviour explains the increase of the sound damping above the value caused by the viscosity mechanism. Comparison of Fig. 3.9(b) with 3.9(a) exhibits the $\gamma > 1$ effects. The value for γ is chosen smaller and χ is chosen bigger than typical for simple liquids like argon, in order to exhibit all features on a common scale for the frequency axis.

The frequency scales for the hydrodynamic-limit results are given by ω_q and θ_q. If t^* denotes the time scale for the dynamics described by $\mathcal{U}_Q(t)$, one gets $\lim_{q\to 0} t^*\theta_q = \lim_{q\to 0} t^*\omega_q = 0$. Thus, the time-scale-separation property is valid in the limit of vanishing wave numbers. The hydrodynamic-limit results are asymptotic formulas for the correlators for small q and small z.

3.3.7 *Visco-elastic features and glassy-dynamics precursors of the density-fluctuation correlators*

In this section, a visco-elasticity model will be considered for the longitudinal stress-fluctuation correlator. This leads to a theory of high-frequency sound. In addition, the stress dynamics causes a central weakly-wave-number-dependent relaxation peak for the density-fluctuation spectra.

Maxwell's assumption of a simple-relaxation law for the stress correlations implies a two-parameter formula for the kernel $k_L(t)$ in Eq. (3.102a). It shall be written as $k_L^M(t) = Cc_{is}^2 \exp[-|t|/\tau_L^M]$. The formula for the correlator in the frequency domain, $k_L^M(z) = -Cc_{is}^2/[z \pm i/\tau_L^M]$, Im $z \gtrless 0$, is the analogue of Eq. (3.87a) for the transverse-stress dynamics. Langevin's theory provides a justification of this formula within the microscopic theory of liquids. This holds, if the time-scale septation, $\tau_L^M >> t^*$, is valid in analogy to what is discussed in connection with Eq. (3.87d). A similar two-parameter formula can be motivated for the heat-fluctuation correlator $k_h(z)$. But, for the reason of mathematical simplicity, this shall not be done here; the hydrodynamic-limit result $\overline{\overline{k}}_h(z)$ shall be used instead. Substituting these formulas into Eq. (3.103a), one gets the simplest model for a generalized-hydrodynamics description of $\overline{K}_{Lq}(z)$. It shall be denoted by $K_{Lq}^M(z)$, i.e.

$$K_{Lq}^M(z)/c_{is}^2 = -\{C/[z + i/\tau_L^M]\} - \{(\gamma - 1)/[z + i\gamma\chi q^2]\}. \tag{3.112a}$$

The formula holds for Im $z > 0$; it provides also the analytical continuation of the kernel onto the complete frequency plane. The continuation exhibits the symmetry from Eq. (3.20c). According to Eq. (3.102b), one gets for the non-negative strength parameter the expression

$$C = (c_L^2/c_{is}^2) - \gamma. \tag{3.112b}$$

The hydrodynamic-limit result (3.104b) is reproduced for such small frequencies that $|z\tau_L^M| \ll 1$. The value for the transport coefficient ν_L is given by a formula analogous to Eq. (3.87c):

$$\nu_L = c_{is}^2 C \tau_L^M. \tag{3.112c}$$

Substitution of Eq. (3.112a) in Eqs. (3.103b–d) yields a model for visco-elastic effects, which was introduced and analysed by Mountain (1966).

Let us consider first the case $\gamma = 1$. The Eq. (3.112a) for the kernel has the same form as noted in Eq. (3.104b), if one translates the parameters in the

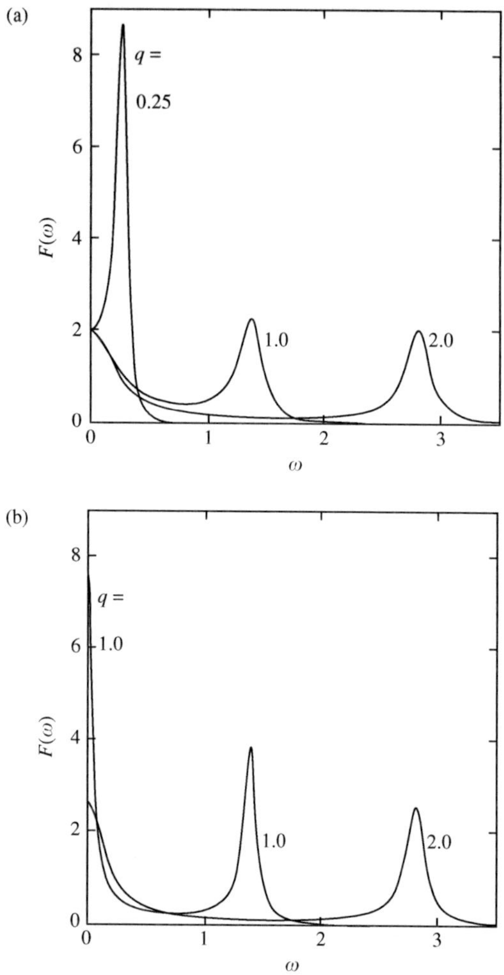

FIG. 3.10. Normalized density-fluctuation spectra $\bar{\phi}_q''(\omega) = F(\omega)$ for various wave numbers q as a function of frequency ω calculated within Mountain's visco-elasticity model (3.112a) for the longitudinal-stress-relaxation correlator $K_{Lq}^M(z)$. The units of time and length are chosen such that the relaxation time and isothermal sound speed have the values $\tau_L^M = 2.0$ and $c_{is} = 1.0$, respectively. The high-frequency-sound speed is chosen as $c_L = \sqrt{2}$ and the thermal diffusivity as $\chi = 0.05$. Panels (a) and (b) exhibit the results for the adiabaticity ratio $\gamma = 1.0$ and $\gamma = 1.5$, respectively.

latter as follows: $\nu_L \to 0, (\gamma - 1) \to C$, and $\gamma\chi q^2 \to 1/\tau_L^M$. Therefore, the implications as formulated in Eqs. (3.108a)–(3.111e) can be translated into the results, which are of interest here. The exception is the inequality (3.110a). It has to be reconsidered, since the q-dependent quantity $\gamma\chi q^2$ is translated to a q-independent one. The case of interest is that there is a doublet of peaks and a central peak. The spectrum is described by Eq. (3.110b) with properly translated parameters. From Eq. (3.111b), one gets $\omega_q = \sqrt{1+C}c_{is}q + O(q^3)$. From Eq. (3.112b), one obtains $(1 + C)c_{is}^2 = c_L^2$. Translation of Eq. (3.111c) yields the damping parameter. One gets sound-wave excitations specified by the speed $c_L > c_{is}$ and a damping, which is independent of the wave number. The small-q asymptotes read

$$\omega_q = c_L q, \quad \gamma_q = \tfrac{1}{2}[C/(1+C)]/\tau_L^M \tag{3.113}$$

Decreasing q, ω_q decreases proportional to q. The width of the peak remains unchanged. This scenario is demonstrated in Fig. 3.10(a) for a change of the dimensionless wave number from 2 to 1; it is the analogue to what is demonstrated in Fig. 3.6 for the transverse-current spectra for a change of q from 3 to 2. There is a critical wave number $q_c = [C/(1+C)]/(2\tau_L^M c_L)$ so that the preceding two formulas for high-frequency sound hold for $q \gg q_c$. For q decreasing to q_c, there occurs the crossover to the hydrodynamic-limit behaviour. Decreasing q further, the peak width shrinks proportional to q^2. There appears the Brillouin doublet of hydrodynamic-sound peaks for $q \ll q_c$, as discussed in the upper panel of Fig. 3.9. In Fig. 3.10(a), such a Brillouin spectrum is exhibited for the dimensionless wave number $q = 0.25$.

The high-frequency-sound spectrum does not exhaust the complete spectrum, but only the fraction a_{1q}, Eq. (3.107b). The remaining spectrum is described by Eq. (3.110c) with properly translated parameters. This spectrum of a simple relaxation process is a qualitatively new feature of visco-elasticity, which is called Mountain's peak. Substituting $K_{Lq}^M(z) = -C/[z + i\tau_L^M]$ into Eq. (3.103b) and neglecting z in comparison to $K_{Lq}^M(z)$, one gets:

$$\phi_q^0(\omega) = f(1/\tau)/[\omega^2 + (1/\tau)^2]. \tag{3.114a}$$

The relaxation rate $1/\tau$ is the smaller the larger C:

$$1/\tau = [1/\tau_L^M]/(1+C). \tag{3.114b}$$

The strength factor $f = a_{0q} = 1 - a_{01}$ increases with C towards unity:

$$f = C/(1+C). \tag{3.114c}$$

The Landau–Placzek ratio a_{0q}/a_{1q} is given by

$$R_{\mathrm{LP}} = f/(1-f) = C. \tag{3.114d}$$

The condition $q \gg q_c$ implies that the high-frequency sound peaks are well separated from Mountain's peak. As a result, the central peak depends on q only weakly, as is demonstrated in Fig. 3.10(a). This q-insensitivity is the signature discriminating Mountain's peak from Rayleigh's peak.

For $\gamma = 1$, Eqs. (3.103a,c) yield $\{z^2 - q^2[-zk_L(z) + c_{is}^2]\}\overline{\phi}_{Lq}(z) = (-z)$. Fourier transformation into the position space leads to a field equation for the Laplace transform of the longitudinal-current-density correlator. It is the analogue of Eq. (3.84b) for the transverse quantities:

$$(-z^2)\overline{\phi}_L(\vec{r}, z) - [G_L(z)/(\mu\rho)]\vec{\partial}\vec{\partial}\,\overline{\phi}_L(\vec{r}, z) = z\delta(\vec{r}). \tag{3.115a}$$

Here, the abbreviation is introduced

$$G_L(z) = (\mu\rho)[c_{is}^2 - zk_L(z)]. \tag{3.115b}$$

Because of Eq. (3.70b), $(\mu\rho)c_{is}^2 = 1/\kappa_{is}$ is the isothermal compression modulus. The low-frequency limit of $k_L(z)$ is determined by the longitudinal kinematic viscosity ν_L. One gets

$$k_L(\omega + i0) = i\nu_L + O(\omega), \tag{3.116a}$$

$$G_L(\omega + i0) = 1/\kappa_{is} - i\omega\nu_L(\mu\rho) + O(\omega^2), \quad \text{liquid.} \tag{3.116b}$$

The function $G_L(z)$ generalizes the isothermal modulus to a frequency dependent function. Therefore, it is called the frequency-dependent compression modulus or frequency-dependent longitudinal modulus. Substitution of the leading and next-to-leading small-ω -expansion terms for the modulus into Eq. (3.115a), one reproduces the Laplace transform of the wave equation (3.105b). From Eq. (3.102c) one gets $\lim_{z\to\infty}[-zk_L(z)] = c_L^2 - c_{is}^2$. Hence

$$\lim_{z\to\infty} G_L(z) = (\mu\rho)c_L^2. \tag{3.116c}$$

In the high-frequency limit, Eq. (3.115a) reduces to the Laplace transform of a wave equation. But the velocity for the waves is the high-frequency sound speed c_L. This is related to the high-frequency modulus in the same manner as explained in Sec. 3.3.4 for the high-frequency transverse-current fields. The difference between the transverse and longitudinal current correlators of a liquid is the following: the static limit of the modulus is zero for the former, Eq. (3.85b), while it is the positive number $1/\kappa_{is}$ for the latter.

It was explained in connection with Eq. (3.86b) that an ideal elastic medium is characterized by a zero-frequency pole of the transverse stress correlator $K_T(z)$. Let us contemplate, whether it makes sense to assume the same property for the longitudinal stress correlator. This means that the high-frequency asymptote of $G_L(z)$ is extrapolated to frequency $z = 0$:

$$k_L(\omega + i0) = -c_{is}^2 C/\omega + O(\omega^0), \quad \text{elastic medium.} \tag{3.117a}$$

Within the generalized hydrodynamics model, this limit is obtained by substituting $1/\tau_L^M = 0$. As a result, Eq. (3.114b) implies $(1/\tau) = 0$. Therefore, the correlator for Mountain's peak degenerates to

$$\phi_q^0(z) = -(f/z), \qquad \text{elastic medium.} \tag{3.117b}$$

The corresponding spectrum is a spike at frequency $\omega = 0 : \phi_q^{0\prime\prime}(\omega) = \pi f \delta(\omega)$. The density correlator has an arrested part f, Eq. (2.85).

According to Eq. (2.95), the preceding formula is equivalent to the following statement. If a density fluctuation average $\langle \delta\rho_{\vec{q}} \rangle_{\mathrm{B}}$ was produced at time $t = 0$ for the system equilibrated in the presence of a perturbing field b, the averaged long-time limit of the perturbation of the freely evolving system will not be zero: $\mathrm{limav}_{t\to\infty} \langle \delta\rho_{\vec{q}}^* \rangle_{b,t} \neq 0$. After switching off the perturbing field, the density fluctuation average will change. But, no matter how large is t, there will always be values of time $t' > t$, where $|\langle \delta\rho_{\vec{q}}^* \rangle_{b,t'}|$ exceeds $|\langle \delta\rho_{\vec{q}} \rangle_{\mathrm{B}} \cdot f|(1-\epsilon)$. Here, ϵ can be any small positive value. This statement is in contradiction to the observations made for a liquid. But, it is in accord with the observations made for a glass, the purest example for an isotropic elastic medium. One can check that Eqs. (3.117a,b) hold also, if $\gamma > 1$ is chosen in Eq. (3.112a) and if a constant $i\nu_L'$ is added there. The limit of Mountain's peak for vanishing width $1/\tau$ is a characterizing feature of the amorphous solid state. A central spectral spike is a signature of the density-fluctuation spectrum, which discriminates an amorphous solid from a liquid.

For $\gamma > 1$, the correlator can be written as ratio of a numerator polynomial $N(z)$ of degree three and a denominator polynomial $D(z)$ of degree four in analogy to what is discussed in Eqs. (3.108a-e). For $C = 0$ and $\gamma - 1 = 0$, there are four non-degenerate zeros of the denominator: $z_\pm = \pm c_{is} q$, $z_0 = -i\chi q^2$, and $z_0' = -i/\tau_L^M$. Since the coefficients of the polynomials are analytic functions of C and $\gamma - 1$, one can continue the discussion as done in the preceding section. For sufficiently small $\gamma - 1$ and C, there remains a pair of poles $z_\pm = \pm\omega_q - i\gamma_q$ describing high-frequency sound for large q. This sound crosses over to hydrodynamic sound for small q. The speed of the high-frequency sound is c_L and the one for hydrodynamic sound is c_{ad} as before. There is a Mountain peak, which is weakly dependent on q; it is caused by a pole z_0' on the negative imaginary axis. For small enough q, there is a diffusion peak on top of Mountain's peak. This is caused by a hydrodynamic pole $z_0 = -i\theta_q$, where θ_q vanishes proportional to q^2. These properties are demonstrated in Fig. 3.10(b) for an ad-hoc choice of parameters.

Levesque *et al.* (1973) have analyzed molecular-dynamics-simulation results for the density-fluctuation spectra of a Lennard-Jones liquid for a state, which is discussed above in Figs. 3.7 and 3.8. The small-q-limit regime cannot be studied because the simulated liquid volume is too small. But the expected pattern of a Brillouin doublet complemented by a central peak was detected and analysed for the two wave numbers $q = 0.62$ and 0.75. In Fig. 3.11, data are reproduced for larger wave numbers. These exhibit a central peak on top of a background.

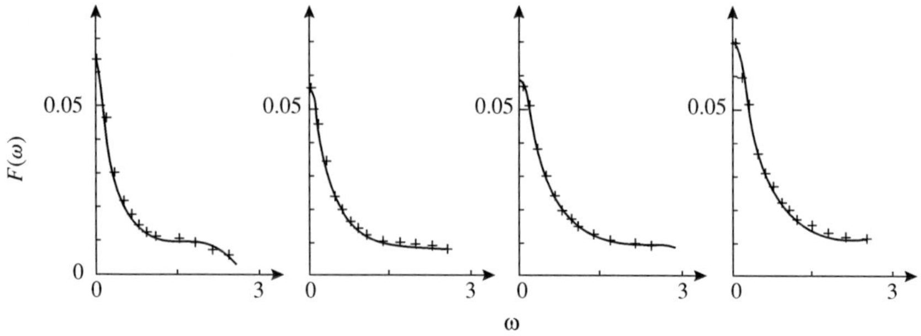

FIG. 3.11. Density-fluctuation spectra $F(\omega) = 2S_q\phi_q''(\omega)/\pi$ as a function of frequency ω for various wave numbers q. The data are results of molecular-dynamics simulations for a Lennard-Jones liquid for the same state as discussed in Fig. 3.7. The units are specified in Eqs. (1.8a–c). From left to right, the wave numbers are 1.93, 2.53, 3.18 and 3.81. The full lines are obtained by using the two-relaxator model, Eq. (3.118a), for the longitudinal kernel $M_{Lq}(z)$ in Eq. (3.90a). Reproduced from Levesque *et al.* (1973).

The peak width and peak hight do not vary proportional to q^2 and $1/q^2$, respectively, as one would expect for a Rayleigh peak. The authors report that their data can be fitted with the visco-elasticity formulas for the longitudinal-stress correlator, Eq. (3.112a,b), provided c_{is}^2 and c_L^2 are restored as their general q-dependent values $(\Omega_q/q)^2$ and $(\Omega_{Lq}/q)^2$, respectively, and if $\tau_L^M, \chi\gamma$ and γ are used as q-dependent fit parameters $\tau_{Lq}^M, \chi_q\gamma_q$ and γ_q, respectively. However, the extrapolation of γ_q to small q suggests a value for the adiabaticity ratio 3.5, which is considerably above the correct value $\gamma \approx 1.9$. This fit appears unplausible. A fit done with the bias $\gamma_q < 2$ is not possible.

The fits mentioned in the preceding paragraph are based on a simple relaxation model for the first term in Eq. (3.99d): $k_{Lq}(t)$ was replaced by $k_{Lq}(t = 0)\exp[-|t|/\tau_{Lq}^M]$. In order to arrive at a more plausible data description, this approximation was generalized by using a sum of two exponentials, specified by two relaxation times, say t_1' and t_2'. A perfect fit could be achieved with a smoothly q-dependent γ_q smaller than 2. The coupling parameter to the heat-fluctuation $(\gamma_q - 1)$ was found to be zero for $q \geq 2$. Thus, the fits, which are shown as full lines in Fig. 3.11, are obtained from the representation (3.90a) of the density-fluctuation correlator in terms of the following two-relaxator model for the longitudinal kernel

$$\tilde{M}_{Lq}(t) = \Omega_q^2\{C_q' \exp[-(t/t_1') + C_q \exp[-(t/t_2')]\}. \tag{3.118a}$$

The correct initial condition for the kernel is imposed $\tilde{M}_{Lq}(t = 0) = M_{Lq}(t = 0)$. Since Eqs. (3.91a,b) require $C_q' + C_q = [\Omega_{Lq}^2 - \Omega_q^2]/\Omega_q^2$, there is only one fit amplitude, say C_q. This model for $M_{Lq}(t)$ is the analogue of Eq. (3.89) for the transverse-stress kernel $K_{Tq}(t)$ or of Eqs. (3.56a,b) for the friction functions.

Also the scale t'_1 for the short-time decay is close to the value $t_1 \approx 0.6$ for the transverse dynamics. The same holds for the q-independent scale t'_2, which is about 7 times larger than t'_1. This data analysis suggests that visco-elasticity effects for the longitudinal-stress dynamics are quite similar to the ones for the transverse-stress dynamics. They cannot be treated within a Langevin theory, because there is no time-scale separation.

The frequency scale for the short-time process in Eq. (3.118a) is about 1 THz, which corresponds to ω near 2 in the units used in Fig. 3.11. This is the scale for the inter-molecular vibrations of van der Waals liquids as was demonstrated above in connection with Figs. 1.1 and 1.2. Spectra in this regime are material specific. General results can be expected only for lower frequencies. With the aim to understand general low-frequency phenomena, let us restrict the preceding findings to the regime of frequencies z, obeying $|zt'_1| \ll 1$. One arrives at the model for the kernel for Im $z \gtrless 0$:

$$\tilde{M}_{Lq}(z) = \pm iq^2\nu_q - \Omega_q^2 C_q/[z \pm i/\tau_{Lq}]. \tag{3.118b}$$

Here, $\nu_q = [\Omega_q^2/q^2]C'_q t'_1$ is the white-noise spectrum produced in the low-frequency regime by the band of normal-liquid excitations. The scale $\tau_{Lq} = t'_2$ quantifies a long-time tail of the longitudinal-stress correlations.

Let us examine the implications of Eq. (3.118b) for such small relaxation rates $1/\tau_{Lq}$ that there is a low frequency regime with the property

$$|z \pm iq^2\nu_q| \ll \Omega_q^2 C_q/|z \pm i/\tau_{Lq}|. \tag{3.119a}$$

In a leading-order calculation, which treats the ratio of left and the right-hand side of Eq. (3.119a) as a small parameter, one can write $z + \tilde{M}_q(z) = -\Omega_q^2 C_q/[z \pm i/\tau_{Lq}]$. As a result, the density correlator (3.90a) describes a simple relaxation process

$$\tilde{\phi}_q(z) = -f_q/[z \pm i/\tau_q], \tag{3.119b}$$

$$1/\tau_q = [1/\tau_{Lq}]/(1 + C_q), \tag{3.119c}$$

$$f_q = C_q/(1 + C_q). \tag{3.119d}$$

The relaxation rate $1/\tau_q$ is smaller than the one for the stress relaxation $1/\tau_{Lq}$, and it tends to zero if C_q tends to infinity. The Landau–Placzek ratio is $f_q/(1 - f_q) = C_q$. These formulas explain qualitatively the central peaks in Fig. 3.11. To describe the low-frequency relaxation process by Eq. (3.119b), frequencies of the order of $1/\tau_q$ are relevant. One concludes from Eq. (3.119a) that the preceding equations are an implication of Eq. (3.118b) if the limit $1/\tau_{Lq} \to 0$ is considered. The reasoning can be reversed. If there is a central peak in the density-fluctuation spectrum, parameterized by a strength factor f_q and a small relaxation rate $1/\tau_q$, a similar peak occurs in the longitudinal-stress spectrum. The parameters C_q and $1/\tau_{Lq}$ follow from f_q and $1/\tau_q$ by inversion of Eqs. (3.119c,d).

Equations (3.119b–d) are generalizations to non-vanishing wave numbers of Eqs. (3.114a–d) for Mountain's peak. It is an import implication of Fig. 3.11 that the central-peak spectrum is observed for wavelengths on the microscopic length scale d of the liquid. According to the discussion of Eq. (3.117b), a vanishing of $1/\tau_q$ is a signature of an amorphous solid. An approach of $1/\tau_q$ towards zero is the signature of a liquid approaching a glass transition. The appearance of a quasi-elastic spectrum as described by Eq. (3.119b) for $1/\tau_q < 1/t^*$ is a glassy-dynamics precursor.

Density-fluctuation spectra $S_q\phi_q''(\omega)$ determine the cross-section for the inelastic scattering of light for the momentum transfer $\hbar\vec{q}$ and energy change $\hbar\omega$ of the photons. This holds under the assumption that the dielectric constant couples to the electric field merely via density changes. In this case, $2\pi/q$ is of the order of the wavelength of the light, i.e., q is of the order of 10^{-3} Å^{-1}. Light-scattering is the ideal tool to measure the Brillouin doublet, the Rayleigh peak, and to study visco-elastic effects. Mountain (1966) developed his theory in order to explain light-scattering data for molecular liquids. He attributed the slow relaxation process, which causes the central peak, to some soft intra-molecular mode of the constituents.

The dynamics exhibited in Fig. 3.11 deals with wavelengths which are several orders of magnitude smaller than those studied by light-scattering. This range of q can be tested by neutron-scattering measurements. It is tempting to consider each particle of a dense liquid as a constituent of a molecule. The particle is connected with its cage-forming neighbours. The large times τ_{Lq} and τ_{Tq} specify the scale for the motion of such clusters. There is the problem of how to relate these times to the microscopic structure of the clusters. In particular, a means has to be found to average over the various clusters expected for an amorphous state. An approach towards the treatment of these problems will be discussed in the following chapter. In Chapter 5, it will be demonstrated how the proposed explanation of the quasi-elastic density-fluctuation spectrum implies one for the glassy dynamics of the shear and the tagged-particle motions.

4

FOUNDATIONS OF THE MODE-COUPLING THEORY FOR THE EVOLUTION OF GLASSY DYNAMICS IN LIQUIDS

This chapter starts with the Zwanzig–Mori relation between density correlators and fluctuating-force correlators. An expression for the latter in terms of the former will be motivated with the intention to treat slow dynamics of liquids in a region where there might be no separation of time scales for the correlation decay of density fluctuations and of force fluctuations (Sec. 4.1). As a result, closed equations of motion for the density-fluctuation correlators are obtained. The coefficients in these equations are determined by the equilibrium-structure functions. The challenge of these equations is to understand the interplay of nonlinearities with retardation effects. It will be shown that these equations provide a model for a statistical description of a dynamics for every choice of the coupling coefficients (Sec. 4.2). The study of the equations is referred to as mode-coupling theory (MCT).

The MCT equations of motion are regular in the sense that the correlators enter as polynomials and as convolution integrals of polynomials. The coupling coefficients appear as coefficients for these polynomials. The long-time limits of the solution for the density correlators exhibit bifurcation singularities. A classification of these singularities is possible (Sec. 4.3). They are equivalent to the bifurcations of the zeros of real polynomials, or degeneracies and combinations thereof. The basic singularity describes a transition from a liquid to an amorphous solid, i.e., it deals with an idealized liquid-to-glass transition. Examples for the simplest bifurcation scenarios will be explained. Details for the liquid–glass transitions for hard-sphere systems and for systems of hard spheres with additional short-range attractions are considered and compared with data measured for some colloids.

In Sec. 4.4, numerical solutions of the MCT equations of motion are discussed, which demonstrate the evolution of the dynamics upon shifting the coupling coefficients towards a liquid-glass-transition point. Some properties of this MCT dynamics are identified in order to explain that the slow dynamics is quite different from that known from conventional models for a nonlinear dynamics. It will be demonstrated also that some features of the MCT-bifurcation dynamics are similar to what is identified in Chapter 1 as a facet of the glassy dynamics of liquids.

4.1 Self-consistent-current-relaxation approaches

In a glassy liquid, the correlation functions for density fluctuations exhibit tails for times t, which exceed by several orders of magnitude the natural scale t_{mic} for condensed-matter motion. These slow processes occur for wave numbers q specifying the microscope length scale of the liquid, i.e., the wavelength $(2\pi/q)$ of the fluctuations is of the order of the particle diameter d. Figures 1.4, 1.5, 1.7, 1.15, 1.20 and 1.21 document these facts. The long-time tails are equivalent to quasi-elastic contributions to the density-fluctuation spectra. There are central peaks within an interval of frequencies ω, which are below the natural scale $1/t_{\text{mic}}$ for particle motion in the system. It is discussed in connection with Eqs. (3.119a–d) that there is a one-to-one correspondence of a quasi-elastic spectrum for density fluctuations with that of the fluctuating-forces or of stress fluctuations. The latter reflect the slow relaxation of the transient localization of the particles in the cages formed by their neighbors. There is no separation of the time scale τ_q for the glassy dynamics of the density fluctuations from the scale τ_{Lq} of the fluctuating forces. A theory of glassy dynamics in liquids has to provide a simultaneous explanation of the autocorrelators for density fluctuations and force-fluctuations, and this for wave numbers extending from macroscopic to microscopic sizes. In the following, closed sets of equations of motion for the liquid dynamics shall be motivated, which are compatible with such request. The considerations shall be restricted to systems of N equal spherical particles.

4.1.1 *The factorization ansatz*

The formalism introduced in Sec. 3.3 for the description of simple-liquid dynamics shall be used also in this section. The normalized density correlators in the time domain, $\phi_q(t)$, are used as defined in Eq. (3.63). The equivalent functions in the domain of complex frequencies z, $\phi_q(z)$, are obtained by Laplace transformation. The density fluctuations $\rho_{\vec{q}}$ of wave vector $\vec{q}$ can vary with time because there is the possibility of a time-dependent current fluctuation of the same wave vector, $\vec{j}_{\vec{q}}$, Eqs. (3.65a,b). The Zwanzig–Mori formula (3.95a) provides a fraction representation of $\phi_q(z)$ in terms of a fluctuating-current correlator $K_q(z)$. A non-trivial current flow is caused by interaction forces between the particles, and the Zwanzig–Mori formula (3.95d) provides a fraction representation of $K_q(z)$ in terms of a fluctuating-force correlator denoted by $M_{Lq}(z) = q^2 K_{Lq}(z)$. Combining these formulas, one gets the double-fraction representation (3.90a) of the density correlator:

$$\phi_q(z) = -1/\Big\{z - \Omega_q^2/\big[z + M_{Lq}(z)\big]\Big\}. \tag{4.1a}$$

The kernel $M_{Lq}(z)$ is the Laplace transform of a function $M_{Lq}(t)$. The latter is a matrix element of the time-translation operator $\mathcal{U}_{\text{red}}(t) = \exp[i\mathcal{L}_{12}t]$, generated by the self-adjoint reduced Liouvillian $\mathcal{L}_{12} = \mathcal{Q}_{12}\mathcal{L}\mathcal{Q}_{12}$:

$$M_{Lq}(t) = (F_{Lq}|\mathcal{U}_{\text{red}}(-t)F_{Lq}). \tag{4.1b}$$

The operator $\mathcal{Q}_{12} = \mathcal{Q}_1\mathcal{Q}_2 = \mathcal{Q}_2\mathcal{Q}_1$ projects perpendicular to the density fluctuations, $\mathcal{Q}_1 = 1-\mathcal{P}_1$, $\mathcal{P}_1 = |\rho_{\vec{q}*})[NS_q]^{-1}(\rho_{\vec{q}*})|$, and to the longitudinal part of the current fluctuations, $\mathcal{Q}_2 = 1 - \mathcal{P}_2$, $\mathcal{P}_2 = |j^3_{\vec{q}*})[Nv_{th}^2]^{-1}(j^3_{\vec{q}*}|$. Here, $\vec{q}^* = (0, 0, q)$ denotes a distinguished wave vector of modulus q. Rewriting Eqs. (3.90d,e) with the aid of Eqs. (3.71b–d), one arrives at an expression for the fluctuating force

$$F_{Lq} = \mathcal{Q}_1\mathcal{L}j^3_{\vec{q}*}/[v_{th}\sqrt{N}]. \tag{4.1c}$$

In this section, an expression for the kernel $M_{Lq}(t)$ as a linear combination of correlator products $\phi_k(t)\phi_p(t)$ shall be derived. Thereby, the long-time tails of the density-fluctuation correlators will be identified as the origin of the long-time tails of the fluctuating-force correlations.

A paragraph with a digression on the generality of the preceding three formulas will be of importance in later sections. Equations (4.1a–c) are valid also if the N-particle system is imbedded as a subsystem in a larger many-particle system. In this case, $\mathcal{L}$ generates the dynamics of the complete system, and also the averages refer to the larger system. Equations (4.1a–c) hold under the assumptions that the complete system is homogeneous, isotropic and free of chirality. These restrictions ensure the validity of the general symmetries of the correlators for density and current fluctuations discussed in Sec. 3.1. The Eqs. (3.95a–c) remain valid with the definitions of the thermal velocity $v_{th} = \sqrt{k_{\mathrm{B}}T/\mu}$, structure factor S_q, Eq. (3.58c), and frequency Ω_q, Eq. (3.67). These equations are implications of the continuity equation, which holds also for the subsystem. Similarly, Eqs. (4.1a–c) are general implications of the Zwanzig–Mori reformulations and the indicated symmetries. The difference between a truly simple N-particle system and a subsystem of a larger one concerns the explicit form of the fluctuating forces. The general equation of motion (3.71a) can be written as $\mathcal{L}\vec{j}_{\vec{q}} = -i[\vec{f}^{(1)}_{\vec{q}} + \vec{f}^{(2)}_{\vec{q}}]/\mu$. The first contribution to the general force fluctuation describes the convection contribution and the interaction contribution due to the potential for the N particles of the subsystem. This contribution is noted in Eqs. (3.71b–d): $\vec{f}^{(1)}_{\vec{q}} = \vec{f}_{\vec{q}}$. The second contribution is due to the interaction forces between the particles of the subsystem with the ones of the remaining part of the complete system. If the interaction potential between a particle κ of the subsystem with a particle ν of the remaining system at position $\vec{r}_\nu$ is denoted by $V_\nu(|\vec{r}_\kappa - \vec{r}_\nu|)$, there holds $\vec{f}^{(2)}_{\vec{q}} = \sum_\nu\sum_\kappa[-\vec{\partial}V_\nu(|\vec{r}_\kappa - \vec{r}_\nu|)]\exp[i\vec{q}\vec{r}_\kappa]$. Therefore, the fluctuating force in Eq. (4.1c) consists of two contributions

$$F_{Lq} = F^{(1)}_{Lq} + F^{(2)}_{Lq}, \tag{4.1d}$$

where $F^{(1,2)}_{Lq} = -i\mathcal{Q}_1 f^{(1,2)3}_{\vec{q}*}/[\mu v_{th}\sqrt{N}]$. The first contribution is due to a part of the Hamiltonian, which is invariant under translation of the subsystem. This part cannot yield a change of the subsystem's total momentum. The force $\vec{f}^{(1)}_{\vec{q}}$ vanishes with q, as explained in connection with Eqs. (3.71c,d). However, because of the interaction of the subsystem with the remaining system, the mentioned invariance is broken. The subsystem's total momentum is not conserved. Momentum can be transferred from the subsystem to the remainder. This momentum change is described by the $q \to 0$ limit of $\vec{f}_{\vec{q}}^{(2)}$. One gets a contribution to the relaxation kernel in the long-wavelength limit

$$\lim_{q\to 0} M_{Lq}(t) = (F_{L0}^{(2)}\,|\,\mathcal{U}_{\rm red}(-t)F_{L0}^{(2)}). \tag{4.1e}$$

Interesting examples for the specified subsystem physics are colloids. In this case, the subsystem is the solute and the remaining particles are the constituents of the solvent.

A possible transient localization of a particle near some position $\vec{r}_1$ is caused by its interaction with other particles at neighbour positions $\vec{r}_2$. The dynamical variable describing the presence of such interacting pairs is $\rho(\vec{r}_1)\rho(\vec{r}_2)$ with $\rho(\vec{r}) = \sum_\kappa \delta(\vec{r}-\vec{r}_\kappa)$ denoting the particle density. Via a sixfold Fourier transform, the pair densities can be expressed as linear combination of fluctuation pairs $\rho_{\vec{k}_1}\rho_{\vec{k}_2}$. The interaction contribution of the forces $\vec{f}_{\vec{q}}$ in Eq. (3.71b) is an integral over the variables $\rho(\vec{r}_1)\rho(\vec{r}_2)$ weighted with potential derivatives and phase factors. Hence, it is a linear combination of the variables $\rho_{\vec{k}_1}\rho_{\vec{k}_2}$. In a first step towards the study of the cage effect, the correlation of the forces shall be expressed by those of the density-fluctuation pairs. Let us anticipate the set of wave vectors $\vec{k}$ to be labelled somehow. Thereby an ordering $\vec{k}_1 < \vec{k}_2$ is introduced in the sense that the label of $\vec{k}_1$ precedes that of $\vec{k}_2$. The label of the pair fluctuations shall be denoted by $k = (\vec{k}_1, \vec{k}_2)$, where $\vec{k}_1 < \vec{k}_2$ is implied. The dynamical variables $A_k = \rho_{\vec{k}_1}\rho_{\vec{k}_2} = \rho_{\vec{k}_2}\rho_{\vec{k}_1}$ span the space of all pair fluctuations. Assuming the A_k to be linearly independent, the projector $\mathcal{P}$ on this subspace is given by Eq. (2.55), where $g_{m,n} = (A_m|A_n) = \langle(\rho_{\vec{m}_1}\rho_{\vec{m}_2})^*\rho_{\vec{n}_1}\rho_{\vec{n}_2}\rangle$ denote the elements of the metric matrix $\boldsymbol{g}$. The inverse of the latter is the solution of the equation

$$\sum_{\vec{k}_1<\vec{k}_2} (\rho_{\vec{m}_1}\rho_{\vec{m}_2}\,|\,\rho_{\vec{k}_1}\rho_{\vec{k}_2})(\boldsymbol{g}^{-1})_{\vec{k}_1\vec{k}_2,\vec{n}_1\vec{n}_2} = \delta_{\vec{m}_1\vec{n}_1}\delta_{\vec{m}_2\vec{n}_2}. \tag{4.2}$$

The projector on the space perpendicular to that of the pair fluctuations is denoted by $\mathcal{Q} = 1 - \mathcal{P}$.

As explained in connection with Eqs. (2.57a–f), the Laplace transform of kernel $M_{Lq}(t)$ is a resolvent matrix element: $M_{Lq}(z) = (F_{Lq}\,|\,\mathcal{R}(z)F_{Lq})$, $\mathcal{R}(z) = [\mathcal{L}_{12} - z]^{-1}$. Formula (2.64) achieves the desired expression, where $\mathcal{L}$ has to be replaced by $\mathcal{L}_{12}$ and $X = Y = F_{Lq}$. The result shall be denoted as

$$M_{Lq}(z) = M_{Lq}^{\rm reg}(z) + \sum_{m,n}\Big[\Omega_{q,m}^L + \delta\Omega_{q,m}^L(z)\Big]\phi_{m,n}(z)\Big[\Omega_{q,n}^L + \delta\Omega_{q,n}^L(z^*)\Big]^*. \tag{4.3a}$$

The first contribution on the right-hand side is given by Eq. (2.63e): $M_{Lq}^{\rm reg}(z) = (\mathcal{Q}F_{Lq}\,|\,\mathcal{R}_{\mathcal{Q}}(z)\mathcal{Q}F_{Lq})$. The reduced resolvent reads $\mathcal{R}_{\mathcal{Q}}(z) = [\mathcal{Q}\mathcal{L}_{12}\mathcal{Q} - z]^{-1}$. It deals with the dynamics generated by the reduced Liouvillian $\mathcal{L}_{\mathcal{Q}} = \mathcal{Q}\mathcal{L}_{12}\mathcal{Q}$. This dynamics treats densities, longitudinal currents and all pair fluctuations A_k as static. Therefore, it seems reasonable to assume that this kernel does not exhibit anymore the quasi-elastic contributions caused by the slow correlations of the pair fluctuations. In this sense, kernel $M^{\text{reg}}_{Lq}(z)$ is considered as regular. The same holds for the functions $\delta\Omega^L_{q,m}(z) = -\sum_k (\mathcal{Q}F_{Lq} \,|\, \mathcal{R}_{\mathcal{Q}}(z)\mathcal{Q}\mathcal{L}_{12}A_k)(\boldsymbol{g}^{-1})_{k,m}$. This formula is a rewriting of Eq. (2.63c) into the present notation. These regular functions $\delta\Omega^L_{q,m}(z)$ renormalize the thermodynamic functions $\Omega^L_{q,m}$. According to Eq. (2.63a), the latter are given by

$$\Omega^L_{q,\vec{m}_1\vec{m}_2} = \sum_{\vec{k}_1 \le \vec{k}_2} (F_{Lq} \,|\, \rho_{\vec{k}_1}\rho_{\vec{k}_2})(\boldsymbol{g}^{-1})_{\vec{k}_1\vec{k}_2,\vec{m}_1\vec{m}_2}. \tag{4.3b}$$

The functions $\phi_{m,n}(z)$ describe the pair correlations in the frequency domain. According to Eq. (4.1b), their dynamics is generated by the Liouvillian $\mathcal{L}_{12}$:

$$\phi_{\vec{m}_1\vec{m}_2,\vec{n}_1\vec{n}_2}(t) = (\rho_{\vec{m}_1}\rho_{\vec{m}_2} \,|\, \mathcal{U}_{\text{red}}(-t)\rho_{\vec{n}_1}\rho_{\vec{n}_2}). \tag{4.3c}$$

The initial value of the relaxation kernel is noted in Eqs. (3.91a,b): $M_{Lq}(t = 0) = -\lim_{z\to\infty}[zM_{Lq}(z)] = \Delta_q^2$. Introducing the initial value of the regular kernel

$$M^{\text{reg}}_{Lq}(t = 0) = m^{(1)}_{Lq}, \tag{4.4a}$$

one gets $m^{(1)}_{Lq} = -\lim_{z\to\infty}[zM^{\text{reg}}_{Lq}(z)] = (\mathcal{Q}F_{Lq} \,|\, \mathcal{Q}F_{Lq})$. Since $\delta\Omega^L_{q,m}(z) = O(1/z)$ and $-\lim_{z\to\infty}[z\phi_{m,n}(z)] = \phi_{m,n}(t = 0)$, the sum in Eq. (4.3a) contributes to the initial value the number

$$\sum_{m,n} \Omega^L_{q,m}\phi_{m,n}(t = 0)\Omega^{L*}_{q,n} = m^{(2)}_{Lq}. \tag{4.4b}$$

Consequently, there holds

$$\Delta_q^2 = m^{(1)}_{Lq} + m^{(2)}_{Lq}. \tag{4.4c}$$

Since $\phi_{m,n}(t = 0) = g_{m,n}$, one obtains $m^{(2)}_{Lq} = (\mathcal{P}F_{Lq} \,|\, \mathcal{P}F_{Lq})$. Both contributions to Δ_q^2 are non-negative: $m^{(1,2)}_{Lq} \ge 0$.

The overlaps of the forces with the density-pair fluctuations read:

$$(F_{Lq} \,|\, \rho_{\vec{k}}\rho_{\vec{p}}) = (F^{(a)}_{q,\vec{k}\vec{p}} - F^{(b)}_{q,\vec{k}\vec{p}})v_{th}\sqrt{N}. \tag{4.5a}$$

Here, one gets $F^{(a)}_{q,\vec{k}\vec{p}}v_{th}\sqrt{N} = (\mathcal{L}j^3_{\vec{q}*} \,|\, \rho_{\vec{k}}\rho_{\vec{p}})/[v_{th}\sqrt{N}]$ and $F^{(b)}_{q,\vec{k}\vec{p}}v_{\text{th}}\sqrt{N} = (\mathcal{P}_1\mathcal{L}j^3_{\vec{q}*} \,|\, \rho_{\vec{k}}\rho_{\vec{p}})/[v_{th}\sqrt{N}]$. The first contribution is reformulated by using Eqs. (2.4b), (2.14b) for the Liouvillian, the connection (2.8) between scalar product and averages, and the continuity equation: $(\mathcal{L}j^3_{\vec{q}*} \,|\, \rho_{\vec{k}}\rho_{\vec{p}}) = (j^3_{\vec{q}*} \,|\, [(\mathcal{L}\rho_{\vec{k}})\rho_{\vec{p}} + \rho_{\vec{k}}(\mathcal{L}\rho_{\vec{p}})]) =$

$\langle j^{3*}_{\vec{q}*}[\vec{k}\vec{j}_{\vec{k}}\rho_{\vec{p}} + \rho_{\vec{k}}\vec{p}\vec{j}_{\vec{p}}]\rangle$. Substituting the explicit expressions (3.57) for the fluctuations of densities and currents, and performing the independent averaging over the velocities, one can continue the reformulation to $v^2_{th}[k^3\langle\rho^*_{\vec{q}*-\vec{k}}\rho_{\vec{p}}\rangle + p^3\langle\rho^*_{\vec{q}*-\vec{p}}\rho_{\vec{k}}\rangle]$. The request (3.13) of homogeneity requires $\vec{q}^* = \vec{k}+\vec{p}$. The remaining averages are given by Eq. (3.58c). Hence,

$$F^{(a)}_{q,\vec{k}\vec{p}} = \delta_{\vec{q}*,\vec{k}+\vec{p}}\big[k^3S_p + p^3S_k\big]. \tag{4.5b}$$

The second contribution is determined by $(\mathcal{P}_1\mathcal{L}j^3_{\vec{q}*} \,|\, \rho_{\vec{k}}\rho_{\vec{p}}) = (\mathcal{L}j^3_{\vec{q}*} \,|\, \rho_{\vec{q}*})[NS_q]^{-1}(\rho_{\vec{q}*} \,|\, \rho_{\vec{k}}\rho_{\vec{p}})$. The hermiticity of $\mathcal{L}$, the continuity equation, and Eq. (3.58a) yield $(\mathcal{L}j^3_{\vec{q}*} \,|\, \rho_{\vec{q}*}) = (j^3_{\vec{q}*} \,|\, \vec{q}^*\vec{j}_{\vec{q}*}) = qv^2_{th}N$. Because of the homogeneity implication (3.13), the average $\langle\rho^*_{\vec{q}*} \,|\, \rho_{\vec{k}}\rho_{\vec{p}}\rangle$ vanishes unless $\vec{q}^* = \vec{k}+\vec{p}$. A rotation $D(\varphi)$ around the third coordinates axis leaves $\vec{q}^*$ fixed, Eq. (B.3c). Rotational invariance implies $\langle\rho_{\vec{q}*}\rho_{D(\varphi)\vec{k}}\rho_{D(\varphi)\vec{p}}\rangle = \langle\rho_{\vec{q}*}\rho_{\vec{k}}\rho_{\vec{p}}\rangle$. Thus, the triple average does not depend on the plane of the triangle formed with $\vec{q}^*, \vec{k}$, and $\vec{p}$; it is given by the lengths of the sides q, k and p. One arrives at

$$F^{(b)}_{q,\vec{k}\vec{p}} = \delta_{\vec{q}*,\vec{k}+\vec{p}}qS_kS_p\big[1 + \rho^2c_3(q,k,p)\big]. \tag{4.5c}$$

Here, the triple-density average is expressed by a quantity called triple-correlation function $c_3(q,k,p)$:

$$\langle\rho^*_{\vec{q}}\rho_{\vec{k}}\rho_{\vec{p}}\rangle = \delta_{\vec{q},\vec{k}+\vec{p}}S_qS_kS_p\big[1 + \rho^2c_3(q,k,p)\big]N. \tag{4.6}$$

Neglecting $\rho^2c_3(q,k,p)$ relative to 1 is known as convolution approximation (Jackson and Feenberg 1962).

It appears plausible that a slow-relaxation contribution to the density-fluctuation correlators is sufficient to cause a slow-relaxation contribution to the correlation functions for the density-fluctuation pairs. Even if there is no correlation of the slow motion of the particles near position $\vec{r}_1$ with that of the particles near position $\vec{r}_2$, the distance and hence the force between the particles changes slowly. The simplest model for the pair correlator $\phi_{m,n}(t)$, which is consistent with this insight, is obtained by the factorization ansatz:

$$\phi_{\vec{m}_1\vec{m}_2,\vec{n}_1\vec{n}_2}(t) \longrightarrow (\rho_{\vec{m}_1}(t) \,|\, \rho_{\vec{n}_1})(\rho_{\vec{m}_2}(t) \,|\, \rho_{\vec{n}_2}). \tag{4.7a}$$

This factorization is the essential step towards the intended discussion of the cage effect. Because of the homogeneity condition (3.11b), only terms with $\vec{m}_1 = \vec{n}_1$ and $\vec{m}_2 = \vec{n}_2$ contribute. For $\vec{m}_1 < \vec{m}_2$ and $\vec{n}_1 < \vec{n}_2$, factorization products $(\rho_{\vec{m}_1}(t)|\rho_{\vec{n}_2})(\rho_{\vec{m}_2}(t)|\rho_{\vec{n}_1})$ cannot occur. Because of Eq. (3.63), the right-hand side of Eq. (4.7a) is given by $\delta_{\vec{m}_1,\vec{n}_1}\delta_{\vec{m}_2,\vec{n}_2}N^2S_{m_1}S_{m_2}\phi_{m_1}(t)\phi_{m_2}(t)$. For $t = 0$, the factorization ansatz (4.7a) implies the replacement of the metric matrix $g_{\vec{m}_1\vec{m}_2,\vec{n}_1\vec{n}_2}$ by $\delta_{\vec{m}_1,\vec{n}_1}\delta_{\vec{m}_2,\vec{n}_2}N^2S_{m_1}S_{m_2}$. Equation (4.2) yields $(\boldsymbol{g}^{-1})_{\vec{k}_1\vec{k}_2,\vec{m}_1\vec{m}_2} =$

$\delta_{\vec{k}_1,\vec{m}_1}\delta_{\vec{k}_2,\vec{m}_2}/[N^2 S_{k_1}S_{k_2}]$. Using Eqs. (4.3b), (4.5a), one finds that the factorization ansatz is equivalent to replacing the term $\sum_{m,n}\Omega^L_{q,m}\phi_{m,n}(t)\Omega^{L*}_{qn}$ by

$$M^{\rm mc}_{Lq}(t) = \frac{1}{2}v^2_{th}\sum_{\vec{k},\vec{p}}\Big\{\big[(F^{(a)}_{q,\vec{k}\vec{p}} - F^{(b)}_{q,\vec{k}\vec{p}})/(S_k S_p)\big]^2\big[S_k S_p \phi_k(t)\phi_p(t)/N\big]\Big\}. \quad (4.7b)$$

Here, the symmetry of the expression with respect to interchanges of $\vec{k}$ and $\vec{p}$ is used to justify the replacement of $\sum_{\vec{k}<\vec{p}}\cdots$ by $\frac{1}{2}\sum_{\vec{k}\neq\vec{p}}\cdots$. In the thermodynamic limit, the restriction $\vec{k}\neq\vec{p}$ can be ignored.

The frequency-dependent renormalizations $\delta\Omega^L_{q,m}(z)$ of the coupling amplitudes shall be neglected for the sake of simplicity:

$$\delta\Omega^L_{q,m}(z) \longrightarrow 0. \quad (4.8)$$

The preceding ad hoc simplification and the ansatz from Eq. (4.7a) lead to the replacement of formula (4.3a) for the fluctuating-force correlator by a function to be indicated by a tilde:

$$M_{Lq}(z) \longrightarrow \tilde{M}_{Lq}(z), \quad (4.9a)$$

$$\tilde{M}_{Lq}(z) = M^{\rm reg}_{Lq}(z) + M^{\rm mc}_{Lq}(z). \quad (4.9b)$$

The second term is the Laplace transform of the function defined in Eq. (4.7b). The function $M^{\rm mc}_{Lq}(t)$ models the coupling of the slow-relaxation part of the fluctuating-force correlator to the slow-relaxation parts of the density correlators. The former deals with the mode of changes of stresses in space and time, and the latter with the one of densities. Expressions like $M^{\rm mc}_{Lq}(t)$ are called mode-coupling contributions. The above described derivation of the mode-coupling formula for $\tilde{M}_{Lq}(z)$ is an expatiate version of the one presented by Götze and Lücke (1976) for the density correlator of liquid helium. The latter work was done within the quantum-mechanical theory of correlation functions; but, it is easy to simplify the formulas to ones for the classical limit.

Five comments shall be added to the preceding results. The first one concerns the factorization ansatz for the function $\phi(t) = (\rho_{\vec{m}_1}\rho_{\vec{m}_2}\,|\,\mathcal{U}_{\rm red}(-t)\rho_{\vec{n}_1}\rho_{\vec{n}_2})$. Function $\phi(t)$ deals with the reduced dynamics generated by $\mathcal{L}_{12}$. This function differs from the correlator of the pair densities $\psi(t) = (\rho_{\vec{m}_1}\rho_{\vec{m}_2}\,|\,\mathcal{U}(-t)\rho_{\vec{n}_1}\rho_{\vec{n}_2})$ dealing with the full dynamics. The latter is generated by the complete Liouvillian; and it contains a single-density-fluctuation contribution due to the superposition of pair fluctuations: $\psi^{(1)}(t) = (\rho_{\vec{m}_1}\rho_{\vec{m}_2}\,|\,\mathcal{P}_1\mathcal{U}(-t)\mathcal{P}_1\rho_{\vec{n}_1}\rho_{\vec{n}_2}) = [(\rho_{\vec{m}_1}\rho_{\vec{m}_2}\,|\,\rho_{\vec{q}*})(\rho_{\vec{q}*}|\,\rho_{\vec{n}_1}\rho_{\vec{n}_2})/(NS_q)]\phi_q(t)$. A factorization ansatz for $\psi(t)$ in analogy to Eq. (4.7a) would be a blunder since, thereby, the contribution $\psi^{(1)}(t)$ would get lost. The single-density-fluctuation dynamics does not occur for function $\phi(t)$. It is the effect of the projectors in $\mathcal{L}_{12} = \mathcal{Q}_1\mathcal{Q}_2\mathcal{L}\mathcal{Q}_2\mathcal{Q}_1$ to eliminate this contribution.

The second comment concerns the initial values of the relaxation kernels discussed in Eqs. (4.4a–c). Replacing $M_{Lq}(t)$ by $\tilde{M}_{Lq}(t)$ implies the replacement of the initial value Δ_q^2 by $\tilde{\Delta}_q^2 = m_{Lq}^{(1)} + m_{Lq}^{(2)\text{mc}}$. Because of the factorization ansatz, $m_{Lq}^{(2)\text{mc}} = M_{Lq}^{\text{mc}}(t=0)$ differs from $m_{Lq}^{(2)}$. For a Lennard-Jones system near its triple point, $m_{Lq}^{(2)\text{mc}}$ can exceed Δ_q^2 seriously. Independent of the model chosen for the correlator $M_{Lq}^{\text{reg}}(t)$, one gets $\tilde{\Delta}_{Lq}^2 > \Delta_{Lq}^2$ for some intermediate-size values of q. The model $\tilde{M}_{Lq}(t)$ is motivated for a description of the long-time correlations; this description implies errors for the treatment of the short-time dynamics.

Third, because of the strong interaction forces in condensed matter, there are important correlations between the particle positions at short distances. These are described by correlation functions like $(\rho(\vec{r}_1)\rho(\vec{r}_2)|\mathcal{U}_{\text{red}}(-t)\rho(\vec{s}_1)\rho(\vec{s}_2))$, in particular by their initial values. The factorization ansatz ignores these correlations. Therefore, a direct factorization of the function $(F_{Lq}|\mathcal{U}_{\text{red}}(-t)F_{Lq})$ leads to divergent results. However, rewriting the fluctuating-force kernel in the form of Eq. (4.3a) has achieved the separation of two problems. One concerns the evaluation of the products of diverging forces and the vanishing probability of finding particle pairs experiencing these forces. The other problem concerns the reasonable description of the long-time behaviour of the pair correlator. The first problem enters in form of the amplitudes $(F_{Lq}|\rho_{\vec{k}_1}\rho_{\vec{k}_2})$ in Eq. (4.3b). These have been expressed in terms of the equilibrium functions S_q and $c_3(q,k,p)$. Hence, the pair correlator in Eq. (4.3a) is multiplied by well behaved functions. The errors of the factorization ansatz do not lead to diverging results anymore.

All formulas used above for a description of a dynamics assume an underlying Hamilton function with sufficiently regular interaction potentials. It is of interest to consider also equations of motions and canonical averages for singular potentials. Such theory can be defined with the aid of limiting procedures for systems with regular interactions so that the structure factor S_q, the direct correlation function c_q, and the triple correlations $c_3(q,k,p)$ are well defined continuous functions of the wave numbers. The coefficients of $\phi_k(t)\phi_p(t)$ in Eq. (4.7b) are explained. The fourth comment is the suggestion, to use Eq. (4.9b) also for a discussion of the density-fluctuation dynamics for a system with the singular interaction potentials mentioned.

The fifth comment concerns the assumed regularity of $M_{Lq}^{\text{reg}}(z)$ and $\delta\Omega_{q,m}^L(z)$. This assumption is likely to be illegitimate. One would expect that the underlying variables have overlaps with density fluctuation products $A'_{\vec{k}_1,\ldots,\vec{k}_\ell} = \rho_{\vec{k}_1}\cdots\rho_{\vec{k}_2}$ for $\ell \geq 3$ so that the functions $\phi'(t) = (A'_{\vec{k}_1,\ldots,\vec{k}_\ell}|\mathcal{U}_{\text{red}}(-t)A'_{\vec{p}_1,\ldots,\vec{p}_\ell})$ exhibit cage-effect-induced tails. In order to proceed, the set of distinguished variables A_m should be extended so that also the variables with $\ell \geq 3$ are included. Using the factorization ansatz as above, one arrives at Eqs. (4.9a,b). However, Eq. (4.7b) has to be extended. Expressions of order $\ell \geq 3$ have to be added, which read $\sum_{\vec{k}_1,\ldots,\vec{k}_\ell} V(\vec{q},\vec{k}_1,\ldots,\vec{k}_\ell)\phi_{k_1}(t)\cdots\phi_{k_\ell}(t)$. The coefficients $V(\vec{q},\vec{k}_1,\ldots,\vec{k}_\ell)$ are determined by averages of products of more than three density fluctuations. So far, such contributions have not been evaluated. However, for the following general mathematical considerations, such higher-order functionals will be included.

It will be shown that $\ell \geq 3$ terms do not lead to qualitative effects, which are not described already by the quadratic expression in Eq. (4.7b).

4.1.2 *Self-consistency equations for density correlators*

The above discussed mode-coupling result for the fluctuating-force correlator shall be used in this section as motivation of a closed set of equations of motion for the density-fluctuation correlation functions of simple amorphous systems.

Let $\tilde{M}_{Lq}(z)$ denote a positive analytic function of the complex frequency z exhibiting the large-frequency behaviour $\lim_{y\to\infty} \tilde{M}_{Lq}(iy)/y = 0$. Such function shall be used to define

$$\tilde{\phi}_q(z) = -1/\{z - \Omega_q^2/[z + \tilde{M}_{Lq}(z)]\}. \tag{4.10a}$$

A function of this form was discussed in connection with Eqs. (3.90a), (3.92a). It was shown there that $\tilde{\phi}_q(z)$ is the Laplace transform of a continuous positive definite function $\tilde{\phi}_q(t)$ obeying the normalization condition $\tilde{\phi}_q(t=0) = 1$. The kernel $\tilde{M}_{Lq}(z)$ shall be specialized further by requesting the representation

$$\tilde{M}_{Lq}(z) = \pm i\nu_q + \hat{M}_{Lq}(z), \quad \text{Im}\, z \gtrless 0. \tag{4.10b}$$

Here $\nu_q \geq 0$ quantifies a possible white-noise spectrum. The kernel $\hat{M}_{Lq}(z)$ is requested to be positive analytic obeying the large-frequency behaviour (3.93b). According to Appendix A.3, this request is equivalent to the possibility to write $\hat{M}_{Lq}(z)$ as a Laplace transform of a continuous positive definite function $\hat{M}_{Lq}(t)$. Equation (4.10a) has the same form as the representation of the density correlator $\phi_q(z)$ in terms of the fluctuating-force correlator $M_{Lq}(z)$ if one replaces the latter by $\tilde{M}_{Lq}(z)$. Such a replacement was motivated by Eqs. (4.9a,b). The formula (4.10b) holds if the regular kernel is approximated by a white-noise term: $M_{Lq}^{\text{reg}}(z) = \pm i\nu_q$, Im $z \gtrless 0$. Since the discussion focusses on the low-frequency dynamics, the white-noise approximation for the regular kernel appears reasonable. The expression for $\hat{M}_{Lq}(z)$ is given by the Laplace transform of the mode-coupling contribution, Eq. (4.7b).

If the simplifications discussed in connection with Eqs. (4.9a,b) should make sense, the low-frequency part of the Laplace transform of $\phi_k(t)\phi_p(t)$ should be close to that of $\tilde{\phi}_k(t)\tilde{\phi}_p(t)$. This observation motivates a formula for $\hat{M}_{Lq}(t)$, which is obtained from Eq. (4.7b) by adding tildes on the correlators of the right-hand side. The result shall be expressed by a dimensionless function $m_q(t)$:

$$\hat{M}_{Lq}(t) = \Omega_q^2 m_q(t), \tag{4.11a}$$

$$m_q(t) = L^{-3} \sum_{\vec{k}+\vec{p}=\vec{q}*} V(\vec{q}^{\,*}, \vec{k}\vec{p})\tilde{\phi}_k(t)\tilde{\phi}_p(t). \tag{4.11b}$$

The frequency Ω_q is defined in Eq. (3.67). It absorbs a factor q from the amplitudes in Eqs. (4.5b,c) and a factor v_{th} from Eq. (4.7b). The inverse functions

S_k^{-1} and S_p^{-1}, which appear in $F^{(a)}_{q,kp}/(S_kS_p)$, are expressed in terms of the direct correlation functions c_k and c_p, respectively, with the aid of Eq. (3.61). The length of the normalization box is denoted by L so that $\rho = N/L^3$ abbreviates the number density of the particles. As a result, one gets

$$V(\vec{q}^*, \vec{k}\vec{p}) = \tfrac{1}{2}\rho S_q S_k S_p \Big\{ \big[(k^3 c_k + p^3 c_p)/q\big] + \rho c_3(q,k,p)\Big\}^2. \qquad (4.11c)$$

Here, $\vec{p}$ is understood as an abbreviation of $\vec{q}^* - \vec{k}$. The sum in Eq. (4.11b) can be converted in an integral: $L^{-3}\sum_{\vec{k}} \cdots = (2\pi)^{-3}\int d^3\vec{k} \cdots$. The latter is discussed conveniently in polar coordinates, using $\vec{q}^*$ as axis. Rotational invariance of the integrand with respect to this axis reduces the problem to a two-dimensional integral. Using k and p as variables, one arrives at:

$$m_q(t) = [\rho/(32\pi^2)] \int_0^\infty dk \int' dp S_q S_k S_p (pk/q^5)$$
$$\Big\{(q^2 + k^2 - p^2)c_k + (q^2 + p^2 - k^2)c_p + 2q^2 \rho c_3(q,k,p)\Big\}^2 \tilde{\phi}_k(t)\tilde{\phi}_p(t). \qquad (4.11d)$$

The prime at the p-integral means that the integration is restricted to the interval $|q-k| \le p \le |q+k|$.

Equation (4.11c) implies $\lim_{q\to 0} V(\vec{q}^*, \vec{k}\vec{p}) = \frac{1}{2}\rho S_0 S_k^2 \{c_k + [(k^3)^2 c'_k/k] + \rho c_3 (0,k,k)\}^2$, where $c'_k = dc_k/dk$. Consequently, Eq. (4.11b) yields for the zero-wave-number limit of $m_q(t)$:

$$m_0(t) = S_0 \int_0^\infty dk v_L(k) \tilde{\phi}_k(t)^2, \qquad (4.12a)$$

$$v_L(k) = [\rho/(4\pi^2]k^2 S_k^2 \Big\{\big[c_k + \rho c_3(0,k,k)\big]^2 + \tfrac{2}{3}k\big[c_k + \rho c_3(0,k,k)\big]c'_k + \tfrac{1}{5}k^2 c'^2_k\Big\}. \qquad (4.12b)$$

From Eq. (4.11a), one obtains for the small-q limit of kernel $\hat{M}_{Lq}(t)$:

$$\lim_{q\to 0} \hat{M}_{Lq}(t)/q^2 = [v_{th}^2/S_0] m_0(t). \qquad (4.12c)$$

The factor q^2 results from the fact that both amplitudes $F^{(a,b)}_{q,\vec{k}\vec{p}}$ in Eq. (4.5a) vanish proportional to q in the long-wavelength limit. This holds for simple systems as well as for those subsystems of larger many-particle ensembles, which are specified in connection with Eq. (4.1d).

Assuming the wave-number dependent functions ν_q, Ω_q and $V(\vec{q}, \vec{k}\vec{p})$ as given, the formulas (4.10a)–(4.11c) specify a closed set of equations to be obeyed by the functions $\tilde{\phi}_q(t)$ and $\hat{M}_q(t)$. The parameters $V(\vec{q}, \vec{k}\vec{p})$ play the role of coupling constants. Notice that neither the temperature T nor the interaction potentials occur explicitly in the cited equations. These quantities occur only implicitly as parameters controlling the functions S_q, c_q and $c_3(q,k,p)$. The latter relate the coupling constants to the equilibrium structure. It will be explained in later sections that the parameters ν_q and Ω_q do not play a subtle role for the long-time dynamics exhibited by the solutions $\tilde{\phi}_q(t)$. The Eqs. (4.10a)–(4.12c) have been formulated first by Bengtzelius *et al.* (1984). There is a series of papers dealing with derivations of these formulas by reasonings different from the one outlined above, as can be inferred from a review by Das (2004).

Equations (4.10a)–(4.11c) formulate the request of a simultaneous evaluation of current-relaxation processes and the density-fluctuation propagation. The force-fluctuation kernel $\hat{M}_{Lq}(z)$ determines the current relaxation. The latter determines the autocorrelation functions for the density fluctuations via the continuity equation. The expression (4.10a) formulates this implication of the microscopic equations of motion as an elementary nonlinear relation between the correlators in the frequency domain. The factorization results (4.11a–c) formulate an elementary nonlinear connection between the density correlators and the force correlators in the time domain. They model the fact that slow density fluctuations imply slow force fluctuations. The subtlety of the self-consistency problem is connected with the requirement that the correlators in the time domain have to be positive definite functions, which are related to the correlators in the frequency domain by an integral transform.

It appears appropriate to close this section with some comments on self-consistent mode-coupling approaches, which have been analysed prior to the one formulated above. The factorization ansatz, which is an essential step in all mode-coupling theories, was introduced by Kawasaki (1966). He extended a study of Fixman (1962) to a general theory for the low-frequency behaviour of transport coefficients near second-order phase transitions. In every particular mode-coupling theory, insight into the dynamics is required in order to decide, which generalized fluctuating force has to be coupled to which correlators. Near critical points for phase transitions, long-range-order effects cause small-wave-number divergencies of susceptibilities. These imply singularities of the mode-coupling coefficients, which are the analogues of $V(\vec{q}^*, \vec{kp})$. The mentioned singularities are taken as input information from the equilibrium theory. As a result, one finds power-law divergencies of the relaxation spectra specified by non-trivial exponents (Kadanoff and Swift 1968; Wegner 1968; Kawasaki 1970). The factorization ansatz cannot handle all long-range-order effects adequately, and, therefore, the calculated exponents are not exact. But, often, the results are satisfactory approximations.

The closed mode-coupling equations studied this book differ in two respects from those studied in the cited papers on critical-point dynamics. First, the equations considered here depend smoothly on control parameters like the temperature T. There is no critical temperature introduced as input quantity. Second, small wave-number contributions to the mode-coupling integrals (4.11d), (4.12a) do not play a distinguished role. These two statements hold for all systems, where glassy dynamics does not interfere with critical-point effects.

Self-consistent mode-coupling equations for a calculation of the complete fluctuation spectra for densities and currents of simple liquids have first been motivated and solved for liquid argon in its normal state (Götze and Lücke 1975; Bosse *et al.* 1978a,b). This and the follow-up work achieved agreement of the theoretical results with the data known from experiment and simulation studies on a 20% accuracy level. But, the mathematical structure of the self-consistency equations used in the cited work differs from the one considered here. Sjögren (1980a) imbedded the factorization ansatz in the kinetic equation approach towards liquid dynamics (Boon and Yip 1980). His theory implies expressions for the quantities $\delta\Omega^L_{q,m}(z)$ and $M^{\mathrm{reg}}_{Lq}(z)$ in Eq. (4.3a). Dropping the former and simplifying the latter, his formulas reduce to Eqs. (4.10a)–(4.11c). The solution of his equations for liquid rubidium (Sjögren 1980b) provide an almost perfect description of the data.

The quantum-mechanical generalization of Eqs. (4.10a)–(4.11c) was studied with the aim to explain the low-frequency excitations of liquid helium at zero temperature (Götze and Lücke 1976). In this work, the regular term ν_q was dropped, and the convolution approximation was made. An iteration procedure was used to solve the self-consistency equations. Thereby, a calculation of the dynamical structure factor $S(q,\omega) = S_q\phi''_q(\omega)$ was performed using the experimental results for the structure factor S_q as the only input information. One goal of this work was the determination of the Landau spectrum for the intermediate-wave-number region $10\ \mathrm{nm}^{-1} < q < 30\ \mathrm{nm}^{-1}$. It had been argued by R.P. Feynman that localized rearrangements of particles are unlikely to occur in the densely packed liquid of strongly repelling helium atoms. A particle can move only together with its cage. Particle motion is accompanied with a flow pattern similar to that around a sphere moving in an incompressible ideal fluid. This implication of the cage effect, which is called back flow, is described by the cited mode-coupling approach.

Another facet of self-consistent current-relaxation approaches had been described in a study of the density correlator $\phi_q(z)$ of a gas of non-interacting electrons moving in an array of impurity atoms. This model deals with an example for a two component system mentioned in connection with Eq. (4.1d). The factorization ansatz for the electrons' fluctuating-force kernel leads to a mode-coupling formula in analogy to Eq. (4.7b). But, because the forces act between the electrons and the impurities, the correlator product $\phi_k(t)\phi_p(t)$ has to be replaced by $\phi_k(t)\phi^{(i)}_p(t)$. Here, $\phi^{(i)}_p(t)$ denotes the impurity-density-fluctuation correlator. Since the impurities are considered as fixed in space, there holds $\phi^{(i)}_p(t) = \phi^{(i)}_p(t=0)$. As a result, the relaxation kernel is a linear function of the density correlators

$$M^{\mathrm{mc}}_q(t) = \sum_k U_{(q,k)}\phi_k(t). \tag{4.13}$$

The non-negative coupling constants $U_{(q,k)}$ depend smoothly on the impurity density $n^{(i)}$, which is the relevant control parameter of the model. There appear two regimes for the dynamics separated by a critical density $n_c^{(i)}$. For weak coupling, $n^{(i)} < n_c^{(i)}$, the density correlations decay to zero, $\phi_q(t \to \infty) = 0$, as anticipated throughout Chapter 3. The hydrodynamic regime deals with diffusion, quantified by a finite positive diffusivity D. However, in the strong-coupling regime, $n^{(i)} > n_c^{(i)}$, the electrons cannot percolate through the system. The diffusivity vanishes. The particles are localized in regions of a finite averaged diameter $r_0 : \phi_q(t \to \infty) = f_q = 1/[1 + (r_0 q)^2]$. Upon approaching the critical density $n_c^{(i)}$, the localization length r_0 diverges according to the law: $(1/r_0)^2 = C\epsilon + O(\epsilon^2)$, where $\epsilon = (n^{(i)} - n_c^{(i)}/n_c^{(i)}$ denotes a distance parameter (Götze 1978). The transition to localization results from a positive feed-back between density-fluctuation propagation and scattering efficiency. The scattering destroys the velocity correlations. As a result, the density-fluctuation propagation is slowed down. This is reflected by a diffusion-spectrum enhancement: $\phi_q''(\omega \to 0) \propto 1/(Dq^2)$, Eq. (3.40b). The increase of $\phi_q''(\omega)$ enhances the scattering efficiency, and this leads to a decrease of D (Götze 1981). If $n^{(i)}$ exceeds $n_c^{(i)}$, no propagation of the particles over long distances is possible. The particles are localized by the cage formed by the neighbouring scattering centres. The cited approach describes a transition from states with vanishing correlator-long-time limits to ones exhibiting arrested parts.

The classical version of the system discussed in the preceding paragraph is known as the Lorentz model. It deals with a single particle, or, equivalently, with a gas of non-interacting particles, moving in an array of scattering centers. The above cited self-consistent current-relaxation theory was applied to the simplest version of the model: a hard sphere of diameter d moves in an array of point scatterers, which are distributed with a density $n^{(i)}$ without any correlations. The relevant control parameter of the model is the dimensionless density $n^* = n^{(i)}d^3$. A transition from a system exhibiting diffusion to one with all particles being localized was obtained for the critical value $n_c^* = 9/4\pi$. For $(n_c^* - n^*)/n_c^* = \epsilon$ decreasing to zero, the localization length diverges as described above, $r_0^2 \propto 1/\epsilon$ (Götze *et al.* 1981b). Results of the approach (Götze *et al.* 1981a) and of its extension to a model with correlated scattering centers (Leutheusser 1983a) agree satisfactorily with the molecular-dynamics-simulation results, which were known at that time.

Edwards and Anderson (1975) suggested to characterize a glass state by the property that some correlation function exhibits an arrested part. They studied a system of fixed particles carrying a spin s_κ, $\kappa = 1, \ldots, N$. The spins experience a random distribution of interactions. The relevant control parameter of the model is the temperature T. The authors identified a critical value T_c separating regions of different dynamical behaviour. For $T > T_c$, the spin correlators $C(t) = \langle s_\kappa(t)s_\kappa\rangle/\langle s_\kappa^2\rangle$ relax to zero. But, for $T < T_c$, there is a non-trivial long-time limit: $C(t \to \infty) = q_{EA}$, $0 < q_{EA} < 1$. The arrested part q_{EA} was calculated by studying the averaging process of the distribution of free energies. The instability of the spin glass upon heating manifests itself by a linear decrease of the

Edwards–Anderson parameter: $q_{EA} \propto \epsilon$, where $\epsilon = (T_c - T)/T_c$. Hence, the spin-glass transition of the Edwards–Anderson theory is characterized by a continuous but singular control-parameter dependence of the correlator's long-time limit.

In this book, the concepts of an ideal glass state and of an ideal glass transition shall be used as suggested by Edwards and Anderson (1975). In a liquid state, the density fluctuation correlators $\phi_q(t)$ relax to zero. In a glass, there is an arrested part: $\phi_q(t \to \infty) = f_q$, $0 < f_q \leq 1$. This definition implies that the dynamical structure factor $S(q, \omega) = S_q \phi_q''(\omega)$ exhibits an elastic contribution $\pi S_q f_q \delta(\omega)$. Such spikes are a well-known signature of crystalline solids, where the analogue of f_q is called the Debye–Waller factor. For the Debye–Waller factor f_q of the amorphous glass, one expects a continuous dependence of f_q on the wave number. The definition of the ideal glass state is consistent with the discussions of Eqs. (3.117b), (3.119b): an ideal glass is a system exhibiting Mountain's peak whose width is degenerated to zero.

The mentioned mode-coupling approaches towards the glass transition (Götze 1978; Götze *et al.* 1981b) exhibit severe shortcomings. For example, interference effects for the scattering of De Broglie waves cause a low-frequency singularity of the velocity spectrum, which is not reproduced. Furthermore, the cited calculations had been done with the approximation of the kernel $M_q^{\text{mc}}(t)$ by its $q = 0$ value. Avoiding this simplification, yields artifacts (Leutheusser 1983b). Finally, it is obvious that the density correlator of the Lorentz system exhibits an arrested part f_q not only for $n^* > n_c^*$ but for all scatterer densities. No matter how small the impurity density might be, there is a non-zero probability for the particle to be trapped in a cluster of scatterers. The cited mode-coupling-theory calculations for the simple Lorentz system can describe only 95% of the decrease of D/D_0 due to the increase of n^*. Here, $D_0 = (dv_{th}/(3\pi n^*)$ denotes the small-n^* limit of the diffusivity D. The theory overestimates the trend to arrest in the sense that the calculated critical value n_c^* is about 17% smaller than the value established by molecular-dynamics-simulation studies (Höfling *et al.* 2006). The exponent ν, which specifies the divergence of the localization length $r_0 \propto 1/|\epsilon|^\nu$ for $|\epsilon| < 0.3$, is about 0.68 rather than the calculated value 1/2. The indicated defects are implications of the notorious problem that mode-coupling theories cannot handle divergent-length-scale problems.

It was pointed out by Sjölander and Turski (1978) and by Geszti (1983) that some self-consistent mode-coupling approach might yield a critical temperature T_c for simple systems separating a liquid phase from an amorphous-solid one. However, no equations have been formulated in these papers, which could be used to identify the conjectured transition or to calculate glassy-dynamics features of correlators. The closed set of equations (4.10a)–(4.11c) has been applied by Bengtzelius *et al.* (1984) to study the hard-sphere system. A liquid-to-glass transition was identified for the critical packing fraction $\varphi = 0.516$. The Debye–Waller factor f_q of the glass decreases monotonically to $f_q^c > 0$ for φ decreasing to φ_c.

Contrary to what is known for the transition from a spin glass to a paramagnetic system or for the transition from a Lorentz glass to a Lorentz liquid, the correlators' long-time limits vary discontinuously. The localization length r_0 does not diverge at the transition but increases to a finite value: $\lim_{\varphi \to \varphi_c +} r_0 = r_0^c \approx 0.07d$. The cited equations do not deal with a divergent-length-scale problem.

A simplified model for the mathematical problem posed by Eqs. (4.10a)–(4.11b) was analysed by Leutheusser (1984) and by Bengtzelius *et al.* (1984). This model deals with a single correlator $\tilde{\phi}(z)$ and a single kernel $\hat{M}(z)$ related by $\tilde{\phi}(z) = -1/\{z - \Omega^2/[z + i\nu + \hat{M}(z)]\}$, $\Omega > 0$, $\nu > 0$. The mode-coupling kernel is written as $\hat{M}(t) = \Omega^2 v \tilde{\phi}(t)^2$. The model exhibits a glass transition for the critical coupling constant $v_c = 4$. At $v = v_c$, there occurs a discontinuous change of the correlator's long-time limit. Increasing v towards v_c, there evolves a low-frequency loss peak albeit one described by Eq. (1.1a).

4.2 A mode-coupling theory

In this section, the basic version of a theory for the evolution of glassy dynamics in liquids shall be formulated as a problem of solving equations of motions for a set of M autocorrelation functions. The mathematical aspects shall be explained without reference to the preceding considerations of the dynamics of liquids. This is done in order to make it evident that the discussions of the solutions of the equations of motion are built on a solid basis. It will be shown that the above derived self-consistent current-relaxation approaches provide special examples of the general theory.

4.2.1 *Equations of motion and fixed-point equations*

A dynamics of a system shall be characterized by a set of M functions of the time denoted by $\phi_q(t)$, $q = 1, \ldots, M$. The requested standard properties are: the functions depend on the time t continuously, they are real, and they are invariant under time reversal. In this section, equations of motion for these functions shall be formulated. A covariance property of these equations will be derived. In order to simplify the notations, arrays of M numbers shall be denoted by bold letters, e.g., $\boldsymbol{x} = (x_1, \ldots, x_M)$. The arrays with all components being zero or one, shall be denoted by $\mathbf{0} = (0, \ldots, 0)$ and $\mathbf{1} = (1, \ldots, 1)$, respectively. Using this notation, the dynamics is a continuous mapping of the time axis in the set of arrays,

$$\boldsymbol{\phi}(t) = (\phi_1(t), \ldots, \phi_M(t)), \tag{4.14}$$

obeying the standard symmetries: $\boldsymbol{\phi}(t) = \boldsymbol{\phi}(t)^* = \boldsymbol{\phi}(-t)$.

The central concept of the following discussions is an array of polynomials $\boldsymbol{\mathcal{F}}[P, \boldsymbol{x}]$. The components shall be written as

$$\mathcal{F}_q[P, \boldsymbol{x}] = \sum_{n=1}^{n_0} \sum_{k_1 \cdots k_n} V^{(n)}_{q,k_1 \cdots k_n} x_{k_1} \cdots x_{k_n}, \quad q = 1, \ldots, M. \tag{4.15a}$$

The coefficients $V^{(n)}_{q,k_1 \cdots k_n}$ are symmetric functions of the n labels $k_1, \ldots, k_n$. There are no terms of zeroth order: $\boldsymbol{\mathcal{F}}[P, \mathbf{0}] = \mathbf{0}$. The N independent coefficients play the role of coupling constants of the theory. They shall be labelled somehow, $V_1, \ldots, V_N$, and considered as coordinates of a point P in an N-dimensional space. The point $P \leftrightarrow (V_1, \ldots, V_N)$ shall be called the state of the system. The coordinates of P, V_α, $\alpha = 1, \ldots, N$, are also called the control parameters. The essential restriction requested is that none of the coefficients is negative:

$$V^{(n)}_{q,k_1 \cdots k_n} \geq 0; \quad n = 1, \ldots, n_0; \quad q, k_1, \ldots, k_n = 1, \ldots, M. \tag{4.15b}$$

Let us note some obvious implications of these inequalities. If $0 \leq x_k$ for all k, there holds

$$0 \leq \partial^n \mathcal{F}_q[P, \boldsymbol{x}] / \partial x_{k_1} \ldots \partial x_{k_n} \tag{4.15c}$$

for all $n = 0, 1, \ldots$ and all $q, k_1, \ldots, k_n = 1, \ldots, M$. Similarly, if there are two arrays $\boldsymbol{x}$ and $\boldsymbol{y}$ obeying $0 \leq x_k \leq y_k$ for all k, there holds for all q:

$$\mathcal{F}_q[P, \boldsymbol{x}] \leq \mathcal{F}_q[P, \boldsymbol{y}]. \tag{4.15d}$$

The polynomials are used to define an array of kernels $\boldsymbol{m}(t) = \boldsymbol{\mathcal{F}}[P, \boldsymbol{\phi}(t)]$. The components

$$m_q(t) = \mathcal{F}_q[P, \boldsymbol{\phi}(t)] \quad q = 1, \ldots, M, \tag{4.16}$$

are continuous and they have the standard symmetries.

There are two versions of the equations of motion. The first one is specified by positive frequencies Ω_q and non-negative frequencies $\nu_q, q = 1, \ldots, M$. It is requested that the functions $\phi_q(t)$ have continuous first and second derivatives for $t \neq 0$ and that there exist finite one-sided limits $\partial_t \phi_q(\pm 0)$ of the first derivatives at the origin of the time axis. For $t > 0$ and $q = 1, \ldots, M$, there holds

$$\partial_t^2 \phi_q(t) + \nu_q \partial_t \phi_q(t) + \Omega_q^2 \big[\phi_q(t) + \int_0^t m_q(t - t') \partial_{t'} \phi_q(t') dt'\big] = 0 \tag{4.17a}$$

together with the initial conditions $\phi_q(+0) = 1$, $\partial_t \phi_q(+0) = 0$. The second version of the equations of motion is specified by an array of positive times τ_q, $q = 1, \ldots, M$. The functions $\phi_q(t)$ are requested to have continuous first derivatives for $t \neq 0$ and finite limits $\partial_t \phi_q(\pm 0)$. For $t > 0$ and $q = 1, \ldots, M$, there holds

$$\tau_q \partial_t \phi_q(t) + \phi_q(t) + \int_0^t m_q(t - t') \partial_{t'} \phi_q(t') dt' = 0 \tag{4.17b}$$

together with the initial conditions $\phi_q(+0) = 1$. Equations for $t < 0$ can be formulated by observing the requested symmetry $\boldsymbol{\phi}(-t) = \boldsymbol{\phi}(t)$.

If all the coefficients in Eq. (4.15b) vanish, all kernels are zero: $\boldsymbol{m}(t) = \mathbf{0}$. In this case, the first version for a dynamics deals with the correlation functions of M uncoupled harmonic oscillators specified by the oscillator frequencies Ω_q and damping constants ν_q. The second version for a dynamics deals with M uncoupled simple relaxation processes specified by the relaxation times τ_q. It shall be assumed in the following that the M polynomials are non-trivial: for each value of q, not all coefficients $V^{(n)}_{q,k_1\cdots k_n}$ must be zero. The kernels $m_q(t)$ depend on the array $\boldsymbol{\phi}(t)$, i.e., the equations of motion are nonlinear. The kernel $m_q(t)$ may depend on $\phi_k(t)$ for $k \neq q$, thereby providing a coupling of $\phi_q(t)$ with $\phi_k(t)$. The formulated equations of motion are nonlinear integrodifferential equations of the Volterra type. The latter property means that the equation for $\phi_q(t)$ is connected with the functions $\phi_k(t')$ for such times t', which precede t. Equations of this type have been studied extensively motivated, e.g., by problems of population dynamics. However, the form of the nonlinearities studied in that context (Gripenberg *et al.* 1990) are different from the one considered here.

Let us assume that there is a solution $\boldsymbol{\phi}(t)$ for the equations of motion, which is defined for $0 \leq t < t_{\max}$. Integrating Eq. (4.17a) twice and observing the requested initial conditions, one gets for times within the specified interval:

$$\begin{aligned}\phi_q(t) =& 1 + \nu_q t - \nu_q \int_0^t \phi_q(t')dt' \qquad (4.18a)\\ &- \Omega_q^2 \int_0^t \left\{ \int_0^{t'} \left[\phi_q(t'') + m_q(t'-t'')\phi_q(t'') - m_q(t'')\right] dt'' \right\} dt'.\end{aligned}$$

Integrating Eq. (4.17b) once and observing $\phi_q(0) = 1$, one obtains

$$\phi_q(t) = 1 - (1/\tau_q) \int_0^t \left[\phi_q(t') + m_q(t-t')\phi_q(t') - m_q(t')\right] dt'. \qquad (4.18b)$$

In order to arrive at the preceding formulas, the convolution integral together with the initial condition $\phi_q(0) = 1$ has been reformulated according to: $\int_0^t m_q(t-t')\partial_{t'}\phi_q(t')dt' = \partial_t \int_0^t m_q(t-t')\phi_q(t')dt' - m_q(t)$. Conversely, if an array $\boldsymbol{\phi}(t)$ obeys Equation (4.18a) or (4.18b), there hold the equation of motion (4.17a) or (4.17b), respectively, together with the corresponding initial conditions. Hence, the equations of motion together with the initial conditions are equivalent to the corresponding integral equations of the Picard type formulated above.

A first implication of the equations of motion is that a possible solution $\boldsymbol{\phi}(t)$ for $0 < t < t_{\max}$ has continuous derivatives of all orders, which have finite limits at the origin: $\boldsymbol{c}^{(\ell)} = \partial^\ell \boldsymbol{\phi}(t \to +0)$, $\ell = 0, 1, \ldots$ For $\ell = 0$ and 1, these properties have been incorporated in the definition of the solution. For larger ℓ, the proof can be done by induction. It is based on the following observations. If the array $\partial^\ell \boldsymbol{\phi}(t)$ is continuous and has finite limits $\boldsymbol{c}^{(\ell)}$ for

$t \to +0$ for $\ell = 0, 1, \ldots, n$, $n \geq 1$, also the polynomials $m_q(t)$ have continuous derivatives up to order n with finite limits $m_q^{(\ell)} = \partial_t^\ell m_q(t \to 0)$, $q = 1, \ldots, M$. Consequently, also the convolution integrals $i_q(t) = \int_0^t m_q(t-t')\partial_{t'}\phi_q(t')dt'$ have continuous derivatives up to order n and finite limits $\boldsymbol{i}^{(\ell)} = \partial_t^\ell \boldsymbol{i}(t \to +0)$. For example, $i_q^{(0)} = 0$, $i_q^{(1)} = m_q^{(0)} c_q^{(1)}$. From Eq. (4.17a), one concludes that $\partial_t^2 \phi_q(t)$ has first and second continuous derivatives with finite limits at the origin. Hence, the theorem holds for $n = 2$ and 3; etc. From Eq. (4.17b) one concludes that $\partial_t \phi_q(t)$ has a continuous derivative with a finite limit at the origin. Thus, the theorem holds for $n = 2$, etc.

The preceding paragraph implies that a possible solution exhibits an asymptotic Taylor series: $\phi_q(t) = 1 + \sum_{\ell=1}^n c_q^{(\ell)} |t|^\ell/\ell! + O(|t|^{n+1})$. Substituting this result into the equations of motion, one can determine recursively the coefficients $c_q^{(n+1)}$ from the coefficients $c_q^{(\ell)}$ for $\ell = 1, \ldots, n$. Let us note the result up to the first contribution, which is influenced by the kernels. These enter through their initial values $m_q^{(0)} = \mathcal{F}_q[P, \mathbf{1}]$. The first version of equation of motion yields

$$\begin{aligned}\phi_q(t) =& 1 - \tfrac{1}{2}(\Omega_q t)^2 + \tfrac{1}{6}(\Omega_q t)^2 \nu_q |t| \\ &+ \tfrac{1}{24}\Big\{\big[1 + m_q^{(0)}\big](\Omega_q t)^2 - (\nu_q t)^2\Big\}(\Omega_q t)^2 + O(|t|^5).\end{aligned} \tag{4.19a}$$

Equation (4.17b) leads to

$$\phi_q(t) = 1 - |t|/\tau_q + \tfrac{1}{2}\big[1 + m_q^{(0)}\big](t/\tau_q)^2 + O(|t|^3). \tag{4.19b}$$

A third implication of Eqs. (4.18a,b) is the uniqueness theorem. If $\boldsymbol{\phi}^{(1,2)}(t)$ are solutions for $0 < t \leq t_{\max}^{(1,2)}$, one gets $\boldsymbol{\phi}^{(1)}(t) = \boldsymbol{\phi}^{(2)}(t)$ for $0 < t < \min(t_{\max}^{(1)}, t_{\max}^{(2)})$. The proof can be copied from that which is done for the Picard equations within the theory of ordinary differential equations. Notice that the results of this paragraph and of the three preceding ones do not make use of the restrictions imposed by Eq. (4.15b).

For the study of the solutions of the equations of motion, it will be important to identify special arrays, to be called fixed points $\boldsymbol{f}^* = (f_1^*, \ldots, f_M^*)$. These are invariant arrays for a mapping $\boldsymbol{\mathcal{T}}[P, \boldsymbol{x}]$. This mapping is defined with the aid of the polynomials $\mathcal{F}_q[P, \boldsymbol{x}]$ for all arrays $\boldsymbol{x}$ with non-negative components.

$$0 \leq x_q : \mathcal{T}_q[P, \boldsymbol{x}] = \mathcal{F}_q[P, \boldsymbol{x}]/\{1 + \mathcal{F}_q[P, \boldsymbol{x}]\}, \quad q = 1, \ldots, M. \tag{4.20a}$$

Since $0 \leq \mathcal{F}_q[P, \boldsymbol{x}]$, there holds

$$0 \leq \mathcal{T}_q[P, \boldsymbol{x}] < 1, \quad q = 1, \ldots, M. \tag{4.20b}$$

This result can be generalized to pairs of arrays $\boldsymbol{x}$ and $\boldsymbol{y}$.

$$0 \leq x_q \leq y_q : \mathcal{T}_q[P, \boldsymbol{x}] \leq \mathcal{T}_q[P, \boldsymbol{y}], \quad q = 1, \ldots, M. \tag{4.20c}$$

A fixed point $\boldsymbol{f}^*$ is an array obeying:

$$0 \leq f_q^* < 1; \quad q = 1, \ldots, M : \boldsymbol{f}^* = \boldsymbol{\mathcal{T}}[P, \boldsymbol{f}^*]. \tag{4.21a}$$

It is an array, whose M components solve the M implicit equations

$$f_q^*/(1 - f_q^*) = \mathcal{F}_q[P, \boldsymbol{f}^*], \quad q = 1, \ldots, M. \tag{4.21b}$$

The fixed points are defined for each state P. If the dependence on P shall be emphasized, it will be indicated by writing $f_q^* = f_q^*(P)$. There is always the trivial fixed point $\boldsymbol{f}^* = \mathbf{0}$. Depending on the state P, there may be one or several non-trivial fixed points.

For a given fixed point $\boldsymbol{f}^*(P)$ for the state P, an invertible linear mapping $\boldsymbol{x} \to \hat{\boldsymbol{x}}$, can be defined by

$$x_q = f_q^*(P) + [1 - f_q^*(P)]\hat{x}_q, \qquad q = 1, \ldots, M. \tag{4.22a}$$

Substituting these formulas into Eq. (4.15a) and using Eq. (4.21b), one gets $\mathcal{F}_q[P, \boldsymbol{x}] = \{f_q^*(P) + \hat{\mathcal{F}}_q[P, \hat{\boldsymbol{x}}]\}/[1 - f_q^*(P)]$. Here, the new polynomials read

$$\hat{\mathcal{F}}_q[P, \hat{\boldsymbol{x}}] = \sum_{n=1}^{n_0} \sum_{k_1 \cdots k_n} \hat{V}_{q,k_1 \cdots k_n}^{(n)} \hat{x}_{k_1} \cdots \hat{x}_{k_n}, \tag{4.22b}$$

$$\begin{aligned} \hat{V}_{q,k_1 \cdots k_n}^{(n)} &= \frac{1}{n!}[1 - f_q^*(P)]\{\partial^n \mathcal{F}_q[P, \boldsymbol{f}^*(P)]/\partial x_{k_1} \cdots \partial x_{k_n}\} \\ &\quad [1 - f_{k_1}^*(P)] \cdots [1 - f_{k_n}^*(P)]. \end{aligned} \tag{4.22c}$$

The new array $\hat{\boldsymbol{\mathcal{F}}}[P, \hat{\boldsymbol{x}}]$ is given by polynomials of degree n_0 without constant terms as was explained for $\boldsymbol{\mathcal{F}}[P, \boldsymbol{x}]$. Also, the new coefficients $\hat{V}_{q,k_1 \cdots k_n}^{(n)}$ obey the condition (4.15b). All conclusions drawn from Eqs. (4.15a,b) hold also for the new quantities, which are denoted by the symbol $\hat{}$.

If one introduces a new mapping $\hat{\boldsymbol{\mathcal{T}}}[P, \hat{\boldsymbol{x}}]$ in analogy to Eq. (4.20a),

$$\hat{\mathcal{T}}_q[P, \hat{\boldsymbol{x}}] = \hat{\mathcal{F}}_q[P, \hat{\boldsymbol{x}}]/\{1 + \hat{\mathcal{F}}_q[P, \hat{\boldsymbol{x}}]\}, \tag{4.23a}$$

one finds the transformation rule $\mathcal{T}_q[P, \boldsymbol{x}] = f_q^*(P) + \hat{\mathcal{T}}_q[P, \hat{\boldsymbol{x}}][1 - f_q^*(P)]$. This means that the equations $\boldsymbol{y} = \boldsymbol{\mathcal{T}}[P, \boldsymbol{x}]$ and $\hat{\boldsymbol{y}} = \hat{\boldsymbol{\mathcal{T}}}[P, \hat{\boldsymbol{x}}]$ are equivalent. Let $\boldsymbol{f}'^*$ denote some fixed point for state P. It may be equal to the fixed point $\boldsymbol{f}^*(P)$ or not. One concludes on the equivalence of the following pair of equation

$$\boldsymbol{f}'^* = \boldsymbol{\mathcal{T}}[P, \boldsymbol{f}'^*], \qquad \hat{\boldsymbol{f}}'^* = \hat{\boldsymbol{\mathcal{T}}}[P, \hat{\boldsymbol{f}}'^*]. \tag{4.23b}$$

There holds the covariance theorem: the fixed-point equation is covariant under the transformation defined in Eqs. (4.22a–c).

Equation (4.22a) can also be used to define an invertible linear mapping between time dependent arrays $\boldsymbol{\phi}(t)$ and $\hat{\boldsymbol{\phi}}(t)$:

$$\phi_q(t) = f_q^*(P) + [1 - f_q^*(P)]\hat{\phi}_q(t); \quad q = 1, \ldots, M. \tag{4.24a}$$

The array $\boldsymbol{\phi}(t)$ obeys the standard conditions and exhibits the initial conditions $\boldsymbol{\phi}(t = 0) = \mathbf{1}$ if and only if the same holds for $\hat{\boldsymbol{\phi}}(t)$. Furthermore, $\boldsymbol{\phi}(t)$ has continuous derivatives for $t > 0$ up to order n with finite limits for $t \to \pm 0$ if and only if this holds for the array $\hat{\boldsymbol{\phi}}(t)$. Similarly, $\partial_t \phi_q(\pm 0) = 0$ holds if and only if $\partial_t \hat{\phi}_q(\pm 0) = 0$. Substituting Eq. (4.24a) into Eq. (4.16) one gets $m_q(t) = \{f_q^*(P) + \hat{m}_q(t)\}/[1 - f_q^*(P)]$, where $\hat{m}_q(t)$ obeys the analogue of Eq. (4.16):

$$\hat{m}_q(t) = \hat{\mathcal{F}}_q[P, \hat{\boldsymbol{\phi}}(t)]. \tag{4.24b}$$

Substituting these results in the equations of motion (4.17a,b), the latter are reproduced with $\boldsymbol{\phi}(t)$ and $\boldsymbol{m}(t)$ replaced by $\hat{\boldsymbol{\phi}}(t)$ and $\hat{\boldsymbol{m}}(t)$, respectively. Furthermore, the frequencies Ω_q, ν_q and the times τ_q are replaced by frequencies $\hat{\Omega}_q, \hat{\nu}_q$ and times $\hat{\tau}_q$, respectively, where

$$\hat{\Omega}_q^2 = \Omega_q^2/[1 - f_q^*(P)], \quad \hat{\nu}_q = \nu_q, \quad \hat{\tau}_q = \tau_q[1 - f_q^*(P)]. \tag{4.24c}$$

The equations of motion do not change their form. The covariance theorem is extended to the equations of motion. All formulas and theorems, which hold for the Eqs. (4.17a,b), hold also for the ones with the transformed quantities (Götze 1991).

4.2.2 *Mode-coupling-theory models*

The general theorems for the above defined equations of motion shall be formulated in this section. Some terminology for the discussion of these equations shall be introduced and a few examples will be specified.

The two basic propositions on the equations of motion concern the existence of a solution and its regularity behaviour. The existence theorem reads: there exists a solution $\boldsymbol{\phi}(t)$, which is defined for all times t. The components $\phi_q(t)$ are positive-definite functions, $q = 1, \ldots, M$. To formulate the regularity theorem, some positive $t_{\max}$ shall be chosen, and the time shall be restricted to the finite closed interval $0 \le t \le t_{\max}$. Some positive $v_{\max}$ shall be chosen, and the N coupling coefficients shall be restricted to the bounded closed regions $0 \le V_\alpha \le v_{\max}$, $\alpha = 1, \ldots, N$. If the first version of equations of motion are considered, three positive frequencies $\Omega_{\min}, \Omega_{\max}$, and $\nu_{\max}$ shall be chosen. The characteristic frequencies shall be restricted to the closed intervals $\Omega_{\min} \le \Omega_q \le \Omega_{\max}$, $0 \le \nu_q \le \nu_{\max}$, $q = 1, \ldots, M$. If the second version for the equations of motion are considered two positive rates $\gamma_{\min}$ and $\gamma_{\max}$ are chosen and the restriction $\gamma_{\min} \le 1/\tau_q \le \gamma_{\max}$, $q = 1, \ldots, M$, shall be imposed. The regularity theorem reads: the M functions $\phi_q(t)$ depend continuously on t, V_α, $\alpha = 1, \ldots, N$, and on $(\Omega_q, \nu_q, \tau_q)$, $q = 1, \ldots, M$, if all these variables are restricted to the specified finite closed intervals. The Laplace transforms $\phi_q(z) = LT[\phi_q(t)](z)$ and $m_q(z) = LT[m_q(t)](z)$ define invertible mappings of the functions and kernels, respectively, in the set of positive-analytic functions. For every $\epsilon > 0$, the functions $|\phi_q(\pm iy)y|$ and $|m_q(\pm iy)y|$ have a common bound for all $y \geq \epsilon$.

This and the following three paragraphs present the proof of the two theorems. The solution $\boldsymbol{\phi}(t)$ is constructed as the limit of a sequence of approximants $\boldsymbol{\phi}^{(r)}(t)$, $r = 0, 1, \ldots$. The sequence shall be defined recursively. By induction, the following properties of the functions $\phi_q^{(r)}(t)$, $r = 0, 1, \cdots$, $q = 1, \ldots, M$, shall be demonstrated. The functions are continuous and positive definite; for $t \neq 0$, they have continuous derivatives of all orders. The functions are normalized, $\phi_q^{(r)}(0) = 1$, so that Eq. (2.27b) implies

$$|\phi_q^{(r)}(t)| \leq 1. \tag{4.25a}$$

The functions $\phi_q^{(r)}(t)$ are elements of a linear function space $\mathbb{V}'$. The set $\mathbb{V}'$ consists of all complex valued functions $F(t)$, which are continuous, which are absolutely integrable, and which have an absolutely integrable spectral function $F''(\omega) = FT[F(t)](\omega)/2$. Hence, $F''(\omega)$ is an element of $\mathbb{V}'$ as well. The space $\mathbb{V}'$ is so restricted that all previously formulated theorems on Fourier transforms and Laplace transforms can be derived within the framework of elementary analysis. There hold the results on spectral representations and Bochner's theorem in a form so that only Riemann integrals occur. The functions $\phi_q^{(r)}(t)$ exhibit standard symmetries. Because of Eqs. (A.11b,c), this is equivalent to the request that the spectra exhibit the standard symmetries. Because of Eqs. (A.5b,c), the standard symmetries are equivalent to the following symmetries of the Laplace transforms. $F(z) = F(z^*)^* = -F(-z)$. Approximants for the kernels are defined by

$$\boldsymbol{m}^{(r)}(t) = \boldsymbol{\mathcal{F}}[P, \boldsymbol{\phi}^{(r)}(t)]. \tag{4.25b}$$

The components of $\boldsymbol{m}^{(r)}(t)$ are real polynomials of the components of $\boldsymbol{\phi}^{(r)}(t)$. Hence, the $m_q^{(r)}(t)$ are elements of $\mathbb{V}'$, which exhibit the standard symmetries. For $t \neq 0$, the functions have continuous derivatives of all orders. According to Appendix A.6, the conditions (4.15b) imply that the functions $m_q^{(r)}(t)$ are positive definite. In particular, there holds the analogue of the inequality (4.25a):

$$|m_q^{(r)}(t)| \leq \mathcal{F}_q[P, \mathbf{1}]. \tag{4.25c}$$

The elementary relaxation functions $\exp[-\Gamma|t|]$, $\Gamma > 0$, are examples for functions exhibiting all properties mentioned for the approximants. They are taken to define the zeroth-order array $\boldsymbol{\phi}^{(0)}(t)$. One chooses some sequence of positive rates Γ_q and writes $\phi_q^{(0)}(t) = \exp[-\Gamma_q|t|]$, $q = 1, \ldots, M$.

The sequence of approximants shall be defined by complementing Eq. (4.25b) by a linear integrodifferential equation. If the first version of equations of motion is considered, $\boldsymbol{\phi}^{(r+1)}(t)$ is the solution of

$$[\partial_t^2 + \nu_q \partial_t + \Omega_q^2]\phi_q^{(r+1)}(t) + \Omega_q^2 \int_0^t m_q^{(r)}(t-t')\partial_{t'}\phi_q^{(r+1)}(t')dt' = 0, \qquad (4.26a)$$

to be evaluated for the initial conditions $\phi_q^{(r+1)}(0) = 1$, $\partial_t \phi_q^{(r+1)}(0) = 0$, $q = 1, \ldots, M$. If the second version is discussed, the array is the solution of

$$[\tau_q \partial_t + 1]\phi_q^{(r+1)}(t) + \int_0^t m_q^{(r)}(t-t')\partial_{t'}\phi_q^{(r+1)}(t')dt' = 0, \qquad (4.26b)$$

to be solved with $\phi_q^{(r+1)}(0) = 1$, $q = 1, \ldots, M$. These equations hold for $t \geq 0$. The results for $t \leq 0$ are not considered explicitly, since $\boldsymbol{\phi}^{(r)}(t) = \boldsymbol{\phi}^{(r)}(-t)$. It follows from the discussion of Eqs. (3.90)–(3.94) that the equations for the first version define uniquely an array $\boldsymbol{\phi}^{(r+1)}(t)$. The functions $\phi_q^{(r+1)}(t)$ are positive-definite elements of $\mathbb{V}'$, which exhibit the standard symmetries. They have continuous first and second derivatives, which obey the inequalities (3.92b) with $\tilde{\phi}_q(t) = \phi_q^{(r+1)}(t)$. The request that there is a solution for the second version, which has a determining function as derivative, is equivalent to the following identity for the functions in the frequency domain: $\phi_q^{(r+1)}(z) = -1/[z + M_q(z)]$, $M_q(z) = -1/[\pm i\tau_q + m_q^{(r)}(z)]$, Im $z \gtrless 0$. Obviously, $M_q(z)$ is positive analytic and it exhibits continuous bounded limits $M_q(\omega \pm i0)$. From Sec. 2.4.2, one concludes that there is a uniquely determined $\phi_q^{(r+1)}(t)$, which is a positive-definite element of $\mathbb{V}'$ and which exhibits the standard symmetries. Summarizing, the sequence of approximants is constructed and all functions $\phi_q^{(r)}(t)$ exhibit the properties requested. The proofs imply that $\phi_q^{(r)}(t)$ depends continuously on all parameters $V_\alpha, \Omega_q, \nu_q$, and τ_q, which enter the equations of motion. The existence of continuous derivatives for $t \neq 0$ can be shown as done in the preceding section for the solutions of Eqs. (4.17a,b).

It is shown in Appendix D.1 that the sequence of approximants converges towards a limit $\boldsymbol{\phi}(t)$. For $q = 1, \ldots, M$, there holds

$$\lim_{r\to\infty} \phi_q^{(r+1)}(t) = \phi_q(t). \qquad (4.26c)$$

In Appendix D.1, it is shown in addition that the limits are uniform for all finite time intervals $0 \leq t \leq t_{\max}$. The uniformity holds also with respect to the dependence of the functions on the parameters $V_\alpha, \Omega_q, \nu_q$, and τ_q, provided they are restricted as specified above. Consequently, the proposition on regularity of the limit function $\boldsymbol{\phi}(t)$ is established. Let us select a sequence of times $t_1 < t_2 < \cdots < t_N$ and a sequence of complex numbers $\xi_1, \ldots, \xi_N$. According to Eq. (2.27a), the numbers $g_q^{(r)} = \sum_{j,k=1}^N \xi_j^* \phi_q^{(r)}(t_j - t_k)\xi_k$ are not negative. Equation (4.26c) implies $\lim_{r\to\infty} g_q^{(r)} = g_q \geq 0$, where $g_q = \sum_{j,k=1}^N \xi_j^* \phi_q(t_j - t_k)\xi_k$.

Because of Eq. (2.27a), this means that the components of $\boldsymbol{\phi}(t)$ are positive-definite functions. Since the $\phi_q^{(r)}(t)$ are continuous, the uniform limit is a continuous function as well. The standard symmetries for the approximants imply the same symmetries for the limits.

The two Eqs. (4.26a,b) are equivalent to a pair of Picard equations. The latter are obtained from Eqs. (4.18a,b) by identifying $m_q(t)$ with $m_q^{(r)}(t)$. For the first version of equations of motion, one gets for $t \geq 0$:

$$\phi_q^{(r+1)}(t) = 1 + \nu_q t - \nu_q \int_0^t \phi^{(r+1)}(t')dt' \tag{4.27a}$$
$$-\Omega_q^2 \int_0^t \Big\{ \int_0^{t'} [\phi_q^{(r+1)}(t'') + m_q^{(r)}(t'-t'')\phi_q^{(r+1)}(t'') - m_q^{(r)}(t'')]dt'' \Big\} dt'.$$

For the second version, one gets for $t \geq 0$:

$$\phi_q^{(r+1)}(t) = 1 - (1/\tau_q) \int_0^t [\phi_q^{(r+1)}(t') + m_q^{(r)}(t-t')\phi_q^{(r+1)}(t') - m_q^{(r)}(t')]dt'. \tag{4.27b}$$

Since the convergence in Eq. (4.26c) is a uniform one for every fixed finite time interval, one can interchange the limit of r tending to infinity with the integrations in the two preceding equations. As a result, one concludes that the limit $\boldsymbol{\phi}(t)$ obeys the Eqs. (4.18a) or (4.18b), respectively. In addition, there holds Eq. (4.16) for the kernels $m_q(t) = \lim_{r\to\infty} m_q^{(r)}(t)$. The limit array provides a solution of the equations of motion; and this finishes the proof of the two propositions.

For the second version of equations of motion, there holds a third basic theorem: the M functions $\phi_q(t)$ are completely monotonic (Götze and Sjögren 1995). Each of these functions obeys the sequence of inequalities (1.2b). They can be represented in the form of Eq. (1.2a) with $\rho(\gamma)d\gamma$ to be meant as abbreviation of a Stieltjes measure for an increasing weight function $d\sigma(\gamma)$. It is shown in Appendix D.2 that the approximants can be written as finite sums of elementary relaxation functions:

$$\phi_q^{(r+1)}(t) = \sum_{k=0}^{k_r} f_{q,k}^{(r+1)} \exp[-\gamma_{q,k}^{(r+1)'}|t|]. \tag{4.28}$$

The amplitudes $f_{q,k}^{(r+1)}$ and the rates $\gamma_{q,k}^{(r+1)'}$ are positive parameters. The weight function $\sigma_q^{(r+1)}(\gamma)$ in the representation (D.10) is a step function. The $(k_r + 1)$ steps are located at $\gamma = \gamma_{q,k}^{(r+1)'}$ and their size is $f_{q,k}^{(r+1)}$: $\rho_q^{(r+1)}(\gamma) = \sum_k f_{q,k}^{(r+1)}\delta(\gamma - \gamma_{q,k}^{(r+1)'})$. The limit of a convergent sequence of completely monotonic functions is completely monotonic (Feller (1971), Sec. XIII.1). Thereby, the third theorem is established.

The sequence of approximants $\boldsymbol{\phi}^{(r)}(t)$ can be used to determine numerically the solution of the equations of motion (Götze and Lücke 1976; Bengtzelius 1986a;

Götze and Sjögren 1988). For the applications of interest in this book, the solutions $\phi(t)$ are stretched over several orders of magnitude increase of (t/t_{mic}). In such cases, the convergence of the approximant sequence can be extremely slow.

The cited regularity conditions for the solutions permit to use Eqs. (A.6a,b) for the evaluation of the Laplace transform of the convolution integrals in Eqs. (4.17a,b). Equations (A.7a,b) can be used to evaluate the Laplace transform of the derivatives occurring in the equations of motion. Thereby, the latter equations are transformed in nonlinear relations between $\phi_q(z)$ and $m_q(z)$. As a result, the first version of the equations of motion together with their initial conditions is found to be equivalent to

$$\phi_q(z) = -1/\Big\{z - \Omega_q^2/\big[z \pm i\nu_q + \Omega_q^2 m_q(z)\big]\Big\}; \quad \operatorname{Im} z \gtrless 0. \tag{4.29a}$$

The second version of the equations of motion together with their initial conditions is equivalent to

$$\phi_q(z) = -1/\Big\{z - 1/\big[\pm i\tau_q + m_q(z)\big]\Big\}; \quad \operatorname{Im} z \gtrless 0. \tag{4.29b}$$

For $\operatorname{Im} z \neq 0$, there is a one-to-one correspondence between the holomorphic function $\phi_q(z)$ and a holomorphic function $\chi_q(z)$, which is defined by $\chi_q(z) = z\phi_q(z) + 1$. The first equation of motion is equivalent to

$$\chi_q(z) = -\Omega_q^2/\Big\{z^2 - \Omega_q^2 + z\big[\pm i\nu_q + \Omega_q^2 m_q(z)\big]\Big\}, \tag{4.30a}$$

while the second one is equivalent to

$$\chi_q(z) = 1/\Big\{1 - z\big[\pm i\tau_q + m_q(z)\big]\Big\}, \quad \operatorname{Im} z \gtrless 0. \tag{4.30b}$$

From a mathematical point of view, continuous dependence on time and positive definiteness are the defining properties of autocorrelation functions, as is explained in Chapter 2. Therefore, all properties discussed in Chapter 2 for autocorrelators are valid for the unique solutions of the equations of motion (4.17a,b). This holds in particular for the formulas on spectral representations. Therefore, the solutions $\phi_q(t)$ and $\phi_q(z)$ shall be called correlator of number q in the time domain and in the frequency domain, respectively. The functions $m_q(t)$ and $m_q(z)$ are referred to as relaxation kernels or fluctuating-force kernels. The equations of motion define a model for a statistical description of a dynamics. A particular model is specified by the number M for the length of the array, by the values for the frequencies Ω_q, ν_q or the times τ_q, and by the N numbers for the coefficients of the polynomials $\mathcal{F}_q[P, \boldsymbol{x}]$. The equations (4.16), (4.17a,b) are regular in the sense that the mentioned parameters enter linearly and that nonlinearities merely occur in form of monomials of the functions $\phi_q(t)$, $q = 1, \ldots, M$.

The calculated results $\phi(t)$ should be related to measurements by proposing the function $\phi_q(t)$ as normalized auto correlator in the time domain of some dynamical variable, say X_q,: $\phi_q(t) = (X_q(t)|X_q)/(X_q|X_q)$. Because of Eq. (2.25), a definite parity under conjugation, $X_{qc} = \pm X_{qc}$, would be sufficient to justify the requested reality of $\phi_q(t)$. A definite time-reversion parity, $\mathcal{S}_{re}X_q = \pm X_q$, would justify the requested time inversion symmetry (3.3c). Because of Eq. (2.82a), the function $\chi_q(z)$ is called the corresponding normalized dynamical susceptibility. It is related to the X_q–X_q dynamical susceptibility by $\chi_{X_q,X_q}(z) = \chi_q(z)\chi_q^T$, where Eq. (2.72b) determines the thermodynamic susceptibility $\chi_q^T = (X_q|X_q)/(k_\mathrm{B}T)$. The regularity theorem guarantees that the qualitative features of $\boldsymbol{\phi}(t)$ on a fixed finite time interval do not depend on fine tuning of the model parameters.

One can define a low-frequency regime for the dynamics by the request

$$|z| \ll \nu_q; \quad q = 1, \ldots, M. \tag{4.31a}$$

This is possible provided that none of the M white-noise parameters ν_q vanishes. In a leading-order approximation for the small parameters $|z|/\nu_q$, one can neglect z relative to $\pm i\nu_q$ in Eq. (4.29a). This leads to Eq. (4.29b) with the time scales given by

$$\tau_q = \nu_q/\Omega_q^2; \quad q = 1, \ldots, M. \tag{4.31b}$$

In this sense, the second version of the equations of motion is a limit case of the first one. Extending to all frequencies z the result of the low-frequency approximation can imply qualitative changes of the correlators for small times. This was explained in Sec. 3.2.2 for the theory of tagged-particle motion. In the present case, the solution of Eq. (4.17b) is fixed by the initial value $\phi_q(t=0) = 1$. This initial value is the same as that requested for Eq. (4.17a). There is no possibility to keep intact the second initial condition $\partial_t\phi_q(t=0) = 0$, requested for Eq. (4.17a). According to Eq. (4.19b), one gets for the second version for the equations of motion $\partial_t\phi_q(\pm 0) = \mp 1/\tau_q$, i.e., the ϕ_q versus t curves have a cusp artifact at $t = 0$. Using the results from Appendix C, one concludes: the coarse-grained solutions of the two versions of the equations of motion agree provided the polynomials $\mathcal{F}_q[P, \boldsymbol{x}]$ are identical, the scales for the short-time dynamics are related according to Eq. (4.31b), and the time scale t^* for the smoothening is not smaller than the minimum of M times $1/\nu_q$, $q = 1, \ldots, M$.

Equation (4.16) for the kernels together with the definitions of the polynomials $\mathcal{F}_q[P, \boldsymbol{x}]$ in Eqs. (4.15a,b) formulate that part of the models which goes beyond rewriting microscopic equations of motion. These equations formulate a coupling of the mode of the time evolution for a selected set of dynamical variables X_q, which is described by the correlators $\phi_q(t)$, and the mode of the dynamics of the fluctuating forces 'causing' the non-trivial features of this evolution. The relevant part of the fluctuating-force correlators are described by the functions $m_q(t)$. Thus, the models defined here are descriptions of a dynamics of the kind studied previously as self-consistent mode-coupling equations. Considering the strong theorems discussed above, it appears adequate to refer to the study of the models for a dynamics specified in Sec. 4.2.1 as mode-coupling theory (MCT). The polynomials $\mathcal{F}_q[P, \boldsymbol{x}]$ will be called mode-coupling functionals or mode-coupling polynomials. The non-negative coefficients $V^{(n)}_{q,k_1\cdots k_n}$ are referred to as mode-coupling coefficients. From a strict point of view, there is only one a priori justification for using Eq. (4.16) as model for fluctuating-force correlators: it is a formula which makes sense and which leads to non-trivial results.

The equations of motion (4.17a,b) have the form of linear differential equations with an inhomogeneity given by $i_q(t) = \int_0^t m_q(t-t')\partial_{t'}\phi_q(t')dt'$. As explained in Sec. 2.4, $i_q(t)$ is the retarded effect of the motion of the non-distinguished variables on the correlations of the distinguished ones. The effect is caused by the coupling to the 2 M distinguished variables $\{X_q, \mathcal{L}X_q, q = 1,\ldots,M\}$ at time t'. Therefore, $i_q(t)$ starts with the initial value zero. It evolves smoothly with increasing time, as it is caused by the short-time decay of the correlators. Therefore, the leading short-time decay of the correlations is not influenced by mode-coupling effects. This is demonstrated by the first lines on the right-hand side of Eqs. (4.19a,b). The second lines show the leading effect of mode-coupling on the short time dynamics. It is determined by the positive initial value of the correlators $m_q(t)$. It describes a delay of the correlator decay compared to that obtained without mode-coupling effects. This initial decay and extensions thereof shall be referred to as transient dynamics.

It is tempting to interpret Eq. (4.17a) as an equation of motion for some oscillator coordinate, which experiences a force $i_q(t)$. Such an interpretation can be very misleading, however. The equations of motion do not deal with the time dependence of coordinates, but with that of correlation functions. For a harmonic oscillator, the equation of motion for a correlator agrees with that for the displacement coordinate. Therefore, in the weak-coupling limit, the solutions can be viewed as the ones for coupled oscillators, and the dynamics can be studied by perturbation expansions. The resulting functions $\phi_q(t)$ decay to zero within time intervals specified by the natural time scales of the system $\Omega_q^{-1}, \nu_q^{-1}$ and τ_q. But, this book deals with features of a long-time dynamics, which are caused by strong-interaction effects. In this dynamical regime of intermediate and strong coupling, there is no obvious relation between the time dependence of the correlators and that of some typical orbit of the many-particle system within the same time interval.

The simplest MCT models deal with a single correlator to be denoted by $\phi(t)$. The label q can be dropped in all formulas. The mode-coupling functional for such one-component model is a polynomial of some degree n_0, $\mathcal{F}[P,x] = \sum_{n=1}^{n_0} v_n x^n$. The non-negative coefficients v_ℓ of this polynomial are the control parameters of the state P. The first version of the equations of motion (4.17a) is specified by a single frequency $\Omega > 0$ and by a friction parameter $\nu \geq 0$:

$$\partial_t^2\phi(t) + \nu\partial_t\phi(t) + \Omega^2[\phi(t) + \int_0^t m(t-t')\partial_{t'}\phi(t')dt'] = 0. \tag{4.32a}$$

The second version, Eq. (4.17b) is obtained by replacing $[\partial_t^2\phi(t) + \nu\partial_t\phi(t)]$ by $[\Omega^2\tau\partial_t\phi(t)]$, where $\tau > 0$ is the characteristic timescale of the system. Two examples shall be considered in detail later. The first one is based on a second-degree polynomial (Götze 1984):

$$m(t) = v_1\phi(t) + v_2\phi(t)^2. \tag{4.32b}$$

The other one deals with the third-order polynomial (Götze and Sjögren 1984):

$$m(t) = v_1\phi(t) + v_3\phi(t)^3. \tag{4.32c}$$

In either case, the state P is specified by a point a quadrant of a control-parameter plane: $(v_1, v_2), v_{1,2} \geq 0$ and $(v_1, v_3), v_{1,3} \geq 0$, respectively.

The simplest two-component model treats the correlators asymmetrically (Sjögren 1986). The dynamics of the first correlator is not influenced by the second one. It shall be denoted by $\phi(t)$, it obeys Eq. (4.32a) with $m(t) = \mathcal{F}[P, \phi(t)]$. The second correlator refers to a normalized variable $A = X_2/\sqrt{(X_2|X_2)}$. The parameters and functions connected with this correlator shall carry an index A. The equation of motion is specified by the two frequencies $\Omega_A > 0$ and $\nu_A \geq 0$:

$$\partial_t^2\phi_A(t) + \nu_A\partial_t\phi_A(t) + \Omega_A^2[\phi_A(t) + \int_0^t m_A(t-t')\partial_{t'}\phi_A(t')dt'] = 0. \tag{4.33a}$$

The coupling of the second correlator to the first one is quantified by a positive control parameter v_A:

$$m_A(t) = v_A\phi(t)\phi_A(t). \tag{4.33b}$$

Correlator $\phi_A(t)$ can exhibit a dynamics outside the transient region only if such dynamics is exhibited by $\phi(t)$. In this sense, the variable A probes the slow dynamics described by $\phi(t)$. Therefore, $\phi_A(t)$ is called the probing correlator.

A two-component model using only quadratic terms for the mode-coupling functionals was introduced by Bosse and Krieger (1987). The state P is specified by three control parameters $v_\ell \geq 0$, $\ell = 1, 2, 3$. The second version for the equations of motion shall be formulated for the correlator array $\boldsymbol{\phi}(t) = (\phi_1(t), \phi_2(t))$:

$$\tau_q\partial_t\phi_q(t) + \phi_q(t) + \int_0^t m_q(t-t')\partial_{t'}\phi_q(t')dt' = 0; \quad q = 1, 2. \tag{4.34a}$$

The mode-coupling kernels read

$$m_1(t) = v_1\phi_1(t)^2 + v_2\phi_2(t)^2, \tag{4.34b}$$

$$m_2(t) = v_3\phi_1(t)\phi_2(t). \tag{4.34c}$$

The model generalizes the one defined in Eqs. (4.33a,b) in the sense that the second correlator influences the changes of the first one.

Models of the kind defined above are called schematic ones. They can be used in order to demonstrate interesting features of the MCT with a minimum of efforts. The simplicity of these models implies artifacts for the solutions, i.e., features, which are not typical for the general models. A discussion on which results of a certain schematic model is typical requires an understanding of the generic results of the general MCT equations of motion. Such understanding will be achieved in the following parts of this book. Thereby, a posteriori motivations will appear for considering the specified schematic models.

A priori motivations for the study of schematic $M = 1$ models have been provided by generalizations of an approach, which was proposed by Sherrington and Kirkpatrick (1975) for the theory of spin-glass transitions. Within this frame, one describes the system by a set of spin variables σ_κ, $\kappa = 1, \ldots, N$. These spins obey a stochastic equation of motion, whose non-trivial part is due to an interaction energy $E_p = \sum_{\kappa_1,\ldots,\kappa_p} V^{(p)}_{\kappa_1\cdots\kappa_p}\sigma_{\kappa_1}\cdots\sigma_{\kappa_p}$, $p \geq 2$. The interaction constants $V^{(p)}_{\kappa_1\cdots\kappa_p}$ between p different spins are independent random variables, whose distribution is a Gaussian. The width parameter for the Gaussian is written as J_p^2/N^{p-1}. The difficult task is the evaluation of the properly averaged spin-correlation functions. The model does not yield correlators for different spins. The spin correlator $C_N(t) = \langle\sigma_\kappa(t)\sigma_\kappa\rangle/\langle\sigma_\kappa\sigma_\kappa\rangle$ is studied for the limit N tending to infinity. The result agrees with the solution of an one-component MCT model for the second version of the equation of motion. The mode-coupling function reads $\mathcal{F}[P, f] = v_n f^n$, $n = p - 1$. The coupling constant v_n is proportional to J_p^2/T^2, where T denotes the temperature (Kirkpatrick and Thirumalai 1987). Generalizing the model to $E = \sum_{p\geq 2} E_p$, one obtains the generalization of the mode-coupling functional to a general polynomial (Ciuchi and Crisanti 2000).

4.2.3 *The basic version of microscopic mode-coupling theories*

The discussions of Sec. 4.1.2 resulted in a closed set of equations for the evaluation of functions $\tilde{\phi}_q(t)$. These functions have been motivated as models for the low-frequency parts of the Laplace transforms of the correlators $\phi_q(t)$ for density fluctuations of wave number q in simple amorphous systems. In this section, it will be shown that the mentioned equations provide a special example for the equations of motion of the mode-coupling theory for an M-component array $\boldsymbol{\phi}(t)$. This holds provided M is chosen so large that the wave-number dependence of structure functions and correlation functions can be represented adequately by their values on a grid of M discrete values.

The set of equations (4.10a,b), (4.11a) is equivalent to the first version of the MCT equations of motion (4.29a). This holds if structure functions like S_q and correlators like $\phi_q(t)$ are replaced by their values on a discrete set of wave numbers. Under this condition, the integrals over k and p in Eq. (4.11d) can be replaced by a double Riemann sum. Thereby, $m_q(t)$ gets the form of Eq. (4.16). The functions $\mathcal{F}_q[P, \boldsymbol{x}]$ are homogeneous polynomials of degree 2 with non-negative coefficients. Hence, $\mathcal{F}_q[P, \boldsymbol{x}]$ is a polynomial of the form specified by Eqs. (4.15a,b). But, there occur only coefficients for $n = 2$. The standard symmetries are to be requested for density correlators according to the discussions of Sec. 3.3.1. Similarly, the initial conditions for $\phi_q(t)$ and its first and second derivative agree with those of the MCT solutions, as one sees by comparing Eq. (3.68b) with Eq. (4.19a). Thereby, the statements of the preceding paragraph are proven. The indicated rewriting of the equations of Bengtzelius *et al.* (1984) was the motivation to formulate the MCT equations in Sec. 4.2.1. The generalization of mode-coupling polynomials in the form of Eq. (4.15a) is done in order to achieve more transparency for the following mathematical considerations and to have a frame for generalizations of the microscopic approach.

To be more explicit, a length scale d of the order of the particle diameter shall be introduced. The spacing of the wave-number grid shall be chosen as some fraction of $1/d$, denoted by h/d. The grid is specified by the mid-points of the intervals on the wave-number axis:

$$q = \hat{q}(h/d), \quad \hat{q} = 1/2, 3/2, \ldots, \ (M - 1/2). \tag{4.35a}$$

The wave-number intervals are limited by the cutoff value $q_{co} = (Mh/d)$. Wave-number integrals are considered as equivalent to their Riemann sums:

$$\int_0^\infty \cdots dq = (h/d) \sum_{\hat{q}} \cdots . \tag{4.35b}$$

The density-fluctuation dynamics is described by the M functions $\tilde{\phi}_{\hat{q}(h/d)}(t)$, which shall be denoted by $\phi_q(t)$. It does not seem necessary to introduce a separate symbol like a tilde in order to avoid confusion with the true density correlators of a simple liquid as studied in Sec. 3.3. As mentioned, there holds the first version of the equations of motion (4.29a) or its equivalent (4.17a) in the time domain. The frequencies Ω_q are given by Eq. (3.67). According to Eqs. (4.11d), (4.16), the mode-coupling polynomials read

$$\mathcal{F}_q[P, \boldsymbol{x}] = \left\{[\rho S_q/(32\pi^2)](h/d)^3\right\} \sum_{\hat{k}} \sum_{\hat{p}}{}' S_k S_p(\hat{k}\hat{p}/\hat{q}^5)$$
$$\times \left\{(\hat{q}^2 + \hat{k}^2 - \hat{p}^2)c_k + (\hat{q}^2 + \hat{p}^2 - \hat{k}^2)c_p + 2\hat{q}^2 \rho c_3(q, k, p)\right\}^2 x_k x_p. \tag{4.36}$$

The prime on the sum indicates the restriction to $|\hat{q} - \hat{k}| + 1/2 \leq \hat{p} \leq \hat{q} + \hat{k} - 1/2$. The other symbols have the same meaning as explained in connection with Eq. (4.11c). The coefficients $V^{(2)}_{q,kp}$ in Eq. (4.15a) are given by the product of the factors in front $x_k x_p$. The functionals $\mathcal{F}_q[P, \boldsymbol{x}]$ are determined by equilibrium quantities. A dependence on the particle mass and an explicit dependence on the temperature enter the theory merely via the frequencies Ω_q and, possibly, via the friction parameters ν_q.

All mode-coupling coefficients $V^{(2)}_{q,kp}$ share the factor ρS_q. The coefficients increase with ρ, since this factor denotes the averaged number density for finding a neighbour of a tagged particle, which can contribute to the interaction force. According to Eq. (3.62b), the wave-number dependence of the factor S_q is that of the compressibility. This q-dependence affects strongly that of the fluctuating-force kernel $m_q(t)$. The two terms $(S_k S_p)$ and (kp) in the sum (4.36) shall be referred to as compressibility factor and phase-space factor, respectively. The first factor describes the expectation value $\langle|\rho_{\vec{k}}|^2\rangle\langle|\rho_{\vec{p}}|^2\rangle$ for finding a density fluctuation with wave vector $\vec{k}$ and one with wave vector $\vec{p}$. An overlap with the fluctuating force in Eq. (4.1c) exists only if $\vec{k}+\vec{p}=\vec{q}^*$. This is the selection rule (3.13), which is valid for fluctuating densities in homogeneous systems. There is invariance of overlaps with respect to rotations around the distinguished vector $\vec{q}^*$. As a result, the six-dimensional integral over $\vec{k}$ and $\vec{p}$ reduces to a two-dimensional one: $\int d^3k \int d^3p\delta(\vec{q}^* - \vec{k} - \vec{p}) \ldots = \int dk \int' dp 2\pi(kp/q) \ldots$. The number of fluctuations with wave number k and wave number p, which contribute, increases linearly with k and with p.

For weak interaction potentials $V(r)$, the direct correlation function is given by $c_q = -\hat{V}_q/(k_\mathrm{B}T)$. Here, $\hat{V}_q$ denotes the Fourier transform of $V(r)$ as defined by Eqs. (3.9a,b). The function $(-k_BT)c_q$ has the meaning of an effective or screened interaction potential (Hansen and McDonald 1986). The effective-interaction factors c_k and c_p are crucial contributions to the curly bracket in Eq. (4.36). The triple-correlation term $c_3(q,k,p)$ results from the projector $\mathcal{Q}_1$ in Eq. (4.1c). The projector accounts for the fact that a contribution to the forces F_{Lq} has to be projected out, since it is already accounted for by the $\Omega_q^2\phi_q(t)$-term in Eq. (4.17a). The forces are not given by the potentials but by their gradients $\vec{\partial}V(r)$. Fourier transformation yields the contributions $k^3c_k = (\vec{k}\vec{q}^*)c_k/q$ and $p^3c_p = (\vec{p}\vec{q}^*)c_p/q$. Transformation to bipolar coordinates yields $\vec{k}\vec{q}^* = (q^2+k^2-p^2)/2$ and $\vec{p}\vec{q}^* = (q^2+p^2-k^2)/2$, respectively. The parenthesis formulate the gradient factors. They increase with the wave numbers k or p.

The specified example of a mode-coupling theory is definitive in the following sense. A set of given structure functions and the thermal velocity, complemented by a set of M friction parameters ν_q, determine the M correlation functions $\phi_q(t)$ uniquely. These MCT models formulate microscopic theories for simple amorphous matter.

The study of small wave-number limits of correlators and kernels for homogeneous systems is of particular interest since it leads to the results for hydrodynamic phenomena. For later reference, the convenient procedure for determining such limits within microscopic MCT models shall be explained already here. The fluctuating-force kernel $m_q(t)$ shall be chosen as representative example. For a fixed cutoff number q_{co}, the limit M increasing to infinity has to be considered in order to allow for the smallest point of the grid, $h/(2d) = q_{co}/(2M)$, to approach zero. But, this limit transforms the sums in Eq. (4.35b) to the integrals used before the discretization. For example, from Eq. (4.36) together with Eq. (4.16), one gets Eq. (4.11d). Here, the limit of vanishing q can be performed. For the example, one gets Eq. (4.12a) for the desired quantity $m_0(t)$.

Discretization according to Eq. (4.35b) leads to an expression in terms of a new mode-coupling functional, which shall be denoted by $\mathcal{F}_L[P, \boldsymbol{x}]$:

$$m_0(t) = \mathcal{F}_L[P, \boldsymbol{\phi}(t)]. \tag{4.37a}$$

This formula is the analogue of Eq. (4.16). It expresses $m_0(t)$ as a polynomial with positive coefficients of the M components of the array $\boldsymbol{\phi}(t)$:

$$\mathcal{F}_L[P, \boldsymbol{x}] = \sum_k V_{Lk} x_k^2, \quad V_{Lk} = S_0(h/d) v_L(k). \tag{4.37b}$$

The MCT equation (4.29a) formulates the correct general structure of the density correlator, which is noted in Eqs. (3.90a,d) with $M_{Lq}(z) = q^2 K_{Lq}(z) = \pm i\nu_q + \hat{M}_{Lq}(z)$ and $\hat{M}_{Lq}(z) = \Omega_q^2 m_q(z)$. In Sec. 3.3, it is discussed that a factor q^2 can be split off from $M_{Lq}(z)$ so that there is a finite zero-wave-number limit of $K_{Lq}(z)$. This factor q^2 reflects the momentum-conservation law of the Newtonian dynamics of the many-particle system with translation-invariant interactions. According to Eq. (4.12c), the mode-coupling kernel $\hat{M}_{Lq}(z)$ reproduces this general result. Hence, also the white-noise term has to split off a factor q^2:

$$\nu_q = q^2 \nu_q^{\mathrm{reg}}. \tag{4.38}$$

Here, $\nu_{q=0}^{\mathrm{reg}} = \nu_0$ is a finite non-negative number. There is no possibility to restrict frequencies to such region that the condition from Eq. (4.31a) is obeyed for all small wave numbers. Therefore, the second version of the MCT equations of motion must not be used as a model for small-q phenomena in homogeneous simple systems.

Let us consider also the case discussed above in connection with Eqs. (4.1d,e). The many-particle system is a subsystem of a larger one. There is a non-trivial dynamics of the total momentum of the subsystem, which is described by the zero-wave-number limit of the fluctuating-force kernel $\lim_{q\to 0} M_{Lq}(z) = M_{L0}(z)$. Because of Eq. (4.12c), the mode-coupling part of the kernel does not contribute to this limit. Thus, the total-momentum relaxation is determined by the white-noise term:

$$\nu_{q=0} = \nu > 0. \tag{4.39a}$$

In this case, a low-frequency regime can be defined according to Eq. (4.31a); and, within this regime, the dynamics is described by the second version of the equations of motion. There hold Eqs. (4.17b), (4.29b) with the time scales given by

$$\tau_q = [S_q/(qd)^2][\nu_q (d/v_{th})^2]. \tag{4.39b}$$

If the density ρ tends to zero, the structure factor S_ρ tends to unity and $m_q(t)$ vanishes. The correlator from Eq. (4.29b) reduces to $\phi_q(z) = -1/[z \pm i D_q q^2]$, with

$D_q = v_{th}^2/\nu_q$. This is the density correlator (3.40a) for a tagged particle with a diffusion constant generalized to a wave-number-dependent function. The latter is related to a wave-number-dependent friction constant $\xi_q = \mu\nu_q$ according to Einstein's law (3.49c). Hence, the long-wave-length dynamics of the dilute system deals with Brownian motion. The MCT model deals with the modifications of this dynamics due to cage-effect dominated interactions among the particles of the subsystem.

The discussions of the preceding two paragraphs suggest that we refer to the second version of the equations of motion as mode-coupling theory for a stochastic dynamics. The MCT based on the first version will be referred to as theory for a Newtonian dynamics. The motivations for the self-consistent-current relaxation approach given in Sec. 4.1 imply the suggestion to use the first and second version of the MCT equations for a description of conventional simple systems and of simple colloidal suspensions, respectively.

Usually, theories on colloid dynamics are based on the Smoluchowski equation for the solute. This is a reduction of Hamilton's equations for the phase-space variables of the complete system to a closed equation of motion for the configuration-space variables of the suspended subsystem. The Smoluchowski equations are obtained by considering the limit of a diverging separation of the time scale for the dynamics of the solvent from that of the solute. A formalism for correlation functions and for their expressions in terms of kernels can be developed in analogy to what is discussed in Chapter 2. Within this frame, it can be shown that the autocorrelators of all configuration-space variables X are completely monotonic. Details can be inferred from a review by Nägele (1996). The mode-coupling theory for a stochastic dynamic reproduces the complete monotony of the correlators. Within the Smoluchowski-equation formalism, Szamel and Löwen (1991) have derived self-consistent mode-coupling equations of motion for the density-fluctuation correlators. Their equations are reproduced by the above specified MCT for simple colloidal suspensions.

Fundamental properties of autocorrelation functions concern connections of the values of the functions at any given time t_1 with the value at any other time t_2. This is exemplified, e.g., by the request of positive definiteness. Equivalently, there is the requirement that the spectra have to be non-negative for all frequencies. A theory which intends not to violate these connections has to deal with the correlators for all times. The MCT equations of motion have been motivated by the desire to treat a cage-effect-induced small-frequency dynamics. But, in order to arrive at a well defined set of equations for correlators, the results are imbedded in a formalism, which yields also a theory for the transient. The leading terms of the transient are given by the first lines in Eqs. (4.19a,b). These terms are not modified by mode-coupling effects. The transient is influenced by the mode-coupling functionals, as is shown by the second lines in the cited formulas. It is evident that the MCT results for the transient do not reproduce the state of the art of normal-liquid theory. For example, there are binary collision effects in conventional liquids and hydrodynamic-interaction effects in colloids, which are not treated.

4.2.4 *An elementary mode-coupling-theory model*

In this section, a one-component model shall be discussed. Its fluctuating-force kernel is a linear function of the correlator, which is quantified by a single control parameter v. The solution of the equations of motion in the frequency domain can be expressed in terms of elementary functions. It describes a glass transition characterized by a continuous but non-analytic dependence of the correlator's long-time limit on v (Götze 1984). The low-frequency dynamics for states near the transition point can be described by a scaling law (Götze and Sjögren 1984). It will be shown that the range of frequencies and times for the validity of the scaling-law description can be reduced by oscillation phenomena of the transient dynamics.

(i) The ideal glass transition

The $M = 1$ model to be studied obeys the equation of motion (4.32a) or its equivalent in the frequency domain. There holds Eq. (4.29a) with the index q dropped for all quantities. The mode-coupling functional is given by Eq. (4.32b), specialized to $v_2 = 0$. The relaxation kernel reads $m(t) = v\phi(t)$, $v > 0$. The formula for the kernel in the frequency domain is as elementary as that in the time domain: $m(z) = v\phi(z)$. This identity exemplifies an essential feature of the MCT in its pure form: there is no separation of the time scale for the correlations of the distinguished variable from that for the fluctuating-force correlations. The former are described by the normalized correlator $\phi(t)/\phi(t = 0)$. The non-trivial part of the latter is described by the normalized function $m(t)/m(t = 0)$. Both functions are identical.

For Im $z > 0$, the equation to be solved reads:

$$\phi(z) = -1/\Big\{z - \Omega^2/\big[z + i\nu + \Omega^2 v\phi(z)\big]\Big\}. \tag{4.40a}$$

The problem for Im $z < 0$ shall not be formulated explicitly, since the results are implied by the symmetry: $\phi(z) = -\phi(-z)$. The equation for a stochastic dynamics is obtained by replacing $z+i\nu$ by $i\nu$ with $\nu = \Omega^2\tau$, $\tau > 0$. The equation defines an algebraic function on two Riemann sheets connected by second-order branch points. To write the result explicitly, the square root of a complex number $\zeta = r\exp[i\varphi]$, $r > 0, -\pi < \varphi \leq \pi$, shall be specified by $\sqrt{\zeta} = \sqrt{r}\exp[i\varphi/2]$. Let us note the result as the formula for the susceptibility $\chi(z) = [z\phi(z) + 1]$. On a first sheet, there holds

$$\begin{aligned}\chi(z) =\Big\{ & - z(z + i\nu) + (1 + v)\Omega^2 \\ & + \sqrt{z(z + i\nu) - \Omega_-^2}\sqrt{z(z + i\nu) - \Omega_+^2}\Big\}\Big/(2v\Omega^2).\end{aligned} \tag{4.40b}$$

Here, two frequencies are introduced by

$$\Omega_\pm = |1 \pm \sqrt{v}|\Omega. \tag{4.40c}$$

The function on the second sheet is obtained by changing the sign in front of the product of the square roots.

The function $\chi(z)$ is holomorphic for all finite z except, possibly, at the zeroes of the two polynomials $z(z + i\nu) - \Omega_\pm^2$. They characterize the correlators of harmonic oscillators, which are specified by the frequencies $\Omega_\pm$ and by the damping constant ν. As is discussed in connection with Eq. (A.40a), the zeroes are located in the half plane given by Im $z \leq 0$. Hence, on both sheets, the function $\chi(z)$ is holomorphic for Im $z > 0$. For the second sheet, one finds the large frequency behaviour: $\chi(z) = -z^2/(v\Omega^2) + O(z^0)$. This implies $\phi(z) = (\chi(z) - 1)/z = -z/(v\Omega^2) + O(1/z)$, which contradicts the asymptotic behaviour of a correlator. Consequently, the unique solution for the model under study is given by $\chi(z)$ noted above for the first sheet.

For $v \neq 1$, one gets $0 < \Omega_- < \Omega_+$. Function $\chi(z)$ is holomorphic in a neighbourhood of the origin of the frequency plane specified by $|z(z+i\nu)| < \Omega_-$. Here, one can expand $\chi(z) = \chi^0 + \chi^1 z(z + i\nu) + O([z(z + i\nu)]^2)$. One finds $\chi^0 = [1 + v - |1 - v|]/(2v)$ and $\chi^1 = [1 + v - |1 - v|]/[2v|1 - v|\Omega^2]$. Writing $f = 1 - \chi^0$, the function $\phi_{\text{reg}}(z) = \phi(z) + (f/z)$ is holomorphic except for the mentioned branch points. For $v = 1$, one gets $\Omega_- = 0$ and $f = 0$. The spectrum $\chi''(\omega)$ may exhibit a $\sqrt{\omega}$ singularity. In this case, $\phi''(\omega)$ has a $1/\sqrt{\omega}$ singularity. Hence, the spectrum $\phi''_{\text{reg}}(\omega)$ is integrable. The back transformation $\phi_{\text{reg}}(t)$ vanishes for large times as follows from the Riemann–Lebeque theorem. The Laplace back transformation of $(-1/z)$ is unity. One can summarize as follows. For coupling constants v not exceeding the critical value $v^c = 1$, the correlator has a vanishing long-time limit:

$$\phi(t \to \infty) = 0; \quad v \leq 1. \tag{4.41a}$$

For coupling parameters exceeding v^c, there is an arrested part:

$$\phi(t \to \infty) = f = (v - 1)/v; \quad v > 1. \tag{4.41b}$$

The arrested part of the correlator increases linearly with the separation of the control parameter v from the critical value: $f = (v - v^c) + O((v - v^c)^2)$. This is the same behaviour as was discussed by Edwards and Anderson (1975) for a spin-glass model. The arrested part f is determined by the mode-coupling functional; it is independent of the parameters Ω and ν for the short-time dynamics.

Figure 4.1 exhibits the evolution of the fluctuation spectrum $\phi''(\omega)$ with changes of the control parameter v. The spectrum for the undamped harmonic oscillator is obtained for $\nu = 0$ and $v = 0$. It consists of two spikes: $\phi''(\omega) = (\pi/2)[\delta(\omega - \Omega) + \delta(\omega + \Omega)]$. The arrested part of the correlator, $-f/z$, yields a spike $\pi f \delta(\omega)$. The spectrum for $\nu = 0.8\,\Omega$ and $v = 0$ is that of an underdamped harmonic oscillator. Because of the correlator normalization, $\phi(t = 0) = 1$, there

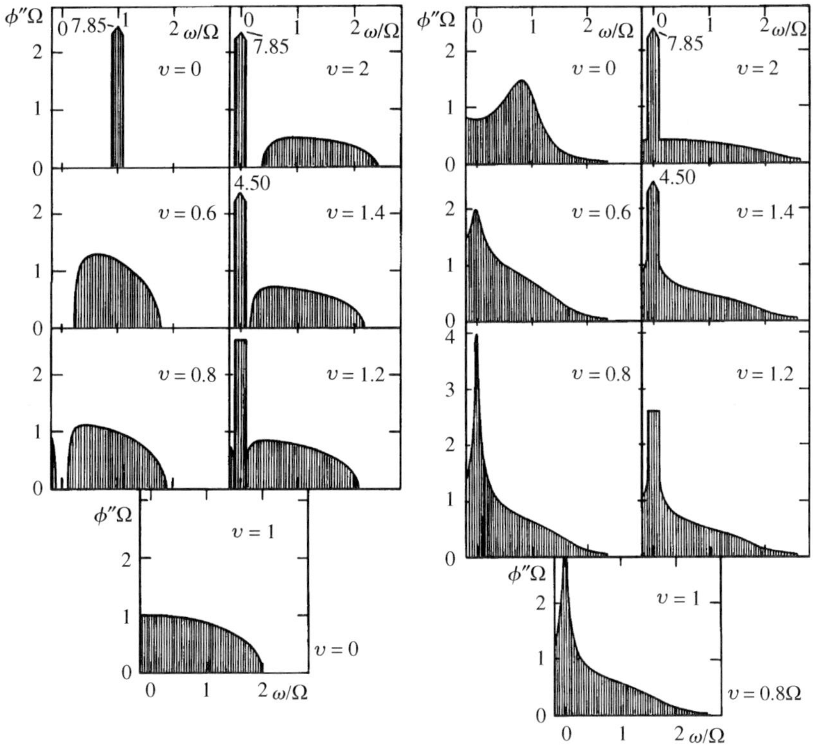

FIG. 4.1. Dimensionless fluctuation spectra $\phi''(\omega)\Omega$ as a function of the dimensionless frequency ω/Ω for various values of the control parameter v calculated for the elementary model defined by Eq. (4.40a). The left panel refers to vanishing damping constant $\nu = 0$, and the right one to $\nu = 0.8\Omega$. Ideal spectral spikes $c\delta(\omega - \omega_0)$ are shown as a square well for $|\omega - \omega_0| \leq 0.1\Omega$ and area c.

holds the sum rule for the regular part of the spectrum: $S = \int_{0+}^{\infty} \phi''(\omega) d\omega = (\pi/2)(1 - f)$. The sum S is independent of v if $v \leq 1$; but, it decreases with increasing coupling within the glass regime: $S = \pi/(2v)$, $v \geq 1$.

Even if there holds $\nu = 0$, the susceptibility spectrum is a continuous function of the frequency provided v is positive. In this case, the spectrum extends between square-root edges on the intervals $-\Omega_+ \leq \omega \leq -\Omega_-$ and $\Omega_- \leq \omega \leq \Omega_+$. Approaching the transition point by increasing v towards v^c or by decreasing it towards v^c, the spectral gap for $-\Omega_- \leq \omega \leq \Omega_-$ decreases. At the critical point, there is a positive spectrum extending between -2Ω and 2Ω.

(ii) The scaling limit

The results for a positive white-noise term ν are most relevant for later discussions. A positive ν leads to a smoothing of the spectra. The remaining discussions of this section shall be restricted to this case. For $v \neq 1$, $\chi''(\omega)$ is an analytic function of the frequency ω. The Taylor expansion of $\chi(z)$ mentioned above yields $\chi''(\omega) = \omega\nu\chi^1 + O(\omega^3)$. Hence, the low-frequency fluctuation spectrum is given by:

$$\phi''(\omega \neq 0) = [\tau/(1-v)] + O(\omega^2), \quad v < 1; \tag{4.42a}$$

$$= [\tau/v(v-1)] + O(\omega^2), \quad v > 1. \tag{4.42b}$$

The time scale τ is the one introduced in Eq. (4.31b) for the discussion of the second version of the equations of motion:

$$\tau = \nu/\Omega^2. \tag{4.42c}$$

The figure demonstrates that the increase of v from 0 to 0.6 implies a change from the spectrum of an underdamped oscillator to one similar to that of an overdamped one. The further increase of v to 0.8 implies a further increase of $\phi''(\omega = 0)$. The feedback between density-fluctuations increase and fluctuating-force-spectrum increase yields a quasi-elastic spectrum. The transition point is characterized by a divergency of the fluctuation spectrum. Equation (4.42b) shows that also the decrease of v towards the glass-instability point manifests itself in an increase of the zero-frequency spectrum towards infinity. For v near v^c, the wings of the quasi-elastic peak for $|\omega/\Omega| > 0.2$ are insensitive with respect to changes of v, as one notices by comparing the figures for $v = 0.8$, 1.0. and 1.2. These wings are precursors of the critical spectrum, i.e., of the spectrum at the transition point. For $v = 1$, one gets $\chi(z) = 1 - \sqrt{-iz\tau} + O(z)$. The transition point manifests itself by a square-root divergence of the fluctuation spectrum:

$$\phi''(\omega) = \sqrt{\tau/(2|\omega|)} + O(\omega^0), \quad v = 1. \tag{4.42d}$$

The correlator $\phi(z)$ depends in a subtle manner on the variable pair (z, ϵ), where ϵ denotes the distance of the control parameter from the critical point;

$$\epsilon = v - 1. \tag{4.43a}$$

For example, Eqs. (4.42a,b) yield $\lim_{\epsilon\to 0}[\lim_{\omega\to 0}\sqrt{\omega}\phi''(\omega)] = 0$, while Eq. (4.42d) leads to $\lim_{\omega\to 0}[\lim_{\epsilon\to 0}\sqrt{\omega}\phi''(\omega)] = \sqrt{\tau/2}$. In order to find a transparent description of the low-frequency dynamics near the transition point, Eq. (4.40b) shall be rewritten as

$$-z\phi(z) = \left\{iz\tau a(z) + \epsilon + \sqrt{\epsilon^2 - iz\tau[a(z)b(\epsilon)]}\sqrt{1 - iz\tau[a(z)/b(\epsilon)]}\right\}\Big/[2(1+\epsilon)]. \tag{4.43b}$$

The two functions

$$a(z) = 1 - [iz/(\Omega^2\tau)], \qquad b(\epsilon) = 2 + 2\sqrt{1+\epsilon} + \epsilon \tag{4.43c}$$

are introduced so that one can write $z(z+i\nu) = [iz(\Omega^2\tau)]a(z)$, $(\Omega_+/\Omega)^2 = b(\epsilon)$, and $(\Omega_-/\Omega)^2 = \epsilon^2/b(\epsilon)$. The function $a(z)$ is holomorphic and $a(z=0)=1$. The function $b(\epsilon)$ is holomorphic for $|\epsilon| < 1$. One concludes the following. For a given pair of constatns c_1 and c_2, which are restricted by $0 < c_{1,2} < 1$, one can find a sufficiently small positive number ϵ^* and a sufficiently large time t^* so that there holds

$$|z/\Omega^2\tau| < c_1, \qquad |z\tau a(z)/b(\epsilon)| < c_2, \tag{4.43d}$$

for all z and ϵ obeying

$$|zt^*| < 1, \qquad |\epsilon/\epsilon^*| < 1. \tag{4.43e}$$

In the following, frequencies z and distance parameters ϵ are restricted according to the preceding inequalities. Then $iz\tau a(z)+\epsilon$, $1/(1+\epsilon)$ as well as the second root in Eq. (4.43b) are holomorphic functions of ϵ and of z. A non-smooth behaviour of the function $-z\phi(z)$ can be caused solely by the factor $[\epsilon^2 - iz\tau a(z)b(\epsilon)]^{1/2}$.

The mentioned square root is qualitatively different depending on whether the small quantity $z\tau$ is small or large compared to the small quantity ϵ^2. To characterize these differences, a correlation scale c_ϵ and a time scale t_ϵ shall be defined by

$$c_\epsilon = |\epsilon|/(1+\epsilon), \qquad t_\epsilon = \tau b(\epsilon)/\epsilon^2. \tag{4.44a}$$

Both scales exhibit a power-law dependence on ϵ for states near the transition point: $c_\epsilon = |\epsilon|(1+O(\epsilon))$, $t_\epsilon = (4\tau/\epsilon^2)(1+O(\epsilon))$. Rescaled times $\hat{t}$, rescaled frequencies $\hat{z}$ and rescaled correlators $\hat{\phi}$ shall be defined by

$$\hat{t} = t/t_\epsilon, \quad \hat{z} = zt_\epsilon, \quad \phi(t) = c_\epsilon\hat{\phi}(\hat{t}), \quad \phi(z) = c_\epsilon t_\epsilon \hat{\phi}(\hat{z}). \tag{4.44b}$$

One concludes from Eq. (A.5a) that $\hat{\phi}(\hat{z}) = LT[\hat{\phi}(\hat{t})](\hat{z})$. For $\epsilon \gtrless 0$, Eq, (4.43b) is equivalent to

$$\hat{\phi}(\hat{z}) = \Big\{ \mp 1 - [i\hat{z}|\epsilon|\hat{a}(\hat{z},\epsilon)/b(\epsilon)] \\ - \sqrt{1 - [i\hat{z}\hat{a}(\hat{z},\epsilon)]}\sqrt{1 - [i\hat{z}\epsilon^2\hat{a}(\hat{z},\epsilon)/b(\epsilon)^2]}\Big\}\Big/(2\hat{z}), \tag{4.44c}$$

$$\hat{a}(\hat{z},\epsilon) = 1 - \big[i\hat{z}\epsilon^2/(\Omega^2\tau^2 b(\epsilon))\big]. \tag{4.44d}$$

For fixed rescaled frequency $\hat{z}$, the rescaled correlator $\hat{\phi}(\hat{z})$ is a smooth function of ϵ. The limit of a vanishing distance parameter reads:

$$\lim_{v\to v^c} \hat{\phi}(\hat{z}) = \varphi(\hat{z}) = \{ \mp 1 - \sqrt{[1 - i\hat{z}]}\}/(2\hat{z}); \quad \epsilon \gtrless 0. \tag{4.45}$$

The function $\overline{\varphi}(\hat{z}) = \{1 - \sqrt{1 - i\hat{z}}\}/(2\hat{z})$ is holomorphic for all $\hat{z}$ except for a square-root branch point for $\hat{z} = -i$. The Laplace back-transform $\overline{\varphi}(\hat{t})$, Eq. (A.4), can be evaluated as contour integral around the square-root cut on the negative imaginary axis:

$$\overline{\varphi}(\hat{t}) = \int_1^\infty \rho(\gamma) \exp[-\gamma \hat{t}] d\gamma, \qquad \rho(\gamma) = \sqrt{(\gamma - 1)}/(2\pi\gamma). \tag{4.46a}$$

One concludes from the discussion of Eqs. (1.2a,b) that $\overline{\varphi}(\hat{t})$ is completely monotonic. The integral can be expressed in terms of the error-function complement (Gradshteyn and Ryzhik (2000), Eq. (3.363)):

$$\overline{\varphi}(\hat{t}) = \left\{ \exp[-\hat{t}]/\sqrt{(\pi\hat{t})} - \mathrm{erfc}[\sqrt{\hat{t}}] \right\}/2. \tag{4.46b}$$

Consequently, the function $\varphi(\hat{z})$ is the Laplace transform of

$$\varphi(\hat{t}) = (1 \pm 1)/2 + \overline{\varphi}(\hat{t}), \qquad \epsilon \gtrless 0. \tag{4.46c}$$

The products $\hat{z}\epsilon$ and $\hat{z}\epsilon^2$ are the variables for the expansion of $\hat{\phi}(\hat{z})$ for small ϵ. One can write

$$\hat{\phi}(\hat{z}) = \varphi(\hat{z}) + \epsilon\hat{z}\varphi_1(\hat{z}) + O((\hat{z}\epsilon)^2, \hat{z}\epsilon^2). \tag{4.47a}$$

Let us prescribe some error margin ϵ^*. There is some constant c^* so that

$$|\hat{\phi}''(\hat{\omega}) - \varphi''(\hat{\omega})| \leq \epsilon^* \varphi''(\hat{\omega}) \tag{4.47b}$$

is valid, provided there holds the restriction

$$|\epsilon\hat{\omega}| \leq c^*. \tag{4.47c}$$

Ignoring a relative error ϵ^* and implying the restriction for $|\hat{\omega}|$ to values below $c^*/|\epsilon|$, one can write

$$\phi(z) = c_\epsilon t_\epsilon \varphi(z t_\epsilon). \tag{4.48}$$

Laplace back-transformation yields the formula

$$\phi(t) = c_\epsilon[\theta(\epsilon) + \overline{\varphi}(t/t_\epsilon)]. \tag{4.49}$$

According to Appendix C, the restrictions underlying the preceding identity mean that the coarse-grained functions agree within some error margin ϵ^{**}. The latter is of order ϵ^*. There holds $|\phi_*(t) - \tilde{\varphi}_*(t)| \leq \epsilon^{**}$, with $\tilde{\varphi}(t) = c_\epsilon[\theta(\epsilon) + \overline{\varphi}(t/t_\epsilon)]$. The scale for the smoothening is restricted by the condition $\hat{t}^* \geq \epsilon/c^*$.

The preceding two formulas are scaling laws. The sensitive dependence on ϵ, which is exhibited by $\phi(z)$ and $\phi(t)$, is described by the control-parameter independent function $\varphi(\hat{t})$ and its Laplace transform $\varphi(\hat{z})$, respectively. The sensitive dependence on ϵ enters via the two scales c_ϵ and t_ϵ. The approach towards the transition point manifests itself by a divergence of the time scale t_ϵ. The functions $\overline{\varphi}(\hat{t})$ and $\varphi(\hat{z})$ are called shape functions in the time domain and frequency domain, respectively. Equation (4.45) is called the scaling limit. It yields the leading term for an asymptotic expansion of the rescaled correlator $\hat{\phi}(\hat{z})$. The function $\epsilon\hat{z}\varphi_1(\hat{z})$ in Eq. (4.47a) is called the leading correction for the expansion. The scaling-law description for $\phi(z)$ of the elementary model under study is equivalent to that considered within the MCT for the percolation transition of a Lorentz model (Götze *et al.* 1981b).

The conditions (4.43d,e) for the applicability of a scaling-law description are equivalent to the request that the spectra for the transient dynamics are white noise ones. For the example exhibited in the right panel of Fig. 4.1, the conditions are valid for $|\omega/\Omega| < 0.2$. Validity of the scaling law means that rescaled spectra $\phi''/(c_\epsilon t_\epsilon)$ are independent of ϵ if they are considered as function of the rescaled frequency $\hat{\omega} = \omega t_\epsilon$. Equivalently, the rescaled correlators ϕ/c_ϵ are independent of ϵ if they are considered as function of the rescaled time $\hat{t} = t/t_\epsilon$. The restriction from Eq. (4.47c) means the following. The interval of rescaled frequencies $\hat{\omega}$ or of rescaled times $\hat{t}$, where the rescaled spectra or rescaled correlators, respectively, coincide for different ϵ, expands upon decreasing $|\epsilon|$.

The formula for the master functions $\varphi(\hat{z})$ is derived as limit of a correlator for small frequencies. In order to calculate the Laplace back transformation $\varphi(\hat{t})$, the formula for $\varphi(\hat{z})$ was used also for frequencies violating the conditions from Eq. (4.43e). For large frequencies ω, $\varphi(\hat{z})$ describes a non-integrable spectrum proportional to $1/\sqrt{\omega}$. This implies an artifact for the short-time behaviour of $\varphi(\hat{t})$. The function $\overline{\varphi}(\hat{t})$ exhibits a short-time divergency: $\overline{\varphi}(\hat{t}) = [1/\sqrt{\pi\hat{t}}](1 + O(\sqrt{\hat{t}}))$. A related phenomenon is discussed in Sec. 3.2.2 in connection with the short-time cusp of the hydrodynamic-limit correlator for the tagged-particle dynamics. The above mentioned coarse-graining procedure eliminates the artifacts.

The time scale t_ϵ separates regions of different dynamical behaviour. For low rescaled frequencies, Eq. (4.45) yields $\varphi(\hat{z}) = \theta(\epsilon)(-1/\hat{z}) + (i/4) + O(\hat{z})$. Substituting this result for the master function in the scaling law (4.48), one finds a frequency independent spectrum, which diverges for vanishing distance parameter $|\epsilon|$:

$$\phi''(\omega) = c_\epsilon t_\epsilon/4, \qquad 0 < \omega \ll 1/t_\epsilon. \tag{4.50a}$$

In a leading-order expansion for small $|\epsilon|$, the formula reproduces Eqs. (4.42a,b). The counterpart of the low-frequency spectrum is a control-parameter sensitive tail of the correlator in the time domain. From Eqs. (4.46b), (4.49), one gets

$$\phi(t) = c_\epsilon\theta(\epsilon) + c_\epsilon(t_\epsilon/t)^{3/2}\exp[-t/t_\epsilon]/4\sqrt{\pi}, \quad t_\epsilon \ll t. \tag{4.50b}$$

For large rescaled frequencies, the master function for the scaling-law asymptote exhibits an algebraic singularity $\varphi(\hat{z}) = -\sqrt{-i}/(2\sqrt{\hat{z}}) + O(1/\hat{z})$. Equation (4.48) leads to a power-law variation of the spectrum, which depends smoothly on ϵ. For ϵ tending to zero, this high-frequency asymptote of the scaling-law spectrum describes the low-frequency fluctuation spectrum (4.42d) for the transition point. In the context of the scaling-law discussion, the high-frequency condition refers to the restriction $\omega \gg 1/t_\epsilon$. In the context of Eq. (4.42d) the low-frequency concept means that the frequencies have to be so small that the transient spectra can be treated as white noise ones. But this restriction is imposed in Eqs. (4.43d,e) also for the derivation of the scaling limit. One gets:

$$\phi''(\omega) = \sqrt{\tau b(\epsilon)/(8v^2\omega)}, \qquad 1/t_\epsilon \ll \omega \ll \min(\nu, 1/\tau). \tag{4.51a}$$

The counterparts of the preceding results are the power-law decay of the correlator, which follows from the master function for short rescaled times $\hat{t}$: $\overline{\varphi}(\hat{t}) = 1/\sqrt{\pi\hat{t}} + O(\hat{t}^0)$. The scaling law (4.49) yields power-law relaxation of the correlator:

$$\phi(t) = \sqrt{\tau b(\epsilon)/(4\pi v^2 t)}, \qquad \max(1/\nu, \tau) \ll t \ll t_\epsilon. \tag{4.51b}$$

The scaling laws describe those qualitative features, which distinguish the dynamics for the states near the transition point from that for normal states. The formulas are results of a leading-order asymptotic expansion. The corrections to these results can be used to quantify the range of applicability for a scaling-law description of the correlators. Including correction terms and accounting for smooth deviations of the various parameters from their small-ϵ asymptotes can extend the range of validity for the asymptotic descriptions to larger values of $|v - v^c|$ and also to larger frequencies and shorter times, respectively.

The shape functions for the asymptotic expansions are determined by the coupling constant v. This holds also for the correlation scale c_ϵ. The transient dynamics influences the formulas for the leading-order results and also for the first correction solely via the parameter τ, which enters the time scale t_ϵ as factor. In particular, the correlators obtained from the equation with a Newtonian dynamics agree with those obtained from the one with a stochastic dynamics. This holds if the correlators are considered as function of $z\tau$ or t/τ, respectively. However, the range of validity of the mentioned results is not determined solely by the size of $z\tau$ or t/τ. There appear the factor $(\Omega\tau)^2$ in Eq. (4.44d); and this factor enters the second correction to the scaling-law result. Since $\nu = (\Omega\tau)^2/\tau$, the factor enters also the alternatives indicated in the result (4.51a) for the upper frequency limit and in the finding (4.51b) for the lower time limit. The necessary conditions for the critical $1/\sqrt{\omega}$ asymptote for the fluctuation spectrum and for the critical $1/\sqrt{t}$ power-law decay of the correlator read as follows:

$$\omega \ll 1/\tau \quad \text{and} \quad t \gg \tau \quad \text{if } (\Omega\tau)^2 > 1; \tag{4.51c}$$

$$\omega \ll (\Omega\tau)^2/\tau \quad \text{and} \quad t \gg \tau/(\Omega\tau)^2 \quad \text{if } (\Omega\tau)^2 < 1. \tag{4.51d}$$

In particular, the results (4.51c) are valid for the equations of motion with a stochastic short-time dynamics, since this model is obtained by the limit $\Omega \to \infty$

for fixed τ. However, if the oscillator frequency Ω decreases below $1/\tau$, the range of validity of the mentioned power laws shrinks. With decreasing $\Omega\tau$, the $1/\sqrt{\omega}$ regime of the fluctuation spectrum decreases to lower frequencies. The onset time of the $1/\sqrt{t}$ law for the correlation decay increases to larger values. In this sense, one concludes that soft oscillations expand the transient regime thereby masking parts of the scaling-law results. The left panel of Fig. 4.1 demonstrates this phenomenon for the extreme situation $\Omega\tau \to 0$, Ω fixed. In this case, there is no $1/\sqrt{\omega}$ spectrum for $\epsilon = 0$.

4.3 Glass-transition singularities

This section starts with the demonstration that the array $\boldsymbol{f} = \boldsymbol{f}(P)$ of the M long-time limits of the MCT correlators is a distinguished solution of the fixed-point equation. The control-parameter space splits in two parts. Within the weak-coupling region, all correlators relax to zero: $\boldsymbol{f} = \boldsymbol{0}$. Within the strong-coupling region, at least one of the limits $f_q, q = 1, \ldots, M$, is positive. The discussion presented at the end of Sec. 4.1.2 on self-consistent current-relaxation approaches provides the motivation to call states with a non-trivial $\boldsymbol{f}$ a glass while states with $\boldsymbol{f} = \boldsymbol{0}$ are referred to as liquid. Liquid–glass transition points are singularities of $\boldsymbol{f}(P)$. Every singularity P^c of the function $\boldsymbol{f}(P)$ has states nearby, which describe a liquid–glass or a glass–glass transition. Therefore, the singularities P^c are called glass-transition singularities.

Singularities, which are exhibited by the solutions of regular equations, are called bifurcations. The glass-transition singularities are special bifurcation states of the fixed-point equation. The coordinates $V_\alpha, \alpha = 1, \ldots, N$, of the state P enter these equations as coefficients of the mode-coupling polynomial. The singularities are caused by the nonlinear dependence of the equations on the components of $\boldsymbol{f}$.

The schematic models introduced in Sec. 4.2.2 exhibit a variety of glass-transition singularities, which can be identified by elementary calculations. It will be shown that all singularities of the general MCT equations are equivalent to the bifurcations exhibited by the roots of real polynomials. The arrested part $\boldsymbol{f}$ for states near the simplest liquid-glass-transition points P^c shall be evaluated by asymptotic solution of the fixed-point equation. The arrest of the density fluctuations in a hard-sphere system and in a system with short-range attraction will be discussed in detail.

4.3.1 *Regular and critical states*

It is explained in Sec. 2.6 that the averaged long-time limits of correlation functions are interesting since they effect drastically the low-frequency behaviour of dynamical susceptibilities. In this section, some general properties of the $\mathrm{limav}_{t\to\infty}\, \phi_q(t) = f_q$, $q = 1, \ldots, M$, shall be specified for the MCT models for a dynamics. In the original MCT literature, the arrested parts f_q are called non-ergodicity parameters or form factors of the glass. If the dependence of some quantity like f_q on the state P shall be emphasized, a more involved notation will be used, e.g., by writing $f_q(P)$ instead of f_q.

According to Eq. (2.85), the array $\boldsymbol{f}(P)$ is given by the residues of possible zero-frequency poles of the correlators in the frequency domain:

$$\boldsymbol{f}(P) = \lim_{\epsilon \to 0}(-i\epsilon)\boldsymbol{\phi}(i\epsilon). \tag{4.52a}$$

The formula (2.86) implies the inequalities $0 \leq f_q(P) \leq 1 = \phi_q(t=0)$. Similarly, the arrested part of the fluctuating-force kernels, $C_q = \lim_{\epsilon\to 0}(-i\epsilon)m_q(i\epsilon)$, obey the inequalities $0 \leq C_q \leq m_q(t=0) < \infty$. Equation (4.29a) yields $-z\phi_q(z) = 1/\{1 + \Omega_q^2/[-z(z \pm i\nu_q) + \Omega_q^2(-zm_q(z))]\}$, Im $z \gtrless 0$; and from Eq. (4.29b) one gets the formula with $z(z \pm i\nu_q)$ replaced by $\pm iz\Omega_q^2\tau_q$. In both cases, one can put $z = i\epsilon$ and perform the limit $\epsilon \to 0$ with the result: $f_q = 1/\{1 + [1/C_q]\}$. Hence, f_q cannot reach unity:

$$0 \leq f_q(P) < 1, \qquad q = 1, \ldots, M. \tag{4.52b}$$

The completely monotonic correlators $\phi_q(t)$ for a stochastic dynamics are positive and they decrease monotonically with increasing time. Hence, there exist the long-time limits $\phi_q(t \to \infty)$. According to Sec. 2.6, these limits are identical with the averaged long-time limits:

$$\lim_{t\to\infty} \boldsymbol{\phi}(t) = \boldsymbol{f}(P). \tag{4.52c}$$

Using this result in Eq. (4.16), one finds $C_q = \mathcal{F}_q[P, \boldsymbol{f}(P)]$. Consequently, there holds Eq. (4.21a) with $\boldsymbol{f}^*$ replaced by $\boldsymbol{f}(P)$. The limit $\boldsymbol{f}(P)$ is a fixed point. The M components solve the M implicit equations (4.21b),

$$f_q(P)/[1 - f_q(P)] = \mathcal{F}_q[P, \boldsymbol{f}(P)], \qquad q = 1, \ldots, M \tag{4.52d}$$

(Bengtzelius *et al.* 1984).

If $\boldsymbol{f}^*(P)$ denotes some fixed point for state P, one can express the correlator array $\boldsymbol{\phi}(t)$ in terms of another array $\hat{\boldsymbol{\phi}}(t)$ as specified by Eq. (4.24a). The long-time limit $\hat{\boldsymbol{f}}$ of $\hat{\boldsymbol{\phi}}(t)$ exists; and there holds $f_q(P) - f_q^*(P) = [1 - f_q^*(P)]\hat{f}_q$, $q = 1, \ldots, M$. Because of the covariance theorem, $\hat{\phi}_q(t)$ has all properties of an autocorrelation function. In particular, there holds: $\hat{f}_q(P) \geq 0$. The arrested parts exhibit the maximum property (Götze 1991):

$$f_q(P) \geq f_q^*(P), \qquad q = 1, \ldots, M. \tag{4.52e}$$

None of the M components of the distinguished fixed point $\boldsymbol{f}(P)$ can be exceeded by the corresponding component of a fixed point $\boldsymbol{f}^*(P)$ for the same state.

Let $\boldsymbol{g}(P)$ denote some fixed point for state P, which obeys the preceding conditions for all fixed points $\boldsymbol{f}^*(P) : g_q(P) \geq f_q^*(P)$, $q = 1, \ldots, M$. Substituting the special fixed point $\boldsymbol{f}^*(P) = \boldsymbol{f}(P)$, one gets $g_q(P) \geq f_q(P)$. On the other hand, one can use $\boldsymbol{f}^*(P) = \boldsymbol{g}(P)$ in formula (4.52e) in order to get $f_q(P) \geq g_q(P)$. Consequently, there holds $\boldsymbol{f}(P) = \boldsymbol{g}(P)$. The maximum property determines $\boldsymbol{f}(P)$ uniquely.

In Sec. 4.2.4(i), the existence of the long-time limit for a MCT correlator was demonstrated for a model with a Newtonian dynamics; and this even for the case of a vanishing friction parameter ν. This suggests that the long-time limits $\boldsymbol{\phi}(t \to \infty)$ exist for all MCT models. The discussions shall be continued with this assumption. But, all conclusions based on Eq. (4.52c) require the reservation that the existence of $\boldsymbol{\phi}(t \to \infty)$ is not proven for the solutions of the first version of the MCT equations of motion.

Let us consider a sequence of arrays $\boldsymbol{f}^{(n)}$ defined recursively by $\boldsymbol{f}^{(0)} = \mathbf{1}$ and

$$\boldsymbol{f}^{(n+1)} = \boldsymbol{\mathcal{T}}[P, \boldsymbol{f}^{(n)}], \qquad n = 0, 1, \ldots. \tag{4.53a}$$

From Eqs. (4.20b,c) one finds that

$$0 \leq f_q^{(n+1)} \leq f_q^{(n)}, \qquad n = 1, 2, \ldots, \qquad q = 1, \ldots, M. \tag{4.53b}$$

The sequence of decreasing non-negative numbers $f_q^{(n+1)}$ has to converge. There exists the array $\boldsymbol{f}^{**}(P) = \lim_{n\to\infty} \boldsymbol{f}^{(n)}$ so that $0 \leq f_q^{**} < 1$ for $q = 1, \ldots, M$. This limit is the maximum fixed point (Götze and Sjögren 1995):

$$\lim_{n\to\infty} \boldsymbol{f}^{(n)} = \boldsymbol{f}(P). \tag{4.53c}$$

The specified iteration procedure is an efficient method to determine the fixed point $\boldsymbol{f}(P)$ numerically. In order to prove Eq.(4.53c), one uses that $\boldsymbol{\mathcal{T}}[P, \boldsymbol{x}]$ is a continuous function of $\boldsymbol{x}$. Equation (4.53a) yields: $\boldsymbol{f}^{**}(P) = \boldsymbol{\mathcal{T}}[P, \boldsymbol{f}^{**}(P)]$, i.e., $\boldsymbol{f}^{**}(P)$ is a fixed point for sate P. The covariance theorem (4.23b) implies $\lim_{n\to\infty} \hat{\boldsymbol{f}}^{(n)} = \hat{\boldsymbol{f}}^{**}(P)$. In particular, there holds $\hat{f}_q^{**}(P) \geq 0$ for all q. Consequently, Eq. (4.22a) leads to $f_q^{**}(P) = f_q^*(P) + (1 - f_q^*(P))\hat{f}_q^{**}(P) \geq f_q^*(P)$. The limit $\boldsymbol{f}^{**}(P)$ obeys the maximum property; and this implies the desired result: $\boldsymbol{f}^{**}(P) = \boldsymbol{f}(P)$.

Let us examine the assumption that there are glass states for arbitrarily small mode-coupling coefficients V_α, $\alpha = 1, \ldots, N$. For every $\epsilon_0 > 0$, there should be a state $P^{(0)}$ with the following properties. The coordinates of this state obey the condition $0 < \sum V_\alpha^{(0)2} < \epsilon_0^2$; and there is an array $\boldsymbol{f}^{(0)}$ whose M components obey the conditions: $f_q^{(0)} = \mathcal{T}_q[P^{(0)}, \boldsymbol{f}^{(0)}]$, $0 \leq f_q^{(0)} \leq 1$, $\eta = \sum_q f_q^{(0)} > 0$. The mode-coupling polynomials $\mathcal{F}_q[P, \boldsymbol{x}]$ vanish for $\boldsymbol{x} = 0$ and the coefficients are the coordinates V_α. Hence, there is a Lipschitz constant L_0 so that $|\mathcal{F}_q[P, \boldsymbol{x}]| \leq L_0 r \sum_q |x_q|$ for $r^2 = \sum_\alpha V_\alpha^2$. The inequalities shall be applied with $\epsilon_0 = 1/(2L_0 M)$, $P = P^{(0)}$, and $\boldsymbol{x} = \boldsymbol{f}^{(0)}$. One gets $\eta = \sum_q \mathcal{T}_q[P^{(0)}, \boldsymbol{f}^{(0)}] \leq \sum_q \mathcal{F}_q[P^{(0)}, \boldsymbol{f}^{(0)}] \leq \epsilon_0 L_0 \sum_q f_q^{(0)} = \epsilon_0 L_0 M \eta = \eta/2$. This wrong inequality shows that the assumption is illegitimate. Hence, there is some $\epsilon > 0$ so that

$$f_q(P) = 0, \qquad q = 1, \ldots, M, \quad \text{whenever} \sum_\alpha V_\alpha^2 < \epsilon^2. \tag{4.54}$$

There is a neighbourhood of the origin of the N-dimensional control-parameter space, where all states are liquids.

A state P^1 with coordinates V_α^1, $\alpha = 1, \ldots, N$, can be used to define an array of states. This is a one-dimensional set of states extending from the liquid region to the region of large couplings. The states are functions of a real number ξ. The coordinates V_α^ξ of P^ξ read

$$V_\alpha^\xi = \xi V_\alpha^1, \qquad \alpha = 1, \ldots, N, \qquad \xi > 0. \tag{4.55a}$$

Introducing $r_q = \mathcal{F}_q[P^1, \mathbf{1}]$, the initial values $m_q(t = 0)$ of the relaxation kernels for the states P^ξ are given by ξr_q. If there would be a number q_0 so that $r_{q_0} = 0$, the kernel $m_{q_0}(t)$ would vanish for all times for all states P^ξ. The correlator $\phi_{q_0}(t)$ would be independent of mode-coupling effects. This trivial case should be eliminated from the discussion either by reducing M or by reducing the number of control parameters. In this sense, it is no restriction of generality to request

$$r_q = \mathcal{F}_q[P^1, \mathbf{1}] > 0, \qquad q = 1, \ldots, M. \tag{4.55b}$$

It shall be proven that there is some positive ξ_0 so that

$$f_q(P^\xi) > 0, \qquad q = 1, \ldots, M \quad \text{if} \qquad \xi > \xi_0. \tag{4.55c}$$

Hence, states with all coupling constants chosen sufficiently large are glasses with all M correlators having an arrested part.

The proof shall be done by construction of a non-trivial fixed point $\boldsymbol{f}^*$ for sufficiently small values of $\epsilon = 1/\xi$. Substituting the ansatz $\boldsymbol{f}^* = \mathbf{1} - \epsilon \boldsymbol{g}$ in Eq. (4.15a), one can write $\mathcal{F}_q[P^\xi, \boldsymbol{f}^*] = \xi \mathcal{F}_q[P^1, \boldsymbol{f}^*] = \xi\{r_q + \mathcal{Q}_q(\epsilon \boldsymbol{g})\}$. Here $\mathcal{Q}_q(\boldsymbol{x})$ is a polynomial of $\boldsymbol{x}$ of degree n_0 with $\mathcal{Q}_q(\mathbf{0}) = 0$. Hence, one can write $\mathcal{Q}_q(\epsilon \boldsymbol{g}) = \epsilon a_q^{(1)}(\boldsymbol{g}) + \epsilon^2 a_q^{(2)}(\boldsymbol{g}) + \cdots \epsilon^{n_0} a_q^{(n_0)}(\boldsymbol{g})$, with $a_q^{(n)}(\boldsymbol{x})$ denoting a homogeneous polynomial of $\boldsymbol{x}$ of degree n. The fixed-point equation (4.21b) is equivalent to $g_q = 1/[r_q + \epsilon + \mathcal{Q}_q(\epsilon \boldsymbol{g})]$. The right hand side is an analytic function of ϵ for all $\boldsymbol{g}$ obeying $|g_q| \leq 1$, $q = 1, \ldots, M$, and $\epsilon < \epsilon_0$. Hence, there is a solution $g_q = r_q^{-1} + O(\epsilon)$. This shows the existence of a fixed point $f_q^* = 1 - (r_q^{-1}/\xi) + O(1/\xi^2)$ for $\xi > (1/\epsilon_0)$. Because of Eq. (4.52e), there holds $f_q(P^\xi) \geq f_q^*$; and this finishes the proof.

From Eqs. (4.15d), (4.20a), one gets $f_q(P^\xi) \leq \mathcal{T}_q[P^\xi, \mathbf{1}] = 1/\{1 + 1/\mathcal{F}_q[P^\xi, \mathbf{1}]\} = 1/\{1 + [\epsilon/r_q]\} \leq 1 - [\epsilon/r_q] \leq 1 - (r_q^{-1}/\xi)$. Therefore, for sufficiently small ϵ_0, the fixed point $\boldsymbol{f}^*$ is the maximum one:

$$f_q(P^\xi) = 1 - [r_q/\xi] + O(1/\xi^2). \tag{4.55d}$$

The M components $f_q^*(P)$ of some fixed point $\boldsymbol{f}^*(P)$ obey the M implicit equations $J_q[P, \boldsymbol{x}] = 0$, $q = 1, \ldots, M$; $J_q[P, \boldsymbol{x}] = x_q - \mathcal{T}_q[P, \boldsymbol{x}]$. Here, Eq. (4.20a) expresses the M functions $\mathcal{T}_q[P, \boldsymbol{x}]$ in terms of the mode-coupling polynomials. For the discussion of the solutions, it is important to understand some properties of the Jacobian matrix, i.e., of the M-by-M matrix of derivatives $\partial J_q[P, \boldsymbol{x}]/\partial x_p$.

One gets $\partial J_q[P, \boldsymbol{f}^*(P)]/\partial x_p = \delta_{q,p} - [1 - f_q^*]^2 \partial \mathcal{F}_q[P, \boldsymbol{f}^*(P)]/\partial x_p$. It will be more convenient to represent the matrix by an equivalent one: $\partial J_q[P, \boldsymbol{f}^*(P)]/\partial x_p = [1 - f_q^*(P)]\{\delta_{q,p} - A_{q,p}[P, \boldsymbol{f}^*(P)]\}[1 - f_p^*(P)]^{-1}$. For all arrays $\boldsymbol{x}$ obeying

$$0 \leq x_q < 1, \quad q = 1, \ldots, M, \tag{4.56a}$$

the M-by-M matrix $A[P, \boldsymbol{x}]$ is defined by its M^2 elements

$$A_{q,p}[P, \boldsymbol{x}] = [1 - x_q]\{\partial \mathcal{F}_q[P, \boldsymbol{x}]/\partial x_p\}[1 - x_p]. \tag{4.56b}$$

Because of Eq. (4.15c), none of the elements can be negative,

$$A_{q,p}[P, \boldsymbol{x}] \geq 0, \quad q, p = 1, \ldots, M. \tag{4.56c}$$

The characteristic polynomial of matrix A, i.e., the determinant of the matrix $e1 - A$, shall be denoted by χ_e.

$$\chi_e[P, \boldsymbol{x}] = det\{\delta_{q,p} e - A_{q,p}[P, \boldsymbol{x}]\}. \tag{4.57a}$$

The M roots of the polynomial shall be denoted by $e_k[P, \boldsymbol{x}]$, $k = 1, \ldots, M$. The coefficients of $\chi_e[P, \boldsymbol{x}]$ are polynomials of the M components of $\boldsymbol{x}$ and of the N coordinates of P. Hence, the roots are continuous functions of $\boldsymbol{x}$ and P. A nondegenerate root is an analytic function of $x_1, \ldots, x_M$ and $V_1, \ldots, V_N$. Matrices obeying the condition (4.56c) are called non-negative. According to a theorem of G. Frobenius and O. Perron (Gantmacher 1974), there exists a maximum eigenvalue determining the spectral radius of matrix A. It is a real non-negative root of the characteristic polynomial, which cannot be exceeded by the moduli of the eigenvalues $e_k[P, \boldsymbol{x}]$. Let us denote this special root by $e_M[P, \boldsymbol{x}] = E[P, \boldsymbol{x}]$. There holds

$$|e_k[P, \boldsymbol{x}]| \leq E[P, \boldsymbol{x}], \quad k = 1, \ldots, M - 1. \tag{4.57b}$$

If none of the roots is unity,

$$e_k[P, \boldsymbol{x}] \neq 1, \quad k = 1, \ldots, M, \tag{4.58a}$$

matrix $1 - A$ can be inverted. Denoting this inverse matrix by I, one can write for $q, p = 1, \ldots, M$:

$$\sum_k I_{q,k}[P, \boldsymbol{x}]\{\delta_{k,p} - A_{k,p}[P, \boldsymbol{x}]\} = \delta_{q,p}. \tag{4.58b}$$

This observation is the basis of the theorem on implicit functions. With the present notation, it states the following. If $\boldsymbol{f}^*(P)$ is a fixed point so that the conditions (4.58a) are valid for $\boldsymbol{x} = \boldsymbol{f}^*(P)$, a neighbourhood of states Q can be defined in the sense that $\sum_\alpha |V_\alpha^P - V_\alpha^Q|^2 < \epsilon^2$. Here, $V_\alpha^{P,Q}$, $\alpha = 1, \ldots, N$, are the coordinates of P and Q, respectively, and $\epsilon > 0$. There exist a fixed point

$\boldsymbol{f}^*(Q)$ for every Q, whose components are analytic functions of $V_\alpha^Q, \ldots, V_N^Q$. For Q tending to P, $\boldsymbol{f}^*(Q)$ approaches $\boldsymbol{f}^*(P)$. For none of the states Q, there is another fixed point, say $\tilde{\boldsymbol{f}}(Q)$, which obeys $|\tilde{f}_q^*(Q) - f_q^*(Q)| < \epsilon$, $q = 1, \ldots, M$. Loosely summarized: if there holds condition(4.58a), the fixed point $\boldsymbol{f}^*(P)$ can be analytically continued in a whole neighborhood, and there is no other fixed point nearby. Differentiating the fixed-point equation with respect to the mode-coupling coefficients, one gets the addendum:

$$\partial f_q^*(P)/\partial V_\alpha = [1 - f_q^*(P)] \sum_p I_{q,p}[P, \boldsymbol{f}^*(P)]\{[1 - f_p^*(P)]\partial \mathcal{F}_p[P, \boldsymbol{f}^*(P)]/\partial V_\alpha\}. \tag{4.58c}$$

Representing matrix A in its Jordan form, one estimates for the asymptotic behaviour of its n-th powers $(A[P, \boldsymbol{x}]^n)_{p,q} = O(n^\ell E[P, \boldsymbol{x}]^n)$, where ℓ is restricted by $0 \leq \ell \leq M$. Hence, if the maximum eigenvalue is smaller than unity,

$$E[P, \boldsymbol{x}] < 1, \tag{4.59a}$$

the elements of A^n decrease exponentially with increasing n. In this case, the Neumann series provides an efficient tool to evaluate the matrix I:

$$I_{q,p}[P, \boldsymbol{x}] = \delta_{q,p} + \sum_{n=1}^{\infty} (A[P, \boldsymbol{x}]^n)_{q,p}. \tag{4.59b}$$

The results formulated above shall be specialized to the maximum fixed point $\boldsymbol{f}(P)$. The corresponding M-by-M matrix $A[P, \boldsymbol{f}(P)]$ shall be called the stability matrix for the state P:

$$A_{q,p}(P) = [1 - f_q(P)]\{\partial \mathcal{F}_q[P, \boldsymbol{f}(P)]/\partial x_p\}[1 - f_p(P)]. \tag{4.60a}$$

Its maximum eigenvalue is called the maximum eigenvalue for state $P : E(P) = E[P, \boldsymbol{f}(P)]$. It is a zero of the characteristic polynomial,

$$\chi_{E(P)}(P, \boldsymbol{f}(P)) = 0. \tag{4.60b}$$

There holds the maximum-eigenvalue theorem:

$$E(P) \leq 1. \tag{4.60c}$$

The proof can be found in Appendix D.3.

The preceding inequality suggests the classification of the states P in regular ones, defined by the condition $E(P) < 1$, and critical ones. The latter shall be denoted by P^c, and they are defined by the condition

$$E(P^c) = 1. \tag{4.61a}$$

The corresponding long-time limit,

$$\boldsymbol{f}^c = \boldsymbol{f}(P^c), \tag{4.61b}$$

is called the critical form factor or the critical arrested part of the correlator array $\boldsymbol{\phi}(t)$. The corresponding matrix (4.60a) is called the critical stability matrix:

$$A^c_{q,p} = [1 - f^c_q]\{\partial \mathcal{F}_q[P^c, \boldsymbol{f}^c]/\partial x_p\}[1 - f^c_p]. \tag{4.61c}$$

For regular states P, all consequences of conditions (4.58a), (4.59a) hold for the special fixed points $\boldsymbol{f}^*(P) = \boldsymbol{f}(P)$. In particular, there is a unique analytical continuation of $\boldsymbol{f}(P)$ in a whole neighbourhood of this state P. This means that the set of regular points is an open one. The matrix $I = (1 - A)^{-1}$ can be evaluated as Neumann series. From Eq. (4.56c) one gets that every power A^n is a non-negative matrix, and hence,

$$I_{q,p} \geq 0, \quad p, q = 1, \ldots, M. \tag{4.62a}$$

From Eq. (4.58c), one concludes that

$$\partial f_q(P)/\partial V_\alpha \geq 0 \tag{4.62b}$$

for all $q = 1, \ldots, M$ and all $\alpha = 1, \ldots, N$. If a mode-coupling coefficient V_α increases, none of the components $f_q(P)$ can decrease.

The set of critical points is closed, since it is the complement of an open set. Let us note that the N coordinates $V^c_1, \ldots V^c_N$ of a critical point P^c and the M components $f^c_1, \ldots, f^c_M$ of $\boldsymbol{f}^c$ are a solution of the $(M + 1)$ analytical equations:

$$x_q = \mathcal{T}_q[P, \boldsymbol{x}], \quad q = 1, \ldots, M \tag{4.63a}$$

$$\chi_1[P, \boldsymbol{x}] = 0. \tag{4.63b}$$

Appeal to the implicit-function theorem leads to the following conclusion. Generically, the critical points are elements of a smooth manifold of codimension 1 in the parameter space. They are points on an $(N-1)$ dimensional smooth manifold of states. The critical points, which cannot be continued in a whole neighborhood of this manifold, are called endpoints. It will be shown below that the function $\boldsymbol{f}(P)$ exhibits a singularity at P^c.

For the microscopic MCT models defined in Sec. 4.2.3, the coordinates of P are determined by the equilibrium structure functions. These are specified by a set of, say, K parameters $\xi_1, \ldots, \xi_K$ to be called the external control parameters. The most common external control parameters are the density ρ and the temperature T. For the sake of simplicity, it shall be assumed that the N control parameters $V_\alpha, \alpha = 1, \ldots, N$, are analytic functions of $\vec{\xi} = (\xi_1, \ldots, \xi_K)$. All quantities, which have been identified as continuous, smooth or analytic functions of the coefficients V_α, are also continuous, smooth or analytic functions of

$\vec{\xi}$, respectively. The purpose of the MCT is the study of the evolution of the correlator array $\boldsymbol{\phi}(t)$ upon changes of $\vec{\xi}$. The simplest and most relevant case deals with a single control parameter ξ only. The description of the state in terms of the external control parameter ξ is an analytic path

$$\xi \longrightarrow P^{\xi} \tag{4.64a}$$

given by N analytic coordinate functions $V_\alpha(\xi)$, $\alpha = 1, \ldots, N$. To be specific, let us assume that P^ξ for small ξ is located in the liquid region and for large ξ within the glass region. There must exist a largest value of ξ, say, ξ^c so that

$$\boldsymbol{f}(P^{\xi}) = 0 \quad \text{for } \xi < \xi^c. \tag{4.64b}$$

Generically, there is some interval of values of ξ above ξ^c so that the states are regular glass states. For one component, say number q_0, there holds

$$f_{q_0}(P^{\xi}) > 0 \qquad \text{for } \xi > \xi^c. \tag{4.64c}$$

There is no analytic array $\boldsymbol{f}(P)$ obeying both preceding conditions, i.e., $P^c = P(\xi^c)$ is a critical point. The liquid–glass transition state P^c is a special critical point.

4.3.2 *Examples for bifurcation diagrams*

The schematic models defined in Sec. 4.2.2 lead to various types of glass-transition singularities. These bifurcation points shall be discussed in this section, and some terminology for their description will be introduced.

Let us start with one-component models explained in connection with Eq. (4.32a). The mode-coupling polynomial $\mathcal{F}[P, x]$ of degree n_0 is specified by not more than n_0 non-negative coefficients $V_1, \ldots, V_N$. Equation (4.21b) for the fixed point f^* is equivalent to the equation for the real roots of the polynomial $Q(x) = x - (1 - x)\mathcal{F}[P, x]$:

$$f^* - (1 - f^*)\mathcal{F}[P, f^*] = 0. \tag{4.65a}$$

Only the roots within the interval $0 \le x < 1$ have to be considered. There is always the trivial root $f^* = 0$. Since $Q(x)$ is of degree $(n_0 + 1)$, the number of fixed points $f_1^*, f_2^* \ldots$, cannot exceed $(n_0 + 1)$. The maximum property identifies the correlator long-time limit f as the largest root:

$$f(P) = \max_{\kappa} f_\kappa^*(P). \tag{4.65b}$$

The one-by-one stability matrix agrees with its maximum eigenvalue: $E[P, f^*] = [1 - f^*(P)]^2 \partial\mathcal{F}[P, f^*(P)]/\partial x$. The condition (4.61a) for the state P^c to be a critical one with the critical arrested part $f^c = f(P^c)$ reads

$$1 - [1 - f^c]^2 \partial\mathcal{F}[P^c, f^c]/\partial x = 0. \tag{4.65c}$$

By induction, one shows for $\ell \geq 2 : \partial^\ell \mathcal{Q}(P^c, f^c)/\partial x^\ell = \ell \partial^{(\ell-1)} \mathcal{F}[P^c, f^c]/\partial x^{(\ell-1)} - [1 - f^c]\partial^\ell \mathcal{F}[P^c, f^c]/\partial x^\ell$.

The preceding equation is equivalent to the statement that the maximum root $f^* = f^c$ obeys the condition $\partial Q[f^c]/\partial x = 0$. Thus, a point P^c is a critical one if and only if the root f^c is degenerate. The non-degenerate roots of a polynomial are analytic functions of the coefficients. This exemplifies the general result that $f(P)$ can be analytically continued in a neighbourhood of a regular point P. The roots are continuous functions of P which exhibit branch points at the critical points P^c. This implies that the critical points are singularities of $f(P)$ as claimed in the preceding section.

According to the terminology used by Arnol'd (1975), a bifurcation is called an A_ℓ singularity, if it is equivalent to the one described by a real root of degeneracy $\ell, \ell = 2, 3, \ldots$ of a real polynomial. The A_2 and A_3 singularity are also called fold singularity and cusp singularity, respectively. They are the generic singularities occurring for the inverse of smooth maps of smooth two-dimensional manifolds on planes (Arnold 1992). The A_4 singularity is also named a swallow-tail bifurcation. An A_3 singularity occurs because two A_2 singularities coalesce. Similarly, an A_4 singularity occurs because of the collapse of two A_3 singularities, and so on. In this sense, the A_2 singularity is the basic one for the class of bifurcations A_ℓ, $\ell = 2, 3, \ldots$ Motivated by the cited terminology, a state P^c shall be called an A_ℓ glass-transition singularity if the corresponding maximum root $f^c = f^*(P^c)$ of $Q(x)$ has a degeneracy ℓ, $\ell \geq 2$. It is defined by $\partial^n Q(P^c, f^c)/\partial x^n = 0$ for $n = 0, 1, \ldots, \ell - 1$, while the ℓ-th derivative is non zero. This means that there holds Eq. (4.65a) with $f^* = f^c$ and $P = P^c$; and, in addition,

$$1 - [1 - f^c]^{n+1}\Big\{\frac{1}{n!}\partial^n \mathcal{F}\big[P^c, f^c\big]\Big/\partial x^n\Big\} = 0, \quad n = 1, \ldots, \ell - 1, \tag{4.66a}$$

while

$$1 - [1 - f^c]^{\ell+1}\Big\{\frac{1}{\ell!}\partial^\ell \mathcal{F}\big[P^c, f^c\big]\Big/\partial x^\ell\Big\} = \mu_\ell^c \tag{4.66b}$$

is non-zero. The number μ_ℓ^c shall be called the singularity parameter. Because of Eq. (4.15c), the second term in the preceding formula cannot be negative; and, thus, $\mu_\ell^c \leq 1$. Taylor's theorem implies $Q[P^c, x] = [\mu_\ell^c/(1-f^c)^\ell](x-f^c)^\ell + O((x-f^c)^{\ell+1})$. If μ_ℓ^c were negative, there would be some $\epsilon > 0$ so that $Q(x) < 0$ for all x obeying $0 < x - f^c < \epsilon$. Since $Q(1) = 1$, there would be a root, $Q(f^*) = 0$, located above $f^c : f^c < f^* < 1$. Because of the maximum property, this is impossible. Consequently, there holds

$$0 < \mu_\ell^c \leq 1. \tag{4.66c}$$

Let us consider the $(N + 1)$-dimensional space spanned by the variables $V_1, \ldots, V_N, f$. An A_ℓ singularity $[P^c, f^c]$ is characterized as such point in this space, whose coordinates $(V_1^c, \ldots, V_N^c, f^c)$ obey the ℓ equations $\partial^n Q(f^c)/\partial x^n = 0$, $n = 0, \ldots, \ell - 1)$. Generically, these ℓ equations for polynomials of the $(N+1)$ variables define a smooth manifold of codimensionality ℓ, i.e., a smooth subset of dimensionality $(N + 1 - \ell)$. Generically, the singularities P^c of type A_ℓ are points on an $(N + 1 - \ell)$-dimensional smooth manifold in the N-dimensional control-parameter space. The singularity parameter μ_ℓ^c is a smooth function of the coordinates of P^c. If P^c converges towards P^{cc} so that $\lim_{P^c \to P^{cc}} \mu_\ell^c(P^c) = 0$, P^{cc} is an endpoint of the manifold. It is an A_k singularity with $k > \ell$.

The sets $(V_1^c, \ldots, V_N^c, f^c)$, which specify some A_ℓ-glass-transition singularity, are not identical to the ones describing the full bifurcation scenario of real polynomials of degree ℓ. Firstly, the maximum theorem restricts the discussion to a part of the manifold only. Second, the coefficients of the admissible polynomials $Q(x) = x - (1-x)\mathcal{F}[P, x]$ form only a subset of all real coefficients. There is no coefficient of order zero, $Q(x = 0) = 0$, and the coefficients of the polynomial $\mathcal{F}[P, x]$ are restricted to non-negative ones.

The one-component model defined by Eq. (4.32b) for a control-parameter quadrant $P \leftrightarrow (v_1, v_2)$ is specified by $Q(x) = x[(1-v_1)+(v_1-v_2)x+v_2x^2]$. Besides $f_0^* = 0$, there are the fixed points $f_\pm^* = \{(v_2-v_1) \pm \sqrt{[(v_2 - v_1)^2 - 4(1 - v_1)v_2]}\}/(2v_2)$. This holds, provided these numbers are real and obey the restriction $0 \leq f_\pm^* < 1$. From Eq. (4.66b) one gets $\mu_2^c = 1 - (1 - f^c)^3 v_2^c$.

The fixed point f_+^* agrees with f_0^* if and only if $v_1 = 1$. In this case $\mu_2^c = 1 - v_2^c$. The conditions (4.65b, 4.66c) imply that there is the line of A_2 singularities, which is characterized by

$$0 \leq \lambda^c < 1,\ \mu_2^c = 1 - \lambda^c,\ v_1^c = 1,\ v_2^c = \lambda^c,\ f^c = 0. \tag{4.67a}$$

Figure 4.2 exhibits this line as dashed. A second transition line is defined by $f_+^* = f_-^*$. It is a part of a parabola. The latter is defined by those points, which lead to a vanishing of the argument of the above noted square root. The condition (4.66c) and the request $v_1 \geq 0$ yield a curve of A_2 singularities specified by

$$1/2 \leq \lambda^c < 1,\ \mu_2^c = 1 - \lambda^c,\ v_1^c = (2\lambda^c - 1)/\lambda^{c2},\ v_2^c = 1/\lambda^{c2},\ f^c = 1 - \lambda^c. \tag{4.67b}$$

This line is shown as the full curve. The only remaining glass-transition singularity for the model is an A_3 singularity with the properties

$$\mu_3^c = 1,\ v_1^{cc} = 1,\ v_2^{cc} = 1,\ f^{cc} = 0. \tag{4.67c}$$

This bifurcation point is marked by an open circle. The identified set of singularities specifies the transition line between liquid states and glass states.

Let us consider an array of states $(v_1^\xi, v_2^\xi) = (v_1^c, v_2^c)\xi$, $0 < \xi$, as discussed in connection with Eq. (4.55a). For $\xi < \xi^c = 1$, the states are liquids, and for $\xi > \xi^c$ they are glasses. For $\xi = \xi^c$, there occurs a liquid–glass transition. The arrested part for non-negative distance parameters $\epsilon = (\xi - \xi^c)$ is determined by f_+^*. One derives for states near the dashed line:

$$f = (1 - \lambda^c)^{-1}[\epsilon + O(\epsilon^2)], \quad \epsilon \geq 0, \quad 0 \leq \lambda^c < 1. \tag{4.68a}$$

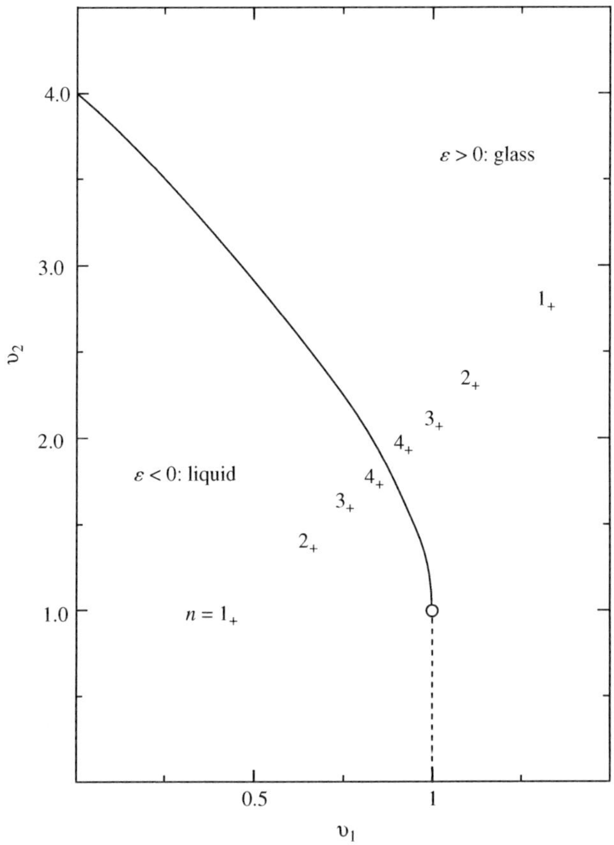

FIG. 4.2. Singularity diagram for the one-component model defined by Eqs. (4.32a,b). The dashed and the full lines are the sets of degenerate and generic A_2-liquid–glass transitions, respectively. The circle marks the common endpoint of these lines, which is a degenerate A_3 singularity. The crosses with labels n mark states $v_{1,2} = v_{1,2}^c(1+\epsilon)$, $\epsilon = \pm 2^{-n}$, $n = 1$–4, with $v_{1,2}^c$ given by Eq. (4.67b) for $\lambda^c = 3/4$.

Crossing this transition line, the long-time limit varies continuously. The glass-form factor increases linearly with increasing distance ϵ from the transition point. The result for $\lambda^c = 0$ reproduces the one discussed in Sec. 4.2.4(i) for the elementary model. For states near the full line, one derives:

$$f - f^c = \lambda^c[\sqrt{\epsilon} + O(\epsilon^{3/2})], \quad \epsilon \geq 0\,, \quad 1/2 \leq \lambda^c < 1. \tag{4.68b}$$

Crossing this transition line, the long-time limit varies discontinuously. With increasing $\epsilon, f - f^c$ increases proportional to the square-root of the distance parameter. This is the normal behaviour for the bifurcation of the root of quadratic equations. The singularities for $f^c > 0$, which describe discontinuous transitions, shall be referred to as generic A_2 glass-transitions. Singularities with $f^c = 0$, shall be called degenerate glass-transition singularities. The degenerate A_2 glass-transition singularities describe continuous transitions of the correlator's long-time limit. In the original MCT literature, degenerate and generic A_2 transitions have been called type-A and type-B transitions, respectively.

The states of the one-component model defined by Eq. (4.32c) are also specified by two control parameters: $P \leftrightarrow (v_1, v_3)$. The mode-coupling polynomial is given by $\mathcal{F}[P, x] = v_1 x + v_3 x^3$. Equation (4.65a) for the fixed points reads $f^*\{1 - (1 - f^*)[v_1 + v_3 f^{*2}]\} = 0$. Equation (4.66b) yields for the A_2-singularity parameter: $\mu_2^c = 1 - 3[1 - f^c]^2 v_3^c f^c$.

The trivial fixed point $f_0^* = 0$ is degenerate if and only if $v_1^c = 1$. In this case, there holds $\mu_2^c = 1$. As a result, the line of degenerate fixed points does not have an endpoint. The upper panel of Fig. 4.3 exhibits this straight line for $0 \le v_3^c < 4$. There may be up to three non-trivial fixed points $f_\kappa^*, \kappa = 1, 2, 3$. Rather than solving the cubic equation for the f_κ^*, the inverse method shall be used. Equation (4.65a) yields the first condition for a generic fixed point $f^* = f^c = x : [v_1^c + v_3^c x^2] = 1/(1 - x)$. The second condition is given by Eq. (4.65c): $[v_1^c + 3v_3^c x^2] = 1/(1 - x)^2$. This is a pair of linear equations, which defines the parameter representation of the singularity line: $v_1 = v_1^c(x)$, $v_3 = v_3^c(x)$, $0 \le x < 1$. A starting point for the line is identified for $v_1^c = 0 : f^c = x = 2/3$, $v_3^c = 27/4$, $\mu_2^c = 1/2$. For $x > 2/3$, there holds $v_1^c(x) < 0$ and the curve is outside the admissible parameter region. With decreasing x, $v_1^c(x)$ increases, while $v_3^c(x)$ and μ_2^c decrease. An endpoint is obtained for

$$v_1^{cc} = 9/8,\ v_3^{cc} = 27/8,\ f^{cc} = 1/3,\ \mu_3^c = 1/3. \tag{4.69a}$$

The specified line and its endpoint are shown in the upper panel of Fig. 4.3 as the full curve and circle, respectively. The tangent vector $(\partial v_1^c(x)/\partial x, \partial v_3^c(x)/\partial x)$ of this branch of the set of singularities vanishes at the endpoint. Here, the curve exhibits a cusp. For x decreasing below 1/3, the curve continues with $v_1^c(x)$ decreasing and $v_3^c(x)$ increasing. But, this branch is irrelevant for the discussion since $\mu_2 < 0$. The inequality (4.66c) is violated for $x < 1/3$; the corresponding solution $f^* = x$ is not a maximum fixed point.

There is a peculiar point on the above specified line of generic A_2 glass-transition singularities, which is called a crossing point. It is characterized by

$$v_1^x = 1,\ v_3^x = 4,\ f_x^c = 1/2,\ \mu_2^x = 1/4. \tag{4.69b}$$

Because of the maximum property, the above identified line of degenerate A_2 singularities cannot reach this point; nor can it extend beyond it.

The boundary of the liquid region is given by the line of generic A_2 glass transition for states with $0 \le v_1^c \le v_1^x$. This line deals with discontinuous liquid–glass transitions. Within the glass, the form factor increases like $f - f^c \propto \sqrt{\epsilon}$, as was discussed above in connection with Eq. (4.68b). The remaining part of

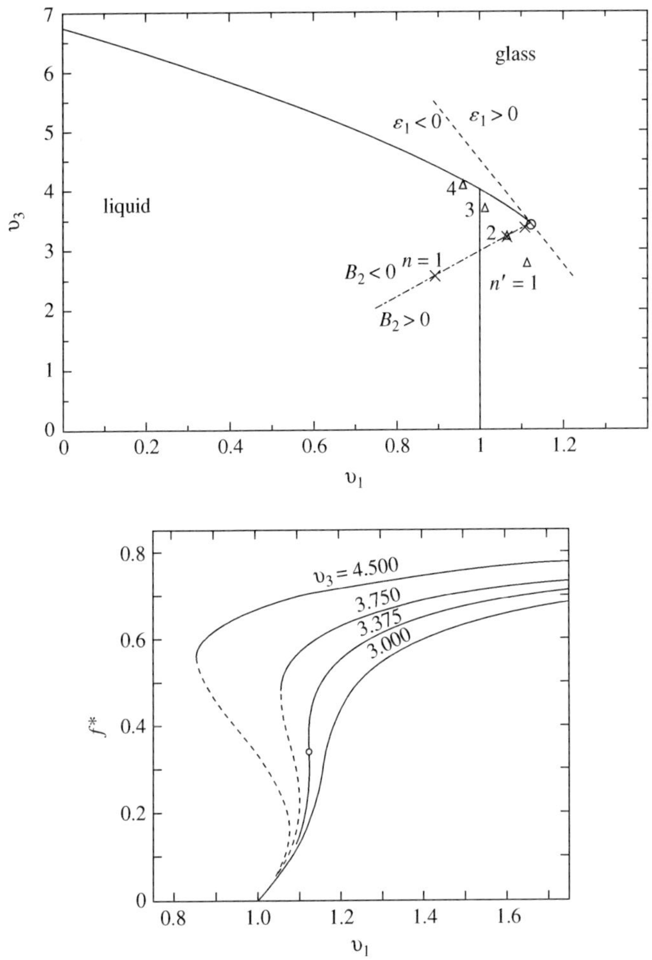

FIG. 4.3. The upper panel shows the singularity diagram for the one-component model defined by Eqs. (4.32a,c). The upper full curve exhibits the line of generic A_2 glass-transition singularities ending in a generic A_3 singularity, which is marked by an open circle. The straight vertical line is the set of degenerate A_2 singularities. The crosses and triangles indicate states discussed in Sec. 6.3.2(iii) and the other lines and symbols are explained there as well. Modified version of a figure from Götze and Sperl (2002).

The lower panel exhibits the f^* versus v_1 curves for the fixed points for values v_3 as indicated. The full-line parts are the maximum fixed points and the dashed ones exhibit fixed points, which are no glass form factors. The open circle marks the A_3 singularity. Reproduced from Götze and Haussmann (1988).

the boundary is given by the line of the degenerate A_2 singularities shown in the figure. This line deals with continuous liquid–glass transitions. Within the glass, the arrested part increases linearly with the distance parameter, $f \propto \epsilon$, as was discussed above in connection with Eq. (4.68a). The boundary exhibits a corner at the crossing point, where the two lines meet transversally.

The line of generic A_2 glass-transition singularities between the crossing point and the endpoint deals with discontinuous transitions from one glass state to another glass state. If this line of glass-glass transitions is crossed by a path as defined in connection with Eqs. (4.55a–c), all points below the transition line are regular. Here, the arrested part $f(P^\xi)$ is positive. Because of Eq. (4.58c), it increases with increasing ξ towards some limit, say, $f_- = \lim_{\xi \to \xi^c - 0} f(P^\xi)$. At the line, the arrested part increases discontinuously to the value $f^c > f_-$. For larger ξ, the arrested part increases in analogy to what was explained in connection with Eq. (4.68b): $f - f_c \propto \sqrt{\epsilon}$.

The lower panel of Fig. 4.3 presents a survey of the evolution of the manifold of non-trivial fixed points f^*. It is obtained by the inverse method. Formula (4.65a) is written as an elementary equation for v_1 as function of f^* for fixed parameter $v_3 : v_1 = [1/(1 - f^*)] - v_3 f^{*2}$. All curves start for $v_1 = 1$ with slope unity: $v_1 = 1 + f^* + (1 - v_3) f^{*2} + O(f^{*3})$. The minima and maxima of the v_1 versus f^* curves determine the fixed points with degeneracy $\ell = 2$. The cusp singularity (4.69a) appears as horizontal inflection point of the curve for $v_3 = v_3^{cc}$. Depending on the value of v_3, there are four qualitatively different curves. First, if $v_3 \geq v_3^x$, the v_1 versus f^* curves exhibit a minimum for $f^* > f_x^c$ and a maximum for $f^* < f_x^c$. The full line for $v_3 = 4.500$ exemplifies the discontinuous increase of the long-time limit upon increasing v_1 through the critical value v_1^c. If $v_3^x > v_3 > v_3^{cc}$, the minimum of the v_1 versus f^* curve is located above $v_1 = 1$. As second possibility, there occurs a continuous transition if v_1 increases above 1 and, then, there occurs a generic A_2 bifurcation to a glass state with a larger arrested part. This is exhibited for $v_3 = 3.750$. The third case deals with the merging of the two extrema of the curves for $v_3 = v_3^{cc}$. The crossing of the A_3 singularity manifests itself by a f^* versus v_1 function, which exhibits an $f^* - f^{cc} \propto (v_1 - v_1^{cc})^{1/3}$ behaviour. Crossing the glass–glass-transition line transversally at the A_3-endpoint singularity, the arrested part varies continuously; but, there is a divergency of the slope of the f^* versus v_1 curve. The fourth case concerns the values $v_3 < v_3^{cc}$, and it is demonstrated for $v_3 = 3.000$. There is a continuous glass transition for $v_1 = 1$ with a smooth increase of f^* towards unity if v_1 increases from unity towards infinity. If $v_3 < 1$, the f^*- versus -v_1 curve is bent downward. This behaviour is qualitatively the same as described by Eq. (4.41b) for the elementary model. The results for the latter are reproduced for $v_3 = 0$. If $1 < v_3 < v_3^{cc}$, the f^* versus v_1 curve exhibits a crossover from upward to downward bending with a maximum of the slope for v_1 near v_1^{cc}. This is a precursor effect for the approach towards the A_3 singularity.

The three control parameters v_1, v_2, and v_A specify the state of the two-component model defined by Eqs. (4.32a,b), (4.33a,b). The two correlators have been denoted by $\phi_1(t) = \phi(t)$ and $\phi_2(t) = \phi_A(t)$. Let us denote the corresponding long-time limits by $f_1 = f$ and $f_2 = f_A$. The two mode-coupling polynomials read $\mathcal{F}_1[P, \boldsymbol{x}] = v_1 x_1 + v_2 x_1^2$ and $\mathcal{F}_2[P, \boldsymbol{x}] = v_A x_1 x_2$. The first fixed-point equation deals with the problem discussed above in connection with Fig. 4.2. Independent of the size of v_A, the cited figure exhibits the region of the liquid also for the model under discussion. And the complementary region deals with glasses characterized by $f(P) > 0$. The second fixed-point equation for glass states P reads $f_A/(1 - f_A) = v f_A$, where $v = f(P) \cdot v_A$. This is the fixed-point equation for the elementary model discussed in Sec. 4.2.4(i). Hence, there are two glass states. If the coupling constant v_A is so large that $v > 1$, the second correlator exhibits a positive arrested part $f_A = (v - 1)/v$. These states shall be referred to as glass 1. States with $v \leq 1$, describe a glass with a second correlator, which relaxes to zero for long times. These states shall be referred to as glass 2. The two glasses are separated by a line of degenerate A_2 glass-transition points. Crossing this line, e.g., by increasing v_A for fixed v_1 and v_2, the first arrested part f remains fixed. The second long-time limit f_A varies continuously, but it changes from zero to a positive value. The transition manifold is specified by the equation $v = v^c = v_A f(P) = 1$ (Sjögren 1986).

Figure 4.4 exhibits the singularity diagram for a cut through the three-dimensional parameter space for the model under discussion. The parameter pair determining the first correlator are chosen on the path $P^\xi \leftrightarrow (v_1^c, v_2^c)\xi$, $0 < \xi$. The values (v_1^c, v_2^c) denote a transition point according to Eq. (4.67b) for $\lambda^c = 3/4$. Some points on this array are exhibited as crosses in Fig. 4.2. In Fig. 4.4, the horizontal line $\xi = \xi^c = 1$ separates the liquid from the glass. If ξ increases from ξ^c to infinity, $f(P^\xi)$ increases from $f^c = 1 - \lambda^c = 1/4$ to unity. For $v_A < 1$, the second correlator relaxes to zero. For $v_A > 1/f^c$, both correlators exhibit an arrested part. The dashed glass–glass-transition line decreases monotonically from $\xi = \infty$ to $\xi = 1$ if v_A increased from 1 to $v_A^c = 1/f^c$. Because of Eq. (4.68b), one gets for small positive $\epsilon = \xi - 1$, for the transition line: $v_A = v_A^c - [\lambda^c/(1 - \lambda^c)^2]\sqrt{\epsilon}(1 + O(\epsilon^{1/2}))$. The dashed transition line merges the full one with a horizontal slope at the point $\xi = \xi^c$, $v_A = v_A^c$.

The bifurcation at the merging point is not equivalent to some A_ℓ bifurcation. The essential difference between this merging singularity and the A_2 bifurcations occurring for the other points of the lines concerns the maximum eigenvalue $E(P^c) = 1$ of the critical stability matrix. For the latter points, the maximum eigenvalue is not degenerate. The second eigenvalue of the two-by-two stability matrix is smaller than unity. For the merging point, the critical eigenvalue $E(P^c) = 1$ is degenerate. The discussion of the preceding paragraph has shown, that this singularity can be described adequately as an iteration of two A_2 singularities. It is a degenerate A_2 singularity for the glass 1–glass 2 transition line. But, its control parameter v is not a smooth function of the control parameters for the mode-coupling functional. Rather, its control parameter depends singularly on the mode-coupling coefficients. This latter $\sqrt{\epsilon}$ behaviour is determined by the A_2 bifurcation for the points of the $\xi = \xi^c$ line.

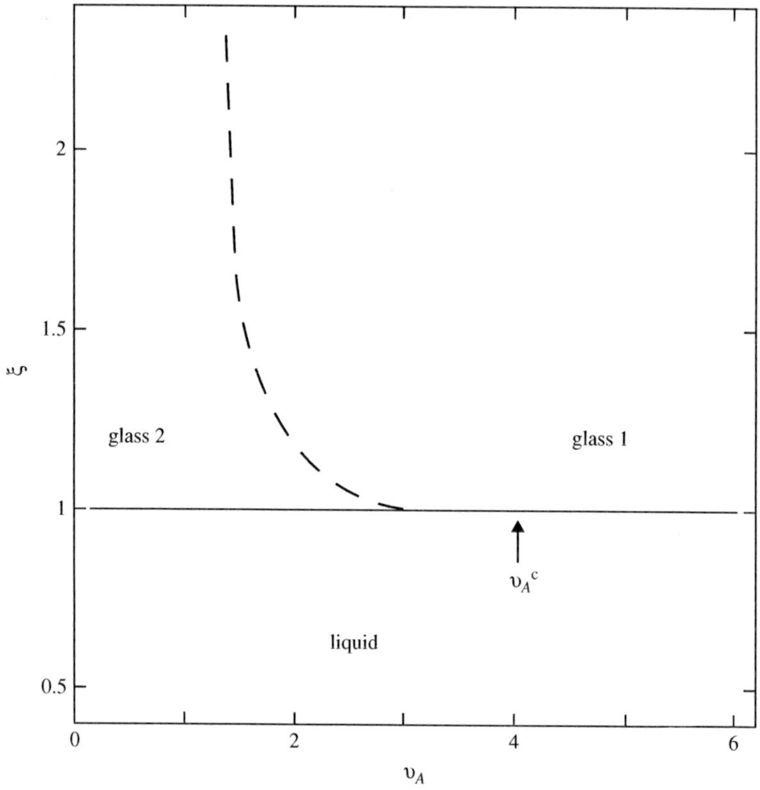

FIG. 4.4. Bifurcation lines for the two-component model defined by Eqs. (4.32a,b), (4.33a,b). The coupling constants v_1 and v_2 increase with parameter ξ according to $(v_1, v_2) = (v_1^c, v_2^c)\xi$. Here, v_1^c and v_2^c are determined by Eq. (4.67b) for $\lambda^c = 3/4$. The horizontal axis shows the third coupling constant v_A. The states below the horizontal line $\xi = 1$ are liquids. The dashed line for $\xi \geq 1$ shows points of continuous transitions between two glasses. For the large-v_A states (glass 1), both correlators exhibit an arrested part: $f > 0, f_A > 0$. For the small-v_A states (glass 2), the first correlator has an arrested part $f > 0$; but the second one relaxes to zero for long times. The two transition lines merge for $\xi = \xi^c = 1$, $v_A = v_A^c = 4$ as indicated by the arrow.

The two-component model defined by Eqs. (4.34a–c) for a three-dimensional space of control parameters, $P \leftrightarrow (v_1, v_2, v_3)$, shall be used to demonstrate a generic A_4-glass-transition singularity (Götze and Haussmann 1988; Götze and Sperl 2002). The second mode-coupling polynomial reads: $\mathcal{F}_2[P, \boldsymbol{x}] = v_3 x_1 x_2$. This is the same formula as discussed for the explanation of Fig. 4.4. One concludes that $f_1 = 0$ implies $f_2 = 0$. Similarly, the fixed-point equation for the second component yields $f_2^* = 1 - (v_3 f_1^*)^{-1}$, provided $v_3 > 1/f_1^*$. For such states, $f_1 > 0$ implies $f_2 > 0$. Let us restrict the discussion to the region of large v_3 where there are only glass states with both correlators exhibiting arrest.

Equation (4.34b) implies: $\mathcal{F}_1[P, \boldsymbol{x}] = v_1 x_1^2 + v_2 x_2^2$. The fixed-point equation for f_1^* can be simplified by expressing f_2^* in terms of f_1^*. Hence, the fixed-point component f_1^* is given as solution of $G[P, f_1^*] = 0$, where $G[P, z] = [z/(1-z)] - \{v_1 z^2 + v_2[1-(v_3 z)^{-1}]^2\}$. A critical state P^c with the arrested parts f_1^c and $f_2^c = 1 - (v_3 f_1^c)^{-1}$ is determined by those f_1^c, which are the degenerate zeros of $G[P, z]$. It is a point on the surface in the parameter space, which is defined by the solutions of the pair of equations $G[P^c, f_1^c] = 0$, $\partial G[P^c, f_1^c]/\partial z = 0$. The solution can be determined by the inverse method. The equations define a pair of linear relations for the two parameters v_1^c and v_2^c. The coefficients and the inhomogeneities are elementary functions of v_3^c and f_1^c. Thus, the solution yields the manifold of singularities as expressions for v_1^c and v_2^c in terms of the two variables v_3^c and f_1^c. To simplify the notation, the coordinates for the manifold shall be denoted by y and z:

$$y = v_3^c, \qquad z = f_1^c. \tag{4.70a}$$

The other two coordinates for the fixed points read:

$$v_1^c = \big[(2+y)z - 3\big] / \big[2(1-z)^2 z(yz-2)\big], \tag{4.70b}$$

$$v_2^c = \big[y^2 z^3(1-2z)\big] / \big[2(1-z)^2(y^2z^2 - 3yz + 2)\big]. \tag{4.70c}$$

The second component of the fixed point is given by

$$f_2^c = 1 - (yz)^{-1}. \tag{4.70d}$$

The singularity parameter has the values

$$\mu_2^c = \big[3y(y+2)z^3 - (y^2+18y+8)z^2 + 6(y+3)z - 6\big] \\ / \big[2y(y+2)z^3 - 12yz^2 + 2(y+2)z\big]. \tag{4.70e}$$

There holds the restriction $0 < z < 1$, and the maximum property has to be used to identify f_1^c in those cases, where there are several fixed points for the same state P.

For fixed y, Eqs. (4.70b,c) describe a curve with z as curve parameter. The curve is the cut $v_3^c = y$ through the singularity manifold. The curves enter the admissible parameter quadrant for $v_1^c = 0$ with z having the minimal value $z_{\text{min}} = 3/(2+y)$. With increasing z, the curves are located in the quadrant $v_1^c > 0$, $v_2^c > 0$, till they leave it for $z = 1/2$ at $v_1^c = 4$, $v_2^c = 0$. There are three types of curves as shown in Fig. 4.5. The curve for $v_3 = 45$ is representative for very large values of y. The cubic polynomial of z, which determines the numerator of Eq. (4.70e), has two zeros, say $z_-(y) < z_+(y)$. These can be expressed as smooth functions of y by elementary formulas. For $z = z_\pm(y)$, the derivatives $\partial v_{1,2}^c(z)/\partial z$ change sign, i.e., the curves exhibit cusps there. For z increasing above z_{min}, $v_1^c(z)$ increases

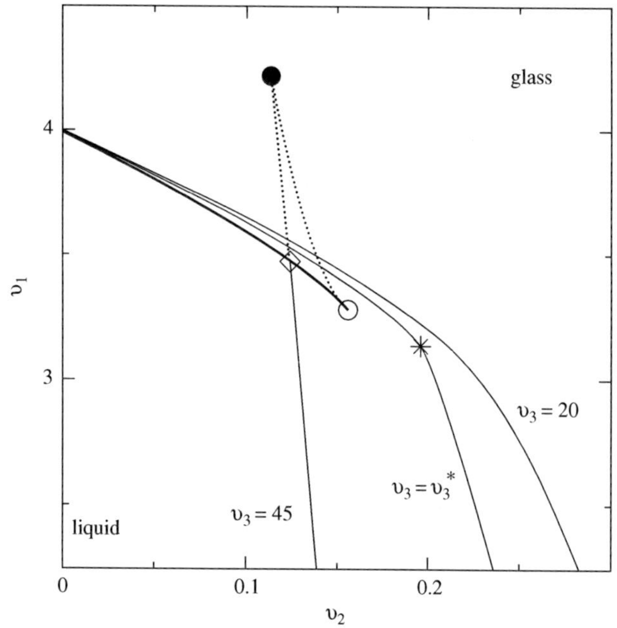

FIG. 4.5. The full lines show points for A_2 glass-transition singularities describing discontinuous liquid–glass and glass–glass transitions for the two-component model defined by Eqs. (4.34a–c). The lines are obtained by cuts through the parameter octant $v_\alpha \geq 0$, $\alpha = 1, 2, 3$, for the three indicated fixed values for v_3. The cut for $v_3 = v_3^* = 24.7\cdots$ contains the A_4 singularity, which is marked by a star. The circle and diamond mark the A_3 singularity and the crossing point, respectively, for the cut $v_3 = 45$; the dotted lines and the filled circle complete this bifurcation manifold, but they refer to non-maximum fixed points. Modified version of a figure from Götze and Sperl (2002).

and $v_2^c(z)$ decreases. At $z = z_-(y)$, an A_3 singularity is reached. Increasing z further, $v_1^c(z)$ decreases and $v_2^c(z)$ increases till another A_3 singularity is reached for $z = z_+(y)$. This singularity is marked by a circle. The branch between $z_-(y)$ and $z_+(y)$ is shown as the dotted curve. These points cannot be maximum fixed points, since $\mu_2^c < 0$. Increasing z above $z_+(y)$ towards $1/2$, $v_1^c(z)$ increases and $v_2^c(z)$ decreases. The two branches $z < z_-(y)$ and $z > z_+(y)$ cross at some point, which is marked by a diamond. Because of the maximum property, the first branch cannot reach this point. The dotted part of the first branch between the crossing point and the first A_3 singularity does not describe glass-transition singularities. The liquid regime is bounded by two smooth lines of A_2-liquid-glass-transition points. These lines form a corner. One line extends as glass–glass-transition line into the glass and it ends there in an A_3 singularity. The scenario is similar to that discussed in Fig. 4.3. But here, all lines deal with generic A_2 singularities, which describe discontinuous transitions.

Substituting $z = z_{\pm}(y)$ in Eqs. (4.70b,c), the triple v_1^c, v_2^c, v_3^c describes a curve in parameter space. These are lines of A_3 bifurcations; the one constructed with $z_+(y)$ is a line of A_3-glass-transition singularities. The line of points constructed with $z_-(y)$ are fixed points, which are below a larger one. Decreasing y, there appears a critical value y^*, where the two lines merge: $z_+(y^*) = z_-(y^*)$. The coalescence of two A_3 singularities is an A_4 singularity. The equation for the degeneracy of two zeros of the numerator polynomial of Eq. (4.70e) is a polynomial of order 6. Two factors $(y^* - 2)$ and $(y^* - 4)$ can be split off, so that the searched-for value can be calculated with the known formulas for the roots of a quartic polynomial: $y^* = 24.779392$, $z^* = f_1^{c*} = 0.24266325$. The A_4 singularity is shown by a star. As opposed to the generic A_3 singularity, the A_4 singularity can be approached arbitrarily closely by liquid states.

For $4 < v_3 < v_3^{c*}$, the liquid–glass transition surface is a smooth manifold of generic A_2 singularities. A typical example for a cut is shown for $v_3^c = 20$.

4.3.3 *Classification of the critical states*

All subtle results of the MCT equations deal with implications of the existence of glass-transition singularities P^c. These implications can be explained qualitatively by the asymptotic solution of the equations of motion for states P approaching the singular ones. In this section, the basic concepts and equations shall be formulated, which underly the asymptotic-solution theory. The general theory will be developed in Chapter 6. Here, the procedure will be started with the aim to calculate the arrested part $\boldsymbol{f}(P)$ of the correlator array. It will be shown that all glass-transition singularities are generic and degenerate A_ℓ bifurcations, $\ell = 2, 3, \ldots$, or iterations thereof.

Let us represent the array of the long-time limits $\boldsymbol{f}(P)$ of the correlators for the state P in terms of another array to be denoted by $\boldsymbol{g}(P)$:

$$f_q(P) = f_q^c + (1 - f_q^c) g_q(P), \quad q = 1, \ldots, M. \tag{4.71a}$$

Here and in the following text, the index c shall indicate that the corresponding quantity refers to a given glass-transition singularity P^c, e.g., $\boldsymbol{f}^c = \boldsymbol{f}(P^c)$. The representation of $\boldsymbol{f}(P)$ in terms of $\boldsymbol{g}(P)$ can be substituted in Eq. (4.15a) in order to get $\boldsymbol{\mathcal{F}}[P, \boldsymbol{f}(P)]$ as polynomial of order n_0 for the components of $\boldsymbol{g}(P)$:

$$\mathcal{F}_q[P, \boldsymbol{f}(P)] = (1 - f_q^c)^{-1} \sum_{n=0}^{n_0} \sum_{k_1 \cdots k_n} A_{q,k_1 \cdots k_n}^{(n)}(P) g_{k_1}(P) \cdots g_{k_n}(P). \tag{4.71b}$$

The coefficients

$$A_{q,k_1 \cdots k_n}^{(n)}(P) = \frac{1}{n!}(1 - f_q^c)\{\partial^n \mathcal{F}_q[P, \boldsymbol{f}^c]/\partial x_{k_1} \cdots \partial x_{k_n}\}(1 - f_{k_1}^c) \cdots (1 - f_{k_n}^c) \tag{4.71c}$$

are linear functions of the original mode-coupling coefficients $V_1, \ldots, V_N$. Since $0 \le f_q^c < 1$ for all q, one concludes from Eq. (4.15c), that

$$A^{(n)}_{q,k_1\cdots k_n}(P) \ge 0 \tag{4.71d}$$

for all P and all $q, k_1, \ldots, k_n = 1, 2, \ldots, M$. One can write

$$A^{(n)}_{q,k_1\cdots k_n}(P) = A^{(n)c}_{q,k_1\cdots k_n} + D^{(n)}_{q,k_1\cdots k_n}(P). \tag{4.71e}$$

The coefficients $A^{(n)c}_{q,k_1\cdots k_n} = A^{(n)}_{q,k_1\cdots k_n}(P^c)$ are determined by the M components of the critical glass form factor $\boldsymbol{f}^c$ and by the values of the N mode-coupling coefficients at the critical point P^c. These are numbers characterizing the glass-transition singularity under consideration. The difference coefficients $D^{(n)}_{q,k_1\cdots k_n}(P)$ are linear combinations of the N control-parameter differences $(V_\alpha - V_\alpha^c)$, $\alpha = 1, \ldots, N$. The functions $D^{(n)}_{q,k_1\cdots k_n}(P)$ quantify the mode-coupling polynomial in a neighbourhood of the singularity; they are used as small parameters for the intended asymptotic expansions. The preceding formulas have similarities to those in Eqs. (4.22a–c). However, $\boldsymbol{f}^c$ in Eq. (4.71c) is control-parameter independent. Its analogue $\boldsymbol{f}^*(P)$ in Eq. (4.22c) depends on $V_1, \ldots, V_N$, and this dependence can be a singular one.

The asymptotic solution shall be searched for under the assumption that there is a neighbourhood of states P so that $|g_q(P)| < 1$ for all q. Then, one can write: $f_q(P)/(1 - f_q(P)) = f_q^c/(1 - f_q^c) + \sum_{n=1}^{\infty} g_q(P)^n/(1 - f_q^c)$. The fixed-point equation (4.52d) gets the form:

$$\sum_k [\delta_{q,k} - A^c_{q,k}] g_k(P) = J_q(P), \tag{4.72a}$$

$$\begin{aligned} J_q(P) = {} & D_q^{(0)}(P) + \sum_k D^{(1)}_{q,k}(P) g_k(P) \\ & + \sum_{n \ge 2} \Big\{ \Big[\sum_{k_1\cdots k_n} A^{(n)}_{q,k_1\cdots k_n}(P) g_{k_1}(P) \cdots g_{k_n}(P) \Big] - g_q(P)^n \Big\}. \end{aligned} \tag{4.72b}$$

Here, Eq. (4.52d) for $P = P^c$ was applied: $f_q^c/(1 - f_q^c) = (1 - f_q^c)^{-1} A_q^{(0)c}$. Formula (4.61c) for the critical stability matrix was used: $A^{(1)c}_{q,p} = A^c_{q,p}$.

Above, the fixed-point equation is rewritten as a set of M linear equations with inhomogeneity $\boldsymbol{J}(P)$. The nonlinearities of the problem are hidden in the nonlinear dependence of $\boldsymbol{J}(P)$ on the array $\boldsymbol{g}(P)$. The inhomogeneity starts with the two terms given by the coefficients $D_q^{(0)}(P)$ and $D^{(1)}_{q,k}(P)$, which are small parameters. This suggests that $\boldsymbol{g}(P)$ is also the smaller the smaller the $(V_\alpha - V_\alpha^c)$, $\alpha = 1, \ldots, N$, so that one can solve the equation iteratively considering $\boldsymbol{g}(P)$ as the relevant small quantity. The procedure requires some care, however, since the M-by-M matrix $(1 - A^c)$ does not have an inverse. To proceed, one has to understand the eigenvectors with eigenvalue zero of this matrix. These are the eigenvectors of the critical stability matrix for the eigenvalue unity. The M eigenvalues of matrix A^c shall be denoted by e_k, $k = 1, \ldots, M$. They are special cases of the ones discussed in connection with Eqs. (4.57a,b): $e_k = e_k[P^c, \boldsymbol{f}^c]$. According to the discussion of Eqs. (4.60), (4.61), there is the maximum eigenvalue unity. It shall be chosen as e_M:

$$|e_k| \leq 1; \quad k = 1, \ldots, M-1; \tag{4.73a}$$

$$e_M = 1. \tag{4.73b}$$

The critical stability matrix A^c is an example for a non-negative matrix. In order to understand further properties of such matrices, this paragraph shall be used to explain the concept of irreducibility. Let us consider a non-trivial splitting of the index numbers in two groups. The first group consists of numbers obeying $1 \leq q \leq L$, and the second one of numbers obeying $L+1 \leq q \leq M$; $1 \leq L < M$. Thereby, the M-by-M matrix A^c is written as a two-by-two block matrix: $A^{(i,j)}$, $i, j = 1, 2$. The diagonal blocks are given by the L-by-L matrix $A^{(1,1)}$, defined by its elements $A^{(1,1)}_{q,p} = A^c_{q,p}$ with $q, p = 1, \ldots, L$, and by the $(M-L)$-by-$(M-L)$ matrix $A^{(2,2)}$, given by its elements $A^{(2,2)}_{q,p} = A^c_{q,p}$ with $q, p = L+1, \ldots, M$. One off-diagonal block is the L-by-$(M-L)$ matrix $A^{(1,2)}$, defined by its elements $A^{(1,2)}_{q,p} = A^c_{q,p}$ with $q = 1, \ldots, L$ and $p = L+1, \ldots, M$. The other off-diagonal block is the $(L-M)$-by-L matrix $A^{(2,1)}$, defined by its elements $A^{(2,1)}_{q,p} = A^c_{q,p}$, $q = L+1, \ldots, M$ and $p = 1, \ldots, L$. If the block $A^{(2,1)}$ is a matrix of zeros, matrix A^c is called reducible. More generally, A^c is called reducible if the above structure is obtained after some permutation of the array labels. If this is not possible, the matrix is called irreducible. The matrix A^c is irreducible if there are not enough vanishing elements in order to transform it by index permutation in a triangular block matrix.

There hold the following theorems due to G. Frobenius and O. Perron (Gantmacher 1974). If a non-negative matrix is irreducible, the maximum eigenvalue is non-degenerate. The left and the right eigenvectors for the maximum eigenvalue can be chosen so that all their components are positive.

If the critical stability matrix is irreducible, there holds:

$$e_k \neq 1, \quad k = 1, \ldots, M-1. \tag{4.74a}$$

The left and right eigenvectors of A^c for the eigenvalue $e_M = 1$ can be chosen so that all M components are positive. These eigenvectors shall be denoted by $\boldsymbol{a}^*$ and $\boldsymbol{a}$, respectively:

$$\sum_p a^*_p A^c_{pk} = a^*_k, \qquad \sum_p A^c_{qp} a_p = a_q, \tag{4.74b}$$

$$a^*_k > 0, \qquad a_q > 0, \qquad k, q = 1, \ldots, M. \tag{4.74c}$$

The eigenvectors shall be determined uniquely by imposing the convention:

$$\sum_q a_q^* a_q = 1, \qquad \sum_q a_q^* a_q a_q = 1. \tag{4.74d}$$

The arrays $\mathbf{a}^*$ and $\mathbf{a}$ are called the distinguished eigenvectors of P^c.

Let us consider the following M linear equations for an array $\boldsymbol{X}$. The equations are specified by an inhomogeneity array $\boldsymbol{I}$:

$$\sum_k [\delta_{q,k} - A^c_{q,k}] X_k = I_q, \quad q = 1, \ldots, M. \tag{4.75a}$$

According to linear-algebra theorems, there is a solution if and only if the inhomogeneity obeys the condition

$$\sum_q a_q^* I_q = 0. \tag{4.75b}$$

The general solution is the sum of a special one, say $\tilde{\boldsymbol{X}}$, and an arbitrary multiple X of the eigenvector $\boldsymbol{a}$:

$$\boldsymbol{X} = X\boldsymbol{a} + \tilde{\boldsymbol{X}}. \tag{4.75c}$$

A special solution $\tilde{\boldsymbol{X}}$ can be constructed as a linear mapping of $\boldsymbol{I}$, specified by an M-by-M matrix R:

$$\tilde{X}_q = \sum_k R_{q,k} I_k, \quad q = 1, \ldots, M. \tag{4.75d}$$

The matrix elements $R_{q,k}$; $q, k = 1, \ldots, M$, can be expressed in terms of the ones for the matrix A^c. They are determined uniquely by imposing the conventions

$$\sum_q a_q^* R_{q,k} = 0 = \sum_q R_{k,q} a_q, \quad k = 1, \ldots, M. \tag{4.75e}$$

The M-by-M matrix R is called the distinguished resolvent of P^c.

A remark shall be added concerning the evaluation of the two eigenvectors and of the matrix R. An irreducible non-negative matrix like A^c is called primitive if there is some number $m, m = 1, 2, \ldots$, so that all elements of the m-th power of the matrix are positive:

$$\sum_{k_1 \cdots k_{m-1}} A^c_{q,k_1} A^c_{k_1,k_2} \cdots A^c_{k_{m-1},p} > 0, \quad q, p = 1, \ldots, M. \tag{4.76a}$$

For a primitive A^c, the $M - 1$ first eigenvalues have a modulus smaller than the maximum eigenvalue (Gantmacher 1974):

$$|e_k| < 1, \quad k = 1, \ldots, M - 1. \tag{4.76b}$$

In this case, it is possible to calculate the eigenvectors $\boldsymbol{a}^*$ and $\boldsymbol{a}$ as well as the matrix R from exponentially convergent power series of matrix A^c. The proof of this statement is given in Appendix D.4. For all MCT models studied so far, the irreducible stability matrix is primitive.

The discussion of Eq. (4.75a) can be applied to Eq. (4.72a). The M components of the array $\boldsymbol{g}(P)$ can be expressed in terms of an amplitude to be called $g(P)$ and another array:

$$g_q(P) = g(P)a_q + \sum_k R_{qk}J_k(P). \tag{4.77a}$$

The convention ensures the uniqueness of the splitting of $\boldsymbol{g}$ in the two terms on the right-hand side of this formula. There holds

$$\sum_q a_q^* g_q(P) = g(P). \tag{4.77b}$$

In this sense, $g(P)$ is the 'projection' of $\boldsymbol{g}(P)$ on the 'dangerous' eigenvector $\boldsymbol{a}$.

Substituting the preceding formulas in Eq. (4.71a), one arrives at formulas for the arrested parts of the correlators:

$$f_q - f_q^c = h_q[g(P) + \tilde{g}_q(P)]. \tag{4.78a}$$

Here, the critical amplitudes are defined by

$$h_q = (1 - f_q^c)a_q. \tag{4.78b}$$

They are properties of the singularity and do not depend on the control-parameter differences $(V_\alpha - V_\alpha^c)$, $\alpha = 1, \ldots, N$. Because of Eqs. (4.52b), (4.74c), the critical amplitudes are positive:

$$h_q > 0, \quad q = 1, \ldots, M. \tag{4.78c}$$

The second term in the bracket of Eq. (4.78a) reads:

$$\tilde{g}_q(P) = \sum_k R_{qk}J_k(P)/a_q. \tag{4.78d}$$

The weight function $w_q = a_q a_q^*$ is positive and normalized because of Eqs. (4.74c,d): $\sum_q w_q = 1$. The amplitude $g(P)$ is the average of $\delta f_q = [f_q - f_q^c]/h_q$ with respect to the specified weight: $\sum_q w_q \delta f_q = g(P)$. The second term in Eq. (4.78a) describes the fluctuations around this average:

$$\sum_q w_q \tilde{g}_q(P) = 0. \tag{4.78e}$$

The problem of determining $\boldsymbol{g}(P)$ is reformulated in Eq. (4.77a) as one of determining the amplitude $g(P)$ and the array $\boldsymbol{J}(P)$. They are connected by the solubility condition (4.75b),

$$\sum_q a_q^* J_q(P) = 0, \tag{4.79}$$

and by the formula (4.72b) for the inhomogeneity $\boldsymbol{J}(P)$. Substituting Eq. (4.77a), one gets

$$J_q(P) = D_q^{(0)}(P) + g(P) D_q^{(1,0)}(P) + \sum_k D_{q,k}^{(0,1)}(P) J_k(P) \\ + \sum_{m+n\geq 2} g(P)^m \sum_{k_1\cdots k_n} \left[A_{q,k_1\cdots k_n}^{(m,n)}(P) J_{k_1}(P)\cdots J_{k_n}(P)\right]. \tag{4.80}$$

The difference coefficient $D_q^{(0)}(P)$ is defined in Eq. (4.71e) for $n=0$. For $n=1$, that formula defines $D_{q,p}^{(1)}(P)$, which determines the two new coefficients

$$D_q^{(1,0)}(P) = \sum_p D_{q,p}^{(1)}(P) a_p, \tag{4.81a}$$

$$D_{q,k}^{(0,1)}(P) = \sum_p D_{q,p}^{(1)}(P) R_{p,k}. \tag{4.81b}$$

The three difference coefficients in the first line of Eq. (4.80) are linear functions of the control-parameter differences $(V_\alpha - V_\alpha^c)$, $\alpha = 1,\ldots,N$. The other coefficients are linear combinations of the coefficients defined in Eq. (4.71c). Thus, they are linear functions of the control parameters $V_\alpha, \alpha = 1,\ldots,N$:

$$A_{q,k_1\cdots k_n}^{(m,n)}(P) = \Big\{\Big[\sum_{p_1\cdots p_m q_1\cdots q_n} A_{q,p_1\cdots p_m q_1\cdots q_n}^{(m+n)}(P) a_{p_1}\cdots a_{p_m} \\ R_{q_1,k_1}\cdots R_{q_n,k_n}\Big] - a_q^m R_{q,k_1}\cdots R_{q,k_n}\Big\} \\ \times [(m+n)!/(m!n!)]. \tag{4.81c}$$

Equation (4.80) is a nonlinear equation to determine $J_q(P)$ as function of $g(P)$ and of the control parameters. Substituting the result in the solubility condition (4.79), one gets the equation to determine $g(P)$.

Let us rewrite Eq. (4.80) as follows:

$$J_q(P) \\ = \left[D_q^{(0)}(P) + g D_q^{(1,0)}(P) + g^2 A_q^{(2,0)}(P) + g^3 A_q^{(3,0)}(P) + g^4 A_q^{(4,0)}(P) + \cdots\right] \\ + \sum_k \left[D_{q,k}^{(0,1)}(P) + g A_{q,k}^{(1,1)}(P) + g^2 A_{q,k}^{(2,1)}(P) + \cdots\right] J_k(P) \\ + \sum_{kp} \left[A_{q,kp}^{(0,2)}(P) + \cdots\right] J_k(P) J_p(P) + \cdots . \tag{4.82a}$$

Here, $g(P)$ was denoted as g for reasons of transparency. An asymptotic solution is constructed by assuming that g is small in the sense that $P \to P^c$ implies $g \to 0$. It is assumed in addition that $|D^{(0)}(P)/g|$, $|D_q^{(1,0)}(P)/g|$, $|D_q^{(0,1)}(P)/g|$ and $|J_q(P)/g|$ remain bounded. Then, one can replace recursively $J_k(P)$ on the right-hand side of Eq. (4.82a) by this expression $J_q(P)$. Given some arbitrary power $r_0 \geq 2$, one arrives at the formula

$$J_q(P) = \sum_{r=0}^{r_0} c_q^{(r)}(P) g^r + O_q, \tag{4.82b}$$

Here, $O_q/g^{r_0} \to 0$ if $g \to 0$.

Introducing the functions

$$c^r(P) = \sum_q a_q^* c_q^{(r)}(P), \tag{4.83a}$$

the remaining condition (4.79) reads

$$0 = \sum_{r=0}^{r_0} c^{(r)}(P) g^r + O. \tag{4.83b}$$

The function $O = \sum_q a_q^* O_q$ obeys the condition $\lim_{g \to 0} O/g^{r_0} = 0$. The coefficients $c_p^{(r)}(P)$ are sums of products of the coefficients $A_{q,k_1 \cdots k_n}^{(m,n)}(P)$ occurring in Eq. (4.82a). Hence, these functions and $c^{(r)}(P)$ depend smoothly on the control parameters. There holds

$$c^{(0)}(P^c) = c^{(1)}(P^c) = 0. \tag{4.83c}$$

$\boldsymbol{J}(P)$ is given as a polynomial of $g(P)$, up to arbitrarily small corrections. According to Eq. (4.83b), $g(P)$ is determined by the roots of a real polynomial, up to arbitrarily small corrections. One concludes that $g(P)$ exhibits an A_ℓ singularity at $P = P^c$ with $g(P^c) = 0$. The order ℓ is determined by the condition:

$$c^{(r)}(P^c) = 0, \qquad r \leq \ell - 1; \qquad c^{(\ell)}(P^c) \neq 0. \tag{4.83d}$$

Increasing r_0 to arbitrary large values, the functions $g(P)$ can be evaluated with arbitrary accuracy for $P \to P^c$. Details for the A_2 singularity will be determined in the following Sec. 4.3.4. The A_3 singularity and the A_4 bifurcation will be studied in Sec. 6.3.1.

The reducibility of the stability matrix A^c is the necessary condition for a bifurcation of the arrested parts $\boldsymbol{f}(P)$, which is not an A_ℓ singularity. If A^c is reducible, a renumbering of the M indices casts A^c in the form of a triangular R-by-R block matrix: $A^{(i,j)}$, $i, j = 1, \ldots, R$, $1 < R \leq M$; $A^{(i,j)} = 0$ for $i > j$. The R diagonal blocks, $A^{(i,i)}$, $i = 1, \ldots, R$, are irreducible. The set of eigenvalues of A^c is composed of the eigenvalues of the R matrices $A^{(i,i)}$. If $E^{(i)}$ denotes the non-degenerate maximum eigenvalue of $A^{(i,i)}$, the maximum eigenvalue E of A^c

is the largest of the numbers $E^{(i)}$, $i = 1, \ldots, R$. Suppose that P^c is a critical point with a non-degenerate eigenvalue $E = 1$. This means that only one of the $E^{(i)}$ is unity, while the other ones are smaller than 1. In this case, the preceding discussions remain valid with one exception. Even though the components of the eigenvectors $\boldsymbol{a}^*$ and $\boldsymbol{a}$ cannot be negative, some of them may be zero. A new type of singularity can occur for points P_0, where the maximum eigenvalue $E = 1$ is degenerate. This means that the maximum eigenvalue of two or more of the diagonal blocks become unity simultaneously at $P = P_0$. Such singularity is exhibited by the merging point of the two transition lines discussed in connection with Fig. 4.4. For this example, the singularity P_0 is described by the degenerate zero of a second-order polynomial, whose coefficients exhibit the singular control-parameter dependence characteristic for an A_2 bifurcation. It will be shown in connection with Fig. 5.23 that this procedure can be used also for the discussion of a microscopic MCT model, which exhibits a degenerate-maximum-eigenvalue singularity. A more detailed discussion of the specified problem is not yet available. It should be emphasized that the appearance of a block of zeros, $A^{(i,j)} = 0$, means that there are no mode couplings of the variables of block i to those of block j. Such a peculiarity must have a special reason like, for example, some symmetry of the microscopic equations of motion.

4.3.4 *Correlation arrest near A_2 singularities*

In this section, a path in control-parameter space shall be considered as described by Eqs. (4.64a–c). The external control parameter ξ shall be shifted to the distance parameter $\epsilon = \xi - \xi^c$. The transition point P^0 shall be a singularity P^c of type A_2. For $|\epsilon|$ smaller than some positive bound, P^c separates the path in two kinds of states:

$$\text{liquid states } P^\epsilon \quad \text{if } \epsilon < 0, \ \text{ glass states } P^\epsilon \quad \text{if } \epsilon > 0. \tag{4.84}$$

Asymptotic expansion formulas for $\boldsymbol{f}(P^\epsilon)$ shall be derived for small positive ϵ.

(i) Generic liquid–glass transitions

The most important case deals with a discontinuous change of all M long-time limits:

$$f_q^c > 0, \quad q = 1, \ldots, M. \tag{4.85}$$

Because of Eq. (4.52d), the functions $A_q^{(0)}(P) = (1 - f_q^c)\mathcal{F}_q[P, \boldsymbol{f}^c]$, $q = 1, \ldots, M$, are positive for all states P in a neighbourhood of P^c. They increase strictly with increasing coupling constant V_α, $\alpha = 1, \ldots, N$. The corresponding difference functions from Eq. (4.71e) are used to define

$$\sigma(P) = \sum_q a_q^* D_q^{(0)}(P). \tag{4.86a}$$

This is a linear function of the control-parameter differences. Its reduction to the path defines a smooth function of ϵ:

$$\sigma_\epsilon = \sigma(P^\epsilon). \tag{4.86b}$$

Since the components of $\boldsymbol{a}^*$ are positive, the function σ_ϵ is positive for $\epsilon > 0$ and negative for $\epsilon < 0$. Requesting $C = d\sigma_\epsilon(\epsilon = 0)/d\epsilon \neq 0$, one gets

$$\sigma_\epsilon = C\epsilon + O(\epsilon^2), \quad C > 0. \tag{4.86c}$$

The function σ_ϵ is called the separation parameter.

Generically, the condition $\sigma(P) = 0$ defines a hyperplane in the control-parameter space. The request $C \neq 0$ means that the path crosses this plane at P^{c} transversally, i.e., its tangent at P_{c} is not in the plane. The parameter C quantifies the position of the tangent of the curve relative to the mentioned plane. Let us imagine that one considers the microscopic MCT for a Lennard-Jones liquid, and uses the temperature as external control parameter. Anticipating that the high-temperature region deals with liquids and the low-temperature one with glasses, one could use $\epsilon = (T_{\mathrm{c}} - T)/T_{\mathrm{c}}$ as distance parameter from the transition point at temperature T_{c}. The constant C would relate the experimental control parameter T to the relevant microscopic parameter σ_ϵ. The constant C for isobaric changes would differ from that for isochoric ones.

Equation (4.86c) can be used to write for the coefficients from Eq. (4.71e):

$$D^{(n)}_{q,k_1\cdots k_n}(P^\epsilon) = d^{(n)}_{q,k_1\cdots k_n}\sigma_\epsilon + O(\epsilon^2). \tag{4.87a}$$

The coefficients $d^{(n)}_{q,k_1\cdots k_n}$ characterize the leading-order changes of the mode-coupling constants at P^c. The preceding remarks on $A_q^{(0)}(P)$ ensure the inequalities

$$d_q^{(0)} > 0, \quad q = 1, \ldots, M, \tag{4.87b}$$

and Eq. (4.86a) implies the normalization $\sum_q a_q^* d_q^{(0)} = 1$. Equations (4.81a,b) yield

$$D_q^{(1,0)}(P^\epsilon) = d_q^{(1,0)}\sigma_\epsilon + O(\epsilon^2); \qquad D_{q,k}^{(0,1)}(P^\epsilon) = d_{q,k}^{(0,1)}\sigma_\epsilon + O(\epsilon^2), \tag{4.87c}$$

with the abbreviations $d_q^{(1,0)} = \sum_p d_{q,p}^{(1)} a_p$ and $d_{q,k}^{(0,1)} = \sum_p d_{q,p}^{(1)} R_{p,k}$.

For later reference, let us also define two linear functions of the control parameters by

$$\lambda(P) = \sum_q \sum_{kp} a_q^* A_{q,kp}^{(2)}(P) a_k a_p, \tag{4.88a}$$

$$\mu_2(P) = 1 - \lambda(P). \tag{4.88b}$$

Using the definition (4.81c) for $A_q^{(2,0)}(P)$ and observing the convention (4.74d), one can write:

$$\mu_2(P) = -\sum_q a_q^* A_q^{(2,0)}(P). \tag{4.88c}$$

The conditions (4.71d) imply $\lambda(P) \geq 0$, and this is equivalent to $\mu_2(P) \leq 1$.

It would be non-generic if the transition point P^c were to obey the equation $\mu_2(P^c) = 0$. The discussion shall be continued with the assumption that $\mu_2(P^c) \neq 0$. Hence, there is a whole neighbourhood of P^c, where $\mu_2(P)$ does not have a zero. Let us restrict the discussion to this neighbourhood. The bound for $|\epsilon|$ shall be reduced so that all P^ϵ are located in this neighbourhood as well.

The leading contribution to $J_q(P^\epsilon)$ is given by $J'_q(P) = D_q^{(0)}(P)+g^2 A_q^{(2,0)}(P)$. Every other term in Eq. (4.82a) is smaller than $D_q^{(0)}(P)$ or $g^2 A_q^{(2,0)}(P)$ if one considers the limit $\epsilon \to 0, g \to 0$. The solubility condition (4.79) reads in leading order $0 = \sum a_q^* J'_q(P) = \sigma(P) - g^2(P)\mu_2(P)$. Since $\mu_2(P) \neq 0$, this equation describes an A_2 bifurcation. The request of a generic transition is equivalent to the request of a generic A_2 singularity. Since the glass side is located for $\sigma(P^\epsilon) > 0$, $\mu_2(P)$ must be positive. Within the neighbourhood specified, there holds

$$0 < \mu_2(P) \leq 1, \quad 0 \leq \lambda(P) < 1. \tag{4.89a}$$

The function $\mu_2(P)$ determines the singularity parameter discussed in Sec. 4.3.2: $\mu_2(P^c) = \mu_2^c$. Let us also write

$$\lambda^c = 1 - \mu_2^c. \tag{4.89b}$$

Because of the maximum property, the leading-order solution reads $g(P^\epsilon) = \sqrt{\sigma_\epsilon/\mu_2^c}$. There holds $J'_q(P^\epsilon) = O(\epsilon)$.

The results of the preceding paragraph motivate the construction of a solution as asymptotic series in powers of the small parameter $s = \sqrt{\epsilon} : g(P) = g^{(1)}(P) + g^{(2)}(P) + \cdots, g^{(r)}(P) = O(s^r)$. The three difference coefficients in Eq. (4.82a) are of order s^2. Hence, the expansion of $\boldsymbol{J}(P)$ starts with terms of order s^2 : $J_q(P) = J_q^{(2)}(P) + J_q^{(3)}(P) + \cdots, J_q^{(r)}(P) = O(s^r)$. Iteration of the equation for the inhomogeneity yields:

$$J_q^{(2)}(P) = D_q^{(0)}(P) + g^2 A_q^{(2,0)}(P), \tag{4.90a}$$

$$J_q^{(3)}(P) = g\Big[D_q^{(1,0)}(P) + \sum_k A_{q,k}^{(1,1)}(P) J_k^{(2)}(P) + g^2 A_q^{(3,0)}(P)\Big]. \tag{4.90b}$$

The solubility condition (4.79) reads

$$\sigma(P) - g^2\mu_2(P) + \sum_q a_q^* J_q^{(3)}(P) + O(s^4) = 0. \tag{4.90c}$$

The preceding equation permits the evaluation of $g^{(1)}(P)$ and $g^{(2)}(P)$. Writing $g = \sqrt{\sigma_\epsilon/\mu_2(P)}[1 + \sqrt{\sigma_\epsilon} \cdot \kappa + O(s^2)]$, one gets $2\sqrt{\mu_2(P)}\kappa = \sum_q a_q^*[J^{(3)}(P)/(g\sigma_\epsilon)]$. The contribution $\tilde{g}_q(P)$ in Eq. (4.78a) is determined from Eq. (4.78d): $\tilde{g}_q(P)a_q = \sum_k R_{qk}J_k^{(2)}(P) + O(s^3)$. One arrives at

$$f_q(P^\epsilon) - f_q^c = h_q\sqrt{\sigma_\epsilon/\mu_2(P^\epsilon)}\Big\{1 + \sqrt{\sigma_\epsilon}k_q(P^\epsilon) + O(\epsilon)\Big\}, \qquad \epsilon \geq 0. \qquad (4.91a)$$

The correction amplitude $k_q(P)$ is split into two parts

$$k_q(P) = \overline{K}_q(P) + \kappa(P). \qquad (4.91b)$$

The first term is due to the contribution $\tilde{g}_q(P)$:

$$\overline{K}_q(P) = \sum_k R_{qk}A_k(P)/a_q, \qquad (4.92a)$$

$$A_k(P) = \Big[d_k^{(0)}\sqrt{\mu_2(P)}\Big] + \Big[A_k^{(2,0)}(P)/\sqrt{\mu_2(P)}\Big]. \qquad (4.92b)$$

As discussed in connection with Eq. (4.78e), the array $\overline{\boldsymbol{K}}$ describes fluctuations around zero:

$$\sum_q w_q\overline{K}_q = 0. \qquad (4.92c)$$

The average of the amplitude k_q is coefficient $\kappa(P) = \sum_q w_q k_q(P)$. It reads

$$\begin{aligned}\kappa(P) = \big[1/(2\sqrt{\mu_2(P)})\big]\sum_q a_q^*\Big\{\Big[d_q^{(1,0)}\Big]\\ + \Big[\sum_k A_{q,k}^{(1,1)}(P)A_k(P)/\sqrt{\mu_2(P)}\Big] + \Big[A_q^{(3,0)}(P)/\mu_2(P)\Big]\Big\}. \qquad (4.92d)\end{aligned}$$

The leading asymptotic contribution to the increase of $f_q(P) - f_q(P^c)$ is given by the first term on the right-hand side of Eq. (4.91a) (Götze 1985). It exhibits a $\sqrt{\epsilon}$ variation. This square-root law is the signature of a generic A_2 bifurcation. The dependence on the correlator index q is determined by the critical amplitude h_q. The leading asymptotic correction is specified by the amplitude $k_q(P)$ (Franosch *et al.* 1997). It yields a term to $f_q(P)$, which varies linearly with the distance parameter ϵ. If one intends to evaluate $f_q(P)$ strictly up to order ϵ, one can choose the functions $\mu_2(P)$ and $k_q(P)$ for their values at the transition state P^c. These functions vary smoothly with ϵ and their changes contribute only terms of order $\sqrt{\epsilon} \cdot \epsilon$. For the same reason, σ_ϵ can be replaced by $C\epsilon$. If the transition point P^c is shifted on the transition hypersurface, μ_2^c can vary. Approaching a higher-order singularity A_ℓ, $\ell \geq 3$, means that μ_2^c approaches zero. As a result, the amplitude k_q increases beyond any bound. Consequently, the ϵ interval for the applicability of the asymptotic expansion approaches zero.

The leading-order result $f_q(P) - f_q^c \propto \sqrt{\epsilon}$ formulates a universal feature of the generic liquid-glass transition. It holds for all models and all correlators. This statement will be corroborated in Chapter 5. Therefore, it is tempting to suggest tests of such results by experiments in order to find out whether or not MCT is relevant for the description of liquids. Figures 1.12, 1.13 and 1.17 exhibit responses to such suggestion. However, a test is not straightforward. One obstacle is that the range of validity of the leading-order formula is not universal. If one defines the range of validity by a margin δ for the relative error, the leading-order result for the range of validity is given by: $\epsilon < [\delta/|k_q(P)|]^2/C$. The range of validity depends on the correlator considered, on the transition point chosen, and on the path P^ϵ through the transition point.

There is the elementary connection (4.29a) between the correlator in the frequency domain and the fluctuating force correlator $m_q(z)$. From an ad hoc point of view, it is irrelevant whether the results are discussed for one quantity or the other. Let us assume that $\phi_q(t)$ is the normalized autocorrelation function of the system's dipole moment. Up to some constant of proportionality, and up to some additive constant, the dynamical dielectric susceptibility $\chi_D(z)$ is given by $z\phi_q(z) : \chi_D(z) \propto z\phi_q(z) + 1$. The negative inverse of $\chi_D(z)$ is the dynamical dielectric modulus. One concludes from Eq. (4.30a): Up to some constant of proportionality, and up to some additive second-order polynomial of the frequency, the dielectric modulus $M_D(z)$ is given by $zm_q(z) : M_D(z) \propto zm_q(z) - 1 \pm i(\tau_q z) + (z/\Omega_q)^2$. Dielectric spectroscopy can provide results for $\chi_D(z)$ as well as for $M_D(z)$. In this example, also from an experimental point of view, it is irrelevant whether the dynamics is discussed for $z\phi_q(z)$ or for $zm_q(z)$. The question is, which of the quantities can formulate most transparently the physical effect under study. The arrested part of the correlators, $f_q = \lim_{\epsilon\to 0}(-i\epsilon)\phi_q(i\epsilon)$, and the arrested part of the relaxation kernels, $f_q^m = \lim_{\epsilon\to 0}(-i\epsilon)m_q(i\epsilon)$, are related by the fixed point equation $f_q/(1-f_q) = f_q^m = \mathcal{F}_q[P, \boldsymbol{f}]$. One gets the asymptotic result for f_q^m from Eq. (4.91a) for f_q:

$$f_q^m(P) - f_q^{mc} = h_q^m \sqrt{\sigma_\epsilon/\mu_2(P)}\Big\{1 + \sqrt{\sigma(P)}k_q^m(P) + O(\epsilon)\Big\}. \tag{4.93a}$$

Here, $f_q^{mc} = f_q^c/(1-f_q^c) = \mathcal{F}_q[P^c, \boldsymbol{f}^c]$ is the discontinuity of the kernel's long-time limit at the transition point. The critical amplitude for the fluctuating-force correlator reads

$$h_q^m = h_q/(1-f_q^c)^2. \tag{4.93b}$$

The correction amplitude is given by

$$k_q^m = k_q + h_q/[(1-f_q^c)\sqrt{\mu_2(P)}]. \tag{4.93c}$$

The last term in the preceding equation is positive. If k_q is positive or zero, there holds $k_q^m > k_q$. In this case, the leading asymptotic law describes f_q better than a formula based on the leading-order result for the kernel. An example is given by the schematic model discussed in connection with Fig. 4.2. For the path chosen, one infers from Eq. (4.68b) that $k = 0$. On the other hand, if $k_q < 0$, there is a trend of cancellation in Eq. (4.93c); $|k_q^m|$ can be smaller than $|k_q|$. Under such conditions, the leading-order result for f_q^m can determine a better approximation for the arrested part $f_q = 1/[1 + (1/f_q^m)]$ than formulated in Eq. (4.91a). This is the case for the arrested part f_A for the second correlator of the model discussed in Fig. (4.4), provided v_A is large.

(ii) Degenerate liquid–glass transitions

The second case of interest deals with a degenerate transition from the liquid to the glass, i.e.,

$$\boldsymbol{f}^c = \mathbf{0}. \tag{4.94}$$

Equation (4.71a) reduces to $\boldsymbol{f}(P) = \boldsymbol{g}(P)$. Similarly, the coefficients in Eq. (4.71c) agree with the mode-coupling coefficients: $A^{(n)}_{q,k_1\cdots k_n}(P) = V^{(n)}_{q,k_1\cdots k_n}$. The mode-coupling polynomial does not have a zeroth-degree term, $\mathcal{F}_q[P, \mathbf{0}] = 0$. Therefore, the difference coefficients of order $n = 0$ vanish:

$$D_q^{(0)}(P) = 0, \quad q = 1, \ldots, M. \tag{4.95a}$$

This is the essential speciality, which requires a modification of the asymptotic analysis of the preceding subsection. The formula (4.61c) for the critical stability matrix simplifies to the coefficient matrix for the linear contributions to the mode-coupling polynomials:

$$A^c_{q,p} = V^{(1)c}_{q,p}. \tag{4.95b}$$

Let us recall that, generically, $A^c_{q,p}$ is an irreducible non-negative matrix with a maximum eigenvalue unity. In order to have a continuous liquid–glass transition, it is necessary that the mode-coupling polynomial has a linear contribution. Such contribution does not exist for the microscopic MCT models discussed in Sec. 4.2.3. Hence, conventional liquid–glass transitions are discontinuous ones. However, if the liquid is immersed in an arrested array of scattering centers, there exists a linear mode-coupling contribution. This is explained in connection with Eq. (4.13). In that case, there may exist parameter regions for continuous transitions.

The difference coefficient from Eq. (4.81a) shall be used to define the separation parameter $\sigma(P)$ for the continuous transition:

$$\sigma(P) = \sum_q a_q^* D_q^{(1,0)}(P). \tag{4.96a}$$

Because of Eqs. (4.74b,d), one can write explicitly

$$\sigma(P) = \left[\sum_{q,p} a_q^* V^{(1)}_{q,p} a_p\right] - 1. \tag{4.96b}$$

Reducing P to a generic path P^ϵ, there holds the result (4.86c) for $\sigma_\epsilon = \sigma(P^\epsilon)$. The formulas (4.87c) hold and one gets the analogue of the conditions (4.87b):

$$d_q^{(1,0)} > 0, \qquad q = 1, \ldots, M. \tag{4.96c}$$

The definitions for the functions $\lambda(P)$ and $\mu_2(P)$ in Eqs. (4.88a,b) will be used here as well; Eq. (4.88c) remains valid.

From Eq. (4.82a), one concludes that the leading contribution to the inhomogeneity $J_q(P)$ is given by $J'_q(P) = gD_q^{(1,0)}(P) + g^2 A_q^{(2,0)}(P)$. In the limit of ϵ and g tending to zero, the other terms can be neglected compared to those noted. The solubility condition (4.79) reads in leading order $0 = \sum_q a_q^* J'_q(P) = g[\sigma(P) - g\mu_2(P)] = 0$. The non-trivial solution reads $g = \sigma(P)/\mu_2(P)$. On the glass side of the transition, there holds $\sigma(P) > 0$. The maximum property implies $g > 0$. Hence, $\mu_2(P) > 0$. There hold Eqs. (4.89a–c); in particular, $\mu_2^c = \mu_2(P^c)$ is the singularity parameter for the degenerate fold bifurcation.

The preceding discussions suggest to solve the equations for the arrested parts by asymptotic expansions in powers of $s = \epsilon$. Using the notations introduced in connection with Eqs. (4.90a–c), one starts the series for $g(P)$ with terms of order ϵ and that for $J_q(P)$ with terms of order ϵ^2. One obtains

$$J_q^{(2)}(P) = gD_q^{(1,0)}(P) + g^2 A_q^{(2,0)}(P), \tag{4.97a}$$

$$J_q^{(3)}(P) = \sum_k \Big[D_{q,k}^{(0,1)}(P) + gA_{q,k}^{(1,1)}(P)\Big] J_k^{(2)}(P) + g^3 A_q^{(3,0)}(P). \tag{4.97b}$$

The solubility equation gets the form:

$$g(P)\sigma(P) - g^2(P)\mu_2(P) + \sum_q a_q^* J_q^{(3)}(P) + O(s^4) = 0. \tag{4.97c}$$

The procedure for solving the equations follows the one explained in the preceding subsection. The result reads:

$$f_q(P^\epsilon) = h_q[\sigma_\epsilon/\mu_2(P)]\Big\{1 + \sigma_\epsilon k_q(P) + O(\epsilon^3)\Big\}, \qquad \epsilon \geq 0. \tag{4.98}$$

The amplitude $k_q(P)$ is the sum of an average κ and a fluctuation $\overline{K}_q(P)$ as noted in Eq. (4.91b). The latter obeys Eq. (4.92c), since it is represented in terms of the matrix $R_{q,k}$ as formulated in Eq. (4.92a). The amplitude in the latter equation reads

$$A_k(P) = \Big[d_k^{(1,0)}\Big] + \Big[A_k^{(2,0)}(P)/\mu_2(P)\Big]. \tag{4.99a}$$

The average κ is determined by

$$\kappa = [1/\mu_2(P)] \sum_q a_q^* \Big\{ \Big[A_q^{(3,0)}(P)/\mu_2(P)\Big] + \sum_k \Big[\big(D_{q,k}^{(0,1)}(P)\mu_2(P)/\sigma(P)\big) + A_{qk}^{(1,1)}(P)\Big] A_k(P) \Big\}. \tag{4.99b}$$

The leading asymptotic contribution to $f_q(P)$ is given by $h_q(C/\mu_2^c)\cdot\epsilon$ (Franosch and Götze 1994). Upon shifting the state into the glass, the arrested parts increase linearly with the distance parameter ϵ. This change of $\sqrt{\epsilon}$-variations to ϵ-variations is the essential difference between a degenerate and a generic A_2 bifurcation. Also for the continuous transition under discussion, the range of validity of the expansion formula (4.98) shrinks to zero if P^c approaches a higher-order singularity. In this case, μ_2^c approaches zero; and this causes a divergency of $k_q(P)$. If one intends to perform a strict expansion in powers of ϵ, one has to notice that the smooth deviations of $\mu_2(P)$ from $\mu_2(P^c)$ and the deviation of σ_ϵ from $C\epsilon$ also lead to ϵ^2 contributions to $f_q(P)$.

(iii) Glass–glass transitions

The preceding discussions in subsections (i) and (ii) deal with a path P^ϵ, which crosses transversally a hypersurface of A_2 singularities at a critical point $P^c = P^{\epsilon=0}$. For $\epsilon < 0$, the correlator long-time limit is the trivial fixed point $\boldsymbol{f}^* = \boldsymbol{0}$, characterizing a liquid. For $\epsilon > 0$, there is a non-trivial long-time limit $\boldsymbol{f}(P^\epsilon)$. Let us complete this discussion with a comment on the general case of a path crossing transversally a hypersurface of A_2 bifurcations. Such a path deals with a glass–glass transition at P^c. Let us denote the fixed point, which describes the correlator long-time limit for $\epsilon < 0$, by $\boldsymbol{f}^*(P)$. At P^c, there occurs the transition to some fixed point $\boldsymbol{f}(P)$. Let this fixed point exhibit a glass-transition singularity at P^c with $\boldsymbol{f}^c = \boldsymbol{f}(P^c)$. Generically, the other fixed point $\boldsymbol{f}^*(P^c)$ will refer to a regular point. From the discussion of Sec. 4.3.1, one concludes the following. There is some positive ϵ_0 so that $\boldsymbol{f}^*(P^c)$ can be continued uniquely as function $\boldsymbol{f}^*(P^\epsilon)$ along the path for $|\epsilon| < \epsilon_0$. The fixed points $\boldsymbol{f}^*(P^\epsilon)$ depend smoothly on ϵ. Because of the maximum property, there holds $f_q^*(P^c) \le f_q(P^c)$, $q = 1, \dots, M$. The path describes the transition:

$$\phi(t \to \infty) = f^*(P^\epsilon), \quad \epsilon < 0, \tag{4.100a}$$

$$= f(P^\epsilon), \quad \epsilon \ge 0. \tag{4.100b}$$

If $f_q^*(P^c) = f_q^c$ for $q = 1, \dots, M$, there occurs a continuous glass–glass transition. The transition from glass 2 to glass 1 in Fig. 4.4 deals with this case. If $f_{q_0}^*(P^c) < f_{q_0}^c$ for some $q_0, 1 \le q_0 \le M$, there is a discontinuous glass–glass transition. This case is demonstrated in the lower panel of Fig. 4.3 for $v_3 = 3.750$. The full line for $v_1 > 1$ shows $f^*(P)$, which can be continued smoothly up to near $v_1 = 1.1$. But for $v_1 = v_1^c \approx 1.06$, there occurs a discontinuous increase to $f^c > 0.4$.

The covariance theorem of MCT can be applied to reduce the discussion of the glass–glass transition to that of a liquid–glass transition. To proceed, one uses the fixed point $\boldsymbol{f}^*(P)$ in Eq. (4.24a) in order to transform the correlator array $\boldsymbol{\phi}(t)$ to the array $\hat{\boldsymbol{\phi}}(t)$. The latter has the long-time limit $\hat{\boldsymbol{f}}(P^\epsilon)$. There hold all MCT results for the transformed problem. The coupling constants $\hat{V}^{(n)}_{q,k_1\cdots k_n}$ in Eq. (4.22c) are complicated functions of $V_1, \dots, V_N$; however, they vary smoothly with changes of the original control parameters. By construction, $\hat{\boldsymbol{f}}(P^\epsilon) = 0$ for $\epsilon < 0$. Hence, $\hat{\boldsymbol{\phi}}(t)$ exhibits a liquid–glass transition and all the preceding formulas are applicable with the parameters replaced by transformed ones. One gets for $q = 1, \ldots, M$:

$$f_q(P^\epsilon) = f_q^*(P^\epsilon) + [1 - f_q^*(P^\epsilon)]\hat{f}_q(P^\epsilon), \quad 0 \leq \epsilon < \epsilon_0. \tag{4.100c}$$

One can substitute Eq. (4.91a) or Eq. (4.98) for the transformed array $\hat{\boldsymbol{f}}(P^\epsilon)$ if the transition is discontinuous or continuous, respectively.

(iv) Some scenarios with a reducible stability matrix

Let us consider the possibility that there is a subgroup of L correlators exhibiting a discontinuous change at the transition; $1 \leq L < M$. After permutation of the labels, there holds

$$f_q^c > 0; \quad q = 1, \ldots, L. \tag{4.101a}$$

For the other correlators, $f_q(P^c) = 0$ is assumed; $q = L+1, \ldots, M$. This means that there are no couplings connecting the long-time limits of the latter correlators with those of the former. Such peculiarity must have a special reason. It is non generic that this reason is present only for the special states P^c. Thus, the generic case requires that

$$f_q(P) = 0, \quad q = L+1, \ldots, M, \tag{4.101b}$$

holds for a whole neighbourhood of states P near P^c. The scenario was explained for the transition from the liquid to the glass 2 states in connection with Fig. 4.4. The transition leads to a reducible stability matrix as explained in Sec. 4.3.3. Equations (4.86)–(4.93) remain valid, albeit for the set of the first L components only.

A further scenario is obtained if Eq. (4.101b) is combined with a continuous transition for the first group of L correlators. It might occur also that the group of singular correlators split in two subgroups, say, $q = 1, \ldots, L_1$ and $q = L_1 + 1, \ldots, L$; $L_1 + L_2 = L$, $1 \leq L_1 < L$. The first group can exhibit a discontinuous transition and the second group a continuous one. The asymptotic expansion has to be done separately for each group.

4.3.5 *Density-fluctuation arrest in hard-sphere-like systems*

The simplest model of matter deals with hard spheres of some diameter d, which form a homogeneous isotropic system with a number density ρ. A single control parameter like the packing fraction $\varphi = (\pi/6)\rho d^3$ specifies the equilibrium structure. The system can be studied experimentally in form of slightly polydisperse colloidal suspensions. It was demonstrated in Sec. 1.7 that measurements of density correlators $\phi_q(t)$ by photon-correlation spectroscopy have identified a liquid–glass transition in the sense defined within the mode-coupling theory. The value for the transition point of the hard-sphere system (HSS) is $\varphi_g = 0.578(\pm 0.7\%)$ (van Megen 1995). In this section, some quantitative MCT results for the density-fluctuation arrest of the HSS shall be considered.

Within the framework of the microscopic version of the MCT, Bengtzelius *et al.* (1984) have evaluated Debye–Waller factors f_q for the hard-sphere glass as function of the packing fraction φ. The convolution approximation was made to simplify the coefficients of the mode-coupling functional, i.e., Eq. (4.11d) was used with $c_3(q,k,p) = 0$. The structure factor S_q was evaluated in the Percus-Yevick approximation (Hansen and McDonald (1986), Appendix B). The critical point for the correlators' long-time limits was predicted to be:

$$\varphi_c = 0.516; \quad \text{Percus–Yevick approximation for } S_q. \tag{4.102a}$$

The results of this work for the form factor at the transition point, f_q^c, and for $(\varphi - \varphi_c)/\varphi_c = 0.066$ are reproduced in Fig. 1.16 as full lines.

The Verlet–Weis approximation formulates a modification of the Percus–Yevick results so that S_q agrees better with the data known from simulation studies for the hard-sphere liquid (Hansen and McDonald (1986), Appendix C). Using this expression for the determination of the mode-coupling functional, one gets for the transition point: $\varphi_c = 0.525$ (Bengtzelius 1986b). Foffi *et al.* (2004) determined the structure factor by molecular-dynamics simulation for packing fractions up to 0.54. The results for S_q vary regularly with changes of φ so that an extrapolation of the data for the structure factor can be made for φ up to 0.55. Using this result in Eq. (4.36), they found:

$$\varphi_c = 0.546; \quad \text{simulation results for } S_q. \tag{4.102b}$$

This value for the critical point does not suffer from errors of data for the system's structure factor. The 6% underestimation of the transition value φ_g appears as a genuine error of the MCT. This conclusion requires the reservation that φ_g refers to slightly polydisperse systems and that the hard-sphere potential is an idealization for the repulsion of the colloid particles.

Voigtmann *et al.* (2004) reported extensive studies of glassy dynamics based on molecular-dynamics-simulation results for a system of $N = 1000$ particles. The diameters of the particles $d_\kappa, \kappa = 1, \ldots, N$, are selected as random numbers from a distribution with constant density extending from $(1-p)d$ to $(1+p)d$. The parameter p quantifies the polydispersity and leads to a packing fraction $\varphi = (\pi/6)\rho d^3(1+p^2)$. The interaction between particles κ and σ with distance r is modelled by the very steep but regular repulsion potential

$$V_{\kappa\sigma}(r) = (k_B T)[(d_\kappa + d_\sigma)/(2r)]^{36}. \tag{4.103}$$

The investigation is done for $p = 0.1$. This parameter is large enough so that no indications of nucleation phenomena could be detected. From the simulation data for the dynamics, the glass-transition value $\varphi_g = 0.594(\pm 0.2\%)$ was deduced. Using the simulation results for the structure factor in order to determine the mode-coupling coefficients, the critical value $\varphi_c = 0.585$ was calculated. In agreement with the report in the preceding paragraphs, the theory overestimates the trend to arrest, albeit by less than 2% only.

The regularity theorems from Sec. 4.2.2 ensure that microscopic MCT models, which are built on similar structure factors, yield similar results for the correlators $\phi_q(t)$. There is no reason to expect loss of insight if one restricts the discussion of models for hard-sphere-like systems by using one or the other alternative for the structure-factor input. Most of the MCT results to be cited in this book for the HSS are based on the Percus–Yevick approximation for the structure factor. This has the advantage that the mode-coupling coefficients in Eq. (4.36) are given by elementary expressions. Specifying the number M of the discretization for the wave numbers and the wave-number cutoff q_{co}, the mode-coupling functional $\boldsymbol{\mathcal{F}}[P, \boldsymbol{x}]$ is defined. Every result mentioned for such model is reproducible within the accuracy provided by the computer, which is used for the numerical work.

Extensive studies of the HSS by Franosch *et al.* (1997) are based on a model with $M = 100$ components using $q_{co} = 40/d$. The critical point for this model is $\varphi_c = 0.51591213$. Such precision for φ_c permits the numerical analysis of the changes of correlators due to changes of the distance parameter

$$\epsilon = (\varphi - \varphi_c)/\varphi_c \tag{4.104a}$$

for values as small as 10^{-7}. This model yields the parameter in Eq. (4.89c):

$$\lambda^c = 0.735. \tag{4.104b}$$

From Eq. (4.92d), one deduces the parameter

$$\kappa^c = 0.961. \tag{4.104c}$$

Equations (4.86a,b) for the separation parameter imply an upward-bending σ_ϵ versus ϵ curve. For $|\epsilon| \leq 0.04$, this curve does not increase by more than 10% above its linear asymptote

$$\sigma_\epsilon = C\epsilon, \qquad C = 1.54. \tag{4.104d}$$

The deviations of σ_ϵ from $C\epsilon$ do not enter the leading and next-to-leading expansion terms of the Debye–Waller factors for small ϵ. But, outside the specified 4% interval of density changes, deviations from the cited asymptotic laws are expected to become of order 10%. Equation (4.91a) yields the formula (1.12) for the leading-order result for the small-ϵ variation of the arrested parts f_q of the density fluctuations. Here, f_q^c denotes the critical Debye–Waller factor and the factor $\tilde{h}_q$ is proportional to the critical amplitude h_q:

$$\tilde{h}_q = h_q\sqrt{C/(1 - \lambda^c)}. \tag{4.104e}$$

Figure 4.6 exhibits some representative results of the cited work.

The dashed lines in Fig. 1.17 show that the measured density dependence of the Debye–Waller factors is consistent with the $\sqrt{\epsilon}$ law. One method to measure f_q^c is the extrapolation of the f_q versus $\sqrt{\epsilon}$ interpolations to $\epsilon = 0$. Figure 1.16 demonstrates comparisons between data and prediction. Determining the slopes of the interpolations is one means to measure $\tilde{h}_q$. van Megen (1995) has shown that the data for $\tilde{h}_q$ agree with the prediction of Eq. (4.104e), provided the number C is chosen as 1.2. Presumably, it is adequate to distribute the 12% discrepancy between calculated and measured $\tilde{h}_q$ equally on errors for C and h_q.

Within the framework of the MCT, it is the change of the structure factor S_q with increasing density which causes the arrest of the HSS. The change due to a shift of φ from 4.7% below to 4.7% above φ_c is demonstrated in Fig. 4.6(a). There are two dominant effects. First, the height of the structure-factor peak for q near $q_1 \approx 2\pi/d$ increases with φ. As discussed in connection with Fig. 3.4, the peak is caused by the intermediate-range order of the particles. Upon increasing φ, the shell structure of the pair distribution becomes more pronounced, and this leads to an increase of the first peak. A large S_q means large fluctuations of the densities $\langle |\rho_{\vec{q}}|^2 \rangle \propto S_q$ and, hence, of large fluctuations of the forces. This is reflected by the factor $S_q S_p S_k$ in Eq. (4.36). Increasing this factor increases the coefficients in the mode-coupling functional.

The second effect is due to the decrease of the free volume caused by increasing φ. The probability to find two particles in close contact increases with φ. The increase of $g(r \approx d)$ leads to an increase of the oscillation amplitude of $S_q - 1 = \rho c_q/(1 - \rho c_q)$, as is demonstrated in Fig. 4.6(a) for $q \sim q_2$ and for $q \sim 13/d$. This is equivalent to an increase of $|c_q|$. The fluctuating force amplitude, and thereby the mode-coupling coefficients, increase with $|c_q|$.

An increase of the density leads to a decrease of the averaged interparticle spacing a, given by $(4\pi/3)\rho a^3 = 1$. This is reflected by the structure-factor-peak position to shift to larger wave numbers with increasing φ. A similar shift occurs for the position of the first minimum of S_q. Consequently, the contributions from the gradient factors for the forces and also those from the phase-space factor increase. In Eq. (4.36), the former term is given by the parenthesis terms like $(\hat{q}^2 \pm \hat{k}^2 \mp \hat{p}^2)$ and the latter one by the factor $\hat{k}\hat{p}$, as is explained in Sec. 4.2.3. These length-scale-changing effects enhance the effects discussed in the two preceding paragraphs. The same statement is true for the increase of the factor ρ in the first bracket of Eq. (4.36).

One concludes that the structural arrest of the HSS is caused predominantly by coupling density fluctuations for intermediate wave numbers, say, wave numbers from the interval $5 \leq qd \leq 15$. If one ignores coupling to fluctuations with $qd > 30$ by a corresponding decrease of the cutoff q_{co}, the values for φ_c and λ^c alter by not more than 1% . If one ignores all fluctuations for $qd \leq 3$, φ_c increases by 0.8%, λ^c increases by 1% and f_q^c varies by less than 0.015 (Sperl 2004a). The existence of a hydrodynamic-fluctuation regime is irrelevant for the MCT explanation of structural arrest in hard-sphere-like systems.

The identified liquid–glass transition is a non-trivial implication of the factorization ansatz (4.7a) for the fluctuating-force correlators. This ansatz is the

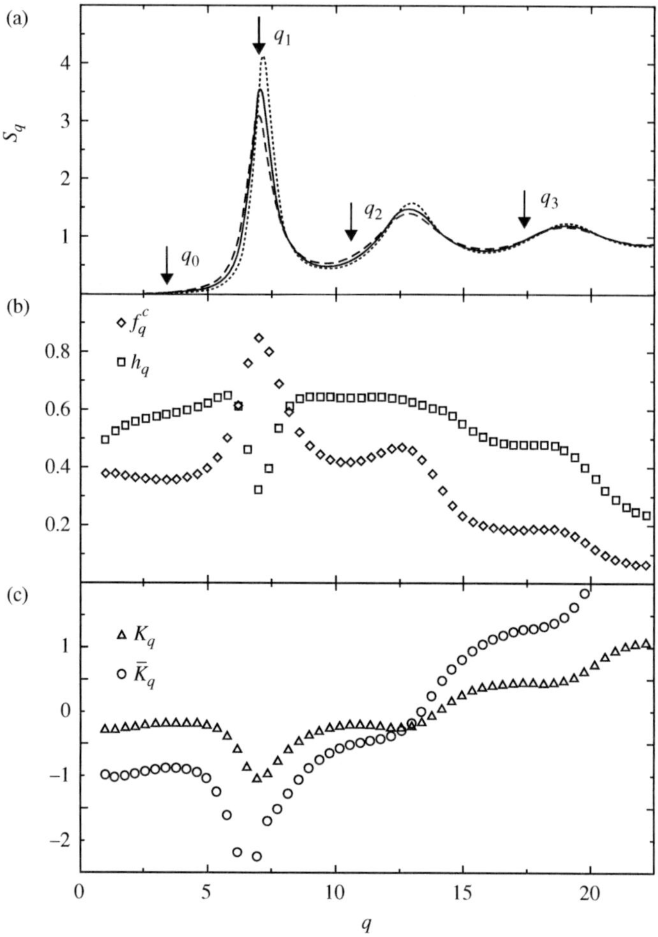

FIG. 4.6. (a) The structure factor S_q of the hard-sphere system calculated in the Percus–Yevick approximation for the packing fractions $\varphi = 0.492$ (dashed line, $\epsilon = -0.047$), $\varphi = 0.516$ (solid line, $\epsilon = 0$), and $\varphi = 0.540$ (dotted line, $\epsilon = 0.047$). The arrows mark the wave numbers $q_0 = 3.4, q_1 = 7.0, q_2 = 10.6$ and $q_3 = 17.4$; they are measured in units of $(1/d)$, with d denoting the particle diameter. (b) The critical Debye–Waller factor f_q^c and the critical amplitude h_q. (c) The correction amplitude $\overline{K}_q = \overline{K}_q(P^c)$, Eq. (4.92a), and the correction amplitude $K_q = K_q(P^c)$, Eq. (6.23). The underlying mode-coupling functional (4.36) is based on the Percus–Yevick structure factor and the convolution approximation for the triple-density averages. A wave-number grid with $M = 100$ values and a cutoff $q_{co} = 40$ is used. Reproduced from Franosch *et al.* (1997).

simplest procedure to formulate a self-consistent treatment of the cage effect for strongly interacting particles. The simultaneous treatment of the particle distributions in configuration space and of the forces determining the particle orbits leads to two alternatives for the long-time limits, i.e., for the equilibrium. For densities ρ smaller than a critical value ρ_c, the system behaves similarly as studied for a long time for liquids, say, argon near its triple point. A particle can move further than the diameter of its cage by pushing ahead one of its neighbours. This neighbour can move, because it can push ahead its neighbour, etc. As a result, the particle moves together with its deforming cage. This cooperative motion is similar to an irrotational flow of a continuum liquid around a sphere. It is not connected with appreciable density fluctuations on macroscopic length scales. However, if ρ exceeds ρ_c, a particle cannot move over distances of order of its diameter d, because its neighbours are arrested near their equilibrium positions. They are arrested, because their neighbours are arrested, etc. The solution for $\rho > \rho_c$ describes an arrested distribution of particles, i.e., an amorphous solid. The average for density fluctuations vanish, but the average of the density-fluctuation squares does not. The distribution of the fluctuations is described by the Debye–Waller factor f_q in analogy to what one knows for crystalline solids. The fluctuations of the particle positions cause fluctuations of forces. The averages of these fluctuations vanish. The averages of the force-fluctuation squares are positive, they are described by the mode-coupling functional $\mathcal{F}_q[P, \boldsymbol{f}] = m_q(t \to \infty)$. The fixed-point equation specifies a procedure for solving the balance condition for particle distributions and force distributions.

It is explained above that the factor S_q in Eq. (4.36) reflects the wave-number dependence of the compressibility and, thereby, of the force fluctuations. This factor dominates the q-dependence of the mode-coupling polynomial $\mathcal{F}_q[P, \boldsymbol{f}]$. Because of Eq. (4.52d), there holds $f_q = \mathcal{F}_q[P, \boldsymbol{f}]/\{1+\mathcal{F}_q[P, \boldsymbol{f}]\}$, i.e., f_q increases with $\mathcal{F}_q[P, \boldsymbol{f}]$. The oscillations of the S_q versus q curves cause in-phase oscillations of the f_q versus q ones. The pronounced peak of S_q shown in Fig. 4.6(a) for q near q_1 implies the peak of f_q^c shown in Fig. 4.6(b) in the same wave-number region.

Within the density-functional theory of freezing, one motivates a formula for the free energy of the N-particle system as function of the particle-density deviations from the homogeneous reference fluid. Like the mode-coupling functional, the density functional is determined by the direct correlation function c_q for the reference fluid. The equilibrium state is derived from the request to minimize the free energy with respect to choices of the density deviations (Oxtoby 1991). Kirkpatrick and Wolynes (1987) have shown that a simplifying ansatz for a trial density field for an amorphous solid yields an arrested state identical to that obtained from an approximate solution of the MCT equation for f_q. This indicates that there is no fundamental difference between the mode-coupling-theory approach towards arrest and the more traditional way of determining the equilibrium state from a minimum principle for the free energy.

In order to make plausible the main feature of the q-dependence of the critical amplitude h_q and of the correction amplitude $k_q = \overline{K}_q + \kappa$, let us consider a reference system. This shall account only for the shell of relevant wave numbers identified above. Within this shell, the wave-number dependence of the correlator shall be ignored. This model deals with a single correlator only and the mode-coupling kernel reads $v_2\phi(t)^2$. The model is defined by Eqs. (4.32a,b) for $v_1 = 0$. In this case, one gets from Eqs. (4.67b), (4.68b): $\lambda^c = f^c = h = 1/2$, $k = 0$. The strong peak of the structure function S_q causes a strong increase of f_q^c above the reference value 1/2. But, $f_q - f_q^c$ is bounded by $1 - f_q^c$. Avoiding an increase above this value can be achieved by a decrease of h_q below the reference value 1/2 and by a negative correction amplitude k_q.

The description of the increase of $f_q - f_q^c$ by the asymptotic formulas for small distance parameters ϵ is demonstrated in Fig. 4.7 for three representative wave numbers. Within a 10% accuracy level, the expansion formula up to the next-to-leading term describes the results for ϵ up to 0.03 or somewhat further. The leading-order result accounts for the increase for q_1 and q_3 only up to $\epsilon = 0.003$ and 0.002, respectively. Hence, including the corrections expands the ϵ-range for the validity of the asymptotic formulas by an order of magnitude. Based on the somewhat arbitrary 10%-accuracy criterion, the range of validity of the formula with corrections describes the Debye–Waller-factor increase for wave number q_2 worse than the leading-order expression. This is due to 'accidental' cancellation effects with terms of higher order.

One infers from Fig. 4.6(c) that the correction amplitude $k_q = 0.961 + \overline{K}_q$ is rather small for $qd \leq 6$. Hence, one expects the leading-order result to describe the data on a 10% accuracy level for ϵ up to about 0.05. This explains that the data in Fig. 1.17 for $qR = qd/2 = 3.05$ and 2.68 exhibit the leading-order law for $\sqrt{\epsilon}$ as large as 0.23. The data for $qR = 3.42$ should be similar to those exhibited in Fig. 4.7 for q_1. Hence, one expects the leading-order description to be valid only for ϵ below about 0.01; and this conclusion is consistent with what is shown in Fig. 1.17 for the colloids.

It was mentioned in Sec. 3.3.4 that the structure factor of a Lennard-Jones liquid for intermediate wave numbers is very close to that of a hard-sphere system. This holds provided the diameter of the equivalent hard spheres is chosen so that it decreases with increasing temperature. Hence, an increase of the temperature T induces similar changes of the relevant mode-coupling coefficients as a properly chosen decrease of the packing fraction φ for the equivalent HSS. Consequently, the T_c versus φ_c line for a liquid–glass transition of the Lennard-Jones system (Bengtzelius 1986b) is an increasing curve in the $T - \varphi$ plane.

4.3.6 *Arrest in systems with short-ranged-attraction*

In this section, the glass-transition singularities shall be discussed for a microscopic model, which describes a system of particles whose interaction potential consists of a hard-core repulsion complemented by a short-ranged attraction. A simple model describes the attraction by a shell of a constant negative potential.

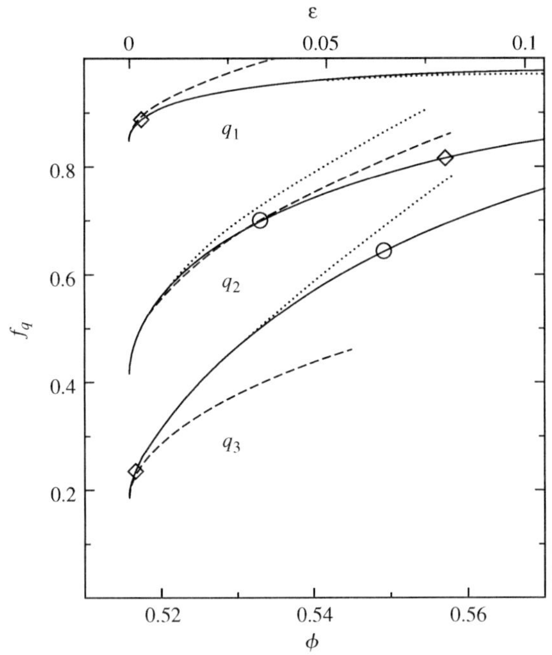

FIG. 4.7. The form factors f_q of the hard-sphere glass (solid lines). The model for the mode-coupling functional and the wave numbers are specified in the caption of Fig. 4.6. The dashed lines exhibit the asymptotic laws (1.12, 4.104e), and the diamonds at $\epsilon = 0.003$, 0.08 and 0.002 mark the densities, where the dashed and solid lines deviate by 10% for q_1, q_2, and q_3, respectively. The dotted lines exhibit the formula (4.91a). They deviate from the full lines by 10% for $\epsilon = 0.141$, 0.033 and 0.064 (from top to bottom). For q_2 and q_3, these densities are marked by circles. Reproduced from Franosch *et al.* (1997).

The interaction of this square-well system (SWS) is quantified by the hard-core diameter d, by the value $-u_0$, $u_0 > 0$, of the attraction potential, and by the width Δ for the interval of distances $r, d < r < d+\Delta$, where there is the negative potential. The equilibrium structure is specified by three control parameters. They can be chosen as the packing fraction of the hard cores, $\varphi = (\pi/6)\rho d^3$, the attraction-range parameter $\delta = \Delta/d$, and the attraction-strength parameter $\Gamma = u_0/(k_B T)$. The inverse of the latter can be viewed as the dimensionless temperature $\theta = 1/\Gamma$.

It is explained in Sec. 1.8 that colloidal suspensions can be prepared with effective interactions similar to the one of the SWS. The solvent contains a certain percentage of non-adsorbing coiled polymers; and this mixture produces the Asakura–Oosawa depletion interaction between the solute particles. Varying the coil diameter, the range parameter δ can be changed in certain limits. The attraction is of entropic origin and, therefore, the dominant temperature dependence of the attraction-potential strength is a factor (k_BT). Hence, the strength parameter Γ depends on T only indirectly via polymer–polymer interactions. The parameter Γ can be varied to some extent by changing the polymer density c_p.

The mean-spherical approximation is an extension of the Percus–Yevick theory for the HSS so that the effect of the particle attraction on the structure factor can be treated (Hansen and McDonald (1986), Chapter 5). For the case of small δ, say, $\delta < 0.2$, the solution of the integral equation for S_q can be obtained by series expansion in δ. This yields S_q and c_q explicitly in terms of elementary functions (Dawson *et al.* 2001). Neglecting the triple correlations in Eq. (4.36), the mode-coupling functional is determined for every state P specified by the coordinate triples $(\varphi, \Gamma, \delta)$. As explained in Sec. 4.2.3, the MCT model is defined by specifying the step size h/d for the wave-number discretization and by the number M of discrete q values considered. The results to be cited are obtained by using $h = 0.4/d$. The cutoff wave number $q_{co} = Mh$ has to be adjusted so that the results do not vary seriously upon a further increase of M. For decreasing δ, finer spatial structures have to be resolved; therefore, q_{co} has to be increased. For $\delta = 0.03$, $M = 500$ correlators have been considered. Figure 4.8 exhibits structure factors S_q and the corresponding pair-distribution functions $g(r)$ for some representative states from the cited work. Figure 4.9 shows cuts through the bifurcation manifold for the SWS, which are specified by a sequence of fixed values for δ.

Let us consider the structure factors for fixed Γ but increasing φ. Figure 4.8 demonstrates the change of S_q due to the change from state 1 to state 2 as well as that from state 4 to state 3. The increase of φ causes an increase of the height of the structure-factor peak and a decrease of the structure factor minimum for $qd \sim 10$. These are the two changes due to the increase of intermediate-range order and the decrease of the free volume, which have been discussed above in connection with Fig. 4.6(a). These changes lead to arrest at sufficiently large φ because of the increasing efficiency of the cage effect. No matter the size of Γ, increasing the density will cause a generic liquid–glass transition at some critical packing fraction $\varphi_c(\Gamma)$, which depends on Γ. For δ below some critical value δ^*, and Γ within a certain interval, Fig. 4.9 demonstrates the occurrence of a further discontinuous A_2 transition from a glass with a form factor $f_q^{(1)}$ to another glass with a larger form factor $f_q^{(2)}$.

The changes of the particle distribution due to an increase of Γ is subtle because there are three effects. An increase of Γ causes an increasing trend to dimerization. This first attraction effect makes the system more inhomogeneous, it disturbs the shell structure of the cage-forming neighbours of a given particle, it reduces the intermediate-range order. Thereby, the attraction counteracts the trend to arrest, it stabilizes the liquid. These effects are reflected by a reduction of the height of the structure-factor peak with increasing Γ. Consider states for $\varphi = 0.50$. From Fig. 4.6(a) one infers a peak height above 3.1 for $\Gamma = 0$;

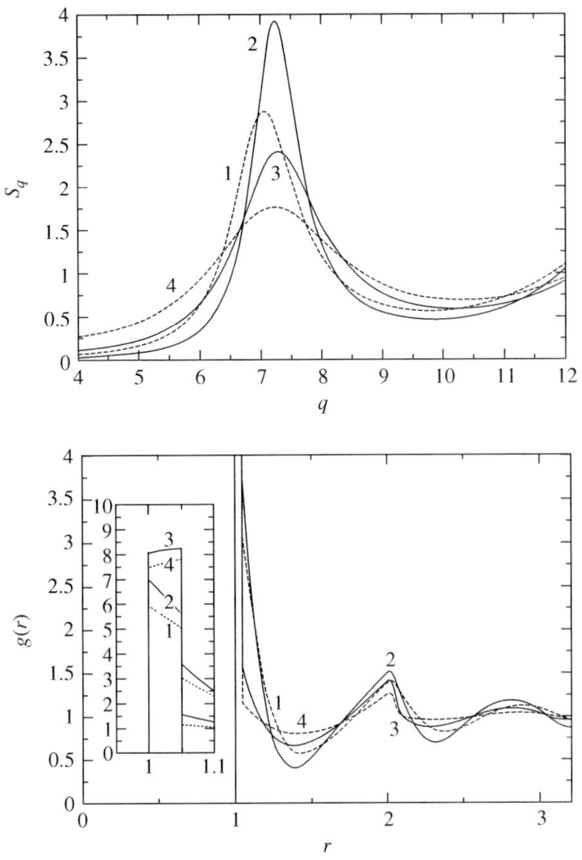

FIG. 4.8. Structure factors S_q and pair-distribution functions $g(r)$ of the square-well system (SWS) with attraction-range parameter $\delta = 0.05$, evaluated in the mean-spherical approximation. The unit of length is the hard-core diameter d. The states 1, 2, 3 and 4 are specified by the pairs $(\varphi, \Gamma) = (0.50, 2.0)$, $(0.55, 2.0)$, $(0.50, 6.7)$ and $(0.40, 6.7)$, respectively. Here, φ denotes the packing fraction and Γ is the attraction-strength parameter. Reproduced from Dawson *et al.* (2001).

and Fig. 4.8 shows for states 1 and 3 the peak heights near 2.9 and below 2.4 for $\Gamma = 2$ and 6.7, respectively. The first attraction effect causes a decrease of the compressibility factor in Eq. (4.36) and, consequently, a reduction of the mode-coupling coefficients. Let us suppose that the identified attraction effect dominates over the other effects to be discussed below. Then, the decrease of the mode-coupling coefficients due to an increase of Γ has to be compensated by an increase of φ in order to reach the glass transition point. Consequently, the critical packing fraction $\varphi_c(\Gamma)$ increases with Γ. The Γ_c versus φ_c curves

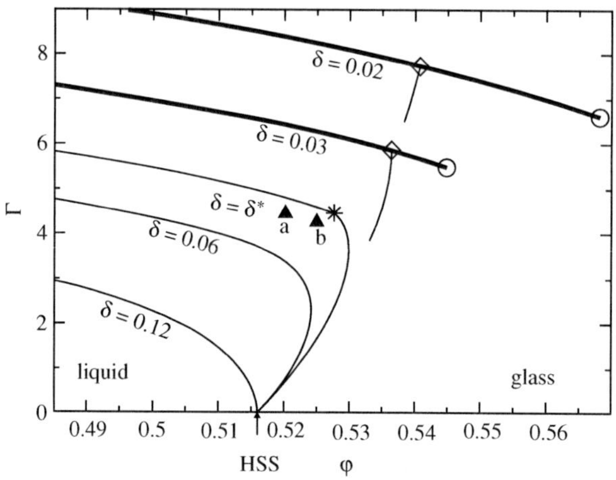

FIG. 4.9. Cuts through the bifurcation manifold for a model of the SWS for constant attraction-range parameters δ as indicated. The results are based on mode-coupling coefficients calculated with the mean-spherical approximation for the structure factor. The arrow marks the critical packing fraction $\varphi_c^{\rm HSS} = 0.516$ of the HSS. The star denotes the position $\varphi = \varphi^* = 0.52768$ and $\Gamma = \Gamma^* = 4.4759$ of the A_4 singularity for $\delta = \delta^* = 0.04381$. The open circles mark A_3 singularities as endpoints of glass–glass-transition lines. The diamonds denote crossing points of generic liquid–glass-transition lines. The filled triangles denote states with coordinates $\varphi = 0.520, \Gamma = 4.46$ (a) and $\varphi = 0.525, \Gamma = 4.27$ (b). Modification of a figure from Sperl (2004b).

start with positive slope at the $\Gamma = 0$ axis at the transition point of the HSS. For sufficiently small δ, Fig. 4.9 shows indeed that there appear liquid states for a packing fraction exceeding the critical value $\varphi_{\rm c}^{\rm HSS}$ of the hard sphere system. These liquid states arrest if the interaction strength parameter Γ is decreased. The increase of the effective temperature $\theta = 1/\Gamma$ leads to freezing of the liquid in an amorphous solid.

In order to explain the above identified $\Gamma_{\rm c}$ versus $\varphi_{\rm c}$ freezing curve as a glass-melting line, let us consider a hard-sphere glass at, say, $\varphi = 0.520$. The particles are localized in cages with a distribution characterized by some length r_s. The dimerization effect due to attraction causes voids in the shell of the cage-forming neighbours provided the range Δ of the attraction is smaller than the localization length r_s. This trend destabilizes the arrest. The effect increases with the increase of Γ and the decrease of δ. If Δ is sufficiently larger than $r_{\rm s}$, the dimerization mechanism becomes inefficient and the increasing part of the $\Gamma_{\rm c}$ versus $\varphi_{\rm c}$ curve disappears. The value 0.117 is the critical one for the existence of liquid states of the SWS for $\varphi > \varphi_{\rm c}^{\rm HSS}$ (Sperl 2004b). This value suggests that r_s is of the order 0.1 d.

The comparison of S_q for states 1 and 3 demonstrates a second attraction effect, namely a shift of the peak position to larger wave numbers. It implies an increase of S_q for $7.5 < qd < 10.0$. The corresponding increase of the factor $S_p S_k$ in Eq. (4.36) causes an increase of the mode-coupling coefficients in the specified range of wave numbers. The effect is enhanced by the increasing importance of the gradient factors $(q^2 \pm k^2 \mp p^2)$ and of the phase-space factor $\hat{k}\hat{p}$. Thus, the second attraction effect, as opposed to the first one, favours arrest. The phenomenon is a length-scale shrinking effect with the same implication as discussed in the preceding section for the density increase of a HSS. The second attraction effect causes a decrease of the $\Gamma_{\rm c}$ versus $\varphi_{\rm c}$ transition lines shown in Fig. 4.9 for all φ below $\varphi_{\rm c}^{\rm HSS}$. If δ increases for fixed density, an increasing part of the attraction potential contributes to the binding energy of clusters. Therefore, the transition points $\Gamma_{\rm c}$ for $\varphi < \varphi_{\rm c}^{\rm HSS}$ decrease with increasing δ.

The small-Γ branch of the liquid–glass transition line, where the first attraction effect destabilizes arrest, and the large-Γ branch, where the second attraction effect supports arrest, can join smoothly in a parameter region, where both effects nearly cancel. For $\delta = 0.06$, Fig. 4.9 demonstrates such boundary of the liquid region.

For small δ and large Γ, there is a third attraction effect, which is called the sticking phenomenon. If Γ increases strongly, there is also a strong increase of the pair distribution within the shell $d < r < d + \Delta$. The effect is demonstrated for the pair of states 1 and 3 in the inset in Fig. 4.8. The resulting rapid variation of $g(r)$ on a length scale Δ causes large contributions to the direct correlation function c_q for wave numbers q increasing up to $1/\Delta$. This large-q region is not important for the other effects discussed above. Baxter (1968) has analysed the sticking effect in the limit of Δ tending to zero and Γ tending to infinity so that the pair distribution gets the contribution $g^{\rm stick}(r) = Ad\delta(r - d)$. The number A quantifies the percentage of particle pairs, which stick together because of reversible bond formation. According to Eq. (3.59a), this sticking contribution is equivalent to a structure factor contribution $S_q^{\rm stick}$, which is given by $(S_q^{\rm stick} - 1)/\rho = (4\pi A d^3)\sin(qd)/(qd)$.

Some side remarks might help to appreciate Baxter's formula. Suppose that $g(r)$ and its first and second derivatives are continuous. One concludes from Eq. (A.14b) for the asymptotes of Fourier transforms that the corresponding structure factor exhibits the large-q behaviour: $(S_q - 1)/\rho = O(1/q^3)$. If one substitutes a discontinuous contribution $g_\alpha(r) = -g_\alpha \theta(d_\alpha - r)$ in Eq. (3.59a), one gets $(S_q^\alpha - 1)/\rho = (4\pi g_\alpha d_\alpha^3)\cos(qd_\alpha)/(qd_\alpha)^2 + O(1/q^3)$. Let $g(r)$ be regular in the following sense. Up to some points $r = d_\alpha$, the functions $g(r)$, $g'(r)$, and $g''(r)$ are continuous. The limits $g_\pm^\alpha = g(d_\alpha \pm 0)$, $g_\pm'^\alpha = g'(d_\alpha \pm 0)$ and $g_\pm''^\alpha = g''(d_\alpha \pm 0)$ exist. Then, $(S_q - 1)/\rho = \sum_\alpha (4\pi g_\alpha d_\alpha^3)\cos(qd_\alpha)/(qd_\alpha)^2 + O(1/q^3)$ with $g_\alpha = (g_+^\alpha - g_-^\alpha)$ denoting the discontinuity of $g(r)$ at $r = d_\alpha$. Substituting the results into Eq. (3.61), one arrives at the following conclusion. A regular pair distribution may have a $1/q^2$ tail for the direct correlation function. The tail amplitude is determined by the discontinuities:

$$c_q^{reg} = \sum_\alpha (4\pi g_\alpha d_\alpha^3)\cos(qd_\alpha)/(qd_\alpha)^2 + O(1/q^3). \tag{4.105a}$$

The Baxter limit of the sticking effect implies a $(1/q)$ tail:

$$c_q^{\text{stick}} = (4\pi A d^3)\sin(qd)/(qd) + O(1/q^2). \tag{4.105b}$$

For large q, the sticking effect causes a huge enhancement of c_q relative to the direct correlation of the hard-sphere system. This enhancement gets a large weight for the mode-coupling coefficients in Eq. (4.36) due to the gradient and phase-space factors. The third attraction effect increases the mode-coupling coefficients for short-wave-length density fluctuations. Sticking favours arrest.

One must not perform the Baxter limit under the integral in Eq. (4.11d) because the limits $q \to \infty$ and $(\delta, \Gamma) \to (0, \infty)$ cannot be interchanged. To identify the qualitative change of c_q due to the sticking effect, one has to notice that the pair-distribution function $g(r)$ of the SWS has two discontinuities g_1 and g_2 at $r_1 = d$ and $r_1 = d + \Delta$, respectively. One gets from Eq. (4.105a)

$$\begin{aligned} c_q^{\text{SWS}} = [4\pi d^3/(qd)^2] \Big\{ & \big[(g_1 + g_2)\cos[qd(1+\delta/2)]\cos[q\Delta/2]\big][1 + O_+(\delta)] \\ & + \big[(g_1 - g_2)\sin[qd(1+\delta/2)]\sin[q\Delta/2]\big][1 + O_-(\delta)]\Big\} \\ & + O(1/q^3). \end{aligned} \tag{4.105c}$$

The corrections $O_\pm(\delta)$ are modifications of $(g_1 + g_2)$ and $(g_1 - g_2)$, respectively, which are unimportant for short-ranged attractions. For the same reason, one can neglect $\delta/2$ compared to unity in the arguments of the trigonometric functions. The first contribution in the curly bracket describes a small modification of the asymptote of the direct correlation function for the HSS. The qualitatively new feature of c_q^{SWS} is given by the second contribution to the curly bracket. For small Δ, there is a large wave-number interval, where $\sin(q\Delta/2)$ can be replaced by $(qd\delta/2)$. For sufficiently large Γ, the factor $g_1 - g_2$ becomes large; and the second term in the curly bracket dominates. One arrives at

$$c_q^{\text{SWS}} = c_q^{\text{stick}} \quad \text{for } q_\ell \ll q \ll q_u. \tag{4.105d}$$

The amplitude in Eq. (4.105b) is given by $A = (g_1 - g_2)\delta/2$. Different approximations for S_q of the SWS differ in the value they produce for A. The lower crossover wave number q_ℓ is to be chosen so that the first term in the curly bracket can be neglected: $q_\ell \approx 10/d$. The upper crossover wave number increases with decreasing $\delta : q_u d \approx 1/\delta$.

Equation (4.105d) is a general result for the sticking phenomenon. Different models differ in the details of the crossover features if q increases from intermediate q-values to above q_ℓ and if q increases above q_u into the region of the

$(1/q^2)$ behaviour. The simplest model uses the Baxter limit for the structure factor within the framework formulated for the microscopic MCT. In this case, the cutoff q_{co} plays the role of q_u.

Let us consider a large-Γ branch of the liquid-glass transition line for such small δ that the sticking effect provides the dominant contribution for the arrest. Within the interval $q_\ell \ll q \ll q_u$, the critical Debye–Waller factor will be much larger than one would get by the repulsion dominated arrest of hard-sphere-like systems. The latter mechanism deals with fluctuation wave numbers q, which obey the condition $q \ll q_\ell$. Hence, a smooth joining of this branch with the small-Γ branch is impossible. The simplest possibility for a non-smooth joining of two branches is a corner. It results from some crossing point as is explained in connection with Figs. 4.3 and 4.5. According to the results of Sec. 4.3.3, this is also the only generic possibility occurring in a two-dimensional parameter manifold. Figure 4.9 demonstrates this scenario for $\delta = 0.03$ and $\delta = 0.02$. In order to corroborate this reasoning, one can eliminate the sticking effect on the mode-coupling functional by introducing a cutoff $q_{co} = 20/d$ in Eq. (4.36). The remaining interplay of the hard-sphere-localization mechanism with the first and second attraction effect yields a smooth liquid-glass transition line also for large Γ and small δ (Dawson *et al.* 2001).

There are two different types of routes from a liquid state close to the corner into the glass, as was explained in connection with Fig. 4.3. For the first type, φ is increased for nearly constant Γ. Such route crosses the small-Γ branch. The transition is caused by the arrest mechanism of hard-sphere-like systems modified by the first and second attraction effect. It is characterized by a critical Debye–Waller factor $f_q^{c(1)}$, similar to that shown in Fig. 4.6(b) and explained in the preceding section. For the second type, Γ is increased for nearly constant φ. The transition is dominated by the sticking effect. The critical Debye–Waller factor $f_q^{c(2)}$ is larger than $f_q^{(1)}$ for, say, $qd = 20$. Because of the inequality (4.52e), there holds $f_q^{c(1)} < f_q^{c(2)}$ for all q. The upper branch continues as glass–glass transition line.

The differences between the glass states obtained on the two mentioned transition routes manifest themselves most clearly by the glass–glass transition. Figure 4.10 exhibits the evolution of the arrested parts f_q of the square-well glass upon increasing Γ for fixed δ and φ. The three lowest curves refer to a parameter region, where the repulsion-dominated effects cause arrest like in the HSS. The second attraction effect dominates over the first one so that an increase of Γ supports arrest. The Debye–Waller factor depends smoothly on Γ; it can be analytically continued beyond the critical value $\Gamma_c = 1.05\ldots \times \Gamma^*$ for the glass–glass transition. The continuation yields a fixed point, but this is not the maximum one. At Γ_c, there occurs a generic A_2 bifurcation. There is a large discontinuous increase from $f_q^{(1)c} = f_q(\Gamma \to \Gamma_c - 0)$ to $f_q^{(2)c} = f_q(\Gamma = \Gamma_c)$. The outstanding signature of the sticking-effect-dominated glass state is a very large arrested part.

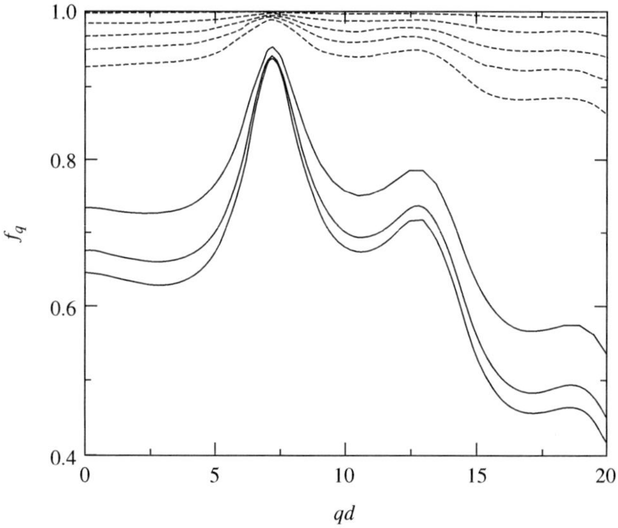

FIG. 4.10. Debye–Waller factors f_q of the square-well glass with packing fraction $\varphi \approx 0.540$, attraction-range parameter $\delta \approx 0.03$, and attraction-strength parameters Γ/Γ^* increasing from bottom to top as 0.846, 0.957, 1.0506 (full lines) and 1.0507, 1.063, 1.091, 1.222, 1.833 (dashed lines). q denotes the wave number and d abbreviated the hard-sphere diameter. $\Gamma^* = 0.9091$ is the attraction-strength parameter of the A_4 singularity. The determination of the mode-coupling coefficients are based on the Percus–Yevick approximation for the structure factor. Reproduced from Dawson *et al.* (2001).

If one shifts the states along the glass–glass-transition line by increasing φ, the arrested parts increase. But, $f_q^{(2)c}$ does not increase very much since the packing fraction enters the bond-formation mechanism only indirectly. On the other hand, the Debye–Waller factor $f_q^{(1)c}$ of the hard-sphere-like glass increases appreciably similar to what is demonstrated in Fig. 4.7. The transition line gets an endpoint near some packing fraction, where $f_q^{c(1)}$ is increased close to $f_q^{c(2)}$. The simplest possibility for such endpoint is an A_3 glass-transition singularity as explained in connection with Fig. 4.3. According to Sec. 4.3.3, this is the only generic possibility for an endpoint in a two-dimensional manifold of control parameters. Figure 4.9 shows these singularities by open circles.

If δ increases, the relevant cutoff q_u decreases and the sticking effect reduces. The endpoint shifts towards the crossing point as is demonstrated in Fig. 4.9 for the change from the cut for $\delta = 0.02$ to the one for $\delta = 0.03$. Increasing δ further, there must occur a critical value δ^*, where the A_3 singularity merges the crossing point. This value marks the connection between two scenarios. There is the scenario with corner for the liquid–glass-transition line for $\delta < \delta^*$. And there is the one with a smooth boundary of the liquid region for $\delta > \delta^*$. According to the discussion of Sec. 4.3.3, the only generic possibility for the merging of an A_3 glass-transition singularity with a crossing point in a three-dimensional parameter manifold is an A_4 singularity. Using the numerical solution of the Percus–Yevick equations for the structure factor of the SWS as input for the mode-coupling coefficients, one gets the parameters for the A_4 glass-transition singularity (Dawson *et al.* 2001):

$$\varphi^* = 0.529, \quad \Gamma^* = 0.909, \quad \delta^* = 0.0429; \quad \text{Percus–Yevick}\, S_q. \tag{4.106}$$

Within the Percus–Yevick approximation, one replaces the Fourier transform of the direct correlation function for $d < r < d + \Delta$ by $c(r) = g(r)[1 - \exp(-u_0/k_{\mathrm{B}}T)]$. The mean-spherical approximation simplifies this expression to $c(r) = u_0/k_{\mathrm{B}}T = \Gamma$. This approximation, which was used to calculate the results in Figs. 4.8 and 4.9, yields the A_4-singularity parameters: $\varphi^* = 0.528$, $\Gamma^* = 4.48$ and $\delta^* = 0.0438$. It is illegitimate to use the simplification of the mean-spherical approximation in the region of large attraction strength parameters Γ. However, for small δ, the simplification concerns only a short interval of r near $r^* = d + \Delta/2$. The replacement of $c(r^*)$ by Γ is equivalent to a rescaling of Γ^*. This explains, why the results of the mean-spherical approximation agree with those of the Percus–Yevick theory on an accuracy level of a few percent, if the attraction strength parameter Γ of the former approach is rescaled by a factor $0.909/4.48$ (Dawson *et al.* 2001).

The A_4 glass-transition singularity organizes a subtle arrest scenario, which is described in connection with Fig. 4.5. The MCT provides a qualitative explanation of the glassification diagram of certain colloidal suspensions, whose properties have been reported in Figs. 1.19 and 1.20. Glass-transition diagrams with reentry regions and A_3 singularities have first been calculated for hard-sphere systems with short-ranged attraction by Bergenholtz and Fuchs (1999) and by Fabbian *et al.* (1999).

Two additions to the preceding discussions can be inferred from Fig. 4.11. This diagram reproduces the identification of various states of colloid–polymer mixtures from the pioneering work on this subject (Poon *et al.* 1993). The polydispersity of the system is so small that metastable states cannot be observed outside the glass region. If the system is not an amorphous solid, it is either a liquid, or a crystal, or a fluid–crystal mixture. If the polydispersity were increased somewhat, all states outside the glass region would be fluid, as is discussed in Sec. 1.7. In this sense, the states shown in Fig. 4.11 by crosses and by squares also have to be understood as liquids. The figure shows arrest lines calculated within the microscopic MCT (Bergenholtz *et al.* 2003). The mode-coupling coefficients in this work are based on a semi-empirical formula for the structure factor S_q. It is derived from the microscopic expression for the Asakura–Oosawa depletion attraction. The density is parameterized by the packing fraction φ, which is rescaled so that the glass-transition point φ_g of the HSS is reproduced. The attraction strength is expressed in terms of the mass density c_p of the polymer. The range parameter ξ is the ratio of the radius of gyration of the

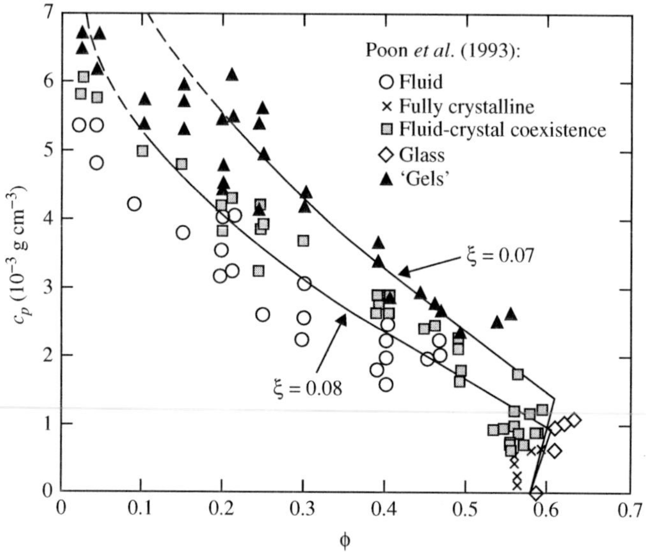

FIG. 4.11. The symbols mark states of a colloid–polymer mixture of packing fraction φ and polymer density c_p as identified in experiments by Poon *et al.* (1993). The ratio ξ of the polymer gyration radius and the colloid-particle radius is 0.08. The curves are the lines of A_2 bifurcations for the arrest of the density-fluctuation correlators calculated within a microscopic MCT model for a system of particles interacting via a hard-core repulsion complemented by an Asakura–Oosawa attraction with range parameters $\xi = 0.08$ and 0.07. Further details can be inferred from the text. Reproduced from Bergenholtz *et al.* (2003).

polymer coils and of that for the hard core. The formulas describe the sticking effect on S_q so that simulation data for the system's structure factor are reproduced for all $qd > 10$ and for packing fractions between 0.1 and 0.4. The figure demonstrates that MCT explains the arrest diagram for the high-density region, say, for $\varphi > 0.5$. The cited calculations imply that the critical value ξ^* of the A_4 -singularity is between 0.08 and 0.10. Hence, the higher-order glass-transition singularity is predicted to occur in a region of parameters, which can be explored by experimental studies.

The second remarkable result documented in Fig. 4.11 is that the calculated transition line continues to describe the data for the arrest line down to very small densities. In the limit of φ tending to zero, the direct correlation function in the cited work reduces to the Fourier transform of the Meyer function for the interaction potential $V(r)$: $\exp[-V(r)/(k_BT)] - 1$. This is the result expected from the perturbation expansion for the equilibrium structure in the limit $\varphi \to 0$ (Hansen and McDonald 1986). In this limit, the MCT results for the bifurcation line can be evaluated analytically yielding the dashed lines in the figure. A hard-sphere system for $\varphi \approx 0.1$ is a model for a dense gas rather than for a liquid. A homogeneous amorphous solid at such low density is called a gel. Hence, the mode-coupling theory describes the first step towards the formation of gels via a self-consistent treatment of the sticking effect. The corner in the boundary of the high-density-liquid region can be viewed as the joining of the gel-transition line with the glass-transition line. The ordering phenomena of a gel state due to sticking-effect-induced cluster formation causes a strong peak of the structure factor for small wave numbers. This fact is not reflected by the cited theory.

4.4 Dynamics near glass-transition singularities

In this section, numerical solutions of the MCT equations of motion shall be considered for states P approaching a liquid–glass-transition point P^c. It will be demonstrated that the evolution of the bifurcation dynamics exhibits features, which are similar to those discussed in chapter 1 as facets of the glassy dynamics of liquids.

4.4.1 *Relaxation through plateaus*

Figure 4.12 exhibits correlators $\phi(t)$ for the one-component model defined by Eqs. (4.32a,b). The upper panel of Fig. 4.13 shows the corresponding susceptibility spectra $\chi''(\omega) = \omega\phi''(\omega)$; the lower one exhibits susceptibility spectra $\chi''_A(\omega) = \omega\phi''_A(\omega)$ of the second correlators of the two-component model defined by Eqs. (4.32a,b), (4.33a,b). The states are specified by the coupling constants v_1 and v_2; and they are chosen on an array through a transition point (v_1^c, v_2^c) given by Eqs. (4.67b) for $\lambda^c = 3/4$. The distance parameters for the states $\epsilon = (v_{1,2} - v_{1,2}^c)/v_{1,2}^c = \pm 2^{-n}$, $n = 1, 2, \ldots$, decrease with increasing n in an geometric sequence. Some of these states are indicated in Fig. 4.2 by crosses.

The decay of the correlators $\phi(t)$ from the initial value 1 to 0.7 occurs for times t increasing up to about $0.8/\Omega$; it can be described by Eq. (4.19a) for the transient dynamics. This regime is followed by a one-decade time interval of oscillatory motion, which depends appreciably on the mode-coupling coefficients. The liquid state with label $n = 1$ deals with coupling constants, which are 1/2 of the critical ones. The corresponding correlator decays to zero if the time is increased to about $10/\Omega$. The glass state with label $n = 1$ refers to coupling constants, which are 3/2 of the critical ones. The corresponding correlator decays to its arrested part f if Ωt is increased to about 10. The specified one-decade time interval deals with the dynamics of a 'normal'-liquid state or of a 'normal'-disordered-solid state. The correlation functions are calculated for a model with a Newtonian short-time dynamics without friction terms. The damping of the oscillations is caused solely by mode-coupling effects. Both panels of Fig. 4.13 show that the transient dynamics causes humps of the loss spectra for $\omega/\Omega > 0.1$. These humps are the 'normal-dynamics' excitation spectra. For $\omega/\Omega < 0.1$, the $n = 1$ spectra $\chi''(\omega)$ for the liquid and for the glass exhibit a linear frequency

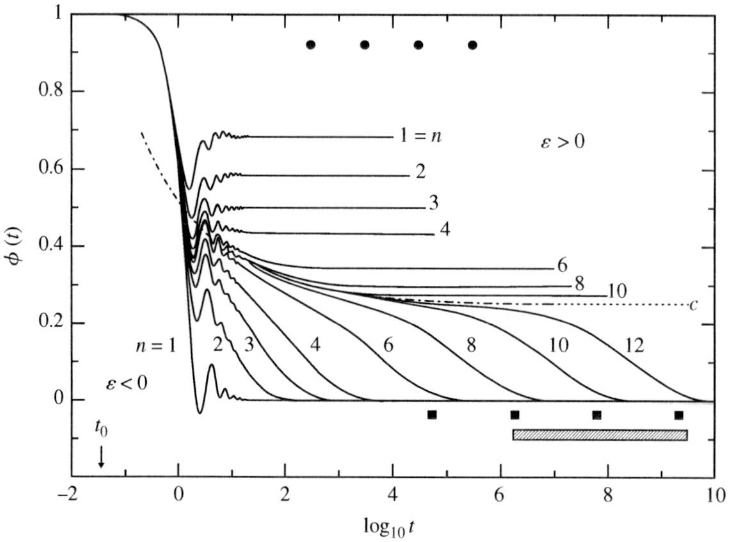

FIG. 4.12. Solutions $\phi(t)$ of the equations of motion (4.32a,b) for $\nu = 0$ and units of time chosen so that $\Omega = 1$. The dotted line labelled c refers to the critical point $P^c = (v_1^c, v_2^c)$, which is specified by Eqs. (4.67b) with $\lambda^c = 3/4$. The states labelled n refer to coupling constants $v_{1,2} = v_{1,2}^c(1+\epsilon)$, $\epsilon = \pm 2^{-n}$. The dashed-dotted line exhibits the asymptote (6.17b) for the critical decay $f^c + h(t_0/t)^a$ with $f^c = 0.25$, $h = 0.75$, $a = 0.305$ and t_0 adjusted as indicated by the arrow. Further details are explained in the text (Voigtmann 2003a).

dependence. The linear ω variation is indicated by the dashed straight lines of slope unity in the double-logarithmic presentation of the loss spectra. Such low-frequency behaviour is induced by white-noise spectra: $\phi''(\omega) \approx \phi''(\omega = 0)$.

The liquid-state correlators $n \geq 2$ exhibit long-time tails for $\Omega t > 10$. The decay extends to times which increase dramatically if the distance parameter $\epsilon = -2^{-n}$ tends to zero. This precursor phenomenon of the arrest has its counterpart in loss spectra, which are located the further below 0.1 Ω, the closer the state is shifted to the arrest point. The glass-state correlators for $n \geq 6$ exhibit a stretched time dependence for $t\Omega > 10$. Again, this long-time tail extends the further the smaller the distance of the state from the critical one. This precursor phenomenon of the glass instability at the transition point has a counter part in the loss spectra for $\omega < 0.1\Omega$. There is a large enhancement of the susceptibility spectra above any reasonably estimated white-noise-induced linear ω dependence for 'normal' liquid dynamics. The glass spectra $\chi''(\omega)$ and $\chi''_A(\omega)$ for $n = 8$ exhibit such enhancement for $10^{-3} < \omega/\Omega < 10^{-1}$. The shown ϵ-sensitive long-time decay processes and the corresponding control-parameter-sensitive low-frequency spectra are non-trivial features of the MCT bifurcation dynamics.

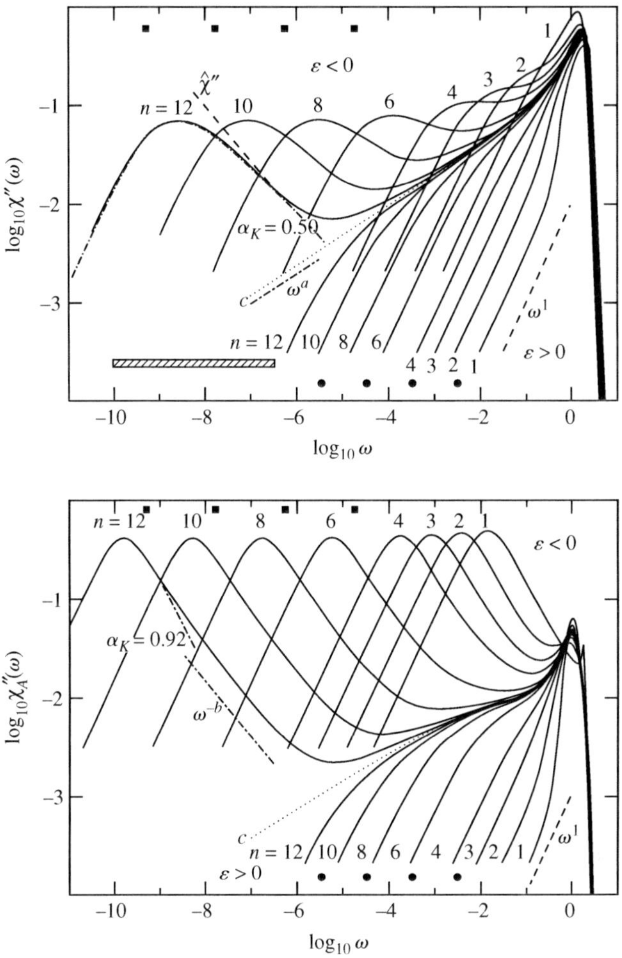

FIG. 4.13. Susceptibility spectra $\chi''(\omega) = \omega\phi''(\omega)$ for the correlators shown in Fig. 4.12 (upper panel), and corresponding spectra $\chi''_A(\omega) = \omega\phi''_A(\omega)$ (lower panel) for the correlators calculated from Eqs. (4.33a,b) with $v_A = 50, \Omega_A = \Omega/5, \nu_A = 0$. The dashed straight lines labelled ω^1 have slope unity. Other details are explained in the text (Voigtmann 2003a).

Let $\phi^c(t)$ denote the critical correlator, i.e., the correlator calculated for the transition point P^c. It is shown as the dotted line in Fig. 4.12. This correlator relaxes to the critical arrested part f^c for t tending to infinity. For every positive error margin δ^*, one can find a time t_- so that there holds

$$|\phi^c(t) - f^c| \leq \delta^*/2 \tag{4.107a}$$

for all times exceeding t_-. According to Sec. 4.2.2, the correlators are continuous functions of the control parameters for times of any given finite time interval. Hence, for every time t_+, which exceeds t_- by an arbitrary large finite value, there exists some positive ϵ_+ so that there holds

$$|\phi^\epsilon(t) - \phi^c(t)| < \delta^*/2 \tag{4.107b}$$

for all distance parameters ϵ, which obey $|\epsilon| \leq \epsilon_+$, and for all times from the interval $t_- \leq t \leq t_+$. Here, $\phi^\epsilon(t)$ denotes the correlator for a state with distance parameter ϵ. Combining the preceding two results, one obtains: $|\phi^\epsilon(t) - f^c| \leq \delta^*$ for $t_- \leq t \leq t_+$ and $|\epsilon| \leq \epsilon_+$. All correlators are arbitrarily close to the critical arrested part f^c within an arbitrarily long time interval, provided the initial point t_- is chosen large enough and provided the distance parameters are not outside a certain interval $-\epsilon_+ \leq \epsilon \leq \epsilon_+$. In this sense, the $\phi(t)$ versus t curves exhibit a plateau f^c of arbitrary length if the states are arbitrarily close to the transition point. Figure 4.12 demonstrates the result for $\delta^* = 0.03$ and $\epsilon_+ = 2^{-12}$ for the two-orders-of-magnitude time increase from $t_- = 10^4/\Omega$ to $t_+ = 10^6/\Omega$. The considerations can be done for every correlator $\phi_q(t)$ of an M-component model. The parameters get a label. One arrives at:

$$|\phi_q^\epsilon(t) - f_q^c| \leq \delta^*, \quad t_- \leq t \leq t_+, \quad |\epsilon| \leq \epsilon_+, \quad q = 1, \ldots, M, \tag{4.108}$$

with $(\delta^*, t_-, t_+, \epsilon_+) = (\min \delta_q^*, \max t_{q-}, \min t_{q+}, \min \epsilon_{q+}; q = 1, \ldots, M)$. The numbers f_q^c denote a plateau for the $\phi_q(t)$ versus $\log t$ curves for states near the critical point. The critical arrested parts f_q^c are the simplest of the general concepts, which characterize the MCT bifurcation dynamics.

The preceding conclusions are demonstrated in Fig. 4.14 for a microscopic MCT model. It exhibits the evolution of a long-time dynamics for two correlators of the hard-sphere system (HSS). The mode-coupling-functional of the underlying $M = 100$ model is explained above in connection with Fig. 4.6. The correlators $\phi_q(t)$ are calculated from the equation of motion (4.17b) for a colloid dynamics. The scales τ_q for the short-time dynamics are modelled by ignoring the wave-number dependence of the friction parameter ν_q in Eq. (4.39b):

$$\tau_q = t_{\text{mic}} S_q/(qd)^2. \tag{4.109a}$$

The curves with label $n = 0$ exhibit the decay without mode-coupling effects: $\phi_q^{(0)}(t) = \exp[-(t/\tau_q)]$. If the wave number is restricted to such small values that one can ignore the q-dependence of the structure factor, this correlator describes particle diffusion. It reduces to the hydrodynamic-limit expression (3.40c) with the diffusivity given by

$$D_s = d^2/(t_{\text{mic}} S_0). \tag{4.109b}$$

Such a correlator describes the long-wave length relaxation for short times, i.e., the leading terms in Eq. (4.19b) for the $q \to 0$ limit. Therefore, D_s is called the

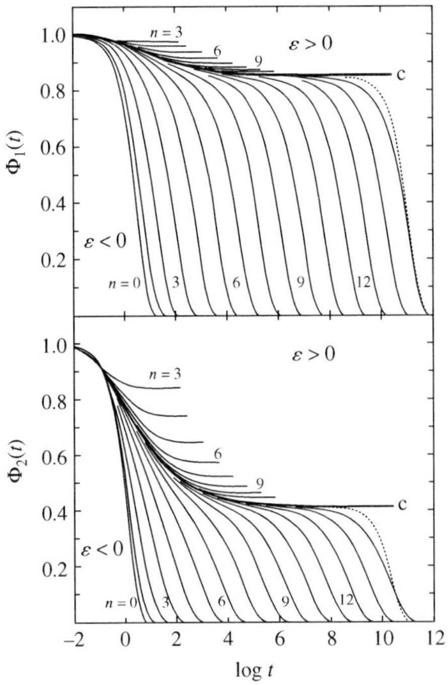

FIG. 4.14. Density correlators $\phi_1(t)$ and $\phi_2(t)$ for a model of a system of hard spheres with diameter d for wave numbers $q_1 = 7.0/d$ and $q_2 = 10.6/d$, respectively. The labels $n = 0, 1, \ldots, 14$ specify the packing fraction by $\varphi = \varphi_c(1 + \epsilon)$, $\epsilon = \pm 10^{-(n/3)}$ with $\varphi_c = 0.5159\ldots$ denoting the critical value. The heavy curves with label c present the correlators for the critical point, $\epsilon = 0$; they relax to the plateaus $f_1^c = 0.849$ and $f_2^c = 0.417$, respectively. The correlators are solutions of a $M = 100$ model. The mode-coupling constants are explained in the caption of Fig. 4.6. The equations of motion (4.17b) are solved for a stochastic short-time dynamics. The time constants are modelled by Eq. (4.109a). The unit of time is chosen so that $t_{\mathrm{mic}} = 160$. The dotted lines exhibit simple-relaxation laws for the below-plateau decay: $f_{1,2}^c \exp[-(t/\tau^{1,2}]$, $\tau^1 = 13.7 \times 10^{10}$, $\tau^2 = 2.83 \times 10^{10}$. Reproduced from Franosch *et al.* (1997).

short-time diffusion constant of the system. The time scale for the presentation in Fig. 4.14 is chosen so that $t_{\mathrm{mic}} = 160$. Thereby, the 'normal' dynamics is restricted to the regime $t \leq 10$. Within this time interval, the $n = 1$ liquid correlators decay to zero and the $n = 3$ deep-glass correlators reach their arrested limits f_q. As explained in Sec. 4.2.2, the correlators are completely monotonic; they do not exhibit oscillation features. All correlators for $n > 3$ exhibit an ϵ-sensitive dynamics for $t > 10$. It extends to the larger times the smaller the distance parameter ϵ. For $n \geq 14$, the correlators agree with the plateaus f_q^c

within an error margin $\delta^* = 0.02$ for the three-decade time interval extending from $t_- = 10^5$ to $t_+ = 10^8$. According to Fig. 4.6, the wave number q_2 for the second correlator is chosen to get a minimal value for the plateau f_q^c within the q-range accessible for light-scattering experiments for colloidal suspensions. The wave number q_1 for the correlator $\phi_1(t)$ is chosen near the structure-factor-peak position; the plateau f_1^c is close to the largest possible value for the HSS.

For liquid states, the correlators $\phi_q(t)$ decay to zero for large enough times. Hence there exists a time for the plateau crossing. It shall be denoted by τ_q^{pc} so that

$$\phi_q(\tau_q^{pc}) = f_q^c, \qquad \phi_q(t > \tau_q^{pc}) < f_q^c, \quad \epsilon < 0. \tag{4.110a}$$

The ϵ dependence of the correlators and of the plateau-crossing times τ_q^{pc} is not indicated explicitly. For large times and small ϵ, $\phi_q(t)$ is arbitrarily close to f_q^c. Hence, τ_q^{pc} diverges for ϵ tending to zero:

$$\lim_{\epsilon \to 0-} \tau_q^{pc} = \infty. \tag{4.110b}$$

τ_q^{pc} quantifies a slowing down of the liquid dynamics upon approaching the arrest point. In Fig. 4.12, the plateau-crossing times τ^{pc} are marked by filled circles for the states with labels $n = 6, 8, 10, 12$. Increasing n by 2, i.e., decreasing $\log|\epsilon|$ by $2\log 2$, $\log \tau^{pc}$ increases by a fixed amount. If this behaviour were to be valid for all values of $|\epsilon|$, one could conclude a power-law divergency of the time scale: $\tau^{pc} \propto 1/|\epsilon|^\delta$. The figure suggests a δ between 1.6 and 1.7.

The frequencies $\omega^{pc} = 1/\tau^{pc}$ for the above specified states are also marked in Fig. 4.13 by filled circles. Up to an ϵ-independent factor, these ϵ-dependent frequencies ω^{pc} mark the scale for a control-parameter sensitive crossover phenomenon of the spectra. For $\omega^{pc} \ll \omega \ll \Omega/10$, the spectra are close to that for the critical correlator. This is the counterpart to the fact that the correlators $\phi(t)$ are close to the critical ones for $10/\Omega \ll t \ll \tau^{pc}$. Within the specified regions of frequencies or times, the spectra or correlators, respectively, of the liquid are close to those for the glass. This holds for sufficiently small distance parameters and reflects the inequality (4.107b). For frequencies far below ω^{pc}, the glass spectra exhibit a white-noise-induced spectrum: $\chi''(\omega \ll \omega^{pc}) \propto \omega$. This reflects the fact that the glass correlators have reached their long-time limit for $t \gg \tau^{pc}$. For small $|\epsilon|$, the liquid spectra for $\omega \ll \omega^{pc}$ exhibit a low-frequency loss peak similar to the one discussed in Fig. 1.3 for glassy spectra of liquids. It corresponds to the decay of $\phi(t)$ from the plateau to zero for $t > \tau^{pc}$, as discussed in connection with Fig. 1.4. Up to an ϵ-independent factor, ω^{pc} marks the position of the control-parameter-sensitive loss minimum of the liquid for states close to the arrest point.

The dotted curve with label c in the upper panel of Fig. 4.13 is a straight line of slope $a \approx 0.30$, provided $\omega < \Omega/50$. A straight line of this slope is shown in the dashed-dotted line labelled ω^a. The straight line exhibits such a power law relaxation spectrum. The critical loss spectrum in the lower panel exhibits the

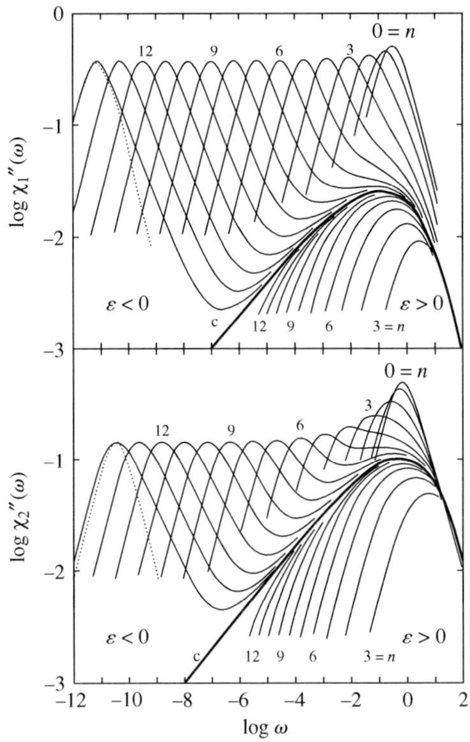

FIG. 4.15. Susceptibility spectra $\chi''_{1,2}(\omega) = \omega\phi''_{1,2}(\omega)$ for the correlators of a HSS exhibited in Fig. 4.14. The dotted lines show Lorentzian spectra $2\chi^{1,2}(\omega\tau^{1,2})/[1 + (\omega\tau^{1,2})^2]$ with relaxation times $\tau^{1,2}$ specified in the caption of Fig. 4.14 and $\chi^1 = 0.370, \chi^2 = 0.147$ chosen to fit the maxima of the low-frequency-loss peaks. Reproduced from Franosch *et al.* (1997).

same behaviour $\chi''_A(\omega) \propto \omega^a$, albeit only for $\omega < \Omega/10^4$. Figure 4.15 exhibits the loss spectra of the density fluctuation with wave numbers q_1 and q_2 for the model of a HSS for states defined in Fig. 4.14. For both critical correlators, a power-law spectrum specified by a common exponent $a \approx 0.3$ is exhibited for $\omega < 10^{-5}$. If the power-law spectra were to extend to $\omega = 0$, one could conclude a power-law decay of the critical correlators towards the plateau: $\phi^c_q(t) - f^c_q \propto 1/t^a$. Such a connection between some t^{-a} long-time decay and an ω^a small-frequency loss spectrum is established by a Tauberian theorem, which is explained in Appendix A.7. This law is exhibited in Fig. 4.12 as the dashed dotted line for $a = 0.30$. It describes the critical correlator perfectly for $t\Omega > 10$. Similar indications of an ω^a-loss spectrum and of a t^{-a}-correlation decay have been found for the glassy-relaxation data discussed in Figs. 1.3 and 1.4, respectively.

A major difference between the loss spectra of the two correlators of the two-component model is demonstrated by Fig. 4.13. It concerns the weight of the low-frequency peaks. According to the discussions made for Eq. (1.11b), this difference is a consequence of the difference of the plateau heights. Since $f^c = 1/4$, the low-frequency loss peak of the first correlator exhausts only 25% of the total weight $\int_{-\infty}^{\infty} \chi''(\omega) d\ln\omega$. As was discussed in connection with Fig. 4.4, the plateau of the second correlator reads $f_A^c = 1 - 1/(v_A f^c)$. For the value of v_A used in Fig. 4.13, there holds $f_A^c = 0.92$. Only 8% of the total weight is due to the spectrum for $\omega > \omega^{pc}$. Therefore, the figure for $\chi_A''(\omega)$ is dominated by the low-frequency loss peak. For the upper panel, the situation is reversed: 75% of the weight are due to frequencies exceeding the loss-minimum position. An analogous conclusion holds for the comparison of the spectra shown in Fig. 4.15 for the HSS. The ratio of the plateaus reads: $f_1^c/f_2^c \approx 2.0$. Since the loss peaks of the first correlator are somewhat narrower than the ones of the second, the ratio of the peak heights exceeds a factor 2: $\chi_1/\chi_2 \approx 2.5$.

Figure 4.15 shows that the positions of the loss minima of the curves referring to the two wave numbers q_1 and q_2 agree for the same density, provided $|\epsilon| \leq 10^{-3}$. If this would be correct for all q, one could conclude that the plateau-crossing times are independent of the correlator index: $\tau_q^{pc} = \tau^{pc}$. Such a result would hold if the factorization property (1.10b) would be valid for the time interval dealing with the plateau crossing. The function $G(t)$ there should be ϵ-dependent and exhibit a zero for the plateau-crossing time: $G(\tau^{pc}) = 0$. The amplitude F_q there has to be identified with f_q^c. If this factorization property would hold also for the glass, one would get $\phi_q(t \to \infty) - f_q^c = f - f_q^c = H_q G(\infty); \epsilon > 0$. For small ϵ, Eq. (4.91a) yields: $H_q G(\infty) = h_q \sqrt{\sigma_\epsilon/\mu_2^c}$. The q-dependence of H_q would be given by that of the critical amplitude h_q.

The considerations of the preceding paragraph provide the motivation for defining functions $\hat{\phi}_q(t)$ and equivalent loss spectra $\hat{\chi}_q(\omega)$:

$$\hat{\phi}_q(t) = [\phi_q(t) - f_q^c]/h_q, \tag{4.111a}$$

$$\hat{\chi}_q(\omega) = \chi_q''(\omega)/h_q. \tag{4.111b}$$

These formulas provide expressions for $\phi_q(t)$ and $\chi_q(\omega)$ in terms of $\hat{\phi}_q(t)$ and $\hat{\chi}_q(\omega)$, respectively. The factorization property is equivalent to the statement that $\hat{\phi}_q(t)$ is independent of the correlator index q. The formula (1.10b) holds with $f_q = f_q^c$, $H_q = h_q$, and $G(t) = \hat{\phi}_q(t)$. Via Fourier transformation, the property is equivalent to the statement that $\hat{\chi}_q(\omega)$ does not depend on q.

Figures 4.16 and 4.17 exhibit rescaled correlators and loss spectra for three distance parameters $|\epsilon|$ for the hard-sphere-colloid model specified above. Four representative wave numbers are chosen as marked by arrows in Fig. 4.6. Within the accuracy of the drawing, the rescaled correlators collapse on a common curve for $|\hat{\phi}_q(t)| \leq 0.1$. Let us denote this q-independent part $G(t)$. Similarly, the rescaled spectra collapse on a common function for $\log \hat{\chi}_q(\omega) \leq -2$ within the accuracy of the drawing. Let us denote this common spectrum by $\hat{\chi}(\omega)$. If one could assume the shown results to hold for all q and if one could extrapolate

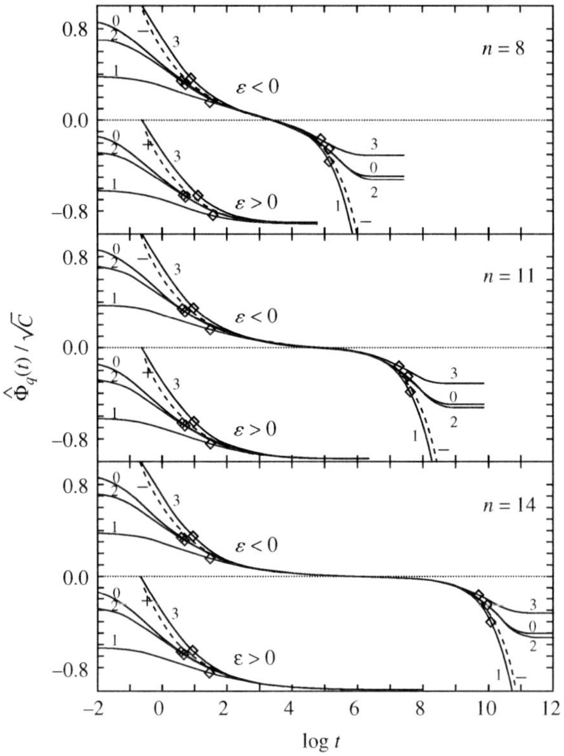

FIG. 4.16. Correlators for the model of a HSS described above for three values of the distance parameter $|\epsilon| = 10^{-n/3}$, $n = 8, 11, 14$, rescaled to functions $\hat{\phi}_q(t)$ according to Eq. (4.111a). The labels 0 to 3 indicate the wave numbers $q_0 = 3.4$, $q_1 = 7.0$, $q_2 = 10.6$, $q_3 = 17.4$. The curves for the glass states are shifted downwards by 1 to avoid overcrowding. The correlators for q_1 and q_2 are shown in Fig. 4.14, where also the model and units are explained. The constant $C = 1.54$ connects σ and ϵ according to Eq. (4.104d).

The dashed lines with label $\pm$ show q-independent first-scaling-law asymptotes for $\epsilon \gtrless 0$ calculated for the right-hand side of Eq. (6.66a); and the diamonds mark the times where $\hat{\phi}_q(t)$ differs from the asymptotes by ± 0.05. Reproduced from Franosch *et al.* (1997).

the obvious trends shown in the figures, one could conclude the following. For every positive error margin δ^{**}, there is some $\epsilon^{**} > 0$ so that

$$|\hat{\phi}_q(t) - G(t)| < \delta^{**} \quad \text{if } |G(t)| < \epsilon^{**}. \tag{4.111c}$$

Similarly, for every positive δ^{***}, there is some $\epsilon^{***} > 0$ so that

$$|\hat{\chi}_q(\omega) - \hat{\chi}(\omega)| < \delta^{***}\hat{\chi}(\omega) \quad \text{if } \hat{\chi}(\omega) < \epsilon^{***}. \tag{4.111d}$$

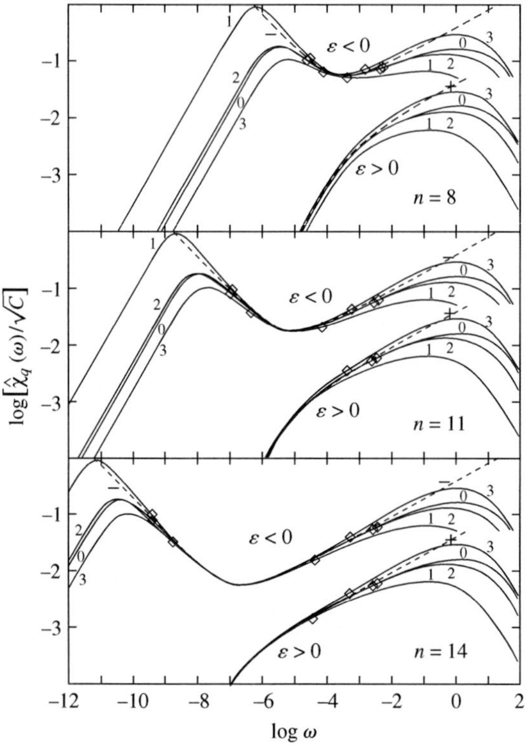

FIG. 4.17. Double-logarithmic presentation of the susceptibility spectra for the correlators of a HSS model specified in the caption of Fig. 4.16. The spectra are rescaled according to Eq. (4.111b); the glass spectra are shifted downwards by one decade to avoid overcrowding. The susceptibility spectra for q_1 and q_2 are shown also in Fig. 4.15.

The dashed lines with labels $\pm$ are q-independent first-scaling-law asymptotes for $\epsilon \gtrless 0$ calculated for the right-hand side of Eq. (6.66b); and the diamonds mark the frequencies where $\hat{\chi}_q(\omega)$ deviates by 10% from the asymptote. Reproduced from Franosch *et al.* (1997).

For the liquid states, the implicit equation $|G(t)| \leq \epsilon^{**}$ defines a finite time interval $t_{\min} \leq t \leq t_{\max}$. Within this interval, there holds Eq. (1.10b) within the error δ^{**}. For times decreasing below $t_{\min}$, the $\hat{\phi}_q(t)$ versus t curves fan out. A similar fanning out is observed for times increasing above $t_{\max}$. The described scenario agrees with that discussed for glassy relaxation of a liquid in connection with Fig. 1.8. The data exhibited there show a time interval with $t_{\max}/t_{\min} \approx 600$; this is similar to what is shown for the upper panel in Fig. 4.16. The function $G(t)$ describes the stretched relaxation towards the plateau and the beginning of the stretched relaxation below the plateau. The remarkable differences for the plateau-crossing shown in Fig. 4.14 for the two wave numbers

q_1 and q_2 are caused solely by the q-dependence of the quantities f_q^c and h_q. The critical amplitude h_q is another general concept characterizing the MCT-bifurcation dynamics.

As mentioned above, τ^{pc} is given by $G(\tau^{pc}) = 0$. Similar to what was derived from Fig. 4.12, one infers from Fig. 4.16 that $\log \tau^{pc}$ increases by the same value δ if one changes from $n = 8$ to $n = 11$ as if one changes from $n = 11$ to $n = 14$. If this would hold for all $n = 3 \cdot \log |\epsilon|$, one could conclude again on a power-law: $\tau^{pc} \propto 1/|\epsilon|^{\delta}$. From the figure, one deduces $\delta \approx 1.6$.

If one defines the end of the plateau τ_{end} by $\hat{\phi}_q(\tau_{\text{end}}) = -0.1$, one concludes in an analogous manner on a power-law scale: $\tau_{\text{end}} \propto 1/|\epsilon|^{\gamma}$. The exponent deduced from the diagrams is larger than $\delta : \gamma \approx 2.5$. The evolution of the plateau crossing upon approaching the critical point is governed by two power-law time scales. The approach towards the plateau occurs on scale τ^{pc}. The decay below the plateau occurs on scale τ_{end}.

Comparing the two panels for $n = 11$ and $n = 14$ of Fig. 4.17 one verifies that the minima of the liquid spectra shift according to the power law $\omega_{\min} \propto |\epsilon|^{\delta} \propto 1/\tau^{pc}$. The result confirms what was deduced above from Figs. 4.12 and 4.13. One checks also that the loss peak maxima decrease with $|\epsilon|$ according to the law $\omega_{\max} \propto |\epsilon|^{\gamma}$. This holds for each of the 4 wave numbers; and this for the change from label $n = 8$ to $n = 11$ as well as for the one from $n = 11$ to $n = 14$. The second scale is one characterizing the loss-peak dynamics. The decrease of $\omega_{\max}/\omega_{\min}$ upon decreasing $|\epsilon|$ is the same feature, which is exhibited in Fig. 1.3 for loss spectra of a glassy liquid upon decreasing the temperature.

The possibility to describe correlators and spectra by some common function $G(t)$ and $\hat{\chi}(\omega)$, respectively, appear as asymptotic properties of the MCT dynamics for control parameters approaching a glass-transition singularity. Figures 4.16 and 4.17 demonstrate the evolution of the asymptotic behaviour upon decrease of ϵ. Let us note an important difference between the two figures, which is of relevance for possible comparisons of the MCT results with data for glassy dynamics obtained from experiment or molecular-dynamics simulations. Using the 'within the accuracy of the drawing' criterion, one infers that the factorization property holds for $t \geq t_{\min} = 10^{2.5}$. The ϵ-independent $t_{\min}$ is a factor 30 larger than the time for the end of the normal-liquid-dynamics regime. There is an appreciable interval for glassy dynamics, $10 < t < 300$, where Eqs. (1.10a,b) do not hold. The lowest panel of Fig. 4.17 does not show that the regime of the asymptotic law $\chi_q''(\omega) = h_q \hat{\chi}(\omega)$ holds for ω smaller than $1/t_{\min}$. Rather, the factorization property is exhibited for $-9 < \log \omega < -4.5$. Let us remember that the normal-dynamics spectra are located for frequencies above 0.1; for normal dynamics, there is a white-noise induced linear loss spectrum for $\omega < 0.1$. Consequently, there is a 3.5-decade regime of frequencies below the band of normal dynamics where there are spectra caused by the bifurcation phenomenon. But, these spectra are not described by the asymptotic law. One has to decrease ω by more than a factor 3000 below 0.1 before the description by the function $\hat{\chi}(\omega)$ is possible.

For the $n = 14$ state, the factorization property holds for times up to $\tau_{\text{end}} \approx 10^{9.5}$. Hence, there is a 7-decade window for time changes, which obey Eq. (1.10b). The spectra collapse for frequencies down to about 10^{-9}. Hence, the factorization formula for the description of the dynamics in the frequency domain holds only for a 4.5-decade window. As noted above, $\log \tau_{\text{end}}$ and $-\log \omega_{\max}$ increase by $\gamma \Delta n/3$ if $\log |\epsilon|$ is increased by $\Delta n/3$. As shown in the upper panel of Fig. 4.16, there is a 2-decade window for time changes left for the validity of the factorization property for the $n = 8$ state. This window reduces to about 0.5 decades for the $n = 6$ state, which has a distance parameter $\epsilon = -0.01$. This is a large enough time interval for demonstrating that the plateau-crossing time τ^{pc} is independent of the correlator index q. Figure 4.17 shows that there is no frequency interval left to demonstrate the factorization property for the description of the minimum of the $n = 8$ state. For the state with $\epsilon = -0.01$, the spectrum for wave number q_1, does not even exhibit a minimum, as can be inferred from Fig. 4.15.

Let us conclude this section with a glimpse on the density-fluctuation dynamics for states near a higher-order glass-transition singularity. Results for two such states P_a and P_b are reproduced in the upper two panels of Fig. 4.18. They are calculated for the square-well system (SWS). The model for the mode-coupling functional is explained in connection with Fig. 4.9 for the singularity diagram. The calculated decay curves exhibit a strongly stretched decay to zero; but, there is no indication of a plateau. In this respect, the $\phi_q(t)$ versus $\log t$ curves are quite different from the ones calculated for the HSS. However, the decay curves are similar to the ones shown in Fig. 1.21 for molecular-dynamics-simulation results for square-well systems.

The two liquid states P_a and P_b have a packing fraction larger than the transition value φ_c^{HSS} of the hard-sphere system. They are partly encircled by the liquid–glass transition line. Shifting the transition point P^c along this line by increasing the attraction strength parameter Γ, the critical form factors f_q^c increase strongly. For Γ between 3 and 4, f_q^c is close to that for an HSS. The particle arrest is dominated by the strong-repulsion-induced cage effect. The Debye–Waller factors and their increase with Γ are similar to the ones shown as full lines in Fig. 4.10. For Γ near 5, the sticking effect is the dominant mechanism for arrest. The critical Debye–Waller factors f_q^c are much larger; they are similar to what is shown by the dotted lines in Fig. 4.10. If P^c is shifted through the A_4 singularity, f_q^c increases very steeply, similar as shown for the $v_3 = 3.375$ curve in the lower panel of Fig. 4.3. There is no distinguished f_q^c, which could cause a plateau for the correlators for the states near P_a or P_b. Rather, there is a distribution of plateaus between the two extreme values mentioned. The reasoning given at the beginning of this section for the slowing down of the liquid dynamics upon approaching a transition point remains valid. But, the explanation for the existence of a plateau does not apply. The states S_1 and S_2 discussed in Fig. 1.21 had been selected carefully close to the suspected position of the A_4 singularity. The states P_a and P_b had been selected in order to demonstrate qualitative agreement of some MCT results with the findings of the cited simulation work.

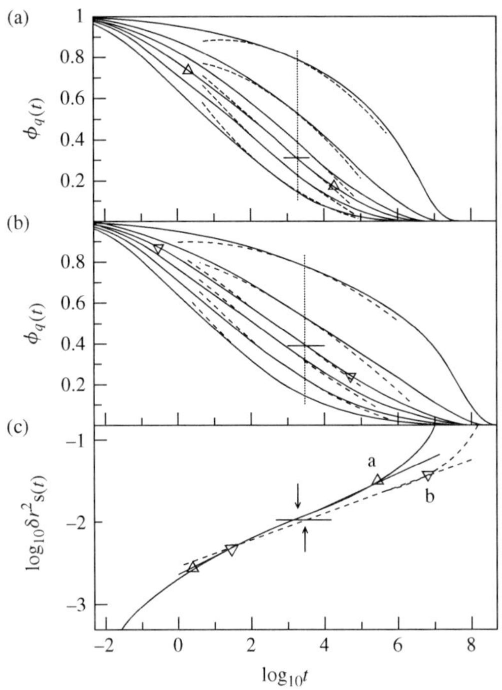

FIG. 4.18. Solutions for a microscopic model for the SWS with stochastic dynamics. The mode-coupling coefficients are based on the mean-spherical approximation for the structure factor. The time scale τ_q is modelled by Eq. (4.109a). The unit of length and time are chosen so that the hard-core diameter d is unity and $t_{\text{mic}} = 160$. The states a and b have the attraction range parameter $\delta^* = 0.04381$ of the A_4 singularity and (φ, Γ) as specified in Fig. 4.9. The full lines in panels (a) and (b) show the density fluctuation correlators for the respective states for wave numbers 4.2, 20.2, 24.2, 27.0, 32.2 and 36.2 (from top to bottom). The dashed lines exhibit the result of Eq. (1.14) with τ fitted as indicated by the vertical dotted lines. The amplitudes are the critical arrested part at the A_4 singularity $F_q = f_q^c$ and $H_q^{(1)} = B_0' h_q$, $H_q^{(2)} = h_q B_2(q)$ are calculated from the mode-coupling functional according to Eqs. (6.182a,c,d). For states a and b, $B_2(q)$ vanishes for $q = 27.0$ and $q = 24.2$, respectively. The short horizontal lines mark f_q^c for $q = 27.0$ and 24.2, respectively. The triangles mark the points, where the correlators differ from the logarithmic-decay function by 5%. Reproduced from Sperl (2003).

In panel (c), the full and dashed line show the mean-squared displacement for states a and b, respectively. The short horizontal line marks $6(r_{sc})^2$ with r_{sc} denoting the critical localization length for the A_4 singularity. The straight lines exhibit the power law $(r_{sc})^2(t/\tau)^{\alpha'}$. The time scales τ, which are noted by arrows, are the same as used in the upper panels. The exponents $\alpha' = 0.201$ and 0.161 for states a and b, respectively, are calculated from Eq. (6.194b). The points of 5% deviation of the straight asymptotes from the solution $\delta r_s^2(t)$ are marked by triangles (Sperl 2004a).

It is demonstrated in Fig. 1.21 that the liquid correlators can be described by Eq. (1.14) within time intervals of 2.5 and 4 decades, respectively. The data points in the upper panel of Fig. 4.19 show the fit amplitudes F_q; and those in the lower one show the amplitudes $H_q^{(1)}$ for the first state together with the rescaled ones for the second state. The data demonstrate two remarkable properties: the amplitudes F_q are the same for both states and the amplitudes $H_q^{(1)}$ are proportional to each other (Sciortino *et al.* 2003). The dashed lines in Fig. 4.18 are also calculated with Eq. (1.14), and they also describe the correlators for intervals of 2.5 and 4 decades, respectively. The amplitude F_q is chosen as the critical arrested part f_q^c at the A_4 singularity. It is shown as full line in the upper panel of Fig. 4.19. The amplitude $H_q^{(1)}$ is chosen proportional to the critical amplitude h_q at the A_4 state; it is shown as line in the lower panel.

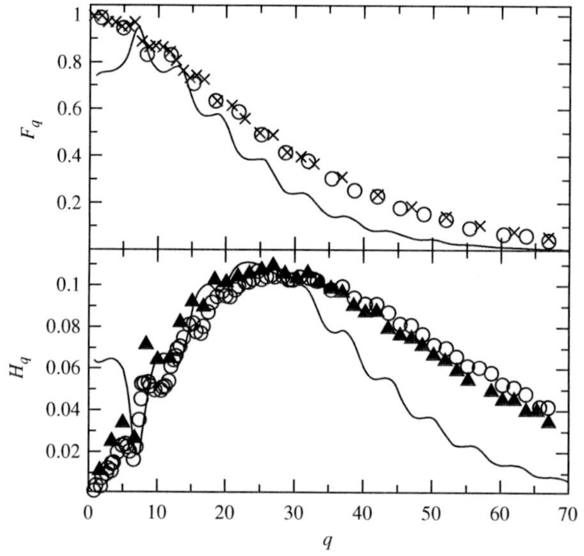

FIG. 4.19. The data points present the fit parameters from Eq. (1.14) for the description of simulation data for the density correlators of square-well systems as presented in Fig. 1.21 and determined by Sciortino *et al.* (2003). The circles and crosses in the upper panel present F_q for systems S_1 and S_2, respectively. The circles in the lower panel present $H_q^{(1)}$ for system S_1 and the triangles exhibit $2.43 \cdot H_q^{(1)}$ for system S_2. The lines are the critical arrested parts f_q^c and 14% of the critical amplitude h_q, respectively, calculated at the A_4 glass-transition singularity of the SWS model specified in the caption of Fig. 4.18. Reproduced from Sperl (2003).

4.4.2 *Below-plateau relaxation*

In this section, some properties shall be discussed of the below-plateau decay of the correlators and of the corresponding low-frequency-loss peaks for liquid states close to the arrest points.

The shaded bar in Fig. 4.12 marks a 3.2-decade time interval. Within this interval, the $n = 12$-liquid-state correlator decays from 0.95 f^c to 0.05 f^c. An increase of the time by a factor larger than 1500 is needed for the specified 90% of the below-plateau-decay process. This time-increase factor is about 27 times larger than that needed for the same 90% of the decay described by the elementary-relaxation process (1.1b). The shaded bar in the upper panel of Fig. 4.13 marks a 3.6-decade frequency interval. Within this interval, the $n = 12$-state low-frequency loss-peak rises above 0.2 $\chi_{\max}$, where $\chi_{\max}$ denotes the height of the peak maximum. A frequency increase by a factor larger than 3500 is needed to scan the loss-peak spectrum above its 20% level. This increase factor is more than 23 times that for the corresponding scanning of the Lorentzian spectrum $\omega\tau/[1+(\omega\tau)^2]$ of the elementary-relaxation process. The dashed-dotted line labelled $\alpha_K = 0.50$ shows the spectrum of the Kohlrausch function (1.3), for the stretching exponent $\beta = \alpha_K$. It describes the loss peak well for $\chi''(\omega) \geq 0.15\chi_{\max}$. The mentioned results for the one-component model exhibit an outstanding feature of the glassy dynamics of liquids, which is discussed in connection with Figs. 1.1, 1.2, 1.6 and 1.7: there is stretching of the dynamics of the slowest relaxation process.

The dashed-dottel line with label $\alpha_K = 0.92$ in the lower panel of Fig. 4.13 shows a fit of the low-frequency loss peak by that of a Kohlrausch function with exponent $\beta = \alpha_K$. The stretching for the below-plateau decay of the second correlator of the two-component model is less pronounced than that for the first correlator. Moreover, the Kohlrausch spectrum fits the peak well only for $\chi''_A(\omega) > 0.3\chi_{A\max}$. The size of the stretching and the quality of a Kohlrausch fit depend on the correlator considered.

The features of the MCT-bifurcation dynamics noted in the preceding two paragraphs are exhibited also by the results for the microscopic MCT model for the hard-sphere system. The dotted lines in Fig. 4.14 exhibit elementary decay processes $f_q^c \exp[-(t/\tau_q)]$ with a relaxation time scale τ_q chosen to fit the 20%-decay level of the $n = 14$ correlators. Stretching manifests itself by the $\phi_q(t)$ versus $\log t$ curves being flatter than the exponential-relaxation curves. The stretching is more pronounced for the correlator ϕ_2 with the low plateau than for the correlator ϕ_1 with its high plateau. Loss spectra corresponding to the Lorentzian (1.1a) are shown in Fig. 4.15 as dotted lines. Stretching manifests itself by the $\chi''_q(\omega)$ versus $\log\omega$ curves describing a much broader peak than is described by the spectra of the elementary-relaxation process. The larger peak shown for the wave number q_1 is less stretched than the smaller peak for wave number q_2. The figure demonstrates also a feature, which is identified for glassy-dynamics spectra of liquids in Figs. 1.3 and 1.6: the loss peaks are skewed. The low-frequency wing exhibits a white-noise induced linear frequency dependence of the spectra, $\chi''_q(\omega\tau_q \ll 1) \propto \omega$. But, the high-frequency wing decays less steeply with increasing $\log\omega$ than that for a Lorentzian. The stretching is a phenomenon for $\omega \gtrsim \omega_{\max}$, if $\omega_{\max}$ denotes the position of the loss maximum. This corresponds to the fact that the stretching of the $n = 14$ curves in Fig. 4.14 occurs mainly for times short compared to τ_q. The major part of the stretching occurs for times, where the correlator is close to the plateau. This suggests that understanding the plateau-crossing is a prerequisite for understanding the stretching of the low-frequency-loss-peak dynamics.

A comprehensive documentation of the below-plateau relaxation and of the low-frequency loss peaks for the HSS can be found in a paper by Fuchs *et al.* (1992); and Fig. 4.20 reproduces some of their findings. The upper panel shows the wave-number dependence of the Kohlrausch exponents β obtained by optimizing the fits to the time dependence of the correlators and to the frequency dependence of the loss peaks, respectively. The exponent is large, i.e., the stretching is not so pronounced, for wave numbers near the structure-factor peak position. For larger wave numbers, the exponents decrease with increasing q. For $qa > 12$, the exponents are similar to those known for various data for van der Waals liquids. The Kohlrausch law does not provide a correct description of the correlators. Therefore, the Kohlrausch exponents depend on whether they result from fits to the correlators or to the spectra. They depend also on the manner one optimizes the fit. The fit values for β should be considered as a practical way to quantify stretching rather than numbers relevant for the understanding of the complex MCT dynamics. The lower panel shows that the q-variation of the relaxation time is similar to that shown in Fig. 4.6(b) of the plateau value f^c_q, but the variation of τ_q is more pronounced. The relaxation time at the structure-factor-peak position rises by more than a factor 4 above that for intermediate-q values off the peak position.

For the one-component model defined by Eqs. (4.32a,b), the stretching depends on the position of the critical point $P^c = (v^c_1, v^c_2)$ on the transition line in the control-parameter plane, which is exhibited in Fig. 4.2. The results shown in Fig. 4.12 and in the upper panel of Fig. 4.13 refer to the critical point characterized by $\lambda^c = 3/4$. If λ^c increases above this value towards unity, the stretching increases and the fit exponent β tends towards zero. If λ^c decreases towards $1/2$, the stretching exponent tends towards unity (Götze and Sjögren 1987a). The schematic model studied in the first papers on the mode-coupling theory for ideal glass transitions deals with the special value $\lambda^c = 1/2$ (Leutheusser 1984; Bengtzelius *et al.* 1984). This model is non generic in the sense that it does not describe stretching of the below-plateau decay. It was demonstrated first by De Raedt and Götze (1986) that the MCT solutions yield a stretched low-frequency loss peak, which can be described very well by a Kohlrausch spectrum. Bengtzelius (1986a) was the first who solved the equations of motion for a microscopic MCT model. He studied a Lennard-Jones system and demonstrated stretching of a below-plateau-decay for a density-fluctuation correlator, which can be described well by a Kohlrausch function with a stretching exponent $\beta = 0.68$.

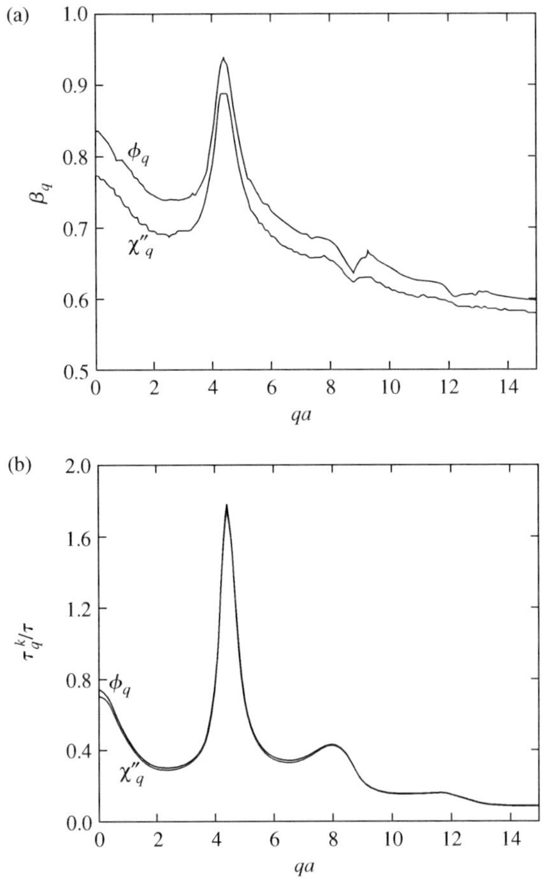

FIG. 4.20. Parameters for fits of the below-plateau-relaxation processes of the MCT correlators $\phi_q(t)$ for the hard-sphere system by Kohlrausch functions $f_q^c \exp[-(t/\tau_q^K)^{\beta_q}]$. The correlators and spectra are calculated for an $M = 300$ model based on the Verlet–Weiss approximation for the structure factor. The averaged interparticle distance $a = (3/(4\pi\rho))^{1/3}$ is used as the unit of the length so that the structure-factor peak position at the critical point is near $4.4/a$. The upper and lower curves in panel (a) are obtained by determining β_q from the inflection point of the $\phi(t)$ versus $\log t$ curves and from the width of the loss peak at half height, respectively. The upper and lower curve in panel (b) for the time scales τ_q^K are obtained by fits to the position of the inflection point and to the loss-peak-maximum position, respectively. The time scale τ is proportional to the plateau-end time τ_{end} introduced in Sec. 4.4.1. Reproduced from Fuchs *et al.* (1992).

It has been tacitly assumed in the preceding paragraphs that the stretching properties are independent of the distance parameters ϵ, provided these are small enough. This assumption appears justified as shall be demonstrated now. The reader can check that Fig. 4.12 exhibits the following property: a major part of the liquid correlators below the plateau for the states with labels $n = 8, 10$ and 12 coincide if they are shifted parallel to the horizontal axis. The same feature is exhibited by the large-n liquid correlators shown in Fig. 4.14; moreover, the shifts from the results for n to the ones for $n \pm 1$ is the same in both panels. A change of $\log t$ by some amount Δ, is equivalent to a change of t to $t \cdot 10^{\Delta}$. If the mentioned shift properties would hold for all Δ and all q, one could represent the correlators for the below-plateau process as

$$\phi_q(t) = \tilde{\phi}_q(\tilde{t}), \qquad \tilde{t} = t/\tau, \quad q = 1, \ldots, M. \tag{4.112a}$$

Here, $\tilde{\phi}_q(\tilde{t})$ would be an ϵ-independent shape function. The strong control-parameter sensitivity of the below-plateau decay, which is demonstrated in Figs. 4.12 and 4.14, would be caused solely by that of the scale τ. Equation (4.112a) formulates a scaling law, which is discussed in Sec. 1.3 as superposition principle. The result formulated here extends Eq. (1.9a) in the sense that the same scale τ appears in all M correlators. The stretching discussed above for the correlation functions in the time domain is a property of the shape function $\tilde{\phi}_q(\tilde{t})$.

The strongly ϵ-dependent scale τ is defined only up to some ϵ-independent factor S. A change from τ to $\tau_S = S\tau$ is equivalent to a change of the shape function $\tilde{\phi}_q(\tilde{t})$ to $\tilde{\phi}_q^S(\tilde{t}) = \tilde{\phi}(S\tilde{t})$. One can define a scale for the below-plateau decay process of correlator $\phi_q(t)$, say τ_q^x, as the time, for which the correlator has decayed to the percentage x of the plateau, $x < 1$:

$$\phi_q(t = \tau_q^x)/f_q^c = x. \tag{4.112b}$$

The scaling law implies

$$\tau_q^x = \tilde{\tau}_q^x \tau. \tag{4.112c}$$

The strongly ϵ-dependent time τ_q^x is given by the control-parameter independent number $\tilde{\tau}_q^x$, defined by the solution of the equation $\tilde{\phi}_q(\tilde{t} = \tilde{\tau}_q^x) = x f_q^c$, and by the time scale τ. The latter is shared by all correlators. The squares in Fig. 4.12 mark the scales τ^x for the schematic-model correlators defined there; $x = 0.1$ is chosen and the states with $\epsilon = -1/2^n$ are considered for $n = 6, 8, 10$ and 12. A change from n to $n + 2$ causes an n-independent addition to $\log \tau^x$. If this were to hold for all ϵ, one could conclude on a power-law divergency of the time scale for ϵ tending to zero: $\tau \propto 1/|\epsilon|^{\gamma}$. From the figure, one infers for this model $\gamma \approx 2.5$. The squares in Fig. 4.13 exhibit the frequencies $1/\tau^x$. The data in both panels suggest $\omega_{\max} \propto 1/\tau$, with an ϵ-independent constant of proportionality. Quite similarly, one infers from Fig. 4.14 for the states with label $n \geq 9$: there

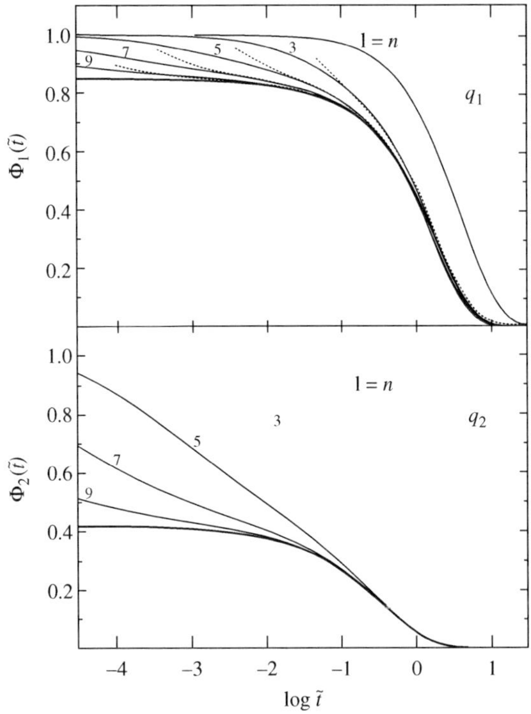

FIG. 4.21. Density correlators from Fig. 4.14 for distance parameters $\epsilon = (\varphi - \varphi_c)/\varphi_c = -10^{-n/3}$, $n = 1, 3, \ldots, 9$ as a function of the rescaled time $\tilde{t} = t/\tau$. The n-dependent scaling time reads $\tau = 0.578/|\sigma|^\gamma$, $\gamma = 2.46$, and σ is the separation parameter $\sigma = C\epsilon$, Eq. (4.104d). The heavy solid lines are the shape functions $\tilde{\phi}_q(\tilde{t})$, Eq. (4.112a).

The dotted lines are the functions $\tilde{\phi}_q(\tilde{t}) + h_q B_1|\sigma|\tilde{t}^{-b}$, with h_q denoting the critical amplitude, $B_1 = 0.431$ and $b = 0.583$, according to Eq. (6.134a). Reproduced from Franosch *et al.* (1997).

holds $\tau_q^x \propto 1/|\epsilon|^\gamma$ with $\gamma \approx 2.5$ for both correlators of the hard-sphere system. Figure 4.15 confirms that $\omega_{\max} \propto |\epsilon|^\gamma$. A similar result was noted in the preceding section for the relation between $\omega_{\max}$ and the plateau-end time, i.e., $\tau \propto \tau_{\text{end}}$. Hence, an understanding of the plateau crossing should imply an understanding of the time scale τ for the complete below-plateau-relaxation process.

The validity of the superposition principle means that correlators for different states coincide if they are considered as function of the time, which is rescaled with the state dependent time τ. Such ϕ_q versus $\log \tilde{t}$ curves are shown in Fig. 4.21 for two correlators of the hard-sphere-system model for some representative values of the distance parameter ϵ. The time τ is not fitted but chosen as $c/|\sigma|^\gamma$. Here, c is an ϵ-independent number, $\gamma = 2.46$, and σ denotes the separation parameter from Eqs. (4.104d). The curves collapse for $\phi_q/f_q^c \leq 0.03$ if $n \geq 3$. This demonstrates the above discussed power law for the times τ_q^x for $x \leq 0.03$. However, the scaling law does not describe the complete below-plateau decay. Rather, the law holds within a chosen positive error margin δ^{sc},

$$|[\phi_q(t) - \tilde{\phi}_q(\tilde{t})]/f_q^c| \leq \delta^{sc}, \tag{4.113a}$$

for the final percentage $x_{\max}$ if ϵ is sufficiently small:

$$|\phi_q(t)/f_q^c| \leq x_{\max}, \quad |\epsilon| \leq \epsilon_{\max}. \tag{4.113b}$$

The figure suggests that the superposition law is an asymptotic property in the following sense. The percentage increases towards its upper limit unity if the states approach the transition point: $\lim_{\epsilon_{\max}\to 0} x_{\max} = 1$. The evolution scenario of the superposition-principle regime upon decreasing the distance parameter ϵ, is similar to that, which is shown in Fig. 1.5 for a density-fluctuation correlator of a glassy liquid upon decreasing the temperature.

The plateau-crossing time τ^{pc} does not vary with changes of ϵ in the same manner as τ. Therefore, the superposition principle cannot hold for times approaching τ^{pc} too closely. From the data, the small-ϵ behaviour $\tau^{pc} \propto |\epsilon|^{-\delta}$ and $\tau \propto |\epsilon|^{-\gamma}$ was deduced. Since $\gamma > \delta$, the rescaled plateau-crossing time $\tilde{\tau}^{pc} = \tau^{pc}/\tau$ tends to zero for ϵ approaching zero. For any given lower bound $\tilde{t}_{\min}$ for the rescaled times, $\log \tilde{\tau}^{pc}$ decreases below $\log \tilde{t}_{\min}$ if ϵ tends to zero. The rescaled times $\tilde{t}$ near $\tilde{\tau}^{pc}$, where there occur the deviations from the superposition principle, shift outside this interval given by $\tilde{t} \geq \tilde{t}_{\min}$. The shape functions $\tilde{\phi}_q(\tilde{t})$, which are shown in Fig. 4.21 as heavy curves for $\log \tilde{t} \geq -4.5 = \log \tilde{t}_{\min}$, can be calculated numerically as solutions of the equations of motion for $|\epsilon| < 10^{-6}$.

Let us consider Eq. (4.112a) as a definition of the correlator $\tilde{\phi}_q(\tilde{t})$ in terms of the correlator $\phi_q(t)$. Using Eq. (A.5a) for Laplace transforms, an equivalent relation between the correlators in the frequency domain is derived:

$$\phi_q(z) = \tau\tilde{\phi}_q(\tilde{z}), \qquad \tilde{z} = \tau z. \tag{4.114a}$$

Introducing dynamical susceptibilities by the equations $\chi_q(\omega) = 1 + \omega\phi_q(\omega + i0)$ and $\tilde{\chi}_q(\tilde{\omega}) = 1 + \tilde{\omega}\tilde{\phi}_q(\tilde{\omega} + i0)$ one can rewrite the result as

$$\chi_q(\omega) = \tilde{\chi}'_q(\tilde{\omega}) + i\tilde{\chi}''_q(\tilde{\omega}), \qquad \tilde{\omega} = \tau\omega. \tag{4.114b}$$

The validity of the superposition principle for all times is equivalent to the statement that $\tilde{\phi}_q(\tilde{z})$ or $\tilde{\chi}'_q(\tilde{\omega})$ or $\tilde{\chi}''_q(\tilde{\omega})$ do not depend on the control parameters. In this case, Eqs. (4.114a,b) formulate scaling laws for the correlators in the frequency domain and for the dynamical susceptibility, respectively. Equation (4.114b) generalizes the superposition principle, which was discussed in Eq. (1.9c), to one, which holds for the reactive as well as for the dissipative part of the dynamical susceptibility. Figure 4.22 demonstrates the evolution of this scaling law for the susceptibility. The curves refer to the hard-sphere-liquid results considered in Fig. 4.21. The master spectra $\tilde{\chi}''_q(\tilde{\omega})$ describe the

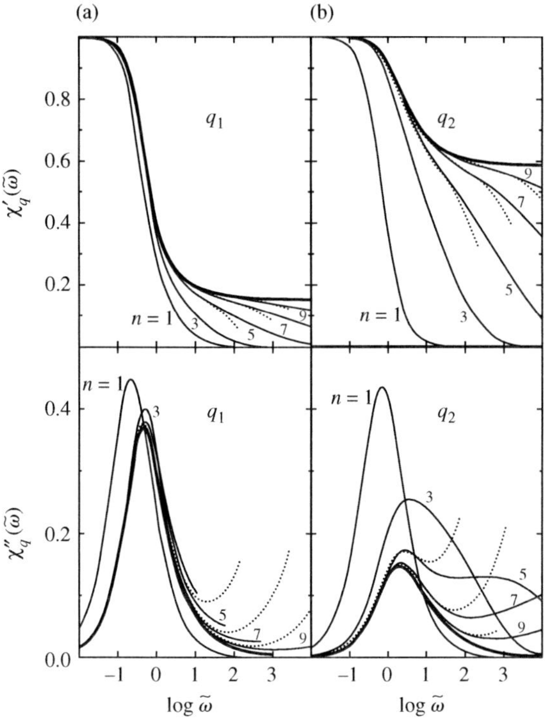

FIG. 4.22. Semilogarithmic presentation of the dynamical susceptibilities $\chi_q(\omega) = \chi'_q(\omega) + i\chi''_q(\omega)$ for the functions from Fig. 4.21. The rescaled frequencies are $\tilde{\omega} = \omega\tau$. Reproduced from Franosch *et al.* (1997).

low-frequency loss peaks in the limit of ϵ tending to zero. For a given upper bound $\tilde{\omega}_{\rm up}$ for the rescaled frequencies, the spectra around the rescaled spectral minimum $\tau\omega_{\rm min} \propto \tau/\tau^{pc}$ shift above $\tilde{\omega}_{\rm up}$ if the state P approaches P^c. The shown numerical results suggest that the scaling-law formula $\tilde{\chi}''_q(\tilde{\omega})$ describes the loss peak in the limit of diverging separation between the low-frequency loss peak and the loss minimum.

The non-degenerate A_2 bifurcation of the MCT-correlator-long-time limits, which describes the transition from a liquid to an ideal glass state, is the simplest generic bifurcation imaginable for a non-linear dynamics. This phenomenon is an example for a catastrophic change of the equilibrium state caused by a smooth change of a single control parameter ϵ. The square-root singularity for the arrest is the general signature for a fold bifurcation. Appeal to regularity properties for the time dependence of a relaxing variable leads to the conclusion that there will develop a plateau, a diverging plateau crossing time τ^{pc}, and a below-plateau relaxation characterized by a diverging time scale τ. These are generic change-of-equilibrium properties for every nonlinear theory. Therefore, it might be worthwhile to close this section with a digression emphasizing more explicitly some general features of the MCT bifurcation dynamics, which are qualitatively different from those of the fold-bifurcation scenario of a conventional dynamics.

To proceed, let us specify universal bifurcation properties of a conventional dynamics by considering the generic A_2 bifurcation of a system with one degree of freedom. The position variable shall be denoted by x and the configuration space shall be restricted to $-1 \leq x \leq 1$. Newton's equation of motion reads $\mu\ddot{x}(t) + \nu\dot{x}(t) = K(t)$, with $\nu > 0$ denoting some friction constant. The force $K(t)$ shall be derived from a potential $u(x) : K(t) = -\partial_x u(x(t))$. In order to focus on the essence of the problem, the limit of strong damping shall be considered so that the inertia term $\mu\ddot{x}(t)$ can be neglected compared to the friction term $\nu\dot{x}(t)$. The unit of time is chosen so that $\nu = 1$, and the initial condition $x(t = 0) = 1$ shall be considered. The potential shall be the polynomial $u(x) = \frac{1}{3}x(x - f^c)^3 - \frac{1}{12}[(x - f^c)^4 - f^{c\,4}] - \frac{1}{2}\epsilon x^2, f^c = 1/4$. Here, ϵ denotes the control parameter; it shall be restricted to $|\epsilon| < 0.05$. Independent of ϵ, the potential has a minimum for $x = 0$, $u(x = 0) = 0$. Since $K(x) = -x[(x-f^c)^2-\epsilon]$, this minimum is the only equilibrium point for $\epsilon < 0$. In this case, $x(t)$ decreases monotonically to zero if the time increases to infinity: $x(t \to \infty) = 0$. For $\epsilon > 0$, there are two further equilibrium points at $f_\pm = f^c \pm \sqrt{\epsilon}$. The value f_- specifies a potential maximum and f_+ a minimum. For $\epsilon \geq 0$, $x(t)$ decreases monotonically till it arrests at a positive long-time limit: $x(t \to \infty) = f_+$. The variable $x(t)$ exhibits the fold bifurcation in analogy to what is shown in Fig. 4.12 for the MCT correlator $\phi(t)$. The inset of Fig. 4.23 demonstrates the smooth changes of the potential for the system due to changes of ϵ. The solution of the equation of motion yields t as elementary function of x; its discussion is left to the reader. Figure 4.23 exhibits the evolution of the bifurcation dynamics upon increasing ϵ from -0.02 to 0.02. The existence of a plateau and the increase of τ^{pc} and τ are demonstrated.

The large-time asymptote for the critical decay reads in leading order: $f^c + 4/t$. For $t > 10^3$, this correlator agrees with the plateau f^c within the accuracy of the drawing, while Fig. 4.12 exhibits a clearly visible stretched critical decay up to $t = 10^4$. The $1/t$ decay leads to a logarithmically-divergent low-frequency spectrum. The corresponding susceptibility spectrum, $\chi''(\omega) \propto -\omega \log \omega$ can hardly be distinguished from a white-noise-induced linear variation. The critical loss spectrum does not have a strong enhancement over a white-noise-induced loss spectrum. These universal results cannot be used to describe the critical correlators in Fig. 4.14 or the critical loss spectra in Fig. 4.15. Nor can they be used to fit the 120 °C spectrum shown in Fig. 1.3 or the correlator decay exhibited in Fig. 1.4 for $-4 \leq \log(t/\tau) \leq -2$.

In the limit of small $|\epsilon|$, the plateau crossing times for the $\epsilon < 0$ curves vary proportional to $1/\sqrt{|\epsilon|}$. The same is true for the times τ^x, specifying the decay to xf^c, Eq. (4.112b). The complete slow dynamics is characterized by a single time scale τ^{pc}, which diverges upon approaching the arrest. Let us consider a value x_{in} close to but below f^c. The variable reaches this value at some time $t_{\text{in}} : x(t_{\text{in}}) = x_{\text{in}}, t_{\text{in}} > \tau^{pc}$. For small $|\epsilon|$, t_{in} increases proportional to τ^{pc}. For

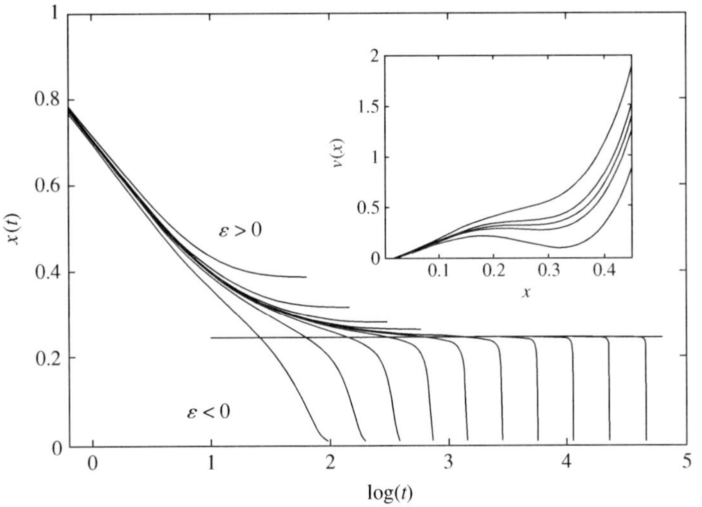

FIG. 4.23. Decay curves for a relaxator $x(t)$ in a potential $u(x) = \frac{1}{3}x\ (x - f^c)^3 - \frac{1}{12}[(x - f^c)^4 - f^{c4}] - \frac{1}{2}\epsilon x^2, f^c = 1/4$. The curves refer to $\epsilon = 0.08/4^n$ for $n = 1, 2, 3, 4$ (from top to bottom) and $\epsilon = -0.08/4^n$ for $n = 1, 2, \ldots, 10$ (from left to right). The horizontal line marks f^c. The inset shows $v(x) = 10^3 u(x)$ for $100\epsilon = -0.500, -0.125, 0.000, 0.125$ and 0.500 (from top to bottom).

increasing $\delta t = t - t_{\text{in}}$, $\delta x = x(t) - x_{\text{in}}$ decreases. The speed is given by the potential derivative for $x < x_{\text{in}}$; it is of the order of the speed of the transient decay, and it depends only smoothly on ϵ. With decreasing ϵ, the δx versus t curve becomes ϵ-independent. The slowing-down of the decay of $x(t)$ below x_{in} is due to the increase of t_{in}. Contrary to what is described by the scaling law (4.112a), the speed $\dot{x}(t) = \delta\dot{x}(t)$ does not decrease with $|\epsilon|$. As a result, the $x(t)$ versus $\log t$ curve approaches a vertical line as demonstrated in the figure. The A_2-bifurcation of a conventional dynamics cannot produce results for the low-frequency-loss peak dynamics, which are similar to the glassy-dynamics data shown in Figs. 1.1 or 1.3 or which are exhibited as solutions of MCT equations of motion in Figs. 4.13 or 4.15.

4.4.3 *Structure and structure relaxation*

In this section, solutions of the MCT equations of motion are compared for models which have the same mode-coupling functional but differ in the details for the short-time dynamics.

The evolution of the complex dynamics for states near a liquid–glass transition is demonstrated in Fig. 4.14 for a microscopic MCT model for the hard-sphere system with stochastic short-time dynamics. The results for the density correlator with wave number $q = 10.6/d$ are reproduced in Fig. 4.24 for four packing fractions φ. In this figure, the time is measured in units of a φ-independent

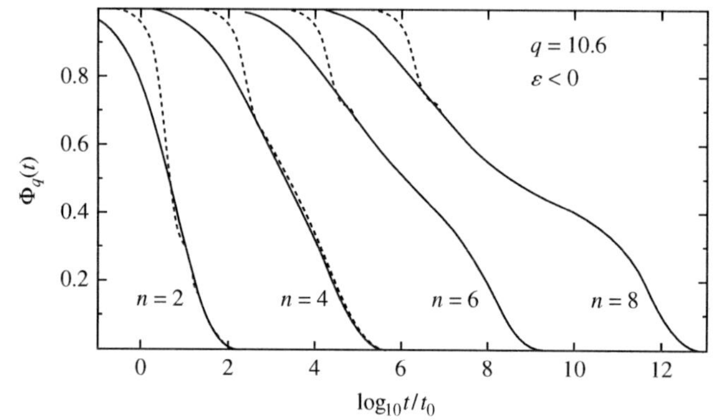

FIG. 4.24. Density correlators obtained for a microscopic MCT model for a system of hard spheres with diameter d. The wave number is $q = 10.6/d$ and the distances of the packing fraction φ from the transition value φ_c are given by $-\epsilon = (\varphi_c - \varphi)/\varphi_c = 10^{-n/3}$, $n = 2, 4, 6, 8$. Curves for successive values of n are shifted horizontally by two decades in order to avoid overcrowding. The model for the mode-coupling functional is explained in connection with Fig. 4.6. The full lines are calculated from Eq. (4.17b) for a stochastic short time dynamics with τ_q given by Eq. (4.109a). The time scale reads $t_0 = 0.425(t_{\text{mic}}/160)$. The correlators are replotted from Fig. 4.14. The dashed lines are obtained as solutions of Eq. (4.17a) for a Newtonian dynamics with $\nu_q = 0$. The time scale reads $t_0 = 0.00944(0.4d/v_{th})$ with v_{th} denoting the thermal velocity. Reproduced from Franosch *et al.* (1998).

scale t_0. The dashed lines reproduce correlators for the same wave number and the same densities calculated from a microscopic MCT model for the same mode-coupling functional. But, Eq. (4.17a) for a Newtonian short-time dynamics is applied with vanishing friction constants, $\nu_q = 0$. Again, the data are represented as function of $\log(t/t_0)$ with some properly chosen density-independent scale t_0. For times outside the region of normal-liquid dynamics, say $t/t_0 \geq 10$, the correlators for the two versions of the theory agree within the accuracy of the drawing for states with label $n \geq 6$. They agree also for all $n \geq 3$, i.e., for $|\epsilon| \leq 0.10$, if a small ϵ-dependence of the scales t_0 is permitted. If one presents the results for the glass states rescaled with the same values for the two times t_0, one arrives at the same conclusion (Franosch *et al.* 1998).

An analogous statement about the agreement of correlators outside the transient regime is demonstrated by the molecular-dynamics simulation data shown in Fig. 1.4. The correlators in that figure exhibit a stretching of the decay from 0.9 to 0.1, which extends over a 5.5-decade time increase. A stretching of a similar size is exhibited by the $n = 8$ correlators in Fig. 4.24. But, in the latter diagram, the wave number is chosen larger in order to get a lower plateau than that exhibited in Fig. 1.4. Thereby, it can be emphasized that the statement on the agreement of the two correlators holds for the stretched decay towards the plateau with the same accuracy as for the stretched below-plateau relaxation. Figure 4.24 shows in addition that the complete density dependence of the correlators for the glassy dynamics is the same. This property of the solutions of the MCT equations has been confirmed by a molecular-dynamics study for a model of a slightly polydisperse hard-sphere liquid (Voigtmann *et al.* 2004).

If the shown numerical results would be valid generally, one could conclude as follows. The solutions of the MCT equations for the dynamics outside the transient can be written as

$$\phi_q(t) = \phi_q^{\mathrm{str}}(t/t_0), \quad t/t_0 \gg 1. \tag{4.115a}$$

The correlators $\phi_q^{\mathrm{str}}(x)$ are determined by the mode coupling functional. The transient dynamics determines the time scale t_0. Since the mode-coupling effects yield smooth variations of the transient dynamics, also t_0 depends smoothly on the mode-coupling coefficients. The slow dynamics of a model for a conventional liquid, which is based on Newton's equations of motion with or without friction effects, is the same as that for a colloid, whose dynamics is described by stochastic equations of motion. This holds up to some time scale t_0, which drifts smoothly with changes of control parameters. For states near the transition points, the scale t_0 can be considered as state-independent.

One can follow the reasoning presented in the preceding section in connection with the superposition principle and motivate relations for the correlators in the frequency domain or for the dynamical susceptibilities. The numerical results for the comparison of the loss spectra demonstrate the following (Franosch *et al.* 1998):

$$\phi_q(z) = t_0 \phi_q^{\mathrm{str}}(z t_0), \quad |z t_0| \ll 1, \tag{4.115b}$$

$$\chi_q(\omega) = \chi_q^{\mathrm{str}}(\omega t_0), \quad |\omega t_0| \ll 1, \tag{4.115c}$$

Here, $\phi_q^{\mathrm{str}}(z)$ is the Laplace transform of $\phi_q^{\mathrm{str}}(t)$ and $\chi_q^{\mathrm{str}}(\omega) = \omega\phi^{\mathrm{str}}(\omega + i0) + 1$. For frequencies ω sufficiently below the band of 'normal' dynamics, the susceptibilities and the density correlators in the frequency domain are determined solely by the mode-coupling functional, provided the frequencies are measured in units of $1/t_0$.

It will be shown in Chapter 6 that the above formulated extrapolations of some numerical results are correct implications of the MCT equations of motion. The conclusion formulates asymptotic-limit results for the long-time and low-frequency correlators for states P approaching the bifurcation points P^c. Anticipating this insight, two further conclusions are obtained. First, according to the results of Sec. 4.2.2, the correlators $\phi_q(t)$ for all MCT models, which are based on a stochastic short-time dynamics, are completely monotonic functions. They are general relaxation processes in the sense of Eq. (1.2a). Consequently, outside the transient, all MCT solution are superpositions of elementary relaxation processes.

Second, according to Sec. 4.2.3, the mode-coupling functional for microscopic models is determined by canonically-defined equilibrium-structure functions. Consequently, up to a time scale t_0, the complex bifurcation dynamics outside the transient regime is determined by the Boltzmann factor $\exp[-(U/k_{\mathrm{B}}T)]$, where U is the total interaction potential of the many-particle system. In general, correlation functions $\phi(t)$ provide statistical information on the orbits of the system in phase space. The MCT solutions for $t/t_0 \gg 1$, provide a model for the statistics of orbits in configuration space. The statistics is determined by the potential landscape, i.e., by the function U of the $3N$ position coordinates of the system. The temperature T defines the weights of the configurations. Via the time t_0, the details of the transient motion merely specify the scale for the exploration of the configuration space (Götze and Sjögren 1992). The slow bifurcation dynamics deals with structure relaxation.

4.4.4 *Descriptions of some glassy-dynamics data*

This section presents some quantitative comparisons of solutions of the MCT equations of motion with data for the glassy dynamics of liquids.

Figure 4.25 shows density-fluctuation correlators ϕ_q determined for six packing fractions φ for nearly-monodisperse colloidal suspensions of nearly-hard spheres. The data are represented as function of the time measured in units of some density dependent scale τ_α. The latter is adjusted to get the curves coinciding for $\phi_q(t) < 0.2$. The original measurements for panels (b) and (c) are shown in Fig. 1.15. The wave number $q = 3.42/R$ is near the structure-factor-peak position q_p, while the results in panels (a) and (c) refer to q values on the wings of the peak of S_q. The lines show the functions $f(q)G(t/\tau_\alpha)$ with $f(q)$ adjusted to the data. The quantities $G_q(\tilde{t}) = \tilde{\phi}_q(\tilde{t})/f_q^c$ are the shape functions of the scaling law (4.112a) normalized by the critical Debye–Waller factor as calculated by Fuchs *et al.* (1992) for a model for the HSS. The figure demonstrates that the evolution of the superposition principle of the hard-sphere colloids is consistent with that discussed in connection with Fig. 4.21. The data confirm the calculated shape functions. The measurements exhibit the increase of the stretching upon increasing q above q_p and upon decreasing q below q_p in accord with the results shown in Fig. 4.20(a). Analogous figures for $qR = 2.68$ and $qR = 3.77$ corroborate this conclusion (van Megen and Underwood 1994). The fit of $f(q)$ is a means to determine the critical arrested parts f_q^c of the glass from data for the dynamics of the liquid. The discrepancies between $f(q)$ and f_q^c are of the size exhibited in Fig. 1.16 for the $\epsilon = 0$ results (van Megen and Underwood 1993).

The crosses in Fig. 4.26 show density correlators measured for a wave number near the peak position of the structure factor for three large packing fractions φ for a nearly monodisperse hard sphere system. Similar to what is exhibited in the

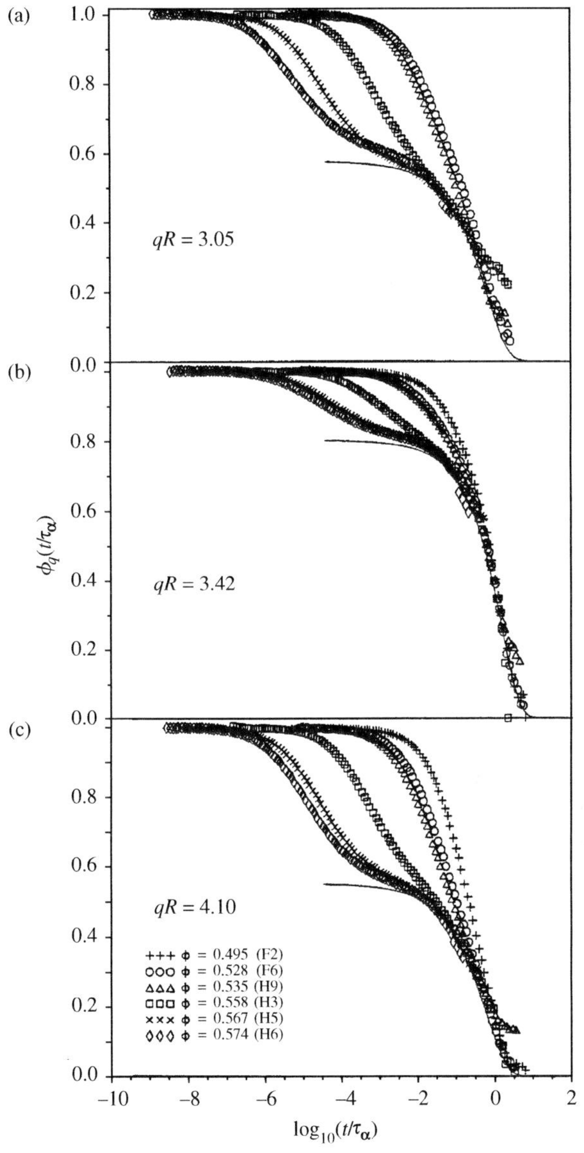

FIG. 4.25. Density-fluctuation correlators $\phi_q(\tilde{t})$, $\tilde{t} = t/\tau_\alpha$, for wave numbers q measured by photon-correlation spectroscopy for hard-sphere colloids of particles with radius R for six packing fractions φ. The time t is rescaled by the time τ_α near inflection point of the curves so that they collapse for large t/τ_α. The lines are MCT shape functions for the superposition-principle description of the below-plateau decay for a hard-sphere model calculated by Fuchs *et al.* (1992); they are explained in the text. Reproduced from van Megen and Underwood (1993).

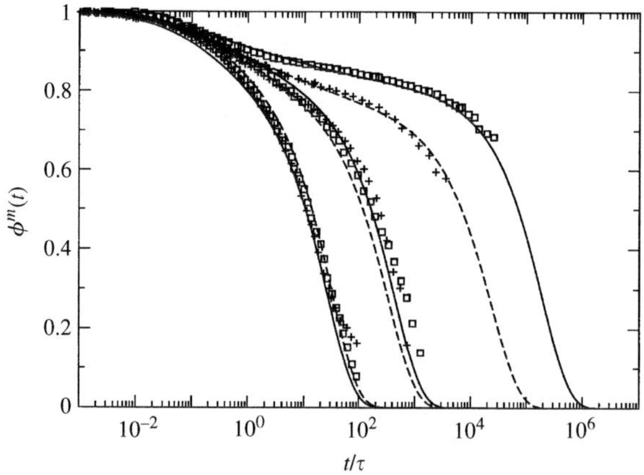

FIG. 4.26. The symbols are normalized photon-correlation spectroscopy results $\phi^m(t)$ for hard-sphere colloids for a scattering wave-number q near the structure-factor-peak position measured by Henderson *et al.* (1996). The three sets of crosses refer to nearly monodisperse systems of spheres of diameter d with packing fractions $\varphi^{\rm exp} = 0.535, 0.558$ and 0.567 (from left to right). The three sets of squares refer to binary mixtures of nearly mono-disperse hard spheres of diameters d and $0.8d$. The packing fractions are $\varphi^{\rm exp} = 0.536, 0.556$ and 0.566 (from left to right). The relative packing fraction of the small particles is $\hat{x}_B = 0.12$. The scale is the Brownian time $\tau = d^2/24D_0$, with D_0 denoting the diffusivity of an isolated particle of diameter d.

The lines are results of microscopic MCT equations for a model with $M = 200$ wave numbers, stochastic short-time dynamics, and mode-coupling coefficients based on the Percus–Yevick approximation for the structure factors. The dashed lines are evaluated for a simple hard-sphere system, $q = 6.6/d$ is chosen and $t_{\rm mic}$ in Eq. (4.109a) is fitted. The packing fractions are $\varphi = 0.485, 0.505, 0.5140$ (from left to right). The full lines are evaluated for a binary hard-sphere mixture. Equations (1.13), (5.100) are used to evaluate $\phi^m(t)$, $q = 7.4/d$ is chosen, and the parameter $t_{\rm mic}$ in Eqs. (5.99a,b) is adjusted. The packing fractions are $\varphi = 0.485, 0.505$ and 0.5145. Reproduced from Voigtmann (2003b). The dashed line for the highest density has been recalculated for a packing fraction, which is 0.1% smaller than that used in the cited publication (Voigtmann 2003a).

upper panel of Fig. 1.15, the elementary-relaxation law $\phi^m(t) = \exp[-(t/\tau')]$, $\tau' \approx \tau$, accounts for the data only for $t/\tau \leq 0.1$. Within the one-decade time interval $0.1 < t/\tau < 1$, 'normal' liquid dynamics is exhibited, which is affected by the interactions. The evolution of glassy dynamics is demonstrated for a four-orders-of-magnitude time increase beyond $t/\tau = 1$. The dashed lines are results of the microscopic MCT equation for a hard-sphere system with stochastic short-time dynamics. There is no obvious feature of the data, which is not described qualitatively on the preceding pages for the MCT bifurcation scenario. The figure documents the first quantitative comparison of results of the microscopic version of the mode-coupling theory with data for measured correlators, which deal with the complete time interval accessible in the experiment.

The MCT solutions for $\phi^m(t)$ are determined by the wave number q, by the time scale τ for the dynamics, and by the packing fraction φ. These parameters are reported by Henderson *et al.* (1996). But, the mode-coupling theory does not provide exact results for the correlators of the N-particle problem. In order to quantify the shortcomings of the theory and to describe data quantitatively, three φ-independent constants have been used as fit parameters. First, $qd = 6.6$ was chosen in order to calculate the dashed lines in Fig. 4.26. This value is slightly smaller than the position q_p of the main peak of the structure factor. One infers from Fig. 4.6 that an adjustment of q in this wave-number region implies adjustments for the critical Debye–Waller factor f_q^c. Thus, the adjustment of q is done in order to correct for the errors of the MCT prediction for the plateau value. The size of these errors can be inferred from Fig. 1.16. The transient dynamics determines the scale t_0 for the complex long-time dynamics as is formulated by Eq. (4.115a). No error in t_0 can be tolerated if the below-plateau relaxation is to be described quantitatively. But, MCT treats the transient dynamics in an oversimplified fashion. For example, as pointed out in Sec. 4.1.1, the short-time behaviour of the fluctuating-force kernel is not correct. In order to minimize the consequences of these defects of the theory, a common time scale t_{mic} is fitted for the three dashed lines.

The third fit parameter is the difference Δ between measured values for the packing fraction, φ^{exp}, and the value φ used to calculate the dashed lines in Fig. 4.26. Near the bifurcation singularity, the MCT solution for large times depends sensitively on the distance parameter $\epsilon = (\varphi - \varphi_c)/\varphi_c$. For example, if the value 0.5140 is changed to 0.5145, the solution for $\phi^m(t)$ exceeds the corresponding data points for the one-decade time interval $t/\tau \geq 400$ (Voigtmann 2003b). Here, as in any other application of a theory for a singularity, one must not compare theory and data for the same value of the control parameter φ, but for the same value of the distance parameter ϵ. This can be done by shifting φ relative to φ^{exp}. The shifts used for the three curves is 0.050, 0.053, and 0.053 (from left to right). According to Sec. 4.4.2, the time scale for the below-plateau decay varies proportional to $1/|\epsilon|^\gamma, \gamma \approx 2.5$. Changing the first noted shift from 0.050 to 0.053, implies a shift of the dashed line to the left by a value about the size of the data symbols. Thus, the shown fit quality can be obtained by fitting the difference Δ to the φ-independent value 0.053. As explained in Sec. 4.3.5, a major part of this 10% correction is required to reduce the errors, which result from those of the Percus–Yevick approximation for S_q.

In Secs. 4.4.1 and 4.4.2, the similarity between the bifurcation scenario for schematic models with that for the evolution of glassy dynamics in liquids has been indicated. This observation shall be corroborated by comparing quantitatively solutions for schematic models with glassy-dynamics data. A minimal request for the use of some schematic model is that it can reproduce the plateau and the stretching of the below-plateau-relaxation process. Therefore, one cannot expect to succeed with the one-component-model correlator $\phi(t)$, defined by Eqs. (4.32a,b). According to Eq. (4.67b), the transition point for this model is fixed by the parameter λ^c. As discussed in Sec. 4.4.2, pronounced stretching requires λ^c to exceed 1/2 sufficiently. But, this leads to a small plateau $f^c = 1 - \lambda^c$. The connection between stretching size and plateau value is an artifact of this model, and so is the upper bound 1/2 for f^c. The two-component model, defined by Eqs. (4.33a,b) for a second correlator $\phi_A(t)$, appears as the simplest model worth trying. The two mode-coupling coefficients v_1 and v_2 determine the glassy dynamics of the first correlator $\phi(t)$. This correlator serves as a caricature of the density-fluctuation dynamics. The parameter λ^c, which fixes the transition point in the control-parameter plane of Fig. 4.2, is mainly used to describe the stretched loss minima. The second correlator is used to describe the dynamics of the probing variable. The third mode-coupling coefficient v_A of the model is fixed mainly by the request to reproduce the plateau for the bifurcation scenario of the measured variable: $f_A^c = 1 - 1/(v_A f^c)$. If one uses the model as a caricature for a microscopic one, one has to request that the mode-coupling constants v_1, v_2, and v_A are smooth functions of the external control parameters, e.g., of the temperature T. The two-component model under discussion specifies the 'normal liquid dynamics' by two frequencies Ω and Ω_A and two friction frequencies ν and ν_A. This is an oversimplification. In the best, it can serve to characterize the low-frequency end of the vibrational spectrum of a molecular liquid. Naturally, the parameters will also drift smoothly with temperature. According to Sec. 4.4.3, these model parameters enter the description of the structural relaxation indirectly via an overall time scale t_0.

The evolution of the optical-Kerr-effect response function $\chi^{\exp}(t)$ of salol upon decreasing the temperature from 340 K to 247 K is shown in Fig. 4.27. The result for $T = 257$ K is discussed in Fig. 1.2 as an example for glassy dynamics stretched over a five-orders-of-magnitude time interval. The dotted lines show the functions $N\chi_A(t)$, where $\chi_A(t) = -\partial_t \phi_A(t)$ is the normalized response function for the probing variable of the two-component model under discussion. For $T = 340$ K, the calculated curve agrees with the measured one for times exceeding 0.5 ps. Lowering the temperature, there appear oscillations in the data for times around 1 ps, which are not described by the calculations. For the lowest measured temperatures, the dotted lines are close to the data for times exceeding 2 ps.

The mentioned two-component model is applied with temperature independent frequencies for the short-time dynamics. The values $\Omega = 2\Omega_A = 7.9$ THz are of the order of magnitude expected for molecular vibrations in van der Waals

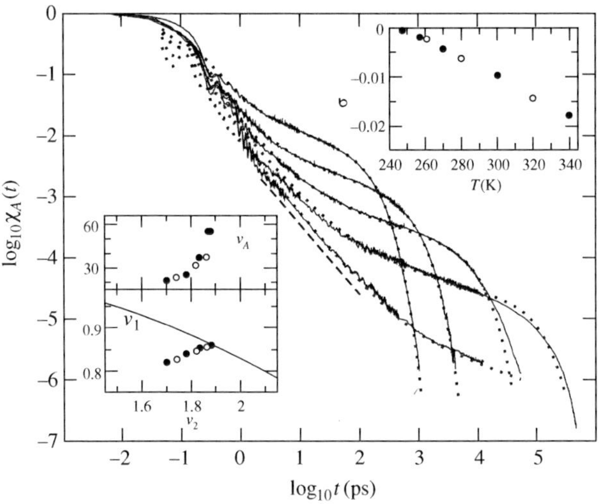

FIG. 4.27. Optical-Kerr-effect response functions of salol measured for temperatures $T/\mathrm{K} = 247, 257, 270, 300$ and 340 (from bottom to top) and normalized to unity for $t = 0.01$ ps (Hinze *et al.* 2000). The dashed line in the double logarithmic presentation has slope -1.15 and represents power-law decay of the correlator $t^{-a'}$, $a' = 0.15$. The dotted lines show $N\chi_A(t)$, with $\chi_A(t) = -\partial_t\phi_A(t)$ denoting the response function for the normalized probing-variable correlator of the two component model defined by Eqs. (4.32a,b), (4.33a,b). The constants N are 25, 37.3, 27.8, 17.9 and 19 (from bottom to top). The parameters are $\Omega = 2\Omega_A = 10\nu_A = 7.9$ THz, $\nu = 0$. The mode-coupling coefficients are shown by the filled circles in the lower inset. The upper inset exhibits the separation parameter σ, Eq. (4.86a). The open circles in the insets show the parameters used in the fits for the data measured for other temperatures; these curves are not exhibited in order to avoid overcrowding. Reproduced from Götze and Sperl (2004b).

systems. The data-normalization at 0.01 ps determines the value of $\chi^{\exp}(t)$ at the beginning of the structural-relaxation, say, for $t = 2$ ps. This value is influenced by the interplay of molecular vibrations and intermolecular-interaction effects. These features cannot be described by the oscillator model for the short-time dynamics. Therefore, one cannot use a T-independent normalization constant N. The fit values for N vary irregularly by up to a factor 2. The major part of the change of $\chi^{\exp}(t = 2 \text{ ps})$ upon decreasing T is due to mode-coupling effects on the transient dynamics. This result is the analogue to the description of the changes shown in Fig. 4.26 for $\phi^m(t/\tau = 1)$. With reservation concerning the factor N, the measured evolution of glassy dynamics is described by the temperature dependence of the three mode-coupling coefficients. The insets show that the coefficients v_1 and v_2 vary linearly with T. They change by about 12% due to the nearly 100 K change of temperature considered. The straight line in $v_1 - v_2$ plane extrapolates to a transition point specified by $\lambda^c = 0.73$. The interpolation of the separation parameter by $\sigma = C(T_c - T)/T_c$ suggests $T_c = 245\,\mathrm{K}$ as temperature for the arrest. It is essential for the data description that the coupling constant v_A for the probing variable to the density fluctuations increases by about a factor 3 if the temperature is lowered. This increase is non linear. It roughly follows an Arrhenius law, $v_A \propto \exp(T_0/T)$. The value $T_0 \approx 1000\,\mathrm{K}$ is not unreasonable for the change of a structure variable for a van der Waals liquid. It does not seem possible to identify a feature of glassy dynamics for the experimental results in Fig. 4.27, which are not reproduced by the solution of the schematic model.

It was discussed in connection with Fig. 1.2 that $\chi^{\exp}(t)$ for $T = 257$ exhibits a von Schweidler-law decay for a time interval of 2.5 decades. One concludes that this power-law decay, Eq. (1.4a), is reproduced by the schematic model as initial part of the below-plateau decay of the probing-variable correlator $\phi_A(t)$. A corresponding power-law wing of the low-frequency-loss peak, Eq. (1.4b), is indicated in the lower panel of Fig. 4.13 by a dashed-dotted straight line denoted ω^{-b}.

Depolarized light-scattering in backward direction yields loss spectra $\chi''_A(\omega)$ for the same probing variable A, whose response function $\chi_A(t)$ is measured by optical-Kerr-effect spectroscopy. Figure 4.28 exhibits a set of such spectra, which shows the evolution of glassy-dynamics upon decreasing the temperature T for orthoterphenyl (OTP). The spectra for $290\,\mathrm{K} \leq T \leq 320\,\mathrm{K}$ exhibit a stretched loss minimum with a high-frequency part, which is strongly enhanced above any reasonable estimation of some white-noise-induced back-ground spectrum. This is the same feature as exhibited by the spectra in Fig. 1.3 for the molten salt CKN. Opposed to the salt spectra, the OTP spectra exhibit a low-frequency loss peak, which is much stronger than the spectra of the microscopic-excitation band. This is typical for all van der Waals liquids. Presumably, this is a consequence of the scattering mechanism, which is due to coupling of light to the reorientational degrees of freedom (Berne and Pecora 1976). These, in turn, are strongly coupled to the translational degrees of freedom, which are essential for the evolution of glassy dynamics. The low-frequency loss peak is stretched less than the one discussed in Figs. 1.3 and 1.6 for CKN. It exhibits the evolution of the superposition principle in qualitative agreement with the one discussed in Fig. 4.22; and the shape function is close to one for a Kohlrausch law with a stretching exponent $\beta = 0.79$ (Cummins *et al.* 1997).

The full lines in Fig. 4.28 exhibit $N\chi''_A(\omega)$ with $\chi''_A(\omega)$ denoting loss spectra calculated for the specified two-component schematic model. The four frequencies for the description of the transient dynamics, the normalization constant N, and the mode-coupling parameter v_1 are chosen as temperature independent. The path for the states, which is shown in the right panel, intersects the transition line at $v_1 = 0.87$; according to Eq. (4.67b), this corresponds to $\lambda^c = 0.735$. The temperature-dependence of v_2 is close to a linear one. It extrapolates to a transition temperature $T_c = 280\,\mathrm{K}$, which is 10 K below the value deduced

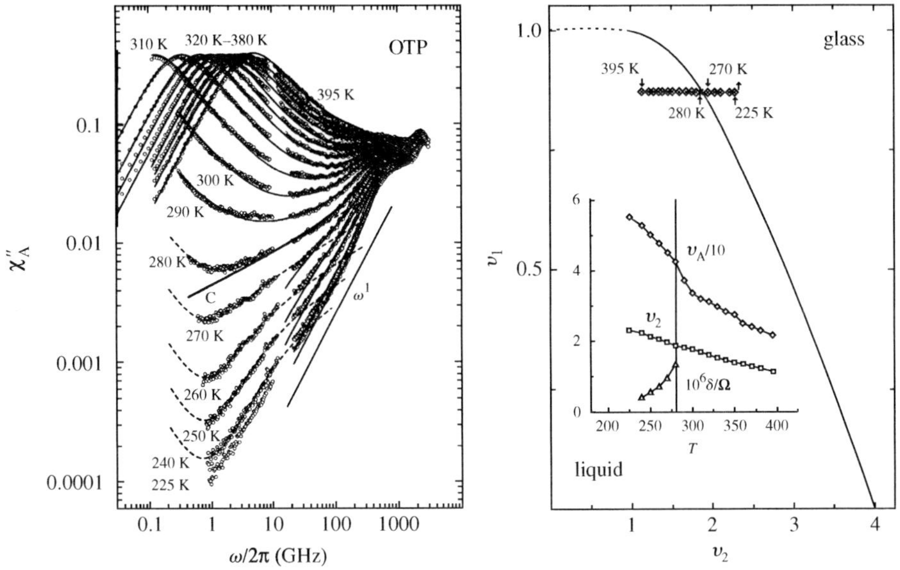

FIG. 4.28. The data in the left panel show loss spectra $\chi''_A(\omega)$ of orthoterphenyl for various temperatures T measured by depolarized light scattering spectroscopy (Cummins *et al.* 1997). The line denoted ω^1 has slope unity and indicates the behaviour of a white-noise-induced loss spectrum $\chi''_A(\omega) \propto \omega$. The full lines exhibit $N\chi''_A(\omega)$ with $\chi''_A(\omega)$ denoting loss spectra for the probing variable calculated for the two-component model defined by Eqs. (4.32a,b), (4.33a,b). The normalization constant N is kept T-independent and the temperature-independent frequencies for the transient dynamics are: $\Omega/2\pi = 1.81$ THz, $\nu = 0, \Omega_A/2\pi = 0.14$ THz, $\nu_A/2\pi = 0.44$ THz. The mode-coupling coefficient $v_1 = 0.87$ is chosen T-independent and the choice of v_2 and v_A is shown in the right panel. The heavy full line with label c exhibits the spectrum for the critical temperature $T_c = 280$ K.

The dashed lines are results of an extended version of the MCT, which is specified by a parameter δ; the model (Götze and Sjögren 1987b) is not discussed in this book. Reproduced from Singh *et al.* (1998).

from the extensive neutron-scattering studies made for this system (Tölle 2001). The inset of the right panel shows that the third mode-coupling coefficient v_A is of a similar size as that used in Fig. 4.27 for salol; it increases also with decreasing T. The schematic MCT model describes the evolution of the glassy-dynamics spectra for OTP for an frequency decrease by a factor 1000 below the band of normal-liquid excitations. This evolution is caused by a decrease of the temperature from a value exceeding the melting temperature T_m by 66 K to temperatures 39 K below T_m.

The spectra of OTP exhibit an appreciable temperature dependence for $\omega/2\pi = 0.3$ THz. This phenomenon is reproduced by the model as result of the interplay of the mode-coupling effects and the transient dynamics. The data exhibit a similar temperature dependence for the frequency band 0.3 THz $< \omega <$3 THz: but the magnitude of the effect decreases with increasing ω. The schematic model is too primitive to describe these dependencies.

The line with label c represents the spectrum for the state on the transition line. The low-frequency part exhibits a power-law behaviour, $\chi''_A(\omega) \propto \omega^a, a \approx 0.31$. The data confirm this result, albeit only for a one-decade interval, 4 GHz $\leq \omega/2\pi \leq$ 40 GHz. For frequencies below 3 GHz, there is a qualitative discrepancy between the data and the critical spectrum. There is a low-frequency loss peak due to a below-plateau decay, while MCT predicts arrest. The spectra measured for 0.8 GHz $< \omega <$ 30 GHz and $T < T_c$ are very small; but, they are considerably larger than spectra calculated for the MCT model. The data for temperatures near and below the critical value T_c demonstrate low-frequency relaxation processes, which are not described by the theory discussed in this book.

If the above formulated interpretation of the two-component model could be taken at its face value, one could argue as follows. Let us describe the density-fluctuation dynamics of a given liquid by the first correlator $\phi(t)$. Upon changing the temperature T, the state moves on a path $(v_1(T), v_2(T))$ in the control-parameter plane. Let us consider also a set of probing variables A_k, $k = 1, 2, \ldots$, and characterize their dynamics by the normalized correlators $\phi_k(t) = (A_k(t)|A_k)/(A_k|A_k)$. The transient dynamics is specified by pairs of frequencies Ω_k, ν_k and by mode-coupling coefficients v_A^k. The latter quantify the coupling of the probing variables to the density fluctuations. The set of correlators for a given temperature T is given as solution of Eqs. (4.33a,b) using the same $\phi(t)$ for all k. Such common analysis of a set of probing variables was suggested and tested with encouraging results for supercooled $Na_{0.5}Li_{0.5}PO_3$ by Rufflé *et al.* (1999). The authors discussed coherent neutron-scattering cross sections for a representative set of wave numbers and, in addition, the longitudinal elastic modulus.

The glassy dynamics of the van der Waals liquid propylene carbonate has been studied by several techniques and a common interpretation of some of the data according to the above described suggestion has been made (Götze and Voigtmann 2000). The smooth parameter path for all states analysed crosses the transition line at a point specified by $\lambda^c = 0.75$. The results shown in Fig. 4.12 and in the upper panel of Fig. 4.13 for states with label $n \geq 4$ are close to the ones used in this study. The separation parameter σ exhibits a linear temperature dependence and yields $T_c = 180\,\mathrm{K}$ as critical temperature for the arrest. Incoherent neutron-scattering spectra for 10 wave numbers between $5\,\mathrm{nm}^{-1}$ and $14\,\mathrm{nm}^{-1}$ as measured by Wuttke *et al.* (2000) are described perfectly by the model. Figure 4.29 reproduces the results for three examples. The coupling coefficients v_q are of the order 20 and increase somewhat with decreasing T. The measured spectra are the ones for tagged-particle density fluctuations. Depolarized

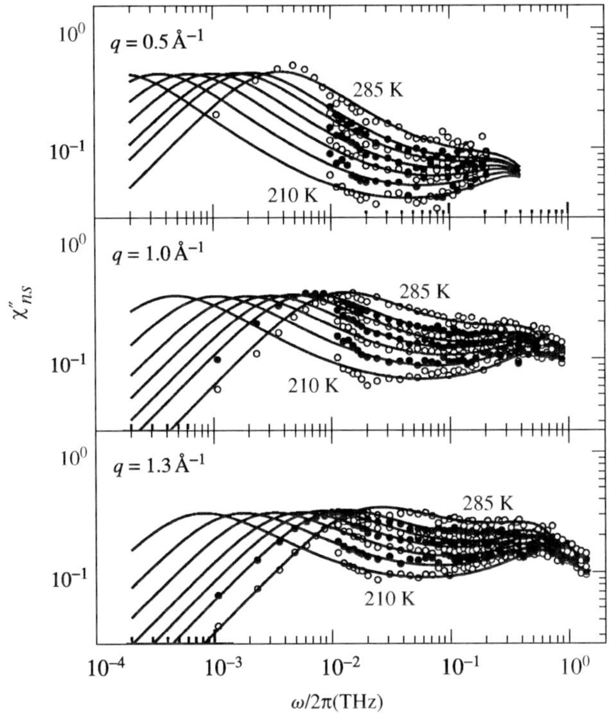

FIG. 4.29. Susceptibility spectra $\chi''_{ns}(\omega)$ for propylene carbonate ($T_{\mathrm{m}} = 218\,\mathrm{K}$, $T_g = 160$ K) measured by Wuttke *et al.* (2000) by incoherent neutron-scattering spectroscopy for temperatures $T/\mathrm{K} = 210, 220, 230, 240, 251, 260$, and 285 (from bottom to top; open and filled symbols are used to help discriminating different data sets). Up to a normalisation constant, the full lines exhibit $\chi''_A(\omega) = \omega\phi''_A(\omega)$ with ϕ_A denoting the second correlator of the schematic model defined by Eqs. (4.32a,b), (4.33a,b). The parameters for the transient dynamics for the first correlator are $\Omega = 1$ THz, $\nu = 0$. Diagrams for the parameters Ω_A, ν_A and v_A can be found in the paper by Götze and Voigtmann (2000), where the figure is reproduced from.

light-scattering spectra (Du *et al.* 1994) have been described by the model for the measured interval 0.3 GHz $< \omega/2\pi <$ 0.5 THz. The coupling constant v_A for this probing variable increased up to 50 for T extrapolating to T_{c}. The lower panel of Fig. 4.13 exhibits an example similar to the one studied. Furthermore, dielectric loss spectra (Schneider *et al.* 1999) for $200\,\mathrm{K} \le T \le 293\,\mathrm{K}$ have been described for the five-orders of magnitude band for frequencies decreasing from 1 THz to below 10 MHz. For $153\,\mathrm{K} \le T \le 193\,\mathrm{K}$, the model describes the data only for frequencies exceeding 10 GHz. The dielectric relaxation experiments exhibit a low-frequency loss peak also for $T \le T_{\mathrm{c}}$, which does not exist for the MCT results for the ideal glass states.

In Fig. 4.15, loss-peak spectra $\chi_1''(\omega)$ and $\chi_2''(\omega)$ are compared for density fluctuations with wave numbers q_1 and q_2 for the HSS. It is demonstrated there that the former peak has a larger intensity, a lower peak position and a smaller stretching than the latter. These properties characterize the shape function $\tilde{\phi}_q(\tilde{t})$ in Eq. (4.112a). They reflect properties of the mode-coupling functional. The properties are quantified by the plateau f_q^c, Fig. 4.6(b), the relaxation time τ_q and the stretching parameter β_q shown in Fig. 4.20. The loss peaks of propylene carbonate for the dielectric response and the one for the depolarized-light-scattering one exhibit a similar relation; the former response is stronger, slower and less stretched than the latter. Within the cited schematic-model data analysis, the correct intensities of the peaks are fitted by adjusting the mode-coupling coefficient v_A. For example, for $T = 220\,\mathrm{K}$, the dielectric loss spectrum is fitted with $v_A \approx 50$ and the light-scattering susceptibility with $v_A \approx 22$. It appears non-trivial that thereby the measured differences in the relaxation times and in the stretching are also reproduced quantitatively (Götze and Voigtmann 2000).

Alba-Simionesco and Krauzman (1995) were the first to interpret glassy-dynamics data of liquids by schematic-model results. They analysed depolarized light-scattering spectra and motivated their model by the assumption that dipole-induced-dipole interaction is the dominant scattering mechanism. In this case, a light-wave-induced dipole moment of a molecule induces a non-rotational-invariant dielectric polarizability of the neighboring molecules; and this causes the scattering with change of the light-wave polarization (Berne and Pecora

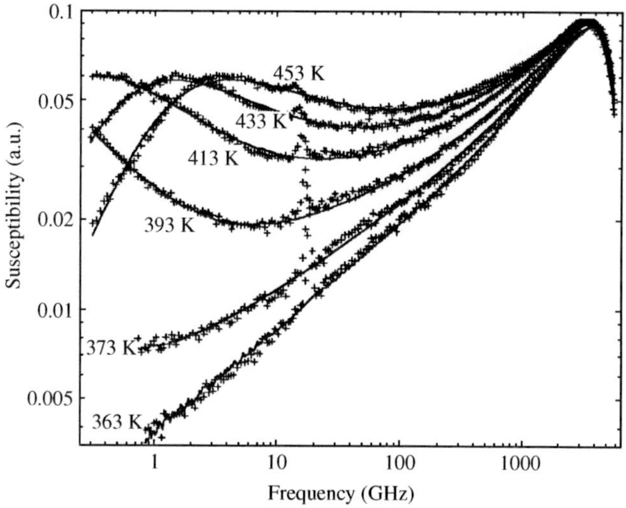

FIG. 4.30. Susceptibility spectra of the molten salt CKN measured by depolarized-light-scattering spectroscopy by Li *et al.* (1992). The full lines are descriptions of the data by results of a schematic model explained in connection with Eq. (4.116). Reproduced from Krakoviack *et al.* (1997).

1976). The measured variable is a sum of pair-fluctuations $A = \sum_{\vec{k}+\vec{p}=\vec{q}} C_{\vec{q},\vec{k}\vec{p}}\, \rho_{\vec{k}} \rho_{\vec{p}}$. The corresponding correlator is of similar form as the one for the kernel $M_{Lq}(t)$ in Eq. (4.1b). Simplifying the correlator of the pairs by a factorization ansatz as discussed in Sec. 4.1.1, one gets the correlator as superposition of products of density correlators in analogy to Eq. (4.7b). Since the wave number q for the light quanta is small compared to the natural scale $2\pi/d$ of the liquid, one can specialize to the limit $q = 0$. Hence, the expression describing the measurements is given by $\phi_A(t) = \sum_{\vec{k}} C_k \phi_k(t)^2$, where the C_k denotes some coupling coefficient. Schematically, this formula is reduced by considering only two terms for the sum:

$$\phi_A(t) = N[\phi(t)^2 + \gamma\phi_1(t)^2]/(1+\gamma). \tag{4.116}$$

The correlator $\phi(t)$ is taken as solution of the one-component model defined by Eqs. (4.32a,b). The kernel $m_1(t)$ for the other correlator $\phi_1(t)$ is written as $rm(t) = (rv_1)\phi(t) + (rv_2)\phi(t)^2$. Figure 4.30 demonstrates that spectra for the molten salt CKN can be described perfectly by the above described model. The four upper curves have been discussed also in Fig. 1.3. The parameters N, $r = 10$, $\gamma = 2.4$ as well as the four frequencies for the transient dynamics of the two correlators are chosen independent of the temperature. The parameter γ is mainly used to adjust the intensity of the low-frequency loss peak. The complete evolution of the shown spectra is caused by a regular drift of the two coupling constants (v_1, v_2) towards the transition line in the parameter plane similar to what is discussed above for the other schematic-model studies.

5

EXTENSIONS OF THE MODE-COUPLING THEORY FOR THE EVOLUTION OF GLASSY DYNAMICS OF LIQUIDS

The hitherto discussed microscopic versions of the mode-coupling theory for the glassy dynamics are restricted to considerations about density-fluctuation correlators of simple systems. It is the goal of the present chapter to broaden the scope of this approach. As before, N particles enclosed in a cubic box of side length L shall be considered and also the conventional assumptions on thermodynamic limits and symmetries will be made. But, correlation functions formed with variables different from density fluctuation shall be calculated, and the theory will be extended to non-simple systems. In Sec. 5.1, shear correlations and tagged-particle motion will be studied for simple systems. Then, the MCT equations will be derived for mixtures of spherical particles (Sec. 5.2). In Sec. 5.3, a theory for molecular systems will be formulated by reducing the equations of motion for the composed particles to ones for the mixture of the spherical constituents of the molecules. The chapter will be closed by specifying some features of the complex dynamics of glass-forming liquids, which are not considered in this book.

The intended extensions of the MCT shall be derived in close analogy to what is explained in Sec. 4.1 for the density correlators. Therefore, the various motivations for the simplifications of the general equations of motion shall not be repeated. Rather, the discussion shall focus on technical details and on considerations of typical results. The derivations proceed in three steps. In the first one, expressions for the correlators in terms of fluctuating-force kernels are derived within the frame work of the Zwanzig–Mori theory. The kernels are correlators for a reduced dynamics, say $M(t) = (X|\mathcal{U}_{\mathrm{red}}(-t)Y)$. The essential request is that the dynamical variables X and Y have some overlaps with density-fluctuation pairs A_m. These pair variables describe the slowly fluctuating forces, which cause the cage effect. As the second step, the kernels $M(t)$ are expressed in terms of the correlators of density-fluctuation pairs $\phi_{m,n}(t) = (A_m|\mathcal{U}_{\mathrm{red}}(-t)A_n)$. The relevant formula (2.64) is simplified by a white-noise approximation for the first term on the right-hand side and by neglecting the kernels from Eq. (2.63c). The remaining task is the evaluation of the overlaps $(F|A_k)$ in order to calculate the amplitudes Ω_m^X in Eq. (2.63a). The crucial third step of the procedure is the factorization ansatz as is explained in connection with Eqs. (4.7a,b). As a result, the non-trivial part of the kernels $M(t)$ are obtained as mode-coupling functionals of the density-fluctuation correlators.

5.1 Extensions of the MCT for simple systems

The shear-response function and the mean-squared displacement are classical quantities for a statistical description of liquid dynamics. In this section, the glassy-dynamics features of these functions shall be discussed for simple systems.

5.1.1 *MCT equations for the glassy shear dynamics*

The most obvious distinction between the macroscopic behaviour of solid matter and of liquid matter is that the former can sustain a static shear stress while the latter cannot. Therefore, the understanding of the low-frequency shear dynamics is an important part of the understanding of the liquid-glass transition. In this section, the shear correlator $K_T(z)$ shall be expressed in terms of a mode-coupling functional of the density-fluctuation correlators. The expression to be derived is the basis for describing glassy-dynamics features of the shear as corollary of those of the density fluctuations.

The shear correlations determine the long-wavelength correlations of the transverse currents. The relevant dynamical variable is the transverse-current fluctuation, say, $A = j^1_{\vec{q}*}/[v_{th}\sqrt{N}]$. Here, $\vec{q}^* = (0, 0, q)$ is a reference wave vector of modulus q. The correlation function $(A(t)|A)$ agrees with the transverse-current correlator $\phi_{Tq}(t)$ from Eqs. (3.64a,c,e). Its general properties are discussed in Secs. 3.3.1–3.3.3. The result of the first step of the procedure outlined above is formulated in Eqs. (3.75a,b). The latter provide the desired expression of the correlator in terms of a fluctuating-force correlator. Introducing the abbreviation $M_{Tq}(z) = q^2 K_{Tq}(z)$, one can write

$$\phi_{Tq}(z) = -1/[z + M_{Tq}(z)], \tag{5.1a}$$

$$M_{Tq}(z) = (F_{Tq}|\mathcal{U}_{\text{red}}(-t)F_{Tq}), \tag{5.1b}$$

$$F_{Tq} = \mathcal{L}j^1_{\vec{q}*}/[v_{th}\sqrt{N}]. \tag{5.1c}$$

The reduced time-evolution operator $\mathcal{U}_{\text{red}}(t) = \exp[i\mathcal{L}_Q t]$ is generated by the reduced Liouville operator $\mathcal{L}_Q = \mathcal{QLQ}$ with $\mathcal{Q}$ denoting the projector perpendicular to A. The projector $\mathcal{Q}$ is simpler than the one entering Eq. (4.1b). In particular $\mathcal{QL}A = \mathcal{L}A$, since A has a definite time-inversion parity. The analogue of projector $\mathcal{Q}_1$ in Eq. (4.1c) is missing in Eq. (5.1c).

The limit function

$$K_T(t) = \lim_{q\to 0}[M_{Tq}(t)/q^2] \tag{5.2}$$

and its Laplace transform $K_T(z)$ are the quantities of main interest. According to the generalized Green–Kubo formula (3.76b), it is the correlator of the shear-stress variable V_T. Because of Eqs. (3.79), (3.80), $K_T(z)$ determines the generalized-hydrodynamics form of the transverse-current correlator.

The second step of the approach starts with expressing kernel $M_{Tq}(z)$ in terms of the density-fluctuation-pair correlators $\phi_{m,n}(t) = \phi_{\vec{m}_1\vec{m}_2,\vec{n}_1\vec{n}_2}(t)$. The result is given by Eqs. (4.3a–c) with label L replaced by label T. The white-noise approximation for the regular kernel $M^{\text{reg}}_{Tq}(z) = q^2 K^{\text{reg}}_{Tq}(z)$ and the neglection of the amplitude corrections $\delta\Omega^T_{q,m}(z)$ results in $M_{Tq}(z) = \pm i\nu^T_q q^2 + \sum_{m,n} \Omega^T_{q,m} \phi_{m,n}(z)\Omega^{T*}_{q,n}$, Im $z \gtrless 0$. The evaluation of the overlaps $(F_{Tq}|\rho_{\vec{k}}\rho_{\vec{p}})$ in Eq. (4.3b) follows strictly the one given for $F^{(a)}_{q,\vec{q}\vec{p}}$ in the paragraph following Eq. (4.5a). One merely has to replace the coordinate index 3 by the index 1. The result is the analogue of Eq. (4.5b):

$$(F_{Tq}|\rho_{\vec{k}}\rho_{\vec{p}}) = \delta_{\vec{q}*,\vec{k}+\vec{p}}\big[k^1 S_p + p^1 S_k\big]. \tag{5.3}$$

Here, as opposed to the results of Sec. 4.1.1, triple correlations do not occur in the coupling amplitudes.

The third step yields a result, which shall be expressed in terms of the Laplace transform $m^T_q(z)$ of a dimensionless function $m^T_q(t)$:

$$M_{Tq}(z) = \pm i\nu^T_q q^2 + \Omega^2_q m^T_q(z), \quad \text{Im } z \gtrless 0. \tag{5.4a}$$

The characteristic frequency Ω_q, determines the initial decay of the density-fluctuation correlations according to Eqs. (3.67), (3.68). The kernel $m^T_q(t)$ is the analogue of kernel $m_q(t)$ in Eq. (4.11b). The factorization ansatz yields a quadratic functional of the density correlators:

$$m^T_q(t) = L^{-3} \sum_{\vec{k}+\vec{p}=\vec{q}*} V^T(\vec{q}^*, \vec{k}\vec{p})\phi_k(t)\phi_p(t). \tag{5.4b}$$

The mode-coupling coefficients read

$$V^T(\vec{q}^*, \vec{k}\vec{p}) = \tfrac{1}{2}\rho S_q S_p S_k \Big\{\big[k^1 c_k + p^1 c_p\big]/q\Big\}^2 \tag{5.4c}$$

with $\vec{p}$ abbreviating $\vec{q}^* - \vec{k}$. The coefficients $V^T(\vec{q}^*, \vec{k}\vec{p})$ are determined by the structure factor S_q and by the direct correlation function c_q. Equations (5.1a), (5.4a–c) (Munakata and Igarashi 1978) determine the transverse-current correlators for all wave numbers q in terms of the density-fluctuation correlators.

Equations (5.4b,c) can be rewritten as double integral in analogy to Eq. (4.11d): $m^T_q(t) = \int_0^\infty dk \int' dp \cdots \phi_k(t)\phi_p(t)$. The integrals can be expressed as Riemann sums over a grid of M wave-vector values of spacing (h/d). This implies a large-q cutoff $q_{co} = Mh/d$, as explained in connection with Eqs. (4.35), (4.36). As a result, the array $\boldsymbol{m}^T(t)$ of M kernels $m^T_q(t), q = 1, \ldots, M$, is given by quadratic mode-coupling polynomials of the correlator arrays $\boldsymbol{\phi}(t) = (\phi_1(t), \ldots, \phi_M(t))$:

$$m^T_q(t) = \mathcal{F}^T_q[\boldsymbol{\phi}(t)], \qquad \mathcal{F}^T_q[\boldsymbol{x}] = \sum_{kp} V^T_{q,kp} x_k x_p. \tag{5.5}$$

The non-negative coefficients $V^T_{q,kp}$ are determined by the equilibrium structure; hence, they are smooth functions of external control parameters like the density.

One gets $\lim_{q\to 0} V^T(\vec{q}^*, \vec{k}\,(\vec{q}^* - \vec{k})) = \rho S_0 S_k^2[k^1 k^3 c'_k/k]^2/2$ with $c'_k = dc_k/dk$ and $S_0 = \lim_{q\to 0} S_q$. Introducing the positive coupling coefficients

$$v_T(q) = [\rho/(60\pi^2)]\left[q^2 c'_q S_q\right]^2, \tag{5.6a}$$

one can write the integral representation for the zero-wave-vector kernel $m_0^T(t) = \lim_{q\to 0} m_q^T(t)$:

$$m_0^T(t) = S_0 \int_0^\infty dq v_T(q)\phi_q(t)^2. \tag{5.6b}$$

This integral shall be considered as equivalent to a sum over the grid of discrete wave numbers. One arrives at the mode-coupling expressions:

$$m_0^T(t) = \mathcal{F}_0^T[\boldsymbol{\phi}(t)], \qquad \mathcal{F}_0^T[\boldsymbol{x}] = (h/d)\sum_q v_T(q)x_q^2. \tag{5.6c}$$

From Eq. (5.4a), one obtains for the shear correlator in the frequency domain:

$$K_T(z) = \pm i\nu^\infty + c_{is}^2 m_0^T(z), \quad \text{Im } z \gtrless 0. \tag{5.7}$$

Here, Eq. (3.70a) was used to express the small-q limit of the dispersion law in terms of the isothermal sound speed c_{is} : $\lim_{q\to 0}[\Omega_q^2/q^2] = c_{is}^2 = v_{th}^2/S_0 = k_B T/(\mu S_0)$. Furthermore, the abbreviation is used: $\lim_{q\to 0}\nu_q^T = \nu^\infty$. This notation is motivated by Eq. (3.85a). The white-noise term ν^∞ is that part of the frequency dependent kinematic viscosity, which remains for large frequencies.

Suppose, the density-fluctuation-correlator array $\boldsymbol{\phi}(t)$ has been calculated. According to Appendix A.6, Eqs. (5.5), (5.6c) determine continuous positive-definite functions $m_q^T(t)$ and $m_0^T(t)$, respectively. They have the symmetries of the density correlators. If the second version of the equations of motion are used, the two kernels are completely monotonic. Equations (5.4a) and (5.7) define positive analytic functions exhibiting the behaviour $|M_{Tq}(z) \mp i\nu_q^T q^2| < C_q/|\text{Im } z|$ and $|K_T(z) \mp i\nu^\infty| < C/|\text{Im } z|$, Im $z \gtrless 0$, respectively. From the discussion of Sec. 2.4.2, one concludes that Eqs. (5.1a) and (5.2) define correlation functions with the correct general properties of the transverse-current correlators and of its generalized-hydrodynamics forms, respectively. The cited equations formulate the mode-coupling theory for the shear dynamics.

Let us rewrite the MCT formulas for the shear correlators in terms of the shear modulus $G_T(z)$, which is defined in Eq. (3.84c). The traditional notation for this quantity reads

$$G_T(z) = G'_T(\omega) - iG''_T(\omega), \quad z = \omega + i0. \tag{5.8a}$$

The real part $G'_T(\omega)$ and the negative imaginary part $G''_T(\omega)$ of $G_T(\omega + i0)$ are referred to as the storage modulus and loss modulus, respectively. The former is an even function of ω and the latter is an odd one. There holds:

$$G'_T(\omega) = -\left[(\rho k_B T)/S_0\right]\omega m_0^{T\prime}(\omega), \tag{5.8b}$$

$$G''_T(\omega) = \omega\eta^\infty + \left[(\rho k_B T)/S_0\right]\omega m_0^{T\prime\prime}(\omega). \tag{5.8c}$$

The factor $(\rho k_{\mathrm{B}}T)$ is the compression modulus of a gas of non-interacting particles of number density ρ and temperature T. Because of Eq. (3.62b,d), $(\rho k_B T)/S_0 = 1/\kappa_{is}$ is the isothermal compression modulus of the liquid specified by the structure factor S_q. Hence, the dimensionless function $[-\omega m_0^{T\prime}(\omega)]$ denotes the storage modulus relative to the thermodynamic modulus of the system. The loss modulus contains a trivial part $\omega\eta^\infty$. Here $\eta^\infty = \mu\rho\nu^\infty$ denotes the shear viscosity of a liquid in the absence of the cage effect. The glassy-dynamics part of $G_T''(\omega)$ in units of the modulus $1/\kappa_{is}$ is the loss spectrum $\omega m_0^{T\prime\prime}(\omega)$ of the kernel $m_0^T(t)$.

A paragraph with a digression shall be added, which explains more explicitly the common features of longitudinal and transverse elastic moduli. As done in Chapter 4, let us assume that heat fluctuations can be ignored for the theory of glassy dynamics. Then, Eqs. (3.99d), (4.1a) imply $\lim_{q\to 0} K_{Lq}(z) = \lim_{q\to 0} [M_{Lq}(z)/\, q^2] = k_L(z)$. Kernel $k_L(t)$ is the analogue of $K_T(t)$. Using Eqs. (4.10b), (4.11a), one arrives at the analogue of Eq. (5.7):

$$k_L(z) = \pm i\nu_L^\infty + c_{is}^2 m_0(z), \quad \mathrm{Im}\, z \gtrless 0. \tag{5.9}$$

As anticipated by the notation, the mode-coupling kernel $m_0(z)$ is the counterpart of $m_0^T(z)$. The white-noise term has a similar meaning as ν^∞; it is a longitudinal kinematic viscosity due to 'normal-liquid dynamics': $\nu_L^\infty = \lim_{q\to 0} \nu_q/q^2$. The longitudinal modulus is given by Eq. (3.115b). It is represented by a storage modulus $G_L'(\omega)$ and a loss modulus $G_L''(\omega)$:

$$G_L'(\omega) = [(\rho k_{\mathrm{B}}T)/S_0][1 - \omega m_0'(\omega)], \tag{5.10a}$$
$$G_L''(\omega) = \omega\eta^\infty + [(\rho k_{\mathrm{B}}T)/S_0]\omega m_0''(\omega). \tag{5.10b}$$

The results for the loss moduli $G_T''(\omega)$ and $G_L''(\omega)$ have a similar structure; the addition to the background is given by the loss spectrum of correlators, which are quadratic mode-coupling functionals. The corresponding reactive parts enter in a similar manner as term $[-\omega m_0^{T\prime}(\omega)/\kappa_{is}]$ and $[-\omega m_0'(\omega)/\kappa_{is}]$, respectively. However, the longitudinal storage modulus has the thermodynamic modulus $[1/\kappa_{is}]$ as an additional term.

The preceding extension of the MCT was formulated for an N-particle system obeying Hamilton's equations of motion. The introduction of the stress anticipates the momentum-conservation law. This law does not apply for the solute subsystem of a colloidal suspension, since momentum can be transferred to the solute. Handling this problem is less obvious for the shear than it is for the density-fluctuation dynamics. The preceding MCT formulas have been applied to dense colloids with the assumption that the indicated complications merely concern the theory of the regular term ν_q^T.

5.1.2 *Glassy-relaxation features of shear correlations*

In this section, it will be shown that the ideal glass transition, which is described by the MCT for the density-fluctuation dynamics, implies a transition to an isotropic medium with a positive static shear modulus. Results for the shear modulus of the hard-sphere system and for the square-well system shall be considered in order to demonstrate transition scenarios for the elastic response.

The formulas (5.5) yield a simple relation between the long-time limit of the density-fluctuation-correlator array $\boldsymbol{f} = \boldsymbol{\phi}(t \to \infty)$ and the arrested parts of the transverse-momentum-flux correlations:

$$m_q^T(t \to \infty) = \mathcal{F}_q^T[\boldsymbol{f}], \quad q = 1, \ldots, M. \tag{5.11a}$$

Generically, the arrest of the density-fluctuation correlations implies the arrest of the transverse-momentum-flux correlations. The generalized shear-relaxation kernel in Eq. (5.4a) exhibits a zero-frequency pole for the glass but not for the liquid:

$$\lim_{z \to 0}[-zM_{Tq}(z)] = \Omega_q^2 \mathcal{F}_q^T[\boldsymbol{f}]. \tag{5.11b}$$

Generically, the reverse statement holds as well. The vanishing of one of the non-negative values $\mathcal{F}_q^T[\boldsymbol{f}]$, $q = 1, \ldots, M$, is possible only if all M components of the array $\boldsymbol{f}$ vanish. Consequently, density-fluctuation arrest occurs if and only if transverse-fluctuating-force correlations arrest.

The alternative between liquid dynamics and glass dynamics can also be reformulated as alternatives for the static elastic moduli $G_T^0 = G_T'(\omega \to 0)$ or $G_L^0 = G_L'(\omega \to 0)$ (Bengtzelius *et al.* 1984). From Eq. (5.6d), one gets $\lim_{\omega \to 0}[-\omega m_0^T(\omega)] = f_T$, where

$$f_T = \mathcal{F}_0^T[\boldsymbol{f}]. \tag{5.12a}$$

Generically, the non-negative number f_T can vanish only if $\boldsymbol{f} = \mathbf{0}$. Similarly, Eq. (4.37a) yields $\lim_{\omega \to 0}[-\omega m_0(\omega)] = \mathcal{F}_L[\boldsymbol{f}] = \lim_{q \to 0} \mathcal{F}_q[\boldsymbol{f}]$. Again, generically, $\mathcal{F}_L[\boldsymbol{f}] > 0$ unless $\boldsymbol{f} = \mathbf{0}$. The fixed-point equation (4.52d) implies $\mathcal{F}_L[\boldsymbol{f}] = f_0/(1 - f_0)$. Here,

$$f_0 = \lim_{q \to 0} f_q \tag{5.12b}$$

denotes the zero-wave-number limit of the Debye–Waller factor. One arrives at

$$G_T^0 = [1/\kappa_{is}] f_T, \tag{5.12c}$$

$$G_L^0 = [1/\kappa_{is}]/(1 - f_0). \tag{5.12d}$$

The system is in a liquid state if and only if the longitudinal static modulus G_L^0 agrees with the thermodynamic modulus $[1/\kappa_{is}]$. This is the case if and only if the static shear modulus G_T^0 vanishes. The system is in an ideal glass state if G_L^0 exceeds $[1/\kappa_{is}]$, and this holds if and only if G_T^0 is positive. Continuing the discussions of Secs. 3.3.4 and 3.3.7 on visco-elastic phenomena, one can summarize: the transition from a liquid to an isotropic elastic medium in the sense discussed by

Maxwell is equivalent to a transition from a continuous low-frequency spectrum for density-fluctuation correlations to a spectrum with a degenerate Mountain's peak. Transverse fluctuating forces arrest if and only if longitudinal fluctuating forces arrest. And the arrest of both types of fluctuating forces is due to the arrest of the density-fluctuations. The latter arrest is caused by a sufficiently strong cage effect.

In the remainder of this section, the shear modulus shall be studied for states P near some generic liquid-glass transition point P^c. A path through P^c shall be considered, which is specified by a distance parameter ϵ as explained in Eqs. (4.84), (4.85). The separation parameter σ and the singularity parameter μ_2^c are used as defined in Sec. 4.3.4. Substituting Eq. (4.91a) into Eqs. (5.6), (5.12a), one gets the leading order asymptotic expansion formula:

$$f_T = f_T^c + h_T \sqrt{C\epsilon/\mu_2^c} + O(\epsilon) \ , \quad \epsilon \geq 0. \tag{5.13a}$$

Shifting P from the liquid through the critical point, the long-time limit $f_T = m_0^T(t \to \infty)$ increases discontinuously from zero to the critical arrested part

$$f_T^c = S_0 \int_0^\infty dq v_T(q) f_q^{c\,2}. \tag{5.13b}$$

The increase of f_T within the glass according to the $\sqrt{\epsilon}$ law is quantified by the positive critical amplitude

$$h_T = 2S_0 \int_0^\infty dq v_T(q) f_q^c h_q. \tag{5.13c}$$

From the discussions of Sec. 4.4.1, one concludes that the bifurcation dynamics of the shear evolves around a plateau of hight f_T^c for the $m_0^T(t)$ versus $\log t$ curves. For liquid states, the plateau is crossed at some time τ_T^{pc}. In analogy to Eq. (4.110b), there holds $\lim_{\epsilon \to 0-} \tau_T^{pc} = \infty$. In Sec. 4.4.3, evidence was presented that the correlators $\phi_q(t)$ decrease monotonically for times exceeding a scale t_{mic} for the transient dynamics. Consequently, $m_0^T(t)$ decreases with increasing time for $t \geq t_{\text{mic}}$. From the Green–Kubo equation (3.77d) one gets $\eta/(\mu\rho) = \nu \geq \int_0^{t_{\text{mic}}} K_T(t)dt + c_{is}^2 \int_{t_{\text{mic}}}^{\tau_T^{pc}} m_0^T(t)dt$. The first integral is a smooth function of ϵ. The second one is larger than $c_{is}^2[\tau_T^{pc} - t_{\text{mic}}] f_T^c$. As a result, one finds that the viscosity diverges if the liquid state P approaches the critical point P^c:

$$\lim_{\epsilon \to 0-} \eta = \infty. \tag{5.14}$$

Numerical evidence was discussed in connection with Eqs. (4.111a,c) that there holds the factorization property $\phi_q(t) = f_q^c + h_q G(t)$ for the plateau-crossing relaxation. Let us assume that this result is correct in a leading-order expansion for small $G(t)$. Substitution into Eq. (5.6b) shows that $G(t)$ describes also the leading-order result for the plateau-crossing dynamics of the shear:

$$m_0^T(t) = f_T^c + h_T G(t). \tag{5.15a}$$

The numerical results discussed in connection with Eqs. (4.111b,d) suggest that the preceding formula can be Laplace transformed. Let us assume this to be the case. Then, $m_0^T(z) = [-f_T^c/z] + h_T LT[G(t)](z)$ describes the shear correlator in the frequency domain for the dynamics outside the 'normal excitation regime' and outside the central part of the low-frequency loss-peak region. From Eqs. (5.8a–c) one gets

$$G_T(\omega + i0) - G_T^0 = -[\rho k_B T/S_0] h_T \omega LT[G(t)](\omega + i0). \tag{5.15b}$$

The regular contribution from Eq. (5.8c) is dropped here. The anticipated asymptotic result implies that the white-noise-induced term $\omega\eta^\infty$ can be neglected compared to term on the right-hand side of the preceding equation.

Let us assume that the below-plateau-decay processes of the density correlators of the liquid states obey the superposition principles, which are discussed in connection with Eqs. (4.112a), (4.114b). From Eq. (5.6b), one obtains the superposition principle for the below-plateau decay of the shear correlator

$$m_0^T(t) = \tilde{m}_0^T(t/\tau). \tag{5.16a}$$

Here, τ is the common time scale of the relaxation processes and

$$\tilde{m}_0^T(\tilde{t}) = S_0^c \int_0^\infty dk v_T^c(q) \tilde{\phi}_q(\tilde{t})^2 \tag{5.16b}$$

denotes the control-parameter-independent shape function for the shear. Laplace transformation yields $m_0^T(z) = \tau\tilde{m}_0^T(z\tau)$ with $\tilde{m}_0^T(\tilde{z}) = LT[\tilde{m}_0^T(\tilde{t})](\tilde{z})$. From Eqs. (5.8a–c), one arrives at the scaling laws for the modulus

$$-G_T'(\omega) + i[G_T''(\omega) - \omega\eta^\infty] = [\rho k_B T/S_0]\tilde{\omega}\tilde{m}_0^T(\tilde{\omega}), \tag{5.16c}$$

with $\tilde{\omega} = \omega\tau$ abbreviating the rescaled frequency. Because of Eq. (3.85b), the low-frequency limit of the imaginary part of this equation implies the formula for the viscosity

$$\eta = \eta^\infty + C_\eta \cdot \tau. \tag{5.17a}$$

Here, C_η denotes the number

$$C_\eta = [\rho k_B T/S_0]\tilde{m}_0^{T\prime\prime}(\tilde{\omega} = 0). \tag{5.17b}$$

The divergency of the viscosity is caused by that of the scale τ.

All formulas (5.13-5.17) have their obvious analogies for the longitudinal modulus.

Figure 5.1 exhibits the MCT results for the evolution of the glassy shear dynamics for the same model of a HSS, which is discussed in connection with Fig. 4.15. The regular viscosity contribution η^∞ is neglected. In units of $(k_B T/d^3)$, there is a critical value for the thermodynamic shear modulus $(1/\kappa_{is})^c \approx 74$, a discontinuity of the static longitudinal modulus $G_L^{0c} - (1/\kappa_{is})^c \approx 56$, and a critical static shear modulus $G_T^{0c} \approx 18$ (Chong 2005). Up to a factor, the upper panel exhibits the loss spectra of the shear correlator $K_T(t)$. The evolution of the superposition principle for the low-frequency loss peaks exhibits the pattern demonstrated in Fig. 4.22 for a density-fluctuation susceptibility $\chi_q(\omega)$ for $q = q_2$. This can be checked by the reader by shifting a replica of the figure parallel to the logarithmic frequency axis so that all peak positions $\omega_{\max}$

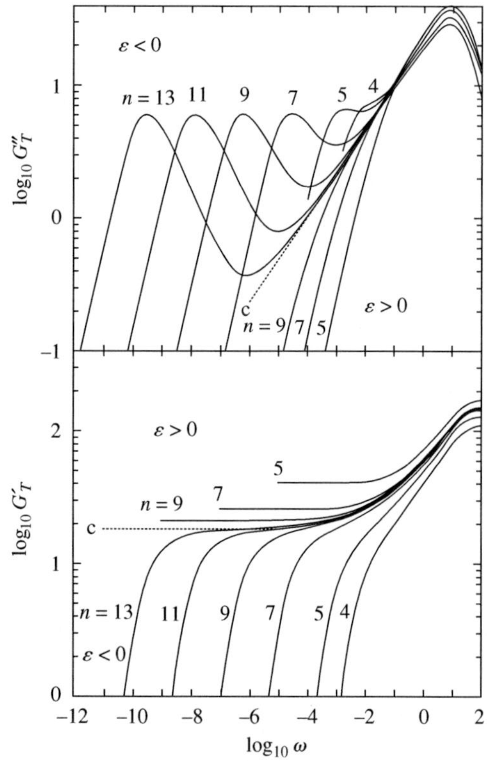

FIG. 5.1. Loss moduli G''_T and storage moduli G'_T as function of frequency ω calculated for a model of the HSS with stochastic short-time dynamics. The unit of the moduli is $(k_B T/d^3)$ with d denoting the sphere diameter. The unit of time and other details of the model are explained in the captions of Figs. 4.6 and 4.14. The dotted lines with label c are the results for the critical packing fraction φ_c and the full ones refer to the distance parameters $\epsilon = (\varphi - \varphi_c)/\varphi_c = \pm 10^{-(n/3)}$ (Chong 2005).

coincide with the position of the $n = 13$ peak. The shape of the loss peaks between the maximum $\chi_{\max}$ and $\chi_{\max}/2$ can be fitted well by a Kohlrausch function with a stretching exponent $\beta = 0.61$ (Fuchs *et al.* 1992). The low-frequency critical density-fluctuation loss spectra shown in Fig. 4.15 exhibit a power-law behaviour: $\chi_q''(\omega) \propto \omega^a$, $a \approx 0.31$. Anticipating the factorization property, one gets for the critical correlator in Eq. (5.15a): $G^c(t) = (t_0/t)^a$. The dotted line with label c in the upper panel exhibits the resulting critical spectrum, $\log G_T''(\omega) = \text{constant} + a \log \omega$, for $\omega \leq 0.002$.

By Laplace transformation, one gets from the anticipated factorization property (5.15a): $-\omega m_0^{Tc\prime}(\omega) - f_T^c \propto \omega^a$. For frequencies tending to zero, $-\omega m_0^{Tc\prime}(\omega)$ decreases to f_T^c. Equation (5.8b) implies that the critical storage modulus $G_T'^c(\omega)$ decreases to the critical static shear modulus G_T^{0c} if $\log \omega$ tends to minus infinity. This is shown by the dotted line with label c in the lower panel of Fig. 5.1. One can follow the discussion of Sec. 4.4.1 in order to discuss the evolution of the low-frequency behaviour of the $G_T'(\omega)$ versus $\log(1/\omega)$ curves in analogy to the evolution of the long-time behaviour of the $\phi(t)$ versus $\log t$ curves. The storage modulus evolves around the plateau G_T^{0c}. From Eqs. (5.12c), (5.13a) one concludes that $G_T^0 = G_T'(\omega \to 0)$ increases according to a $\sqrt{\epsilon}$ law if φ increases above φ_c. The $n = 5$ curve demonstrates that G_T^0 increases by more than a factor 2 if φ increases by about 2% above φ_c. For liquid states, the $G_T'(\omega)$ versus $\log \omega$ curves cross the plateau at some frequency ω^{pc}; and $\lim_{\epsilon \to 0-} \log \omega^{pc} = -\infty$. The plateau extends to frequencies near the position $\omega_{\max} \propto 1/\tau$ of the low-frequency loss peak. The figure suggests that ω^{pc} decreases with $|\epsilon|$ proportional to the position $\omega_{\min}$ of the loss minimum.

Within the regime of hydrodynamic behaviour of the liquid, which is characterized by $\omega \ll \omega_{\max}$, the loss modulus varies linearly with the frequency: $G_T''(\omega) = \eta\omega + O(\omega^3)$. The storage modulus is a regular even function of ω, i.e., $G_T' = O(\omega^2)$. Hence, it is small compared to the loss modulus. Within the regime of 'normal liquid dynamics', the storage modulus is of a similar size as the loss modulus, as is demonstrated in Fig. 5.1 for $\omega \approx 1$. The existence of a plateau for the storage modulus is an outstanding feature of the glassy dynamics. This plateau occurs for frequencies above the position $\omega_{\max}$ of the low-frequency-loss peak and below the regime of normal liquid dynamics. Within this plateau region, the storage modulus decreases only slightly with decreasing frequency and it is much larger than the loss modulus. The figure exemplifies the phenomenon for the interval $\omega_{\max} \leq \omega \leq 0.01$, where G_T' exceeds G_T'' by more than a factor 5.

The preceding formulas for the shear shall be used to demonstrate some further features of the MCT dynamics. The first one concerns the influence of the transient dynamics on the long-time correlations. The lower full line and the dashed line in Fig. 5.2 exhibit critical correlators calculated for the mode-coupling functional of a HSS with $M = 100$ components and $q_{c0} = 40/d$ as studied above. The full line refers to stochastic short time motion; its Laplace transform determines the dotted curves with label c in Fig. 5.1. The dashed line is obtained as solution of the equations of motion (4.17a) for a Newtonian dynamics without friction terms. But this correlator is plotted for a time scale,

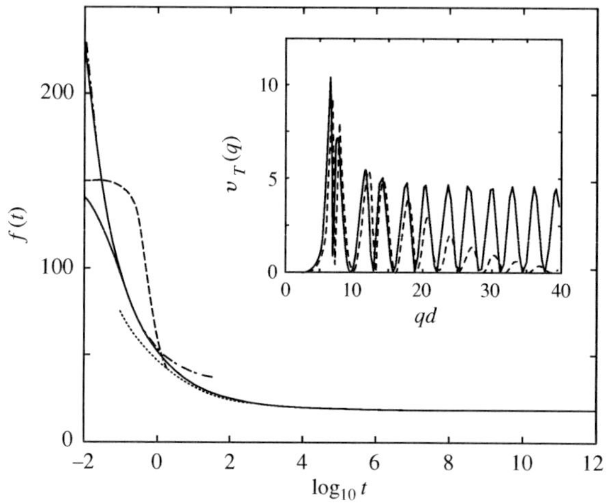

FIG. 5.2. Up to some arbitrary common scale factor, the functions $f(t)$ present the contribution $m_0^{Tc}(t)$ from Eqs. (5.6a,b) for the shear correlator $K_T(t)$ calculated for microscopic models of the HSS at the critical packing fraction. The convolution approximation is used and the structure factor is calculated within the Percus–Yevick theory. The two full lines are obtained for models with a stochastic short-time dynamics. The unit of time is chosen as in Figs. 4.14 and 5.1. The lower line is obtained for a model with $M = 100$ points on the wave-number grid and a cutoff value $q_{\text{co}} = 40/d$. The upper line refers to a model with an extended grid of $M = 250$ points and $q_{\text{co}} = 100/d$. The dashed-dotted line is a fit of a small-time part of the upper full line by a function $c_1 + c_2/\sqrt{t}$. The dotted line shows a power-law (6.18) of the long-time parts: $f_T^c + h_T(t_0/t)^a$, $a = 0.31$, $t_0 = 0.425$; and f_T^c, h_T are evaluated from Eqs. (5.13b,c). The dashed line refers to the $M = 100$ model; but the stochastic dynamics is replaced by a Newtonian one with vanishing friction parameter. Furthermore, the time is rescaled so that dashed and dotted lines cannot be distinguished for $\log t > 8$. The inset shows the coefficients $v_T(q)$ from Eq. (5.6a) calculated for Percus–Yevick structure factors at the critical points. The full line is obtained for the HSS. The dashed one is calculated for a Lennard-Jones system for a density close to that for argon at its triple point. The length σ in Eq. (1.8a) is adjusted such that the structure-factor-peak position is close to that for the HSS (Chong 2005).

which is chosen so that the full and dashed lines agree for $\log t > 8$ within the accuracy of the drawing. Within this accuracy, the two $K_T^c(t)$ versus $\log t$ curves coincide for $\log t > 0.5$. In agreement with the results suggested by Fig. 4.24, one observes that the $K_T^c(t)$ versus $\log(t/t_0)$ curves are independent of the transient dynamics for large t/t_0.

The following three paragraphs deal with properties of a liquid of hard spheres, which are absent in systems with regular interaction potentials. The strongest singularity of the pair-distribution function $g(r)$ of the HSS is a discontinuous change from zero to the positive contact value $g_+ = g(r \to d+)$ if the distance r increases from values below the sphere diameter d to above. According to Eq. (4.105a), this discontinuity implies that the derivative of the direct correlation function exhibits a $(1/q^2)$ large-wave-number tail: $c'_q = dc_q/dq = [4\pi g_+ d^3][-d\sin(qd)]/(qd)^2 + O(1/q^3)$. Hence, the mode-coupling coefficient $v_T(q)$ in Eq. (5.6a) has a tail, which oscillates with a constant amplitude:

$$v_T(q) = B[1 - \cos(2qd)] \cdot d + O(1/q), \qquad B = [2\rho d^3/15]g_+^2. \tag{5.18a}$$

The full line in the inset of Fig. 5.2 exhibits this tail for $qd > 20$. The tail renders the integral (5.6b) divergent for vanishing times:

$$m_0^T(t \to 0) = \infty, \qquad \text{HSS.} \tag{5.18b}$$

Suppose, the interaction is given by a regular repulsion potential $V(r) = \epsilon(\sigma/r)^n$, $n > 3$. One can show that the characteristic frequencies Ω_{Tq} in Eq. (3.74b) are finite. In particular, the small-q limits of Ω_{Tq}^2/q^2 are finite numbers. Because of Eq. (3.74e), the latter are the numbers c_T^2. These determine the initial values of $K_T(t)$ via Eqs. (4.76) and thereby the values $m_0^T(t \to 0)$ in Eq. (5.7). For n tending to infinity, the equilibrium structure converges to one for the HSS and one gets $\lim_{n\to\infty} \Omega_{Tq}^2/q^2 = \infty$ (Branka and Heyes 2004). One concludes that the result (5.18b) reproduces a peculiarity of the structure of a HSS. It is elementary, to demonstrate the analogous result for the longitudinal kernel of MCT,

$$m_0(t \to 0) = \infty, \qquad \text{HSS,} \tag{5.18c}$$

or, more generally, for all MCT expressions for the fluctuating-force correlations:

$$M_{Lq}(t \to 0) = M_{Tq}(t \to 0) = \infty, \qquad \text{HSS.} \tag{5.19}$$

For small times, the density correlators for the system with stochastic short time dynamics are given by $\phi_q^{(0)}(t) = \exp[-t/\tau_q]$. The small-$t$ divergency of the integral (5.6b) is caused by the large-q terms. Here, one can replace $v_T(q)$ by the bracket term in Eq. (5.18a); and Eq. (4.109a) yields $\tau_q = t_{\text{mic}}/(qd)^2$. Hence, the dominant contribution to the integral reads $By \int_0^\infty dx[1-\cos(2yx)]\exp[-x^2]$ with $y = \sqrt{t_{\text{mic}}/(2t)}$. The Fourier integral of the Gaussian vanishes exponentially for y tending to infinity. Hence, one gets a $1/\sqrt{t}$ divergence of the kernel:

$$m_0^T(t) \sim BS_0\sqrt{\pi t_{\text{mic}}/(8t)}\,, t \to 0; \qquad \text{HSS, stochastic dyn.} \tag{5.20a}$$

Using the Tauberian theorem formulas (A.64a,b) one arrives at the expressions for the moduli

$$G_T^{\prime,\prime\prime}(\omega) \sim [\rho k_{\text{B}}T](\pi B/4)\sqrt{\omega t_{\text{mic}}}\,, \quad \omega \to \infty; \quad \text{HSS, stochastic dyn.} \tag{5.20b}$$

For dilute colloids, the $1/\sqrt{t}$ law can be derived by systematic perturbation theory (Cichocki and Felderhof 1991). The origin of this result is the $\sqrt{t}$ law for the probability density for two diffusing particles to meet in a scattering volume, Eq. (3.42a). This mechanism should be relevant also for the high-density colloids of interest in this book; and the preceding MCT formulas are consistent with such expectation.

A MCT model, which is specified by some cutoff q_{c0}, yields finite values for the initial values of the mentioned kernels; for example, Eq. (5.6b) leads to $m_0^T(t=0) = S_0 \int_0^{q_{c0}} v_T(q)dq$. But, this initial value diverges proportional to q_{c0} if the cutoff tends to infinity. For $t > 0$, $\int_0^{q_{c0}} v_T(q)\phi_q(t)^2 dq$ tends to $m_q(t)$ if q_{c0} diverges. One concludes the following. For every error margin $\delta > 0$ and every cutoff q_{c0}, there is a time t_{c0} so that $m_0^T(t)$ for $t \geq t_{c0}$ does not change beyond δ if the cutoff is increased beyond q_{c0}. Loosely summarized: for $t \geq t_{c0}$, the correlator $m_T^0(t)$ is cutoff independent; for such large times, it reflects a property of the HSS. For $t \leq t_{c0}$, the correlator may change with changing q_{c0}. For such small times, it may deal with cutoff artifacts. The two solid lines in Fig. 5.2 show the results for $m_0^{Tc}(t)$, which are calculated for $q_{c0}d = 40$ and 100. One concludes the following. Within the accuracy of the drawing, the results of the model with $q_{c0} = 40/d$ are cutoff independent for $t \geq t_{c0} = 0.1$. Similar results can be formulated for the correlators in the frequency domain. Within the accuracy of the drawing, the results in Fig. 5.1 for $\omega \leq \omega_{c0} \approx 0.25$ do not change if q_{c0} is increased. The moduli for $\omega > \omega_{c0}$ change if q_{c0} increases above $40/d$ (Chong 2005).

The dashed line in the inset of Fig. 5.2 demonstrates a feature, which has been discussed already in the preceding chapter in the context of the solution of the fixed-point equation for the Debye–Waller factor. This line exhibits the mode-coupling function $v_T(q)$ for a Lennard-Jones system at the glass transition temperature T_c for a density close to that of the triple point. The length scale σ in Eq. (1.8a) is adjusted so that the structure-factor-peak position agrees with that of a HSS with diameter d. Because of the factor q^4 in Eq. (5.6a), $v_T(q)$ is very small for small wave numbers. Fluctuations for $q < q_- \approx 4/d$ do not contribute appreciably to the mode-coupling contribution to the fluctuation-force kernel $m_0^T(t)$. The factor q^4 results from two factors q^2. One is the phase-space factor due to the number of fluctuations with wave-vector modulus q. The other is due to the factor $[qc'_q]^2$, where the gradient factor q reflects the smallness of the derivatives of the effective interaction for long-wavelength changes. For the regular potential, $v_T(q)$ decreases with increasing q so rapidly, that no changes of the MCT solution for the correlators can be exhibited if q_{c0} increases above $40/d$. The contributions for $q > q_+ \approx 20/d$ do not influence the results appreciably. One concludes that the glassy dynamics is determined by the fluctuations within the region $q_- \leq q \leq q_+$. A necessary condition for understanding the glassy dynamics of the shear in van der Waals-like liquids is the understanding of the density-fluctuation-correlators for the specified intermediate-wave-number region. This was pointed out first by Geszti (1983). The fluctuations within this region are determined by the structure factor S_q for $q_- \leq q \leq q_+$; and this S_q is close to that of a HSS. For $q > q_+$, the mode-coupling coefficient $v_T(q)$ of the HSS differs from that for the Lennard-Jones liquid for the corresponding state. However, this difference does not influence the glassy dynamics but only the correlations in the transient region. The glassy shear dynamics for the specified class of liquids is that of the HSS; it is an entropy-driven phenomenon.

The glassy dynamics of systems of particles, which interact via a strong repulsion and an additional short ranged attraction, is more subtle than that of van der Waals-like systems. The equilibrium structure is controlled by three parameters: the packing fraction φ, the attraction strength parameter Γ, and the attraction-range parameter δ. There are non-trivial lines for liquid-glass transitions of the system in the $\varphi - \Gamma$ plane of control parameter points for the system specified by a fixed δ. This is explained for the square-well system (SWS) in connection with Fig. 4.9. There appears an A_4-bifurcation singularity specified by a parameter triple $(\varphi^*, \Gamma^*, \delta^*)$. For $\delta < \delta^*$, the liquid boundary consists of two parts. There is a low-Γ branch, where the arrest is caused predominantly by the hard repulsion. The critical Debye–Waller factor $f_q^{(1)c}$ is close to that of a HSS. There is a large-Γ branch, where the arrest is dominated by bond formation. The critical Debye–Waller factor $f_q^{(2)c}$ has a trend to be larger on the second branch than on the first one. There is a crossing point, where $f_q^{(2)c} > f_q^{(1)c}$ for all values of q. From Eq. (5.13b), one infers that the arrested parts of the shear correlations change discontinuously at the crossing point: $f_T^{(2)c} > f_T^{(1)c}$. The regularity properties of the MCT solutions imply that this holds also for states near the crossing points. Bond formation increases the shear stiffness, i.e., G_T^{0c} is larger on the large-Γ branch than on the small-Γ one. The same conclusion holds for the critical longitudinal modulus G_L^{0c}.

The above explained properties are demonstrated in Fig. 5.3. The full lines exhibit the change of the critical static moduli if the state is shifted along the transition line for $\delta = \delta^*$. The thermodynamic compression modulus $1/\kappa_{is}$ decreases with increasing Γ because the states shift towards the spinodal line. For $\Gamma < 0.9\Gamma^*$, the static shear modulus G_T^{0c} increases with Γ only weakly. For the repulsion-dominated glass states, the bond formation has only a small effect on the stiffness. The increase of Γ/Γ^* from 0.9 to 1.1, however, causes an increase of G_T^{0c} by more than a factor 3. The attraction-dominated glass states have a drastically increased shear stiffness compared to the repulsion-dominated ones. The crossing of the A_4 singularity manifests itself by a vertical tangent of the G_T^{0c} versus Γ curve for $\Gamma = \Gamma^*$. Increasing Γ further, a maximum of G_T^{0c} is reached for Γ/Γ^* near 1.4. For even larger Γ, the stiffness of the glass decreases with increasing attraction since the effect of the accompanying decrease of the critical packing fraction overwhelms the effect of the increasing bond strength. The $(G_L^{0c} - 1/\kappa_{is})$ versus Γ curve has a similar behaviour as described for the shear. The dotted lines in the figure exhibit results for a model describing the attraction by a Yukawa potential. They are similar to the ones for the SWS.

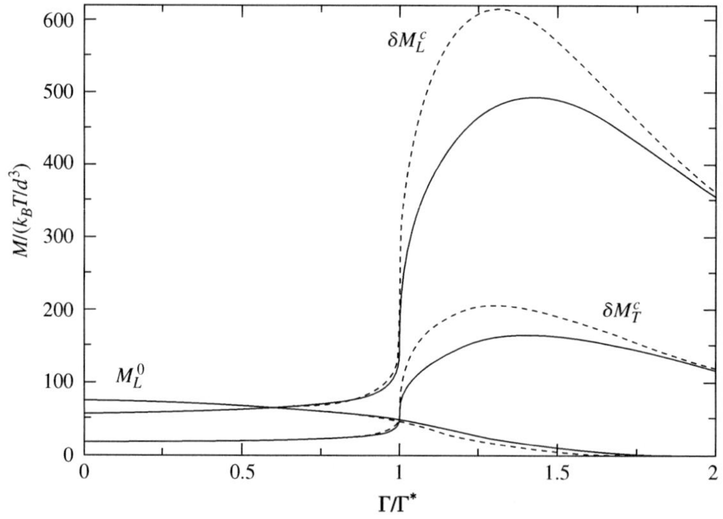

FIG. 5.3. The thermodynamic compression modulus $(1/\kappa_{is}^c) = M_L^0$, the discontinuity of the static longitudinal modulus $G_L^{0c} - (\kappa_{is}^c) = \delta M_L^c$, and the static shear modulus $G_T^{0c} = \delta M_T^c$ for the liquid–glass transition points of a system of hard spheres with diameter d and an additional short-range attraction presented as a function of the attraction-strength parameter Γ. The transition manifold refers to the attraction range δ^* of the A_4 singularity, which is specified by an attraction strength Γ^*. The curves are calculated for a mode-coupling functional using the convolution approximation and the mean-spherical approximation for the structure factor. The full and dashed lines refer to models, which describe the attraction by a square well and a Yukawa potential, respectively. Reproduced from Götze and Sperl (2003).

This section shall be closed with remarks on similarities of MCT results for the shear dynamics with some data. Let us consider first a Lennard-Jones liquid for a state near the triple point or a HSS for a packing fraction near the freezing point. For such states, the density correlators for wave number q near the structure-factor-peak position can be approximated reasonably by a simple-relaxation law: $\phi_q(t) \approx \phi_q^{(0)}(t) = \exp[-\omega_q t]$. A microscopic-approximation formula for the rates ω_q has been derived by de Schepper and Cohen (1982). Correlators of the type mentioned are shown in the upper panel of Fig. 4.14 for label $n = 2$ and 3. The figure exhibits results for a stochastic dynamics; but, the long-time parts of the correlators calculated for a Newtonian dynamics are quite similar (Götze and Mayr 2000). From Eq. (5.6b), one obtains a lowest-order approximation for the shear-relaxation kernel: $m_0^T(t) \approx m_0^{T(0)}(t) = S_0 \int_0^\infty dq v_T(q)$ $\phi_q^{(0)}(t)^2$. This mode-coupling formula describes modifications of the 'normal liquid dynamics' by the cage effect in a parameter regime, where the request of full consistency between density-correlation decay and fluctuating-force-correlation decay is not yet important. The function $m_0^{T(0)}(t)$ exhibits a long-time tail extending to times of order, say, $4t_{\rm mic}$. It was demonstrated by Kirkpatrick (1984) that $m_0^{T(0)}(t)$ reproduces data of the kind shown in Fig. 3.8. Thus, the puzzling tails, which were discovered in the cited molecular-dynamics studies of $K_T(t)$, are explained as precursors of the MCT-structural-relaxation phenomena.

Verberg *et al.* (1997) used the above mentioned mode-coupling formula for $m_0^{T(0)}(t)$ in order to explain dynamical viscosities $\eta(\omega) = [G_T''(\omega) + iG_T'(\omega)]/\omega$, which were measured for hard-sphere colloidal suspensions by van der Werff *et al.* (1989). But, their formula for the mode-coupling function $v_T(q)$ is more sophisticated than Eq. (5.6a). The measured dynamic viscosities exhibit the $1/\sqrt{\omega}$ tails for large frequencies in qualitative agreement with Eq. (5.20b). Within the range of packing fractions analyed, the static viscosity $\eta = \eta(\omega \to 0)$ varies by about a factor 6. This is what one would expect from the change of the scale for the long-time decay of the correlators in Fig. 4.14 for an increase of label n from 1 to about 2 or 3. In agreement with the measurements, the calculated loss moduli $G_T^{(0)\prime\prime}(\omega) = \omega\eta'$ are larger than the storage moduli $G_T^{(0)\prime}(\omega) = \omega\eta''$. The cited data and the corresponding theoretical results deal with subtleties of the dynamics of hard spheres in a regime, where glassy-dynamics features are of a minor significance.

Mason and Weitz (1995) measured the shear moduli of hard-sphere colloidal suspensions for packing fractions φ increasing from 0.50 to 0.56. According to Fig. 1.15, this is the control-parameter region, where the density correlators $\phi_q(t)$ exhibit huge deviations from a simple relaxation and where the time scale for correlation decay increases by more than two orders of magnitude. For φ near 0.5, the measured moduli keep on increasing with increasing ω in qualitative agreement with Eq. (5.20b). As is typical for 'normal liquid dynamics', the storage modulus in this high-frequency regime is comparable in size or smaller than the loss modulus. With increasing φ, there appears a low-frequency region, where the loss modulus develops a minimum and where the storage modulus exhibits a plateau. Most remarkably, the storage modulus can exceed the loss modulus by factors larger than 5. The scenario for the evolution of the glassy dynamics has been fitted by a superposition of a t^{-a}-to-t^b cross over according to Eqs. (1.7a,b) and the $1/\sqrt{t}$ law for the transient dynamics of a HSS. The complex dynamics of micellar systems has been studied also by viscosimetry (Mallamace *et al.* 2004). The measured evolution patterns for the loss moduli are in qualitative agreement with the one shown in Fig. 5.1.

Puertas *et al.* (2005) have analysed molecular-dynamics-simulation results for the shear for two models. The first one is the hard-sphere-like system explained in connection with Eq. (4.103). The studied increase of φ from 0.50 to 0.58 implies an increase of the scale τ for the long-time decay of $m_0^T(t)$ by a factor of about 100. These increases correspond to the ones exhibited in Fig. 4.14 for a change of label n from about 4 to about 7. The second model is obtained by adding a short-range attraction to the interaction potential. A long-range repulsion is included also in order to suppress phase-separation phenomena. The evolution of glassy dynamics is studied for increasing attraction strength Γ for fixed $\varphi = 0.40$. In this case, the time scale τ exhibits an increase by about a factor 1000. The evolution of the superposition principle for the shear correlators and for the density-fluctuation correlators follows the pattern predicted by MCT and the scales are coupled. The master function $\tilde{m}_0^T(\tilde{t})$ in Eq. (5.16a) is proportional to the square of that for the density fluctuations, $\tilde{\phi}_q(\tilde{t})^2$, for $q = 8/d$ and $q = 22/d$ for the first and second model, respectively. This is consistent with the expectation that the main contribution to the force fluctuations of the HSS is due to wave numbers near the structure-factor-peak position. It supports also the prediction that the arrest for the second model is due to the large-wave-number fluctuations, which describe the sticking effect. Plateaus for the $m_0^T(t)$ versus $\log t$ curves cannot be identified in the data. The large increase of $m_0^T(t)$ for small times renders it difficult to evaluate the moduli. With this reservation, the results show that the storage moduli for the states near the arrest points exhibit plateaus, which exceed the values of the loss moduli.

5.1.3 *MCT equations for the tagged-particle dynamics*

A demand for a statistical description of liquid dynamics concerns the characterization of the positions $\vec{r}_s(t)$ of a tagged particle. The tagged-particle-density correlator $\phi_q^s(t)$ is the function of primary importance. It is the autocorrelation function formed with the dynamical variable $\rho_{\vec{q}}^s = \exp[i\vec{q}\vec{r}^{\,s}]$. Other quantities of interest are the mean-squared displacement $\delta r_s^2(t) = \langle[\vec{r}_s(t) - \vec{r}_s(t=0)]^2\rangle = 6\Delta_s(t)$ and the velocity correlator $K_s(t) = \langle\vec{v}_s(t)\vec{v}_s\rangle/3$. In this section, mode-coupling-theory equations of motion for these functions shall be derived.

(i) A self-consistent treatment of tagged-particle-current relaxation

The first step of the procedure, which is outlined at the beginning of this chapter, starts with the representation (3.35a) of the density correlator in the frequency domain in terms of a kernel $K_q^s(z)$. Equation (3.35b) reads $K_q^s(z)/(v_{th}^s)^2 = (A'|\mathcal{R}_1(z)A')$ with $A' = j_{\vec{q}*}^{s3}/v_{th}^s$ denoting the normalized tagged-particle-current-density fluctuation parallel to the distinguished wave vector $\vec{q}^{\,*} = (0,0,q)$. The resolvent $\mathcal{R}_1(z) = [\mathcal{L}_1 - z]^{-1}$ is formed with the reduced Liouvillian $\mathcal{L}_1 = \mathcal{Q}_1\mathcal{L}\mathcal{Q}_1$. Here, $\mathcal{Q}_1 = 1 - \mathcal{P}_1$ with $\mathcal{P}_1 = |\rho_{\vec{q}*}^s)(\rho_{\vec{q}*}^s|$ denoting the projector perpendicular to the density-fluctuation considered. Then, the Zwanzig–Mori reduction is done for the single distinguished variable A'. One gets $K_q^s(z)/(v_{th}^s)^2 = -1/[z + M_{sq}(z)]$. The detailed reasoning follows that for the derivation of Eqs. (4.1a–c). As a result, one obtains the double-fraction representation:

$$\phi_q^s(z) = -1/\Big\{z - \Omega_{sq}^2/\big[z + M_{sq}(z)\big]\Big\}, \tag{5.21a}$$

$$M_{sq}(z) = (F_{sq}|\mathcal{U}_{\text{red}}(-t)F_{sq}), \tag{5.21b}$$

$$F_{sq} = \mathcal{Q}_1\mathcal{L}j_{\vec{q}*}^{s3}/v_{th}^s. \tag{5.21c}$$

The frequency $\Omega_{sq} = v^s_{th} q$ vanishes proportional to q. This is a manifestation of the conservation law for the tagged-particle density. The time evolution is generated by the reduced Liouvillian $\mathcal{L}_\mathcal{Q} = \mathcal{Q}_2 \mathcal{L}_1 \mathcal{Q}_2 : \mathcal{U}_{\text{red}}(t) = \exp[i\mathcal{L}_\mathcal{Q} t]$. Operator $\mathcal{Q}_2 = 1 - |A')(A'|$ is the projector perpendicular to the longitudinal-current fluctuation. The fluctuating force F_{sq} contains the time-derivative of the tagged-particle momentum density $\mu_s \vec{j}^{\,s}_{\vec{q}*}$ and, hence, the cage-forming forces between the tagged particle and its neighbours. Consequently, Eqs. (5.21a–c) provide the desired expression of the correlator in terms of a fluctuating-force kernel $M(t) = M_{sq}(t)$. According to Eq. (3.36b), the velocity correlator $K_s(z)$ in the frequency domain is the zero-wave-number limit of $K^s_q(z)$. One arrives at the fraction representation (3.48a) with the formula for the fluctuating-force correlator

$$\lim_{q \to 0} M_{sq}(z) = X_s(z)/\mu_s. \tag{5.21d}$$

The dynamical variables describing the cage-forming forces are combinations of $\rho(\vec{r}_1)\rho^s(\vec{r}_2)$. They determine the probability density for finding a particle of the system at $\vec{r}_1$ and the tagged particle at $\vec{r}_2$. The coefficients for the combinations are derivatives of the interaction potential. Transforming from the space of positions $\vec{r}$ to the one of wave vectors $\vec{k}$, one understands that the relevant pair fluctuations read $A_k = \rho_{\vec{k}_1}\rho^s_{\vec{k}_2}$; here, $k = (\vec{k}_1, \vec{k}_2)$ denotes the label composed of two labels of wave vectors. The quantities $\rho_{\vec{k}_1}\rho^s_{\vec{k}_2}$ and $\rho_{\vec{k}_2}\rho^s_{\vec{k}_1}$ are different variables; summations over k are meant as unrestricted sums over $\vec{k}_1$ and $\vec{k}_2$.

The second step towards a derivation of a mode-coupling kernel starts with a formula for $M_{sq}(z)$, which is analogous to Eq. (4.3a). It is simplified as explained in Secs. 4.1.1 and 4.1.2: $M_{sq}(z) \to \pm i\nu^s_q + M^{mc}_{sq}(z)$, Im $z \gtrless 0$. Here, the non-negative numbers ν^s_q denote a white-noise background contribution. The non-trivial part is the Laplace transform of the function

$$M^{mc}_{sq}(t) = \sum_{m,n} \Omega^s_{qm} \phi^s_{m,n}(t) \Omega^{s*}_{qn}. \tag{5.22a}$$

The pair correlator reads $\phi^s_{\vec{m}_1\vec{m}_2,\vec{n}_1\vec{n}_2}(t) = (\rho_{\vec{m}_1}\rho^s_{\vec{m}_2}|\mathcal{U}_{\text{red}}(-t)\rho_{\vec{n}_1}\rho^s_{\vec{n}_2})$. Its initial value is the metric matrix for the pair variables: $\phi_{m,n}(t=0) = g_{m,n}$. The coupling amplitudes are given by a formula similar to Eq. (4.3b):

$$\Omega^s_{q,\vec{m}_1\vec{m}_2} = \sum_{\vec{k}_1\vec{k}_2} (F_{sq}|\rho_{\vec{k}_1}\rho^s_{\vec{k}_2})(\boldsymbol{g}^{-1})_{\vec{k}_1\vec{k}_2,\vec{m}_1\vec{m}_2}. \tag{5.22b}$$

Writing $\mathcal{Q}_1 = 1 - \mathcal{P}_1$, Eq. (5.21c) yields two contributions to the coupling amplitudes: $(F_{sq}|\rho_{\vec{k}}\rho^s_{\vec{p}}) = (F^{s(a)}_{q,\vec{k}\vec{p}} - F^{s(b)}_{q,\vec{k}\vec{p}})v^s_{th}$. The first one can be evaluated as explained for the derivation of Eq. (4.5b): $F^{s(a)}_{q,\vec{k}\vec{p}} = (\mathcal{L}j^{s3}_{\vec{q}*}|\rho_{\vec{k}}\rho^s_{\vec{p}})/(v^s_{th})^2 = \langle j^{s3*}_{\vec{q}*}[\vec{k}\vec{j}_{\vec{k}}\rho^s_{\vec{p}} + \vec{p}\rho_{\vec{k}}\vec{j}^{\,s}_{\vec{p}}]\rangle/(v^s_{th})^2 = p^3\langle\rho^{s*}_{\vec{q}*-\vec{p}}\rho_{\vec{k}}\rangle$. The second contribution yields: $F^{s(b)}_{q,\vec{k}\vec{p}} = (\mathcal{P}_1\mathcal{L}j^{s3}_{\vec{q}*}|$

$\rho_{\vec{k}}\rho^s_{\vec{p}})/(v^s_{th})^2 = (\mathcal{L}\rho^s_{\vec{q}*}|j^{s3}_{\vec{q}})\langle\rho^{s*}_{\vec{q}*}\rho_{\vec{k}}\rho^s_{\vec{p}}\rangle/(v^s_{th})^2 = q\langle\rho^{s*}_{\vec{q}*-\vec{p}}\rho_{\vec{k}}\rangle$. The desired overlaps are

$$(F_{sq}|\rho_{\vec{k}}\rho^s_{\vec{p}}) = \delta_{\vec{q}*,\vec{k}+\vec{p}}v^s_{th}(-k^3)(\rho^s_{\vec{k}}|\rho_{\vec{k}}). \tag{5.22c}$$

A digression shall be made with the intention to write the average $\langle\rho^{s*}_{\vec{k}}\rho_{\vec{k}}\rangle$ in a more conventional manner. Expressing the fluctuations in terms of the particle positions $\vec{r}_s$ and $\vec{r}_\kappa$, $\kappa = 1, \ldots, N$, one gets $\langle\rho^{s*}_{\vec{q}}\rho_{\vec{q}}\rangle = \sum_\kappa\langle\exp[i\vec{q}(\vec{r}_\kappa - \vec{r}_s)]\rangle = \int\exp[i\vec{q}\vec{r}]\rho g^s(r)d^3\vec{r}$. Here, the probability density for finding a system particle in a distance r from the tagged particle is defined in analogy to Eq. (3.59b):

$$\rho g^s(r) = \sum_\kappa\langle\delta(\vec{r} - \vec{r}_\kappa + \vec{r}_s)\rangle. \tag{5.23a}$$

One can copy the derivation of Eq. (3.60a) in order to show: $\lim_{r\to\infty} g^s(r) = 1$. Anticipating that $g^s(r)-1$ is an absolutely integrable function, one can introduce a real continuous function of wave number q by Fourier transformation:

$$\tilde{h}^s_q = \int\exp[i\vec{q}\vec{r}][g^s(r) - 1]d^3\vec{r}. \tag{5.23b}$$

There holds

$$\langle\rho^{s*}_{\vec{q}}\rho_{\vec{q}}\rangle = \rho\tilde{h}^s_q + (2\pi)^3\rho\delta(\vec{q}). \tag{5.23c}$$

Suppose, the interaction of the tagged particle with the N particles of the system is the same as the one among the system particles. Then, one gets $g^s(r) = g(r)$. This holds provided the thermodynamic limit is performed. In this case, $\rho\tilde{h}^s_q = \rho\tilde{h}_q = S_q - 1 = \rho c_q S_q$. Here, the Ornstein–Zernike relation (3.61) is used in order to introduce the direct correlation function c_q. It is the practice to generalize the concept of a direct correlation function to the case of an arbitrary interaction potential between the tagged particle and the system particles by writing $c^s = h^s_q/S_q$. Hence, one gets for the overlap in Eq. (5.22c)

$$(\rho^s_{\vec{q}}|\rho_{\vec{q}}) = \rho c^s_q S_q. \tag{5.23d}$$

In the third step of the derivation, the correlations of the density-fluctuation pairs are written as products of the correlations between density-fluctuations of the system and those between the tagged-particle densities: $\phi_{\vec{m}_1\vec{m}_2,\vec{n}_1\vec{n}_2}(t) \to \delta_{\vec{m}_1,\vec{n}_1}\delta_{\vec{m}_2\vec{n}_2}NS_{m_1}\phi_{m_1}(t)\phi^s_{m_2}(t)$. This factorization ansatz implies for the elements of the inverse of the metric matrix $(\boldsymbol{g}^{-1})_{\vec{k}_1\vec{k}_2,\vec{m}_1\vec{m}_2}$ the replacement by $\delta_{\vec{k}_1,\vec{m}_1}\delta_{\vec{k}_2,\vec{m}_2}/[NS_{k_1}]$. Notice that cross terms for the factorization $(\rho_{\vec{m}_1}(t)\rho^s_{\vec{n}_2})$ $(\rho^s_{\vec{m}_2}(t)|\rho_{\vec{n}_1})$ are of order N^0 because of Eq. (5.23d). These terms can be neglected in the thermodynamic limit. Combining all items, one gets the mode-coupling contribution to the relaxation kernel (Sjögren 1980b):

$$M^{mc}_{sq}(t) = (v^s_{th}\rho)^2\sum_{\vec{k}+\vec{p}=\vec{q}*} S_k[c^s_k k^3]^2\phi_k(t)\phi^s_p(t)/N. \tag{5.24a}$$

It will be more convenient to express this kernel in terms of a dimensionless function $m_q^s(t)$ by writing $M_{sq}^{mc}(t) = \Omega_{sq}^2 m_q^s(t)$. The kernel $m_q^s(t)$ is given as mode-coupling functional of the correlators. The mode-coupling coefficients are expressed in terms of equilibrium structure functions:

$$m_q^s(t) = L^{-3} \sum_{\vec{k}+\vec{p}=\vec{q}*} V^s(\vec{q}^*, \vec{k}\vec{p})\phi_k(t)\phi_p^s(t), \tag{5.24b}$$

$$V^s(\vec{q}^*, \vec{k}\vec{p}) = \rho S_k[c_k^s k^3/q]^2. \tag{5.24c}$$

Here, $\vec{p}$ is an abbreviation of $\vec{q}^* - \vec{k}$. As explained for the derivation of Eq. (4.11d), the thermodynamic limit can be considered in order to write the kernel as a double integral

$$m_q^s(t) = [\rho/(16\pi^2)] \int_0^\infty dk \int' dp S_k(pk/q^5) \\ \left\{(q^2 + k^2 - p^2)c_k^s\right\}^2 \phi_k(t)\phi_p^s(t). \tag{5.25}$$

The prime indicates that the p-integral is restricted by $|q - k| \leq p \leq |q + k$.

Substitution of the preceding formulas into Eq. (5.21a), one gets for the tagged-particle-density correlator in the frequency domain:

$$\phi_q^s(z) = -1\Big/\Big\{z - \Omega_{sq}^2\Big/\big[z \pm i\nu_q^s + \Omega_{sq}^2 m_q^s(z)\big]\Big\}; \quad \text{Im } z \gtrless 0. \tag{5.26a}$$

If one restricts the discussion to the regime of such small frequencies that $|z|$ can be neglected compared to ν_q^s, the representation simplifies to

$$\phi_q^s(z) = -1\Big/\Big\{z - 1\Big/\big[\pm i\tau_q^s + m_q^s(z)\big]\Big\}; \quad \text{Im } z \gtrless 0. \tag{5.26b}$$

Here, a time scale $\tau_q^s = \nu_q^s/\Omega_{sq}^2$ has been introduced. These formulas together with the expression (5.25) for the kernel provide equations for the functions $\phi_q^s(t)$. Let us assume that these equations have solutions with the standard properties of density-fluctuation correlators. Then, one can rewrite the fractions as integrodifferential equations of the Volterra type as explained in Sec. 4.2 for the density correlators $\phi_q(t)$. One gets from the first fraction

$$\partial_t^2\phi_q^s(t) + \nu_q^s\partial_t\phi_q^s(t) + \Omega_{sq}^2\left[\phi_q^s(t) + \int_0^t m_q^s(t-t')\partial_{t'}\phi_q^s(t')dt'\right] = 0. \tag{5.27a}$$

This first version of the equation of motion has to be solved for the initial conditions $\phi_q^s(0) = 1$, $\partial_t\phi_q^s(0) = 0$. The second fraction is equivalent to the second version of an equation of motion

$$\tau_q^s\partial_t\phi_q^s(t) + \phi_q^s(t) + \int_0^t m_q^s(t-t')\partial_{t'}\phi_q^s(t')dt' = 0, \tag{5.27b}$$

which has to be solved with the initial condition $\phi_q^s(0) = 1$. It shall be demonstrated below that the assumption is justified provided models are considered with finite discrete sets for the wave numbers.

The above formulated microscopic equations for the tagged-particle dynamics deal with a self-consistent-current-relaxation approach very similar to that discussed in Sec. 4.1. The tagged-particle density-fluctuation correlator can exhibit long-time tails if and only if this is true for the fluctuating-force correlator. A necessary condition for a slow fluctuation of forces is that of pair densities $\rho(\vec{r}_1)\rho^s(\vec{r}_2)$. A sufficient condition for the slow fluctuation of the pair densities is the slow fluctuation of $\rho(\vec{r}_1)$ and $\rho^s(\vec{r}_2)$. The simplest model which accounts for these facts is obtained by the factorization ansatz. The resulting nonlinear equations of motion formulate the task to determine the correlators $\phi_q^s(t)$ so that the resulting fluctuating-force kernels $m_q^s(t)$ reproduce the Zwanzig–Mori expressions for the correlators. The thermodynamic-limit results for the density correlators $\phi_q(t)$ are not influenced by the motion of a single tagged particle. However, the dynamics of the tagged particle in its cage depends on the dynamics of the cage-forming neighbours. The motion of these neighbours is characterized by the density correlators $\phi_q(t)$. Knowledge of the functions $\phi_q(t)$ is a prerequisite for specifying and solving the equations of motion for the tagged-particle-density-correlation functions $\phi_q^s(t)$.

The pair of equations (5.26a), (5.27a) has the same structure as the pair (4.29a), (4.17a) for the density correlators. As before, the equations derived here are considered as the MCT equations of motion for the tagged-particle-density-fluctuation correlators for a simple system of particles moving according to Newton's equations of motion. The Eqs. (5.26b) and (5.27b) have the same form as Eqs. (4.29b) and (4.17b), respectively. These are the MCT equations for particles moving according stochastic equations of motion. These equations are proposed for a discussion of glassy colloidal suspensions. It is anticipated that hydrodynamic-interaction effects enter merely indirectly via the time scales τ_q^s.

(ii) Equations for the velocity correlations and mean-squared displacements

In preceding sections, the glassy generalized-hydrodynamics correlations have been characterized by relaxation kernels $m_0(t)$ and $m_0^T(t)$. They describe the cage-effect-induced contributions to the elastic moduli. With the same goal, let us also characterize the glassy generalized-hydrodynamics correlations of the tagged-particle motion by a kernel to be denoted by $m_0^s(t)$. This kernel shall specify the leading Taylor coefficient for small q for the mode-coupling contribution to the kernel $M_{sq}(t)$. Formula (5.24a) yields $\lim_{q\to 0} M_{sq}^{mc}(t) = (v_{th}^s)^2 \rho L^{-3} \sum_{\vec{k}} S_k [c_k^s k^3]^2 \phi_k(t)\phi_k^s(t)$. Thus, there exists the kernel

$$m_0^s(t) = \lim_{q\to 0} (qd)^2 m_q^s(t). \tag{5.28a}$$

A length d has been introduced so that $m_0^s(t)$ is dimensionless; it shall characterize the particle diameter. One arrives at a mode-coupling formula

$$m_0^s(t) = \int_0^\infty dq v_s(q)\phi_q(t)\phi_q^s(t)\ , \qquad v_s(q) = [\rho/6\pi^2]S_q[q^2 c_q^s d]^2. \tag{5.28b}$$

Because of Eqs. (3.49a), (5.21d), the small-q limit ν_0^s of the background spectrum determines that contribution $\xi_s^0 = \mu_s \nu_0^s$ to the friction coefficient, which is not influenced by mode-coupling effects. Equation (3.49c) can be used to relate ξ_s^0 to a diffusivity $D_0^s = (k_B T)/\xi_s^0$. This is a diffusivity expected for 'normal-liquid dynamics'.

$$\lim_{q\to 0} \nu_q^s/(v_{th}^s)^2 = \lim_{q\to 0} q^2 \tau_q^s = 1/D_0^s. \tag{5.28c}$$

Since $M_{sq}(z) = \pm i\nu_q^s + (v_{th}^s)^2 q^2 m_q^s(z)$, Eq. (5.21d) implies the following formula for the friction kernel for Im $z \gtrless 0$:

$$X_s(z) = \Big\{ \pm \left[i/D_0^s\right] + \left[m_0^s(z)/d^2\right]\Big\}(k_{\mathrm{B}}T). \tag{5.28d}$$

Substituting the preceding formula into (3.48a), one gets the velocity correlator in the frequency domain in terms of $m_0^s(z) = LT[m_0^s(t)](z)$:

$$K_s(z) = -1/\Big\{\left[z/(v_{th}^s)^2\right] \pm \left[i/D_0^s\right] + \left[m_0^s(z)/d^2\right]\Big\}, \quad \text{Im } z \gtrless 0. \tag{5.29a}$$

The equations for the stochastic motion have been obtained by restricting the region of frequencies to such small values that z can be dropped in comparison to $\nu_0^s = (v_{th}^s)^2/D_0^s$. In this regime, the velocity correlator simplifies to

$$K_s(z) = -1/\Big\{\pm \left[i/D_0^s\right] + \left[m_0^s(z)/d^2\right]\Big\}, \quad \text{Im } z \gtrless 0. \tag{5.29b}$$

Let us assume that $m_0^s(z)$ is a positive analytic function and that $zm_0^s(z)$ is finite for Im z tending to infinity. The assumption will be justified in the following subsection. In this case, $K_s(z)$ is positive analytic. From Eq. (5.29a), one derives $K_s(z) = -1/[z/(v_{th}^2)]\,\{1 + O(1/z)\}$. As shown in Sec. 2.4.2, $K_s(z)$ is the Laplace transform of a continuous positive-definite function with $K_s(t=0) = (v_{th}^s)^2$. Hence, $K_s(t)$ has the general properties of a correlation function; it has the standard symmetries, and it exhibits the initial condition of the velocity correlator. For $t \geq 0$, it obeys an equation of motion:

$$\partial_t K_s(t) + \left[(v_{th}^s)^2/D_0^s\right]K_s(t) + \left[v_{th}^s/d\right]^2 \int_0^t m_0^s(t-t')K_s(t')dt' = 0. \tag{5.30}$$

Equation (3.32e) can be used to derive formulas for the Laplace transform $\Delta_s(z)$ of the mean-squared-displacement function. For the second version of the equations of motion, one gets

$$\Delta_s(z) = 1\Big/\Big\{z^2\left[(\pm i/D_0^s) + (m_0^s(z)/d^2)\right]\Big\}, \quad \text{Im } z \gtrless 0. \tag{5.31}$$

The first version of the equations of motion yields $[-iz\Delta_s(z)/(v_{th}^s)^2] \pm [\Delta_s(z)D_0^s] - i[m_0^s(z)\Delta_s(z)/d^2] = -i/z^2$. Since $\Delta_s(t=0)=0$, Laplace-back transformation leads to the equation of motion for $t \geq 0$:

$$\partial_t \Delta_s(t)/(v_{th}^s)^2 + \Delta_s(t)/D_0^s + \int_0^t \left[m_0^s(t-t')\Delta_s(t')/d^2 \right] dt' = t. \tag{5.32a}$$

For the theory based on stochastic motion of the particles, the inertia term has to be dropped. One gets an integral equation of the Volterra type:

$$\Delta_s(t) + (D_0^s/d^2) \int_0^t m_0^s(t-t')\Delta_s(t')dt' = D_0^s t. \tag{5.32b}$$

As is explained in Sec. 4.2.1 for a similar problem, one can show that the solutions $\Delta_s(t)$ of the preceding two equations have continuous derivatives of all orders for $t > 0$. Moreover, these derivatives approach finite limits if the positive t tends to zero.

For small times, the equations for $\Delta_s(t)$ can be solved by asymptotic power-law expansion. For the first version of the equations of motion, there holds

$$\Delta_s(t) = \tfrac{1}{2}(v_{th}^s t)^2 \Big\{ 1 - \tfrac{1}{3}\left[(v_{th}^s)^2/D_0^s\right]t$$
$$- \tfrac{1}{12}\left[(m_0^s(t=0)/d^2) - (v_{th}^s/D_0^s)^2\right](v_{th}^s t)^2 + O(t^3) \Big\}. \tag{5.33}$$

A possible friction term in the equations of motion is caused by a positive value for $1/D_0^s$. Such a term leads to a suppression of the ballistic-law increase of $\Delta_s(t)$ with increasing time. The cage effect in dense systems leads to a suppression of the $\Delta_s(t)$ increase. For small times, this effect enters as the positive initial value $m_0^s(t=0)$ of the mode-coupling kernel.

For the second version of the equations of motion, one derives:

$$\Delta_s(t) = D_0^s|t| \left\{ 1 - \tfrac{1}{2} m_0^s(t=0)\left[D_0^s|t|/d^2\right] + O(t^2) \right\}. \tag{5.34a}$$

For small times, the mean-squared displacement increases with t according to Einstein's law (3.40e). Therefore, the parameter D_0^s is called the short-time diffusivity. Again, the increase of $\Delta_s(t)$ is suppressed due to the cage effect, which enters via the initial value of the mode-coupling function. Motivated by Eq. (3.32a) one can define a velocity 'correlator' $\overline{K}_s(t) = \partial_t^2 \Delta_s(t)$. The Laplace transform of this function is $K_s(z)$ from Eq. (5.29b). The latter function is positive analytic. But, it is not a correlation function in the frequency domain since the high-frequency property (2.30) is violated. An auto correlator is a continuous positive-definite function. In particular, its initial value is positive. Contrary to this property, function $\overline{K}_s(t)$ has a negative initial value. From Eq. (5.34a), one infers

$$\overline{K}_s(t \to +0) = -m_0^s(t=0)\left[D_0^s/d^2\right]. \tag{5.34b}$$

Within the framework of the correlation-function theory developed in Chapter 2, a velocity correlator cannot be defined for systems with stochastic particle dynamics.

An addendum to the preceding conclusion might be worthwhile. The second version of the equations of motion has been obtained by extrapolating the small-z behaviour of the correlators onto the complete frequency plane. Such a procedure is discussed in Sec. 3.2.2 in connection with the derivation of the hydrodynamic description of diffusion. The procedure is equivalent to some coarse-graining, as is discussed in Appendix C. The function $\overline{K}_s(t)$ is such coarse-grained correlator derived from $K_s(t)$. Because of the cage effect, $K_s(t)$ decreases from the positive initial value to zero if the time increases to some value of the order of the scale t_{mic} for the normal-liquid motion. A further increase of t, leads to negative values of $K_s(t)$. This behaviour has been discussed in connection with the upper panel of Fig. 3.2; and it is exhibited by the numerical solutions of Eq. (5.30) for the HSS (Chong *et al.* 2001a). For short times, a coarse graining of $K_s(t)$ on scale t_{mic} implies a cancellation of the positive terms by the negative ones, and this leads to a reduction of $\overline{K}_s(t=0)$ compared to $K_s(t=0)$. The MCT formula (5.34b) states that the negative terms overwhelm the positive ones. The appearance of a negative long-time tail of $K_s(t)$ is an outstanding feature of the glassy dynamics described by MCT. The negative tail exhibited by the upper panel of Fig. 3.2 is a precursor of the glassy dynamics. The coarse-graining on scale t_{mic} does not change the correlator far outside the transient. This is confirmed by comparisons of the numerical results for $\Delta_s(t)$ versus t/t_0 curves and for the second derivatives of these curves, calculated for the HSS for the two versions of the equations of motion (Chong *et al.* 2001a). The details of the models and the values of the scales are specified in the caption of Fig. 4.24. The preceding reasoning and the cited numerical work suggests that $\overline{K}_s(t)$ is the velocity correlator for the model for stochastic dynamics for times outside some transient regime. The details of the transient merely determine the scale t_0 for the glassy dynamics.

(iii) MCT equations of motion

The preceding derivations have to be considered as motivation of closed equations of motion for tagged-particle dynamics. In this subsection, it shall be explained in which sense the above derived equations define models for the evaluation of correlation functions $\phi_q^s(t)$ and for the functions $K_s(t)$ and $\Delta_s(t)$. It shall be demonstrated that these equations have unique solutions for the mentioned functions, which exhibit the general properties discussed in Sec. 4.2 for the basic version of MCT.

To proceed, the wave vector moduli $q, k, \ldots$ used in the equations from above shall be considered as labels for arrays of M elements. All integrals shall be considered as abbreviations of corresponding Riemann sums as explained in connection with Eqs. (4.35a,b). It is the primary goal to calculate an array of functions $\boldsymbol{\phi}^s(t) = (\phi_1^s(t), \ldots, \phi_M^s(t))$. Equations (5.27a,b) together with their initial conditions formulate two versions of equations of motion for these arrays. The first version is specified by an array of positive frequencies $\boldsymbol{\Omega}_s = (\Omega_{s1}, \ldots, \Omega_{sM})$ and by another array of M non-negative frequencies $\boldsymbol{\nu}^s = (\nu_1^s, \ldots, \nu_M^s)$. These two arrays quantify the short-time dynamics in leading order. Similarly, the array of positive times $\boldsymbol{\tau}^s = (\tau_1^s, \ldots, \tau_M^s)$ specifies the short time asymptote of the correlator array for the second version of the dynamics. All general properties to be requested here for the arrays $\boldsymbol{\phi}^s(t)$ are the same as those requested in Sec. 4.2.1 for the arrays $\boldsymbol{\phi}(t)$. The central piece in the equations of motion is the array of kernels $\boldsymbol{m}^s(t) = (m_1^s(t), \ldots, m_M^s(t))$. According to Eq. (5.25), the kernels are given as bilinear polynomials of the arrays $\boldsymbol{\phi}(t)$ and $\boldsymbol{\phi}^s(t)$:

$$\boldsymbol{m}^s(t) = \boldsymbol{\mathcal{F}}^s[P^s, \boldsymbol{\phi}(t), \boldsymbol{\phi}^s(t)]. \tag{5.35a}$$

The mode-coupling polynomials can be denoted in the form

$$\mathcal{F}_q^s[P^s, \boldsymbol{x}, \boldsymbol{y}] = \sum_{kp} V_{q,kp}^s x_k y_p. \tag{5.35b}$$

The explicit values of the coefficients $V_{q,kp}^s$ can be inferred from the following formula, where the same abbreviations are used as in Eq. (4.36):

$$\mathcal{F}_q^s[P^s, \boldsymbol{x}, \boldsymbol{y}] = \Big\{ \big[\rho/(16\pi^2)\big](h/d)^3 \Big\} \sum_{\hat{k}} \sum_{\hat{p}}{}' S_k(\hat{k}\hat{p}/\hat{q}^5) \Big\{ (\hat{q}^2 + \hat{k}^2 - \hat{p}^2) c_k^s \Big\}^2 x_k y_p. \tag{5.35c}$$

The important mathematical property of the mode-coupling polynomials are the restrictions $V_{q,kp}^s \geq 0$ for the mode-coupling coefficients and the equation $\boldsymbol{\mathcal{F}}^s[P^s, \boldsymbol{0}, \boldsymbol{y}] = 0$. The latter condition formulates the request that the dynamics of $\boldsymbol{\phi}^s(t)$ is coupled to that of $\boldsymbol{\phi}(t)$. As was the case for the basic version of MCT, the coefficients $V_{q,kp}^s$ are determined by the structure functions of the equilibrium. The symbol P^s shall be used to denote the set of coupling constants $V_{q,kp}^s$. The latter play the role of general control parameters. The interaction of the tagged particle with the liquid particles might introduce external control parameters, which do not enter P. This happens, for example, if the tagged-particle's diameter d^s is considered as control parameter, which can be different from the diameter d of the liquid constituents.

The preceding two paragraphs formulate the problem of solving a set of M coupled nonlinear integrodifferential equations. The array $\boldsymbol{\phi}(t)$ enters the equations as input information about coefficients. In the present paragraph, it shall be shown that this problem can be considered as a special case of the problem studied in Sec. 4.2. To demonstrate this, arrays of $2M$ components shall be composed of two M-component arrays. They shall be denoted by bold letters with a tilde, for example, $\tilde{\boldsymbol{x}} = (\boldsymbol{x}, \boldsymbol{y}) = (x_1, \ldots, x_M, y_1, \ldots, y_M)$. The theory under discussion deals with the array $\tilde{\boldsymbol{\phi}}(t) = (\boldsymbol{\phi}(t), \boldsymbol{\phi}^s(t))$. The first version of equations of motion is specified by two arrays of frequencies $\tilde{\boldsymbol{\Omega}} = (\boldsymbol{\Omega}, \boldsymbol{\Omega}_s)$ and $\tilde{\boldsymbol{\nu}}^s = (\boldsymbol{\nu}, \boldsymbol{\nu}^s)$; the second one by an array of times $\tilde{\boldsymbol{\tau}}^s = (\boldsymbol{\tau}, \boldsymbol{\tau}^s)$. The retardation integrals in the equations of motion are determined by an array of kernels $\tilde{\boldsymbol{m}}(t) = (\boldsymbol{m}(t), \boldsymbol{m}^s(t))$. The latter are given as functionals of the correlators specified by a pair of coupling-constant sets P and P^s: $\tilde{\boldsymbol{m}}(t) = \tilde{\boldsymbol{\mathcal{F}}}[P, P^s, \tilde{\boldsymbol{\phi}}(t)]$. The general equations of motion for the $2M$-component arrays (4.17a) or (4.17b), respectively, split in M equations for the array $\boldsymbol{\phi}(t)$ and in M equations for the array $\boldsymbol{\phi}^s(t)$. The first set agrees with the equations of motion studied in Sec. 4.2.3 for the microscopic MCT for simple systems. This holds if the first M components of functional $\tilde{\boldsymbol{\mathcal{F}}}$ agree with the functional from Eq. (4.36):

$$\tilde{\mathcal{F}}_q[P, P^s, \tilde{\boldsymbol{x}}] = \mathcal{F}_q[P, \boldsymbol{x}], \quad q = 1, \ldots, M. \tag{5.36a}$$

The second set of the equations of motion for $\tilde{\boldsymbol{\phi}}(t)$ agrees with the set of M equations (5.27a) or (5.27b), respectively, for $\boldsymbol{\phi}^s(t)$. This holds provided one expresses the second group of components of $\tilde{\boldsymbol{\mathcal{F}}}$ in terms of the functionals from Eq. (5.35c):

$$\tilde{\mathcal{F}}_{M+q}[P, P^s, \tilde{\boldsymbol{x}}] = \mathcal{F}_q^s[P^s, \boldsymbol{x}, \boldsymbol{y}], \quad q = 1, \ldots, M. \tag{5.36b}$$

It is left to the reader to check that the above formulated model for a dynamics for the $2M$-component array $\tilde{\boldsymbol{\phi}}(t)$ obeys all the conditions, which are specified in Sec. 4.2.1 for an MCT model.

The observations from above permit to cite all general results from Secs. 4.2 and 4.3 as valid for the components $\phi_q^s(t)$, $q = 1, \ldots, M$ of the general array $\tilde{\boldsymbol{\phi}}(t)$. There hold the existence theorem and the uniqueness theorem for the solutions of the equations of motion for $\boldsymbol{\phi}^s(t)$. For $t \neq 0$, the solutions have continuous derivatives of all orders. The solutions are positive-definite functions. If the second version of the equations of motion are used, the $\phi_q^s(t)$, $q = 1, \ldots, M$, are completely monotonic. The Laplace transforms $\phi_q^s(z)$ exist; they are positive analytic and obey the condition $\lim_{y\to\infty} \phi_q^s(iy)iy = -1$. The analogous properties are exhibited by the kernels $m_q^s(t)$ with $\lim_{y\to\infty} m_q^s(iy)iy = -m_q^s(t=0) = -\mathcal{F}[P^s, \mathbf{1}, \mathbf{1}]$. The double-fraction representations (5.26a) and (5.26b) of $\phi_q^s(z)$ in terms of $m_q^s(z)$ are equivalent to the equations of motion (5.27a) and (5.27b), respectively. For all finite time intervals, the functions $\phi_q^s(t)$ and $m_q^s(t)$, $q = 1, \ldots, M$, depend continuously on the non-negative mode-coupling constants and on the parameters, which specify the transient. In summary, a model for a statistical description of a dynamics is formulated. It is an extension of the microscopic MCT explained in Sec. 4.2.3. The functions $\phi_q^s(t)$ and $m_q^s(t)$ shall be referred to as tagged-particle-density-fluctuation correlators and tagged-particle-fluctuating-force correlators for wave number q, respectively.

The mean-squared-displacement function $\Delta_s(t)$ and the equivalent velocity correlator $K_s(t)$ specify the small-wave-vector limit of $[\phi_q^s(t) - 1]/q^2$. Such limit is not defined, if the wave-number interval is replaced by a grid of M values. Assuming that zero-wave-number limits can be performed in the originally motivated mode-coupling expressions, a kernel $m_0^s(t)$ can be obtained as mode-coupling functional (5.28b). Replacing the integral by a Riemann sum, the kernel is given as polynomial $m_0^s(t) = (h/d)\sum_{\hat{q}} v_s(q)\phi_q(t)\phi_q^s(t)$. This motivates the MCT formula (5.29a) for $K_s(z)$. The kernel $m_0^s(t)$ has all the properties required to derive the Eqs. (5.29b)–(5.34b). Therefore, $\Delta_s(t)$ and $K_s(t)$ are the model functions for the corresponding quantities introduced in Sec. 3.2.1.

The arrested parts of the density-fluctuation correlations play an important role for the discussion of glassy dynamics. Let us consider these quantities for the tagged-particle motion:

$$\lim_{t\to\infty} \boldsymbol{\phi}^s(t) = \boldsymbol{f}^s(P, P^s). \tag{5.37a}$$

The arrested part for the $2M$-component theory defined above reads $\tilde{\boldsymbol{f}}(P, P^s) = (\boldsymbol{f}(P), \boldsymbol{f}^s(P, P^s))$. The first M components of the fixed-point equation for the $2M$-component model reproduce Eq. (4.52d) for the arrested parts $\boldsymbol{f}(P)$ for the density correlators. The second group of M components shows that $\boldsymbol{f}^s(P, P^s)$ is a solution of the fixed-point equation

$$f_q^s(P, P^s)/[1 - f_q^s(P, P^s)] = \mathcal{F}_q^s[P^s, \boldsymbol{f}(P), \boldsymbol{f}^s(P, P^s)]; \quad q = 1, \ldots, M \tag{5.37b}$$

(Bengtzelius *et al.* 1984). The arrested part $\boldsymbol{f}^s(P, P^s)$ is distinguished from other fixed points $\boldsymbol{f}^{s*}(P, P^s)$ by the maximum property (4.52e): $f_q^s(P, P^s) \geq f_q^{s*}(P, P^s)$, $q = 1, \ldots, M$. This distinguished solution of Eq. (5.37b) can be obtained by the iteration procedure explained in Eqs. (4.53a–c). In solid-state physics, $\boldsymbol{f}^s(P, P^s)$ is known as Lamb–Mössbauer factor. It determines the cross section for elastic incoherent scattering of neutrons (Lovesey 1984).

The above formulated $2M$-component MCT model exhibits a peculiarity. And this requires reexamination of those results of Sec. 4.3, which have been derived by appeal to generic behaviour. To specify this peculiarity for the most relevant case, the stability matrix of the theory shall be considered. According to Eq. (4.60a), this is the matrix $\tilde{A}$ formed with the $2M$ derivatives of the $2M$ mode-coupling polynomials: $\tilde{A}_{qp}(P, P^s) = [1 - \tilde{f}_q(P, P^s)]\{\partial\tilde{\mathcal{F}}_q[P, P^s, \tilde{\boldsymbol{f}}(P, P^s)] /\partial\tilde{x}_p\}[1 - \tilde{f}_p(P, P^s)]$, $q, p = 1, \ldots, 2M$. This matrix can be considered as a 2-by-2 block matrix formed with four M-by-M blocks. According to Eq. (5.36a), the upper left block is identical to the stability matrix considered in Sec. 4.3 for the M-component MCT:

$$\tilde{A}_{q,p}(P, P^s) = A_{q,p}(P); \qquad q, p = 1, \ldots, M. \tag{5.38a}$$

This block does not depend on the control parameters P^s. The $\boldsymbol{y}$-independence of the functions $\tilde{\mathcal{F}}_q$, $[P, P^s, \tilde{\boldsymbol{x}}]$, $\tilde{\boldsymbol{x}} = (\boldsymbol{x}, \boldsymbol{y})$, $q = 1, \ldots, M$, is the essential peculiarity; it implies a zero matrix for the upper right block

$$\tilde{A}_{q,p}(P, P^s) = 0; \quad q = 1, \ldots, M, \quad p = M+1, \ldots, 2M. \tag{5.38b}$$

The lower left block shall be specified by a matrix B^s:

$$\tilde{A}_{M+q,p}(P,P^s) = B^s_{q,p}(P,P^s), \quad q,p = 1,\ldots,M, \tag{5.38c}$$

$$B^s_{q,p}(P,P^s) = [1 - f^s_q(P,P^s)] \sum_k V^s_{q,pk} f^s_k(P,P^s)[1 - f_p(P)]. \tag{5.38d}$$

Finally, there is the lower right block matrix $A^s(P,P^s)$, specified by:

$$\tilde{A}_{M+q,M+p} = A^s_{q,p}(P,P^s); \quad q,p = 1,\ldots,M, \tag{5.38e}$$

$$A^s_{q,p}(P,P^s) = [1 - f^s_q(P,P^s)] \sum_k V^s_{q,kp} f_k(P)[1 - f^s_p(P,P^s)]. \tag{5.38f}$$

In the same manner as explained in connection with Eqs. (4.56a–c), one shows that the Jacobian matrix of the set of M implicit equations (5.37b) for the M variables $f^s_1,\ldots,f^s_M$ is equivalent to an M-by-M matrix with elements $\delta_{qp} - A^s_{q,p}(P,P^s)$; $q,p = 1,\ldots,M$. For the equation for fixed point $\boldsymbol{f}^s$, the components of $\boldsymbol{f}(P)$ enter as linear coefficients in the coupling constants. The dependence of the solution $\boldsymbol{f}^s(P,P^s)$ on P enters only indirectly via the P dependence of $\boldsymbol{f}(P)$. The arrested part $\boldsymbol{f}(P)$ influences the arrested part $\boldsymbol{f}^s(P,P^s)$. The peculiarity of the $2M$-component model is that there is no influence of $\boldsymbol{f}^s(P,P^s)$ on the arrested part $\boldsymbol{f}(P)$. As a result, the stability matrix has a triangular form; it does not exhibit the irreducibility property.

5.1.4 *Idealized transitions from diffusion to localization*

It shall be discussed in this section how the generic liquid–glass transition causes a transition from diffusive to localized motion for a tagged particle. If the tagged particle differs from the other constituents of the system, there may exist two types of glasses. In one type, the tagged particle is localized; and in the other one, it can percolate through the amorphous solid.

(i) The diffusivity

Let P denote a liquid state, i.e., there is no arrested part of the density fluctuation correlations: $\boldsymbol{f}(P) = \boldsymbol{0}$. The right-hand side of Eq. (5.37b) vanishes and, hence, there is no arrested part of the tagged-particle correlations either: $\boldsymbol{f}^s(P,P^s) = 0$. This holds independently of the size of the mode-coupling coefficients $V^s_{q,kp}$. If the density fluctuations of the environment of the tagged particle cannot arrest, also the fluctuating-force correlation function have to relax to zero for long times: $m^s_q(t \to \infty) = 0$. As a result, there are no arrested parts of the tagged-particle-density correlations: $\phi^s_q(t \to \infty) = 0$. The tagged-particle probability distribution spreads out over the whole space. The probability for finding the tagged particle in a fixed bounded region becomes arbitrarily small if the time becomes arbitrarily large.

Let us follow the reasonings of Sec. 3.2 and assume that $m^s_0(t)$ is absolutely integrable. In this case, the kernel in the frequency domain has continuous boundary values $m^s_0(\omega \pm i0)$. Because of Eqs. (5.29a,b), there is a continuous velocity spectrum. All results from Sec. 3.2 can be used. In particular, one concludes that the dynamics of the tagged-particle motion within the hydrodynamics-limit regime is described by the diffusion equation. The result is quantified by the diffusivity D_s. Equations (3.33a) and (5.29a) yield the expression for this transport coefficient of the liquid:

$$D_s = 1/\left[(1/D_s^0) + (m_0^{s\prime\prime}(\omega = 0)/d^2)\right]. \tag{5.39a}$$

There holds Einstein's law (3.33c) for the mean-squared displacement

$$\lim_{t\to\infty} \delta r_s^2(t)/t = 6D_s. \tag{5.39b}$$

If the second version of the equations of motion is considered, one can introduce the function $f(t) = \Delta_s(t)/t$ and rewrite Eq. (5.32b) to $f(t)/D_s^0 = 1 - \int_0^\infty \{f(t - t')[1-(t'/t)]\theta(t-t')m_0^s(t')/d^2\}dt'$. Proceeding as for the derivation of Eq. (3.33c), one arrives at $\lim_{t\to\infty} f(t)/D_s^0 = D_s/D_s^0$ with D_s given by Eq. (5.39a). If the distinction between D_s and D_s^0 shall be emphasized, D_s is called the long-time diffusivity. The zero-frequency spectrum of the kernel $m_0^s(t)$ determines the suppression of the diffusivity due to the cage effect: $[(1/D_s) - (1/D_s^0)]d^2 = m_0^{s\prime\prime}(\omega = 0) > 0$.

Three paragraphs with degressions shall be added to the preceding results on liquid dynamics. The first one concerns the number $1/D_s^0$. This coefficient exhibits a non-trivial dependence on control parameters like the density ρ. In a hard-sphere liquid, for example, D_s^0 varies with ρ because of vortex-flow effects (Hansen and McDonald 1986). In colloids D_s^0 changes with ρ also because of hydrodynamic interactions (Pusey 1991). Within the parameter regime of interest, the changes of $1/D_s^0$ are not larger than, say, a factor 3. The theory used in this book does not contribute to the understanding of the indicated variations of $1/D_s^0$ due to 'normal liquid state' effects.

The above made assumption on the integrability of $m_0^s(t)$ has to be considered as a conjecture. The motivation for the conjecture is that the mentioned regularity property can be proven within the theory, which is based on the second version for the equations of motion. In this case, the following theorem is valid (Götze and Sjögren 1995). For all regular states, there is some positive rate γ_0 so that the representation (D.10a) of Appendix D.2 reads

$$\phi_q(t) = f_q + \int_{\gamma_0}^\infty \exp[-\gamma|t|]d\sigma_q^\phi(\gamma), \qquad q = 1, \ldots, M. \tag{5.40a}$$

The weight function $\sigma_q^\phi(\gamma)$ increases with γ from 0 to $1 - f_q$. The proof is too involved to be reproduced in this book. Laplace transformation yields the representation of the correlators in the frequency domain:

$$\phi_q(z) = [-f_q/z] + \int_{\gamma_0}^\infty \{-1/[z + i\gamma]\}d\sigma_q^\phi(\gamma). \tag{5.40b}$$

For regular states, there is a gap for the rate distribution of a size not smaller than γ_0. Consequently, the correlators approach exponentially their long-time limits: $\phi_q(t) = f_q + O(\exp[-\eta t])$, $0 < \eta < \gamma_0$. Up to a possible pole $[-f/z]$, the correlators in the frequency domain are holomorphic for Im $z > -\eta$. As explained above, the theory for the tagged-particle quantities can be considered as a special case of the general formalism. Hence, the preceding formulas are valid also with indices s added in the appropriate places. The specified results for the correlators imply corresponding ones for mode-coupling polynomials like $m_q(t)$, $m_0^T(t)$ or $m_q^s(t)$. In particular, for liquid states, $m_0^s(t)$ decreases exponentially for diverging t. Hence, it is absolutely integrable.

For liquid states, $m_0^s(z)$ is holomorphic for Im $z > -\eta$. One can write $m_0^s(z) = ia_0 - a_1 z + z^2 g_1(z)$ with $g_1(z)$ being holomorphic in the specified region. The integral representation for $m_0^s(z)$ corresponding to Eq. (5.40b) shows: $a_0 = m_0^{s\prime\prime}(0)$ is the positive fluctuation spectrum at $\omega = 0$ and $a_1 = -\partial m_0^s(z = 0)/\partial z$ is the negative slope of the kernel's reactive part at the frequency origin, i.e.,

$$a_0 = \int_{\gamma_0}^{\infty} \gamma^{-1} d\sigma_0^m(\gamma), \qquad a_1 = \int_{\gamma_0}^{\infty} \gamma^{-2} d\sigma_0^m(\gamma). \tag{5.40c}$$

Substitution of the formula for $m_0^s(z)$ into Eq. (5.31) yields $\Delta_s(z) = [-iD_s/z^2] - [D_s^2 a_1/(d^2 z)] + g_2(z)$. Here, $g_2(z)$ is holomorphic for Im $z > -\eta$. Hence, its Laplace back transform is bounded by a multiple of $\exp[-\eta|t|]$ (Widder 1946). One obtains the extension of Eq. (5.39b):

$$\Delta_s(t) = D_s|t|\Big\{1 + \big[(a_1/d^2)D_s/|t|\big]\Big\} + O(\exp[-\eta|t|]). \tag{5.40d}$$

Because of the cage effect, the linear long-time asymptote of the $\Delta_s(t)$ versus t curve is approached from above. This result is opposite to what is predicted by the Langevin-theory formula (3.52); and it explains qualitatively the long-time behaviour of the data exhibited in Fig. 3.1 for a dense high-temperature Lennard-Jones mixture.

(ii) The localization length

The other alternative to be discussed are the ideal glass states. Generically, such states P as well as the critical states P^c are characterized by positive Debye–Waller factors $f_q(P) > 0$, $q = 1, \ldots, M$. Let us express the density-fluctuation correlators $\phi_q(t)$ in terms of other normalized correlation functions $\hat{\phi}_q(t)$ according to Eq. (4.24a):

$$\phi_q(t) = f_q(P) + (1 - f_q(P))\hat{\phi}_q(t), \quad q = 1, \ldots, M. \tag{5.41a}$$

The correlator array $\hat{\boldsymbol{\phi}}(t)$ describes the density-fluctuation dynamics relative to the arrested background. These correlations do not exhibit arrested parts:

$$\hat{\boldsymbol{\phi}}(t \to \infty) = 0. \tag{5.41b}$$

The fluctuating-force correlator for the tagged-particle dynamics splits into two contributions:

$$\boldsymbol{m}^s(t) = \boldsymbol{m}^{s(1)}(t) + \boldsymbol{m}^{s(2)}(t), \tag{5.42a}$$

$$m_q^{s(1)}(t) = \sum_p U_{q,p}^{(1)} \phi_p^s(t), \qquad m_q^{s(2)}(t) = \sum_{kp} U_{q,kp}^{(2)} \hat{\phi}_k(t) \phi_p^s(t). \tag{5.42b}$$

The new mode-coupling coefficients are obtained as modifications of the original ones from Eq. (5.35b):

$$U_{q,p}^{(1)} = \sum_k V_{q,kp}^s f_k(P), \qquad U_{q,kp}^{(2)} = V_{q,kp}^{(s)}[1 - f_k(P)]. \tag{5.42c}$$

The fixed-point equation (5.37b) for the long-time limits of the tagged-particle-density correlations is determined by the coefficients $U_{q,p}^{(1)}$:

$$f_q^s(P, P^s)/[1 - f_q^s(P, P^s)] = \sum_p U_{q,p}^{(1)} f_p^s(P, P^s). \tag{5.43a}$$

The stability matrix for this linear mode-coupling functional is identical with the one noted in Eq. (5.38f):

$$A_{q,p}^s = (1 - f_q^s) U_{qp}^{(1)} (1 - f_p^s). \tag{5.43b}$$

For reasons of transparency, the reference to the control parameters P and P^s is suppressed in the preceding formulas for the coefficients $U^{(1)}$ and $U^{(2)}$.

According to the discussion of Sec. 4.2.1, there are two generic possibilities. The corresponding states shall be called glass 1 and glass 2. The former are characterized by positive long-time limits of the density correlators $\phi_q^s(t)$ and latter by vanishing ones. There holds for $q = 1, \ldots, M$:

$$\text{glass 1}: \quad f_q^s(P, P^s) > 0, \tag{5.44a}$$

$$\text{glass 2}: \quad f_q^s(P, P^s) = 0. \tag{5.44b}$$

If the M^2 mode-coupling constants $U_{q,p}^{(1)}$ are sufficiently small in the sense of Eq. (4.54), there holds $\boldsymbol{f}^s(P, P^s) = 0$. The constant $U_{q,p}^{(1)}$ is a linear combination of the M values $V_{q,kp}^s$, $k = 1, \ldots, M$, formed with the bounded coefficients $f_k(P)$. Consequently, the system is a glass 2 if all the M^3 coupling constants $V_{q,kp}^s$ are sufficiently small. As specified in Eqs. (4.55a–d), the system is in glass 1 provided the $V_{q,kp}^s$ are sufficiently large. Glass 1 states and glass 2 states are separated by transition points. Generically, these are A_2 bifurcations. For all regular glass 2 states, there hold Eqs. (5.39–5.40). For such states, the tagged particle diffuses through the system as explained above for the liquid.

Let (P, P^s) denote a regular glass 1 state. From Eq. (5.28b) one concludes that the kernel $m_0^s(t)$ exhibits a positive finite arrested part. It shall be expressed in terms of a length, which is denoted by $r_s, 0 < r_s < \infty$:

$$m_0^s(t \to \infty) = (d/r_s)^2, \tag{5.45a}$$

$$(r_s/d)^2 = 1 \Big/ \left[\int_0^\infty dq v_s(q) f_q f_q^s \right]. \tag{5.45b}$$

This length quantifies a zero-frequency pole of the kernel in the frequency domain. One can write $m_0^s(z) = -[(d/r_s)^2/z] + g_1(z)$, with $g_1(z)$ being the Laplace transform of a positive-definite continuous function without arrested part. Substituting either Eq. (5.29a) or Eq. (5.29b) into Eq. (3.32e), one gets $\Delta_s(z) = -(r_s^2/z)/[1 + g_2(z)]$. The function $g_2(z)$ is holomorphic for Im $z > 0$, and $g_2(z \to 0) = 0$. Consequently, $\lim_{z \to 0}[-z\Delta_s(z)] = r_s^2 = \Delta_s(t \to \infty)$. The mean-squared displacement approaches a finite long-time limit:

$$\lim_{t \to \infty} \delta r_s^2(t) = 6r_s^2. \tag{5.46a}$$

From Eq. (3.30), one infers that the length r_s determines the small-wave-number limit of the Lamb–Mössbauer factor:

$$f_q^s = 1 - (qr_s)^2 + O(q^4). \tag{5.46b}$$

If the second version is used for the equations of motion, one can modify the reasoning given above and show that $\Delta_s(z) = -(r_s^2/z) + g_3(z)$. Here, $g_3(z)$ is holomorphic for Im $z > -\eta$. This implies that the long-time limit is approached exponentially: $\Delta_s(t) = r_s^2 + O(\exp[-\eta|t|])$.

Equation (5.46b) shows that there is perfect arrest of the correlations of the tagged-particle-density fluctuations in the long-wave-length limit: $\lim_{q \to 0} f_q^s = 1$. Details of the glass structure do not enter the limit. This feature is an implication of the small-q divergency of the mode-coupling coefficients for the fluctuating-force correlator $m_q^s(t)$, which is specified in Eqs. (5.28a–c). Using Eq. (5.45a), one obtains for small values of (qd) : $m_q^s(t \to \infty) = m_0^s(t \to \infty)/(qd)^2 = 1/(qr_s)^2$. Equations (5.35), (5.37) yield the expression for the arrested part: $f_q^s = m_q^s(t \to \infty)/\,[1 + m_q^s(t \to \infty)]$. As a result, one gets the formula for the Lamb–Mössbauer factor:

$$f_q^s = 1/\left[1 + (qr_s)^2\right], \quad qd \ll 1. \tag{5.46c}$$

The generalized-hydrodynamics expression for the tagged-particle correlator is obtained from Eqs. (3.35a), (3.36b): $\phi_q^s(z) = -1/[z + q^2 K_s(z)]$, $qd \ll 1$. This formula yields for the arrested parts: $f_q^s = \lim_{z \to 0}[-z\phi_q^s(z)] = 1/[1 + (qR)^2]$, $R^2 = \lim_{z \to 0}[K_s(z)/z]$. Equation (5.45a) is equivalent to $\lim_{z \to 0}[-zm_0^s(z)/d^2] = 1/r_s^2$. Consequently, one derives from Eq. (5.29a) or from Eq. (5.29b): $R^2 = r_s^2$. Equation (5.46c) (Götze 1978) is the generalized-hydrodynamics result for the arrested part.

The Fourier back transform $\phi^s(r, t)$ of $\phi_q^s(t)$ has a simple meaning, as is discussed in Sec. 3.2.1. It is the probability density for finding the tagged particle at time t at a position $\vec{r}$ with modulus r if it is located at time $t = 0$

at the origin $\vec{r} = \vec{0}$ of the position space. For the states under consideration, this probability density approaches a stationary one: $f^s(r) = \phi^s(r, t \to \infty) = \int \exp[-i\vec{q}\vec{r}] f_q^s d^3\vec{q}/(2\pi)^3$. The tagged particle is localized, and $f^s(r)$ is its stationary distribution. The parameter r_s is a measure for the extend of the distribution: $r_s^2 = \int r^2 f^s(r) d^3\vec{r}/6$. Hence, it is called the localization length. The arrested structure of the glass causes a random array of arrested cages. The frozen cage-forming forces are determined by the interaction between the tagged particle and the neighbour particles, by the frozen structure of the glass, and by the arrested distribution of the tagged particle. These three factors enter Eq. (5.45b) as $v_s(q), f_q$ and f_q^s, respectively. The localization length presents a result for the averaging over the cage distribution. The transitions from the liquid to glass 1 and also the one from glass 2 to glass 1 are transitions from a diffusive dynamics for the tagged particle to a localized one.

(iii) Arrest near the transition from glass 1 to liquid

In order to study the arrest near the liquid–glass transition, a shift of state P through a fold-bifurcation point P^c shall be considered as specified in Eqs. (4.84a–c). The notations and results for the separation parameter σ, the distance parameter ϵ, the parameters λ and μ_2 etc. shall be used as explained in Sec. 4.3.4. These quantities keep their meaning also if the transition is considered as one for the $2M$-component model defined in the preceding section. The coordinates of P^s do not enter these parameters. For the generic transition from the liquid to the glass 2, there holds $\boldsymbol{f}^s = \boldsymbol{0}$ also for sufficiently small positive ϵ. Nothing remains to be done in this case. Therefore, the following discussions refer to transitions from liquid states to glass 1 states.

Generically, the discontinuous change of $\boldsymbol{\phi}(t \to \infty)$ at the transition causes a similar discontinuous change of $\boldsymbol{\phi}^s(t \to \infty)$. The discontinuity depends on P^s and shall be denoted by $\boldsymbol{f}^{sc}(P^s) = \boldsymbol{f}^s(P^c, P^s)$. The $2M$-component model exhibits a generic A_2 bifurcation. It is characterized by a critical arrested part $\tilde{\boldsymbol{f}}^c$ with $2M$ positive components:

$$f_q^c > 0, \quad f_q^{sc}(P^s) > 0, \quad q = 1, \ldots, M. \tag{5.47a}$$

The critical arrested parts for the tagged-particle density correlations are to be calculated from the specialization of Eq. (5.37b) to $P = P^c$.

$$f_q^{sc}(P^s)/[1 - f_q^{sc}(P^s)] = \mathcal{F}_q^s[P^s, \boldsymbol{f}^c, \boldsymbol{f}^{sc}(P^s)]. \tag{5.47b}$$

All results from Sec. 4.3.4 remain valid. For example, Eq. (4.91a) implies the square-root law for the Lamb–Mössbauer factor

$$\lim_{\epsilon \to +0} [\boldsymbol{f}^s(P, P^s) - \boldsymbol{f}^{sc}(P^s)]/\sqrt{\epsilon} = \boldsymbol{h}^s(P^s)\sqrt{C/\mu_2^c}. \tag{5.48a}$$

Equation (5.45b) implies a similar law for the square of the localization length:

$$r_s^2(P, P^s) - r_{sc}^2(P^s) = -h_{\text{MSD}}(P^s)\sqrt{C\epsilon/\mu_2^c} + O(\epsilon). \tag{5.48b}$$

Here, $h_{\mathrm{MSD}}(P^s)$ and $h^s_q(P^s)$, $q = 1, \ldots, M$, denote positive critical amplitudes; r_{sc} is the critical locatization length. These results and their extensions will be of relevance for the discussions in chapter 6. Therefore, some details shall be added.

It is shown in the preceding section that the stability matrix for the $2M$-component model is a triangular 2-by-2 block matrix. Because of Eq. (5.38a), the upper left block is the stability matrix for the basic version of MCT. This matrix has a non-degenerate maximum eigenvalue $E(P)$, which obeys $0 < E(P) \leq 1$. The other $(M-1)$ eigenvalues have a modulus not larger than $E(P)$. Because of Eqs. (5.38e, f), (5.43a,b), the lower right block is the stability matrix $A^s(P, P^s)$ of a generic M-component MCT model. Consequently, also this matrix has a non-degenerate maximum eigenvalue, to be denoted by $E^s(P, P^s)$. Since it is an eigenvalue of the stability matrix $\tilde{A}(P, P^s)$ of the $2M$-component model, there holds: $E^s(P, P^s) \leq 1$. A transition point from the liquid to the amorphous solid is a critical point P^c. It is characterized by $E(P^c) = 1$. It would be non generic if also $E^s(P^c, P^s)$ would be unity. Hence, a generic transition point is characterized by the inequality $E^s(P^c, P^s) < 1$. The eigenvalue $E^s(P, P^s)$ depends continuously on the matrix elements. Consequently, for every generic transition point, there is a neighbourhood of points $P^{s\prime}$ near P^s and a neighbourhood of glass states P near P^c so that

$$E^s(P, P^{s\prime}) < 1. \tag{5.49}$$

The following discussion shall be restricted to these regions of P and P^s.

Since the spectral radius of the matrix $A^s(P, P^s)$ is smaller than unity, there exists the inverse $I^s(P, P^s)$ of the M-by-M matrix with elements $[\delta_{q,p} - A^s_{q,p}(P, P^s)]$. It shall be specified by the elements $I^s_{q,p}(P, P^s)$ so that

$$\sum_{\ell} I_{q,\ell}(P, P^s)[\delta_{\ell,k} - A^s_{\ell,k}(P, P^s)] = \delta_{q,k}; \quad q, k = 1, \ldots, M. \tag{5.50a}$$

As discussed in connection with Eq. (4.59b), matrix $I^s(P, P^s)$ can be calculated as Neumann series. This series representation implies the inequalities

$$I^s_{q,p}(P, P^s) \geq 0; \quad q, p = 1, \ldots, M. \tag{5.50b}$$

The preceding insight shall be applied for the calculation of the change of the critical Lamb–Mössbauer factor $\boldsymbol{f}^{sc}(P^s) = \boldsymbol{f}^s(P^c, P^s)$ due to changes of P^s. To proceed, let $V^{s0}_{q,kp}$ denote the coefficients of the functional (5.35b) for a state P^s_0. The mode-coupling coefficients for the state P^s shall be characterized by difference coefficients $D^s_{q,kp}$, which are defined by:

$$V^s_{q,kp} = V^{s0}_{q,kp} + D^s_{q,kp}. \tag{5.51a}$$

The deviation of the critical arrested part for state P^s from the one for state P^s_0 shall be quantified by an array $\boldsymbol{g}^{s0}$ so that

$$f^{sc}_q(P^s) = f^{sc}_q(P^s_0) + [1 - f^{sc}_q(P^s_0)]g^{s0}_q; \quad q = 1, \ldots, M. \tag{5.51b}$$

Using these formulas, Eq. (5.47b) gets the form: $\sum_p [\delta_{q,p} - A^s_{q,p}(P^c, P^s_0)] g^{sc}_p = Y^{s0}_q + \delta Y^{s0}_q$. Here, the abbreviations are introduced:

$$Y^{s0}_q = [1 - f^{sc}_q(P^s_0)] \sum_{kp} D^s_{q,kp} f^c_k f^{sc}_p(P^s_0), \tag{5.51c}$$

$$\delta Y^{s0}_q = [1 - f^{sc}_q(P^s_0)] \sum_{kp} D^s_{q,kp} f^c_k [1 - f^{sc}_p(P^s_0)] g^{s0}_p - (g^{s0}_q)^2/[1 - g^{s0}_q]. \tag{5.51d}$$

With the aid of Eq. (5.50a), the fixed-point equation can be rewritten as the set of equations for the components of $\boldsymbol{g}^{s0}$:

$$g^{s0}_q = \sum_\ell I_{q,\ell}(P^c, P^s_0)\{Y^{s0}_\ell + \delta Y^{s0}_\ell\}; \quad q = 1, \dots, M. \tag{5.51e}$$

Iteration of these identities yields an asymptotic series for $\boldsymbol{g}^{s0}$ in powers of the small parameters $D^s_{q,kp}$. The leading contribution is of first order; it is obtained by dropping δY^{s0}_ℓ in the preceding formula.

The reasoning from above can be used also to determine the small-ϵ expansion of $\boldsymbol{f}^s(P, P^s)$ for glass states P near P^c. The states are located on a path P^ϵ, $\epsilon > 0$, as discussed in Sec. 4.3.4. It is the goal to derive the analogue of Eq. (4.91a) for the Lamb–Mössbauer factor. To proceed, the deviations of the arrested parts from their critical values shall be expressed in terms of arrays $\boldsymbol{g}(P)$ and $\boldsymbol{g}^s$. For $q = 1, \dots, M$, there holds:

$$f_q(P) = f^c_q + [1 - f^c_q] g_q(P)\,,\, f^s_q(P, P^s) = f^{sc}_q(P^s) + [1 - f^c_q(P^s)] g^s_q. \tag{5.52a}$$

Substitution of these expressions, the fixed-point equations (5.47b) can be rewritten in a form similar to Eq. (5.51e). Using the coefficients from Eq. (5.38c) and introducing the abbreviations

$$\delta Y^s_q = [1 - f^{sc}_q(P^s)] \sum_{kp} V^s_{q,kp} [1 - f^c_k][1 - f^{sc}_p(P^s)] g_k(P) g^s_p - (g^s_q)^2/[1 - g^s_q], \tag{5.52b}$$

one gets the identities for $q = 1, \dots, M$:

$$g^s_q = \sum_\ell I_{q\ell}(P, P^s) \Big\{ \sum_p B_{\ell p}(P^c, P^s) g_p(P) + \delta Y^s_\ell \Big\}. \tag{5.52c}$$

Iteration of these formulas provides an asymptotic series for $\boldsymbol{g}^s$. The small expansion parameters are the components of $\boldsymbol{g}(P)$. The expansion coefficients are smooth functions of the coordinates of P and of P^s. It is shown in Sec. 4.3.4 that $\boldsymbol{g}(P)$ can be represented as series of powers of $\sqrt{\epsilon}$. Consequently, Eq. (5.52c) leads to the analogous series for $\boldsymbol{g}^s$. The coefficients for the latter can be expressed recursively in terms of those for the former. Equation (4.91a) yields $g_p(P) = [f_p(P) - f_p^c]/[1 - f_p^c]$ up to order ϵ. The leading-order terms reproduce Eq. (5.48a) with the formula for the critical amplitudes:

$$h_q^s(P^s) = [1 - f_q^{sc}(P^s)] \sum_{\ell p} I_{q\ell}(P, P^s) B_{\ell p}(P^c, P^s) h_p / [1 - f_p^c]. \tag{5.53}$$

The leading correction shall be specified by some amplitude $k_q^s(P, P^s)$. One gets for $\epsilon \geq 0$

$$f_q^s(P^\epsilon, P^s) = f_q^{sc}(P^s) + h_q^s(P^s)\sqrt{\sigma_\epsilon/\mu_2(P^\epsilon)}\Big\{1 + \sqrt{\sigma_\epsilon} k_q^s(P^\epsilon, P^s) + O(\epsilon)\Big\}. \tag{5.54}$$

The correction amplitude $k_q^s(P, P^s)$ is obtained by using two simplifications in Eq. (5.52c). First, $\boldsymbol{g}(P)$ is replaced by its leading-correction term. Second, Eq. (5.52b) is evaluated with the leading approximations for $\boldsymbol{g}(P)$ and $\boldsymbol{g}^s$.

A change of an external control parameter ϵ might also induce a change of the coupling coefficients $V_{q,kp}^s$, i.e., P^s will depend on ϵ. Let P_0^s denote the point for $\epsilon = 0$. The difference coefficients in Eq. (5.51a) shall be expressed by their leading expansion terms:

$$D_{q,kp}^s = d_{q,kp}^s \sigma_\epsilon + O(\epsilon^2). \tag{5.55a}$$

The formulas (5.51) yield expressions for the leading-order change of the Lamb–Mössbauer factor.

$$f_q^{sc}(P^s) = f_q^{sc}(P_0^s) + F_q \sigma_\epsilon + O(\epsilon^2), \tag{5.55b}$$

$$F_q = \big[1 - f_q^{sc}(P_0^s)\big] \sum_{\ell kp} I_{q,\ell}\big[P^c, P_0^s\big]\big[1 - f_\ell^{sc}(P_0^s)\big] d_{\ell,kp}^s f_k^c f_p^{sc}(P_0^s). \tag{5.55c}$$

Substitution of formula (5.55b) into Eq. (5.54) one gets the result in a form used in the original work by Fuchs *et al.* (1998):

$$f_q^s = f_q^{sc} + h_q^s \sqrt{\sigma_\epsilon/\mu_2^c}\Big\{1 + \sqrt{\sigma_\epsilon}[\kappa + \overline{K}_q^s] + O(\epsilon)]\Big\}. \tag{5.56}$$

Here, $\boldsymbol{f}^{sc} = \boldsymbol{f}^s(P^c, P_0^s)$ and $\boldsymbol{h}^s = \boldsymbol{h}^s(P_0^s)$ denote the arrested parts and critical amplitude at the transition point of the path. $\overline{K}_q^s = k_q^s(P, P^s) + [F_q\sqrt{\mu_2(P)}/h_q^s(P^s)] - \kappa$, with κ being defined in Eq. (4.92d). In all smooth functions of P and P^s, the variables can be restricted to P^c and P_0^s, respectively.

One can substitute the expansions for $\boldsymbol{f}$ and $\boldsymbol{f}^s$ into Eq. (5.45b) in order to get a corresponding expansion for the square of the localization length. The leading-order contribution reproduces Eq. (5.48b) with

$$r_{sc}^2 = d^2 \Big/ \int_0^\infty dq v_s^c(q) f_q^c f_q^{sc} \tag{5.57a}$$

$$h_{\mathrm{MSD}} = (r_{sc}^4/d^2) \int_0^\infty dq v_s^c(q)[f_q^c h_q^s + h_q f_q^{sc}]. \tag{5.57b}$$

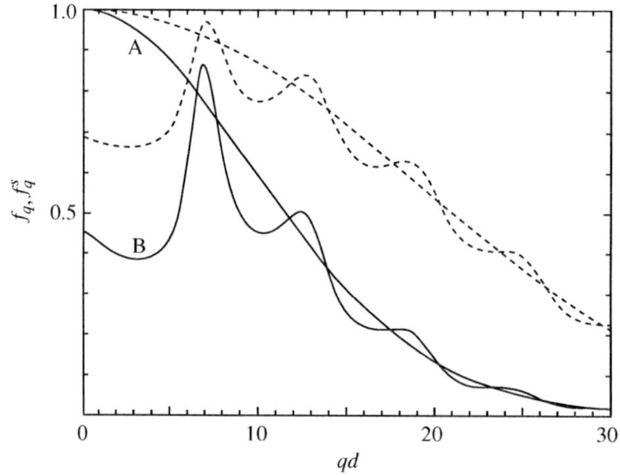

FIG. 5.4. The full lines A and B exhibit the Lamb–Mössbauer factor f_q^s and Debye–Waller factor f_q, respectively, for a system of hard spheres of diameter d, calculated for the critical packing fraction φ_c. The dashed lines are the corresponding pair of curves calculated for a packing fraction $\varphi = \varphi_c + 6.6\%$. The MCT model is based on the Percus–Yevick approximation for the structure factor and the convolution approximation. Reproduced from Bengtzelius *et al.* (1984).

The full line A in Fig. 5.4 shows the critical arrested part f_q^{sc} for the hard-sphere system. The tagged particle is also a hard sphere with the same diameter as that of the surrounding particles. The f_q^{sc} versus q curve exhibits the following features (Bengtzelius *et al.* 1984). It can be described very well by a Gaussian approximation: $f_q^{sc} \approx \exp[-(qr_{sc})^2]$. For $qd > 10$, the curve interpolates that for the critical Debye–Waller factor f_q^c. This finding is demonstrated also for the dashed lines, which are calculated for a distance parameter $\epsilon = 0.066$. Upon decreasing the packing fraction φ of the glass towards the instability point φ_c, the localization length increases strongly. At the critical point, it reaches a finite value r_{sc}. As expected from the Lindemann melting criterion, this critical value is of the order of 10% of the diameter d.

For the $M = 100$-component model for a HSS, which is explained in connection with Fig. 4.6, one obtains for the critical localization length and the critical amplitude for the mean-squared displacement (Fuchs *et al.* 1998):

$$r_{sc}/d = 0.0746\,, \qquad h_{\mathrm{MSD}}/d^2 = 0.0116. \tag{5.58}$$

Figure 5.5 demonstrates that the critical amplitude h_q^s and the correction amplitude $\overline{K}_q^s$ interpolate the corresponding amplitudes h_q and $\bar{K}_q$ from Fig. 4.6 provided qd exceeds about 10. For smaller wave numbers, the critical amplitudes h_q and h_q^s differ drastically. The former approaches a value around 0.4. The latter vanishes in the small-q limit: $h_q^s = q^2 h_{\mathrm{MSD}} + O(q^4)$. This formula follows, if the

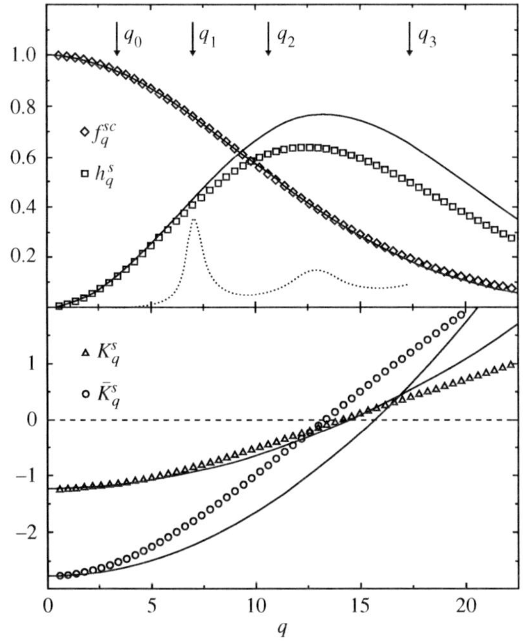

FIG. 5.5. The critical Lamb–Mössbauer factor f_q^{sc}, the critical amplitude h_q^s and the correction amplitude $\overline{K}_q^s$, which determine the expansion formula (5.56) for the arrested part f_q^s of the density correlator of a tagged particle of a HSS. Here, q denotes the wave number in units of the inverse of the particle diameter d. The correction amplitude K_q^s is explained in Eq. (6.33). The full lines present the results for the Gaussian approximation $f_q^s = \exp[-(r_s q)^2]$, with r_s denoting the localization length. The dotted line exhibits one tenth of the structure factors S_q at the critical packing fraction. The MCT model and the marked wave numbers are the same as used for Fig. 4.6; $\kappa = 0.961$. Reproduced from Fuchs *et al.* (1998).

expansion formulas (5.56) and (5.48b) are substituted in Eq. (5.46b). Therefore, the h_q^s versus q curve exhibits a broad maximum. Similarly, there is a large difference between the correction amplitudes k_q and k_q^s for $qd < 5$. The former is nearly zero, while the latter is below -1.5. Therefore, the leading asymptotic formula (5.48b) leads to negative values for r_s^2 for ϵ exceeding about 0.05. The $\sqrt{\epsilon}$-law describes well the strong decrease of r_s^2 with increasing ϵ only for $\epsilon < 0.02$ (Chong *et al.* 2001a).

The structure and the structural arrest of a hard-sphere system with additional short-ranged attraction forces is governed by two length scales, namely the hard-core diameter d and the attraction range Δ. This is discussed in Sec. 4.3.6. The representative bifurcation diagram in Fig. 4.9 exhibits two arrest mechanism provided the attraction-range parameter $\delta = \Delta/d$ is smaller than about

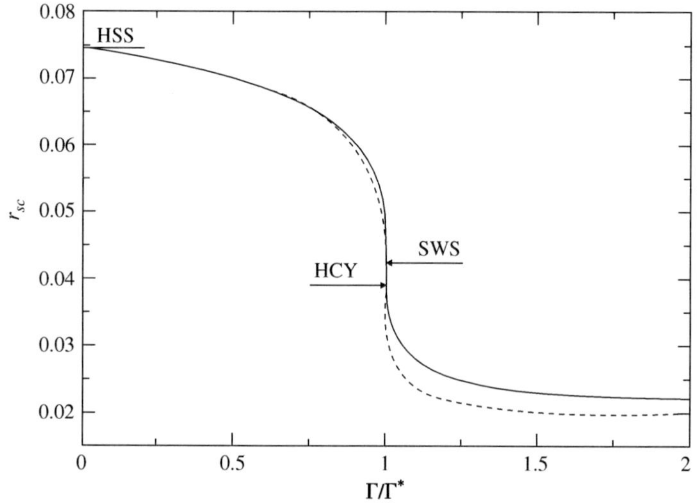

FIG. 5.6. The localization length r_{sc} in units of d of a tagged particle of a system of hard spheres of diameter d with a short-range attraction at the liquid–glass transition points as a function of the attraction-strength parameter Γ. The full and dashed lines refer to systems with square-well attraction and Yukawa-potential attraction, respectively. The arrows with labels SWS and HCY mark the localization length 0.0425 d and 0.0390 d, respectively, for the systems at the A_4 singularity. The line labelled HSS marks the critical localization length 0.0746 d for the hard sphere system. The other details are the same as explained for Fig. 5.3. Reproduced from Götze and Sperl (2003).

0.1. For small attraction strength Γ, there is a transition to a glass whose arrest is caused primarily by the strong repulsion. The localization length r_s will be of the order 0.1 d, as for the HSS. With increasing Γ, r_s decreases, partly because attraction favours localization and partly because the transition shifts to larger packing fractions. For large Γ, the arrest is dominated by the sticking phenomenon. Hence, the localization length will be of the order Δ. The full line in Fig. 5.6 exhibits the decrease of the critical localization length r_{sc} with increasing Γ. The states are calculated for a square-well system with an attraction parameter δ^* for the A_4 singularity. The localization length decreases by about a factor 3.3 if Γ/Γ^* increases from 0 to above 1.5. The dashed line is calculated for a model which describes the attraction by a Yukawa potential. The MCT are robust under changes of the details of the model. The figure deals with the same states, which are considered in Fig. 5.3 for the elastic moduli. The curves corroborate the previous reasoning that bond formation is the cause for the increase of the elastic stiffness as well as for the increase of the Debye–Waller factors.

Let us consider a hard sphere of diameter d^s moving in a hard-sphere system. The diameter ratio d^s/d is a control parameter, which influences the coefficients

P^s for the mode-coupling functional $\boldsymbol{\mathcal{F}}^s$. But, this parameter does not effect the functional $\boldsymbol{\mathcal{F}}$ determining the density correlators $\boldsymbol{\phi}(t)$. The tagged particle gets more free volume for displacements in the arrested cages if (d^s/d) decreases. Hence, the localization length r_s increases with decreasing d^s. The half width of the f_q^s versus q curve shrinks. Similarly, the position of the maximum of the h_q^s versus q curve and the position of the zero of the $\overline{K}_q^s$ versus q curve shift to smaller wave numbers. The dynamics on short-distance scales is effected less than that on long-distance ones. Hence, the suppression of f_q^s is less pronounced for large q than for small ones. As a result, the Gaussian approximation $\exp[-(qr_{sc})^2]$ underestimates the large-q tails of the f_q^{sc} versus q curves the more the more (d^s/d) decreases below unity (Fuchs *et al.* 1998).

(iv) Arrest near the transition from glass 1 to glass 2

Let $\hat{\xi} \to P^s(\hat{\xi})$ denote a path, which leads from the glass 2 region to the glass 1 region. The critical value $\hat{\xi}^c$ shall mark a generic transition point: $\hat{\xi}^c \to P^{sc}$. A distance parameter $\hat{\epsilon}$ shall be used for the characterization of the states P^s near P^{sc}, as explained in Eqs. (4.84a–c). As shown in Sec. 4.3.1, P^{sc} is a bifurcation point of the fixed-point equation (5.43a). The M^2 non-negative coupling coefficients $U_{q,p}^{(1)}$ are smooth functions of $\hat{\epsilon}$. The mode-coupling functional is linear in the M components of $\boldsymbol{f}^s(P, P^s)$. This speciality implies

$$f_q^s(P, P^{sc}) = 0, \quad q = 1, \ldots, M. \tag{5.59a}$$

The generic bifurcation from glass 1 to glass 2 describes a continuous transition. It deals with a degenerate A_2 bifurcation. In order to prove this result, Eq. (5.43a) shall be considered for some specific label, say, $q = q_0$. With $x = f_{q_0}^s$, the left-hand side reads $F_1(x) = x/(1-x)$. Writing $v = U_{q_0,q_0}^{(1)}$ and $v_0 = \sum_{p \neq q_0} U_{q_0,p}^{(1)} f_p^s$, the right-hand side is the linear function $F_2(x) = vx + v_0$. There holds $v \geq 0$ and $v_0 \geq 0$. Within the interval $0 < x < 1$, the $F_2(x)$ versus x curve intersects the $F_1(x)$ versus x curve transversally. Hence, the only degenerate solution occurs for $x = 0$, $v_0 = 0$. Since q_0 is arbitrary, there holds Eq. (5.59a).

Equation (4.98) describes the asymptotic behaviour of the arrested part in glass 1 for states near the degenerate transition to glass 2. In analogy to the derivation of Eq. (4.86c), one can write the separation parameter as

$$\hat{\sigma} = \hat{C}\hat{\epsilon} + O(\hat{\epsilon}^2), \quad \hat{C} > 0. \tag{5.59b}$$

Since the functional in Eq. (5.43a) is linear, Eq. (4.96c) yields $\hat{\lambda} = 0$. The singularity parameter $\hat{\mu}_2$ is unity. There holds:

$$f_q^s(P, P^s) = \hat{h}_q(P)\hat{C}\hat{\epsilon} + O(\hat{\epsilon}^2), \quad \hat{\epsilon} \geq 0. \tag{5.59c}$$

According to Eq. (4.78b), the positive components $\hat{h}_q(P)$ of the critical amplitude agree with the components of the right eigenvector a_q of the critical stability matrix $A^c_{q,p} = U^{(1)}_{q,p}$. Here, the latter matrix elements have to be evaluated for a state P of the glass and $P^s = P^{sc}$.

The preceding results hold for a general $2M$-component MCT model, which yields a reducible stability matrix as specified by Eq. (5.38b). The integrals in Eqs. (5.45b), (5.57a,b) have been meant as abbreviation of Riemann sums of M terms extending over the grid (4.35a). The schematic model defined in Eqs. (4.33a,b) is the simplest example. In the latter case, the arrays $\boldsymbol{\phi}(t)$ and $\boldsymbol{\phi}^s(t)$ are given by the single correlator $\phi(t)$ and $\phi_A(t)$, respectively. The derivation of Fig. 4.4 exemplifies all general features of the theory. The generic transition from glass 1 to glass 2 is characterized by the approach to unity for the eigenvalue $E^s(P, P^s)$, which is discussed in connection with Eq. (5.49).

If one would use carelessly Eq. (5.59c) in order to guess the small-q-limit behaviour of the Lamb–Mössbauer factor, one would arrive at an erroneous result. Using $h^s_q = h_{\mathrm{MSD}}q^2 + O(q^4)$, one would get $f^s_q = [\hat{C}h_{\mathrm{MSD}}\,\hat{\epsilon}]q^2$; and this formula contradicts Eq. (5.46b). Substitution of Eq. (5.59d) into Eq. (5.45b), one gets $(r_s/d)^2 = \hat{\epsilon}^{-1}/[\int_0^\infty dq v_s(q) f_q \hat{C}\hat{h}_q + O(\hat{\epsilon})]$. If the state P^s in glass 1 approaches the transition point, the localization length diverges:

$$\lim_{\hat{\epsilon}\to+0} r_s = \infty. \tag{5.60}$$

The appearance of a diverging length causes problems. For example, the limit $\hat{\epsilon}\to 0$ cannot be interchanged with the limit $q \to 0$ if the arrested part is described by Eq. (5.46c). Handling the continuous transition of the tagged-particle motion from the states with diffusive dynamics to the states with localization within a mode-coupling theory, would require the derivation of equations of motion, which can handle the identified small-wave-number subtleties.

The tagged particle motion in a glass can be discussed most transparently on the basis of Eqs. (5.41)–(5.43). The interaction forces between background and tagged particle consist of two contributions. the first one is due to the static potential, which is produced by the arrested density fluctuations of the glass. This force is time dependent only because of the motion of the tagged particle. Factorization of the density-fluctuation pairs, which determine the force, yields two factors. One is due to the frozen background and the other due to the tagged particle. They are described by the Debye–Waller factor f_k and by the correlator $\phi^s_p(t)$, respectively. As a result, one gets the kernel $m^{s(1)}_q(t)$ in Eq. (5.42b). The relaxation problem described by this kernel is similar to that studied in solid-state physics for the motion of an electron or an ion in a crystal with defects. The latter destroy the translational invariance, thereby leading to momentum relaxation. The standard weak-coupling theory deals with a model for a glass 2, where the arrested fluctuating-force field is considered as given input. The current-relaxation rate increases proportional to the intensity of the force fluctuation. The second contribution $m^{s(2)}_q(t)$ is analogous to the electron–phonon-scattering contribution. It describes momentum relaxation caused by absorption and emission of density-fluctuations of the system.

Because of the scattering of the tagged particle by the defects, it gets the possibility to scatter again within a given fixed time interval. Thereby, the momentum-relaxation rate is increased; and this is reflected by a suppression of the diffusivity below its weak-coupling result. The request to handle this mechanism self consistently leads to a mode-coupling theory for the electron motion in a random potential (Götze 1978) or for the motion of a hard sphere in an uncorrelated array of point scatterers (Götze *et al.* 1981a,b). These theories imply a transition from diffusion to localization. The essential features of these approaches are reproduced by the schematic models, which are discussed in Sec. 4.2.4 for vanishing and non-vanishing white noise damping term, respectively. The divergence of the localization length in these models is described by a Curie–Weiss law, $r_s^2 \propto 1/\hat{\epsilon}$. The cited self-consistent-current-relaxation-theory work uses the ad-hoc simplification for the mode-coupling kernel: $m_q^s(t)(qd)^2 \to m_0^s(t)$. This simplification is illegitimate since the limits $\epsilon \to 0$ and $q \to 0$ cannot be interchanged in the mode-coupling integrals. This unjustifiable simplification step is the mathematical reason for the short comings of the theory for the transition from a glass 1 to a glass 2 state of the Lorentz model, which have been mentioned at the end of Sec. 4.1.2.

5.1.5 *Glassy-dynamics features of tagged-particle motions*

In this section, features of the MCT solutions for the tagged-particle dynamics shall be discussed. Also, some quantitative comparisons shall be considered between calculated results and data obtained from experiments and from molecular-dynamics-simulation studies.

The transition from the liquid to glass 1 is a generic fold bifurcation for the $2M$-component model. Hence, the discussions in Secs. 4.4.1 and 4.4.2 remain valid also for the tagged-particle correlators. There evolves a plateau of height f_q^{sc} for the $\phi_q^s(t)$ versus $\log t$ curves. The plateau is defined the better the smaller the distance parameter ϵ. For the liquid states, there is a plateau-crossing time τ^{pc}, which diverges for ϵ tending to zero. The numerical solutions for the models of a HSS suggest that there holds the factorization formula for the near-plateau correlators: $\phi_q^s(t) = f_q^{sc} + h_q^s G(t)$. The below-plateau decay for the liquid states develops the superposition principle:

$$\phi_q^s(t) = \tilde{\phi}_q^s(t/\tau), \quad \phi_q^s(t) < f_q^{sc}. \tag{5.61a}$$

Here, the function $G(t)$ and the scales τ^{pc}, τ, are the same quantities, which have been discussed for the density-fluctuation correlators $\phi_q(t)$.

Up to a subtraction term and a factor $(1/q^2)$, the mean-squared-displacement function agrees with the tagged-particle density correlator for small q: $[1 - \phi_q^s(t)]/q^2 = \Delta_s(t) + O(q^2)$. Consequently, there evolves also a plateau for the $\Delta_s(t)$ versus $\log t$ curve. Its hight is r_{sc}^2, with r_{sc} denoting the critical localization length. The numerical evidence discussed in Chapter 4 suggests the validity of the factorization property for the near-plateau process, $\delta r_s^2(t)/6 = r_{sc}^2 - h_{\rm MSD} G(t)$.

From the observed evolution of the superposition principle for $\phi_q^s(t)$, one concludes that the above-plateau increase of the $\Delta_s(t)$ versus $\log t$ curves exhibits a corresponding evolution of the superposition principle:

$$\Delta_s(t) = \tilde{\Delta}_s(t/\tau), \qquad \Delta_s(t) > r_{sc}^2. \tag{5.61b}$$

The large-time asymptote is the diffusion law from Eq. (5.39b). Hence, the long-time diffusivity reads

$$D_s = C_s/\tau, \qquad \epsilon \to -0. \tag{5.61c}$$

The constant C_s specifies a limit of the control-parameter independent shape function: $C_s = \lim_{\tilde{t}\to\infty} \tilde{\Delta}_s(\tilde{t})/\tilde{t}$. Upon approaching the freezing point, the diffusivity decreases to zero proportional to the rate $(1/\tau)$ for the below-plateau decay processes. The indicated results are demonstrated by Fuchs *et al.* (1998) for the model of a HSS with stochastic short-time dynamics, which is defined in connection with 4.14. In particular, it is shown there that the Eq. (5.61c) holds for $|(\varphi - \varphi_c)/\varphi_c| < 0.1$ within an error smaller than 5%.

Figure 5.7 demonstrates the evolution of the glassy dynamics for $\Delta_s(t)$ for an $M = 100$-component model for the HSS. The equations of motion (4.17a), (5.27a), (5.32a) for a Newtonian dynamics are used with vanishing friction terms $\boldsymbol{\nu}$ and $\boldsymbol{\nu}^s$. For $\log t < -1.5$, the $\log \Delta_s$ versus $\log t$ curves exhibit the ballistic law (3.31), which is shown as dashed straight line of slope 2. The curves for the liquid states exhibit Einstein's law: $\Delta_s(t \to \infty) = D_s t$ for $\Delta_s(t)/d^2 \geq 0.1$. One checks that the diffusivity decreases proportional to $|\epsilon|^\gamma$ for $|\epsilon| \leq 0.1$. The exponent $\gamma \approx 2.45 \pm 0.02$ is consistent with the one used in Fig. 4.21 for the demonstration of the superposition principle for the density correlators. Thereby, the validity of Eq. (5.61c) is demonstrated. Replotting the results as function of $\tilde{t} = t/\tau \propto t|\epsilon|^\gamma \propto tD_s$ demonstrates the evolution of the superposition principle in analogy to what is shown in Fig. 4.21. The figure explains qualitatively the data shown in Fig. 1.9 for a binary Lennard-Jones mixture. In particular, both figures show that the diffusion asymptote is reached if $\delta r_s^2(t)$ has increased to about 20 times the plateau value. The oscillation features within the transient region $0.1 < t < 1$ are more pronounced for the MCT model than for the mentioned data. This artificial outcome of the theory is a result of neglecting the friction terms $\boldsymbol{\nu}$ and $\boldsymbol{\nu}^s$. In order to compare the evolution of the tagged-particle dynamics for the two versions of equations of motion, one can represent the results for $\Delta_s(t)$ as function of $\log(t/t_0)$. Choosing the t_0 for the two models for the short-time dynamics as specified in the caption of Fig. 4.24, the results for the two models agree within the accuracy of the drawing for $|\epsilon| \leq 0.1$ and $t/t_0 \geq 20$. The transient dynamics merely influences the time scale t_0 for the glassy dynamics (Chong *et al.* 2001a).

Let us consider liquid states P. The long-time diffusivity D_s is a continuous function of the control parameters, which vanishes for P tending to some transition point P^c. Let D_0^s denote another continuous function of the control

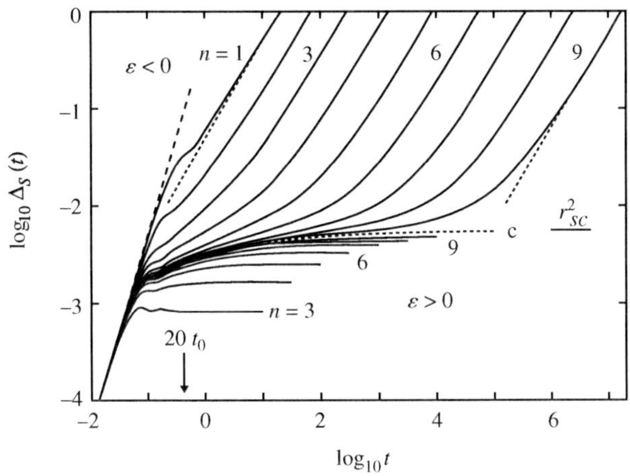

FIG. 5.7. The mean-squared-displacement function $\Delta_s(t) = \delta r_s^2(t)/6$ for a tagged particle of a HSS calculated for a 100-component MCT model with cutoff wave number $q_{\rm co} = 40/d$. The mode-coupling constants are based on the Percus–Yevick structure factor and the convolution approximation. The first version of equations of motion (4.17a), (5.27a) and (5.32a) are used with $\nu_q = \nu_q^s = 1/D_0^s = 0$. The dashed line exhibits the ballistic asymptote $(v_{th}t)^2/2$. The dotted lines exhibit the diffusion asymptotes $D_s t$ for the states $n = 1$ and $n = 9$. The dotted line with label c is the result for the critical packing fraction φ_c, and the full lines refer to $\epsilon = (\varphi - \varphi_c)/\varphi_c = \pm 10^{-n/3}$, $n = 1, 2, \ldots$. The horizontal line marks the square of the critical localization length $r_{sc} = 0.0746$. The time $t_0 = 0.00944\,(0.4d/v_{th})$ is introduced in connection with Fig. 4.24; it defines the arrow for $20t_0$. The units of length and time are chosen so that the particle diameter d and the thermal velocity $v_{\rm th}$ are unity. Reproduced from Chong *et al.* (2001a).

parameters, which is positive for all states near and at P^c. For a fixed and sufficiently large positive constant C, the solutions of the equation

$$D_0^s/D_s = C \tag{5.62}$$

define a surface in parameter space. Sets of such isodiffusivity surfaces quantify the liquid neighbourhood of a transition surface. For systems with a Newtonian dynamics, one can use $D_0^s = dv_{th}$ with d denoting some scale for the particle diameter and v_{th} abbreviating the thermal velocity. Thereby, one eliminates the changes of D_s due to the trivial part of the transient dynamics. Choosing D_0^s as the short-time diffusivity, one can achieve the same goal for systems with stochastic short-time motion. The above defined procedure for characterizing transition manifolds has been introduced by Foffi *et al.* (2002). Molecular-dynamics data have been analyzed for a square-well system with the

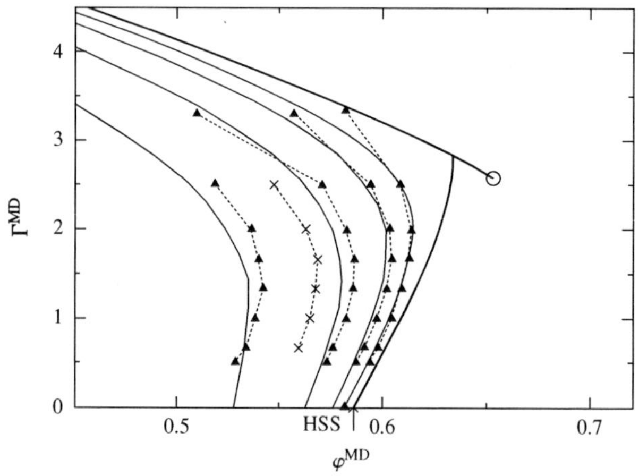

FIG. 5.8. The crosses, which are jointed by dotted lines as guides to the eye, reproduce molecular-dynamics-simulation data of equal diffusivity ratio $D_0^s/D_s = C$, $C = 240$, $D_0^s = dv_{th}^s$. They are obtained for a SWS with attraction-range parameter $\delta = 0.03$ and various values for the packing fraction φ^{MD} and attraction-strength parameter Γ^{MD} (Foffi *et al.* 2002). The triangles reproduce corresponding results for $C = 2 \times 10^2$, 2×10^3, 2×10^4, and 2×10^5 (from left to right) obtained for a binary SWS (Zaccarelli *et al.* 2002). The light full lines are iso-diffusivity curves for the same sequence of four values for C, which are calculated for a model for the SWS. The heavy full lines and the circle show the transition lines and the endpoint singularity, respectively, for this model. Further details are explained in the text. Reproduced from Sperl (2004b).

small attraction-range parameter $\delta = 0.03$. The system and the concepts for its description are explained in connection with Fig. 4.9. The crosses in Fig. 5.8 reproduce data for $C = 240$. The triangles reproduce data for a binary SWS (Zaccarelli *et al.* 2002). This complication of the system is introduced in order to bypass crystallization at high values for the packing fraction φ. The binary mixture exhibits a critical packing fraction for the hard-sphere system $\varphi_c^{\mathrm{HSS}} = 0.585$, which is consistent with the one measured for slightly polydisperse hard-sphere colloids. The data demonstrate liquid states for $\varphi > \varphi_c^{\mathrm{HSS}}$. The isodiffusivity lines consist of a small-Γ branch and a large-Γ one. On the former, the Γ versus φ isodiffusivity curves increase with φ and on the latter they decrease. This behaviour is documented even for states with a packing fraction, which is 10% below φ_c^{HSS}. The data demonstrate that there are two types of glasses. These appear out of liquids, which exhibit qualitatively different arrest mechanisms. Shifting a state along the isodiffusivity line for $C = 2 \cdot 10^5$, the simulation data for the δr_s^2 versus $\log t$ curves change strongly. Upon increasing Γ, the curves deviate from the ballistic short-time asymptote the earlier the larger Γ, since the increasing attraction favors sticking. For the same reason, the plateau decreases by about a factor 20. This is in semiquantitative agreement with what one would expect from the MCT results for the critical localization length, which are shown in Fig. 5.6 for some larger value of the attraction-range parameter δ. For large Γ, the plateau is poorly defined by the data.

The transition lines and the isodiffusivity lines in Fig. 5.8 are MCT results for the SWS with $\delta = 0.03$. They are obtained from solutions for the equations of motion with stochastic short-time dynamics. The mode-coupling coefficients are evaluated with structure factors, which are obtained as numerical solutions of the Percus–Yevick theory. In order to achieve the demonstrated similiarity between the calculated isodiffusivity curves and the data, a linear mapping is used as connection of the control-parameter pairs (φ, Γ) and $(\varphi^{\mathrm{MD}}, \Gamma^{\mathrm{MD}})$: $\varphi^{\mathrm{MD}} = 2.25\varphi - 0.5747$ and $\Gamma^{\mathrm{MD}} = 2.85\Gamma$. The figure demonstrates the relevance of MCT concepts and results for the description of the cited glassy-dynamics data.

The preceding conclusion can be corroborated by a comparison of molecular-dynamics-simulation results for the mean-squared displacements with numerical results of the MCT-equations of motion for $\log \delta r_s^2(t)$ versus $\log t$ curves. Figure 1.22 demonstrates that the interplay of the cage effect with the sticking phenomenon can eliminate the plateau of the curves. Instead, power-law variation is exhibited for times outside the region of transient dynamics and prior to the region of diffusion. Within MCT, the elimination of the plateau is explained as result of the broad distribution of critical arrested parts $6r_{sc}^2$. According to Fig. 5.6, this distribution is a feature of the liquid boundary near an A_4 bifurcation singularity. The lower panel of Fig. 4.18 demonstrates that control-parameter-sensitive sublinear power-law increase is the MCT result for the increase of $\delta r_s^2(t)$ through $6(r_{sc}^*)^2$, where r_{sc}^* is the localization length for the glass-transition singularity P^* of type A_4.

Let us assume that correlators $\phi_q^s(t)$ have been measured by some incoherent scattering experiments for a set of small wave numbers q. A data analysis can be based on the formula $\phi_q^s(t) = \exp[-(q^2\Delta_s(t))]\{1 + 0.5 \cdot [q^2\Delta_s(t)]^2\alpha(t) + O(q^4)\}$. The factor in front of the curly bracket presents the Gaussian approximation for the correlators. It reproduces the leading-order expansion term in Eq. (3.30). The coefficient $\alpha(t)$ is called the non-Gaussian parameter. The expansion formula for the correlator can be rewritten in the form:

$$-[\ln \phi_q^s(t)]/q^2 = \Delta_s(t) - 0.5 \cdot [\alpha(t)\Delta_s(t)^2]q^2 + O(q^4). \tag{5.63}$$

A linear interpolation of a $[\ln \phi_q^s(t)]/q^2$ versus q^2 presentation of the data yields $\Delta_s(t)$ and $\alpha(t)$. The indicated procedure has been used by van Megen *et al.* (1998) in order to determine the mean-squared displacements and to estimate the non-Gaussian parameters for nearly monodisperse colloidal suspensions of nearly hard spheres of radius R. About 2% of the particles were replaced by hard spheres of nearly the same diameter $2R$. But, these particles were tagged by giving them a different refraction-index profile. The incoherent scattering function was measured by photon-correlation spectroscopy. The cited paper presents a

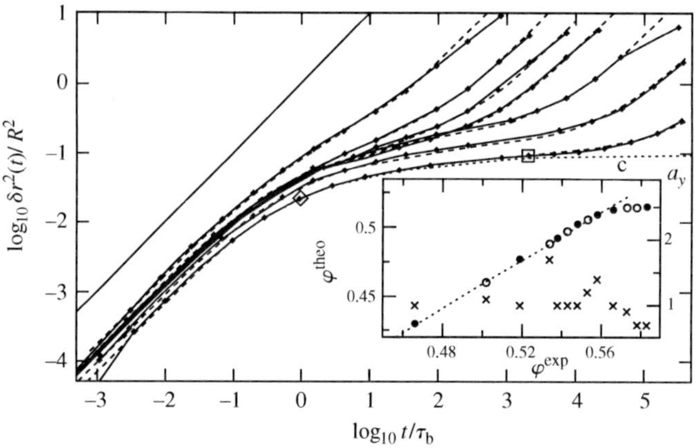

FIG. 5.9. The filled diamonds, which are connected by full lines as guide to the eye, reproduce mean-squared displacements $\delta r^2(t) = 6\Delta_s(t)$ measured by van Megen *et al.* (1998) for colloidal suspensions of hard spheres of radius R. The data refer to the packing fractions $\varphi^{\text{exp}} = 0.466$, (0.502), 0.519, (0.534), 0.538, (0.543), 0.548, (0.553), 0.558, 0.566, (0.573), (0.578), 0.583 (from left to right). Results for $\delta r^2(t)$ for the packing fractions noted in parentheses have not been reproduced in order to avoid overcrowding. The dashed lines exhibit $a_y \delta r_s^2(t)$ with mean-squared displacements calculated for a model for a HSS with stochastic short-time dynamics. The inset exhibits as circles the fit values $\varphi = \varphi^{\text{theo}}$ and as crosses the a_y used in the presentation; the filled dots refer to the curves shown in the main frame. Further details are explained in the text. The dotted line with label c exhibits the result for $\varphi = \varphi^c$, rescaled with the same parameters t_{mic} and a_y used for the dashed line for $\varphi^{\text{exp}} = 0.583$. The open square marks the point, where the dashed line for $\varphi = 0.5145$ and the dotted line deviate by about 10%. The open diamond marks the point of 10% deviation between the dotted line and its description by Eqs. (6.48a–c). Reproduced from Sperl (2005).

lucid description of the complex set-up for this experiment. Data for $\delta r_s^2(t)$ have been obtained for 13 packing fractions between 0.466 and 0.583. The long-time parts for the results for the three data sets with $\varphi > 0.566$ must be considered with reservation, since they exhibit aging effects. Figure 5.9 reproduces data for the seven packing fractions, which are represented in the inset by filled dots. The straight line of slope unity exhibits the diffusion law $\delta r_s^2(t) = 6D_0 t$ for a single colloid particle, which performs Brownian motion in the solvent. The diffusivity D_0 has been measured for the system in the limit of small φ. It defines the Brownian time $\tau_b = R^2/(6D_0)$, which is the natural scale for the short-time dynamics. For $t/\tau_b < 0.02$, there holds the law for the short-time diffusion $\delta r_s^2(t) = 6D_0^s t$, which is quantified by the short-time diffusion constant D_0^s. If φ increases from 0.466 to 0.583, the diffusivity ratio D_0^s/D_0 decreases by about a factor 2.6. The

φ-dependent suppression of D_0^s/D_0 below unity represents the density dependence of the transient dynamics as caused by the interaction between the colloid particles.

The system with the smallest analysed packing fraction has a density, which is about 6% below that for the freezing point. The $\log \delta r_s^2$ versus $\log t$ curve describes a crossover from the short-time diffusion to the long-time diffusion. The curve exhibits an inflection point as glassy-dynamics precursor. This feature of the curve is quite similar to the one, which is discussed in Fig. 3.1 for a van der Waals liquid for $T = 2$. The cage effect is already strong enough to suppress D_s/D_0^s by more than an order of magnitude. The corresponding slowing down of the dynamics is a bit larger than the one which can be inferred from the upper panel of Fig. 1.15 by comparing the dashed line with the line interpolating the crosses. Increasing φ from 0.466 to 0.566, the diffusivity decreases by more than a factor 1000; and the $\delta r_s^2(t)$ versus $\log t$ curve develops some plateau. This change is similar to that demonstrated in Fig. 5.7 for the change of the curve with label $n = 2$ to that with label $n = 6$. The liquid-glass transition occurs for the packing fraction $\varphi_g = 0.572 \pm 0.001$. There is a 1% deviation of this value from that cited in Sec. 4.3.5.

The data for the long-time decay of the density-fluctuation correlators $\phi_q(t)$, which are measured for the colloids, exhibit the superposition principle. The strong dependence of the below-plateau decay of these functions on the distance parameter $\epsilon = (\varphi - \varphi_g)/\varphi_g$ can be expressed in terms of the ϵ-dependence of the common time scale, to be denoted here as τ_α. This is discussed above in connection with Fig. 4.25. A similar discussion can be found in the cited paper for the tagged-particle correlators $\phi_q^s(t)$. Let us denote the common time scale for these correlators by τ_α^s. The measured diffusivity can be used to define a time scale $\tau_D = d^2/D_s$. It characterizes the long-time increase of the mean-squared displacement. A plot of the logarithm of the three times as function of $\log(-\epsilon)$ consists of three parallel straight lines. This means that there are three constants a_α, a_α^s, and a_D so that

$$a_\alpha/\tau_\alpha = a_\alpha^s/\tau_\alpha^s = a_D/\tau_D = |\epsilon|^\gamma, \quad -2 \leq \log|\epsilon| \leq -0.8. \tag{5.64}$$

The slope of the lines is measured as $\gamma = 2.7 \pm 0.3$ (van Megen *et al.* 1998). The proportionality of the scales is called scale coupling for the long-time decay in the glassy liquid. The cited experiments establish the coupling of the diffusivity scale to the scale for the fluctuations of the density and for the tagged-particle density. The scale coupling is reproduced by the numerical results for the MCT equations, as was emphasized above in connection with the discussion of the superposition principles. The MCT scales for the HSS exhibit also the power-law dependence specified by an exponent γ, which is compatible with that measured for the colloids.

The cited measurements do not permit a reliable determination of the non-Gaussian parameter. But, the data are compatible with the conclusion $|\alpha(t)| < 0.2$. This is consistent with the MCT result by Fuchs *et al.* (1998):

$$|\alpha(t)| \leq 0.35. \tag{5.65}$$

The dashed lines in Fig. 5.9 exhibit the solutions for a microscopic MCT model for a HSS. An $M = 100$-component model based on the Percus–Yevick structure factor is used. The equations of motion for a stochastic short-time dynamics are applied as explained in connection with Fig. 4.14. To obtain the MCT curves, three model parameters are adjusted for each of the packing fractions φ^{exp}. First, the time scale t_{mic} for the 'normal liquid' dynamics is adjusted so that the measured short-time diffusivity D_0^s is reproduced. Second, in order to reproduce correctly the distance parameter $\epsilon = (\varphi^{\text{exp}} - \varphi_g)/\varphi_g$ by the MCT quantity $(\varphi - \varphi_c)/\varphi_c$, the theoretical packing fraction $\varphi = \varphi^{\text{theo}}$ is adjusted as shown in the inset of the figure. For the values free of aging problems, the 10% discrepancies between φ^{exp} and φ can be described by a linear parameter mapping. This fit curve reads $\varphi = 0.87\varphi^{\text{exp}} + 0.023$; it is shown as the dotted straight line in the inset. The calculated plateaus agree with the ones implied by the data for 6 out of the 15 measured samples. For the other seven samples analysed, there are some erratic deviations. To account for these deviations, a rescaling factor a_y in front of the calculated mean-squared displacement is used. The values of this third fit parameter are shown in the inset. Notice that the measured $\log \delta r_s^2(t)$ versus $\log t$ curves for $\varphi^{\text{exp}} = 0.548$ and $\varphi^{\text{exp}} = 0.558$ intersect. Deviations of a_y from unity are needed to reproduce such effects. The preceding discussion indicates that MCT has relevance for the explanation of the glassy tagged-particle dynamics of hard-sphere colloids.

The dotted line in the main frame of Fig. 5.9 exhibits the mean-squared displacement calculated for the critical packing fraction φ_c and rescaled with the same values for t_{mic} and a_y, which have been used to fit the data for $\varphi^{\text{exp}} = 0.583$. The experimental data for this high packing fraction agree with the calculated critical curve within a 10% error margin up to a time $t_+ = 2000\tau_b$. Within the large interval $\tau_b \leq t \leq t_+$, the measured $\delta r^2(t)$ and the calculated one exhibit an increase by about a factor 3.8. Such stretched approach-to-the-plateau process is not exhibited by the molecular-dynamics-simulation data shown in Fig. 1.9 and 1.11.

Figure 5.10 shows data for the evolution of the glassy dynamics for correlators $\phi_q^s(t)$ as determined by molecular-dynamics simulations for a nearly monodisperse system of particles with nearly hard-sphere interactions. These findings are compared with results evaluated for a microscopic MCT model. The system for the computer experiment, which is a model for a HSS, has been explained in connection with Eq. (4.103). The simulation is done for a Newtonian dynamics for particles of equal mass. The interaction forces are complemented by a white noise random force with a Gaussian distribution. Introducing such random force is equivalent to complementing the fluctuating-force correlator $M_{Lq}(z)$ in Eq. (4.1a) and the kernel $M_{sq}(z)$ in Eq. (5.21a) by a term $\pm i\nu_q$ and $\pm i\nu_q^s$, Im $z \gtrless 0$, respectively. A model was used with $\nu_q = \nu_q^s = \nu = 30v_{th}/(\sqrt{3}d)$; here, d denotes the averaged particle diameter and v_{th} abbreviates the thermal velocity. If $\nu = 0$ is used, the model produces correlators $\phi_q^{sND}(t)$ for a conventional liquid. The model $\nu > 0$ assumes that there is some interaction with a background

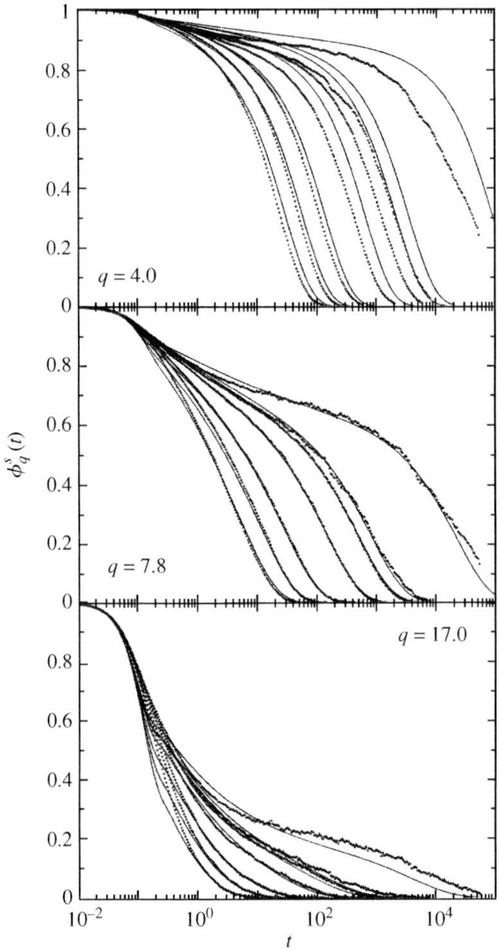

FIG. 5.10. Correlation functions $\phi_q^s(t)$ for tagged-particle-density fluctuations for three wave numbers q obtained by molecular-dynamics simulations for a slightly polydisperse system of particles interacting with a repulsion, which is close to that of hard spheres with diameter d. The packing fractions are $\varphi^{\exp} = 0.50, 0.53, 0.55, 0.57, 0.58, 0.585$, and 0.59 (from left to right). The full lines are MCT results for a HSS calculated for packing fractions $\varphi = 0.445$, 0.470, 0.484, 0.499, 0.505, 0.508, and 0.5135. The data for $q = 4,0$, 7.8 and 17.0 are compared with MCT results for wave numbers $q_{\mathrm{MCT}} = 5.0$, 9.13, 18.3, respectively. The units of length, temperature, and time are chosen so that $d = 1$, $k_{\mathrm{B}}T = 1/3$, $v_{th}^2 = 1/3$. Further details are explained in the text. Reproduced from Voigtmann *et al.* (2004).

system so that the total momentum is not conserved. This is discussed above in connection with Eqs. (4.1d,e). A harmonic oscillator, which is specified by the cited value ν for a friction and by frequencies Ω_q or Ω_q^s, performs overdamped motions for small and intermediate wave numbers. But, the value used for ν is not large enough for the system to be considered as a model for a colloid, since there is no regime for short-time diffusion. The simulation data for $\phi_q^s(t)$ agree with those for $\phi_q^{sND}(t \cdot t^*)$ for $t \geq 10$. This holds if the rescaling parameter t^* is chosen as slightly increasing if the packing fraction φ increases from 0.50 to 0.57. For larger φ, the parameter t^* fluctuates somewhat. If one could ignore the fluctuations, one could conclude that the dynamics for $t > 10$ is independent of the transient motion. The data exhibit increases of the time scale for the glassy relaxation, which extend over nearly four orders of magnitude.

The shown MCT results have been evaluated from Eqs. (4.17a), (5.27a) for a monodisperse hard-sphere system. The Percus–Yevick theory is used to determine the structure factors S_q, and the convolution approximation was applied. The wave numbers are used with a cutoff $q_{c0} = 40/d$, and the discretization for the functions $\phi_q(t)$ and $\phi_q^s(t)$ is done with $M = 300$ values. A fit parameter for an adjustment of the short-time dynamics is not introduced.

A quantitative description of data by results of a microscopic MCT model requires the adjustment of the calculated parameters for the description of the bifurcation singularity to those exhibited by the data. Voigtmann *et al.* (2004) proceed in their analysis by using a linear mapping of the theoretical to the experimental control parameters: $\varphi = 0.810\varphi^{\text{exp}} + 0.037$. This mapping is similar to the one underlying the data analysis of Fig. 5.9. The error of MCT can be quantified by the error of the two numbers specifying the separation parameter, namely the coefficient C in Eq. (4.104d) and the value for the critical point φ_c. Instead of the MCT value $C = 1.54$, the data analysis yields a value 1.2. As discussed in Sec. 4.3.5, there is a cancellation of the errors of MCT against the errors due to those of the structure factor. The genuine error for the critical packing fraction is about 2%. The MCT values for the critical Lamb–Mössbauer factors f_q^{sc} are somewhat larger than the plateau heights of the $\phi_q^s(t)$ versus $\log t$ curves. A manner to quantify this error is to fit data for wave number q by a MCT result for a somewhat larger wave number q_{MCT}. Data fits of the quality demonstrated in the middle and lower panel of Fig. 5.10 have been shown in the cited paper for $qd = 7.8$, 9.0, 13.8, and 17.0 by using $q_{\text{MCT}}d = 9.13$, 10.3, 15.1, and 18.3, respectively. The discrepancies between q and q_{MCT} reflect an underestimation of the critical localization length r_{sc} of about 5%.

The preceding discussion suggests that MCT describes the evolution of the glassy dynamics of the tagged-particle-density fluctuations for wave numbers around and above the structure-factor-peak position. This holds also for the density dependence of the transient dynamics for $0.1 \leq t \leq 10$. The remarkable difference of the relaxation patterns shown for $qd = 7.8$ and 17.0 are explained by the q-dependence of the critical arrested part f_q^{sc}, discussed in Figs. 5.4 and 5.5. For $q = 7.8/d$, the data for $t > 2 \cdot 10^4$ are larger than the MCT results. This implies a violation of the superposition principle in the sense that the relaxation is slower than described by the theory. The effect increases with increasing wave number as is shown for the results for $q = 17.0/d$.

There appear increasing deviations between the simulation data and MCT results for wave numbers decreasing towards zero. As is demonstrated in Fig. 5.10 for $qd = 4.0$, the theory overestimates the trend to arrest in two aspects. The calculated critical arrested parts f_q^{sc} are so much above the measured plateau that a shift of the fit value q_{MCD} above q cannot account for the deviation. Furthermore, the time scale for the below-plateau decay is shorter than calculated. These discrepancies between data and theory increase with packing fraction. Let us characterize the below-plateau decay of the $qd = 4.0$ correlator by a time scale τ_α^s, defined by $\phi_q^s(\tau_\alpha^s) = 0.5$. For the highest measured packing fraction, the calculated value for τ_α^s is about three times larger than the value found in the simulation study. Increasing φ from 0.50 to 0.59, the data exhibit an increase of $\log \tau_\alpha^s$ by about 3. The MCT calculations yield an increase of $\log \tau_\alpha^s$, which is about 20% larger. The small-q correlations are described by the mean-squared displacement. Its long-time part is quantified by the diffusivity D_s. The latter defines a time scale τ_D, as discussed above in connection with Eq. (5.64). The MCT model under discussion describes the $\log \delta r_s^2(t)$ versus $\log t$ curves well for $\varphi \leq 0.55$. But, for larger packing fractions, the above-plateau parts of the curves fall below the simulation data. For the highest φ studied, the data for $\delta r_s^2(t)$ for $t = 10^4$ are about three times larger than the MCT results. The theory underestimates the long-time diffusivity or, equivalently, it overestimates the diffusivity scale τ_D by a factor 3. The simulation results for the non-Gaussian parameters $\alpha(t)$ exceed the calculated one by more than a factor 7 (Voigtmann *et al.* 2004).

The feature, which is described in the preceding paragraph, is referred to as decoupling of the diffusivity scale τ_D from the scale of the density fluctuations for intermediate and large wave numbers. Scale coupling as discussed in connection with Eq. (5.64) means that $D_s \tau_\alpha^s$ is a weakly varying function of the control parameters even though D_s and τ_α^s can vary by several orders of magnitude. The MCT result for the above specified model is exhibited by the lower curve in Fig. 5.11; the product varies by less than 20% within the density range studied. The data, however, demonstrate an increase of $D_s \tau_\alpha^s$ by a factor 3 if the transition point is approached. The specified decoupling phenomenon was demonstrated first by Kob and Andersen (1994). They analysed simulation data for the binary Lennard-Jones mixture, which is explained in connection with Fig. 1.9. A plot of $[-\log \tau_\alpha^s]$ versus $[\log(T - T_c)]$ exhibits a straight line of slope $\gamma \sim 2.4$. Here τ_α^s is a scale for the long-time decay of density fluctuation at intermediate wave number; and $T_c = 0.435$ is the critical temperature estimated by these authors. The lines for the big and small particles are parallel, and this demonstrates coupling of the scales for both species. The $\log D_s$ versus $[\log(T - T_c)]$ curves also exhibit straight lines, albeit with slopes 1.9 and 1.7 for the big and small particles, respectively. Thus, the diffusivities decrease less upon approaching the critical point than

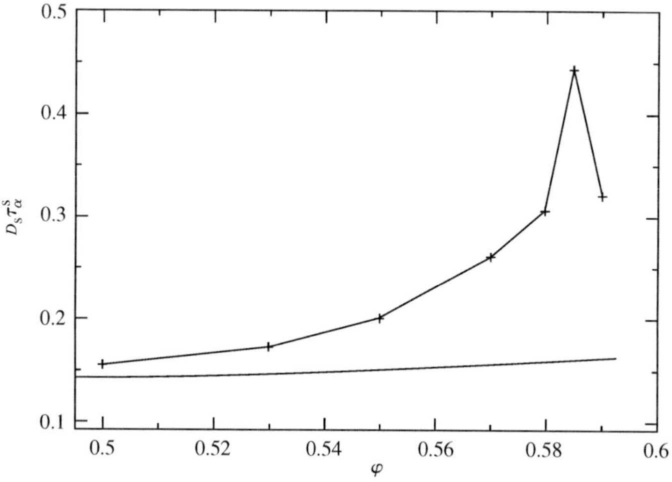

FIG. 5.11. The crosses exhibit the product of the diffusivity D_s and the time scale τ_α^s for the below-plateau decay of tagged-particle density correlators. The data are obtained by molecular-dynamics simulation for the model considered in Fig. 5.10. The relaxation times τ_α^s refer to the correlators of the middle panel of the preceding figure. The lower line is the result of the MCT model discussed in the text. Reproduced from Voigtmann *et al.* (2004).

expected by scale coupling. The $D_s\tau_\alpha^s$ versus $\log(T - T_c)$ curves demonstrate an increase by about a factor 3 upon decreasing T (Kob 2003).

The above cited molecular-dynamics studies lead to the following conclusions. If the temperature of the liquid is decreased or if its density is increased, the tagged-particle-density fluctuations develop long-time correlations, which are not described by the MCT equation for the correlators $\phi_q^s(t)$. Contrary to the MCT results, the simulation data exhibit a decoupling of the diffusivity scale τ_D and the evolution of a large non-Gaussian parameter. In this respect, the simulation data differ qualitatively also from the data measured by photon-correlation spectroscopy for hard-sphere colloids.

5.2 A mode-coupling theory for mixtures of spherical particles

Simple liquids cannot be cooled or compressed sufficiently in order to exhibit fully developed glassy dynamics. Conceptionally, properly chosen mixtures are the simplest systems, for which crystallization can be avoided upon driving the parameters towards their values for the glass transition. This fact provides the motivation for this section, which deals with the extension of the MCT to mixtures of spherical particles. The identification of subtle mixture effects for the complex dynamics is a further goal. Such qualitative features are of particular interest for tests of the relevance of MCT for explaining observations. The theory for mixtures is more involved than that for simple systems. But, the following exposition shall be kept to the minimum, which is necessary as access to the original literature.

5.2.1 *The equations of motion*

The N-particle system to be studied shall consist of n species of identical particles. The species shall be labelled by Greek letters $\alpha, \beta, \ldots$. There are N_α particles with mass μ_α for species α, $\alpha = 1, \ldots, n$. The number concentration and thermal velocity of species α shall be denoted by $x_\alpha = N_\alpha/N$ and $v_{th}^\alpha = \sqrt{k_\mathrm{B}T/\mu_\alpha}$, respectively. The particle label κ used on the preceding pages has to be interpreted as composed of two labels: $\kappa = (\alpha k), \alpha = 1, \ldots, n$, $k = 1, \ldots, N_\alpha$. The vectors for the positions and for the velocities are denoted by $\vec{r}_{\alpha k}$ and $\vec{v}_{\alpha k}$, respectively. Velocity averages are done as explained in Eqs. (3.1a,b): $(v_{\alpha k}^m | v_{\beta p}^n) = \delta^{mn}\delta_{\alpha\beta}\delta_{kp}(v_{th}^\alpha)^2$. The assumptions on the thermodynamic limit and on symmetries are made as specified in Sec. 3.1.1.

The motion of the particles is governed by the forces acting between pairs of them. The forces depend on the species involved. The densities $\rho^\alpha(\vec{r}) = \sum_{k=1}^{N_\alpha} \delta(\vec{r} - \vec{r}_{\alpha k})$ are the generalized dynamical variables characterizing the position of the particles of species α. The generalized dynamical variable for the current density of species α reads: $\vec{j}^\alpha(\vec{r}) = \sum_{k=1}^{N_\alpha} \vec{v}_{\alpha k}\delta(\vec{r} - \vec{r}_{\alpha k})$. The density fluctuations are equivalent quantities, which are obtained by Fourier transform from the space of position vectors $\vec{r}$ to the space of wave vectors $\vec{q}$:

$$\rho_{\vec{q}}^\alpha = \sum_{k=1}^{N_\alpha} \exp[i\vec{q}\vec{r}_{\alpha k}], \qquad \vec{j}_{\vec{q}}^\alpha = \sum_{k=1}^{N_\alpha} \vec{v}_{\alpha k} \exp[i\vec{q}\vec{r}_{\alpha k}]. \tag{5.66a}$$

These dynamical variables are connected by the conservation law for the number of particles of species α:

$$-i\partial_t \rho_{\vec{q}}^\alpha = \mathcal{L}\rho_{\vec{q}}^\alpha = \vec{q}\vec{j}_{\vec{q}}^\alpha. \tag{5.66b}$$

The current-density fluctuations determine the fundamental variables for the equations of motion, namely the momentum-density fluctuations $\vec{\pi}_{\vec{q}}^\alpha = \mu_\alpha \vec{j}_{\vec{q}}^\alpha$. As is explained in Sec. 3.1.2, the $\rho_{\vec{q}}^\alpha$ are scalar-density fluctuations of even parity with respect to inversions of space and time. The $\vec{j}_{\vec{q}}^\alpha$ are vector-density fluctuations of odd parity with respect to the mentioned discrete symmetry operations. Both fields have even conjugation parity. The symmetries imply that the overlaps of the density fluctuations are real functions of the wave number $q = |\vec{q}|$:

$$(\rho_{\vec{q}}^\alpha | \rho_{\vec{q}}^\beta) = N S_q^{\alpha\beta}; \quad \alpha, \beta = 1, \ldots, n. \tag{5.67a}$$

The quantities $S_q^{\alpha\beta} = S_q^{\beta\alpha}$ are called partial structure factors. They provide the simplest information on the equilibrium structure. The fluctuations of densities and current densities are orthogonal:

$$(\rho_{\vec{q}}^\alpha | \vec{j}_{\vec{q}}^\beta) = 0. \tag{5.67b}$$

The current overlaps shall be described by wave-number independent quantities $J^{\alpha\beta}$, which combine the information on concentrations and thermal velocities for the species:

$$J^{\alpha\beta} = x_\alpha (v_{th}^\alpha)^2 \delta^{\alpha\beta}, \tag{5.67c}$$

$$(j_{\vec{q}}^{\alpha k} | j_{\vec{q}}^{\beta \ell}) = N\delta^{k\ell} J^{\alpha\beta}; \quad k, \ell = 1, 2, 3; \quad \alpha, \beta = 1, \ldots, n. \tag{5.67d}$$

The simplest correlation functions for the description of structure dynamics are the ones formed with the density fluctuations. These density correlators are defined by

$$\phi_q^{\alpha\beta}(t) = (\rho_{\vec{q}}^\alpha(t) | \rho_{\vec{q}}^\beta)/N; \quad \alpha, \beta = 1, \ldots, n. \tag{5.68a}$$

Using the notations of Eq. (3.17a), one can write $\phi_q^{\alpha\beta}(t) = \sum_{kp} F_q^{(0)\alpha k\beta p}(t)$. As anticipated by the notation, the functions depend on the wave vector $\vec{q}$ merely via the modulus q. From the symmetry properties (3.18a,c,e), one concludes that $\phi_q^{\alpha\beta}(t) = \phi_q^{\beta\alpha}(t)$, that these functions are real, and that they are invariant under change of t to $-t$. The corresponding symmetries for the correlators in the frequency domain, $\phi_q^{\alpha\beta}(z)$, are formulated in Eqs. (3.20a,b,c). The mentioned symmetries shall be referred to as standard symmetries. The functions $\phi_q^{\alpha\beta}(t)$ generalize the concept of a density correlator $\phi_q(t)$. But, contrary to the latter function, the former are not normalized; one gets $\phi_q^{\alpha\beta}(t=0) = S_q^{\alpha\beta}$. There holds $-i\partial_t \phi_q^{\alpha\beta}(t) = -(\mathcal{L}\rho_{\vec{q}}^\alpha(t)|\rho_{\vec{q}}^\beta) = -(\rho_{\vec{q}}^\alpha(t)|\mathcal{L}\rho_{\vec{q}}^\beta)$ and $(-i\partial_t)^2 \phi_q^{\alpha\beta}(t) = (\mathcal{L}\rho_{\vec{q}}^\alpha(t)|\mathcal{L}\rho_{\vec{q}}^\beta)$. The first Taylor coefficient of the correlators vanishes for $t = 0$ because of Eq. (5.67b). The second one can be reformulated with Eq. (5.67d). Assuming the existence of bounded third derivatives, there holds the short-time expansion formula

$$\phi_q^{\alpha\beta}(t) = S_q^{\alpha\beta} - \tfrac{1}{2} q^2 J^{\alpha\beta} t^2 + O(|t|^3). \tag{5.68b}$$

In the remainder of this chapter, bold letters shall be used to indicate n-by-n matrices. For example, $\boldsymbol{S}_q$, $\boldsymbol{J}$, and $\boldsymbol{\phi}_q(t)$ shall denote the real symmetric matrices, whose elements are defined in Eqs. (5.67a), (5.67c), and (5.68a), respectively. The matrix $\boldsymbol{S}_q$ is the metric matrix for the set of variables $\rho_{\vec{q}}^\alpha$, $\alpha = 1, \ldots, n$. Let us assume that the density fluctuations are linear independent. Then, $\boldsymbol{S}_q$ has an inverse $\boldsymbol{S}_q^{-1}$, and both matrices are positive definite.

This paragraph shall be devoted to the definition of the real symmetric matrix $\boldsymbol{c}_q$ of direct correlation functions. The conventions by Hansen and McDonald (1986) shall be adopted. To relate $S_q^{\alpha\beta}$ to the pair-distribution function $g_{\alpha\beta}(r)$, one writes Eq. (5.67a) in the form:

$$S_q^{\alpha\beta} = x_\alpha \delta^{\alpha\beta} + \rho x_\alpha x_\beta \int \exp[i\vec{q}\vec{r}] g_{\alpha\beta}(r) d^3\vec{r}, \tag{5.69a}$$

$$\rho g_{\alpha\beta}(r) = [N x_\alpha x_\beta]^{-1} \sum_{kp}{}' \langle \delta(\vec{r} + \vec{r}_{\alpha k} - \vec{r}_{\beta p}) \rangle. \tag{5.69b}$$

The prime indicates that, for $\alpha = \beta$, the N_α contributions $k = p$ have to be left out from the double summation. One can copy the proof of Eq. (3.60a) in order to show

$$\lim_{r\to\infty} g_{\alpha\beta}(r) = 1. \tag{5.69c}$$

A pair-correlation function $h_{\alpha\beta}(r)$ is defined by generalizing Eq. (3.60c):

$$g_{\alpha\beta}(r) = 1 + h_{\alpha\beta}(r). \tag{5.69d}$$

The request of the absolute integrability of $h_{\alpha\beta}(r)$ shall be included in the definition of the amorphous state to be discussed. Hence, by Fourier transformation, the functions $\tilde{h}_q^{\alpha\beta}$ can be defined. They depend on the wave number continuously. There holds for $\alpha, \beta = 1, \ldots, n$:

$$S_q^{\alpha\beta} = x_\alpha \delta^{\alpha\beta} + \rho x_\alpha x_\beta \tilde{h}_q^{\alpha\beta} + (2\pi)^3 \rho x_\alpha x_\beta \delta(\vec{q}). \tag{5.69e}$$

For $\vec{q} \neq 0$, one generalizes the Ornstein–Zernike relation (3.61) to mixtures:

$$(\boldsymbol{S}_q^{-1})^{\alpha\beta} = [1/x_\alpha]\delta^{\alpha\beta} - \rho c_q^{\alpha\beta}. \tag{5.70}$$

The goal of the first step of the procedure towards MCT equations is an expression of the set of density-fluctuation correlators $\phi_q^{\alpha\beta}(t)$ in terms of fluctuating-force correlation functions. The result shall be achieved in two substeps. The first one formulates the Zwanzig–Mori equation (2.62a) for the set of n variables $A_\alpha = \rho_{\vec{q}*}^\alpha/\sqrt{N}$, $\alpha = 1, \ldots, n$, $\vec{q}^* = (0, 0, q)$. The metric matrix is identical with $\boldsymbol{S}_q$. The projector on the n-dimensional subspace spanned by these variables reads $\mathcal{P}_1 = \sum_{\alpha\beta} |\rho_{\vec{q}*}^\alpha)(\boldsymbol{S}_q^{-1})^{\alpha\beta}(\rho_{\vec{q}*}^\beta|/N$. The time-translation operator $\mathcal{U}_1(t)$ for the reduced dynamics is generated by $\mathcal{L}_1 = \mathcal{Q}_1\mathcal{L}\mathcal{Q}_1$, with $\mathcal{Q}_1 = 1 - \mathcal{P}_1$. Since A_α and $\mathcal{L}A_\beta$ are perpendicular, the frequency matrix (2.60a) vanishes and the projector $\mathcal{Q}_1$ can be dropped in Eq. (2.61a). The continuity equations can be used to write $\dot{A}_\alpha^{\mathcal{Q}_1} = iqj_{\vec{q}*}^{\alpha 3}/\sqrt{N}$. One arrives at

$$\{z\boldsymbol{1} + q^2 \boldsymbol{G}_q(z)\boldsymbol{S}_q^{-1}\}\boldsymbol{\phi}_q(z) = -\boldsymbol{S}_q. \tag{5.71a}$$

Here, the elements of matrix $\boldsymbol{G}_q(t)$ read: $G_q^{\alpha\beta}(t) = (j_{\vec{q}*}^{\alpha 3}|\mathcal{U}_1(-t)j_{\vec{q}*}^{\beta 3})/N$. The formula is a matrix generalization of Eq. (3.95a) with $\boldsymbol{G}_q(z)\boldsymbol{S}_q^{-1}$ being the generalization of $K_q(z)$.

The second substep formulates Eq. (2.62a) for the set of n variables $A_\alpha = j_{\vec{q}*}^{\alpha 3}/\sqrt{N}$, whose dynamics is generated by $\mathcal{L}_1$. The metric matrix is the diagonal matrix $\boldsymbol{J}$. The projector on the subspace spanned by the variables reads: $\mathcal{P}_2 = \sum_\alpha |j_{\vec{q}*}^{\alpha 3})[x_\alpha(v_{th}^\alpha)^2]^{-1}(j_{\vec{q}*}^{\alpha 3}|/N$. It is perpendicular to $\mathcal{P}_1$ so that $\mathcal{Q} = \mathcal{Q}_1(1-\mathcal{P}_2) = 1 - \mathcal{P}_1 - \mathcal{P}_2$. Again, the frequency matrix (2.60a) vanishes and the projector $\mathcal{Q}_2 = 1 - \mathcal{P}_2$ can be dropped on the right-hand side of Eq. (2.61a): $\dot{A}_\alpha^{\mathcal{Q}_2} = i\mathcal{L}_1 j_{\vec{q}*}^{\alpha 3}/\sqrt{N} = i\mathcal{Q}_1\mathcal{L}j_{\vec{q}*}^{\alpha 3}/\sqrt{N}$. One gets

$$[z\boldsymbol{1} + \boldsymbol{J}\boldsymbol{N}_q(z)]\boldsymbol{G}_q(z) = -\boldsymbol{J}. \tag{5.71b}$$

Here, the elements of a fluctuating-force-correlator matrix $\boldsymbol{N}_q(z)$ are the Laplace transforms of

$$N_q^{\alpha\beta}(t) = (F_q^\alpha | \mathcal{U}_{\rm red}(-t) F_q^\beta). \tag{5.71c}$$

The dynamical variables are the fluctuating forces

$$F_q^\alpha(t) = \mathcal{Q}_1 \mathcal{L} j_{\vec{q}*}^{\alpha 3} / [x_\alpha (v_{th}^\alpha)^2 \sqrt{N}]. \tag{5.71d}$$

The time evolution $\mathcal{U}_{\rm red}(t)$ is generated by the reduced Liouvillian $\mathcal{L}_\mathcal{Q} = \mathcal{Q}_1 \mathcal{Q}_2 \mathcal{L} \mathcal{Q}_2 \mathcal{Q}_1$. After multiplication of Eq. (5.71b) by $q^2 \boldsymbol{S}_q^{-1}$ from the right, one recognizes the result as matrix generalization of Eq. (3.95d). The matrix $\boldsymbol{J}\boldsymbol{N}_q(z)$ is the generalization of $M_{Lq}(z) = q^2 K_{Lq}(z)$; and the forces F_q^α, $\alpha = 1, \ldots, n$, are the generalizations of the force F_{Lq} from Eq. (4.1c).

The density pairs $\rho^{\alpha_1}(\vec{r}_1)\rho^{\alpha_2}(\vec{r}_2)$ are the dynamical variables, whose linear superpositions determine the force between a particle of species α_1 at position $\vec{r}_1$ and a particle of species α_2 at position $\vec{r}_2$. They are linear combinations of the pair fluctuations $A_m = \rho_{\vec{m}_1}^{\alpha_1} \rho_{\vec{m}_2}^{\alpha_2}$. Here, the label m abbreviates the composed set of labels $(\vec{m}_1 \alpha_1 \vec{m}_2 \alpha_2)$. An ordering is anticipated to avoid double counting, as is explained in Sec. 4.1.1. The A_m are the relevant pair variables chosen for the second step of the procedure towards closed equations of motion. Equation (2.64) is applied with $X = F_q^\alpha$ and $Y = F_q^\beta$. The first kernel is approximated by its white noise limit, $M_{X,Y}(z) \to M_{X,Y}(\pm i0) = \pm i n_q^{\alpha\beta}$. The renormalizations are dropped: $M_m^X(z) \to 0$, $M_n^Y(z) \to 0$. For Im $z \gtrless 0$, the reduction of the kernel to its two-density-fluctuation contribution reads

$$N_q^{\alpha\beta}(z) = \pm i n_q^{\alpha\beta} + \sum_{m,n} \Omega_{qm}^\alpha \phi_{m,n}(z) \Omega_{qn}^{\beta *}. \tag{5.72a}$$

Equation (2.63a) provides the formula for the coupling amplitudes:

$$\Omega_{q\vec{m}_1\alpha_1\vec{m}_2\alpha_2}^\alpha = {\sum_{\vec{k}_1\sigma_1\vec{k}_2\sigma_2}}' (F_q^\alpha | \rho_{\vec{k}_1}^{\sigma_1} \rho_{\vec{k}_2}^{\sigma_2}) (\boldsymbol{g}^{-1})_{\vec{k}_1\sigma_1\vec{k}_2\sigma_2, \vec{m}_1\alpha_1\vec{m}_2\alpha_2}. \tag{5.72b}$$

The prime indicates the request to observe the mentioned ordering of the labels.

The evaluation of the overlaps of the variables F_q^α with the pair fluctuations $\rho_{\vec{k}_1}^{\sigma_1} \rho_{\vec{k}_2}^{\sigma_2}$ follows the one explained in the discussion of Eqs. (4.5a–c). One merely has to add labels for the species in order to arrive at

$$\begin{aligned}(F_q^\alpha | \rho_{\vec{k}_1}^{\sigma_1} \rho_{\vec{k}_2}^{\sigma_2}) = \delta_{\vec{q}*, \vec{k}_1 + \vec{k}_2} (\sqrt{N}/x_\alpha) \Big\{ & \delta^{\alpha\sigma_1} k_1^3 S_{k_2}^{\alpha\sigma_2} + \delta^{\alpha\sigma_2} k_2^3 S_{k_1}^{\alpha\sigma_1} \\ & - q x_\alpha \sum_{\epsilon\tau} S_{k_1}^{\sigma_1\epsilon} S_{k_2}^{\sigma_2\tau} [(\delta^{\alpha\epsilon}\delta^{\alpha\tau}/(x_\epsilon x_\tau)) + \rho^2 c_3^{\alpha\epsilon\tau}(q, k_1, k_2)] \Big\}.\end{aligned} \tag{5.73}$$

Equation (4.6) for the expression of the average of a triple-density fluctuation in terms of a triple-correlation function is generalized by the formula:

$$\langle \rho_{\vec{q}}^{\sigma *} \rho_{\vec{k}}^{\alpha} \rho_{\vec{p}}^{\beta} \rangle = N \delta_{\vec{q},\vec{k}+\vec{p}} \sum_{\eta\epsilon\tau} S_q^{\sigma\eta} S_k^{\alpha\epsilon} S_p^{\beta\tau} \left[(\delta^{\eta\epsilon}\delta^{\eta\tau}/(x_\epsilon x_\tau)) + \rho^2 c_3^{\eta\epsilon\tau}(qkp) \right]. \quad (5.74)$$

Here, $\vec{p}$ is meant as an abbreviation of $\vec{q} - \vec{k}$. Dropping the triple-correlation function is the generalization of the convolution approximation to mixtures.

The third step of the procedure formulates the factorization ansatz (4.7a). In the present context, it amounts to the replacement of $\phi_{m,n}(t) = (\rho_{\vec{m}_1}^{\alpha_1} \rho_{\vec{m}_2}^{\alpha_2} | \mathcal{U}_{\text{red}}(-t) \rho_{\vec{n}_1}^{\beta_1} \rho_{\vec{n}_2}^{\beta_2})$ by $N^2 \delta_{\vec{m}_1,\vec{n}_1} \delta_{\vec{m}_2,\vec{n}_2} \phi_{m_1}^{\alpha_1\beta_1}(t) \phi_{n_1}^{\alpha_2\beta_2}(t)$. For $t = 0$, this implies a factorization for the metric matrix, which leads to the replacement of $\boldsymbol{g}^{-1}_{\vec{k}_1\sigma_1\vec{k}_2\sigma_2,\vec{m}_1\alpha_1\vec{m}_2\alpha_2}$ by $\delta_{\vec{k}_1,\vec{m}_1} \delta_{\vec{k}_2,\vec{m}_2} (\boldsymbol{S}^{-1})_{k_1}^{\sigma_1\alpha_1} (\boldsymbol{S}^{-1})_{k_2}^{\sigma_2\alpha_2} / N^2$. Collecting all formulas, one gets the relaxation kernels (Barrat and Latz 1990):

$$\boldsymbol{N}_q(z) = \pm i \boldsymbol{n}_q + \boldsymbol{N}_q^{mc}(z), \qquad \text{Im } z \gtrless 0, \quad (5.75a)$$

$$N_q^{mc\alpha\beta}(t) = \frac{1}{2} [\rho/(x_\alpha x_\beta)] L^{-3} \sum_{\vec{k}} \sum_{\alpha_1\alpha_2\beta_1\beta_2} \left\{ v_{\alpha,\alpha_1\alpha_2}(\vec{q}^*, \vec{k}\vec{p}) \phi_k^{\alpha_1\beta_1}(t) \phi_p^{\alpha_2\beta_2}(t) v_{\beta,\beta_1\beta_2}(\vec{q}^*, \vec{k}\vec{p}) \right\}, \quad (5.75b)$$

$$v_{\alpha,\beta\gamma}(\vec{q}^*, \vec{k}\vec{p}) = \left[k^3 c_k^{\alpha\beta} \delta_{\alpha\gamma} + p^3 c_p^{\alpha\gamma} \delta_{\alpha\beta} + q x_\alpha \rho c_3^{\alpha\beta\gamma}(qkp) \right] / q. \quad (5.75c)$$

Here, $\vec{p}$ abbreviates $\vec{q}^* - \vec{k}$.

Before combining the preceding formulas, some abbreviations shall be introduced. Let us start by remembering the following theorem about real symmetric n-by-n matrices, say, $\boldsymbol{P}$ and $\boldsymbol{Q}$. If $\boldsymbol{P}$ is positive definite, there is some equivalence transformation, which maps simultaneously $\boldsymbol{P}$ and $\boldsymbol{Q}$ on diagonal form. Application to $\boldsymbol{P} = \boldsymbol{J}$ and $\boldsymbol{Q} = \boldsymbol{S}_q^{-1}$ shows that $\boldsymbol{J}\boldsymbol{S}_q^{-1}$ is equivalent to a diagonal matrix. Since both $\boldsymbol{J}$ and $\boldsymbol{S}_q^{-1}$ are positive definite, all eigenvalues are positive and can be written as squares of positive numbers. Consequently, there is a real matrix $\boldsymbol{\Omega}_q$ with n positive eigenvalues so that

$$\boldsymbol{\Omega}_q^2 = q^2 \boldsymbol{J} \boldsymbol{S}_q^{-1}. \quad (5.76a)$$

The matrices $\boldsymbol{\Omega}_q^2 \boldsymbol{S}_q$ and $\boldsymbol{J}^{-1}\boldsymbol{\Omega}_q^2$ are real, symmetric, and positive definite. Let us assume the validity of the standard symmetries of the n-by-n matrix, which is constructed above from the kernels $M_{X,Y}(z)$. Then, $\boldsymbol{n}_q$ is the corresponding zero-frequency spectrum. It is a real, symmetric positive semidefinite matrix. The matrix

$$\boldsymbol{\nu}_q = \boldsymbol{J} \boldsymbol{n}_q \quad (5.76b)$$

is real. Using the above cited theorem with $\boldsymbol{P} = \boldsymbol{J}$ and $\boldsymbol{Q} = \boldsymbol{n}_q$, one concludes that $\boldsymbol{\nu}_q$ is equivalent to a matrix with n non-negative diagonal elements

$\nu_q^1, \ldots, \nu_q^n$. If $\boldsymbol{n}_q$ is positive definite, the n eigenvalues are positive. By construction, the matrices $\boldsymbol{\Omega}_q$ and $\boldsymbol{\nu}_q$ have the dimension of a frequency. A real n-by-n matrix of times can be defined by

$$\boldsymbol{\tau}_q = \boldsymbol{S}_q \boldsymbol{n}_q / q^2 = \boldsymbol{\Omega}_q^{-2} \boldsymbol{\nu}_q. \tag{5.76c}$$

The matrices $\boldsymbol{S}_q^{-1}\boldsymbol{\tau}_q$ and $\boldsymbol{\tau}_q \boldsymbol{S}_q$ are symmetric and positive semidefinite. They are positive definite if and only if this is true for $\boldsymbol{n}_q$.

Using the preceding definitions and introducing the abbreviation $\boldsymbol{m}_q(z) = \boldsymbol{S}_q \boldsymbol{N}_q^{mc}(z)/q^2$, one can combine Eqs. (5.71a,b) to a double-fraction representation for the matrix $\boldsymbol{\phi}_q(z)$ in terms of the matrix $\boldsymbol{m}_q(z)$:

$$\boldsymbol{\phi}_q(z) = -\Big\{ z\mathbf{1} - \big[z\mathbf{1} \pm i\boldsymbol{\nu}_q + \boldsymbol{\Omega}_q^2 \boldsymbol{m}_q(z) \big]^{-1} \boldsymbol{\Omega}_q^2 \Big\}^{-1} \boldsymbol{S}_q; \quad \mathrm{Im}\, z \gtrless 0. \tag{5.77a}$$

If there exists a positive lower bound ν for the eigenvalues $\nu_q^1, \ldots, \nu_q^n$, one can define a low-frequency regime for the dynamics by the request $|z| \ll \nu$. Within this regime, one can replace $z \pm i\nu_q$ by $\pm i\nu_q$. One arrives at the second version for a double-fraction representation of the correlator matrix:

$$\boldsymbol{\phi}_q(z) = -\Big\{ z\mathbf{1} - \big[\pm i\boldsymbol{\tau}_q + \boldsymbol{m}_q(z) \big]^{-1} \Big\}^{-1} \boldsymbol{S}_q; \quad \mathrm{Im}\, z \gtrless 0. \tag{5.77b}$$

Let us assume that there are correlation-function matrices $\boldsymbol{\phi}_q(t)$ and $\boldsymbol{m}_q(t)$ so that their Laplace transforms are connected by the fractions formulated above. Let us assume in addition that the first and second derivatives of $\boldsymbol{\phi}_q(t)$ belong to the extended set of determining functions. Then, one can use Eqs. (A.6a)–(A.7b) to rewrite the first fraction in a first version for equations of motion for $t \gtrless 0$:

$$\partial_t^2 \boldsymbol{\phi}_q(t) \pm \boldsymbol{\nu}_q \partial_t \boldsymbol{\phi}_q(t) + \boldsymbol{\Omega}_q^2 \big[\boldsymbol{\phi}_q(t) + \int_0^t \boldsymbol{m}_q(t-t') \partial_{t'} \boldsymbol{\phi}_q(t') dt' \big] = \mathbf{0}. \tag{5.78a}$$

This equation has to be solved with the initial conditions $\boldsymbol{\phi}_q(t=0) = \boldsymbol{S}_q$ and $\partial_t \boldsymbol{\phi}_q(t=0) = \mathbf{0}$. To rewrite the second fraction into the second version for equations of motion, it is sufficient to assume that the first derivative of $\boldsymbol{\phi}_q(t)$ is a determining matrix function. One gets for $t \gtrless 0$:

$$\pm \boldsymbol{\tau}_q \partial_t \boldsymbol{\phi}_q(t) + \boldsymbol{\phi}_q(t) + \int_0^t \boldsymbol{m}_q(t-t') \partial_{t'} \boldsymbol{\phi}_q(t') dt' = \mathbf{0}. \tag{5.78b}$$

This equation has to be solved with the initial condition $\boldsymbol{\phi}_q(t=0) = \boldsymbol{S}_q$. The preceding fraction representations are matrix generalization of Eqs. (4.29a,b). The equations of motion generalize Eqs. (4.17a,b).

The essential piece of the derivation is the representation of the kernel $\boldsymbol{S}_q^{-1} \boldsymbol{m}_q(t)$ as the bilinear mode-coupling functional $\boldsymbol{N}_q^{mc}(t)/q^2 = \boldsymbol{\mathcal{F}}_q[\boldsymbol{\phi}_k(t), \boldsymbol{\phi}_p(t)]$.

As explained in connection with Eq. (4.11d), one can appeal to the thermodynamic limit in order to transform the summation over $\vec{k}$ in Eq. (5.75b) into a double integral over the wave numbers k and p. One obtains

$$\mathcal{F}_q^{\alpha\beta}[\boldsymbol{X}_k, \boldsymbol{Y}_p] = [\rho/(32\pi^2)] \int_0^\infty dk \int' dp \sum_{\alpha_1\alpha_2\beta_1\beta_2} (pk/q^5) \\ v_{\alpha\alpha_1\alpha_2}(qkp)[X_k^{\alpha_1\beta_1} Y_p^{\alpha_2\beta_2}] v_{\beta\beta_1\beta_2}(qkp), \tag{5.79a}$$

with coupling amplitudes

$$v_{\alpha\alpha_1\alpha_2}(qkp) = \Big\{(q^2 + k^2 - p^2)c_k^{\alpha\alpha_1}\delta_{\alpha\alpha_2} + (q^2 + p^2 - k^2)c_p^{\alpha\alpha_2}\delta_{\alpha\alpha_1} \\ + 2q^2\rho x_\alpha c_3^{\alpha\alpha_1\alpha_2}(qkp)\Big\}\Big/x_\alpha. \tag{5.79b}$$

The prime indicates the restriction $|q - k| \leq p \leq q + k$. The kernel is given as matrix product:

$$\boldsymbol{m}_q(t) = \boldsymbol{S}_q \boldsymbol{\mathcal{F}}_q[\boldsymbol{\phi}_k(t), \boldsymbol{\phi}_p(t)]. \tag{5.80}$$

As explained in connection with Eqs. (4.35), (4.36), a MCT model is specified by a grid of discrete wave numbers. It consists of M steps of equal size (h/d) and it characterizes an interval from zero to a cutoff wave number q_{co}. All integrals are meant as abbreviations of their Riemann sums constructed with this grid.

The nonlinearities of the derived closed set of equations of motion are due to the dependence of the kernels $m_q^{\alpha\beta}(t)$ on the correlator matrix $\boldsymbol{\phi}_k(t)$. This dependence is specified by the mode-coupling polynomials $\mathcal{F}_q^{\alpha\beta}$. These are defined solely in terms of parameters, which quantify the equilibrium structure: the total number density ρ, the number concentrations $x_\alpha, \alpha = 1, \ldots, n$, the direct correlation functions $\boldsymbol{c}_q$, and the triple correlation functions $c_3^{\alpha\beta\gamma}(qkp)$. The temperature and the interaction potentials enter only indirectly via the specified structure functions. These remarks are generalizations of those made in Chapter 4 for the theory of simple systems. This holds also for the extension to the treatment of singular potentials as, e.g., for hard-sphere systems. Similarly, the motivation remains valid for proposing the first version of equations of motion as theory for the description of glassy dynamics in conventional systems. The second version is suggested for a description of colloids. For the latter system, the MCT equations can be derived also within the Smoluchowski-equation approach towards the dynamics (Nägele *et al.* 1999).

If the equations of motion have solutions, one can copy the considereations in Chapter 4 for simple systems in order to show that the correlators $\boldsymbol{\phi}_q(t)$ and $\boldsymbol{m}_q(t)$ have continuous derivatives of all order for $t \neq 0$. These derivatives have finite limits for $t \to \pm 0$. For a system with Newtonian short-time dynamics there holds

$$\boldsymbol{\phi}_q(t) = \{1 - \tfrac{1}{2}\boldsymbol{\Omega}_q^2 t^2 + \tfrac{1}{6}\boldsymbol{\nu}_q\boldsymbol{\Omega}_q^2|t|^3 + O(t^4)\}\, \boldsymbol{S}_q. \tag{5.81a}$$

For the system with stochastic short-time dynamics, one gets

$$\boldsymbol{\phi}_q(t) = \{1 - \boldsymbol{\tau}_q^{-1}|t| + O(t^2)\}\boldsymbol{S}_q. \tag{5.81b}$$

The theorems of Sec. 4.2.2 can be extended to deal with matrix correlators. The first proof, which is given in Appendix A.6, concerns the following statement. If $\boldsymbol{\phi}_q^{(r)}(t)$, $q = 1, \ldots, M$, denotes a set of real continuous positive-definite matrix functions or a set of real completely monotonic matrix functions, the corresponding properties are valid for the M kernels $\boldsymbol{S}_q^{-1}\boldsymbol{m}_q^{(r)}(t)$, $q = 1, \ldots, M$, which are defined by

$$\boldsymbol{m}_q^{(r)}(t) = \boldsymbol{S}_q\boldsymbol{\mathcal{F}}_q[\boldsymbol{\phi}_q^{(r)}(t), \boldsymbol{\phi}_p^{(r)}(t)]. \tag{5.82}$$

This holds for arbitrary real amplitudes $v_{\alpha\beta\gamma}(qkp)$, which are used in Eq. (5.79a) for the definition of the polynomials $\boldsymbol{\mathcal{F}}$. These kernels $\boldsymbol{m}_q^{(r)}(t)$ are substituted in Eqs. (5.78a) or (5.78b), respectively, as replacement of $\boldsymbol{m}_q(t)$. The second proof, which is given in Sec. 2.4.2, establishes the result that the thereby defined equations of motion have solutions $\boldsymbol{\phi}_q^{(r+1)}(t)$, which are positive definite or completely monotonic, respectively. Starting with $\boldsymbol{\phi}_q^{(0)}(t) = \boldsymbol{S}_q \exp[-i\Gamma_q(t)]$, $\Gamma_q > 0$, one obtains a sequence of approximants $\boldsymbol{\phi}_q^{(r)}(t)$, $r = 1, 2, \ldots$. As a third step, one proves the uniform convergence of this sequence to a solution $\boldsymbol{\phi}_q(t)$, $q = 1, \ldots, M$, of the equations of motion, and the uniqueness of this solution. The mentioned proofs have been given by Franosch and Voigtmann (2002) for the second version of the equations of motion. But, their proofs can be used also to extend the proofs of Haussmann (1990) for the first version.

The cited work establishes the following conclusions. The MCT equations of motion define a unique array of continuous positive definite real matrix functions $\boldsymbol{\phi}_q(t)$, $q = 1, \ldots, M$. If the second version of equations of motion are used, the functions $\boldsymbol{\phi}_q(t)$ and the kernels $\boldsymbol{S}_q^{-1}\boldsymbol{m}_q(t)$ are completely monotonic for $q = 1, \ldots, M$. The equations of motion (5.78a) and (5.78b) are equivalent to the equations (5.77a) and (5.77b), respectively, for the Laplace transforms $\boldsymbol{\phi}_q(z)$ and $\boldsymbol{m}_q(z)$. For every finite time interval, the solutions $\phi_q^{\alpha\beta}(t)$ are continuous functions of $\rho, x_1, \ldots, x_n$, of the amplitudes $v_{\alpha\beta\gamma}(qkp)$, and of the elements of the matrices $\boldsymbol{\Omega}_q, \boldsymbol{\nu}_q$, and $\boldsymbol{\tau}_q$. In summary: the MCT equations define a model for a statistical description of a dynamics.

No new problem appears for the discussion of the tagged-particle dynamics in a mixture, which has not been discussed already in Sec. 5.1.3 for that in simple systems. The Eqs. (5.21a)–(d) formulate the first step in the procedure for motivating MCT equations. These formulas are based on the general symmetries and on the conservation law for the tagged-particle density, which hold for mixtures as they do for simple systems. The second step leads to a splitting of the kernel $M_{sq}(t)$ in a white noise term ν_q^s and the mode-coupling contribution $M_{sq}^{mc}(t)$. Writing the latter as $\Omega_{sq}^2 m_q^s(t)$ is a mere convention done for later convenience. There hold the equations of motion (5.26a,b) in the frequency domain or the Eqs. (5.27a,b) in the time domain. The equations for the velocity correlator and for the mean-squared displacement remain unaltered. The task left is the specification of the mode-coupling expression for kernel $m_q^s(t)$.

A particle of species α at some position $\vec{r}_1$ may interact with the tagged particle, which is situated at some position $\vec{r}_2$. The interaction potential $V_{\alpha s}(|\vec{r}_1 - \vec{r}_2|)$ may depend on the species chosen. The relevant pair fluctuations are $A_k = \rho^\alpha_{\vec{k}_1}\rho^s_{\vec{k}_2}$. The generalization from simple system to mixtures amounts to incorporating the label α in the label for the pair variables: $k = (\alpha\vec{k}_1, \vec{k}_2)$. The derivation of Eq. (5.22c) can be copied with the result

$$\left(F_{sq}|\rho^\alpha_{\vec{k}}\rho^s_{\vec{p}}\right) = \delta_{\vec{q}*,\vec{k}+\vec{p}}(v^s_{th})^2(-k^3)\left(\rho^s_{\vec{k}}|\rho^\alpha_{\vec{k}}\right). \tag{5.83a}$$

The formula (5.23d) for the density-fluctuation-pair average is to be generalized to

$$\left(\rho^s_{\vec{q}}|\rho^\alpha_{\vec{q}}\right) = \rho\sum_\beta c^s_{\beta q}S^{\beta\alpha}_q. \tag{5.83b}$$

The third step in the derivation simplifies $M^{mc}_{sq}(t) = \sum_{\vec{m}_1\alpha\vec{m}_2}\sum_{\vec{n}_1\beta\vec{n}_2} \Omega^s_{q\vec{m}_1\alpha\vec{m}_2}\phi_{\vec{m}_1\alpha\vec{m}_2,\vec{n}_1\beta\vec{n}_2}(t)\Omega^{s*}_{q\vec{n}_1\beta\vec{n}_2}$ by the factorization ansatz for the pair correlator $\phi_{\vec{m}_1\alpha\vec{m}_2,\vec{n}_1\beta\vec{n}_2}(t) = (\rho^\alpha_{\vec{m}_1}\rho^s_{\vec{m}_2}|\mathcal{U}_{\text{red}}(-t)\rho^\beta_{\vec{n}_1}\rho^s_{\vec{n}_2})$. The latter is replaced by $N\delta_{\vec{m}_1,\vec{n}_1}\delta_{\vec{m}_2,\vec{n}_2}\phi^{\alpha\beta}_{m_1}(t)\phi^s_{n_1}(t)$. Combining all formulas, one obtains (Bosse and Thakur 1987):

$$M^{mc}_{sq}(t) = (v^s_{th}\rho)^2\sum_{\vec{k}+\vec{p}=\vec{q}*}\sum_{\alpha\beta}(k^3)^2 c^s_{k\alpha}c^s_{p\beta}\phi^{\alpha\beta}_k(t)\phi^s_p(t)/N. \tag{5.84}$$

The preceding formula can be converted into a double-integral representation for the kernel $m^s_q(t)$. One arrives at a generalization of Eq. (5.25):

$$m^s_q(t) = \mathcal{F}^s_q[\boldsymbol{\phi}_k(t), \phi^s_p(t)], \tag{5.85a}$$

$$\mathcal{F}^s_q[\boldsymbol{X}_k, Y_p] = [\rho/(16\pi^2)]\int_0^\infty dk\int' dp(pk/q^5)$$
$$\sum_{\alpha\beta}(q^2+k^2-p^2)^2 c^s_{k\alpha}c^s_{p\beta}X^{\alpha\beta}_k Y_p. \tag{5.85b}$$

Again, the prime indicates the restriction of the integration to $|q-k| \leq p \leq q+k$.

For a discussion of the mean-squared displacement, the kernel $m^s_0(t)$ from Eq. (5.28a) is needed. From Eq. (5.84), one gets $\lim_{q\to 0} M^{mc}_{sq}(t) = (v^s_{th})^2\rho L^{-3}\sum_{\vec{k}} \sum_{\alpha\beta}(k^3)^2 c^s_{k\alpha}c^s_{k\beta}\phi^{\alpha\beta}_k(t)\phi^s_k(t)$. Consequently,

$$m^s_0(t) = \int_0^\infty dq\sum_{\alpha\beta}v^s_\alpha(q)\phi^{\alpha\beta}_q(t)v^s_\beta(q)\phi^s_q(t), \tag{5.86a}$$

$$v^s_\alpha(q) = (\rho/6\pi^2)^{1/2}c^s_{q\alpha}(q^2 d). \tag{5.86b}$$

For general mathematical discussions as well as for numerical calculations, the integrals in Eqs. (5.85b), (5.86a) are to be considered as abbreviations of the Riemann sums for the above specified grid of M values for the wave numbers.

The equations of motion (5.27a) or (5.27b) together with Eqs. (5.85a,b) are closed. The transient dynamics for the first or second versions for the equations is specified by the $2M$ numbers $\Omega_{sq}^2 > 0$ and $\nu_q^s \geq 0$ or the M numbers $\tau_q^s > 0$, respectively, $q = 1, \ldots, M$. The mode-coupling polynomials are given in terms of the equilibrium structure functions $c_{q\alpha}^s$, $q = 1, \ldots, M$, $\alpha = 1, \ldots, n$ and by the total density ρ. Finally, the density-fluctuation-correlator matrices $\boldsymbol{\phi}_q(t)$, $q = 1, \ldots, M$, are needed. This reflects the fact that the dynamics of the tagged particle is influenced by that of the surrounding. One can imitate the trick discussed in connection with Eqs. (5.36a,b) in order to map the problem of a simultaneous evaluation of the M matrix correlators $\boldsymbol{\phi}_q(t)$ and the M correlators $\phi_q^s(t)$ on the problem of evaluating $2M$ matrix correlators $\tilde{\boldsymbol{\phi}}_q(t)$. Here, one uses $\tilde{\phi}_q^{\alpha\beta}(t) = \phi_q^{\alpha\beta}(t)$ and $\tilde{\phi}_{M+q}^{\alpha\beta}(t) = \delta^{\alpha\beta}\phi_q^s(t)$, $q = 1, \ldots, M$. Appeal to the theorems proven by Franosch and Voigtmann (2002) establishes the existence and uniqueness of the solution for the tagged-particle functions. The latter exhibit the standard symmetries and regularity properties.

5.2.2 *Density-fluctuation arrest*

In this section, the equations for the arrest of density-fluctuation correlations in mixtures shall be formulated. Results will be discussed for a binary Lennard-Jones mixture and for a model of silica.

The solutions $\boldsymbol{\phi}_q(t)$ of the second version of the equations of motion are decreasing monotonically and they are bounded. Therefore, the long-time limits exist. The matrix of arrested parts shall be denoted by $\boldsymbol{F}_q$:

$$\lim_{t\to\infty} \boldsymbol{\phi}_q(t) = \boldsymbol{F}_q, \quad q = 1, \ldots, M. \tag{5.87a}$$

It shall be assumed that these limits exist also for the solutions of the first version of the equations of motion. According to Sec. 2.6, the limits quantify a possible zero-frequency singularity for the correlators in the frequency domain:

$$\lim_{\epsilon\to 0}(-i\epsilon)\boldsymbol{\phi}_q(i\epsilon) = \boldsymbol{F}_q. \tag{5.87b}$$

Equations (2.91b,c) ensure that $\boldsymbol{F}_q$ and $\boldsymbol{S}_q - \boldsymbol{F}_q$ are positive semidefinite, i.e.,

$$\boldsymbol{S}_q \geq \boldsymbol{F}_q \geq \boldsymbol{0}. \tag{5.87c}$$

Above and in the following, the concept of semi-ordering of hermitian matrices is used as explained in Appendix A.5. Equation (5.80) yields $\boldsymbol{S}_q^{-1}\boldsymbol{m}_q(t \to \infty)$ as polynomial $\boldsymbol{\mathcal{F}}_q$ of the limits $\boldsymbol{F}_k$. This implies $\lim_{\epsilon\to 0}(-i\epsilon)\boldsymbol{m}_q(i\epsilon) = \boldsymbol{S}_q\boldsymbol{\mathcal{F}}_q[\boldsymbol{F}_k, \boldsymbol{F}_p]$. Specializing Eq. (5.77a) or Eq. (5.77b) to $z = i\epsilon$, one gets a relation between $(-i\epsilon)\boldsymbol{\phi}_q(i\epsilon)$ and $(-i\epsilon)\boldsymbol{m}_q(i\epsilon)$. The limit $\epsilon \to 0$ leads to a generalization of the fixed-point equation (4.21a) (Barrat and Latz 1990):

$$\boldsymbol{F}_q = \left\{ \boldsymbol{S}_q^{-1} + \boldsymbol{\mathcal{F}}_q[\boldsymbol{F}_k, \boldsymbol{F}_p] \right\}^{-1} \boldsymbol{\mathcal{F}}_q[\boldsymbol{F}_k, \boldsymbol{F}_p] \boldsymbol{S}_q, \quad q = 1, \ldots, M. \tag{5.88}$$

These are $M' = Mn(n+1)/2$ nonlinear implicit equations, which are solved by the M' long-time limits $F_q^{\alpha\beta}$, $q = 1, \ldots, M$, $\alpha \leq \beta = 1, \ldots, n$.

In Sec. 4.3.1, several theorems have been discussed for the fixed-point equation, which characterizes the arrested parts for the MCT correlators for simple system. Franosch and Voigtmann (2002) have proven that these theorems have their counterpart in the theory for mixtures. For example, the arrested parts exhibit the maximum property. Let $\boldsymbol{F}_q^*$, $q = 1, \ldots, M$, denote some set of real symmetric matrices, which solve Eq. (5.88) and which obey $\boldsymbol{S}_q \geq \boldsymbol{F}_q^* \geq 0$. Then, $\boldsymbol{F}_q - \boldsymbol{F}_q^*$ is a positive-semidefinite matrix:

$$\boldsymbol{F}_q \geq \boldsymbol{F}_q^*, \quad q = 1, \ldots, M. \tag{5.89}$$

The second theorem concerns the construction of the M matrices $\boldsymbol{F}_q$ as the limit of a sequence of approximants $\boldsymbol{F}_q^{(r)}$, $r = 0, 1, \ldots$. This sequence is constructed recursively. There holds $\boldsymbol{F}_q^{(0)} = \boldsymbol{S}_q$ and $\boldsymbol{F}_q^{(r+1)}$ is determined by

$$\boldsymbol{F}_q^{(r+1)} = \left\{ \boldsymbol{S}_q^{-1} + \boldsymbol{\mathcal{F}}_q[\boldsymbol{F}_k^{(r)}, \boldsymbol{F}_p^{(r)}] \right\}^{-1} \boldsymbol{\mathcal{F}}_q[\boldsymbol{F}_k^{(r)}, \boldsymbol{F}_p^{(r)}] \boldsymbol{S}_q; \quad q = 1, \ldots, M. \tag{5.90a}$$

The sequence of positive-semidefinite matrices is decreasing: $\boldsymbol{F}_q^{(r)} \geq \boldsymbol{F}_q^{(r+1)}$ holds for all r and all q. The limit is the arrested part:

$$\lim_{r \to \infty} \boldsymbol{F}_q^{(r)} = \boldsymbol{F}_q, \quad q = 1, \ldots, M. \tag{5.90b}$$

The terminology introduced in Chapter 4 for the discussion of the solutions of MCT equations will be used also here for the discussion of mixtures. The state of the system is called a liquid if all long-time limits of the correlators vanish: $\boldsymbol{\phi}_q(t \to \infty) = \boldsymbol{0}$, $q = 1, \ldots, M$. A state is called a glass if one of the limits $F_q^{\alpha\beta}$ is non-zero. The arrested parts $F_q^{\alpha\beta}$ are called form factors of the glass. As before, one can show that there are liquid states if all mode-coupling coefficients are sufficiently small. In the sense explained in connection with Eq. (4.55d), there are glass states if a certain number of the coupling coefficients are sufficiently large.

Let us assume that $\sigma(q, \omega)$ denotes the cross-section for the scattering of a neutron by the mixture with momentum transfer $\hbar q$ and energy change $\hbar\omega$. The interaction of the neutron with the nuclei of the particles of species α is described by the scattering length f_α, $\alpha = 1, \ldots, n$. There holds (Lovesey 1984)

$$\sigma(q, \omega) = C \sum_{\alpha\beta} f_\alpha \phi_q''^{\alpha\beta}(\omega) f_\beta. \tag{5.91a}$$

For an ideal glass state, there is a contribution for elastic scattering quantified by the cross-section σ^{el} : $\sigma(q,\omega) = C[\pi\delta(\omega)\sigma_q^{el} +$ regular terms]. This elastic cross-section is expressed in terms of the glass form factors by the formula

$$\sigma_q^{el} = \sum_{\alpha\beta} f_\alpha F_q^{\alpha\beta} f_\beta. \tag{5.91b}$$

If one can determine σ_q^{el} for $n(n+1)/2$ properly chosen sets of scattering lengths, one can invert the set of $n(n+1)/2$ Eqs. (5.91b). As a result, one gets the $F_q^{\alpha\beta}$, $\alpha \leq \beta = 1, \ldots, n$. In this sense, the glass form factors provide measurable informations on the glass.

The form factors $F_q^{\alpha\beta}$ are determined by the mode-coupling functions $\boldsymbol{\mathcal{F}}_q$ and by the structure factors $\boldsymbol{S}_q$. Hence, they are determined by the functions, which characterize the equilibrium structure. Neither the values for the frequencies $\boldsymbol{\Omega}_q$ and $\boldsymbol{\nu}_q$ nor of the times $\boldsymbol{\tau}_q$ have an effect on the arrested parts. In particular, the values of the particle masses $\mu_1, \ldots, \mu_n$ have no influence on the location of possible liquid–glass transition points nor on the values of the form factors $\boldsymbol{F}_q$.

The discussion of the arrest of tagged-particle-density-fluctuation correlations does not require new considerations. Equation (5.37b) for the Lamb–Mössbauer factor $f_q^s = \phi_q^s(t \to \infty)$ remains valid, but there enter the arrested parts $\boldsymbol{F}_k$ in the mode-coupling polynomial from Eq. (5.85a): $f_q^s/(1 - f_q^s) = \mathcal{F}_q^s[\boldsymbol{F}_k, f_p^s]$ (Bosse and Thakur 1987). In analogy to Eqs. (5.42c), (5.43a), one can formulate a fixed-point equation for f_q^s. This equation has the form discussed in Sec. 4.3. All theorems studied there remain valid. There is the speciality, that the mode-coupling functional is linear:

$$f_q^s/(1 - f_q^s) = \sum_p U_{q,p} f_p^s, \tag{5.92a}$$

$$U_{q,p} = [\rho/(16\pi^2)] \int_0^{\infty\prime} dk (pk/q^5)(q^2 + k^2 - p^2) \sum_{\alpha\beta} c_{k\alpha}^s F_k^{\alpha\beta} c_{p\beta}^s. \tag{5.92b}$$

The glass form factors $\boldsymbol{F}_k$ enter linearly in the mode coupling coefficients $U_{q,p}$.

Let γ_r, $r = 1, \ldots, M'$, denote the eigenvalues of the Jacobian matrix of the set of implicit equations (5.88). These can be written as $\gamma_r = 1 - E_r$. The most profound piece of the work by Franosch and Voigtmann (2002) concerns the proof of Eq. (4.60c) and of the accompanying formulas. There is a maximum eigenvalue E, so that $|\gamma_r| \leq E$, $r = 1, \ldots, M'$. It obeys the inequality $E \leq 1$. If the Jacobian matrix is not reducible, the maximum eigenvalue is not degenerate.

As is done in Sec. 4.3.1, states with $E < 1$ shall be referred to as regular ones. The implicit-function theorem ensures, that the set of regular parameter points is open. Near a regular state, the arrested parts $F_q^{\alpha\beta}$ are smooth functions of all mode coupling coefficients. Critical states or singular points are defined by the condition $E^c = 1$. If the Jacobian matrix is not reducible, singular points are bifurcation points of the type A_ℓ, $\ell = 2, 3, \ldots$. As explained in Chapter 4, reducible matrices occur only in exceptional cases; and then, the bifurcation can be described by an iteration of A_ℓ singularities. Even though MCT is extended severely by introducing matrix correlators, there do not appear bifurcation singularities in addition to those occurring within the basic version of this theory. In particular, the generic liquid-glass transition is a fold singularity A_2.

The methods for the asymptotic expansion for the arrested parts, which have been explained in Sec. 4.3.4 for the basic version of MCT, can be generalized for the extended theory under discussion. In analogy to Eq. (4.91a), one gets

$$\boldsymbol{F}_q = \boldsymbol{F}_q^c + \boldsymbol{H}_q\sqrt{\sigma/\mu_2^c}[1 + \sqrt{\sigma}\boldsymbol{k}_q + O(\epsilon)], \quad \sigma \geq 0. \tag{5.93}$$

Here $\sigma = C\epsilon$, $C > 0$, denotes the separation parameter, and $\mu_2^c = 1 - \lambda^c$ abbreviates the positive singularity parameter for the fold bifurcation. $\boldsymbol{F}_q^c$ denotes the positive-definite matrix of the critical glass form factors. It is the discontinuity of the long-time limit $\boldsymbol{\phi}_q(t \to \infty)$ upon shifting ϵ from the negative values for the liquid states to the positive ones for the glass states. The positive-definite matrices $\boldsymbol{H}_q$ are called the critical amplitudes, and $\boldsymbol{k}_q$ denotes the correction amplitudes. The formulas for $\sigma, \lambda, \boldsymbol{H}_q$ and $\boldsymbol{k}_q$ in terms of the mode-coupling coefficients have been derived by Voigtmann (2003c).

The binary Lennard-Jones mixture is a model for a glass-forming van der Waals system. It is defined in connection with Eqs. (1.8a–c), where also the conventional units for its description are explained. Nauroth and Kob (1997) solved Eqs. (5.88) for the arrested parts $\boldsymbol{F}_q$ for this system. The convolution approximation was used. The structure factor matrix $\boldsymbol{S}_q$, which is needed for the determination of the mode-coupling coefficients, has been taken from simulation results for the averages $\langle \rho_{\vec{q}}^{\alpha *} \rho_{\vec{q}}^{\beta} \rangle$. The results are shown in Fig. 5.12 for three temperatures. The structure factors for the other temperatures are obtained by quadratic interpolation. The structure factor for the big majority particles is similar to that shown in Fig. 4.6 for the dense hard-sphere system. But the height of the first sharp diffraction peak of the normalized function S_q^{AA}/x_A is lower than the peak for the HSS for packing fractions near the critical value φ^c. This reflects the fact that the presence of the minority particles weakens the intermediate-range order. The peak height of the normalized structure factor S_q^{BB}/x_B for the minority particles is only about 1.2. The value for $q = 0$ is about 0.4. This structure factor is similar to that of a dense gas rather than to the one of a liquid. The calculated critical temperature is

$$T_c = 0.92 \quad \text{MCT}, \quad \text{LJM}. \tag{5.94a}$$

The simulation results for the density-fluctuation correlators $F_q^{\alpha\beta}(t)$ and for the tagged-particle-density correlation functions $\phi_q^s(t)$ for intermediate wave numbers exhibit the evolution of the time-temperature superposition principle. There is scale coupling and the common scale τ exhibits power-law variation as discussed in connection with Eq. (5.64). A common fit of the data with the two-parameter law $\tau \propto (T - T_c)^\gamma$ is possible. It yields the value $\gamma = 2.7$ and provides the result for the critical point (Kob and Andersen 1994, 1995a):

$$T_c = 0.435 \pm 0.003; \quad \text{simulation}, \quad \text{LJM}. \tag{5.94b}$$

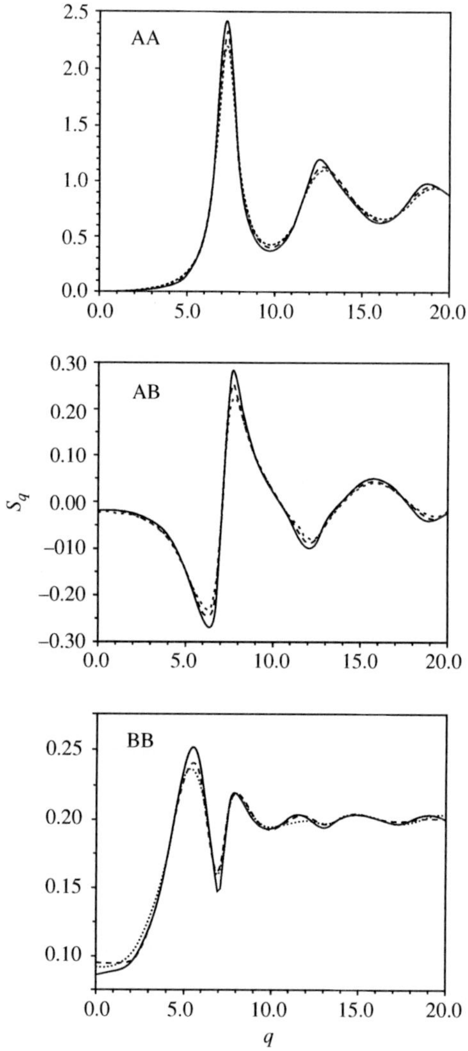

FIG. 5.12. Molecular-dynamics-simulation results for the partial structure factors $S_q^{\alpha\beta}$, $\alpha, \beta = A, B$, of the binary Lennard-Jones mixture for the temperatures $T = 1.0$ (dotted), $T = 0.8$ (dashed), and $T = 0.6$ (solid). The system and the units are explained in connection with Eq. (1.8a–c). Reproduced from Nauroth and Kob (1997).

The structure factor of a Lennard-Jones system is quite similar to that of a HSS. If one would replace the Lennard-Jones potential by its repulsive part

$$V(r) = 4\epsilon(\sigma/r)^{12}, \tag{5.95a}$$

the Boltzmann factors $\exp[-V(r)/(k_B T)]$ would be a function of $\epsilon\sigma^{12}k_B/(r^{12}T)$. A change of the length scale by a factor x would be equivalent to a change of the temperature by a factor $x^{1/12}$. Hence, the equilibrium structure functions would depend on temperature T and number density ρ only via the dimensionless combination

$$\Gamma = (\rho/T^{1/4})[\sigma^3/(k_B/\epsilon)^{1/4}]. \tag{5.95b}$$

The ratio 2.1 between the calculated and measured T_c, would be equivalent to about a ratio 0.83 between the corresponding critical densities ρ_c. One concludes that MCT overestimates the trend to arrest. The about 15% error for the relevant coupling constant Γ for the LJM is of similar size as the error obtained for the critical packing fraction calculated for the HSS on the basis of the Percus–Yevick-structure-factor approximation.

A determination of the critical arrested parts $\boldsymbol{F}_q^c$ requires some extrapolation of the simulation data for the liquid state to ones for the critical point. Such extrapolation should start by examining the evolution of the superposition-principle regime. As mentioned repeatedly, the shape function can often be described satisfactorily by a Kohlrausch law. Indeed, data fits for the correlators for the LJM by $\phi_q^{\sigma\nu}(t) = F_q^{c\sigma\nu}\exp\{-[t/\tau_q^{\sigma\nu}]^{\beta_q^{\sigma\nu}}\}$ yield reasonable results. The MCT values for the critical arrested parts of the correlators and the fit values for $F_q^{c\sigma\tau}$ agree well for $q \leq 10$. But, for larger wave numbers, there appear systematic discrepancies (Nauroth and Kob 1997).

A fit of a part of the data by von Schweidler's law, $\boldsymbol{F}_q(t) = \boldsymbol{F}_q^c - \boldsymbol{H}_q'(t/\tau)^b$, does not lead to reliable results. The fit parameters depend too much on the time interval chosen for the fit. If one assumes that the law represents the leading contributions of a series in powers of the small parameter $(t/\tau)^b$, one would suggest to extend von Schweidler's law for the description of the shape function by the formula:

$$\tilde{\boldsymbol{\phi}}_q(\tilde{t}) = \boldsymbol{F}_q^c - \boldsymbol{H}_q'\tilde{t}^b + \boldsymbol{H}_q''\tilde{t}^{2b}, \quad \tilde{t} = t/\tau. \tag{5.96a}$$

The first two terms are consistent with the factorization property for the near-plateau decay, as discussed in Eqs. (4.111a,c). The third term would describe corrections to this formula, which are relevant if $\boldsymbol{\phi}_q(t)$ deviates too much from the plateau $\boldsymbol{F}_q^c$. The formula was applied first by Sciortino *et al.* (1996) for the analysis of molecular-dynamics-simulation data for supercooled water. The authors demonstrated that a variety of correlation functions could be described adequately by Eq. (5.96a) using a common von Schweidler exponent b. The time interval for the applicability of the fit did not exhibit strong dependencies on the type of correlator chosen.

The discussion of Eqs. (1.7b),(1.10b) has identified von Schweidler's law as large time limit of the factorization formula for the near-plateau decay. This suggests we consider Eq. (5.96a) as a special limit of

$$\boldsymbol{\phi}_q(t) = \boldsymbol{F}_q^c + \boldsymbol{H}_q' G(t) + \boldsymbol{H}_q''(t/\tau)^{2b}. \tag{5.96b}$$

Application of this formula for data fits has the advantage that the plateau values $F_q^{c\alpha\beta}$ are determined by interpolation rather than by extrapolation. The formula was applied first by Gleim *et al.* (1998) for the analysis of the simulation data for the LJM. The results for their critical arrested parts are shown as dots in

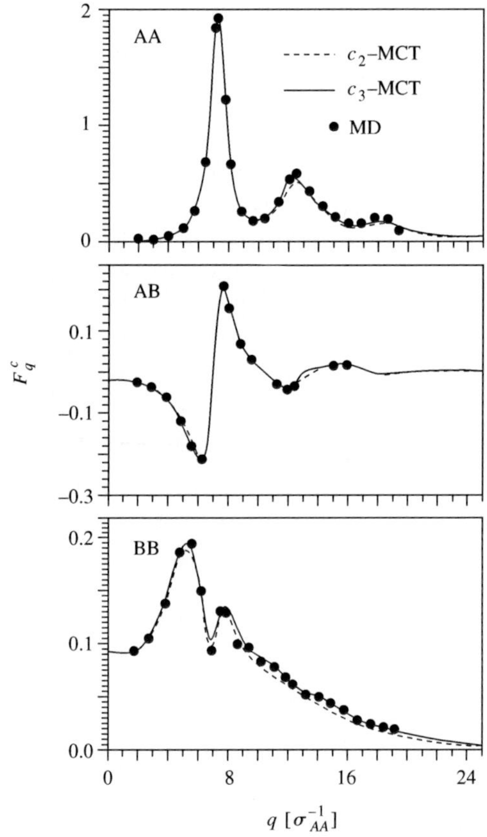

FIG. 5.13. The dots are molecular-dynamics simulation results for the plateaus of the density-fluctuation correlators for the LJM (Gleim *et al.* 1998). The full lines are the MCT results for the critical glass form factors. The dashed lines are MCT results obtained by using the convolution approximation for the evaluation of the mode-coupling coefficients. The calculations are based on molecular-dynamics-simulation results for the structure functions $c_q^{\alpha\beta}$ and $c_3^{\alpha\beta\gamma}(qkp)$. Reproduced from Sciortino and Kob (2001).

Fig. 5.13. The dotted line in Fig. 1.4 exhibits an example for a fit of a tagged-particle correlator by Eq. (5.96b).

The plateau values $F_q^{c\alpha\beta}$ are concepts of MCT for the characterization of the singularities described by that theory. The theory has to provide the formulas for the extraction of the plateau values from the data. Getting these formulas requires an understanding of the relaxation laws for states near the liquid–glass transition. It will be shown in Sec. 6.1.4(ii) that Eqs. (5.96a,b) are asymptotic solutions for the MCT bifurcation. Anticipating this result, one concludes that data analysis with Eq. (5.96a) or with Eq. (5.96b) is legitimate.

Let us compare the data for the $F_q^{c\alpha\beta}$ shown in Fig. 5.13 with the ones for the $S_q^{\alpha\beta}$ shown in Fig. 5.12. The plateau heights oscillate in phase with the partial structure factors. The positions of the first sharp defraction peaks of the $S_q^{\alpha\alpha}$ versus q curves are close to the positions of maxima of the $F_q^{c\alpha\alpha}$ versus q curves, namely about 7 and 5.5 for $\alpha = A$ and $\alpha = B$, respectively. Similar to what is shown in Fig. 4.6 for the simple HSS, the critical arrested parts reach about 80% of the value for the structure factor in those maxima. Similarly, the positions of the second maxima for S_q^{AA} and S_q^{BB} are close to those for the second maxima of the critical glass form factors. Also, the pronounced negative minimum and an equally pronounced maximum of the S_q^{AB} versus q curves are reflected by corresponding extrema for the glass-form factor F_q^{cAB}. In all four mentioned positions for extrema, the critical arrested parts reach about 50% of the values for the corresponding structure factors. For the smallest wave numbers, F_q^{cBB} and F_q^{cAB} approach the values S_q^{BB} and S_q^{AB}, respectively. Similar to what is exhibited in Fig. 1.16 for a simple system, the ratios $F_q^{c\alpha\beta}/S_q^{\alpha\beta}$ exhibit a non-trivial dependence on the wave number q.

The dashed lines in Fig. 5.13 show that the cited MCT calculations for the critical arrested parts describe the plateau values well. Only the data for the B-B correlations are underestimated slightly for $q > 10$.

The significance of the triple correlations for the mode-coupling coefficients has been analysed by Sciortino and Kob (2001). A large effort is necessary to extract from simulation data the functions $c_3^{\eta\epsilon\tau}(qkp)$ in Eq. (5.74) for a temperature near the one for arrest. The result has been included in Eq. (5.79b) without considering its temperature dependence. In the spirit of a leading correction to the theory used with a convolution approximation, this procedure accounts for the modifications caused by the triple correlations. For the LJM, modifications of the fluctuating-force kernels have been found, which are of the order of 15%. But, depending on the wave number, the sign of the modification fluctuates. As a result, the critical temperature is modified by about 1% only. Figure 5.13 demonstrates that the inclusion of the triple correlations does not have a noticeable effect on the MCT results for F_q^{cAA} and F_q^{cAB}. But, the modifications bring F_q^{cBB} in accord with the data also for $q > 10$. The LJM deals with a closely-packed distribution of spherical particles of nearly equal diameters. One cannot imagine three-particle clusters with an arrangement, which cannot be described by the pair-distribution functions. This might be the explanation why the superposition approximation does not lead to appreciable modifications of the MCT results for this system.

SiO_2 crystallizes in an array of tetrahedras for the silicon ions. The oxygen ions are located between two nearest silicon particles. Melting destroys the long-range order, but the intermediate-range order is dominated by clusters similar to the ones in the crystal. Silica is a paradigm for a network liquid. The strong forces between the ions imply a high value for the glass transformation temperature ($T_{\rm g} \approx 1450\,{\rm K}$) and for the melting point ($T_{\rm m} \approx 2000\,{\rm K}$). The open arrangement of the particles with their low coordination numbers renders the properties of silica quite different from those of van der Waals systems. It is a remarkable discovery that the properties of the crystal phase can be deduced from a pair potential $V_{\alpha\beta}(r) = [q_\alpha q_\beta/r] + [A_{\alpha\beta}\exp(-B_\alpha q\beta r)] - c_{\alpha\beta}/r^6$ (van Beest *et al.* 1990); here $\alpha = 1$ and 2 denote Si and O, respectively. Molecular-dynamics studies have shown that a binary system with the cited interaction reproduces satisfactorily the properties of liquid silica. In particular, the complicated q-dependence of the measured total neutron-scattering cross-section $\sum_{\alpha\beta} f_\alpha S_q^{\alpha\beta} f_\beta$ is reproduced very well (Horbach and Kob 1999a). Decreasing the temperature T from 6100 K towards 3500 K, the simulation data for the mean-squared displacements exhibit the familiar pattern for the evolution of a plateau for the glassy dynamics. A power-law extrapolation for the decrease of the diffusivity with temperature suggests a transition value

$$T_c = 3330\,{\rm K}, \quad \text{simulation,} \quad \text{silica.} \tag{5.97a}$$

The analysis of the simulation results for density-fluctuation correlators confirms the estimation of T_c (Horbach and Kob 2001). It provides the plateau values, which are shown as dots in Fig. 5.14. Again, the plateau heights oscillate in phase with the corresponding structure factors. The main peaks of the results for the Si-Si and O-O correlations and the negative minimum for the Si-O ones are located near $2.8\,\text{Å}^{-1}$. They reflect the shell structure for the neighbour positions, and are the analogues of the corresponding extrema in Fig. 5.13. The outstanding speciality of the network system is the appearance of a prepeak located near $1.6\,\text{Å}^{-1}$. It reflects the new length scale of the system's intermediate-range order, which is given by the distance between the tetrahedra centres. The full lines in Fig. 5.14 exhibit the MCT results for the critical arrested parts. They reproduce the data well with one exception. The critical glass form factor for the O-O correlation overestimates the plateau for $4\,\text{Å}^{-1} < q < 6\,\text{Å}^{-1}$. The cited MCT calculations are based on structure factors and triple correlations taken from the molecular-dynamics simulations. Again, this theoretical work overestimates the trend to arrest, since it leads to a critical temperature, which exceeds the value in Eq. (5.97a) by about a factor 1.40:

$$T_c = 4676\,{\rm K}, \quad \text{MCT,} \quad \text{silica.} \tag{5.97b}$$

As expected for the open network structure of liquid silica, the triple correlations have a noticeable effect on the MCT results. Dropping these correlations, i.e., using the convolution approximation, the calculated critical temperature decreases by about 15% to 3962 K (Sciortino and Kob 2001). Figure 5.14

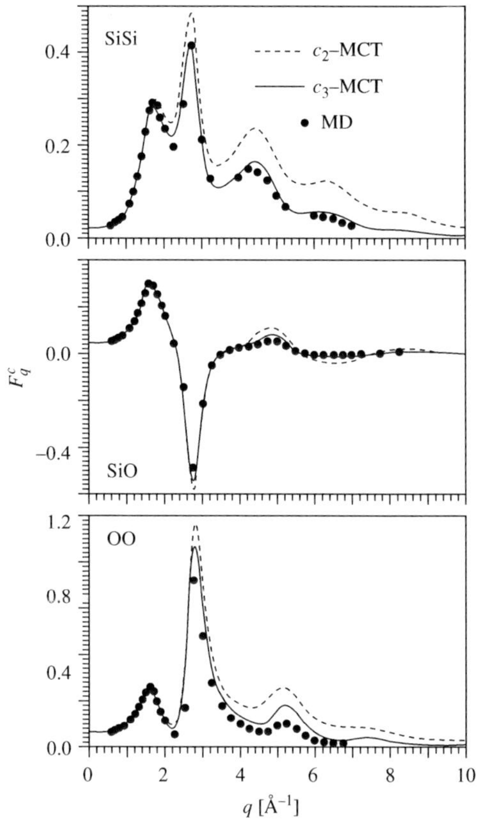

FIG. 5.14. The dots are data for the plateaus of the density-fluctuation correlators deduced from molecular-dynamics simulation results for a model of silica (Horbach and Kob 2001). The full lines are the MCT results for the critical arrested parts. The dashed lines are obtained by simplifying the mode-coupling coefficients by the convolution approximation. The calculations are based on molecular-dynamics-simulation results for the partial structure functions and the triple correlations. Reproduced from Sciortino and Kob (2001).

documents that the simplified version of MCT reproduces the plateau values qualitatively for $q > 2.5\,\text{Å}^{-1}$; but there are systematic errors for the O-O-correlation and the Si–Si ones. The results for the prepeak region are described well, even if the convolution approximation is used.

5.2.3 *Hard-sphere mixtures*

In this section, some results for binary hard-sphere systems shall be discussed in order to demonstrate subtle mixing effects for the glassy dynamics of van der Waals-like liquids.

The structure functions of a hard-sphere mixture are determined by the total number density ρ, by the diameters d_α and number concentrations x_α of the n species. The partial packing fractions read $\varphi_\alpha = (\pi/6)(\rho x_\alpha)d_\alpha^3$. For the MCT calculations to be cited, the convolution approximation is used and the structure-factor matrix $\boldsymbol{S}_q$ is calculated in the Percus–Yevick approximation (Lebowitz and Rowlinson 1964; Baxter 1970). As a result, the coefficients in Eq. (5.79b) can be expressed in terms of elementary functions.

For binary mixtures, big and small particles shall be referred to by labels $1 = A$ and $2 = B$, respectively. In this case, there are three control parameters; they shall be chosen as the packing fraction φ, the relative packing fraction for the small spheres $\hat{x}_B$, and the size ratio $\delta \leq 1$:

$$\varphi = \varphi_A + \varphi_B, \qquad \hat{x}_B = \varphi_B/\varphi, \qquad \delta = d_B/d_A. \tag{5.98}$$

The monotonic function $x_B = (\hat{x}_B/\delta^3)/[1 - \hat{x}_B + (\hat{x}_B/\delta^3)]$ provides the mapping of the interval of relative packing fractions, $0 \leq \hat{x}_B \leq 1$, on the one of number fractions, $0 \leq x_B \leq 1$. If a cutoff value q_{co} for the discretization is used so that the structures on length scale d_A can be resolved, a cutoff q_{co}/δ has to be used to achieve the same goal for the resolution of structures on the second relevant length scale d_B. Restricting the discussion to $\delta \geq 0.5$, a discretization with $M = 200$ components and $q_{co} = 80/d_A$ is chosen. The second version for the equations of motion is used. The regular relaxation kernel in Eq. (5.76c) is approximated by the one for non-interacting colloid particles:

$$n_q^{\alpha\beta} = \delta^{\alpha\beta}/D_\alpha^0. \tag{5.99a}$$

Here D_α^0 denotes the diffusivity of a single particle of species α. The species dependence is described by a Stokes–Einstein law for the friction coefficient:

$$D_\alpha^0 = d^3/(t_{\text{mic}} d_\alpha). \tag{5.99b}$$

The parameter t_{mic} quantifies the overall time scale of the system.

Figure 5.15 exhibits the φ^c versus $\hat{x}_B$ liquid–glass-transition lines for some representative values for the size ratio δ. For small size disparity, say $\delta > 0.75$, mixing stabilizes the glass state. The critical packing fraction φ^c decreases if x_B or x_A increase from zero to small values. The φ^c versus $\hat{x}_B$ line exhibits a minimum as is demonstrated for $\delta = 0.8$. For large size disparity, say $\delta < 0.65$, there occurs the opposite phenomenon, namely, mixing stabilizes the liquid. The critical packing fraction increases if x_B or x_A increase from zero to small values. The resulting transition line exhibits a maximum as is demonstrated for $\delta = 0.6$ and $\delta = 0.5$. These two mixing effects have been predicted originally by Barrat and Latz (1990). In their study of the density-fluctuation arrest, a binary mixture of soft spheres has been considered, whose interaction potential is noted in Eq. (5.95a). There is a range of values for δ, where the transition lines exhibit

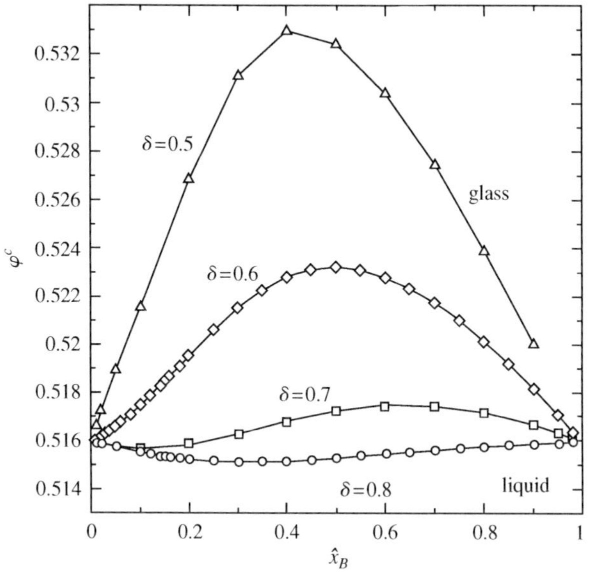

FIG. 5.15. liquid–glass-transition points for a binary hard-sphere mixture for a size ratio $\delta = 0.5$ (triangles), $\delta = 0.6$ (diamonds), $\delta = 0.7$ (squares), and $\delta = 0.8$ (circles) presented as critical total packing fraction φ^c versus relative packing fraction $\hat{x}_B$ of the smaller species. The lines are guides to the eyes. Details for the mode-coupling functional are explained in the text. Reproduced from Götze and Voigtmann (2003).

a minimum and a maximum. For small x_B, mixing stabilizes the glass, and for large x_B, it stabilizes the liquid. This case is exhibited for $\delta = 0.7$.

The calculated changes of φ^c due to mixing are small. For $\delta = 0.8$ and 0.6 they do not exceed 0.2% and 1.5%, respectively. However, a small shift of the position of a bifurcation singularity can cause a large change of the solution of the equation under study. This holds provided the state is close to the singularity. Let us consider a packing fraction for a distance parameter $\epsilon = -0.002 = (\varphi - \varphi^c)/\varphi^c$, with φ^c denoting the critical point for the simple hard-sphere system. A corresponding decay curve is shown for the label $n = 8$ in Fig. 4.14. Increasing $\hat{x}_B$ from 0.0 to 0.1 shifts the state 'closer' to the transition line for $\delta = 0.8$ but 'further away' from it for $\delta = 0.6$. One expects that the time scale for the below-plateau decay increases for the first case but that it decreases for the second one. For the system with the small size disparity, one expects a slowing down of the long-time decay. For the system with the large size disparity, the opposite is expected: the long-time decay of the correlations shall occur faster. The expectation is correct as is demonstrated in Fig. 5.16. The specified increase of $\hat{x}_B$ causes an increase of about a factor 10 and a decrease of about a factor 5 of the long-time-decay scale for $\delta = 0.8$ and $\delta = 0.6$, respectively.

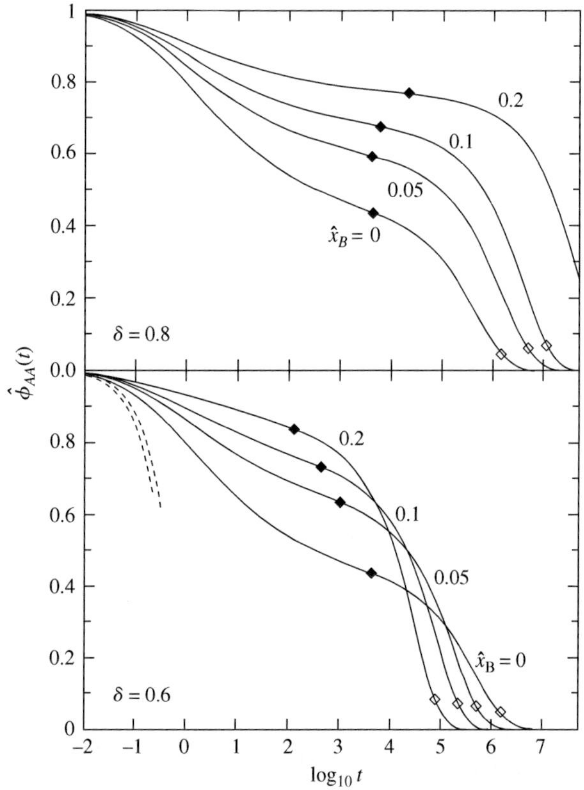

FIG. 5.16. Normalized density-fluctuation correlators of the big particles $\hat{\phi}_{AA}(t) = \phi_q^{11}(t)/S_q^{11}$ for wave number $q = 5.4/d_A$ calculated for the microscopic model for a binary hard-sphere mixture defined in the text. The packing fraction $\varphi = 0.515$ is 0.19% below the critical value φ^c for the simple hard-sphere system. The relative packing fractions for the small particles are $\hat{x}_B = 0.00$, 0.05, 0.10, and 0.20. The size ratio is $\delta = 0.8$ (upper panel) and $\delta = 0.6$ (lower panel). The unit of time is chosen so that $D_\alpha^0 d_\alpha = 0.01$. The filled diamonds mark the points where the correlators cross the value F_q^{c11}/S_q^{11}. The open diamonds mark the time τ, where the correlator has decayed to 10% of its 'plateau' value: $\phi_q^{11}(t = \tau) = F_c^{c11}/10$. The left and right dashed lines in the lower panel exhibit the short-time asymptotes $\hat{\phi}_{AA}(t) = 1 - t/\hat{\tau}_{AA}$ for $\hat{x}_B = 0$ and $\hat{x}_B = 0.2$, respectively, with time scales $\hat{\tau}_{AA}$ derived from Eq. (5.81b). Reproduced from Götze and Voigtmann (2003).

Molecular-dynamics-simulation studies for binary hard-sphere mixtures have confirmed qualitatively the above specified two mixing effects (Foffi *et al.* 2003). A system with $\delta = 0.83$ and $\varphi = 0.582$ exhibits an increase of the below-plateau-decay scale by about a factor 4 if x_B increases from 0.276 to 0.376. A system with $\delta = 0.60$ and $\varphi = 0.60$ demonstrates a decrease of the time scale by about a factor 3 if x_B increases from 0.10 to 0.20.

The appearance of two mixing effects for the long-time-decay scale can be explained as result of the interplay of two mixing effects for the equilibrium structure. Reducing the number of big spheres by a small amount $1 - x_A$ and increasing the number of small spheres by a small percentage $x_B > 1 - x_A$ so that φ remains constant leads to a change of the microscopic length scale of the system from d_A to some smaller averaged value. This implies a decrease of the radii for the nearest and next-to-nearest neighbour shells for the big particles. This length-scale changing effect is demonstrated by the Percus–Yevick results for the pair-distribution functions $g_{AA}(r)$ and $g_{AB}(r)$ (Götze and Voigtmann 2003). Hence, the diameter of the cages, which are formed by the big particles, shrinks. This effect favours arrest, i.e., a decrease of φ^c with increasing of $\hat{x}_B$. The length-scale shrinking causes a shift of the maxima positions of the S_q^{AA} versus q curves to larger values of the wave vectors. The phase-space factor for the relevant couplings of the forces to density fluctuations increases. Thereby, the important mode-coupling-contributions increase. This effect is discussed also in Sec. 4.3.6 in a different context. The explained mechanism dominates for small deviations of δ from unity.

An opposite effect dominates for small δ. An increase of x_B for fixed φ reduces the packing fraction φ_A for the big spheres. The number of channels for the motion of the small particles increases. These 'freely' moving small particles do not contribute efficiently to the cage-forming forces. Hence, they support a trend for the destabilization of the arrested structure. The MCT formulas reflect this effect in a manner, which also has been discussed in Sec. 4.3.6 for a different situation: increasing x_B leads to a smearing-out of the intermediate-range order. This leads to a reduction of the effective interaction, as is reflected by the decrease of the fluctuation amplitudes of $(\boldsymbol{S}_q - \boldsymbol{1}) \approx \boldsymbol{c}_q$.

Figure 5.16 demonstrates two further mixing effects, which occur for small as well as for large δ. First, the length-scale shrinking effect leads to a decrease of the localization length of the tagged particles. This implies an increase of the Lamb-Mössbauer factors f_q^A and f_q^B. Since the Debye–Waller factors $f_q^{\alpha\alpha}/S_q^{\alpha\alpha}$ oscillate around the arrested parts f_q^{α}, the plateaus $f_q^{c\alpha\alpha}/S_q^{\alpha\alpha}$ increase with $\hat{x}_B$. A further mechanism is relevant for wave numbers q below the structure-factor-peak position. Mixing weakens the coherence effects, which are responsible for the strong suppression of the Debye–Waller factor below the Lamb–Mössbauer factor. Second, the $\phi_q^{AA}(t)$ versus $\log t$ curves for the initial part of the structural relaxation become flatter with increasing $\hat{x}_B$. This slowing down of the relaxation towards the plateau due to mixing can be explained on the basis of understanding the general asymptotic laws for the bifurcation dynamics (Götze and Voigtmann 2003). These laws will be discussed in Chapter 6. The changes of the structure factors upon mixing also cause a change of the transient dynamics. As shown by the dashed lines in the lower panel of Fig. 5.16, this is a very small effect only.

Equation (1.13) formulates the correlation function, which is measured in a light-scattering experiments for a colloidal suspension. If one assumes that the particles are optically homogeneous, the scattering amplitudes read (Pusey 1991):

$$b_\alpha(q) = Bd_\alpha^3[\sin(qd_\alpha/2) - \tfrac{1}{2}(qd_\alpha)\cos(qd_\alpha/2)]/(qd_\alpha)^3. \qquad (5.100)$$

The constant B is proportional to the difference of the refraction indices of solvent and solute. The scattering function $\phi_q^m(t)$ exhibits the same mixing effects as demonstrated in Fig. 5.16 for the large particles. However the increase of the plateau of the $\phi_q^m(t)$ versus $\log t$ curves with $\hat{x}_B$ is not as strong as shown in the figure (Götze and Voigtmann 2003).

The lower panel of Fig. 5.16 and the accompanying discussion provide the qualitative explanation of the three mixing phenomena demonstrated by the data in Fig. 1.18 for $\delta = 0.6$. The peculiar crossings of the decay curves for different $\hat{x}_B$ result from the interplay of the slowing down of the towards-the-plateau relaxation due to mixing and the speeding up of the below-plateau decay. The latter phenomenon is typical for sufficiently large size disparities of the two species in the mixture. The melting of the glass state of the simple hard-sphere system due to mixing, which is shown in Fig. 1.18 for $\varphi = 0.58$, reflects the increase of the φ^c versus $\hat{x}_B$ line with the increase of the number concentration of the small species.

Density-fluctuation correlators calculated for the above specified MCT model for a mixture with $\hat{x}_B = 0.20$ and $\delta = 0.60$ are compared in Fig. 5.17 with data measured for a corresponding binary colloidal suspension. The time is measured in units of the Brownian time $\hat{\tau} = d_A^2/(24D_A^0)$. For the calculations, the time scale t_{mic} in Eq. (5.99b) is treated as a fit parameter since MCT cannot describe the renormalization of the single particle diffusivity to the short-time diffusion constant. The theory curves in the figure are calculated for the 10% smaller wave vector $q = 5.4/d_A$ in order to compensate for the 10% error for the MCT prediction of the plateau heights. The two specified fit numbers are the same for all 15 curves shown. The MCT calculations produce an error for the critical point. This is corrected for by describing the system for the five measured packing fractions φ^{exp} by MCT solutions with packing fractions φ^{theo}, which are about 10% smaller. There holds $\varphi^{\text{exp}} = \varphi^{\text{theo}} + 0.05 \pm 0.005$. The results calculated for the highest packing fraction describe all three density correlators. These correlators exhibit a decay from 90% to 10% of the initial value, which is stretched over five orders of magnitude time increase. The evolution of the structural relaxation pattern is described well for the autocorrelator for the big majority particles. The examined density change causes an increase of the long-time-decay scale by a factor 2000. Also, the evolution pattern for the $A - B$ correlations is reproduced reasonably. However, for all but the largest density, the data for the autocorrelators of the small minority particles are considerably below the calculated values. Indeed, these data imply a violation of the condition $\phi_q^{AA}(t)\phi_q^{BB}(t) > \phi_q^{AB}(t)\phi_q^{BA}(t)$ (Voigtmann 2003b), which holds for all correlation-function matrices for variables obeying a Smoluchowski dynamics. In

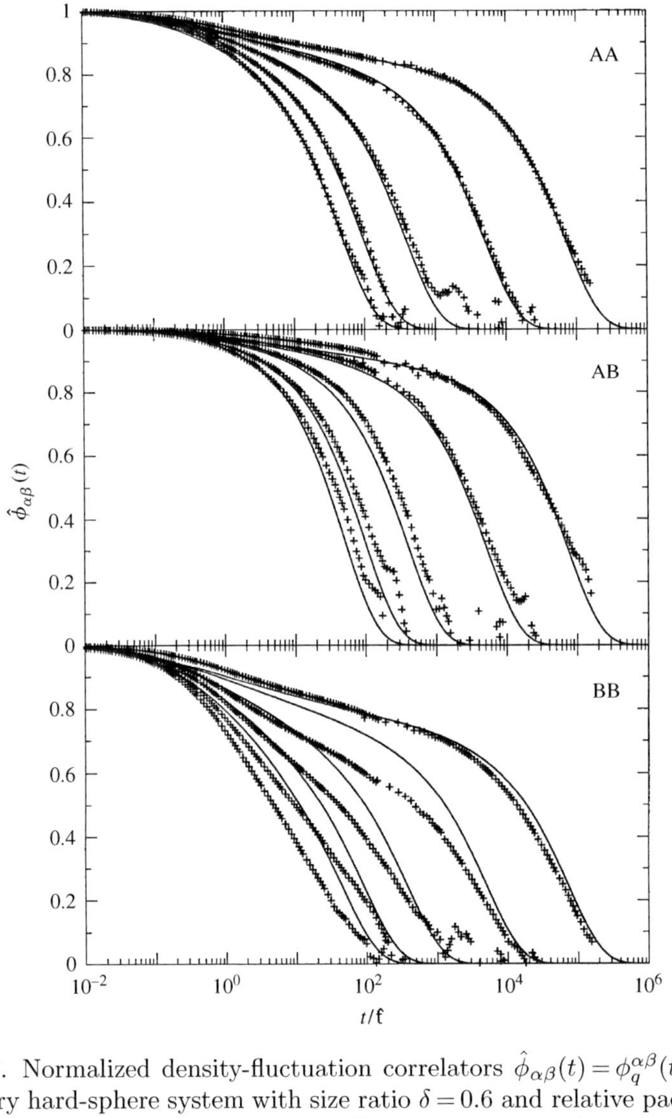

FIG. 5.17. Normalized density-fluctuation correlators $\hat{\phi}_{\alpha\beta}(t) = \phi_q^{\alpha\beta}(t)/S_q^{\alpha\beta}$ for a binary hard-sphere system with size ratio $\delta = 0.6$ and relative packing fraction of the small species $\hat{x}_B = 0.20$. The crosses are results measured for a colloidal suspension with photon-correlation spectroscopy for a wave numbers q obeying $qd_A = 6.0$ (Williams and van Megen 2001). The packing fractions measured are $\varphi^{\text{exp}} = 0.51$, 0.53, 0.55, 0.57, and 0.58 (from left to right). The solid lines are solutions of the MCT model defined in the text for $qd_A = 5.4$ and packing fractions $\varphi^{\text{theo}} = 0.460$, 0.475, 0.497, 0.510, and 0.516 (from left to right). Reproduced from Voigtmann (2003b).

order to judge the relevance of this finding, one would need some understanding on how the experimental errors of the original data for $\boldsymbol{S}_q$ and $\phi_q^m(t)$ influence the curves exhibited in Fig. 5.17 for the minority particles.

Henderson *et al.* (1996) have measured normalized density-fluctuation decay functions $\phi_q^m(t)$, Eq. (1.13), for some colloidal suspensions of nearly hard spheres. Data for systems with small polydispersity are reproduced in Fig. 4.26 as crosses. Data for a binary mixture of nearly monodisperse components are reproduced as squares. The mixtures deal with a size ratio $\delta = 0.80$ and a relative packing fraction of the small species $\hat{x}_B = 0.12$. The data for the two smallest packing fractions refer to distance parameters $\epsilon = (\varphi - \varphi_g)/\varphi_g$ of about -0.07 and -0.03, respectively. The decay curves exhibit structure relaxation, which is stretched over a time increase of about three orders of magnitude. These data do not exhibit mixing effects. This is what one would deduce from Fig. 5.15. The shift of φ_c due to mixing is too small to modify noticeable the distance parameter ϵ for the packing fractions cited. The situation is different, however, for the highest packing fraction studied. The measured packing fraction $\varphi = 0.566$ for the mixture is slightly smaller than the value $\varphi = 0.567$ measured for the $\hat{x}_B = 0$ system. Would the glass transition value φ_g be independent of $\hat{x}_B$ or would it increase with $\hat{x}_B$, $|\epsilon|$ would be larger for the mixture than for the simple hard-sphere liquid. Hence, one would expect the time scale for the below-plateau decay for the $\hat{x}_B = 0.12$ system to be not larger than that for the $\hat{x}_B = 0$ system. Comparison of the squares with the crosses for the largest packing fractions demonstrates the opposite effect. In agreement with the conclusion to be drawn from the transition diagram for $\delta = 0.80$ in Fig. 5.15 or from the curves shown in the upper panel of Fig. 5.16, the data exhibit a considerable slowing down of the long-time decay upon mixing.

The dashed lines in Fig. 4.26 exhibit MCT solutions for the density correlator $\phi_q(t)$ for a model for simple hard-sphere colloids. Details have been discussed in Chapter 4. The full lines represent results for $\phi_q^m(t)$ calculated with Eqs. (1.13), (5.100) from the correlator matrix $\boldsymbol{\phi}_q(t)$ of the model specified above. The wave number for the calculated correlators and the packing fractions used differ on a 10% level from the corresponding parameters measured for the cited data. These changes compensate for the errors of MCT for the plateau values of the correlators and for the error for critical packing fraction of the hard-sphere system. The MCT results reproduce the observed absence of mixing effects for the two smallest packing fractions with $|\epsilon| \geq 0.03$. The observed slowing down of the below-plateau decay by more than an order of magnitude due to mixing for the largest packing fraction is also reproduced by the theory. The slowing down is explained by the addition of two effects. One is caused by a 0.1% decrease of φ^c as shown in Fig. 5.15 for the small-size-disparity scenario. The other contribution results from the assumption that the highest packing fraction for the mixture is 0.1% larger than that for the system with $\hat{x}_B = 0$. This deviation of φ from the value determined in the cited experiment is within the uncertainties for the determination of the solute density of the colloid.

5.2.4 *Sodium-disilicate melts*

Cooling liquid mixtures of silica and alkali oxides, one obtains glasses of great interest for geology and material science. Molecular-dynamics-simulation studies of the glassy dynamics for such liquids have been started by Horbach and Kob (1999b). The system is described as a three-component mixture of spherical particles. The pair potential has the same form of a sum of Coulomb interactions, short-range repulsions, and van der Waals attractions, which is mentioned above for silica. The parameters for the potentials have been derived by Kramer *et al.* (1991). The simulation work is done for a mixture of 1792 Si-ions, 1792 Na-ions, and 4480 O-ions. Two figures for sodium-disilicate shall be considered, which demonstrate an outstanding peculiarity of the glassy relaxation occurring in this class of network-forming liquids.

Adding sodium to silica, Si-O-Si bonds have to be broken. This interruption of the tetrahedral network does neither influence strongly the structure factors nor the density-fluctuation correlators for intermediate and large wave numbers q of the two-component silica subsystem. The sodium ions move within this interrupted silica network. Upon cooling, the autocorrelators of the density fluctuations for all three ions develop the familiar relaxation pattern with some plateau. The left panel of Fig. 5.18 exhibits examples for Na-Na-correlators for three wave numbers. The data demonstrate structure relaxation for an increase

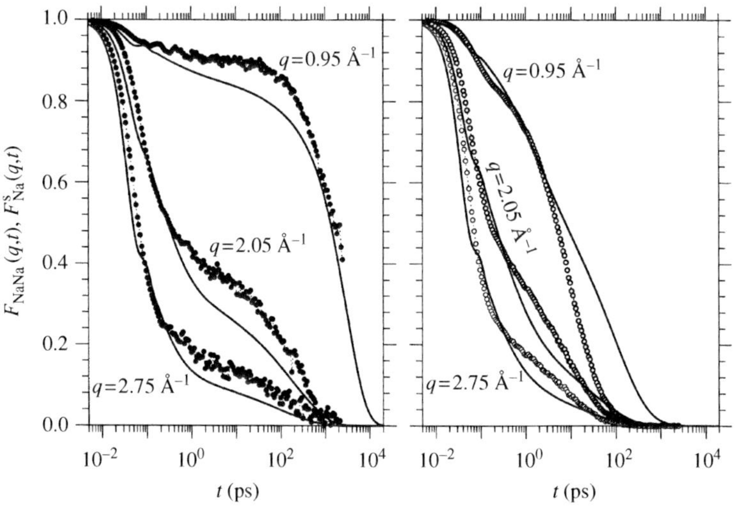

FIG. 5.18. Normalized sodium–sodium density-fluctuation correlators $F_{\mathrm{NaNa}}(q,t) = \phi_q^{\mathrm{NaNa}}(t)/S_q^{\mathrm{NaNa}}$ (left panel) and tagged-sodium-density-fluctuation correlation functions (right panel) for a model of sodium disilicate. The symbols are results of a molecular-dynamics simulation for $T = 2100$ K and the lines are solutions of MCT equations of motion for $T = 3410$ K. Details are explained in the text. Reproduced from Voigtmann and Horbach (2006).

of t by more than four orders of magnitude. The beginning of the below-plateau decay can be described well by the extended von Schweidler law, Eq. (5.96a). The von Schweidler exponent $b = 0.47$ is the same as that found for the pure silica system (Horbach *et al.* 2002). The complete below-plateau processes can be fitted by Kohlrausch functions: $\phi_q^{\alpha\alpha}(t) = A_\alpha(q)\exp[-(t/\tau_\alpha(q))^{\beta_\alpha(q)}]$, $\alpha = 1, 2, 3$, The parameters $A_\alpha(q)$ and $\beta_\alpha(q) < 1$ are temperature insensitive. Thereby, the time-temperature superposition principle is demonstrated. One can write $\tau_\alpha(q) = c_\alpha(q)\tau$ with a temperature insensitive factor $c_\alpha(q)$. This demonstrates scale coupling for the mentioned correlation functions; the strong T-dependence of the below-plateau decay for the three correlators is described by a common scale τ. As discussed in connection with Eq. (5.64), the scale exhibits power-law variation $\tau \propto 1/(T - T_c)^\gamma$. These fits (Horbach and Kob 2002) suggest $\gamma \approx 2.9$ and

$$T_c = 2000\,\mathrm{K}, \quad \text{simulation}, \quad (\mathrm{Na_2O})(\mathrm{SiO_2})_2. \tag{5.101a}$$

The factors $c_\alpha(q)$ are of order unity, i.e., the three scales $\tau_\alpha(q)$ for the long-time decay are of similar size. This is demonstrated in Fig. 5.19 by the filled symbols for the O-O and Na-Na correlators. The features described above are quite similar to those for the Lennard-Jones mixture.

The structure factor of the sodium subsystem exhibits its main peak at a wave number near $2.1\,\text{Å}^{-1}$. It reflects the intermediate-range order, which is enforced by the silica network. But, there is a feature of the structure, which cannot result from a uniform random distribution of the Na ions within the disrupted silica network, namely, a pronounced prepeak of the Na-Na partial structure factor located at a wave number near $0.95\,\text{Å}^{-1}$. A detailed analysis of the particle distributions (Horbach *et al.* 2002) and a look at a snapshot of the system (Meyer *et al.* 2004) shows that the sodium ions are located in channels, which percolate throughout the network. The typical distance of about 6–7 Å between the channels introduces a new length scale for intermediate-range order. These increased averages of the sodium-density fluctuations for q near $0.95\ \text{Å}^{-1}$ or the increased compressibility of sodium for the wave-numbers in the prepeak region cause similar anomalies of the other partial structure factors.

Within MCT, one finds that the plateau heights of the correlators vary with q in phase with the structure factor. A similar statement holds for the variation of the relaxation time $\tau_\alpha(q)$. This is discussed for the hard-sphere system in connection with Fig. 4.6 and Fig. 4.20, respectively. Hence, the theory provides a qualitative explanation of the large plateau height shown in the left panel of Fig. 5.18 for $q = 0.95\ \text{Å}^{-1}$. Similarly, the increased relaxation times for the wave number in the prepeak region, which are shown by the filled symbols in Fig. 5.19, are explained qualitatively.

The quasi-elastic neutron-scattering cross section measured for sodium silicate melts exhibits a shoulder for wave-numbers q near $0.9\,\text{Å}^{-1}$. It has a peculiar dependence on temperature and on composition. This shoulder has been explained on the basis of Eqs. (5.91a,b), using the plateau heights $F_q^{\alpha\beta}$, $\alpha \leq$

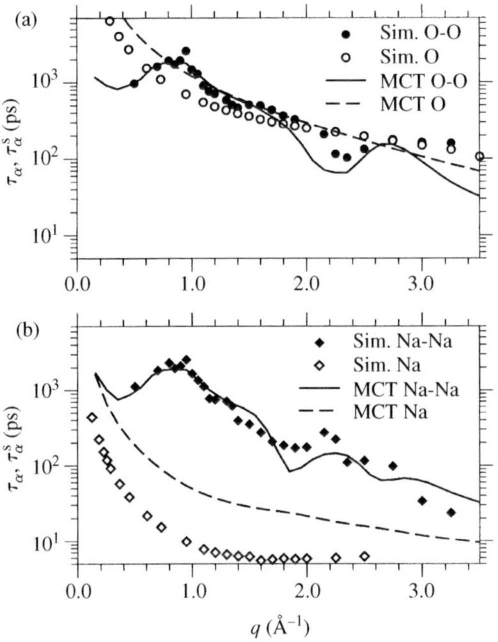

FIG. 5.19. Scales for the description of the long-time decay of correlation functions of sodium disilicate by Kohlrausch functions. The symbols are based on molecular-dynamics-simulation results and the lines are calculated from fits to solutions of MCT equations of motion for the model discussed in Fig. 5.18. The upper panel shows results for the oxygen–oxygen density-fluctuation correlators and for the tagged-oxygen dynamics. The lower panel shows the analogous results for sodium. Reproduced from Voigtmann and Horbach (2006).

$\beta = 1, 2, 3$, from the simulation data (Meyer *et al.* 2002, 2004). These discoveries demonstrate that the model reproduces subtle features of glassy dymamics. They demonstrate also the power of neutron-scattering studies for the identification of length-scale effects of the complex dynamics in glass-forming liquids.

For simple mixtures, the intermediate-wave-number correlators for the tagged-particle-density fluctuations are similar to those for the complete-density fluctuations. This is true also for the motion of the tagged constituents of the silica network of the mixture under discussion. The upper panel of Fig. 5.19 shows, for example, that the time scale τ_α^s for the below-plateau decay of the tagged-oxygen correlations is rather close to the scale τ_α for the decay of the oxygen–oxygen-density correlations. The most important peculiarity of the sodium disilicate melt is that the tagged-sodium-density-fluctuation correlators differ utterly from all other correlation functions under consideration. This can be inferred by comparing the $\phi_q^{\mathrm{Na}}(t)$ versus $\log t$ curves from the right panel of

Fig. 5.18 with the $\phi_q^{\text{NaNa}}(t)$ versus $\log t$ curves from the left one. The lower panel of Fig. 5.19 shows that the long-time-decay scale τ_q^{Na} is 10 to 100 times shorter than the one for the Na-Na-correlation decay. The cage effect is much less efficient for the motion of sodium along channels than one would expect by averaging the paths with equal weight for all directions. The system exhibits extreme scale decoupling.

Voigtmann and Horbach (2006) have solved the MCT equations of motion for the 3-by-3-matrix correlator $\boldsymbol{\phi}_q(t)$ and for the three tagged-particle correlation functions $\phi_q^\alpha(t)$, $\alpha = \text{Si, Na, O}$. The regular friction contributions $\boldsymbol{\nu}_q$ and ν_q^α have been dropped. The structure-factor matrix has been taken from the simulation data for a set of representative temperatures. An interpolation procedure provides these quantities and the corresponding direct-correlation functions for all temperatures of interest. The convolution approximation has been made, i.e., the triple correlations have been neglected in Eq. (5.79b) for the mode-coupling coefficients. The main effect of modifying the silica network by alloying is reproduced by the calculations, namely the strong decrease of the critical temperature relative to the value (5.97b) for pure silica:

$$T_c = 3105\,\text{K}, \qquad \text{MCT with convolution approximation.} \tag{5.101b}$$

The calculated critical temperature is about a factor 1.6 larger than the value noted in Eq. (5.101a).

The temperature is the only fit parameter entering the quantitative comparison of the cited MCT calculation for the correlation functions with the corresponding simulation data. This control parameter has to be adjusted since the structural relaxation under study is governed by the separation parameter $\sigma = C(T_c - T)/T_c$ rather than by T itself. The comparison of the $T = 2100\,\text{K}$ data shown in the two figures is made with curves calculated for $T = 3410\,\text{K}$. Figure 5.18 demonstrates that MCT reproduces qualitatively the mixing effects of sodium disilicate. There is a very high plateau of the Na-Na correlator for the prepeak-wave-number $q = 0.95\,\text{Å}^{-1}$; and there is a strong acceleration of the decay of the tagged-sodium density correlations compared to the long-time decay of the correlations of the total density-fluctuations. Figure 5.19 shows that the non-trivial wave-number dependence of the time scale for the below-plateau decay is reproduced reasonably for the O-O and Na-Na correlations and also for the tagged-oxygen motion. In the latter case, there appears a systematic overestimation of the time for small wave numbers similar to what was discussed above in Fig. 5.11. The scale for the long-time decay of the tagged-Na correlations is overestimated; it differs from the data by up to a factor 10.

5.3 A mode-coupling theory for molecular liquids

A derivation of a mode-coupling theory for molecular liquids was started by Schilling and Scheidsteger (1997). Their work deals with rigid linear molecules whose positions are specified by three coordinates for some reference point and two angular variables characterizing the orientation. The basic variables for a description of the structure are tensor-density fluctuations. The infinite matrix of correlation functions formed with these variables is used for a statistical description of the structure relaxation. The equations of motion should be formulated so that they are covariant with respect to changes of the coordinates for the description of the molecules' positions, in particular, with respect to changes of the reference point. Schilling (2002) derived such equations by introducing three relaxation kernels for fluctuating-force correlations. A factorization ansatz yields the kernels as mode-coupling polynomials of the tensor-density correlators. The coefficients are determined by equilibrium-structure functions. The arrested parts of the solutions of the MCT equations obey a set of nonlinear equations. Fold bifurcations have been identified, which describe liquid-to-glass transitions. The critical arrested parts, which were calculated for a model of water, describe results for correlation-function plateaus exhibited by molecular-dynamics-simulation data. Truncated forms of the equations of motion have been solved for states near the glass-transition point for the reorientational dynamics of a dumbbell moving in a hard-sphere liquid. The results suggest that the bifurcation dynamics is similar to that known from the MCT for simple systems. A list of references to the work indicated can be found in the cited paper.

The form of the mentioned equations of the molecular mode-coupling theory is different from the one discussed in preceding sections. So far, the general properties of these equations are not understood. For example, there is no criterion available to identify the arrested parts within the set of solutions of the generalized fixed-point equation for the same state. None of the asymptotic laws for the bifurcation dynamics, which are discussed below in Chapter 6, has been derived from the equations for the tensor-density correlators. Therefore, a discussion of this theory is beyond the scope of this book.

Within the framework of classical mechanics, the position of a molecule can be described by the positions of the nuclei of the n atoms forming the composed particles. In this context, the positions are referred to as interaction sites since, often, groups of atoms are combined to one entity. Thus, a system of molecules can be viewed as a mixture of n species of interaction sites. Studies of the glassy dynamics of molecular liquids with the intention to evaluate the n-by-n matrix of site-density-fluctuation correlators have been started by Chong and Hirata (1998). In Sec. 5.3.1, MCT equations shall be motivated for the mentioned matrix correlator. Then, some results for a high-density liquid of symmetric hard-sphere dumbbells shall be discussed (Sec. 5.3.2). An extension of this work is used in Sec. 5.3.3 to explain some glassy-dynamics features of the Rouse modes in dense polymer melts.

5.3.1 *A theory for interaction-site-density correlators*

In this section, a mode-coupling theory shall be derived for a system of identical molecules. The results will be obtained for the thermodynamic limit of N

composed particles enclosed in a cubic box of side length L. The number density N/L^3 of the molecules is denoted by ρ. Symmetry properties are assumed as explained in Sec. 3.1. Some results of the theory shall be discussed for a model of orthoterphenyl.

(i) MCT equations of motion

The molecules shall be labelled by Latin letters $k, \ell, \ldots = 1, \ldots, N$, and the sites on the molecules by Greek ones: $\alpha, \beta, \ldots = 1, \ldots, n$. The mass of site α is denoted by μ_α. The labels of the $N_{\text{tot}} = N \cdot n$ sites are composed ones, for example, $\kappa = (\alpha k)$. The system can be viewed as a mixture of n species of number concentrations $x_\alpha = 1/n$. But, for the comparison of formulas of this section with the ones on mixtures, one has to be aware that in Sec. 5.2 N_{tot} and $\rho_{\text{tot}} = \rho n$ have been denoted by N and ρ, respectively.

The position and velocity of site κ are denoted by $\vec{r}_{\alpha k}$ and $\vec{v}_{\alpha k}$, respectively. The density fluctuation for wave vector $\vec{q}$ of site α of a tagged molecule s reads $\rho^\alpha_{s\vec{q}} = \exp[i\vec{q}\vec{r}_{\alpha s}]$, and the density fluctuation for site α is $\rho^\alpha_{\vec{q}} = \sum_k \exp[i\vec{q}\vec{r}_{\alpha k}]$. The corresponding current-density fluctuations are given by $\vec{j}^\alpha_{s\vec{q}} = \vec{v}_{\alpha s} \exp[i\vec{q}\vec{r}_{\alpha s}]$ and $\vec{j}^\alpha_{\vec{q}} = \sum_k \vec{v}_{\alpha k} \exp[i\vec{q}\vec{r}_{\alpha k}]$, respectively. The continuity equations have the conventional form: $-i\partial_t \rho^\alpha_{s\vec{q}} = \mathcal{L}\rho^\alpha_{s\vec{q}} = \vec{q}\vec{j}^\alpha_{s\vec{q}}$, $-i\partial_t \rho^\alpha_{\vec{q}} = \mathcal{L}\rho^\alpha_{\vec{q}} = \vec{q}\vec{j}^\alpha_{\vec{q}}$. The overlaps of the tagged-site-density fluctuations define the intramolecular structure factors

$$(\rho^\alpha_{s\vec{q}}|\rho^\beta_{s\vec{q}}) = w^{\alpha\beta}_q. \tag{5.102a}$$

The partial structure factors for the sites are defined by:

$$(\rho^\alpha_{\vec{q}}|\rho^\beta_{\vec{q}})/N = S^{\alpha\beta}_q. \tag{5.102b}$$

In Eq. (5.67a), the symbol $S^{\alpha\beta}_q$ is used to abbreviate $(\rho^\alpha_{\vec{q}}|\rho^\beta_{\vec{q}})/N_{\text{tot}}$. Consequently, the partial structure factors used in the present section are n times of the ones used in the preceding section on mixtures. The usual symmetry considerations show that the structure factors are real quantities, which depend on the wave vector merely via its modulus $q = |\vec{q}|$. As done in Sec. 5.2, n-by-n matrices shall be denoted by bold letters. The preceding equations define the real symmetric metric matrices $\boldsymbol{w}_q$ and $\boldsymbol{S}_q$. Let us assume that they are non-degenerate. Hence, these matrices are positive definite; the inverses $\boldsymbol{w}_q^{-1}$ and $\boldsymbol{S}_q^{-1}$ exist and are positive definite as well.

The discussion of Eqs. (5.69a–e) for the structure-factor matrix has to be modified since the distance between sites on a tagged molecule is bounded. One writes for $\vec{q} \neq 0$ $S^{\alpha\beta}_q = w^{\alpha\beta}_q + \rho\tilde{h}^{\alpha\beta}_q$. Hence, the inter-chain contribution reads: $\rho\tilde{h}^{\alpha\beta}_q = \int \exp i\vec{q}\vec{r}\langle \sum_{k\neq p} \delta(\vec{r} + \vec{r}_{\alpha k} - \vec{r}_{\beta p})\rangle/N$. The Ornstein–Zernike relation (5.70) for mixtures has to be modified to

$$\boldsymbol{S}_q = [1 - \rho \boldsymbol{w}_q \boldsymbol{c}_q]^{-1} \boldsymbol{w}_q. \tag{5.103}$$

The real symmetric matrix $\boldsymbol{c}_q$ is called matrix of direct correlation functions. The element $c_q^{\alpha\beta}$ characterizes an effective interaction between sites α and β, which are located on different molecules (Hansen and McDonald 1986).

Let us introduce the symmetric matrix $\boldsymbol{J}_q$ for the overlaps of the longitudinal-current-fluctuations for the tagged molecule

$$(\vec{q}\vec{j}_{s\vec{q}}^{\,\alpha}|\vec{q}\vec{j}_{s\vec{q}}^{\,\beta})/q^2 = J_q^{\alpha\beta}. \tag{5.104a}$$

The usual symmetry considerations show that these elements define real symmetric matrices, which depend on $\vec{q}$ via the modulus q only. Velocities of different molecules can be averaged independently. Therefore, the matrix $\boldsymbol{J}_q$ determines also the overlaps of the total-current fluctuations:

$$(\vec{q}\vec{j}_{\vec{q}}^{\,\alpha}|\vec{q}\vec{j}_{\vec{q}}^{\,\beta})/(Nq^2) = J_q^{\alpha\beta}. \tag{5.104b}$$

Suppose, a molecule could be described by the laws of classical mechanics as a complex bound by intramolecular potentials. Then, velocities could be averaged independently from positions. In this case, one would get $J_q^{\alpha\beta} = \delta^{\alpha\beta}(v_{th}^{\alpha})^2$ with $v_{th}^{\alpha} = \sqrt{k_{\mathrm{B}}T/\mu_\alpha}$ denoting the thermal velocity of site α. A q-dependence of $\boldsymbol{J}_q$ occurs due to the reduction of the number of degrees of freedom of the molecule below $3n$ as described by intramolecular constraints.

Correlation functions for the fluctuations of the tagged-site densities and for the partial site densities, respectively, shall be introduced with the conventions:

$$F_{sq}^{\alpha\beta}(t) = (\rho_{s\vec{q}}^{\alpha}(t)|\rho_{s\vec{q}}^{\beta}), \quad F_q^{\alpha\beta}(t) = (\rho_{\vec{q}}^{\alpha}(t)|\rho_{\vec{q}}^{\beta})/N; \quad \alpha, \beta = 1, \ldots, n. \tag{5.105a}$$

Within the site representation, these are the simplest functions for a statistical description of the structure dynamics of the system. The matrix $\boldsymbol{F}_q$ differs by a trivial factor from the corresponding one defined for mixtures in Eq. (5.68a): $\boldsymbol{F}_q(t) = n\boldsymbol{\phi}(t)$. The functions $F_{sq}^{\alpha\beta}(t)$ and $F_q^{\alpha\beta}(t)$ exhibit the standard symmetries as specified in Sec. 5.2.1. The initial conditions are: $\boldsymbol{F}_{sq}(0) = \boldsymbol{w}_q$, $\partial_t \boldsymbol{F}_{sq}(0) = 0$ and $\boldsymbol{F}_q(0) = \boldsymbol{S}_q$, $\partial_t \boldsymbol{F}_q(0) = 0$. The derivation of Eq. (5.68b) can be repeated in order to obtain the short-time expansions:

$$\boldsymbol{F}_{sq}(t) = \boldsymbol{w}_q - \tfrac{1}{2}q^2\boldsymbol{J}_q t^2 + O(|t|^3), \tag{5.105b}$$

$$\boldsymbol{F}_q(t) = \boldsymbol{S}_q - \tfrac{1}{2}q^2\boldsymbol{J}_q t^2 + O(|t|^3). \tag{5.105c}$$

It is explained in connection with Eq. (5.76a) that the coefficients define real frequency matrices $\boldsymbol{\Omega}_{sq}$ and $\boldsymbol{\Omega}_q$, which have n non-negative eigenvalues:

$$\boldsymbol{\Omega}_{sq}^2 = q^2\boldsymbol{J}_q\boldsymbol{w}_q^{-1}, \qquad \boldsymbol{\Omega}_q^2 = q^2\boldsymbol{J}_q\boldsymbol{S}_q^{-1}. \tag{5.106}$$

The first step towards the motivation of MCT equations of motion follows strictly the one explained in connection with Eqs. (5.71a–d). Applying the Zwanzig–Mori procedure twice in succession, the n-by-n matrix correlator in the frequency domain, $\boldsymbol{F}_q(z)$, is given as double-fraction representation in terms of a fluctuating-force matrix $\boldsymbol{N}_q(z)$. There holds Eq. (5.71c) with the expression $F_q^\alpha = \sum_\gamma \mathcal{Q}_1 \mathcal{L} j_{\vec{q}*}^{\gamma 3} (\boldsymbol{J}_q^{-1})^{\gamma\alpha}/\sqrt{N}$. The matrix function $\boldsymbol{F}_{sq}(z)$ can be treated similarly. In this case, $\boldsymbol{S}_q$ has to be replaced by $\boldsymbol{w}_q$ and kernel $\boldsymbol{N}_{sq}(z)$ is formed with $F_{sq}^\alpha = \sum_\gamma \mathcal{Q}_1 \mathcal{L} j_{s\vec{q}*}^{\gamma 3} (\boldsymbol{J}_q^{-1})^{\gamma\alpha}$. Introducing dimensionless kernels $\boldsymbol{m}_q(t) = \boldsymbol{S}_q \boldsymbol{N}_q(t)/q^2$ and $\boldsymbol{m}_{sq}(t) = \boldsymbol{w}_q \boldsymbol{N}_{sq}(t)/q^2$, the fractions are equivalent to the identities:

$$\partial_t^2 \boldsymbol{F}_{sq}(t) + \boldsymbol{\Omega}_{sq}^2 \big[\boldsymbol{F}_{sq}(t) + \int_0^t \boldsymbol{m}_{sq}(t-t')\partial_{t'} \boldsymbol{F}_{sq}(t')dt'\big] = 0, \tag{5.107a}$$

$$\partial_t^2 \boldsymbol{F}_q(t) + \boldsymbol{\Omega}_q^2 \big[\boldsymbol{F}_q(t) + \int_0^t \boldsymbol{m}_q(t-t')\partial_{t'} \boldsymbol{F}_q(t')dt'\big] = 0. \tag{5.107b}$$

These are the equations of motion, which have to be solved with the initial conditions specified in the preceding paragraph. Let us anticipate that the further steps for a motivation of MCT equations have been made, as explained in connection with Eqs. (5.72)–(5.75). The preceding formulas are the MCT equations in analogy to Eq. (5.78a). Friction coefficients similar to $\boldsymbol{\nu}_q$ have been dropped here for the sake of simplicity.

Let us formulate the factorization ansatz as done in Sec. 5.2.1 for mixtures. The set of relevant pair fluctuations is spanned by variables $\rho_{\vec{m}_1}^{\alpha_1}\rho_{s\vec{m}_2}^{\alpha_2}$ and $\rho_{\vec{m}_1}^{\alpha_1}\rho_{\vec{m}_2}^{\alpha_2}$, respectively. One gets a representation of the kernels as quadratic mode-coupling polynomials as explained in connection with Eqs. (5.79), (5.80):

$$\boldsymbol{m}_{sq}(t) = \boldsymbol{w}_q \boldsymbol{\mathcal{F}}_{sq}[\boldsymbol{F}_k(t)\,,\,\boldsymbol{F}_{sp}(t)], \qquad \boldsymbol{m}_q(t) = \boldsymbol{S}_q \boldsymbol{\mathcal{F}}_q[\boldsymbol{F}_k(t)\,,\,\boldsymbol{F}_p(t)], \tag{5.108}$$

$$\mathcal{F}_{sq}^{\alpha\beta}[\boldsymbol{X}_k, \boldsymbol{Y}_p] = (\rho/8\pi^3)\int d^3\vec{k} \sum_{\alpha_1\alpha_2\beta_1\beta_2} \Big\{\hat{v}_{\alpha,\alpha_1\alpha_2}^s(\vec{q}^*, \vec{k}\vec{p}) X_k^{\alpha_1\beta_1} Y_p^{\alpha_2\beta_2} \hat{v}_{\beta,\beta_1\beta_2}^s(\vec{q}^*, \vec{k}\vec{p})\Big\}, \tag{5.109a}$$

$$\mathcal{F}_{q}^{\alpha\beta}[\boldsymbol{X}_k, \boldsymbol{Y}_p] = (\rho/16\pi^3)\int d^3\vec{k} \sum_{\alpha_1\alpha_2\beta_1\beta_2} \Big\{\hat{v}_{\alpha,\alpha_1\alpha_2}(\vec{q}^*, \vec{k}\vec{p}) X_k^{\alpha_1\beta_1} Y_p^{\alpha_2\beta_2} \hat{v}_{\beta,\beta_1\beta_2}(\vec{q}^*, \vec{k}\vec{p})\Big\}. \tag{5.109b}$$

The amplitudes $\hat{v}_{\alpha,\alpha_1\alpha_2}^s(\vec{q}^*, \vec{k}\vec{p})$ and $\hat{v}_{\alpha,\alpha_1\alpha_2}(\vec{q}^*, \vec{k}\vec{p})$ are real, and the latter are symmetric with respect to interchanges of $(\vec{k}\alpha_1)$ and $(\vec{p}\alpha_2)$. As before, $\vec{p}$ is an abbreviation of $\vec{q}^* - \vec{k}$. The integral over $\vec{k}$ can be converted in a double integral over $k = |\vec{k}|$ and $p = |\vec{p}|$, and these can be approximated by sums over finite wave-number grids as explained in Secs. 4.1.2 and 4.2.3.

For given matrices $\boldsymbol{\Omega}_q$ and $\boldsymbol{S}_q$ and given amplitudes $\hat{v}_{\alpha,\alpha_1\alpha_2}(\vec{q}^*, \vec{k}\vec{p})$, Eqs. (5.107b), (5.108), (5.109b) specify a closed set of integrodifferential equations, which have the same form as those discussed for mixtures in Sec. 5.2.1. Hence, all general theorems discussed previously for matrix-MCT models are valid here as well. The statement remains correct also if the tagged-particle correlators are included. From a mathematical point of view, the equations discussed in this section are special cases of those discussed in Sec. 5.2.

In order to determine the coupling amplitudes $\hat{v}_{\alpha,\alpha_1\alpha_2}(\vec{q}^*, \vec{k}\vec{p})$, one has to calculate the overlaps of the forces F_q^α with the pair variables $\rho_{\vec{k}_1}^{\sigma_1}\rho_{\vec{k}_2}^{\sigma_2}$. These overlaps read $(F_q^\alpha|\rho_{\vec{k}_1}^{\sigma_1}\rho_{\vec{k}_2}^{\sigma_2}) = \sum_\gamma (J_q^{-1})^{\alpha\gamma}\{(\mathcal{L}j_{\vec{q}*}^{\gamma 3}|\rho_{\vec{k}_1}^{\sigma_1}\rho_{\vec{k}_2}^{\sigma_2}) - (\mathcal{P}_1\mathcal{L}j_{\vec{q}*}^{\gamma 3}|\rho_{\vec{k}_1}^{\sigma_1}\rho_{\vec{k}_2}^{\sigma_2})\}/\sqrt{N}$. Here, $\mathcal{P}_1 = \sum |\rho_{\vec{q}*}^\alpha)(\boldsymbol{S}_q^{-1})^{\alpha\beta}(\rho_{\vec{q}*}^\beta|/N$ is the projector on the space spanned by the density fluctuations. The second matrix element is evaluated using the continuity equation and Eq. (5.104b): $(\rho_{\vec{q}*}^\beta|\mathcal{L}j_{\vec{q}*}^{\gamma 3}) = (\mathcal{L}\rho_{\vec{q}*}^\beta|j_{\vec{q}*}^{\gamma 3}) = qJ_q^{\gamma\beta}N$. The first matrix element can be rewritten in the form $(\mathcal{L}j_{\vec{q}*}^{\gamma 3}|\rho_{\vec{k}_1}^{\sigma_1}\rho_{\vec{k}_2}^{\sigma_2}) = \langle j_{\vec{q}*}^{\gamma 3 *}[\vec{k}_1\vec{j}_{\vec{k}_1}^{\sigma_1}\rho_{\vec{k}_2}^{\sigma_2} + \rho_{\vec{k}_1}^{\sigma_1}\vec{k}_2\vec{j}_{\vec{k}_2}^{\sigma_2}]\rangle$. It is not obvious how this expression can be formulated in terms of conventional structure factors. To proceed, it shall be evaluated with the approximation that velocities can be averaged independently from positions; e.g., $\langle j_{\vec{q}*}^{\gamma 3*}\vec{k}_1\vec{j}_{\vec{k}_1}^{\sigma_1}\rho_{\vec{k}_2}^{\sigma_2}\rangle = \delta^{\gamma\sigma_1}k_1^3(v_{th}^\gamma)^2(\rho_{\vec{q}*-\vec{k}_1}^\gamma|\rho_{\vec{k}_2}^{\sigma_2})$. It seems a request of consistency to perform the same approximation for the current overlaps: $(J_q^{-1})^{\alpha\gamma} = \delta^{\alpha\gamma}/(v_{th}^\gamma)^2$. One arrives at a formula analogous to Eq. (5.73):

$$(F_q^\alpha|\rho_{\vec{k}_1}^{\sigma_1}\rho_{\vec{k}_2}^{\sigma_2}) = \delta_{\vec{q}*,\vec{k}_1+\vec{k}_2}\sqrt{N}\Big\{k_1^3\delta^{\alpha\sigma_1}S_{k_2}^{\alpha\sigma_2} + k_2^3 S_{k_1}^{\alpha\sigma_1}\delta^{\alpha\sigma_2} - q\sum_\beta(\boldsymbol{S}_q^{-1})^{\alpha\beta}\langle\rho_{\vec{q}*}^{\beta *}\rho_{\vec{k}_1}^{\sigma_1}\rho_{\vec{k}_2}^{\sigma_2}\rangle/N\Big\}. \tag{5.110}$$

This result implies that the mode-coupling amplitudes are mere equilibrium-structure functions. In particular, they are independent of the n masses of the molecules' constituents. The assumption leading to this result is valid if the constraints for the motion are replaced by intramolecular potentials. Such model has been used, for example, in the molecular-dynamics studies for a decamer liquid, which is cited in connection with Figs. 1.5 and 1.11.

As explained for the treatment of Eq. (5.71c), the factorization ansatz yields the formula for the kernel: $N_q^{\alpha\beta}(t) = [1/(2N^2)]\sum_{\vec{k}_1\vec{k}_2}(F_q^\alpha|\rho_{\vec{k}_1}^{\sigma_1}\rho_{\vec{k}_2}^{\sigma_2})(\boldsymbol{S}_{k_1}^{-1})^{\sigma_1\alpha_1}$ $(\boldsymbol{S}_{k_2}^{-1})^{\sigma_2\alpha_2}F_{k_1}^{\alpha_1\beta_1}(t)F^{\alpha_2\beta_2}(t)(\boldsymbol{S}_{k_1}^{-1})^{\beta_1\tau_1}(\boldsymbol{S}_{k_2}^{-1})^{\beta_2\tau_2}(\rho_{\vec{k}_1}^{\tau_1}\rho_{\vec{k}_2}^{\tau_2}|F_q^\beta)$. Here, summation over repeated Greek indices is implied. Substitution of Eqs. (5.103), (5.110) yields Eqs. (5.108), (5.109b) with the expression for the amplitudes

$$\hat{v}_{\alpha,\alpha_1\alpha_2}(\vec{q}^*, \vec{k}\vec{p}) = v_{\alpha,\alpha_1\alpha_2}(\vec{q}^*, \vec{k}\vec{p}) + \delta v_{\alpha,\alpha_1\alpha_2}(\vec{q}^*, \vec{k}\vec{p}), \tag{5.111a}$$

$$v_{\alpha,\beta\gamma}(\vec{q}^*, \vec{k}\vec{p}) = [k^3 c_k^{\alpha\beta}\delta^{\alpha\gamma} + p^3 c_p^{\alpha\gamma}\delta^{\alpha\beta}]/q, \tag{5.111b}$$

$$\delta v_{\alpha,\beta\gamma}(\vec{q}^*, \vec{k}\vec{p}) = \Big[q\sum_{\lambda\mu\nu}(S_q^{-1})^{\alpha\lambda}(S_k^{-1})^{\beta\mu}(S_p^{-1})^{\gamma\nu}(\rho_{\vec{q}*}^\lambda|\rho_{\vec{k}}^\mu\rho_{\vec{p}}^\nu)/N - k^3(\boldsymbol{w}_k^{-1})^{\alpha\beta}\delta^{\alpha\gamma} - p^3(\boldsymbol{w}_p^{-1})^{\alpha\gamma}\delta^{\alpha\beta}\Big]\Big/(q\rho). \tag{5.111c}$$

Up to now, there is no information available for the triple-site-density average. Dropping this average does not appear meaningful since there is some cancellation of it against the terms obtained by the last two terms in the bracket of Eq. (5.111c). It has been suggested by Chong and Götze (2002a) to start studies of molecular liquids by dropping $\delta v_{\alpha,\beta\gamma}(\vec{q}^*, \vec{kp})$ and also the corresponding term for the tagged-molecule correlators. Within these studies, the amplitudes in Eq. (5.109b) are replaced by $v_{\alpha,\beta\gamma}(\vec{q}^*, \vec{kp})$ and those in Eq. (5.109a) by

$$v^s_{\alpha,\beta\gamma}(\vec{q}^*, \vec{kp}) = k^3 c_k^{\alpha\beta} \delta^{\alpha\gamma}/q. \tag{5.112}$$

All quantitative work done so far within the framework of the microscopic MCT for the site-density-fluctuation correlators of glassy molecular systems is based on the specified ad hoc simplifications for the mode-coupling amplitudes. This means that the mode-coupling functionals are used in the same form as for a mixture of n particles simplified by the superposition approximation. The difference between conventional mixtures and molecular systems is accounted for merely by the form of the partial structure factors, by the matrix $\boldsymbol{w}_q$ occurring in Eq. (5.108), and by the modification of the Ornstein–Zernike equation.

(ii) Results for a model of orthoterphenyl

Informative molecular-dynamics studies of the glassy relaxation of a molecular van der Waals liquid have been made for a model introduced by Lewis and Wahnström (1994). The system consists of rigid molecules composed of three sites of equal mass μ. The sites are located at the corners of isosceles triangles. The two short sides have a length $\sigma = 0.483$ nm. They are joined at site $\alpha = 1$ and form a 75° angle there. The interactions between sites of different molecules are derived from identical Lennard-Jones potentials. The interaction parameters σ and ϵ in Eq. (1.8a) are chosen so that the simulation data reproduce well the equilibrium density ρ of orthoterphenyl for temperatures T near 300 K. The interaction sites of the Lewis–Wahnström system (LWS) represent the phenyl rings of the molecules.

Figure 5.20 reproduces data for the normalized correlator $\phi_q^N(t) = (\rho_{\vec{q}}(t)|\rho_{\vec{q}})/(\rho_{\vec{q}}|\rho_{\vec{q}})$ of total-site-density fluctuation $\rho_{\vec{q}} = \sum_{\alpha=1}^{3} \rho_{\vec{q}}^{\alpha}$. The lines exhibit corresponding data calculated for a modified LWS and represented as function of a rescaled time, namely $\phi_q^N(t/s)$ for $s = 0.71$. The modified system agrees with the LWS except for the values of the masses. The latter are chosen as $\mu_2 = \mu_3 = \mu/6$, $\mu_1 = 8\mu/3$. Both systems have the same equilibrium-structure functions. Within MCT, they are described by the same functionals for the relaxation kernels. The equations of motions for the correlators differ only in the values for the frequencies $\boldsymbol{\Omega}_q$. The results for the two sets of functions agree for $t \geq 2$ ps. The change of the mass ratios leads to a change of the oscillation features for $t \leq 1$ ps. An analogous phenomenon is demonstrated in Fig. 4.24 for the MCT results for a HSS. The findings indicate that the dynamics outside the transient regime is influenced by the details of the short-time dynamics merely via

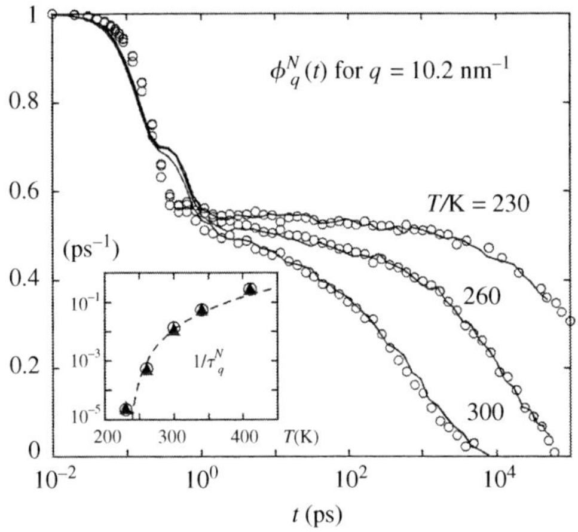

FIG. 5.20. The open dots exhibit molecular-dynamics-simulation results for the normalized correlators for the total site-density fluctuations of the Lewis–Wahnström system. The lines are corresponding simulation results for the same system with modified masses of the sites and times rescaled by a factor s. Details are explained in the text. The inset shows the inverse of the long-time-relaxation scale τ_q^N defined by $\phi_q^N(t = \tau_q^N) = 1/e$; and the dashed line exhibits a fit $\tau_q^N \propto 1/(T - T_c)^\gamma$ with $T_c = 234$ K and $\gamma = 2.76$. Reproduced from Chong and Sciortino (2003).

a temperature-independent time scale. It has been discussed in connection with Eq. (4.115a) that such result would be expected if one could write $\phi_q^N(t) = \phi_q^{\rm str}(t/t_0)$ for $t \geq 2$ ps. Here $\phi_q^{\rm str}$ denotes a correlator, which is determined completely by the equilibrium-structure functions.

The cited simulation data for $\phi_q^N(t)$ demonstrate an increase of the time scale τ_q^N for the below-plateau decay upon the decrease of T by about three orders of magnitude. This variation can be described by a power law, $\tau_q^N \propto 1/(T - T_c)^\gamma$, similar to what is exhibited by the MCT results for a HSS. The fit, which is shown in the inset of Fig. 5.20, suggests a critical temperature $T_c = 234\,$K for the arrest of the LWS. The cited power law also describes the decrease of the diffusivity D, which is observed for the decrease of the temperature from 410 K to 260 K (Mossa *et al.* 2002). Within this T interval, there is coupling of the scale for the diffusivity and the one for the long-time decay of the density fluctuations for $q = 10.2$ nm^{-1}. This feature of glassy dynamics has been discussed for MCT solutions in connection with Eq. (5.64). However, for T decreasing below 260 K, there is decoupling of the mentioned time scales. The diffusivity does not decrease as strongly as does $1/\tau_q^N$. The two decade decrease of D, which is observed for the LWS for $T \leq 280\,$K, can be fitted by an Arrhenius law.

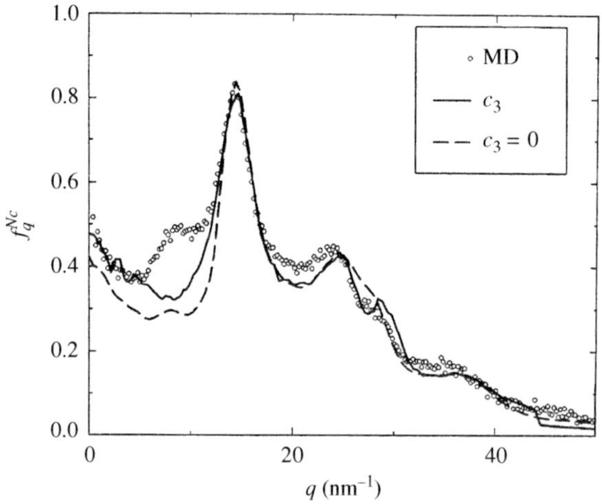

FIG. 5.21. The dots show plateau values f_q^{Nc} determined from molecular-dynamics simulation results for the correlator $\phi_q^N(t)$ for the total-site-density fluctuations of the LWS. Treating this system as a liquid of simple identical particles, the microscopic MCT equations for $\phi_q(t)$ yield the full or the dashed lines for the critical arrested parts f_q^c depending on whether the triple correlations are included in the formulas for the mode-coupling constants or whether the convolution approximation is used, respectively. Reproduced from Rinaldi *et al.* (2001).

The simulation data for various correlators of the LWS demonstrate the evolution of the time-temperature-superposition principle for T decreasing from 346 K to 275 K (Rinaldi *et al.* 2001). The scenario is the same as demonstrated in Figs. 1.5 and 4.21. Thereby, a large part of the shape function for the scaling-law description of the below-plateau decay process can be identified. Fits to the extended von Schweidler law (5.96a) can be used to identify the correlation plateau. The dots in Fig. 5.21 exhibit the plateau values f_q^{Nc} for the correlator $\phi_q^N(t)$. The time scale τ_q^N varies as function of q in phase with the f_q^{Nc} versus q curve. This behaviour is similar to the one demonstrated for the MCT results for the HSS in Figs. 4.6 and 4.20.

Rinaldi *et al.* (2001) have examined, to which extend the glassy dynamics of the LWS can be described by MCT results for a simple system. To proceed, they use the formulas of the microscopic MCT from chapter 4 identifying $\phi_q(t)$ with the correlator $\phi_q^N(t)$. The equilibrium structure function $S_q = \langle \rho_{\vec{q}}^* \rho_{\vec{q}} \rangle / (3N)$ is obtained by interpolating and extrapolating results from the simulation work. The dashed line in Fig. 5.21 exhibits the critical arrested parts f_q^c calculated by using the convolution approximation for the triple-density averages $\langle \rho_{\vec{q}}^* \rho_{\vec{k}} \rho_{\vec{p}} \rangle$. There is good agreement between data and calculations for the region of intermediate and large wave numbers, say, $q \geq 12$ nm^{-1}. If the triple-correlation functions are included by their values determined for the lowest temperature studied, the full lines are obtained. They extend the region of agreement with the data by the additional wave-number interval $q \leq 5\ \mathrm{nm}^{-1}$. The authors have shown for some representative cases that also the MCT shape functions in Eq. (4.112a) are in agreement with the data. The f_q^{Nc} versus q curve exhibits a prepeak for q near $9\ \mathrm{nm}^{-1}$. A corresponding prepeak is detected for the τ_q^N versus q curve. These prepeaks reflect slow relaxation of the below-plateau correlations for wave lengths, which are close to the average distance between the centres of the molecules. For this wave-number regime, there is a large Debye–Waller factor. The described features are not reproduced by the cited MCT calculations. Calculations done for the matrix correlator $\phi_q^{\alpha\beta}(t)$ within the MCT for the site-site functions yield results in close agreement with the dashed line in Fig. 5.21 (Chong and Sciortino 2004).

5.3.2 *Systems of symmetric dumbbells*

Possible symmetries of the molecules cause relations between structure functions; and these imply relations between mode-coupling coefficients. Consequently, there may occur features of the glassy MCT dynamics, which deviate from the ones derived for generic models. Such features offer the possibility for testing the relevance of the theory for a description of liquid dynamics by searching for the indicated symmetry-induced peculiarities. In this section, such phenomena will be demonstrated for a system of symmetric dumbbell molecules. It will be shown that the system has two types of glass states. This implies three different scenarios for the evolution of glassy dynamics depending on whether there is strong, weak, or very weak steric hindrance for reorientational motion.

The interaction sites of the rigid two-centre molecules to be considered shall be labelled as $\alpha = 1 = A$ and $\alpha = 2 = B$. The molecule centre of particle number k is $\vec{r}_{Ck} = (\vec{r}_{Ak} + \vec{r}_{Bk})/2$; the molecule axis is $\vec{e}_k = (\vec{r}_{Ak} - \vec{r}_{Bk})/L_D$. Here $L_D = |\vec{r}_{Ak} - \vec{r}_{Bk}|$ denotes the distance between the interaction sites. If $\mu = \mu_A = \mu_B$ denotes the mass attributed to the interaction centres, the two inertia parameters of the molecules are the total mass 2μ and the moment of inertia $\mu L_D^2/2$. The thermal velocity for the molecules is $v_{th} = \sqrt{k_{\mathrm{B}}T/(2\mu)}$, and the thermal angular velocity is the frequency $\omega_{th} = 2v_{th}/L_D$. The three interaction potentials between the pairs $A - A$, $A - B$, and $B - B$ of sites on different molecules are assumed to be identical. The extension of the repulsive part of the interaction shall be characterized by a length d. The elongation $\zeta = L_D/d$ shall be used to characterize the deviation of the dumbbell from a spherical shape. The restriction $\zeta \leq 1$ will be imposed. A quantity characterizing the space filling by the molecules is the packing fraction $\varphi = (\pi\rho d^3/6)[1 + (3\zeta/2) - (\zeta^3/2)]$. It is the part of the available volume L^3, which is filled by the $N = \rho L^3$ dumbbells under the assumption that the molecular volume is that of two fussed spheres of diameter d. For $\zeta = 0$, the system consists of simple particles and φ is the packing fraction of a hard-sphere system of density ρ. For $\zeta = 1$, the molecule is a dimer of two simple particles sticking together. In this case, φ is the packing fraction of a hard-sphere system of density 2ρ. The quantities from Eqs. (5.102a), (5.104a), which characterize the intramolecular structure, are

$$w_q^{\alpha\beta} = \delta^{\alpha\beta} + (1-\delta^{\alpha\beta})j_0(qL_D), \tag{5.113a}$$

$$J_q^{\alpha\beta} = \frac{1}{3}v_{th}^2\Big\{5\delta^{\alpha\beta} + (1-\delta^{\alpha\beta})\big[j_0(qL_D) - 2j_2(qL_D)\big]\Big\}. \tag{5.113b}$$

Here, $j_0(x) = \sin x/x$ and $j_2(x) = [(3-x^2)\sin x - 3x\cos x]/x^3$ denote spherical Bessel functions of index $\ell = 0$ and $\ell = 2$, respectively.

(i) Symmetry properties

The discussion shall be started by demonstrating that the symmetry of the molecules implies symmetry operators for the dynamics. As a result, the correlators for the interaction-site-density fluctuations can be expressed in terms of auto correlators for eigenvectors of the symmetry operators.

Extending the discussion of Sec. 3.1.1, the phase-space coordinates shall be denoted by $(\boldsymbol{p}, \boldsymbol{q}) = (\pi_1, \ldots, \pi_N)$. Here, $\pi_k = (\vec{p}_{Ak}, \vec{r}_{Ak}, \vec{p}_{Bk}, \vec{r}_{Bk})$ abbreviates the phase-space coordinates of the molecule number $k; \vec{p}_{\alpha k} = \mu\vec{v}_{\alpha k}$. Let $\pi'_k = (\vec{p}_{Bk}, \vec{r}_{Bk}, \vec{p}_{Ak}, \vec{r}_{Ak})$ denote the point in the molecule's phase space, where the two interaction sites are exchanged. The point in phase pace $(\boldsymbol{p}^s, \boldsymbol{q}^s) = (\pi_1, \ldots, \pi'_s, \ldots \pi_n)$ describes the state of the system with exchanged constituents of the tagged molecule number s. A mapping $\mathcal{S}_s$ of the space of dynamical variables shall be defined by: $[\mathcal{S}_s X](\boldsymbol{p}, \boldsymbol{q}) = X(\boldsymbol{p}^s, \boldsymbol{q}^s)$. $\mathcal{S}_s$ is a local exchange operator. The dynamical variable $\mathcal{S}_s X$ is obtained from X by an exchange of the sites of the molecule number s. The mapping $\mathcal{S} = \mathcal{S}_1 \cdots \mathcal{S}_N$ is a global exchange operator: for all N molecules, the two interactions sites are exchanged. The set of all operators and their products form an Abelian transformation group. There holds

$$\mathcal{S}_s^2 = 1, \qquad \mathcal{S}^2 = 1. \tag{5.114}$$

The operator $\mathcal{S}$ exchanges the interaction-site-density fluctuations:

$$\mathcal{S}\rho_{\vec{q}}^{A,B} = \rho_{\vec{q}}^{B,A}. \tag{5.115a}$$

This identity remains valid if $\rho_{\vec{q}}^{\alpha}$ is replaced by $\rho_{s\vec{q}}^{\alpha}$. There holds:

$$\mathcal{S}_s\rho_{s\vec{q}}^{A,B} = \rho_{s\vec{q}}^{B,A}; \quad \mathcal{S}_s\rho_{r\vec{q}}^{A,B} = \rho_{r\vec{q}}^{A,B} \quad \text{if} \quad r \neq s. \tag{5.115b}$$

Let us introduce a 2-by-2 matrix $\boldsymbol{P}$ with elements: $P^{11} = P^{12} = P^{21} = -P^{22} = 1/\sqrt{2}$. This matrix is unitary,

$$\boldsymbol{P}^{-1} = \boldsymbol{P}^{\dagger}, \tag{5.116a}$$

and it obeys

$$\boldsymbol{P}^2 = 1. \tag{5.116b}$$

It shall be used to transform invertibly the pair $\rho_{\vec{q}}^{\alpha}$, $\alpha = A, B$, on another pair to be denoted by $\overline{\rho}_{\vec{q}}^{x}$, $x = +, -$:

$$\sum_{\alpha} \rho_{\vec{q}}^{\alpha} P^{\alpha x} = \overline{\rho}_{\vec{q}}^{x}, \quad \rho_{\vec{q}}^{\beta} = \sum_{x} P^{\beta x *} \overline{\rho}_{\vec{q}}^{x}. \tag{5.116c}$$

In a similar manner, variables $\rho_{s\vec{q}}^{\alpha}$ and $\overline{\rho}_{s\vec{q}}^{x}$ shall be related. From Eqs. (5.115a,b) one concludes

$$\mathcal{S}\overline{\rho}_{\vec{q}}^{x} = x\overline{\rho}_{\vec{q}}^{x}, \qquad \mathcal{S}\overline{\rho}_{s\vec{q}}^{x} = x\overline{\rho}_{s\vec{q}}^{x}, \quad x = \pm \tag{5.116d}$$

The dynamical variables $\overline{\rho}_{\vec{q}}^{x}$ and $\overline{\rho}_{s\vec{q}}^{x}$ are real scalar field fluctuations, which are eigenvectors of the operator $\mathcal{S}$ for eigenvalue x. They have even or odd exchange parity, respectively. They are symmetry-adapted density fluctuations. Similar statements hold for product fluctuations, for example,

$$\mathcal{S}\overline{\rho}_{\vec{k}}^{x}\overline{\rho}_{\vec{p}}^{y} = xy\overline{\rho}_{\vec{k}}^{x}\overline{\rho}_{\vec{p}}^{y}, \quad \mathcal{S}\overline{\rho}_{\vec{k}}^{x}\overline{\rho}_{s\vec{p}}^{y} = xy\overline{\rho}_{\vec{k}}^{x}\overline{\rho}_{s\vec{p}}^{y}. \tag{5.116e}$$

The tagged-molecule fluctuations are also eigenvectors of the local exchange operators:

$$\mathcal{S}_s\overline{\rho}_{s\vec{q}}^{\pm} = \pm\overline{\rho}_{s\vec{q}}^{\pm}, \qquad \mathcal{S}_s\overline{\rho}_{r\vec{q}}^{\pm} = \overline{\rho}_{r\vec{q}}^{\pm}, \quad \text{if } r \neq s. \tag{5.117}$$

The metric matrix for the symmetry-adapted fluctuations is related to that for the interaction-site fluctuations by an equivalence transformation. The latter is created by $\boldsymbol{P}$: $(\overline{\rho}_{\vec{q}}^{x}|\overline{\rho}_{\vec{q}}^{y}) = \sum_{\alpha\beta} P^{\alpha x *}(\rho_{\vec{q}}^{\alpha}|\rho_{\vec{q}}^{\beta})P^{\beta y}$. If $\boldsymbol{M}$ denotes some 2-by-2 matrix, the $\boldsymbol{P}$-created equivalent matrix shall be indicated by a bar:

$$\overline{\boldsymbol{M}} = \boldsymbol{P}^{-1}\boldsymbol{M}\boldsymbol{P}. \tag{5.118}$$

Using this convention, the transformed structure factors are those formed with the density-fluctuations of definite exchange symmetry.

$$\overline{S}_q^{xy} = (\overline{\rho}_{\vec{q}}^{x}|\overline{\rho}_{\vec{q}}^{y})/N, \qquad \overline{w}_q^{xy} = (\overline{\rho}_{s\vec{q}}^{x}|\overline{\rho}_{s\vec{q}}^{y}). \tag{5.119a}$$

Similarly, the equivalence transformation of the matrices $\boldsymbol{F}_q(t)$ and $\boldsymbol{F}_{sq}(t)$ yields the matrices of the correlators for the fluctuations of definite exchange parity:

$$\overline{F}_q^{xy}(t) = (\overline{\rho}_{\vec{q}}^{x}(t)|\overline{\rho}_{\vec{q}}^{y})/N, \qquad \overline{F}_{sq}^{xy}(t) = (\overline{\rho}_{s\vec{q}}^{x}(t)|\overline{\rho}_{s\vec{q}}^{y}). \tag{5.119b}$$

Since the elements of $\boldsymbol{P}$ are real, these are real functions of the time.

Three paragraphs shall be added on the above specified transformation. Noting Eq. (5.118) and its reversion $\boldsymbol{M} = \boldsymbol{P}\overline{\boldsymbol{M}}\boldsymbol{P}^{-1}$ explicitly for the four matrix elements, one recognizes: matrix $\boldsymbol{M}$ exhibits the two symmetry properties

$$M^{AA} = M^{BB} \quad \text{and} \quad M^{AB} = M^{BA} \tag{5.120a}$$

if and only if $\boldsymbol{P}$ transforms $\boldsymbol{M}$ on diagonal form, i.e.,

$$\overline{M}^{xy} = M^x \delta^{xy}, \ x, y = \pm. \tag{5.120b}$$

The two eigenvalues of $\boldsymbol{M}$ are given by

$$M^{\pm} = M^{AA} \pm M^{AB}. \tag{5.120c}$$

They determine the matrix by: $M^{AA} = (M^+ + M^-)/2\,, M^{AB} = (M^+ - M^-)/2$. Matrix $\boldsymbol{M}$ is real if and only if this is true for the two eigenvalues $M^{\pm}$.

According to Eqs. (5.113a,b), the matrices $\boldsymbol{w}_q$ and $\boldsymbol{J}_q$ exhibit the symmetries (5.120a). One gets for the eigenvalues:

$$w_q^{\pm} = 1 \pm j_0(qL_D), \qquad J_q^{\pm} = \tfrac{1}{3} v_{th}^2 \Big\{ 5 \pm \big[j_0(qL_D) - 2 j_2(qL_D) \big] \Big\}. \tag{5.121a}$$

The equivalence transformation also diagonalizes the matrix $\boldsymbol{\Omega}_{sq}^2$ from Eq. (5.106). The two eigenvalues shall be noted as squares of positive frequencies $\Omega_{sq}^{\pm}$:

$$(\Omega_{sq}^{\pm})^2 = q^2 J_q^{\pm} / w_q^{\pm}. \tag{5.121b}$$

One frequency exhibits a linear small-q dispersion as formulated in Eq. (3.26b):

$$\lim_{q \to 0} [\Omega_{sq}^{+}/q] = v_{th}. \tag{5.121c}$$

The other one approaches a positive $q = 0$ limit:

$$\lim_{q \to 0} [\Omega_{sq}^{-} L_D] = \sqrt{8} v_{th}. \tag{5.121d}$$

Let $\boldsymbol{K}$ and $\boldsymbol{L}$ denote matrices, which exhibit the symmetries (5.120a). By

$$M^{\alpha\beta} = K^{\alpha\beta} L^{\alpha\beta}, \qquad \alpha, \beta = A, B, \tag{5.122a}$$

a third matrix $\boldsymbol{M}$ with the mentioned symmetry property is defined.The eigenvalues of $\boldsymbol{M}$ are determined by those of $\boldsymbol{K}$ and $\boldsymbol{L}$:

$$M^{\pm} = (K^+ L^{\pm} + K^- L^{\mp})/2. \tag{5.122b}$$

The transformation of the phase-space coordinates $(\boldsymbol{p}, \boldsymbol{q}) \to (\boldsymbol{p}^s, \boldsymbol{q}^s)$ does not alter the integration measure in Eq. (2.5b). The Hamilton function remains invariant and, thus, the Boltzmann weight $\rho(\boldsymbol{p}, \boldsymbol{q})$ does not change either. Hence, there holds $(\mathcal{S}_s X | \mathcal{S}_s Y) = (X|Y)$ for every pair of dynamical variables X and Y. Consequently, the operators $\mathcal{S}_s$ are unitary mappings of the space of dynamical variables. Hamilton's equations of motion are covariant if the sites A and B are interchanged for a tagged molecule, since $H(\boldsymbol{p}, \boldsymbol{q})$ does not change. If some phase-space orbit connects the initial point $(\boldsymbol{p}, \boldsymbol{q})$ with $(\boldsymbol{p}(t), \boldsymbol{q}(t))$, the orbit with initial point $(\boldsymbol{p}^s, \boldsymbol{q}^s)$ leads to $(\boldsymbol{p}(t)^s, \boldsymbol{q}(t)^s)$. Hence, there hold Eqs. (3.6a,b).

The exchange operators $\mathcal{S}_s$ are unitary transformations, which commute with the unitary operator $\mathcal{U}(t)$ for the time evolution; in particular, they commute with its generator $\mathcal{L}$. They are symmetries for the dynamics. This holds for all products of the local exchange operators and, consequently, for the global exchange $\mathcal{S}$.

From Eqs. (3.6a), (5.116d), one gets $\overline{F}_q^{xy}(t) = ([\mathcal{S}\overline{\rho}_{\vec{q}}^x(t)]|\mathcal{S}\overline{\rho}_{\vec{q}}^x y)/N = xy\overline{F}_q^{xy}(t)$ and, similarly, $\overline{F}_{sq}^{xy}(t) = xy\overline{F}_{sq}^{xy}(t)$. Equivalently, there holds

$$\overline{F}_q^{+-}(t) = \overline{F}_q^{-+}(t) = 0, \qquad \overline{F}_{sq}^{+-}(t) = \overline{F}_{sq}^{-+}(t) = 0. \tag{5.123}$$

More generally, one concludes that there are no correlations between variables of different exchange parity. In particular, one gets for $t = 0 : \overline{S}_q^{+-} = \overline{S}_q^{-+} = 0$. The discussion of Eqs. (5.120a–d) can be applied for the matrices $\boldsymbol{S}_q, \boldsymbol{F}_q(t)$ and $\boldsymbol{F}_{sq}$. The specified matrices can be expressed in terms of the autocorrelation functions for the fluctuations of definite exchange parity, which shall be denoted by $S_q^{\pm} = (\rho_{\vec{q}}^{\pm}|\rho_{\vec{q}}^{\pm})/N$, $F_q^{\pm}(t) = \overline{F}_q^{\pm\pm}(t)$, and $F_{sq}^{\pm}(t) = \overline{F}_{sq}^{\pm\pm}(t)$. Let us also introduce normalized correlators:

$$\phi_q^{\pm}(t) = F_q^{\pm}(t)/S_q^{\pm}, \qquad \phi_{sq}^{\pm}(t) = F_{sq}^{\pm}(t)/w_q^{\pm}. \tag{5.124a}$$

The short-time expansions (5.105b,c) are transformed to

$$\phi_q^{\pm}(t) = 1 - \tfrac{1}{2}\left[\Omega_q^{\pm}t\right]^2 + O(|t|^3), \quad \phi_{sq}^{\pm}(t) = 1 - \tfrac{1}{2}\left[\Omega_{sq}^{\pm}t\right]^2 + O(|t|^3). \tag{5.124b}$$

The characteristic frequencies for the short-time dynamics of the tagged-molecule are noted in Eq. (5.121b). The ones for the total-density fluctuations are given by

$$(\Omega_q^{\pm})^2 = q^2 J_q^{\pm}/S_q^{\pm}. \tag{5.124c}$$

The equation of motion (5.107b) together with its initial conditions is equivalent to $\boldsymbol{F}_q(z) = -\{z - [z + \boldsymbol{\Omega}_q^2 \boldsymbol{m}_q(z)]^{-1}\boldsymbol{\Omega}_q^2\}^{-1}\boldsymbol{S}_q$. This identity expresses the correlator matrix in the frequency domain in terms of the fluctuating-force kernel in the frequency domain. It is equivalent to the expression of matrix $\boldsymbol{m}_q(z)$ in terms of matrix $\boldsymbol{F}_q(z) : \boldsymbol{m}_q(z) = [\boldsymbol{S}_q\boldsymbol{F}_q(z)^{-1} + z]^{-1} - z\boldsymbol{\Omega}_q^{-2}$. Because $[\boldsymbol{P}^{-1}\boldsymbol{M}\boldsymbol{P}]^{-1} = [\boldsymbol{P}^{-1}\boldsymbol{M}^{-1}\boldsymbol{P}]$, the preceding two algebraic relations between $\boldsymbol{F}_q(z)$ and $\boldsymbol{m}_q(z)$ are equivalent to ones, where all matrices get a bar. Since $\overline{\boldsymbol{\Omega}}_q^2$ and $\overline{\boldsymbol{S}}_q$ are diagonal, one concludes: matrix $\overline{\boldsymbol{m}}_q(z)$ is diagonal if and only if this is true for $\overline{\boldsymbol{F}}_q(z)$. The correlator matrix exhibits the symmetries (5.120a) if and only if the fluctuating-force kernel exhibits these properties. In this case, the matrix fraction for $\boldsymbol{F}_q(z)$ is equivalent to two fractions for the diagonal elements: $\phi_q^x(z) = F_q^x(z)/S_q^x = -\{z - \Omega_q^{x2}/[z + \Omega_q^{x2} m_q^x(z)]\}^{-1}$, $x = \pm$. These equations for the normalized autocorrelation functions $\phi_q^x(t)$ for the density-fluctuations with definite parity for the global exchange symmetry are equivalent to the two equations of motion,

$$\partial_t^2 \phi_q^{\pm}(t) + \Omega_q^{\pm 2}[\phi_q^{\pm}(t) + \int_0^t m_q^{\pm}(t-t')\partial_{t'}\phi_q^{\pm}(t')dt'] = 0, \tag{5.125a}$$

together with the usual initial conditions $\phi_q^{\pm}(0) = 1$, $\partial_t \phi_q^{\pm}(0) = 0$. In the same manner, one derives the two equations of motion for the tagged-molecule correlators,

$$\partial_t^2 \phi_{sq}^{\pm}(t) + \Omega_{sq}^{\pm 2}[\phi_{sq}^{\pm}(t) + \int_0^t m_{sq}^{\pm}(t-t')\partial_{t'}\phi_{sq}^{\pm}(t')dt'] = 0, \tag{5.125b}$$

together with the usual initial conditions.

In addition to $\mathcal{S}$, the operators $\mathcal{S}_s$, $s = 1, \ldots, N$, are symmetries for the symmetric-dumbbell system. They imply further relations between the density-fluctuation correlators. From Eq. (5.117), one gets for $s \neq r$: $(\overline{\rho}_{s\vec{q}}^{-}(t)|\overline{\rho}_{r\vec{q}}^{-}) = ([\mathcal{S}_s\overline{\rho}_{s\vec{q}}^{-}](t)|\mathcal{S}_s\overline{\rho}_{r\vec{q}}^{-}) = -(\overline{\rho}_{s\vec{q}}^{-}(t)|\overline{\rho}_{r\vec{q}}^{-})$. Hence, there are no correlations between the tagged particle density fluctuations of different molecules if these have an odd exchange parity. Consequently, $NF_q^{-}(t) = \sum_s(\overline{\rho}_{s\vec{q}}^{-}(t)|\overline{\rho}_{s\vec{q}}^{-}) + \sum_{s\neq r}(\overline{\rho}_{s\vec{q}}^{-}(t)|\overline{\rho}_{r\vec{q}}^{-}) = NF_{sq}^{-}(t)$. The autocorrelator of $\overline{\rho}_{\vec{q}}^{-}/\sqrt{N}$ is identical to the one of $\overline{\rho}_{s\vec{q}}^{-}$:

$$F_q^{-}(t) = F_{sq}^{-}(t). \tag{5.126}$$

Specializing the result to $t = 0$, one concludes that

$$S_q^{-} = w_q^{-}. \tag{5.127a}$$

The Ornstein–Zernike equation (5.103) implies $\boldsymbol{S}_q^{-1} - \boldsymbol{w}_q^{-1} = -\rho\boldsymbol{c}_q$. Since the equivalence transformation diagonalizes $\boldsymbol{S}_q$ and $\boldsymbol{w}_q$, it does the same for $\boldsymbol{c}_q$. There holds $(1/S_q^{\pm}) - (1/w_q^{\pm}) = -\rho c_q^{\pm}$. The preceding equation implies

$$c_q^{-} = 0, \qquad c_q^{AA} = c_q^{AB}. \tag{5.127b}$$

According to Eq. (5.120c), this formula is equivalent to:

$$c_q^{+} = 2c_q^{AA}. \tag{5.127c}$$

A paragraph with a digression shall be added on possible generalizations of the discussions of this section 5.3.2. Suppose that the interaction between the molecules is generalized so that the $A - B$ potential differs from the $A - A$ one. But, the $A - A$ interaction shall remain the same as the $B - B$ one. In this case, the potential for a tagged molecule number s, which is produced by its fixed neighbours, changes if its axis $\vec{e}_s$ is reversed. The potential part of the Hamiltonian $H(\boldsymbol{p}, \boldsymbol{q})$ differs from that of $H(\boldsymbol{p}^s, \boldsymbol{q}^s)$. Therefore, $\mathcal{S}_s$ is not unitary for this model; the local exchange operators do not define a symmetry for this generalized dumbbell system. However, if the axis $\vec{e}_k$, $k = 1, \ldots, N$, of all molecules is reversed, the Hamiltonian remains the same. One can follow the reasoning from above and show that $\mathcal{S}$ is a symmetry operator for the dynamics. All reasonings of this section, which merely exploit the global symmetry $\mathcal{S}$, remain valid also for the generalized dumbbell system. This holds, for example, for Eqs. (5.119)–(5.125). But, Eqs. (5.126)–(5.127) need not be valid for the generalized system.

The system of symmetric electric dipoles is an example for the specified generalization. It is obtained from the symmetric dumbbell system by adding the Coulomb potentials produced by opposite point charges $\pm Q$, which are located at the positions $\vec{r}_{Ck} \pm L'_D \vec{e}_k$.

(ii) MCT equations of motion

In this subsection it will be shown that the MCT equations of motion preserve the above formulated symmetry properties. Consequently, the equations for the matrix correlators can be rewritten as equations for correlator arrays, which are equivalent to those discussed in Chapter 4 for simple systems.

Within the theory under study, the relaxation kernels are replaced by mode-coupling functionals. The symmetry properties of these functionals remain to be examined. Functional $\boldsymbol{\mathcal{F}}_{sq}$ is given by Eq. (5.109a) and by formula (5.112) for the coefficients. In order to understand the symmetries, one has to examine a matrix $\boldsymbol{\psi}_s$ with elements

$$\psi_s^{\alpha\beta} = \sum_{\alpha_1\alpha_2\beta_1\beta_2} \Big\{ [c_k^{\alpha\alpha_1} \delta^{\alpha\alpha_2}] X_k^{\alpha_1\beta_1} Y_p^{\alpha_2\beta_2} [c_k^{\beta\beta_1} \delta^{\beta\beta_2}] \Big\}. \tag{5.128a}$$

There holds $\psi_s^{\alpha\beta} = K^{\alpha\beta} L^{\alpha\beta}$ with $\boldsymbol{K} = \boldsymbol{c}_k \boldsymbol{X}_k \boldsymbol{c}_k$ and $\boldsymbol{L} = \boldsymbol{Y}_p$. Matrix $\boldsymbol{c}_k$ has the symmetries (5.120a). If $\boldsymbol{X}_k$ has these symmetries as well, one concludes from Eqs. (5.120a,b) that also $\boldsymbol{K}$ exhibits these symmetries and $K^\pm = c_k^\pm X_k^\pm c_k^\pm$. If also $\boldsymbol{Y}_p$ exhibits the symmetries, one concludes from Eqs. (5.122a,b) that matrix $\boldsymbol{\psi}_s$ has the symmetries, and its eigenvalues are

$$\psi_s^\pm = \Big\{ c_k^+ X_k^+ c_k^+ Y_p^\pm + c_k^- X_k^- c_k^- Y_p^\mp \Big\} / 2. \tag{5.128b}$$

Since $\boldsymbol{w}_q$ exhibits the symmetries, one can summarize as follows. If the matrices $\boldsymbol{F}_k(t)$ and $\boldsymbol{F}_{sp}(t)$ exhibit the symmetries, which are implied by the global exchange operator $\mathcal{S}$, Eqs. (5.108), (5.109a), (5.112) define a matrix $\boldsymbol{m}_{sq}(t)$, which exhibits the same symmetries. The eigenvalues $m_{sq}^\pm(t)$ of this kernel are given as linear combination of $c_k^+ S_k^+ \phi_k^+(t) c_k^+ w_p^\pm \phi_{sp}^\pm(t) + c_k^- S_k^- \phi_k^-(t) c_k^- w_p^\mp \phi_{sp}^\mp(t)$. For symmetric dumbbells, the condition (5.127b) for the local exchange symmetry eliminates half of the contributions, and one arrives at

$$m_{sq}^\pm(t) = \mathcal{F}_{sq}^\pm[\phi_k^+(t)\,,\, \phi_{sp}^\pm(t)]. \tag{5.129a}$$

Here, mode-coupling functionals are introduced by

$$\mathcal{F}_{sq}^\pm[x_k, y_p] = [w_q^\pm/q^2][\rho/(16\pi^3)] \int d^3\vec{k} (k^3 c_k^+)^2 S_k^+ w_p^\pm x_k y_p, \tag{5.129b}$$

with $p = |\vec{q}^* - \vec{k}|$ and $k^3 = (\vec{q}^* \vec{k})/q$.

In order to examine the symmetries of the functional $\boldsymbol{\mathcal{F}}_q$, one has to substitute the amplitudes (5.111b) into Eq. (5.109b). One arrives at the problem to examine the symmetries of a matrix $\boldsymbol{\psi}$ with elements

$$\psi^{\alpha\beta} = \sum_{\alpha_1\alpha_2\beta_1\beta_2} \Big\{ [k^3 c_k^{\alpha\alpha_1} \delta^{\alpha\alpha_2} + p^3 c_p^{\alpha\alpha_2} \delta^{\alpha\alpha_1}] \\ X_k^{\alpha_1\beta_1} Y_p^{\alpha_2\beta_2} [k^3 c_k^{\beta\beta_1} \delta^{\beta\beta_2} + p^3 c_p^{\beta\beta_2} \delta^{\beta\beta_1}] \Big\}. \quad (5.130a)$$

This matrix can be written as $\boldsymbol{\psi}^{\alpha\beta} = k^3k^3[\boldsymbol{c}_k\boldsymbol{X}_k\boldsymbol{c}_k]^{\alpha\beta}Y_p^{\alpha\beta} + k^3p^3[\boldsymbol{c}_k\boldsymbol{X}_k]^{\alpha\beta}$ $[\boldsymbol{Y}_p\boldsymbol{c}_p]^{\alpha\beta} + p^3k^3[\boldsymbol{X}_k\boldsymbol{c}_k]^{\alpha\beta}[\boldsymbol{c}_p\boldsymbol{Y}_p]^{\alpha\beta} + p^3p^3X_k^{\alpha\beta}[\boldsymbol{c}_p\boldsymbol{Y}_p\boldsymbol{c}_p]^{\alpha\beta}$. If $\boldsymbol{X}_k$ and $\boldsymbol{Y}_p$ obey the symmetries (5.120a), one can discuss each of the four terms as done in the preceding paragraph. Matrix $\boldsymbol{\psi}$ has the same symmetries and the eigenvalues read

$$\psi^{\pm} = \Big\{ \big[(k^3c_k^+ + p^3c_p^{\pm})^2 X_k^+ Y_p^{\pm}\big] + \big[(k^3c_k^- + p^3c_p^{\mp})^2 X_k^- Y_p^{\mp}\big] \Big\}/2. \quad (5.130b)$$

One concludes that Eqs. (5.108), (5.109b), (5.111b) define a matrix $\boldsymbol{m}_q(t)$, which exhibits the symmetries (5.120a) provided the functions $\boldsymbol{F}_k(t)$ have these symmetry properties. Observing Eq. (5.127b), one gets

$$m_q^+(t) = \mathcal{F}_q^+[\phi_k^+(t)] \quad (5.131a)$$

with a mode-coupling functional defined by

$$\mathcal{F}_q^+[x_k] = [\rho S_q^+/(32\pi^3 q^2)] \int d^3\vec{k}(k^3c_k^+ + p^3c_p^+)^2 S_k^+ S_p^+ x_k x_p. \quad (5.131b)$$

Here, $\vec{p}$ abbreviates $\vec{q}^* - \vec{k}$ and $k^3 = (\vec{q}^*/\vec{k}/q)$, $p^3 = (\vec{q}^*/\vec{p}/q)$. If $\boldsymbol{F}_q^- = \boldsymbol{F}_{sq}^-$, the second kernel is given by that for the tagged-molecule dynamics

$$\boldsymbol{m}_q^-(t) = \boldsymbol{m}_{sq}^-(t). \quad (5.132)$$

In the following discussions, it shall be anticipated that wave numbers are chosen on grids of M values. Integrals are considered as equivalent to the corresponding Riemann sums as was explained before. In particular, mode-coupling functionals are used as abbreviations of polynomials. One concludes that Eqs. (5.125a), (5.131a,b) define a microscopic MCT model for an array $\phi_q^+(t)$, $q = 1, \ldots, M$, as it is discussed for simple systems in Chapter 4. Hence, these equations define unique solutions $\phi_q^+(t)$; and all general theorems on MCT solutions are valid for this array. Knowing $\phi_q^+(t)\, q = 1, \ldots, M$, Eqs. (5.125b), (5.129b) are closed; they have the same structure as those discussed in Sec. 5.1.3 for the tagged-particle dynamics. Hence, these equations determine uniquely arrays $\phi_{sq}^{\pm}(t)$, $q = 1, \ldots, M$, which have all the general properties of MCT solutions. The specified solutions can be used to define correlator matrices $\boldsymbol{F}_q(t)$ and $\boldsymbol{F}_{sq}(t)$ via Eqs. (5.124a), (5.120d). Since the derivations of the preceding paragraphs can be reversed, one concludes that these matrices are solutions of the MCT equations from Sec. 5.3.1. Since these solutions are determined uniquely, they are the MCT results for the symmetric dumbbell system. The factorization ansatz, which leads to the equations of Sec. 5.3.1, is consistent with the general implications of the symmetry group generated by $\mathcal{S}$ and $\mathcal{S}_s$, $s = 1, \ldots, N$. The equations of motion (5.125a,b) together with the formulas for the mode-coupling polynomials $\mathcal{F}_{sq}$ and $\mathcal{F}_q^{\pm}$ define the MCT for systems of symmetric two-centre molecules (Chong and Götze 2002a).

The generalized-hydrodynamics results for the tagged-molecule motion are determined by the mean-squared-displacement function $\Delta_C(t)$ of the molecules' centres and by the dipole correlators $C_1(t)$ of the particles. These concepts are of interest for the interpretation of the solutions of the above derived equations. Let us complete the reformulations of the equations for symmetric-dumbbell systems by deriving the equations, which are relevant for the evaluation of the two mentioned functions (Chong *et al.* 2001b).

The intramolecular structure factor w_q^+ has a positive zero-wave-number limit: $w_q^+ = 2 + O(q^2)$. Therefore, the functional $\mathcal{F}_{sq}^+$ exhibits a $(1/q^2)$ divergency for small q, which shall be quantified by a functional $\mathcal{F}_s^+$:

$$\lim_{q\to 0} (qd)^2 \mathcal{F}_{sq}^+[x_k, y_p] = \mathcal{F}_s^+[x_k, y_p], \tag{5.133a}$$

$$\mathcal{F}_s^+[x_k, y_p] = [\rho d^2/(6\pi^2)] \int_0^\infty dk k^4 c_k^{+2} S_k^+ w_k^+ x_k y_k. \tag{5.133b}$$

The kernel $m_{sq}^+(t)$ exhibits a $(1/q^2)$ divergency for q tending to zero, which shall be specified by a dimensionless function $m_s^+(t)$:

$$\lim_{q\to 0} (qd)^2 m_{sq}^+(t) = m_s^+(t), \qquad m_s^+(t) = \mathcal{F}_s^+[\phi_k^+(t), \phi_{sp}^+(t)]. \tag{5.133c}$$

The structure factor w_q^- vanishes with $q : w_q^- = (qL_D)^2/6 + O(q^4)$. Hence, the factor in front of the integral for $\mathcal{F}_{sq}^-$ in Eq. (5.129b) remains finite for vanishing wave number. The zero-q limit of $\mathcal{F}_{sq}^-$ exists and shall be denoted by $\mathcal{F}_s^-$:

$$\lim_{q\to 0} \mathcal{F}_{sq}^-[x_k, y_p] = \mathcal{F}_s^-[x_k, y_p], \tag{5.134a}$$

$$\mathcal{F}_s^-[x_k, y_p] = [\rho d^2/(72\pi^2)]\zeta^2 \int_0^\infty dk k^4 c_k^{+2} S_k^+ w_k^- x_k y_k. \tag{5.134b}$$

The zero-wave-number limit of kernel $m_{sq}^-(t)$ exists and is determined as mode-coupling functional:

$$\lim_{q\to 0} m_{sq}^-(t) = m_s^-(t), \qquad m_s^-(t) = \mathcal{F}_s^-[\phi_k^+(t), \phi_{sp}^-(t)]. \tag{5.134c}$$

The kernels $m_s^{\pm}(t)$ are correlation functions of fluctuating forces of diverging wavelengths, which act on the tagged molecule. These forces are produced by fluctuations of the distances between the interaction sites of the tagged molecule and those of all its neighbours. Therefore, it is necessary to know the correlators $\phi_q^+(t)$ and $\phi_{sq}^{\pm}(t)$ for all wave numbers q in order to evaluate the kernels $m_s^{\pm}(t)$.

The small-q limit of $F_{sq}^-(t) = (\rho_{s\vec{q}}^-(t)|\rho_{s\vec{q}}^-)$ is obtained by expanding the exponentials for $\rho_{sq}^- = \{\exp[i\vec{q}\vec{r}_{As}] - \exp[i\vec{q}\vec{r}_{Bs}]\}/\sqrt{2}$: $F_{s\vec{q}}^-(t) = L_D^2(\vec{q}\vec{e}_s(t)|\vec{q}\vec{e}_s)/2 + O(q^4) = (qL_D)^2(\vec{e}_s(t)\vec{e}_s)/6 + O(q^4)$. For the normalized correlator of the density fluctuations of odd exchange parity, one obtains

$$\phi_{sq}^- = C_1(t) + O(q^2). \tag{5.135a}$$

The leading term is given by the normalized dipole correlator

$$C_1(t) = (\vec{e}_s(t)|\vec{e}_s). \tag{5.135b}$$

This correlator is real. It describes the average of the projection of the molecule's axis on its initial value. $C_1(t) = \langle \cos \vartheta(t) \rangle$, where $\vartheta(t)$ denotes the angle between $\vec{e}(t)$ and $\vec{e}(t = 0)$. This correlator is an even function of the time because this is true for $\phi_{sq}^-(t)$. Assuming that time derivatives can be interchanged with the small-q limit, one derives from Eq. (5.125b) the equation of motion:

$$\partial_t^2 C_1(t) + \Omega_1^2\big[C_1(t) + \int_0^t m_s^-(t-t')\partial_{t'}C_1(t')dt'\big] = 0. \tag{5.136a}$$

The characteristic frequency is determined by the thermal angular velocity

$$\Omega_1^2 = 2\omega_{th}^2. \tag{5.136b}$$

The initial conditions are $C_1(0) = 1$, $\partial_t C_1(0) = 0$. Hence, one obtains the short-time behaviour $C_1(t) = 1 - \frac{1}{2}[\Omega_1 t]^2 + O(|t|^3)$.

The mathematical structure of the preceding formulas is the same as that discussed in connection with Eq. (4.32a). It is the structure obtained by the Zwanzig–Mori procedure for the correlator of a dynamical variable of definite time-inversion parity. The equation of motion with its initial conditions is equivalent to a double-fraction representation for the correlator in the frequency domain in terms of the Laplace transform of the kernel: $C_1(z) = -1\{z - \Omega_1^2/[z + \Omega_1^2 m_s^-(z)]\}$. Equation (2.82a) yields the equivalent expression for the normalized dynamical reorientational susceptibility $\chi_1(z) = (zC_1(z) + 1)$. It has the form of the response function for a harmonic oscillator, characterized by a resonance frequency Ω_1. A damping constant is generalized to a correlation function $m_s^-(z)$:

$$\chi_1(z) = -\Omega_1^2/[z^2 - \Omega_1^2 + z\Omega_1^2 m_s^-(z)]. \tag{5.137a}$$

For the discussion of glassy dynamics, this expression can be restricted to the low-frequency regime:

$$\chi_1(z) = 1/[1 - z m_s^-(z)], \qquad |z| \ll \Omega_1. \tag{5.137b}$$

This formula describes the result of coarse graining the dipole correlations on a time scale t^*, which is large compared to $1/\Omega_1$.

In the preceding two paragraphs and in the following text, the concept of a dipole is used with a mere geometrical meaning. If models of molecules with electrical dipole moments are considered, there appear two complications. The first one results from the violation of the local permutation symmetry. As explained above, this problem can be handled by using Eqs. (5.128b), (5.130b) in their full generality. The second problem is caused by the long range of the Coulomb interactions. As a result, the small-q limits of, for example, the current correlators deal with a singularity. One has to distinguish between longitudinal and transverse susceptibilities.

There are two kinds of mean-squared displacements characterizing the motion of a tagged symmetric dumbbell molecule. One refers to the constituents:

$$\delta r_s^2(t) = \langle [\vec{r}_{\alpha s}(t) - \vec{r}_{\alpha s}(0)]^2 \rangle. \tag{5.138a}$$

The other refers to the molecules' centres:

$$\delta r_C^2(t) = \langle [\vec{r}_{Cs}(t) - \vec{r}_{Cs}(0)]^2 \rangle. \tag{5.138b}$$

Abbreviating $\vec{\delta}_\alpha = [\vec{r}_{\alpha s}(t) - \vec{r}_{\alpha s}(0)]$, one can write $4[\vec{r}_{Cs}(t) - \vec{r}_{Cs}(0)]^2 = 2\vec{\delta}_A^2 + 2\vec{\delta}_B^2 - [\vec{\delta}_A - \vec{\delta}_B]^2$. Since $[\vec{\delta}_A - \vec{\delta}_B]^2 = L_D^2[\vec{e}_s(t) - \vec{e}_s(0)]^2 = 2L_D^2[1 - \vec{e}_s(t)\vec{e}(0)]$, one gets

$$\delta r_s^2(t) = \delta r_C^2(t) + \tfrac{1}{2} L_D^2 [1 - C_1(t)]. \tag{5.138c}$$

The mean-squared displacement of the interaction sites would be equal to that of the molecule's centre if the dipole correlator would be time independent. If the dipole correlations decay, $\delta r_s^2(t)$ exceeds $\delta r_C^2(t)$ by the contribution due to rotation of the molecule's axis.

One infers from Eqs. (3.29c), (3.30) that $\delta r_s^2(t)$ describes the small-q asymptote of the autocorrelators for the tagged-constituents-density fluctuations:

$$F_{sq}^{\alpha\alpha}(t) = 1 - q^2 \Delta_s(t) + O(q^4), \qquad \Delta_s(t) = \delta r_s^2(t)/6. \tag{5.139a}$$

In order to determine the small-q asymptote of the normalized correlator for the total density fluctuations $\rho_{\vec{q}}^A + \rho_{\vec{q}}^B = \rho_{\vec{q}}^+ \sqrt{2}$, one can start with Eq. (5.120d) and write: $\phi_{sq}^+(t) = [2F_{sq}^{AA}(t) - F_{sq}^-(t)]/w_q^+$. Using Eq. (5.135a), one gets $F_{sq}^-(t) = \phi_{sq}^-(t) w_q^- = (q^2 L_D^2/6) C_1(t) + O(q^4)$. Substituting $w_q^+ = 2 - (q^2 L_D^2/6) + O(q^4)$, one arrives at

$$\phi_{sq}^+(t) = 1 - q^2 \Delta_C(t) + O(q^4), \qquad \Delta_C(t) = \delta r_C^2(t)/6. \tag{5.139b}$$

The mean-squared-displacement function $\Delta_C(t)$ determines the small-wave-number limit of the normalized autocorrelator $\phi_{sq}^+(t)$ of the total density fluctuations.

Substituting the formulas (5.121d), (5.133c), (5.139b) into the equation of motion (5.125b) and assuming that the q-to-zero limit can be commuted with the differentiation operation, one obtains

$$\partial_t^2 \Delta_C(t) - v_{th}^2 + (v_{th}/d)^2 \int_0^t m_s^+(t-t')\partial_{t'}\Delta_C(t')dt'] = 0. \tag{5.140a}$$

This equation of motion has to be solved with the initial conditions $\Delta_C(0) = \partial_t \Delta_C(0) = 0$. The leading order behaviour is obtained in analogy to Eq. (3.31):

$$\Delta_C(t) = \tfrac{1}{2}(v_{th}t)^2 + O(|t|^3). \tag{5.140b}$$

Since $\partial_t \int_0^t m^+(t')\Delta_C(t-t')dt' = \int_0^t m^+(t-t')\partial_{t'}\Delta_C(t')dt' + m^+(t)\Delta_C(0)$, the integrodifferential equation from above together with its initial conditions leads to the alternative form for the equation of motion

$$\partial_t \Delta_C(t) + (v_{th}/d)^2 \int_0^t m_s^+(t-t')\Delta_C(t')dt'] = (v_{th})^2 t. \tag{5.140c}$$

This result agrees with Eq. (5.32a) for the mean-squared displacements in simple systems provided the contribution due to friction is dropped there.

(iii) Arrest in systems of symmetric hard dumbbells

Quantitative implications of the preceding MCT equations have been analysed for a system of dumbbells consisting of two fussed hard spheres of diameter d (Chong and Götze 2002a,b). Those calculations are based on a wave-number discretization for a grid of $M = 100$ values extending up to the cutoff $q_{co} = 40/d$. The structure factors and the direct correlation functions are calculated numerically within the reference-interaction-site-model theory (Hansen and McDonald 1986). For $\zeta = 0$, the results differ slightly from the ones using the analytical solutions of the Percus–Yevick equations for the hard-sphere system (HSS). In this subsection, the arrest scenario shall be discussed.

The equation of motion (5.125a) for $\phi_q^+(t)$ has the same form as Eq. (4.17a) for $\phi_q(t)$ if the regular-damping terms ν_q in the latter is specialized to zero. Also, the expressions (5.131a,b) for $m_q^+(t)$ are the same as Eqs. (4.11b,c) for $m_q(t)$, provided one drops c_3 in the latter and identifies S_q^+ and c_q^+ with S_q and c_q, respectively. Therefore, the discussion of the fixed-point equation for $f_q^+ = \phi_q^+(t \to \infty)$ can be done in analogy to that performed in Chapter 4 for simple systems. Within the high-density regime, there is pronounced intermediate-range order for the particle packing. This causes a first peak of the structure factor S_q^+ for q near $2\pi/d$. The system has a low compressibility, and this implies small values for S_q^+ for small wave numbers. The regime $qd < 3$ is of minor importance for the discussion of the mode-coupling effects. The excluded-volume effect causes oscillations of the $(S_q^+ - 1)$ versus q curves. Figure 5.22 demonstrates the specified structure properties for some representative elongations ζ. Increasing the packing fraction φ for fixed ζ, the structure-factor-peak height and the amplitudes of the

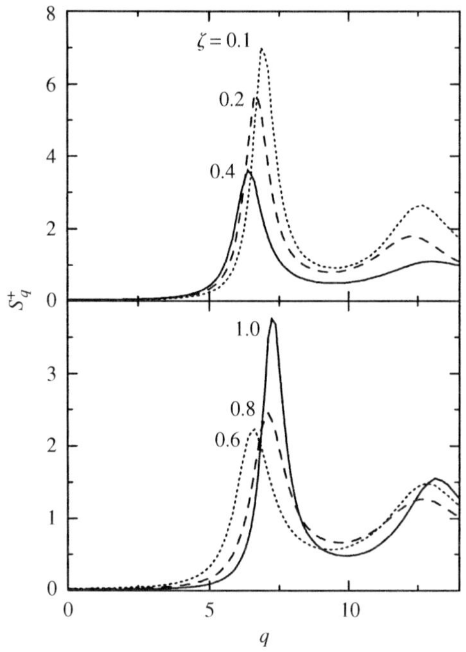

FIG. 5.22. Structure factors S_q^+ for the total-density fluctuations calculated within the reference-interaction-site-model theory for a system of symmetric hard-sphere dumbbells for packing fraction $\varphi = 0.56$ and elongations ζ as indicated. The unit for the wave number q is the inverse of the sphere diameters d. Reproduced from Chong and Götze (2002a).

$(S_q - 1)$-oscillations increase. This leads to an increase of the most relevant mode-coupling coefficients, as explained in more detail in Chapter 4 for the HSS. There occurs an A_2 bifurcation at a critical packing fraction $\varphi_c(\zeta)$. The critical arrested parts f_q^{+c} and the critical amplitude h_q^+ exhibit a similar q-dependence as shown in Fig. 4.6 for the HSS, and this for the same reasons as is explained in Sec. 4.3.5 for the simple system.

Since $\lim_{\zeta \to 0} w_q^+ = 2$, one gets $\lim_{\zeta \to 0} S_q^+ = \lim_{\zeta \to 0} w_q^+ / [1 - \rho w_q^+ c_q^+] = 2/[1 - \rho c_q] = 2S_q$. Here, S_q is the structure factor of the HSS for the packing fraction $\varphi = \pi(\rho d^3)/6$; and $c_q = 2 \lim_{\zeta \to 0} c_q^+$ denotes the corresponding direct correlation function. One checks that the mode-coupling functionals $\mathcal{F}_q^+$ reduce to those for the HSS. Consequently, for $\zeta = 0$, the φ_c versus ζ curve starts at the critical packing fraction φ_c^{HSS} of the hard-sphere system.

In a linear approximation for small ζ, the dumbbell volume increases with the elongation proportional to $[1 + 3\zeta/2]d^3$. To keep the packing fraction φ fixed, the density has to decrease proportional to $[1 - 3\zeta/2]/d^3$. As a result, there appear larger voids. The particles nearly have the shape of ellipsoids with two equal small axis of length d. Thus, a just-arrested structure gets destabilized. To compensate

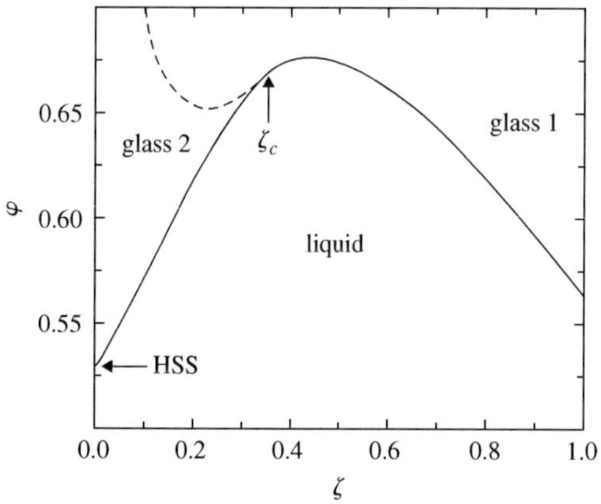

FIG. 5.23. Lines of glass-transition singularities for a system of symmetric hard-sphere dumbbells (see text for details). The solid line shows A_2-bifurcation points $\varphi = \varphi_c(\zeta)$ for the long-time limits of the total-density-fluctuation correlator $\phi_q^+(t)$ describing discontinuous liquid–glass transitions. The dashed line shows degenerate A_2-bifurcation points $\varphi = \varphi_c'(\zeta)$ describing the continuous transitions between glass 1 and glass 2. The two lines merge for the critical elongation $\zeta_c = 0.345$ as indicated by the vertical arrow. The horizontal arrow marks the critical packing fraction φ_c^{HSS} for the hard-sphere system. Reproduced from Chong and Götze (2002a).

this effect, the critical packing fraction has to increase with ζ. This explains the trend, which is exhibited in Fig. 5.23 for $\zeta < 0.3$. The preceding reasoning is reflected by the following properties of the mode-coupling coefficients. Upon dilution of the system of nearly spherical particles, the average distance between the particles increases. The effect of packing restrictions decreases. The shell structure for the pair distribution gets washed out. Hence, the hight of the first-sharp-diffraction peak decreases and its position shifts to smaller wave numbers. Both effects are demonstrated in the upper panel of Fig. 5.22. The first effect decreases the value for important mode-coupling coefficients and the second one reduces their phase-space and gradient factors.

For $\zeta = 1$, the system consists of dimers formed of hard spheres in touch. Compared to the system of unbonded hard spheres of the same packing fraction, the dumbbell system is more inhomogeneous. There are larger voids. The dimerization destabilizes the cages, which localize the particles for $\varphi = \varphi_c^{\mathrm{HSS}}$. In order to compensate this effect, the critical packing fraction has to increase, i.e., $\varphi_c(\zeta = 1) > \varphi_c^{\mathrm{HSS}}$. The destabilization of the arrest due to dimerization has been discussed in more detail for the square-well system in Sec. 4.3.6. In the latter case, bonding is a transient phenomenon caused by short-ranged attraction. For the molecular system with $\zeta = 1$, bonding is a permanent phenomenon caused by intramolecular forces. The critical packing fraction $\varphi_c(\zeta = 1)$ exceeds φ_c^{HSS}

by about 6%. This increase of φ_c due to bonding is much larger than the one demonstrated in Fig. 4.9. The arrest line exhibits an asymmetry in the sense that $\varphi_c(\zeta \approx 0) < \varphi_c(\zeta \approx 1)$.

Writing $\zeta = 1-\delta\zeta$, $\delta\zeta \geq 0$, one gets for the packing fraction $\varphi = (2\pi\rho d^3/6)[1-3\delta\zeta^2/4 + \delta\zeta^3/4]$. If φ is kept fixed, there is no leading-order change of ρ caused by a small decrease $\delta\zeta$ of the elongation. Decreasing ζ for fixed average distance between the centres of the molecules implies that the average distance between the interaction sites of different molecules increases. Hence, the voids between the particles increase. This destabilizes just-arrested structures. In order to compensate this effect, $\varphi_c(\zeta)$ has to increase with increasing $\delta\zeta$. This explains the behaviour, which is exhibited in Fig. 5.23 for $0.6 < \zeta$. The lower panel of Fig. 5.22 demonstrates how the mode-coupling functional reflects the reasoning from above. Increasing the average distance between interaction sites leads to a shift of the diffraction-peak position to smaller wave numbers. The increase of the voids weakens intermediate range order, thereby decreasing the peak height. Both effects increase with increasing $\delta\zeta$. These effects weaken the important mode-coupling coefficients the more the larger $\delta\zeta$.

The preceding discussions explain that the $\varphi_c(\zeta)$ versus ζ curve does not interpolate monotonicly between the values $\varphi_c(\zeta = 0)$ and $\varphi_c(\zeta = 1)$. Rather, the arrest curve exhibits an extremal point for $\zeta = \zeta_{\text{ex}} \approx 0.43$. Here, the system remains liquid for φ exceeding φ_c^{HSS} by more than 25%. For liquids with packing fractions between $\varphi_c(\zeta = 1)$ and $\varphi_c(\zeta = \zeta_{\text{ex}})$, there occurs a reentry phenomenon. Increasing ζ above zero, there occurs a transition from a glass formed by weakly elongated molecules to a liquid. Increasing further, the system reenters a glass; but this one is formed by strongly elongated particles.

To complete the discussion of the arrest scenarios, also the correlations for the fluctuations with negative exchange parity shall be considered. Because of Eq. (5.126), this is equivalent to a study of the correlations of the tagged-particle variables $\rho^-_{s\vec{q}}$. Let us discuss the correlations for both density fluctuations $\rho^\pm_{s\vec{q}}$. From Eqs. (5.125b), (5.129a,b), one derives the fixed-point equations for the long-time limits $f^\pm_{sq} = \phi^\pm_{sq}(t \to \infty)$: $f^\pm_{sq}/(1 - f^\pm_{sq}) = \mathcal{F}^\pm_{sq}[f^+_k, f^\pm_{sp}]$. Both equations have the same structure as the fixed-point equation (5.37b) for the arrest of the tagged particle dynamics in simple systems. To emphasize the specialities of these equations, they shall be noted in the form discussed in connection with Eqs. (5.42c), (5.43a):

$$f^\pm_{sq}/(1 - f^\pm_{sq}) = \sum_p U^\pm_{q,p} f^\pm_{sp}. \tag{5.141a}$$

Here, the coefficients $U^\pm_{q,p}$ of the linear mode-coupling polynomial are obtained by the discretization of the right-hand side of the following identity:

$$\sum_p U^\pm_{q,p} y_p = [w^\pm_q/q^2][\rho/(16\pi^3)] \int d^3\vec{k}(k^3 c^+_k)^2 S^+_k f^+_k w^\pm_p y_p. \tag{5.141b}$$

The solutions exhibit the maximum property, and there hold the other formulas, which are discussed in Chapter 4 for the fixed-point equation. If $f_k^+ = 0$, $k = 1, \ldots, M$, all coefficients $U_{q,p}^{\pm}$ vanish. In this case $f_{sq}^{\pm} = 0$, $q = 1, \ldots, M$. All correlations of density fluctuations vanish in the long-time limit. The states for $\varphi < \varphi_c(\zeta)$ are liquids.

For the states above the full line in Fig. 5.23, the coefficients $U_{q,p}^+$ are so large that the correlations for the total-density fluctuations ρ_{sq}^+ arrest:

$$f_{sq}^+ > 0, \qquad \varphi \geq \varphi_c(\zeta), \quad q = 1, \ldots, M. \tag{5.142a}$$

From Eq. (5.133a), one derives

$$f_{sq}^+ = 1 - (qr_C)^2 + O(q^4), \qquad (r_C d)^2 = 1/\mathcal{F}_s^+[f_k^+, f_{sp}^+]. \tag{5.142b}$$

Because of Eq. (5.139b), the length r_C determines the long-time limit of the mean-squared-displacement function for the molecules' centres

$$\Delta_C(t \to \infty) = r_C^2. \tag{5.142c}$$

The preceding formulas are the strict analogues of those discussed in Sec. 5.1.4 for simple systems. The molecules are localized in the random distribution of cages formed by the arrested density fluctuations. The form factors f_{sq}^+ are the Lamb–Mössbauer factors for the particles centres. The parameter r_C^2 is the mean-squared radius of the probability distribution of the localized molecules. It determines the arrested part for the centres' mean-squared displacements: $\delta r_C^2(t \to \infty) = 6r_C^2$.

The f_{sq}^+ versus q curves are bell shaped with a maximum unity for $q = 0$. With increasing elongation, there appear some wiggles for the curves. These are caused by intramolecular interference effects of the density fluctuations around the two interaction centres. These effects are described by the oscillating function w_q^+, which modulates the mode-coupling coefficients $U_{q,p}^+$.

For $\varphi \geq \varphi_c(\zeta)$ and sufficiently large ζ, also the mode-coupling coefficients $U_{q,p}^-$ are so big that the fixed-point equation for f_{sq}^- exhibits a non-trivial solution. These states shall be referred to as glass 1. Generically, there holds:

$$f_{sq}^- > 0, \quad q = 1, \ldots, M, \qquad \text{glass 1}. \tag{5.143a}$$

Also, the f_{sq}^- versus q curves are bell shaped with a maximum located at $q = 0$. The oscillations of the intramolecular structure factors w_q^-, which modulate the coefficients $U_{q,p}^-$, cause wiggles for sufficiently large ζ. But, there is a qualitative difference between the two form factors f_{sq}^+ and f_{sq}^-, which concerns the values for $q = 0$. The small-q divergency of $\mathcal{F}_{sq}^+$ causes f_{sq}^+ to exhibit the maximal possible value unity for $q \to 0$. This limit is independent of the parameters φ and ζ. According to Eqs. (5.134a,b), the functionals $\mathcal{F}_{sq}^-$ behave regularly for small q. Hence, the form factor f_{sq}^- has a zero-wave-number limit f_1, which is smaller than unity. The value depends on the control parameters φ and ζ:

$$f_{sq}^- = f_1 + O(q^2) \tag{5.143b}$$

$$f_1 = \mathcal{F}_s^-[f_k^+, f_{sp}^-]/\left\{1 + \mathcal{F}_s^-[f_k^+, f_{sp}^-]\right\}. \tag{5.143c}$$

Because of Eq. (5.135a), f_1 denotes the arrested part of the dipole correlations

$$C_1(t \to \infty) = f_1. \tag{5.143d}$$

According to Eq. (2.88b), one concludes that the static reorientational susceptibility $\chi_1^0 = \lim_{\epsilon\to 0} \chi_1(i\epsilon) = \lim_{\omega\to 0} \chi_1'(\omega)$ is smaller than the thermodynamic one. The latter is normalized to unity.

$$\chi_1^0 = 1 - f_1. \tag{5.144a}$$

From Eq. (5.138c), one gets for the arrested part of the mean-squared displacements for the interaction sites

$$\delta r_s^2(t \to \infty) = 6r_C^2 + \tfrac{1}{2}L_D^2(1 - f_1). \tag{5.144b}$$

The long-time limits of the interaction-site-density correlators $F_q^{\alpha\beta} = F_q^{\alpha\beta}(t \to \infty)$ form matrices, which are equivalent to the diagonal matrices with the elements $f_q^{xy} = \delta^{xy}\phi_q^x(t \to \infty)S_q^x$. Consequently, the matrix $\boldsymbol{F}_q$ is positive definite; it has the two positive eigenvalues $f_q^{\pm}S_q^{\pm}$. The analogues statements hold for the matrix $\boldsymbol{F}_{sq} = \boldsymbol{F}_{sq}(t \to \infty)$, which has the two eigenvalues $f_{sq}^{\pm}w_q^{\pm}$.

Equation (5.141b) implies that $U_{q,p}^-$ vanishes with ζ proportional to ζ^4. The integral in this formula abbreviates a sum of a finite number of terms; and each term contains the two factors w_q^- and w_p^-. These factors vanish with ζ proportional to ζ^2. For decreasing elongation, all M^2 coefficients $U_{q,p}^-$ decrease to zero smoothly. From the discussion of Eq. (4.54), one concludes the following. For every fixed φ, there exists some positive $\zeta_0 = \zeta_0(\varphi)$ so that $f_{sq}^- = 0$ for all q and all elongations obeying $0 \leq \zeta < \zeta_0$. Such states for $\varphi \geq \varphi_c(\zeta)$ shall be called glass 2. These small-ζ states are characterized by the fact that the correlations for density fluctuations of negative exchange symmetry vanish for long times:

$$\phi_{sq}^-(t \to \infty) = 0, \qquad q = 1, \ldots, M, \quad \text{glass 2}. \tag{5.145a}$$

In particular, the dipole correlator relaxes to zero:

$$C_1(t \to \infty) = 0. \tag{5.145b}$$

The centres of the molecules arrest in an amorphous array, but the orientational correlations of the molecule vanish in the long time limit. The system is the amorphous analogue of a plastic crystal. The static dipole susceptibility equals the thermodynamic one,

$$\chi_1^0 = 1. \tag{5.145c}$$

The mean-squared displacements of the constituents of the molecule exceed r_C^2 by the amount due to equal partition on a sphere of diameter L_D:

$$\delta r_s^2(t \to \infty) = 6r_C^2 + L_D^2/2. \tag{5.145d}$$

The long-time limits of the site-density correlators are the elements of non-vanishing matrices $\boldsymbol{F}_q$ and $\boldsymbol{F}_{sq}$, which are degenerate. One eigenvalue of the matrices is zero and the other one is the positive number $f_q^+ S_q^+$ and $f_{sq}^+ w_q^+$, respectively.

There is some transition line, $\varphi = \varphi_c'(\zeta)$, which separates glass 1 from glass 2. Let ζ_c denote the elongation of the merging singularity of this line with the liquid-glass transition line, i.e., $\varphi_c'(\zeta_c) = \varphi_c(\zeta_c)$. The dashed line in Fig. 5.23 exhibits the result for the hard-dumbbell model specified above. The critical elongation $\zeta_c = 0.345$ is somewhat below ζ_{ex}. According to the discussion of a mathematically equivalent problem in connection with Eqs. (5.59a–d), the arrested parts f_q^- vanish for all q for $\varphi = \varphi_c'(\zeta)$. Generically, the glass 1-to-glass 2 transition is described by a degenerate A_2 bifurcation. Crossing the line from the glass 1 side to the glass 2 one, f_q^- varies continuously as described by Eq. (5.59c).

Let us characterize the glass states in the neighbourhood of the merging point of the two transition lines by the two small coordinates $\hat{\epsilon} = (\zeta - \zeta_c)/\zeta_c$ and $\epsilon = (\varphi - \varphi_c(\zeta))/\varphi_c(\zeta)$, $\epsilon \geq 0$. Because of Eq. (4.91a), one gets $f_k^+ - f_k^{+c} = \tilde{h}_k \sqrt{\epsilon} + O(\epsilon)$. Hence, the coupling coefficients $U_{q,p}^-$ are smooth functions of the two variables $\hat{\epsilon}$ and $\sqrt{\epsilon}$. The non-degenerate maximum eigenvalue E of the stability matrix $A_{q,p} = U_{q,p}^-$ is a smooth function of the M^2 matrix elements. The condition $E = 1$ for the line of degenerate-A_2-bifurcation points reads $a\hat{\epsilon} + b\sqrt{\epsilon} = 0$. Here, a and b are smooth functions of $\hat{\epsilon}$ and $\sqrt{\epsilon}$. Generically, they are non-zero. The transition line is given by $\hat{\epsilon} = -(b/a)\sqrt{\epsilon}$. In leading order, the line is half of a parabola in the $\hat{\epsilon} - \epsilon$ plane. The two transition lines merge with the same tangent but different curvature. For the hard-sphere-dumbbell model, the lines approach each other for $\hat{\epsilon} < 0$. This holds because a decrease of ζ can be compensated by an increase of φ so that the arrest is not destroyed.

There are two types of glasses for $\varphi > \varphi_c(\zeta)$. Consequently, there are also two types of transitions from the liquid to the glass if the distance parameter $\epsilon = (\varphi - \varphi_c(\zeta))/\varphi_c(\zeta)$ increases from negative to positive values. Let us specialize some results from Sec. 4.3.4 in order to summarize the preceding discussions. As explained in Sec. 5.1.3, the density-fluctuation dynamics can be described by a $2M$-component array of correlators $(\phi_1^+(t), \dots \phi_M^+(t), \phi_1^-(t), \dots, \phi_M^-(t))$. This is equivalent to describing the dynamics by an M-component array of 2-by-2 matrices $(\boldsymbol{F}_1(t), \dots, \boldsymbol{F}_M(t))$. There are recipes for evaluating the separation parameter $\sigma = C\epsilon + O(\epsilon^2)$ and a parameter λ^c for the characterization of the transition. Like the critical packing fraction φ_c, the positive numbers C and λ^c depend on ζ. It is a speciality of the problem that the Eqs. (5.125a), (5.131a,b) formulate a MCT model for the M-component array $(\phi_1^+(t), \dots, \phi_M^+(t))$. Therefore, the parameters $\varphi_c(\zeta)$, σ and λ^c are determined solely by the mode-coupling polynomials $\mathcal{F}_q^+$.

The transition from liquid to glass 1 exhibits the generic properties of a fold bifurcation. The critical arrested parts of the correlators are positive

$$f_q^{+c} > 0, \, f_q^{-c} > 0, \tag{5.146a}$$

and there are positive critical amplitudes

$$h_q^{+c} > 0,\ h_q^{-c} > 0; \quad q = 1, \dots, M, \quad \zeta > \zeta_c. \tag{5.146b}$$

A square-root law describes the leading-order increase of the arrested parts within the glass according to Eq. (4.91a):

$$f_q^{\pm} = f_q^{\pm c} + h_q^{\pm}\sqrt{C\epsilon/(1-\lambda^c)} + O(\epsilon). \tag{5.146c}$$

Analogous formulas hold for the tagged-molecule-density correlators. Expansion for small wave numbers according to Eqs. (5.142b) yields for the localization length of the molecules' centres

$$(r_C)^2 = (r_C^c)^2 - h_{\rm MSD}\sqrt{C\epsilon/(1-\lambda^c)} + O(\epsilon). \tag{5.146d}$$

From Eq. (5.142d), one gets $(r_C^c d)^2 = 1/\mathcal{F}_s^+[f_k^{+c}, f_{sp}^{+c}]$ and an expression for the positive critical amplitude $h_{\rm MSD}$ analogous to formula (5.57b). Similarly, Eqs. (5.143b,c) yield the formulas for the arrested dipole correlations

$$f_1 = f_1^c + h_1\sqrt{C\epsilon/(1-\lambda^c)} + O(\epsilon). \tag{5.146e}$$

Here, $f_1^c = 1/\{1 + 1/\mathcal{F}_s^-[f_k^{+c}, f_{sp}^{-c}]\}$; and the positive critical amplitude reads

$$h_1 = (1 - f_1^c)^2\{\mathcal{F}_s^-[h_k^+, f_{sp}^{-c}] + \mathcal{F}_s^-[f_k^{+c}, h_{sp}^-]\}. \tag{5.146f}$$

For the transition from the liquid to glass 2, the density-fluctuations with even exchange parity exhibit the generic A_2-bifurcation scenario. But, there is no arrest for the variables with odd parity.

$$f_q^{+c} > 0,\ f_q^{-c} = 0, \quad q = 1, \dots, M, \quad \zeta < \zeta_c. \tag{5.147a}$$

There are positive critical amplitudes h_q^+, h_{sq}^+, and $h_{\rm MSD}$ so that the above noted asymptotic formulas for the variation of f_q^+, f_{sq}^+, and r_C remain valid. However, close to the transition line, the correlations for the odd-parity variables relax to zero like in a liquid

$$\phi_q^-(t \to \infty) = \phi_{sq}^-(t \to \infty) = C_1(t \to \infty) = 0, \quad \varphi_c(\zeta) \le \varphi \le \varphi_c'(\zeta). \tag{5.147b}$$

Considered as function of the elongation ζ, the critical arrested parts f_{sq}^{+c} and the critical amplitude h_{sq}^+ exhibit some wiggles. This holds, provided q is not too small. These wiggles are caused by the oscillations of the intramolecular structure factors w_q^+, as mentioned above in a similar context.

The f_q^{-c} versus ζ curves and the h_q^- versus ζ ones are exhibited in Fig. 5.24 for a representative set of wave numbers. There are three regimes, which exhibit qualitatively different behaviour. The regime $\zeta < \zeta_c$ deals with very weak steric hindrance for reorientational motion. The molecules are localized in cages. However, since the shape of the particles deviates little from a sphere, the torsional

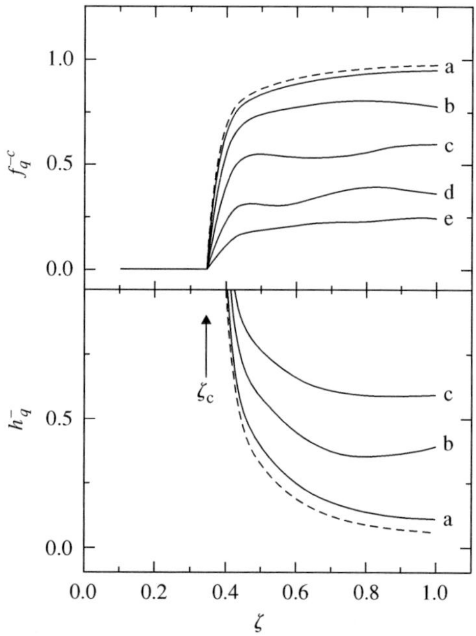

FIG. 5.24. Critical arrested parts f_q^{-c} and critical amplitudes h_q^- as a function of the elongation ζ, $\zeta \geq \zeta_c = 0.345$, for packing fractions $\varphi = \varphi_c(\zeta)$. The dashed lines exhibit the $q = 0$ limits f_1^c and h_1, respectively. The full lines refer to $qd = 3.4$ (a), 7.0 (b), 10.6 (c), 14.2 (d), and 17.4 (e). The results have been calculated for the hard-sphere-dumbbell model, which is explained in the text. Reproduced from Chong and Götze (2002a).

forces are too small to enforce arrest. The dipole correlator relaxes to zero for two reasons. First, the static forces detune the characteristic frequency for the rotation. Averaging over the distribution of frequencies, which is caused by the distribution of cages, yields relaxation to zero via a dephasing mechanism. Second, the dynamics of the solid matrix implies time dependent torques, which destroy the correlations of the dipole orientations.

Both mentioned relaxation mechanisms are of relevance also for $\zeta > \zeta_c$. But, in this case, the static fluctuations are strong enough for producing arrested parts: $f_q^{-c} > 0$, $f_1 > 0$. The second regime deals with large elongation, say, $\zeta > 0.6$. There is strong steric hindrance for reorientational motion. In a simple hard-sphere system, a particle has a critical localization length $r_s^c \approx 0.1d$. A motion of the interaction sites of a $\zeta = 1$ dumbbell over the mentioned length alters $\cos\vartheta$ by less than 0.01. Hence, the arrested part f_1 for $\zeta \approx 1$ is close to unity. The regime is characterized by a large value of f_1, say, larger than 90%. With increasing wave number, f_q^{-c} approaches f_q^{+c} and it exhibits the wiggling ζ-dependence mentioned above.

Third, there is the regime of weak steric hindrance for reorientational motion, say, $0.35 \leq \zeta \leq 0.45$. Here, the dynamics is influenced strongly by the precursor phenomena of the transition to glass 2. The arrest parameters f_q^{-c} and f_1^c decrease with ζ as described by Eq. (4.98). Simultaneously, the critical amplitudes get large values. For the hard-dumbbell system, a less-than-20% increase of ζ above ζ_c implies an increase of f_1 from 0 to 0.75.

This subsection shall be closed with six paragraphs of addenda. First of all, the explanations given above for the general features of the arrest diagram are not restricted to the model of fused-hard-sphere molecules. All systems of linear molecules with top-down symmetry and strong-repulsive-interaction parts exhibit similar results. Indeed, the pronounced maximum of the φ_c versus ζ curve and the existence of a continuous glass 1-to-glass 2 transition have first been predicted by Letz *et al.* (2000) for a system of elongated two-axial hard ellipsoids. This work is done within the framework of the molecular mode-coupling theory mentioned at the beginning of this section 5.3.

Molecular-dynamics-simulation results have been obtained for a binary mixture of 1000 symmetric dumbbells (Chong *et al.* 2005; Moreno *et al.* 2005). A mixture rather than a one-component system has been analysed in order to bypass nucleation phenomena. States have been studied for fixed packing fraction so that the temperature T and the common elongation ζ of the molecules fix the thermodynamic state. The mixture consists of 80% big interaction sites and 20% small ones. The Lennard-Jones potential from Eq. (1.8a–c) is used to describe the interactions between the sites of different molecules. All sites have the same mass μ. For $\zeta = 0$, the system is identical with the binary Lennard-Jones mixture discussed in Figs. 1.4 and 1.9, except for a reduction of the thermal velocity by a factor $1/\sqrt{2}$. The parameter ζ has been varied between 0.125 and 0.85. For the $\zeta - T$ plane of states, six iso-diffusivity lines have been constructed for diffusivities decreasing by almost three orders of magnitude. The data suggest an arrest line $T_c = T_c(\zeta)$ with a pronounced extremal point for $\zeta_{\text{ex}} \approx 0.55$ and an asymmetry in the sense that $T_c(\zeta \approx 0.8) < T_c(\zeta \approx 0.2)$. This asymmetry and the value $T_c(\zeta \approx \zeta_{\text{ex}})/T_c(\zeta \approx 0.8)$ are somewhat larger than one would estimate from Fig. 5.23 under the assumption that an increase of $\varphi_c(\zeta)$ is equivalent to a decrease of $T_c(\zeta)^{1/4}$.

There are two reasons for a subtle ζ-dependence of the dipole dynamics. First, there is the elongation dependence of the density-fluctuation correlators. It is described by the ζ-dependence of the functions $\phi_q^+(t)$. This dependence explains the non-trivial full line in Fig. 5.23 for the liquid–glass transitions. Second, there is the ζ-dependence of the rotation-translation coupling, which is described by the elongation dependence of the functionals $\mathcal{F}_q^-$. The latter dependence explains the existence of the dashed line in Fig. 5.23 and its non-trivial merging at $\zeta = \zeta_c$. One can disentangle qualitatively the two reasons if one considers also a model for a single dumbbell molecule moving in a simple system. The equations for the tagged-molecule functions $\phi_{sq}^{\pm}(t)$, $\Delta_C(t)$, and $C_1(t)$ remain the same as formulated above. But the functions $\phi_k^+(t)$ and f_k^+ for the total-density-fluctuation dynamics have to be replaced by the density correlators $\phi_k(t)$ and the Debye–Waller factors f_k, respectively, of the simple solvent. The correlators of the cage-forming solvent particles are independent of the control parameter ζ. But, the coupling of the dipole fluctuations with the density fluctuations depends on the solute's elongation. This ζ-dependence enters via that of the functions w_q^-, w_p^- and c_k^+ in Eq. (5.129b) for the mode-coupling functional. Results for this model have been analysed for a dumbbell of fused hard spheres of diameter d moving in a system of hard spheres with the same size (Chong *et al* 2001a,b). For this model, the transition function $\varphi = \varphi_c(\zeta)$ is replaced by the constant $\varphi_c^{\rm HSS}$. The function $\varphi = \varphi_c'(\zeta)$ is decreasing monotonically. The arrest mechanism for reorientations is less effective than in a dumbbell liquid and, therefore, the critical elongation $\zeta_c = 0.380$ is somewhat larger than demonstrated in Fig. 5.23. The f_1 versus ζ curve increases less steeply than shown in Fig. 5.24. The elongation ζ has to exceed the critical value ζ_c by about 50% before f_1^c reaches 0.75.

The preceding MCT equations for the $2M$-component arrays $\tilde{\boldsymbol{\phi}}(t) = (\boldsymbol{\phi}^+(t), \boldsymbol{\phi}^-(t))$ are specified by fluctuating-force kernels $\tilde{\boldsymbol{m}}(t) = \tilde{\boldsymbol{\mathcal{F}}}[\boldsymbol{\phi}^+(t), \boldsymbol{\phi}^-(t)]$. In this paragraph bold letters and bold letters with tilde are used as defined in Chapter 4 and Sec. 5.1.3, respectively. It has been a great simplification of the discussion that the function $\boldsymbol{\mathcal{F}}^+[\boldsymbol{x}, \boldsymbol{y}]$ does not depend on $\boldsymbol{y}$. The array $\boldsymbol{\phi}^-(t)$ for the fluctuations with odd exchange parity does not enter the equation for the array $\boldsymbol{\phi}^+(t)$, which describes even-parity fluctuations. However, this simplification is not essential for the existence of the two glass states nor for the continuous transition from one to the other. Rather, the essential condition is the validity of the equation

$$\mathcal{F}_q^-[\boldsymbol{x}, \boldsymbol{0}] = 0, \quad q = 1, \ldots, M, \tag{5.148}$$

for all arrays $\boldsymbol{x}$. For quadratic mode-coupling functionals, the essential condition is that the fluctuating force kernels $m_q^-(t)$ do not contain couplings to pairs $\phi_k^+(t)\,\phi_p^+(t)$. If this is true, there are fixed points $\tilde{\boldsymbol{f}}^* = (\boldsymbol{f}^+, \boldsymbol{0})$ with a non-trivial $\boldsymbol{f}^+$, provided the coupling coefficients for the functional $\boldsymbol{\mathcal{F}}^+$ are sufficiently large. If the coefficients of $\boldsymbol{\mathcal{F}}^-$ are sufficiently small, this is the fixed point, which characterizes a glass 2. If the coefficients of $\boldsymbol{\mathcal{F}}^-$ are sufficiently large, $\tilde{\boldsymbol{f}}^*$ is not the arrested part of $\tilde{\boldsymbol{\phi}}(t)$; there is a glass 1 with the long-time limits of all density-fluctuation correlators being positive. If $\boldsymbol{\mathcal{F}}^+[\boldsymbol{x}, \boldsymbol{y}]$ does not depend on $\boldsymbol{y}$, $\boldsymbol{f}^+$ behaves smoothly at the transition from glass 1 to glass 2. If the kernel $\boldsymbol{m}^+(t)$ contains contributions from $\boldsymbol{\phi}^-(t)$, also $\boldsymbol{f}^+$ exhibits singularities at the transition points between the two glasses (Franosch and Götze 1994).

Let us consider a dumbbell system, which exhibits the global but not the local exchange symmetry. In this case, the direct correlation functions c_q^- do not vanish. Equation (5.130b) shows that there are $X_k^- Y_p^-$ contributions to ψ^+. Hence, there are contributions to the kernel $\boldsymbol{m}^+(t)$, which are proportional to $\phi_k^-(t)\phi_p^-(t)$. But there are no $X_k^+ Y_p^+$ contributions to ψ^-. Hence, there holds

Eq. (5.148); two different glasses are possible. The schematic two-component model, which is defined by Eqs. (4.34a–c), exhibits the above described possibilities. One has to identify $\phi_1(t)$ with $\phi_q^+(t)$ and $\phi_2(t)$ with $\phi_q^-(t)$. There are glass 2 states for $v_3 < 1$ (Götze and Haussmann 1988).

The two-component model defined by Eqs. (4.32), (4.33) provides a simple example, which demonstrates the bifurcation diagram for the symmetric dumbbell system or for the mentioned hard-dumbbell solute in a hard-sphere solvent. The first correlator $\phi(t)$ mimics the array $\boldsymbol{\phi}^+(t)$ of density-fluctuation correlators. The second correlator $\phi_A(t)$ mimics the array $\boldsymbol{\phi}^-(t)$, or its $q = 0$ limit $C_1(t)$. The mode-coupling functional $\mathcal{F}^+[x] = v_1 x + v_2 x^2$ is independent of the second correlator as formulated in Eqs. (5.131a,b). The second mode-coupling functional $\mathcal{F}^-[x, y] = v_A xy$ has the form requested by Eqs. (5.129a,b) with v_A characterizing the strength of the rotation–translation coupling. Let us write $v_{1,2} = v_{1,2}^c \xi$, with $v_{1,2}^c$ denoting the transition point for $\lambda^c = 3/4$ as explained in connection with Fig. 4.12. Hence, ξ mimics φ/φ_c. The critical Debye–Waller factor is $f^c = 1/4$. The discussion from above suggest that v_A should vary like ζ^4 for small elongations and like ζ^2 for large ones. The bifurcation diagram is shown in Fig. 4.4. The critical value for the rotation–translation coupling reads: $v_A^c = 4$. The region $v_A < v_A^c$ mimics the one for very weak steric hindrance for reorientations. For $v_A < 4$, there occur liquid to-glass 2 transitions. For $v_A > 4$, there occur transitions from a liquid to glass 1. The critical arrested part for the second correlator is given by:

$$f_A^c = 1 - (4/v_A). \tag{5.149}$$

This number mimics the critical arrested part f_1^c for the dipole dynamics. In order to reproduce f_1^c, which is shown in Fig. 5.24 for $\zeta = 1$, v_A has to exceed 80. The regime for weak steric hindrance is mimicked for $4 < v_A \leq 10$.

(iv) Glassy-dynamics features

In this subsection, it shall be shown that the glassy dipole relaxation of the liquid has qualitatively different features depending on whether there is very weak, weak, or strong steric hindrance for reorientational motion. The general trends are also exhibited by the above considered two-component model.

The discussion of the arrest scenario suggests that the evolution of the glassy-relaxation parts of the correlators $\phi_q^+(t)$ and $\phi_{sq}^+(t)$ for the total-density fluctuations is similar to that exhibited by the hard-sphere system. Figure 5.25 confirms this suggestion for the hard-sphere-dumbbell liquid for $\zeta = 1$. There is a stretched decay of the correlators towards their plateaus f_q^{+c} and f_{sq}^{+c}, respectively. For the same small negative distance parameter ϵ, all correlators cross their plateaus at nearly the same time τ^{pc}. For $t > \tau^{pc}$, there occurs a stretched decay to zero. In this subsection, a characteristic scale for the below-plateau decay shall be defined as the time needed for the 50% decay; for example $\phi_q^+(\tau_q^+)/f_q^{+c} = \phi_{sq}^+(\tau_{sq}^+)/f_{sq}^{+c} = C_1(\tau_1)/f_1^c = 1/2$. The figure demonstrates scale coupling: there is a common ϵ-sensitive scale τ so that τ_q^+/τ and τ_{sq}^+/τ are nearly

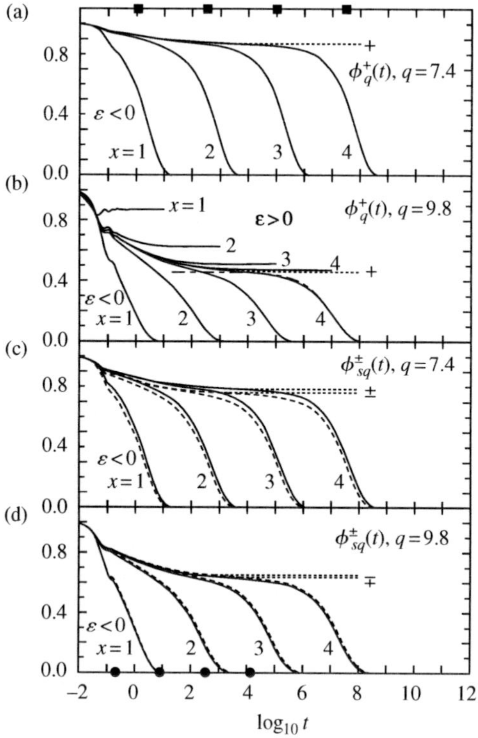

FIG. 5.25. Density-fluctuation correlators calculated within a microscopic MCT for a system of symmetric hard-sphere dumbbells with elongation $\zeta = 1$. The unit of length and time are chosen so that the sphere diameter d and the thermal velocity v_{th} are unity. Further details of the model are specified in the text. Panels (a) and (b) exhibit $\phi_q^+(t)$ for $q = 7.4$ and $q = 9.8$, respectively. The dotted lines refer to the critical point $\varphi = \varphi_c$ and the full ones to distance parameters $\epsilon = (\varphi - \varphi_c)/\varphi_c = \pm 10^{-x}$. The dashed-dotted lines in panel (b) shows a Kohlrausch-law fit to the $x = 4$ liquid correlator: $f_q^{+c} \exp[-(t/\tau)^{\beta}]$, $\beta = 0.68$. Panels (c) and (d) show for $q = 7.4$ and $q = 9.8$, respectively, liquid correlators $\phi_{sq}^+(t)$ as full lines and $\phi_{sq}^-(t)$ as dashed ones. The dotted lines show the results $\varphi = \varphi_c$.

The filled circles and squares mark the time scales for the first-scaling-law description, $t_\sigma = t_0/|\sigma|^{\delta}$ and $t'_\sigma = t_0 B^{-1/b}/|\sigma|^{\gamma}$, respectively. They are explained in Sec. 6.1.3(ii). The exponent parameter is $\lambda = 0.739$ and the various other parameters read: $\sigma = C\epsilon$ with $C = 1.90$ and $\varphi_c = 0.565$; $t_0 = 0.0139$, $\delta = 1/2a$ with $a = 0.310$; $B = 0.857$, $b = 0.576$ and $\gamma = (1/2a) + (1/2b) = 2.48$. Reproduced from Chong and Götze (2002b).

ϵ independent. There holds the superposition principle (4.112a) for percentages of the decay, which are the larger, the smaller is ϵ. The pattern described is exhibited also by $\phi^-_{sq}(t)$; for the large elongation considered and for large and intermediate wave numbers, $\phi^+_{sq}(t)$ is close to $\phi^-_{sq}(t)$.

The numerical solutions obtained for $\zeta = 0.4$ confirm the mentioned properties also for the weak steric-hindrance regime, albeit with two exceptions. First, since f_q^{-c} decreases to zero proportional to $\zeta - \zeta_c$, $\phi^-_{sq}(t)$ is below $\phi^+_{sq}(t)$. Second, the correlators $\phi^-_{sq}(t)$ exhibit severe deviations from the superposition principle (Chong and Götze 2002b).

All features mentioned in the preceding two paragraphs are exhibited also by the solutions calculated in the cited work on the dynamics of a single hard-sphere dumbbell moving in a hard-sphere liquid. This means that the indicated scenario for the reorientational dynamics is an implication of the rotation–translation coupling rather than of the ζ-dependence of the density correlators $\phi^+_q(t)$. This observation is a motivation to discuss the possibilities to describe the glassy dynamics of $C_1(t)$ by the results of the two-component model mentioned in connection with Eq. (5.149). The dashed line in Fig. 5.26 shows the first correlator of this model, which mimics $\phi^+_q(t)$. It reproduces the curve with label $n = 8$ from Fig. 4.12. The full lines show the second correlator of the schematic model. This correlator $\phi_A(t)$ mimics C_1. Increasing the coupling v_A, the states shift along a line in Fig. 4.4, which is parallel to the one for liquid–glass transitions. This scenario mimics the shift of the liquid states along an isodiffusivity curve. The values $v_A = v_A^c = 4$ mimics the value $\zeta = \zeta_c$.

The following discussion shall include also a qualitative comparison of MCT results with data obtained by the above cited molecular-dynamics study. Figure 5.27 reproduces correlators from this work. The time interval up to $10\,t_{\text{mic}} \approx 0.5$ deals with the transient dynamics. Glassy dynamics for $\phi^+_q(t)$ is documented for a time increase by about 4 orders of magnitude. For elongations 0.2 and 0.3, the density correlators $\phi^+_q(t)$ decay to 20% of their initial values for the same time close to 10^4. The long-time decay is isochronic for these two states. In this sense, the states in the two lower panels refer to the same distance parameter $(T - T_c(\zeta))/T_c(\zeta)$. But, the increase of ζ from 0.2 to 0.3 is accompanied by a decrease of T by about 40%. This elongation dependence of the critical points $T_c(\zeta)$ is in qualitative agreement with that exhibited in Fig. 5.23 for the small ζ system.

For $\zeta = 0.2$, Fig. 5.27 demonstrates that $C_1(t)$ relaxes to zero within the transient regime $t_{\text{mic}} < t < 10\,t_{\text{mic}}$. For this normal-liquid-state dynamics, there are no coherence effects; and, therefore, $C_1(t)$ is close to $\phi^-_q(t)$. For $\zeta = 0.3$, there is an appreciable rotation–translation coupling. The dipole correlations relax to zero quickly in comparison to the ones for density fluctuations. However, the cage-forming neighbours exhibit slow relaxation producing the towards-plateau decay of $\phi^+_q(t)$ within the interval $10\,t_{\text{mic}} < t < 1000\,t_{\text{mic}}$. This dynamics is probed by the dipole and it causes a glassy-relaxation tail for $10\,t_{\text{mic}} < t < 100\,t_{\text{mic}}$. The correlators $\phi_A(t)$, which are exhibited in Fig. 5.26,

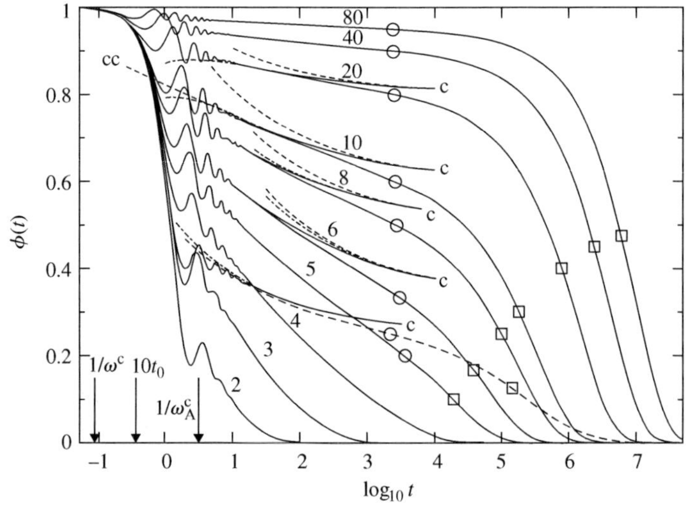

FIG. 5.26. Correlators calculated for the two-component model defined by Eqs. (4.32a,b), (4.33a,b) for $\Omega = \Omega_A = 1$ and $\nu = \nu_A = 0$. The dashed line shows correlator $\phi(t)$ for a state, which is explained in Fig. 4.12 for $n = 8$. The full lines exhibit correlator $\phi_A(t)$ for coupling constants $v_A = 2, 3, \ldots, 80$ as indicated. The circles mark the points where the correlators cross the value for their critical arrested parts $f^c = 1/4$ and $f_A^c = 1 - 4/v_A$, respectively. The squares mark the points, where the correlators decay to half of the critical arrested parts $f^c/2$ and $f_A^c/2$, respectively. The critical corelators $\phi^c(t)$ and $\phi_A^c(t)$ for $v_A = 6, 8, 10$ and 20 are shown as full lines with label c.

The corresponding asymptotes (6.21a) and the asymptotes including the leading corrections according to Eq. (6.30c) are shown as dashed and dashed-dotted lines, respectively. The dashed line with label cc shows the Cole–Cole correlator $\phi_A^{cc}(t)$ for $v_A = 10$, as determined by Eqs. (6.43b), (6.44a). The arrows mark the times $10\,t_0$, $1/\omega^c$ and $1/\omega_A^c$, which specify the critical decay and the characteristic frequencies of the Cole–Cole law (Sperl 2004a).

reproduce qualitatively the above identified features for the coupling constants $v_A = 2$ and $v_A = 3$, respectively. The microscopic time scale for the schematic model is $t_{\text{mic}} \approx 0.5$. The scenario demonstrated for the data and for the schematic model are the precursor phenomena for a transition from the liquid to a glass 2. These features of the dipole relaxation for very weak steric hindrance of reorientational motion are peculiarities, which are caused by the symmetry-induced specialities of the mode-coupling coefficients.

Suppose, the rotation–translation couplings are increased close to the critical values for the merging singularity. If one ignores the time dependence of the density-fluctuation correlator for $10t_{\text{mic}} < t < \tau_{pc}$, the mode-coupling polynomials

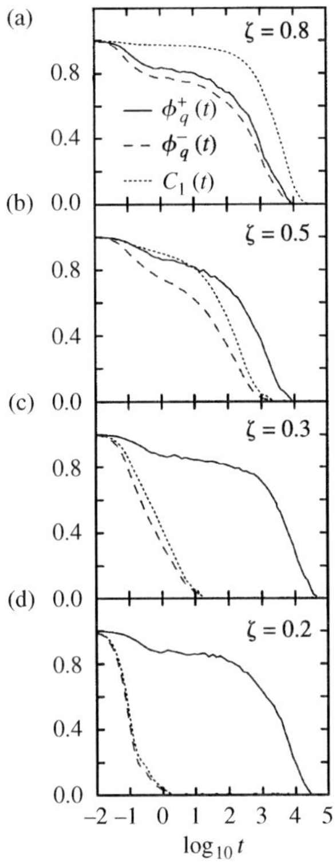

FIG. 5.27. Molecular-dynamics-simulation results for the correlators $\phi_q^{\pm}(t)$ of density fluctuations with a wave number close to the position of the first peak of the structure factor S_q^+ and for the dipole correlator $C_1(t)$. The study is done for binary mixtures of symmetric dumbbells at fixed packing fraction. The system and the units are explained in the text. The states in the four panels refer to the following pairs of elongation ζ and temperature T (from top to bottom): $\zeta = 0.8$, $T = 1.06$; $\zeta = 0.5$, $T = 0.75$; $\zeta = 0.3$, $T = 1.13$; $\zeta = 0.2$, $T = 1.85$. The figure is constructed (Chong 2005) from data obtained by Chong *et al.* (2005).

for the reorientational correlator become linear functions. For the schematic model, there holds $m_A(t) = v\phi_A(t), v = v_A f^c$. The equation of motion describes the degenerate A_2 bifurcation,which is discussed in Sec. 4.2.4. According to Eq. (4.51c), the correlator exhibits a $1/\sqrt{t}$ long-time tail. The transition between the regimes of very weak steric hindrance for reorientations to the one of weak steric hindrance is signalized by an upward-bent $C_1(t)$ versus $\log t$ curve, which stretches out up to about $t = \tau^{pc}$. This feature is shown in Fig. 5.26 for $v_A = 4$.

Increasing the rotation–translation coupling so that the state is located in the regime of weak steric hindrance for rotations, the MCT results for the intermediate-wave-number correlators $\phi_q^-(t)$ exhibit a plateau f_q^{-c}, which is below the plateau f_q^{+c} for the density fluctuations of even exchange parity. From Fig. 5.24, one expects a plateau f_1^c for the dipole correlator above the specified f_q^{-c}. However, the curves in Fig. 5.26 for $v_A = 5$ and 6 and the ones in Fig. 5.28 for $\zeta = 0.4$ demonstrate that the plateau f_1^c is not exhibited clearly. The curves do not have an inflection point for $C_1(t)$ somewhat above f_1^c. The specified MCT results are exhibited also by the simulation data, which are shown in Fig. 5.27 for $\zeta = 0.5$. Figure 5.26 shows that the time scales τ_A of the below-plateau decay of the second correlator for $v_A = 5$ and $v_A = 6$ are considerably smaller than the corresponding scale τ for the first correlator. This is a characteristic feature of the weak-coupling scenario, since the below-plateau decay develops continuously out of the above described correlator at the critical point v_A^c. Similarly, the time scales τ_q^- for intermediate-wave-number fluctuations of a dumbbell liquid with $\zeta = 0.4$ are more than 10 times smaller than the scales τ_q^+ (Chong and Götze 2002b). The results are similar to those for a single dumbbell with $\zeta = 0.43$ immersed in a hard-sphere liquid (Chong *et al.* 2001b). The data shown in Fig. 5.27 for $\zeta = 0.5$ exhibit this property as well.

A typical scenario for the evolution of the superposition principle in simple systems is shown in Fig. 4.21. For a distance parameter ϵ of about 10^{-3}, more than 90% of the below-plateau decay can be described by the rescaled shape function. Within the weak-steric hindrance regime, the asymptotic formulas for the normal A_2 bifurcation are strongly disturbed by the presence of a further glass-transition singularity. Figure 5.29 shows for $\zeta = 0.4$ and $\epsilon = -10^{-3}$ that only less than 70% of the below-plateau decay of $C_1(t)$ can be described by the master function $\tilde{C}_1(\tilde{t})$. For distance parameters relevant for data analysis, the scaling law is valid only for a minor part of the decay of the dipole correlations.

The two correlators $\phi_q^+(t)$ in panels (a) and (b) of Fig. 5.27 exhibit iso-chronic long-time decay. This suggests the same distance parameters $(T - T_c(\zeta))/T_c(\zeta)$. The reasoning is similar as done above for the pair of panels (c) and (d). But here, a decrease of ζ from 0.8 to 0.5 is accompanied by an decrease of T. This result is in qualitative agreement with the expectation from the decreasing φ_c versus ζ curve shown in Fig. 5.23 for the large elongation branch.

According to Fig. 5.25, MCT predicts for the strong steric-hindrance regime that intermediate-wave-number correlators $\phi_q^-(t)$ are close to but below correlators $\phi_q^+(t)$. This feature is exhibited in Fig. 5.27 by the simulation data for $\zeta = 0.8$. As explained above, the critical arrested part f_1^c for the large-ζ dipole correlations is close to unity. This is demonstrated explicitly in the lower panel of Fig. 5.28. This figure shows also that $C_1(t)$ is close to f_1^c for times exceeding the plateau-crossing time τ^{pc} by an order of magnitude. A plateau value cannot be deduced for the dipole correlator since the $C_1(t)$ versus $\log t$ curves do not exhibit an inflection point for $t < \tau_1$. These MCT features for the dipole relaxation are exhibited also for the schematic-model results in Fig. 5.26 for $v_A \geq 40$. And the simulation data for $\zeta = 0.8$ demonstrate the features as well.

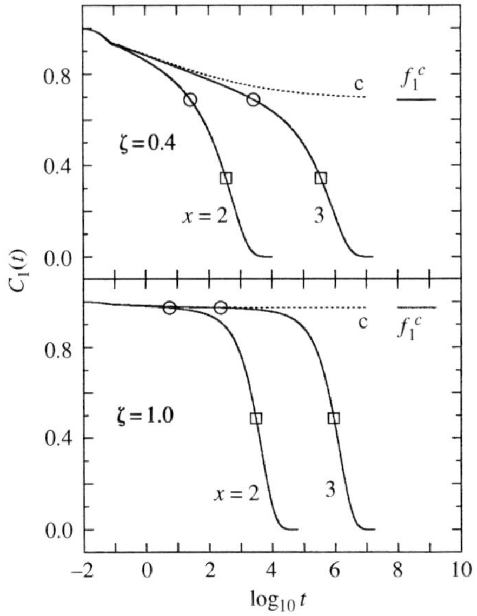

FIG. 5.28. Dipole correlators $C_1(t)$ calculated for a system of symmetric hard-sphere dumbbells. The MCT model and the units are defined in the text. The dotted lines marked with c, refer to the critical packing fraction φ_c, which is 0.675 and 0.565 for $\zeta = 0.4$ and 1.0, respectively. The full lines are liquid curves for distance parameters $\epsilon = (\varphi - \varphi_c)/\varphi_c = -10^{-x}$, $x = 2, 3$. The open circles mark the points of crossing of the critical arrested parts: $C_1(\tau^{pc}) = f_1^c$. The open squares mark the points of decay to half of the critical-arrested-part value: $C_1(\tau_1) = f_1^c/2$. The values $f_1^c = 0.69$ and 0.97 for $\zeta = 0.4$ and 1.0, respectively, are indicated by a horizontal line marked f_1^c. Reproduced from Chong and Götze (2002b).

From the lower panel of Fig. 5.28, one deduces $\tau_1 = 10^6$ as time scale for the below-plateau decay of the dipole correlations for $\epsilon = -10^{-3}$. From Fig. 5.25, one deduces for the scales τ_q^+ of the corresponding processes for the density fluctuations about 2×10^5 and 2×10^4 for $q = 7.4/d$ and $q = 9.8/d$, respectively. Figure 5.26 demonstrates for $v_A \geq 40$ that τ_A exceeds τ by more than an order of magnitude. Also the simulation data in Fig. 5.27(a) demonstrate that τ_1 is more than a factor 10 larger than τ_q^+. This separation of the time scale for the dipole relaxation from that for the relevant density fluctuations is an outstanding feature of the strong steric-hindrance scenario. Opposite to what is shown in Fig. 5.29 for the weak steric-hindrance scenario, the figure demonstrates also the prediction of a large range of validity of the superposition principle for the dipole correlator for the strong steric-hindrance scenario.

Stretching of the below-plateau decay can be quantified by the factor for the time increase, which is necessary for the decay from 90% to 10% of the plateau

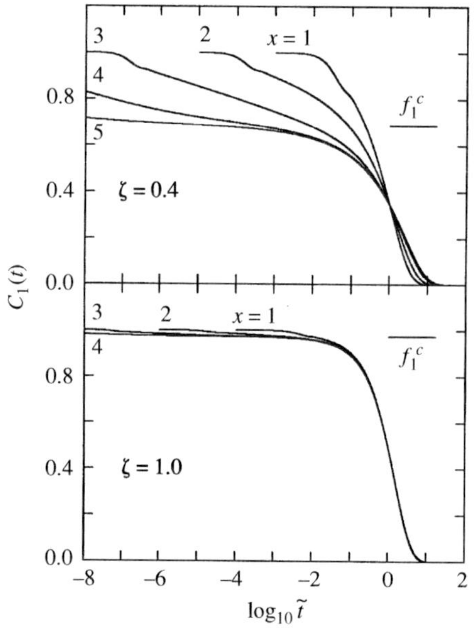

FIG. 5.29. Dipole correlators C_1 for the system of symmetric hard-sphere dumbbells with elongation ζ for liquid states with distance parameters $\epsilon = (\varphi - \varphi_c)/\varphi_c = -10^{-x}$, $x = 1-4$, presented as function of $\log \tilde{t}$. The rescaled time $\tilde{t} = t/\tau_1$ is formed with the decay time τ_1, which is specified by: $C_1(\tau_1) = f_1^c/2$. The critical arrested parts are shown by short horizontal lines marked f_1^c. Reproduced from Chong and Götze (2002b).

value. The simple relaxation process (1.1b) requires an increase of about 1.3 decades. Such increase is exhibited for the $\zeta = 1$ results in Fig. 5.28. These curves are very close to normalized exponentials $\exp[-\gamma t]$. The correlator, which is shown in Fig. 5.26 for $v_A = 80$, and also the simulation result for $C_1(t)$, which is shown in Fig. 5.27 for $\zeta = 0.8$, require a time increase of about 1.6 decades for the specified 80% of the long-time decay. These two examples exhibit stretching, albeit a very small one. The below-plateau decay is equivalent to a low-frequency-loss peak. As discussed before, stretching can be quantified by the exponent β of the Kohlrausch function (1.3), whose loss spectrum is fitted to the upper part of the peak of the susceptibility spectrum. For the cited calculations for a hard-sphere dumbbell moving in a hard-sphere liquid (Chong *et al.* 2001b), such fits yield $\beta = 0.97$ and 0.79 for the elongation $\zeta = 0.80$ and 0.43, respectively. While the small-ζ spectrum exhibits normal stretching, the large-ζ one can hardly be distinguished from the Lorentzian $\gamma\omega/[1 + (\gamma\omega)^2]$ spectrum of the elementary relaxation process. Small stretching of the below-plateau decay of the dipole correlator is another outstanding signature of strong steric-hindrance scenario for reorientational motion.

Let us recognize that the suppression of stretching for strong rotation– translation coupling is an implication of the above noticed separation of time scales. According to Eqs. (5.134a–d), the rate ω^- for the decay of the fluctuating-force correlations $m_s^-(t)$ can be estimated by $1/\tau_1 + 1/\tau_q^+$. For the relevant wave numbers, there holds $\tau_1 \gg \tau_q^+$. Consequently, $1/\tau_1 \ll \omega^-$. The loss-peak position near $1/\tau_1$ for the dipole response is located considerably below the position ω^- of the low-frequency peak of the $\omega m_s^{-\prime\prime}(\omega)$ versus $\log \omega$ curve. Within the frequency regime around $1/\tau_1$, which is of interest for the study of the dipole dynamics, the fluctuating-force spectrum can be treated as a white-noise one: $m_s^{-\prime\prime}(\omega) \approx \tau$. This means $m_s^-(z) \approx \pm i\tau$, Im $z \gtrless 0$. Substitution into Eq. (5.137b) yields the dynamical susceptibility $\chi_1(z) = 1/[1 \mp iz\tau]$. The time-scale separation property implies a Langevin-theory result as discussed in Eqs. (3.50a–f); and the validity of the superposition principle is a trivial implication.

The equations of motion for the correlators formed with tagged-particle-density fluctuations of even exchange parity are very similar to those for the density correlators of simple systems. Hence, the mean-squared-displacement functions $\Delta_C(t)$ for the hard-dumbbell liquid shares essential features with the function $\Delta_s(t)$ of the hard-sphere liquid. For example, the critical arrested parts of these functions differ by less than 10%. And both functions reach the diffusion asymptote at an onset time t_{on}, where they have a value of order $0.1d^2$. In all cases, the time t_{on} exceeds the time τ_q^+ for the below-plateau decay of the density fluctuations considerably. For example, the lower curve in the lower panel of Fig. 5.30 exhibit t_{on} near 10^6 for the state $\zeta = 1$, $\epsilon = -10^{-3}$. For the same state, one infers from Fig. 5.25 values for τ_q^+ near 10^5 and $10^{4.2}$ if $qd = 7.4$ and 9.8, respectively. The diffusion law is the hydrodynamic-limit result for the tagged molecule motion. In order to describe the mean-squared displacement by this law, the time has to exceed the coarse-graining scale τ_q^+.

For weak and very weak steric hindrance for the reorientational dynamics, the scale τ_1 for the long-time decay of the dipole correlator is shorter than τ_q^+ for the relevant intermediate wave numbers. Consequently, for times near and above t_{on}, one gets $C_1(t) = 0$. For $t \geq t_{\text{on}}$, Eq. (5.138c) yields $\Delta_s(t) = Dt + (\zeta d)^2/12$. The offset $(\zeta d)^2/12$ delays the approach of the $\Delta_s(t)/t$ versus $\log t$ curve towards its diffusion asymptote D. For $\zeta = 0.4$, Fig. 5.30 demonstrates this systematic difference between the above-plateau increase of the two functions $\Delta_s(t)$ and $\Delta_C(t)$.

The time-scale-separation property for the strongly-hindered reorientational motion implies a novel feature of the function $\Delta_s(t)$. As explained above, and as is demonstrated by a comparison of the lower panels of Figs. 5.28 and 5.30, the dipole-relaxation scale τ_1 is close to the onset time t_{on}. Consequently, for $t \geq t_{\text{on}}$, there holds: $\Delta_s(t) = Dt + (\zeta d)^2[1 - C_1(t)]/12$. If t increases through τ_1, the bracket provides a factor, which increases. This delays the approach of the $\Delta_s(t)/t$ versus $\log t$ curve towards its asymptote D. For $\zeta = 0.4$, the curve exhibits an onset time t_{on}^s for the diffusion behaviour at about $2\,t_{\text{on}}$. However, the $\zeta = 1.0$ result shows $t_{\text{on}}^s \approx 40\,t_{\text{on}}$. Prior to the diffusion regime,

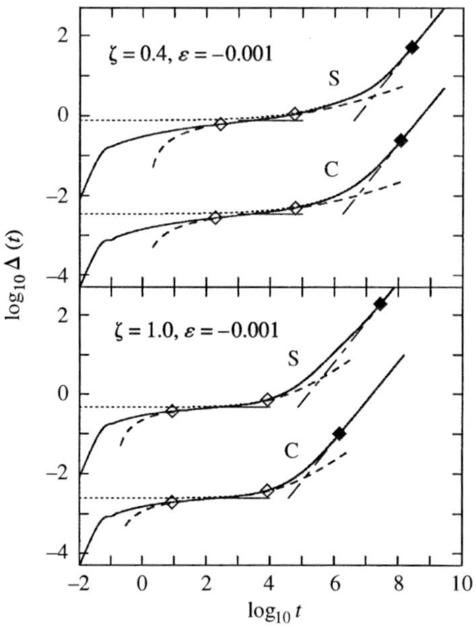

FIG. 5.30. Double-logarithmic presentation of the mean-squared-displacement function $\Delta_S(t)$ for an interaction site and $\Delta_C(t)$ for the molecule centre evaluated for a distance parameter $\epsilon = -10^{-3}$ for a liquid of hard-sphere dumbbells. The system and the units are explained in the text. The results for $\Delta_S(t)$ are shifted upwards by two decades in order to avoid overcrowding. The straight dashed-dotted lines of slope unity exhibit the diffusion asymptotes $\Delta(t) \sim Dt$, and the filled diamonds mark the diffusion-onset points where the curves deviate by 5% from this long-time asymptote. The dotted lines exhibit the description of the functions by the superposition law $\Delta(t) = \tilde{\Delta}(t/\tau)$. The dashed lines exhibit the results of the first-scaling-law formulas (6.54c), (6.56a,b); and the open diamonds mark the points where the functions deviate by 5% from this leading-order asymptotic description of the plateau-crossing process. Reproduced from Chong and Götze (2002b).

there occurs a time interval, where the mean-squared displacement of the constituents can be approximated by a sublinear power-law increase: $\delta r_s^2(t) \propto t^\xi$, $\xi = 0.7$.

The possibilities to approximate $C_1(t)$ by a normalized exponential and $\delta r_s^2(t)$ by a t^ξ variation are implications of the time-separation property. Figure 5.26 suggests that this separation increases with the coupling strength between the reorientational degrees of freedom to those degrees of freedom, which describe primarily the density-fluctuations for intermediate wave numbers. The symmetry properties for the system under study are not of great importance in this context; they merely permit a precise distinction between the group of variables causing glassy dynamics and the group used to probe it. Understanding the separation properties requires an understanding of the close-to-the plateau relaxation. All subtleties of the glassy dynamics originate from this process. However, it is difficult to identify this process for the $C_1(t)$ versus $\log t$ curves for the strong-coupling scenario, because the plateau value f_1^c is so close to its upper limit value unity. In this case, the major part of the dipole decay is described by the elementary relaxation law, which causes a Lorentzian loss peak of the susceptibility spectrum. Those stretching phenomena are absent, which have been discussed in connection with Figs. 1.1 and 1.6. Glassy dynamics manifests itself merely via the appearance of a control-parameter sensitive time scale τ_1, which exceeds the time scale $t_{\rm mic}$ of microscopic liquid dynamics.

A digression shall be added on the theory for non-symmetric dumbbells. The existence of the three scenarios for the dipole dynamics is caused by the degenerate A_2 singularities at some critical elongations ζ_c, which describe transitions from glass 1 to glass 2. The two types of glasses exist if there are sufficiently many relations between the mode-coupling coefficients so that the Eqs. (5.148) are valid. The regularity properties of the MCT equations imply the following. If the coefficients of the polynomials $\mathcal{F}_q^-[\boldsymbol{x},\mathbf{0}]$ are 'small', the solutions of the MCT equations of motion are 'close' to the ones discussed for the symmetric-dumbbell system. Consequently, there appear the same three scenarios for the liquid dynamics as discussed above. It is possible to extract the results for arrest of nearly symmetric dumbbell systems from the general theory outlined in Sec. 4.3.4. The leading-order results for states close to the glass 1-to-glass 2 transition line is quantified by a single small positive parameter σ_0. One gets for the arrested parts f_q^- of the correlator $F_q^{--}(t)$ (Franosch and Götze 1994)

$$f_q^- = h_q^- \left\{\sigma + \sqrt{\sigma^2 + \sigma_0^2}\right\} /(2\mu_2). \tag{5.150}$$

Here, the positive critical amplitudes h_q^- and the positive singularity parameter μ_2 are identical with the corresponding quantities in Eq. (4.98). The same holds for the separation parameter $\sigma = C\epsilon$, $C > 0$, $\epsilon = (\zeta - \zeta_c)/\zeta_c$. For $\sigma_0 = 0$, the previously discussed transition from the plastic glass to the generic glass is reproduced. In this case, there holds: $f_q^- = 0$ for $\epsilon \leq 0$, and $f_q^- = h_q^-[\sigma/\mu_2]$ for $\epsilon \geq 0$. For $\sigma_0 > 0$, f_q^- is positive throughout; it increases smoothly with σ. For $-\sigma \gg \sigma_0$, the arrested part is very small: $f_q^- = h_q^-[\sigma_0(\sigma_0/|\sigma|)/(4\mu_2)]$. For $\sigma \gg \sigma_0$, the arrested part is very close to the one of the symmetric-dumbbell glass: $f_q^- = h_q^-(\sigma/\mu_2)[1 + (\sigma_0/\sigma)^2/(4\mu_2)]$.

5.3.3 *Glassy Rouse dynamics*

The implications of symmetry properties have been demonstrated above for the simplest case. In this section, the studies shall be generalized in order to simplify the MCT equations for systems of cyclic molecules, which consist of n equivalent sites, $n \geq 3$. For large n, such systems serve as models for polymer melts. The models ignore that monomers at the chain ends exhibit a dynamics different from that exhibited by those in the chain centre.

The labels $\alpha, \beta, \ldots = 1, 2, \ldots$ shall be used with the understanding that α and $\alpha + kn$, $k = 0, \pm 1, \pm 2, \ldots$, refer to the same site. Generalizing the considerations done above for $n = 2$, one derives the following. The renumeration $\alpha \to \alpha + 1$ of the sites of molecule number s defines a symmetry operator $\mathcal{S}_s$. It generates the cyclic group of n elements of local permutations $\mathcal{S}_s, \mathcal{S}_s^2, \ldots, \mathcal{S}_s^{n-1}$, $\mathcal{S}_s^n = 1$. In addition, the inversion of the numeration, $\alpha \to n - \alpha + 1$, defines a symmetry $\mathcal{S}'_s$ with $\mathcal{S}'_s\mathcal{S}'_s = 1$. Corresponding global symmetries are obtained from $\mathcal{S} = \mathcal{S}_1 \cdots \mathcal{S}_N$ and $\mathcal{S}' = \mathcal{S}'_1 \cdots \mathcal{S}'_N$. There holds: $\mathcal{S}_s \rho_{s\vec{q}}^{\alpha} = \rho_{s\vec{q}}^{\alpha+1}$ and $\mathcal{S}'_s \rho_{s\vec{q}}^{\alpha} = \rho_{s\vec{q}}^{n-\alpha+1}$. Analogues equations hold also if the local operators are replaced by the global ones. The resulting equations remain valid if the tagged-molecule fluctuations $\rho_{s\vec{q}}^{\alpha}$ are replaced by $\rho_{\vec{q}}^{\alpha}$. If $r \neq s$, there holds $\mathcal{S}_s \rho_{r\vec{q}}^{\alpha} = \rho_{r\vec{q}}^{\alpha}$.

Let us introduce labels $x, y, \ldots = 0, 1, \ldots$ again with the convention that x and $x + kn$, $k = 0, \pm 1, \ldots$ are equivalent. These labels are used in order to define an n-by-n matrix $\boldsymbol{P}$ by its elements $P^{\alpha x} = \exp[i\alpha x(2\pi/n)]/\sqrt{n}$; $\alpha = 1, \ldots, n$, $x = 0, \ldots, n-1$. There holds Eq. (5.116a), i.e., the matrix is unitary. The first part of Eq. (5.116c) can be used to introduce dynamical variables $\overline{\rho}_{\vec{q}}^{x}$. One verifies the validity of the second part of Eq. (5.116c) so that a one-to-one mapping is established of the set $\rho_{\vec{q}}^{\alpha}$, $\alpha = 1, \ldots, n$ on the set $\overline{\rho}_{\vec{q}}^{x}$, $x = 0, \ldots, n-1$. All variables are scalar fluctuations with wave vector $\vec{q}$, which have even parity with respect to inversions of the space and the time. One checks that the new variables are eigenvectors of the global permutation symmetries:

$$\mathcal{S}\overline{\rho}_{\vec{q}}^{x} = \exp[-ix(2\pi/n)]\overline{\rho}_{\vec{q}}^{x}. \tag{5.151a}$$

There holds also

$$\mathcal{S}'\overline{\rho}_{\vec{q}}^{x} = \exp[ix(2\pi/n)]\overline{\rho}_{\vec{q}}^{-x}. \tag{5.151b}$$

In the preceding equations, the fluctuations can be replaced by the ones referring to a tagged molecule of number s. As discussed in connection with Eq. (5.118), the matrix $\boldsymbol{P}$ establishes an equivalence transformation of the matrix correlators formed with the site-density fluctuations and the matrix correlators formed with the new variables:

$$(\boldsymbol{P}^{-1}\boldsymbol{F}_q(t)\boldsymbol{P})^{xy} = (\overline{\rho}_{\vec{q}}^{x}(t)|\overline{\rho}_{\vec{q}}^{y})/N; \quad (\boldsymbol{P}^{-1}\boldsymbol{F}_{sq}(t)\boldsymbol{P})^{xy} = (\overline{\rho}_{s\vec{q}}^{x}(t)|\overline{\rho}_{s\vec{q}}^{y}). \tag{5.152}$$

Combining Eqs. (3.6), (5.151a), one gets $(\overline{\rho}_{\vec{q}}^{x}(t)|\overline{\rho}_{\vec{q}}^{y}) = \exp[i(x-y)(2\pi/n)](\overline{\rho}_{\vec{q}}^{x}(t)|\overline{\rho}_{\vec{q}}^{y})$. A corresponding equation holds for the tagged-variable correlators. Because of the global permutation symmetry, there are no correlations between variables of different index x and y. One can write

$$(\overline{\rho}_{\vec{q}}^{x}(t)|\overline{\rho}_{\vec{q}}^{y})/N = \delta^{xy}F_q^{x}(t), \qquad (\overline{\rho}_{s\vec{q}}^{x}(t)|\overline{\rho}_{s\vec{q}}^{y}) = \delta^{xy}F_{sq}^{x}(t). \tag{5.153a}$$

Matrix $\boldsymbol{P}$ transforms the correlator matrices on diagonal ones. The diagonal elements $F_q^{x}(t)$ and $F_{sq}^{x}(t)$ are autocorrelation functions formed with symmetry-adapted density-fluctuation variables. They are eigenvalues of the matrices $\boldsymbol{F}_q(t)$

and $\boldsymbol{F}_{sq}(t)$, respectively. These are real symmetric functions of the time. Similar formulas are obtained for the structure factors and the characteristic frequencies for the Zwanzig–Mori equations of motion if one specializes the preceding relations to the small-time asymptotic expansions. As explained for the $n = 2$ case, also the relaxation kernels are diagonalized by the specified equivalence transformation. Observing the symmetry $\mathcal{S}$, one can reduce the Zwanzig–Mori equations of motion for the n-by-n matrix correlators $\boldsymbol{F}_q(t)$ and $\boldsymbol{F}_{sq}(t)$ to equations for the eigenvalues $F_q^x(t)$ and $F_{sq}(t)$, respectively. Observing the symmetry $\mathcal{S}'$, one can reduce the problem further, since

$$F_q^x(t) = F_q^{-x}(t), \quad F_{sq}^x(t) = F_{sq}^{-x}(t). \tag{5.153b}$$

Exploiting the local symmetries as done for the derivation of Eq. (5.126), one finds:

$$F_q^x(t) = F_{sq}^x(t), \quad x = 1, \ldots, n-1. \tag{5.154}$$

The correlator $F_{sq}^0(t)$ describes the dynamics of the total-density fluctuation of the tagged molecule for wave vector $\vec{q}^* = (0,0,q)$: $\overline{\rho}_{s\vec{q}*}^0 = \sum_\alpha \exp[iqr_{\alpha s}^3]/\sqrt{n}$. For small q, one gets $\overline{\rho}_{s\vec{q}*}^0 = \{1 + iqr_{Cs}^3 - (q^2/2n)\sum_\alpha (r_{\alpha s}^3)^2 + O(q^3)\}\sqrt{n}$. Here, $\vec{r}_{Cs} = (\sum_\alpha \vec{r}_{\alpha s})/n$ denotes the centre of the molecule. There holds $\langle r_{Cs}^3(t) r_{Cs}^3\rangle = -\Delta_C(t) + 2\langle (r_{Cs}^3)^2\rangle$ with $\Delta_C(t) = \langle [r_{Cs}^3(t) - r_{Cs}^3(0)]^2\rangle/2$ denoting the mean-squared-displacement function for the centre of the molecule. Combining all formulas, one gets $F_{sq}^0(t) = \{1 - q^2\Delta_C(t) + q^2 a + O(q^4)\}n$. Here, a abbreviates some time independent number. One concludes that $\Delta_C(t)$ quantifies the spread of the probability density of the molecules' centres in analogy to what is formulated by Eq. (3.30) or (5.139b):

$$F_{sq}^0(t)/F_{sq}^0(t=0) = 1 - q^2\Delta_C(t) + O(q^4). \tag{5.155}$$

For $x \neq 0$, there holds $\overline{\rho}_{s\vec{q}*}^x = iqZ_x + O(q^2)$. The conformational variables

$$Z_x = \sum_\alpha \exp[ix\alpha(2\pi/n)] r_{\alpha s}^3/\sqrt{n}, \quad x = 1, \ldots, n-1, \tag{5.156a}$$

describe changes of the interaction-site positions for a fixed centre. The variables Z_x are eigenvectors of $\mathcal{S}_s$ for pairwise different eigenvalues $\exp[-ix(2\pi/n)]$. Hence, there are no correlations between variables of different index: $(Z_x(t)|Z_y) = 0$ for $x \neq y$. Since $S_s \vec{r}_{Cs} = \vec{r}_{Cs}$, there are neither correlations between $\vec{r}_{Cs}$ and the Z_x : $(\vec{r}_{Cs}(t)|Z_x) = 0$. The autocorrelators

$$C_x(t) = (Z_x(t)|Z_x) \tag{5.156b}$$

determine the small-q limits of the tagged-molecule-density correlators for $x \neq 0$:

$$F_{sq}^x(t) = q^2 C_x(t) + O(q^4), \quad x = 1, \ldots, n-1. \tag{5.156c}$$

The above specified symmetries for $F_{sq}^{x}(t)$ imply that the correlators $C_x(t)$ are real even function of t, which obey $C_x(t) = C_{-x}(t)$. The functions are multiples of those quantities, which are known as Rouse correlators in polymer physics (Doi and Edwards 1986).

According to the discussion of Eq. (3.30), the function $\Delta_s(t) = \langle [r_{\alpha s}^3(t) - r_{\alpha s}^3(0)]^2 \rangle /2$ characterizes the tagged-interaction-site probability distribution: $F_{sq}^{\alpha\alpha}(t) = 1 - q^2\Delta_s(t) + O(q^4)$. The inversion of Eqs. (5.152), (5.153a) yields $F_{sq}^{\alpha\alpha}(t) = \sum_x F_{sq}^{x}(t)/n$. Substituting the above derived small-q expansions for $F_{sq}^{x}(t)$, one gets $\Delta_s(t) = \Delta_C(t) - \sum_{x=1}^{n-1} C_x(t)/n - b$. The condition $\Delta_s(0) = \Delta_C(0) = 0$ determines the constant b. One arrives at the generalization of Eq. (5.138c):

$$\Delta_s(t) = \Delta_C(t) + \sum_{x=1}^{n-1} [C_x(t=0) - C_x(t)]/n. \tag{5.157}$$

A first study of glassy polymer dynamics within the framework of the microscopic MCT has been presented by Chong and Fuchs (2002). Chains of hard spheres with diameter d are considered. The n correlators $F_q^x(t)$ have been approximated by $F_q^0(t)$. The equation of motion for $F_q^0(t)$ is the same as that for simple systems. The input information on the structure factor and direct correlation function is evaluated within established approximation schemes for polymer physics. The only additional input information is the matrix $\boldsymbol{w}_q$ of intramolecular structure factors. This matrix is calculated from the assumption of a Gaussian-chain-distribution for the monomers. The length parameter is chosen so that the average distance between adjacent monomers is close to the sphere diameter d.

The cited work focusses on the Rouse-mode dynamics and on the mean-squared displacements. The Rouse-mode with the largest x describes out-of-phase motion of neighbouring monomers. This is similar to the flip motion discussed for the dumbbell system with elongation ζ near unity. For this motion with strong steric hindrance, there is a time-scale separation as discussed above for the dimers. Rouse modes with smaller x deal with the motion of complexes of neighbouring monomers. Hence, the steric hindrance is more effective and the time scale τ_x increases with decreasing x. As explained for the dumbbell system, one expects the Rouse correlators to be close to exponentials. The cited numerical work confirms this expectation. The x-dependence of τ_x is close to that found in the Rouse theory for dilute polymer solutions.. The MCT calculations extend the Rouse theory to the case of dense polymer melts. The strong control-parameter dependence of the scales τ_x is coupled to that of the density fluctuations; it is caused by the cage effect for the monomer dynamics.

Figure 5.31 shows the mean-squared displacements $g(t) = 6\Delta(t)$ for a liquid with distance parameter $\epsilon = (\varphi - \varphi_c)/\varphi_c = -0.01$. The two panels refer to molecules with $n = 10$ and $n = 100$ sites, respectively. With increasing time, the functions $\Delta_s(t)$ and $\Delta_C(t)$ approach and cross their plateaus $(r_s^c)^2$ and $(r_C^c)^2$,

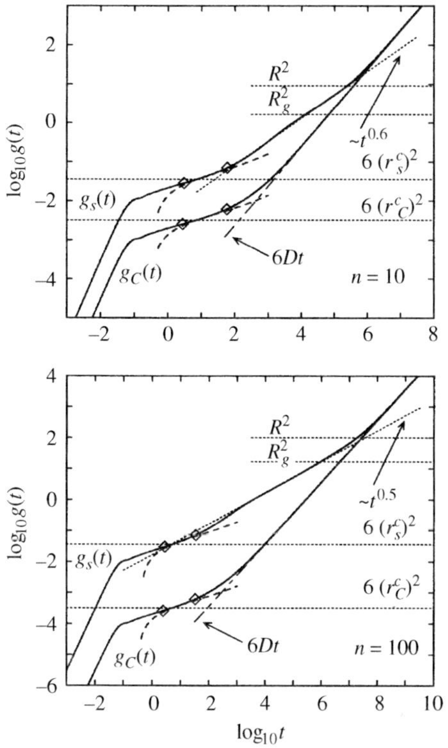

FIG. 5.31. Mean-squared displacements $g_{s,C}(t) = 6\Delta_{s,C}(t)$ evaluated for the sites and the centres, respectively, for the microscopic model for a polymer liquid described in the text. The units are chosen so that the site diameters, the thermal velocities, and the site mass are one. The packing fraction is 1% below the critical value. The molecules have $n = 10$ and $n = 100$ monomers, respectively. The short dotted horizontal lines denote the squares R^2 and R_g^2 for the end-to-end distance and the gyration radius, respectively. The long dotted horizontal lines denote $6(r_s^c)^2$ and $6(r_C^c)^2$ with r_s^c and r_C^c denoting the critical localization length for the sites and the centres, respectively. The dashed straight lines denoted $6Dt$ show the diffusion asymptotes. The dotted straight lines denoted $t^{0.6}$ and $t^{0.5}$ represent power-law variations proportional to t^ξ with $\xi = 0.6$ and 0.5, respectively.

The dashed curves exhibit the scaling-law asymptotes $6[(r_{s,C}^c)^2 - h_{\mathrm{MSD}}^{s,C} g_-(t/t_\sigma)c_\sigma]$ according to Eqs. (6.54c), (6.56a,b). The diamonds indicate the points of 10% deviation between the full and dashed curves. Reproduced from Chong and Fuchs (2002).

respectively, quite similarly to what is demonstrated in Fig. 5.7 for the hard-sphere liquid. The critical localization length r_s^c for the monomers characterize the largest possible cage size for the arrested structure. Therefore, the value r_s^c depends on n only weakly. For independent monomer motion, one would get a decrease of r_C^c proportional to $1/\sqrt{n}$ if n increases to large values. This explains that the plateau value $(r_C^c)^2$ of the $\Delta_C(t)$ versus $\log t$ curves decreases strongly with increasing n. The onset time t_{on} for the diffusion law $\Delta_C(t) = Dt$ is determined by the time needed for the density-fluctuation correlations of intermediate wave number to vanish. Hence, t_{on} is insensitive to changes of n.

The coupling of the conformational dynamics to the translational motion of the molecules' centres causes a delay of the onset time t_{on}^s of the diffusion law for the function $\Delta_s(t)$ relative to t_{on}. This is explained in connection with Fig. 5.30 for the dumbbell. The effect is described by the sum of the Rouse-mode contributions in Eq. (5.157). The delay increases with n since the time scale for the largest relaxation time τ_x diverges with n. For the limit $n \to \infty$, the theory reproduces the result of the Rouse theory: there is no diffusion law, rather $\Delta_s(t)$ increases proportional to $t^{1/2}$ for $t \gg t_{\text{on}}$. Assuming monotonic variation with the chain length, one expects for the MCT results the following. There is a time interval $t_{\text{on}} < t < t_{\text{on}}^s$, where the monomer exhibits approximately a sublinear power-law increase: $\Delta_s(t) \propto t^{\xi}$. The monomers exhibit the diffusion asymptote $\Delta_s = Dt$ for $t > t_{\text{on}}^s$. Increasing n from 2 towards infinity, t_{on}^s tends to infinity and ξ decreases from about 0.7 to 0.5. Figure 5.31 shows that the $n = 100$ model exhibits the Rouse law for a time increase of about four orders of magnitude. For the $n = 10$ model, an exponent $x = 0.6$ is exhibited for a time increase of about a factor 10^3.

The MCT results explain qualitatively the simulation data for a dense decamer liquid, which are discussed in connection with Fig. 1.11. The $t^{0.6}$ variation of the $\Delta_s(t)$ versus $\log t$ curve results from the superposition of elementary relaxation processes for nine conformational degrees of freedom of the molecules. Their scales for the long-time decay exceed the ones for the decay of density fluctuations of intermediate wave number by up to a factor 100. This is caused by strong steric hindrance for conformational changes. From the point of view of polymer physics, the t^{ξ}-behaviour for the decamer liquid is a precursor phenomenon of the Rouse result. The increase of the effective exponent ξ above the Rouse value $1/2$ is caused by finite size effects. For $n = 10$, one cannot replace the sum in Eq. (5.157) by an integral over a density of states of an infinite chain.

5.4 Some addenda

In this section, extensions of MCT studies shall be indicated, which are not discussed in this book. The first subject concerns the vibrational excitations of glassy liquids and of the glass states. A shift of an external control parameter like the temperature T through its critical value T_c is accompanied by a qualitative change of the elastic moduli. There occurs a discontinuous change of the speed of longitudinal hydrodynamic sound. The amorphous solid has a non-vanishing static shear modulus and, as a result, there are hydrodynamic shear waves. The MCT equations provide a quantitative description of these changes. In particular, they provide results for the dynamics, which describe the precursors of the transition due to cooling the liquid towards its freezing point and due to heating the amorphous solid towards its melting point. A comprehensive discussion of the longitudinal and transverse excitations for T near and below T_c is presented by Chong (2006) for a simple van der Waals system. This work deals also with a theory of that part of the dynamical structure factors, which is caused by harmonic vibrations of the particles in the random array of transiently arrested cages. It is demonstrated that some of the results offer a qualitative explanation of the properties of high-frequency sound, which have been observed by inelastic-X-ray-scattering spectroscopy for glass-forming liquids. Schirmacher and Sinn (2008) have solved the microscopic MCT equations of motion with parameters adequate for the normal metallic liquids Li, Na, K, Al, In and Ti. The friction term in Eq. (4.17a) has been neglected. Their results reproduce well the longitudinal-current spectra, which were measured by neutron-scattering experiments for a large range of wave numbers. The results have been described by Eq. (3.118a) for the relaxation kernels. Thereby, it is shown that the theory explains quantitatively the long-time tails of the longitudinal-fluctuating-force correlations of some normal liquids. The tails are precursors of the arrest at the basic glass-transition singularity.

The frequencies of the above mentioned vibrational motions are located within the band of normal-condensed-matter excitations and somewhat below of it. But, the MCT equations of motion discussed on preceding pages are motivated for frequencies far below this band. Therefore, one cannot expect that the MCT results for the vibrational dynamics provide a quantitative description of the data for the measured cross sections for the scattering of X-rays and neutron beams. It remains a challenge, for example, to extend the MCT formulas for the velocity correlator so that the transverse-sound contribution to the density of states for the glass is reproduced.

The second extension concerns the complete description of the correlators of experimental interest for small-wave-number fluctuations. Long-wavelength fluctuations of conserved densities have low frequencies. These slow hydrodynamic fluctuations interfere with low-frequency structure-relaxation processes. A complete theory of the slow dynamics has to handle both types of low-frequency excitations simultaneously. This request is ignored by the derivations of the MCT equations, which are discussed in this book. It has been anticipated that the hydrodynamic-excitation parts to the fluctuating-force kernels are negligible compared to the parts due to the cage-effect-inducing fluctuations of intermediate-size wave numbers. It has been shown a posteriori that it does not matter for the results on glassy dynamics whether hydrodynamic sound waves are incorporated or not. It has been anticipated also that it is irrelevant for the glassy-dynamics parts of the fluctuating-force kernels whether or not heat fluctuations are accounted for.

Light-scattering experiments provide correlation spectra for fluctuations with wave numbers in a region, where hydrodynamic fluctuations are fully developed.

The experimental results are of relevance for the study of glassy dynamics, since the accessible frequency region includes the frequencies adjacent to the band of normal-condensed-matter excitations. For a general set-up of the experiment, the data analysis requires an understanding of the interplay of the hydrodynamic excitations with the structural relaxation processes. It is not sufficient to merely use the MCT generalizations of the shear viscosity and the longitudinal viscosity. Rather, one has to extend the generalized-hydrodynamics description by the introduction of concepts like a frequency-dependent specific heat and dynamical opto-elastic constants. The Zwanzig–Mori formalism offers a framework for deriving formulas for the measurable quantities in terms of kernels so that an exact hydrodynamic-limit result is reproduced for small enough frequencies. The kernels can be treated by the factorization ansatz. This is discussed by Pick *et al.* (2004), by Franosch and Pick (2005), and in the references cited in these papers. The results of optical-Kerr-effect spectroscopy and of depolarized-light-scattering measurements in backward direction are distinguished by the fact that there is no direct coupling of the probing variable with the fluctuations of the conserved densities.

The third problem to be mentioned concerns an extension of MCT to a special non-equilibrium phenomenon. A homogeneous system is considered with boundary conditions, which establish a stationary linear velocity profile. In addition to the relative distance $\epsilon = (T_c - T)/T_c$ or $(n - n_c)/n_c$ of the equilibrium state from the glass-transition singularity, the shear rate $\dot{\gamma}$ enters as control parameter. For $\dot{\gamma} \neq 0$, there is a non-equilibrium contribution to the shear tensor, which is quantified by a parameter σ_s called shear stress. The quantity of main interest is the function $\sigma_s = G(\epsilon, \dot{\gamma})$. It depends on the two control parameters, which determines the stress. For liquid states and small shear rates, there holds $\sigma = \eta\dot{\gamma}$. Since the viscosity η diverges for vanishing ϵ, function $G(\epsilon, \dot{\gamma})$ exhibits a subtle singularity if both control parameters ϵ and $\dot{\gamma}$ tend to zero.

Fuchs and Cates (2002) have analysed the specified problem within the theory for colloidal suspensions. They justified a statistical description of the dynamics in terms of a generalization of correlation functions. Treating the cage effect by a factorization ansatz, they arrive at closed equations of motion for the density-fluctuation dynamics. The solutions can be used to evaluate the stress σ_s outside the linear response regime. The equations formulate a mechanism, which eliminates arrest no matter how small is the non-vanishing shear rate. An asymptotic analysis of $G(\epsilon, \dot{\gamma})$ has been worked out by extending the procedure, which is explained in Sec. 6.1.3 for the MCT for equilibrium systems. As a result, the essential features of $G(\epsilon, \dot{\gamma})$ have been explained by analytical calculations. The new findings for shear melting and yield stress can be reproduced by a one-component schematic model. The latter complements the conventional Zwanzig–Mori equation of motion by a kernel, which modifies Eq. (4.32b) by a cutoff correlator: $m(t) = [v_1\phi(t) + v_2\phi(t)^2]/[1 + (\dot{\gamma}t)^2]$. The cited theory suggest tests by experiments for colloids or by molecular-dynamics simulations. Since the correlators relax to zero for all values of the control parameters, a study of the evolution of the cage-effect-induced glass-transition singularity should be possible not only for states with $T > T_c$ or $n < n_c$ but also for the shear-melted states with $T < T_c$ or $n > n_c$. In particular, it should be possible to identify the critical parameters T_c or n_c by extrapolating results for $|\epsilon| \to 0$ not only for $\epsilon < 0$ but also for $\epsilon > 0$. Preliminary studies support the indicated theory for a new class of glassy liquids as can be inferred from a report by Varnik and Henrich (2006) and the references quoted there.

The MCT equations of motion have been motivated as the simplest model for a treatment of the self-consistency problem for the current relaxation in amorphous systems of strongly interacting particles. The result for the transient localization of a tagged particle in the cage formed by its neighbours implies that ideal arrest can occur only if an infinite number of shells of neighbours is involved. There should be a length characterizing the extension of the clusters of particles, which are participating in the backflow around a moving particle. The approach towards arrest must be reflected by a divergence of this length. But, those correlation functions, which have been analysed within MCT so far, do not exhibit the expected divergent-length-scale phenomenon.

The fourth subject to be mentioned concerns considerations by Biroli and Bouchaud (2004), which demonstrate that the searched-for length scale is exhibited by that part of a four-density-fluctuation average, which cannot be described by the factorization ansatz. This part of a higher-order correlation function depends singularly on three small parameters: the distance parameter like $\epsilon = (T_c - T)/T_c$, the frequency z, and the modulus q of some wave vector. The cited discovery opens two fascinating perspectives. There will be results depending sensitively on the length scale. These might suggest new types of tests of the relevance of MCT for explaining data on the glassy dynamics of liquids. Substituting the length-scale sensitive contribution to the four-density correlator into the formula for the longitudinal-fluctuating-force correlator, one might be able to calculate an important correction term to its mode-coupling-functional contribution. Such a calculation would provide an estimation for the range of validity of the present MCT treatment of the cage effect.

The last issue to be discussed concerns the relaxation for temperatures below the critical one. Let us consider Fig. 4.27 as an example for a description of glassy-dynamics data by MCT results. In the present context, it is the essential point that a smooth drift of coupling constants for regular equations of motion can cause large and subtle changes of the measured response functions. This statement will be corroborated by the asymptotic-analysis results in the following chapter. A linear temperature variation of the separation parameter, $\sigma = C(T_c - T)/T_c$, is the only input necessary to describe the evolution of the relaxation patterns. This holds, provided T exceeds T_c by not more than about 5%. The cited data analysis for salol implies an estimation $T_c = 245\,\mathrm{K}$ with an uncertainty not larger than 1%. The data description is based on the MCT result that there is arrest of the correlations for $T \leq T_c$. However, such arrest is not observed in experiments. Dielectric-loss spectroscopy for salol demonstrates a scenario similar to that exhibited in Fig. 1.14 for a different liquid. Also for $T \leq T_c$, there is a low-frequency loss peak. It's position $\omega_{\max}$ defines a time scale $\tau = 1/\omega_{\max}$ for the slowest relaxation process. The time τ increases with decreasing T. For T near the glass-transformation temperature $T_g = 218\,\mathrm{K}$, τ reaches a value of the order of 100s (Dixon *et al.* 1990). The MCT results in Fig. 4.27 explain values of τ, which exceed the natural scale t_{mic} of the liquid dynamics by up to five orders of magnitude. But, the theory cannot explain the observed increase of τ from about $10^5 t_{\mathrm{mic}}$ to $10^{14} t_{\mathrm{mic}}$.

For temperatures at and below the critical one, τ/t_{mic} is large. Hence, for $T \lesssim T_c$, it is difficult to determine correlation functions of the equilibrium system by molecular-dynamics-simulation studies. But, the simulation results by Horbach and Kob (1999a) for equilibrated liquid silica demonstrate the same shortcomings of MCT, which are described in the preceding paragraph. If T decreases towards T_c, the relaxation scale exhibits a power-law increase similar to that demonstrated by Fig. 4.21 for MCT results: $\tau \propto (T - T_c)^{-\gamma}$, $\gamma \approx 2$. Near T_c, there occurs a crossover to an Arrhenius low-temperature behaviour: $\tau \propto \exp[E/(k_B T)]$. The activation energy E is close to that observed in experiments for silica for T near T_g.

The present status of information is consistent with the following summary (Taborek *et al.* 1986). The MCT value T_c is a significant temperature for the discussion of liquid dynamics. It marks the boundary between two regimes. For $T > T_c$, the motion occurs cooperatively as discussed within the mode-coupling theory for the cage effect. This mechanism implies arrest at $T = T_c$, provided other relaxation mechanisms are ignored. For $T < T_c$, there is transport by mechanisms, which are discussed in the theory for crystalline solids. Local rearrangements of particle complexes permit motions of the constituents over distances larger than their diameter. This leads to diffusion. Since there is no long-range order, the diffusion causes relaxation to zero also for the density correlators. The MCT describes that statistical information on the transient arrest for $T \leq T_c$, which can be related to the Debye–Waller factor. Possibly, it can contribute to the understanding of the dynamics for $T \leq T_c$, which occurs on time scales small compared to the scale τ for the particle diffusion. Goldstein (1969) was the first who suggested the existence of a characteristic temperature T_c for the crossover of the two specified relaxation mechanisms in liquids.

A complete theory of the complex dynamics in glass-forming liquids requires the solution of two problems. First, a theory for correlation functions has to be formulated, which explains the glassy relaxation in the regime $T < T_c$. Second, this theory has to be unified with a correlation-function theory for the dynamics, which is dominated by the cage effect. Starting with papers by Das and Mazenko (1986) and Götze and Sjögren (1987b), a series of proposals has been made for extending MCT so that the two specified problems are solved. Up to now, none of these attempts has matured to a microscopic theory.

6

ASYMPTOTIC RELAXATION LAWS

In the preceding two chapters, mode-coupling equations of motion are considered with the aim to formulate a statistical description of the dynamics for amorphous condensed matter. These nonlinear equations exhibit fold bifurcations and higher-order singularities of the long-time limits of the density-correlation functions. The approach of control parameters towards some bifurcation point implies the evolution of slow relaxation phenomena. The qualitative features of these processes differ drastically from those known from discussions of conventional dynamical systems. Nor are the low-frequency spectra similar to the ones, which are known from studies of second-order phase transitions. These observations indicate that the bifurcations deal with novel paradigms for the evolution of low-frequency excitations. It is demonstrated on the preceding pages that some of the puzzling features of the slow MCT dynamics are similar to those observed for the complex dynamics of glass-forming liquids. Hence, there is the perspective that MCT provides an understanding of some aspects of the glassy dynamics occurring in cooled or compressed liquids, in dense polymer melts, and in colloidal suspensions.

For a discussion of the mentioned perspective, one needs answers to questions like the following ones. Which laws for the time dependence of the correlators are shared by the results for a schematic model shown in Fig. 4.27 and the ones for a microscopic-model study shown in Fig. 4.26? What concepts are adequate to quantify the difference between various correlators calculated for a given model and for correlators obtained from different models? In which sense is the description of the correlators in Fig. 4.18 by the elementary expression (1.14) more than an application of Taylor's theorem?

In this chapter, answers to the indicated questions shall be obtained by deriving laws for the asymptotic properties of the MCT correlators for small frequencies and small separations of the states from the bifurcation points. The leading-order results for the fold bifurcation shall be formulated in terms of two scaling laws, which deal with the interplay of two power-law spectra and two diverging time scales. These results are the essence of the dynamics, which is caused by the simplest singularity under study. The mentioned laws can be demonstrated for the one-component model defined by Eqs. (4.32a,b). The mathematical apparatus developed in Sec. 4.3.3 can be used to derive the leading-order-asymptotic formulas for general models (Götze 1985, 1987).

The range of validity of the leading-order asymptotic description of the glassy dynamics depends on the details of the model under study. For a given model, it depends on the correlator considered. In particular, the range of validity can be so small that the results are irrelevant for an analysis of data from experiments or from molecular-dynamics-simulation studies. The range of validity of the asymptotic-expansion theory can be extended by incorporating the leading corrections to the scaling laws. An understanding of the qualitative features and general trends of the corrections is necessary if one intends to provide a compelling scaling-law analysis of data. Discussion of the correction formulas for the basic version of MCT (Franosch *et al.* 1997; Fuchs *et al.* 1998) is a major goal of this chapter.

There are two regimes for the asymptotic-expansion theory of the MCT solutions for states near generic liquid-glass transition points. The first-scaling-law regime is discussed in Sec. 6.1. It deals with the near-plateau-relaxation phenomena. In the original MCT literature, these phenomena are referred to as beta relaxation. Section 6.2 presents the theory for the second-scaling-law regime. It deals with the below-plateau decay processes, which are also called alpha processes. In Sec. 6.3, results for states near higher-order glass-transition singularities are discussed. Throughout this chapter, it is anticipated that the state P of the system is close to a critical point P^c, which is an A_ℓ singularity, $\ell = 2, 3, \ldots$. It will be the main task to evaluate analytically the asymptotic behaviour of the array $\boldsymbol{\phi}(t) = (\phi_1(t), \ldots, \phi_M(t))$ of density correlators for the basic version of the MCT.

6.1 Dynamics of the first-scaling-law regime

The regularity properties of the equations of motion imply the following. If the state P approaches a critical point P^c, there evolves an interval for the times t of increasing length, within which the M components of the correlator array $\boldsymbol{\phi}(t)$ are close to the corresponding components of the array $\boldsymbol{f}^c$ of critical form factors. It is the goal of this section to evaluate analytically the solutions for states near A_2 glass-transition singularities for times, where the functions $\phi_q(t) - f_q^c, q = 1, \ldots, M$, are small quantities.

In Sec. 6.1.1, the equations of motion are reformulated as closed relations for the mentioned M small functions. As simplest application of these reformulations, power-law decay for the long-time behaviour is derived for A_2-glass-transition singularities P^c (Sec. 6.1.2). Section 6.1.3 starts with a loosely formulated derivation of a scaling law. The main object is a discussion of the implications of this law. Section 6.1.4 presents an asymptotic-solution approach so that the scaling law is obtained as leading-order result. The extension of these result due to the leading asymptotic corrections are discussed in Sec. 6.1.5.

6.1.1 *Reformulation of the MCT equations of motion*

The difference of the array $\boldsymbol{\phi}(t)$ for state P from the critical arrested part $\boldsymbol{f}^c$ shall be characterized by another array to be called $\boldsymbol{g}(t)$. It is defined by

$$\phi_q(t) = f_q^c + (1 - f_q^c) g_q(t), \quad q = 1, \ldots, M. \tag{6.1a}$$

In the domain of complex frequencies z, the relations read:

$$ST[\phi_q(t)](z) = f_q^c + (1 - f_q^c)ST[g_q(t)](z), \quad q = 1, \ldots, M. \tag{6.1b}$$

Here and in the following, it is convenient to use S-transforms. These are defined as elementary modification of Laplace transforms of determining functions $F(t)$:

$$ST[F(t)](z) = -zLT[F(t)](z). \tag{6.2}$$

These transformations define a linear one-to-one mapping of the set of determining functions in the set of functions, which are holomorphic for non-real frequencies z. The S-transform of a constant is the same constant. The formulas (6.1a,b) generalize Eq. (4.71a).

The equation of motion (4.30a) provides an expression of the correlator's S-transform in terms of the kernel's S-transform:

$$1 - ST[\phi_q(t)](z) = 1/\Big\{1 + ST[m_q(t)](z) - \big[z(z + i\nu_q)/\Omega_q^2\big]\Big\}. \tag{6.3a}$$

Equivalently, one can express the kernel in terms of the correlator:

$$ST[m_q(t)](z) - \big[z(z + i\nu_q)/\Omega_q^2\big] = ST[\phi_q(t)](z)/\{1 - ST[\phi_q(t)](z)\}. \tag{6.3b}$$

Here, and in the following parts of this chapter, equations are formulated for $t > 0$ and Im $z > 0$. The results for $t < 0$ and Im $z < 0$ can be obtained by exploiting the standard symmetries. Equation (6.1a) can be substituted into Eq. (4.16) in order to express the kernels as mode-coupling polynomials of the components of $\boldsymbol{g}(t)$:

$$\begin{aligned} m_q(t) = \{A_q^{(0)}(P) &+ \sum_k A_{q,k}^{(1)}(P) g_k(t) \\ &+ \sum_{n=2}^{n_0} \sum_{k_1 \cdots k_n} A_{q,k_1 \cdots k_n}^{(n)}(P) g_{k_1}(t) \cdots g_{k_n}(t)\}/(1 - f_q^c). \end{aligned} \tag{6.3c}$$

The expansion coefficients are defined in Eq. (4.71c).

The first two contributions to $m_q(t)$ shall be reformulated by introduction of difference coefficients (4.71e). The fixed-point equation (4.52d) yields $A_q^{(0)c} = f_q^c$. Formula (4.61c) for the critical stability matrix implies $A_{q,k}^{(1)c} = A_{q,k}^c$. The equations of motion (6.3b) can be written in a form which generalizes Eq. (4.72a):

$$\sum_k [\delta_{q,k} - A_{q,k}^c] ST[g_k(t)](z) = J_q(z), \quad q = 1, \ldots, M. \tag{6.4a}$$

The inhomogeneity consists of contributions of different significance:

$$J_q(z) = J_q^{\text{reg}}(z) + J_q^{\sigma}(z) + \sum_{n=2}^{n_0+1} J_q^{(n)}(z). \tag{6.4b}$$

The regular term formulates a holomorphic frequency dependence: $J_q^{\mathrm{reg}}(z) = -z[z + i\nu_q](1 - f_q^c)/\Omega_q^2$. The second contribution vanishes with vanishing separation parameter σ:

$$J_q^\sigma(z) = D_q^{(0)}(P) + \sum_k D_{q,k}^{(1)}(P) ST[g_k(t)](z). \tag{6.4c}$$

The nonlinear contributions are determined by $J_q^{(n)}(z)$ with $2 \le n \le n_0$,

$$J_q^{(n)}(z) = \Big[\sum_{k_1 \cdots k_n} A_{q,k_1 \cdots k_n}^{(n)}(P) ST[g_{k_1}(t) \cdots g_{k_n}(t)](z) \Big] - ST[g_q(t)](z)^n, \tag{6.4d}$$

and by $J_q^{(n_0+1)}(z) = -ST[g_q(t)](z)^{n_0+1}/\{1 - ST[g_q(t)](z)\}$.

From the discussion of Eqs. (4.74), (4.75), one infers the following. The critical stability matrix defines two M-component arrays with elements a_q^* and a_k as well as an M-by-M matrix with elements $R_{q,k}; q, k = 1, \ldots, M$. The arrays $\boldsymbol{a}^*$ and $\boldsymbol{a}$ are left and right eigenvectors of the critical stability matrix A^c for the eigenvalue unity. They are defined uniquely by imposing the convention (4.74c,d). The matrix R is also determined uniquely in terms of the elements of A^c by imposing the convention (4.75e). The unique solution of Eq. (6.4a) is specified by a function $g(t)$ and an array $\tilde{\boldsymbol{g}}(t)$:

$$g_q(t) = a_q g(t) + \tilde{g}_q(t). \tag{6.5a}$$

There hold the equations

$$ST[\tilde{g}_q(t)](z) = \sum_k R_{q,k} J_k(z), \quad q = 1, \ldots, M, \tag{6.5b}$$

and the solubility condition

$$\sum_q a_q^* J_q(z) = 0. \tag{6.5c}$$

The preceding equations (6.4), (6.5) constitute the desired reformulation of the equations of motion. Notice that the splitting of $\boldsymbol{g}(t)$ as sum of two arrays is unique. Since $\sum_q a_q^* \tilde{g}_q(t) = 0$, there holds $g(t) = \sum_q a_q^* g_q(t)$. Substitution the expression (6.5a) into Eq. (6.1a) and using the definition (4.78b) for the critical amplitudes h_q, one gets

$$\phi_q(t) = f_q^c + h_q g(t) + (1 - f_q^c)\tilde{g}_q(t), \quad q = 1, \ldots, M. \tag{6.5d}$$

Within the microscopic MCT for simple systems, the functions $\phi_q(t)$ denote density-fluctuation correlators for wave numbers q. There is also interest in the correlators $\phi_q^s(t)$ for tagged-particle-density fluctuations. It is explained in connection with Eqs. (5.36)–(5.38) that the simultaneous discussion of the two arrays

$\phi(t)$ and $\phi^s(t)$ can be done within the frame of the basic version of MCT for arrays of $2M$ components. But, it seems worthwhile to write the results for the preceding reformulation of the equations of motion explicitly in order to emphasize simplifying specialities. The mode-coupling kernel for the tagged-particle dynamics shall be used as the quadratic polynomial (5.35b). As done in Sec. 5.1.3, the coupling coefficients of this polynomial are indicated by the symbol P^s. In analogy to Eq. (6.1a), one writes the correlators in the form

$$\phi_q^s(t) = f_q^{sc}(P^s) + [1 - f_q^{sc}(P^s)]g_q^s(t), \quad q = 1, \ldots, M. \tag{6.6a}$$

One gets for the relaxation kernels

$$m_q^s(t) = \Big\{ f_q^{sc}(P^s) + \sum_p A_{q,p}(P^s) g_p^s(t) + \sum_k B_{q,k}(P^s) g_k(t) + \sum_{k,p} C_{q,kp}(P^s) g_k(t) g_p^s(t) \Big\} \Big/ [1 - f_q^{sc}(P^s)]. \tag{6.6b}$$

Here $f_q^{sc}(P^s) = f_q^s(P^c, P^s)$ is the maximum solution of the fixed-point equation (5.37b) at the transition point P^c. The coefficients of the linear contributions to the kernel are the specializations of the functions from Eqs. (5.38f) and (5.38d):

$$A_{q,p}(P^s) = A_{q,p}^s(P^c, P^s) \qquad B_{q,k}(P^s) = B_{q,k}^s(P^c, P^s). \tag{6.6c}$$

The coefficients of the quadratic contributions read:

$$C_{q,kp}(P^s) = [1 - f_q^{sc}(P^s)] V_{q,kp}^s (1 - f_k^c)[1 - f_p^{sc}(P^s)]. \tag{6.6d}$$

There holds the equation of motion (6.3b) with all quantities carrying a label s. One can proceed as above in order to derive the analogue of Eq. (6.4a): $\sum_p [\delta_{q,p} - A_{q,p}(P^s)] ST[g_p^s(t)](z) = J_q^s(z)$. The inhomogeneity reads

$$\begin{aligned} J_q^s(z) = {} & \sum_k B_{q,k}(P^s) ST[g_k(t)](z) \\ & + \Big[\sum_{k,p} C_{q,kp}(P^s) ST[g_k(t) g_p^s(t)](z)\Big] - ST[g_q^s(t)](z)^2 \\ & - \Big[ST[g_q^s(t)](z)^3 / \{1 - ST[g_q^s(t)](z)\}\Big] - z[z + i\nu_q^s][1 - f_q^{sc}(P^s)]/\Omega_{sq}^2. \end{aligned} \tag{6.7a}$$

According to the discussion of Eqs. (5.50a,b), the resolvent matrix with elements $R_{q,k}(P^s) = I_{q,k}^s(P^c, P^s)$ is defined uniquely by the equations:

$$\sum_k R_{q,k}^s(P^s)[\delta_{k,p} - A_{k,p}(P^s)] = \delta_{q,p}; \quad q, p = 1 \ldots, M. \tag{6.7b}$$

The resolvent can be evaluated as Neumann series. Since none of the M^2 matrix elements $A_{k,p}(P^s)$ is negative, the same is true for the M^2 matrix elements

$R^s_{q,k}(P^s)$. One arrives at

$$ST[g_q^s(t)](z) = \sum_k R^s_{q,k}(P^s) J_k^s(z), \quad q = 1, \ldots, M. \tag{6.7c}$$

The preceding three formulas constitute the intended reformulation of the equations of motion for the tagged-particle dynamics.

It is shown in Secs. 5.3.2 and 5.3.3 that the microscopic-MCT models for systems of symmetric dumbbells and for cyclic molecules have the same structure as discussed above for the dynamics of the pairs of arrays $\boldsymbol{\phi}(t)$ and $\boldsymbol{\phi}^s(t)$. Consequently, all results to be derived below for these correlator pairs can be used directly for a discussion of the mentioned molecular systems.

This section 6.1 deals with a generalization of the approach, which is discussed in Sec. 4.3.4 for the asymptotic solution of the fixed-point equation. The above specified general equations of motion will be solved in a regime, where the components of $\boldsymbol{g}(t)$ and their S-transforms can be treated as small:

$$|g_q(t)| \ll 1, \tag{6.8a}$$

$$|ST[g_q(t)](z)| \ll 1, \quad q = 1, \ldots, M. \tag{6.8b}$$

These condition indicate implicitly the set of states P in the neighbourhood of the critical point P^c, the interval of times t, and the region of frequency z, for which the solution will be obtained.

There are two obstacles for the asymptotic-solution theory, which have no counterpart within the theory for the fixed-point equation. The first one concerns the transient dynamics. Equation (6.1a) implies that the M functions $g_q(t)$ are close to unity for times t, which do not exceed the scale t_{mic} for the normal-liquid dynamics. For these times, condition (6.8a) is violated. Typical normalized loss spectra $\chi''_q(\omega) = -\,\text{Im}\, ST[g_q(t)](\omega + i0)(1 - f_q^c)$ exhibit humps of order unity for frequencies ω around $1/t_{\text{mic}}$. Figure 4.15 demonstrates such normal-liquid-dynamics spectra for representative cases. Within the indicated frequency regime, condition (6.8b) is violated. Formulas, which are derived under the condition (6.8a,b), cannot be expected to provide reasonable approximations for the MCT dynamics, unless there hold the restrictions

$$|t_{\text{mic}}/t| \ll 1, \tag{6.9a}$$

$$|t_{\text{mic}} z| \ll 1. \tag{6.9b}$$

The second obstacle concerns liquid states close to some discontinuous liquid-glass transition point. According to the discussions of Secs. 4.4.1 and 4.4.2, the functions $g_q(t)$ are zero near some plateau-crossing time τ^{pc}. The below-plateau decay is specified by some time scale $1/\omega_{\text{max}} > \tau^{pc}$. For $t\omega_{\text{max}} \gg 1, g_q(t)$ decays to its long-time asymptote $-f_q^c/(1-f_q^c)$. For these large times, condition (6.8a) is violated. For frequencies ω near $\omega_{\max}$, the loss spectra exhibit some peak of order unity. Also this feature is demonstrated in Fig. 4.15. Condition (6.8b) is violated for frequencies near and below the position $\omega_{\max}$ of the low-frequency-loss peaks. Results, which are derived by assuming the condition (6.8a,b), are expected to be wrong, unless there holds

$$|\omega_{\max} t| \ll 1, \tag{6.9c}$$

$$|\omega_{\max}/z| \ll 1. \tag{6.9d}$$

The preceding reformulations and the following solutions of the equations of motion are discussed in the frequency domain. The important retardation effects of the original MCT equations appear in form of the integral transformations $ST[g_{k_1}(t)\cdots g_{k_n}(t)](z) = (-iz)\int \theta(t)\exp[izt]g_{k_1}(t)\cdots g_{k_n}(t)dt$. For every value of z, the integrals are effected by the values of the functions $g_q(t)$ for all times. It is the goal to demonstrate that specific quantitative results for the solutions can be obtained, even though no general quantitative results for the correlators are available outside the regimes indicated by the conditions (6.9a–d). Simplifications of the equations for correlators in the region of small frequencies z are discussed in Chapter 3 in connection with the derivation of hydrodynamic-limit results. The corresponding functions in the time domain deal with coarse-grained information. Extrapolating of the results outside their domain of validity yields qualitative deviations from the correct behaviour. There appear artifacts like short-time cusps and large-frequency spectral tails. Similar artifacts will appear also for the results to be derived in this chapter. It will be shown that these artifacts do not falsify the conclusions drawn from the mentioned integral transforms. It is the task to replace the above used undefined symbols '$\ll$' by a statement about limit properties.

6.1.2 *The critical dynamics*

In this section, the MCT solutions shall be analyzed for A_2-bifurcation states P^c. It will be shown that the components of the critical density-fluctuation-correlator arrays $\boldsymbol{\phi}^c(t)$ exhibit power-law variation for long times. This decay of the functions $\phi_q^c(t)$ to the critical arrested parts f_q^c causes power-law divergencies of the critical fluctuation spectra $\phi_q^{c\prime\prime}(\omega)$ for small frequencies ω. Corresponding power-law behaviour is exhibited also by the correlations for the tagged-particle-density fluctuations, for the fluctuating-forces, and by the mean-squared-displacement functions. The power laws are specified by a common exponent $a, 0 < a \leq 1/2$. This critical exponent is determined by the mode-coupling polynomial for the density correlators for coupling coefficients of the critical point P^c. The value for a depends on the model considered. For a given model, it depends on the position of the glass-transition singularity P^c on the transition surface in the control-parameter manifold.

(i) The leading-order asymptotic behaviour

Let us start with three paragraphs, which demonstrate that power-counting 'argumentation' yields the asymptotic forms for the critical correlators $\phi_q^c(t)$. For critical states, the difference coefficients in Eq. (6.4c) vanish. There holds $J_q^c(z) = J_q^{\rm reg}(z) + J_q^{(2)c}(z) + O(g^3)$. Equation (6.5b) yields $ST[\tilde{g}_q^c(t)](z) = O(g^2) + O(z)$. The ansatz $\lim_{z\to 0} z/ST[g_q(t)](z) = 0$ implies that the second contribution in Eq. (6.5a) can be neglected compared to the first one. The latter is quantified by a single function to be called $g_0^c(t)$. The leading-order result has the form $g_q^c(t) = a_q g_0^c(t)$. The inhomogeneity simplifies to: $J_q^c(z) = J_q^{\rm reg}(z) + [\sum_{k_1 k_2} A_{q,k_1 k_2}^{(2)c} a_{k_1} a_{k_2}] ST[g_0^c(t)^2](z) - a_q^2 ST[g_0^c(t)](z)^2 + O(g^3)$. Substitution of this result into Eq. (6.5c), using definition (4.88a) and observing the conventions (4.74d) casts the equation of motion in the form

$$\lambda^c ST[g_0^c(t)^2](z) - ST[g_0^c(t)](z)^2 = -\sum_q a_q^* [J_q^{\rm reg}(z) + O(g^3)]. \tag{6.10}$$

Equations (A.59a,b) in Appendix A.7 show that S-transforms of power laws are power laws. This suggest to solve the preceding equation with the ansatz

$$g_0^c(t) = (t_0/t)^a, \tag{6.11a}$$

which is equivalent to

$$ST[g_0^c(t)] = \Gamma(1-a)(-izt_0)^a. \tag{6.11b}$$

The positive number t_0 denotes some time scale. The exponent is restricted by $0 < a < 1$. The restriction is consistent with the ansatz made above for the derivation of Eq. (6.10). If one requires $a < 1/2$, one gets $ST[g_0^c(t)^2](z) = \Gamma(1-2a)(-izt_0)^{2a}$. Consequently, the function $g_0^c(t)$ obeys the identity

$$\lambda^c ST[g_0^c(t)^2](z) - ST[g_0^c(t)](z)^2 = 0, \tag{6.11c}$$

provided the exponent a is a solution of the equation

$$\Gamma(1-a)^2/\Gamma(1-2a) = \lambda^c, \qquad 0 < a < 1/2. \tag{6.11d}$$

Figure 6.1 for the graph of function $\lambda(x) = \Gamma(1-x)^2/\Gamma(1-2x)$ shows that Eq. (6.11d) establishes a one-to-one relation between λ^c and a. If λ^c increases from 0 to 1, a decreases from 1/2 to 0. The numbers λ^c and a are called the exponent parameter and the critical exponent, respectively, of the A_2 singularity P^c. These are two general concepts characterizing the fold-singularity dynamics. The numbers are determined by the coefficients of the mode-coupling polynomial at the critical point. If there would hold $\lambda^c = 1$, the singularity parameter μ_2^c in Eq. (4.89c) would vanish. In this case, P^c would be a higher-order singularity $A_\ell, \ell \geq 3$. Such possibility is not considered in this section.

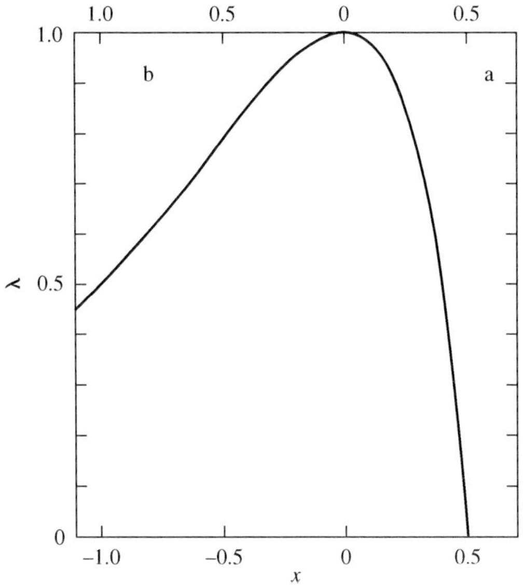

FIG. 6.1. Function $\lambda(x) = \Gamma(1-x)^2/\Gamma(1-2x)$. The upper horizontal axis exhibits the $a = x$ for $x > 0$ and $b = -x$ for $x < 0$.

For $a < 1/2$, there holds $\lim_{z\to 0} J_q^{\mathrm{reg}}(z)/ST[g_0^c(t)](z)^2 = 0$. Both terms on the right-hand-side of Eq. (6.10) can be neglected compared to the two terms on the left-hand one. Because of Eqs. (6.11c,d), the function $g_0^c(t)$ provides a solution of the leading part of the equations of motion for the low-frequency limit.

However, the preceding reasoning is illegitimate. The contribution $\tilde{g}_q(t)$ to Eq. (6.5a) contains terms proportional to $g_0^c(t)^2$. Iterating Eqs. (6.4b), (6.5a,b) one recognizes the following. For every $n = 3, 4, \ldots$, there are contributions to $J_q^c(z)$, which are proportional to $ST[g_0^c(t)^n](z)$. For $n \geq 1/a$, these functions are not defined, since the Laplace–transform deals with divergent integrals. In order to understand that Eqs. (6.11a–d) formulate the germ of an asymptotic-solution approach, the small-time divergence of $g_0^c(t)$ has to be eliminated.

A positive regularization parameter r shall be used to define the function

$$C_r(t) = 1 - \exp[-t/(rt_0)]. \tag{6.12a}$$

Regularized power laws shall be introduced by

$$g_r^c(t) = C_r(t) g_0^c(t). \tag{6.12b}$$

These continuous bounded functions vanish for $t = 0$. For large times, $g_0^c(t)$ approximates $g_r^c(t)$ with a relative error, which vanishes faster than any power of t : $\lim_{t\to\infty}[g_r^c(t) - g_0^c(t)]t^k = 0; k = 1, 2, \ldots$. For $t \geq t_{\mathrm{mic}}$, the functions $g_0^c(t)$ and $g_r^c(t)$ are practically identical provided r is chosen sufficiently below t_{mic}/t_0. From Eq. (A.60b), one infers: $LT[g_r^c(t) - g_0^c(t)](z) = \Gamma(1-a)(-it_0 r^{1-a})/$

$[1 - izrt_0]^{1-a}$. The Laplace transforms of the two functions differ by a function, which is holomorphic for $|zrt_0| < 1$. Within the low-frequency regime of interest, the S-transforms differ by a white-noise-induced function:

$$ST[g_r^c(t)](z) - ST[g_0^c(t)](z) = iK_1(zt_0)[1 + O(zrt_0)]. \tag{6.12c}$$

A constant shall be defined by: $K_1 = \Gamma(1-a)r^{1-a}$. Similar calculations demonstrate that $LT[g_r^c(t)^2 - g_0^c(t)^2](z)$ is holomorphic for $|zrt_0| < 1$. One obtains

$$ST[g_r^c(t)^2](z) - ST[g_0^c(t)^2](z) = iK_2(zt_0)[1 + O(zrt_0)]. \tag{6.12d}$$

Another constant shall be defined by: $K_2 = \Gamma(1-2a)r^{1-2a}(2-2^{2a-1})$. Combining the preceding formulas, one arrives at

$$\lambda^c ST[g_r^c(t)^2](z) - ST[g_r^c(t)](z)^2 = i\lambda^c K_2(zt_0)\left[1 + O((zt_0)^a)\right]. \tag{6.13}$$

For small frequencies, the regularized power-law function $g_r^c(t)$ obeys Eq. (6.11c) except for some background term, which is caused by a white-noise-induced spectrum.

Equations (6.12c,d) present examples for relations between long-time asymptotes of determining functions and the small-frequency asymptotes of their S-transforms. The concepts are explained in Appendix A.7. The asymptotic long-time equality between $g_r^c(t)$ and $g_0^c(t)$, for example, means that the relative difference $[g_r^c(t) - g_0^c(t)]/g_0^c(t)$ vanishes in the limit of diverging t. This property is denoted by: $g_r^c(t) \sim g_0^c(t), t \to \infty$. The asymptotic small-frequency equality of their S-transforms is denoted by: $ST[g_r^c(t)](z) \sim ST[g_0^c(t)](z), z \to 0$. It means $\lim_{z\to 0}\{ST[g_r^c(t)](z) - ST[g_0^c(t)](z)\}/ST[g_0^c(t)](z) = 0$. A Tauberian theorem, which is discussed in connection with Eqs. (A.65a,b), establishes the equivalence of the two limit properties:

$$g_r(t) \sim (t_0/t)^a, \qquad a < 1, \quad t \to \infty, \tag{6.14a}$$

$$ST[g_r(t)](z) \sim \Gamma(1-a)(-izt_0)^a, \quad z \to 0. \tag{6.14b}$$

This holds under some regularity conditions for $g_r(t)$, which are specified in Appendix A.7. These conditions are obeyed by $g_r^c(t)^n, n = 1, 2, \ldots$.

For the discussion of the equations of motion, one needs the S-transforms of the powers $g_r(t)^n, n = 2, 3, \ldots$. From Eqs. (A.65), (A.67), one infers for the small-frequency asymptotes:

$$ST[g_r(t)^n](z) \sim \Gamma(1-na)(-izt_0)^{na}, \quad na < 1, \tag{6.15a}$$

$$\sim (-izt_0)\ell n[i/(zt_0)], \quad na = 1. \tag{6.15b}$$

If $g_r^n(t)$ is absolutely integrable, the zero-frequency limit of the Laplace transform is a finite number. It shall be denoted by $it_0\nu_n$:

$$\nu_n = \int_0^\infty g_r(t)^n dt/t_0, \quad na > 1. \tag{6.15c}$$

Equations (A.66,A.68) imply:

$$ST[g_r(t)^n](z) + (izt_0)\nu_n p(zt_0) \sim \Gamma(1-na)(-izt_0)^{na}, \ na \neq 2,3,\ldots, \qquad (6.15d)$$
$$\sim -(izt_0)^{na}\ell n[i/(zt_0)], \ na = 2,3,\ldots. \qquad (6.15e)$$

Here, $p(\zeta)$ denotes a polynomial of degree $(n-1)$ with normalization $p(0) = 1$.

Restricting the exponent a to values below 1/2, the preceding formulas imply the following proposition. The ansatz

$$g_q(t) = a_q g_r^c(t), \quad q = 1,\ldots,M, \qquad (6.16a)$$

reduces the low-frequency asymptote of the equations of motion to a combination of terms proportional to η^2, where the abbreviation

$$\eta = (-izt_0)^a \qquad (6.16b)$$

is used. The equations are solved in order η^2 if exponent a is related to the exponent parameter λ^c by Eq. (6.11d). To prove the proposition, one uses the formula $\lim_{z\to0} z/\eta^2 = 0$ in order to show: $\lim_{z\to0} J_q^{\mathrm{reg}}(z)/\eta^2 = 0$. Similarly, one concludes from Eqs. (6.15a–e) that $\lim_{z\to0} J_q^{(n)c}(z)/\eta^2 = 0$ for $n \geq 3$. Equations (6.14b), (6.15a) imply that the small-z asymptote of $J_q^{(2)c}(z)$ is a combination of terms proportional to η^2. Consequently,

$$J_q^c(z) \sim \Big[\sum_{k_1k_2} A_{q,k_1k_2}^{(2)c} a_{k_1}a_{k_2}\Big] ST[g_r^c(t)^2](z) - a_q^2 ST[g_r^c(t)](z)^2, \quad z \to 0. \qquad (6.16c)$$

As a result, the equation of motion (6.5c) reduces to Eq. (6.11c). The proposition holds for every choice of t_0.

The discussion shall be continued with the following assumption. There is a time scale t_0 so that the above constructed solution of the low-frequency limit of the equations of motion determines the low-frequency asymptote of the critical correlators. A proof of this assumption is not available.

One can summarize as follows. For a properly chosen t_0, there holds

$$g_q^c(t) \sim a_q g_r^c(t) \sim a_q g_0^c(t), \quad t \to \infty, \quad q = 1,\ldots,M. \qquad (6.17a)$$

Because of Eq. (6.5d), the result for the M functions $g_q(t)$ is equivalent to the statement for the correlators:

$$\phi_q^c(t) - f_q^c \sim h_q(t_0/t)^a, \quad t \to \infty, \quad q = 1,\ldots,M. \qquad (6.17b)$$

Substituting the formulas (6.17a) into Eq. (6.3c), one gets $m_q^c(t) - [A_q^{(0)c}/(1 - f_q^c)] \sim \sum_k A_{q,k}^c a_k g_0^c(t)/(1-f_q^c) = a_q g_0^c(t)/(1-f_q^c)$. The fluctuating-force kernels exhibit an asymptotic behaviour in analogy to the one for the correlators:

$$m_q^c(t) - f_q^{mc} \sim h_q^m(t_0/t)^a, \quad t \to \infty, \quad q = 1,\ldots,M. \qquad (6.17c)$$

The critical arrested parts of correlators and kernels are related by the fixed-point equation $f_q^{mc} = f_q^c/(1-f_q^c)$. The critical amplitudes of the kernels are given by

$$h_q^m = h_q/(1-f_q^c)^2. \tag{6.17d}$$

Within the microscopic MCT models for simple systems, the shear correlator is expressed in terms of a mode-coupling kernel $m_0^T(t)$ as discussed in Eqs. (5.6), (5.7). In the low-frequency limit, a regular term ν^∞/c_{is}^2 can be neglected compared to $m_0^T(z)$. Substitution of Eq. (6.17b) into the mode-coupling functional (5.6b), one gets the analogue of Eq. (6.17c):

$$K_T^c(t)/c_{is}^2 - f_T^c \sim h_T(t_0/t)^a, \quad t \to \infty. \tag{6.18}$$

The critical arrested part f_T^c and the critical amplitude h_T are determined by Eqs. (5.13b,c).

Equations (6.7a,c) show that $ST[g_q^s(t)](z)$ vanishes with z proportional to η. Consequently, one gets $g_q^s(t) \sim \sum_{pk} R_{qp}^s(P^s)B_{pk}(P^s)a_k g_0^c(t)$. Formula (5.53) for the critical amplitude h_q^s and Eq. (6.6a) yield the asymptotic formula for the critical tagged-particle-density correlators:

$$\phi_q^{sc}(t) - f_q^{sc} \sim h_q^s(t_0/t)^a, \quad t \to \infty, \quad q = 1, \ldots, M. \tag{6.19}$$

Two paragraphs shall be added in order to show that the results (6.17)–(6.19) remain valid also for the so-far excluded case $a = 1/2$. The discussion of Fig. 6.1 shows that this case corresponds to a vanishing exponent parameter: $\lambda^c = 0$. The simplest possibility for such case deals with a linear mode-coupling polynomial. If there holds $A_{q,k_1\cdots k_n}^{(n)} = 0$ for $n \geq 2$, one gets $J_q^{(n)}(z) = -ST[g_q(t)](z)^n$ for all contributions to Eq. (6.4b) with $n \geq 2$. There is no need to discuss small-time divergencies of integrands. The solubility condition (6.10) gets the form of an algebraic equation:

$$ST[g_0^c(t)](z)^2 = \sum_q a_q^* J_q^{\text{reg}}(z) + O(ST[g_0^c(t)](z)^3). \tag{6.20a}$$

Let us restrict the analysis to the most important case of non-vanishing rates: $\nu_q > 0, q = 1$. One obtains $ST[g_0^c(t)](z) = \Gamma(1/2)(-izt_0)^{1/2} + O(z)$, with

$$t_0 = \sum_q a_q^* \nu_q (1-f_q^c)/[\Omega_q^2 \pi]. \tag{6.20b}$$

For a one-component model, the preceding formulas reproduce Eq. (4.51c).

The general possibility for $\lambda^c = 0$ accounts for nonlinear mode-coupling polynomials. An example is formulated by Eqs. (4.32a,c) for a one-component model. Because of the ν_n-contributions in Eqs. (6.15c,d), the right-hand side of Eq. (6.10) contains terms, which renormalize the coefficients ν_q in $J_q^{\text{reg}}(z)$. The general results for the asymptote of the correlators remain valid; but, the value for t_0 cannot be calculated analytically.

Equations (4.74c), (4.88a) imply the following conclusion. For generic models, λ^c can be zero only if all second-order mode-coupling coefficients $A^{(2)c}_{q,k_1k_2}$ vanish. For the microscopic theory of liquids, such models are of no relevance. Therefore, in the following, $\lambda^c > 0$ and $a < 1/2$ shall be assumed throughout.

The formulas (6.17b,c), (6.18), (6.19) for the long-time relaxation of various correlators have identical forms. The time dependence is described by the same power-law-function $g_0^c(t) = (t_0/t)^a$. The critical arrested part like f_q^c and the critical amplitude like h_q are general concepts for parameters, which quantify the distinction of the description of the bifurcation dynamics given by different correlation functions. These parameters are determined by the relevant mode-coupling functionals for coefficients chosen at the critical point P^c. Within the microscopic MCT, these quantities are determined by the equilibrium-structure functions. Details of the dynamics for times of order t_{mic} enter the asymptotic-relaxation law via a single number, namely, via the scale t_0. Different choices for the $2M$ frequencies Ω_q and ν_q in Eq. (4.17a) merely result in different values for t_0. The difference between the two versions (4.17a) and (4.17b) for the equations of motion are reflected only by different numbers t_0. The value t_0 is influenced by the mode-coupling effects on the transient dynamics, which cannot be calculated analytically. In comparisons of results from MCT equations with those of asymptotic-expansions, the value t_0 is used as fit parameter.

The result (6.17b) shall be referred to by the statement that the formula

$$\phi_q^c(t) = f_q^c + h_q(t_0/t)^a \tag{6.21a}$$

represents the leading-asymptotic expression for the critical correlators. Similar statements are used for other functions. This formula identifies the first generic source for the stretching of the dynamics. For example, the exponent parameter $\lambda^c = 3/4$ yields a critical exponent $a = 0.305\ldots$. In this case, a reduction of $\phi_q^c(t) - f_q^c$ by a factor 10 requires an increase of the time by more than three orders of magnitude. Figure 4.12 demonstrates this result for a schematic model. The exponent $a = 0.312$ describes the critical decay for the microscopic MCT model of a hard-sphere system, which is discussed in Chapter 4. The upper panel of Fig. 6.2 reproduces from Fig. 4.14 the two critical correlators for this system. For $t > 10^3$, they cannot be distinguished from their leading-order description. The scale t_0 is of order of the scale t_{mic} for the transient motion.

The following feature of the bifurcation dynamics is explained in Sec. 4.4.1. For states P near P^c, there is an interval of times between t_- and t_+, where the correlators $\phi_q(t)$ for states P are close to the critical ones. The time t_- depends on P insensitively; but, t_+ increases beyond any bound if P tends towards P^c. Consequently, the result (6.21a) provides the explanation for the stretching of the relaxation of $\phi_q(t)$ towards f_q^c, which occurs for times obeying $t_- < t < t_+$.

Within the regions of normal liquid dynamics, the low-frequency-correlation spectra $\phi_q''(\omega)$ for density fluctuations of wave numbers of intermediate size

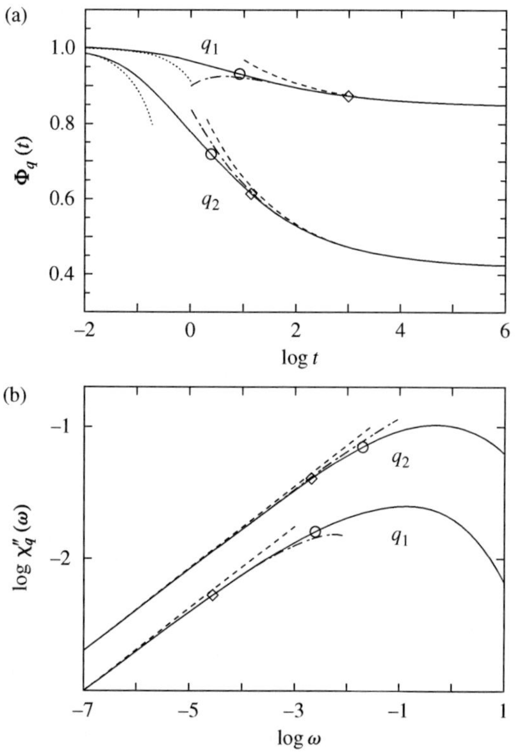

FIG. 6.2. The solid lines exhibit the critical correlators $\phi_q^c(t)$ and critical loss spectra $\chi_q^{c\prime\prime}(\omega)$, respectively, for a hard-sphere-system model. The wave numbers are $q_1 = 7.0$ and $q_2 = 10.6$. The dashed lines are the leading-order asymptotic expressions (6.21a,c) evaluated for $t_0 = 0.425$. The diamonds mark the points t_q^* and ω_q^*, respectively, of 10% deviations between the full and dashed lines. The dashed-dotted lines exhibit the results of Eqs. (6.30c), (6.31a); the circles mark the points of 10% deviation from the full lines. The dotted lines in the upper panel are the short-time asymptotes $(1 - t/\tau_q)$.

The model and the units are explained in connection with Figs. 4.6, 4.14 and 4.15. The amplitudes (f_q^c, h_q, K_q) are (0.849, 0.323, −1.02) and (0.417, 0.647, −0.183) for q_1 and q_2, respectively. The other parameters are: $\lambda^c = 0.735, a = 0.312, b = 0.583; \xi = 0.0422, \zeta = 0.256, \kappa(a) = -0.00165, \kappa(-b) = 0.569, k_a = 3.16, k_b = 1.48$. Reproduced from Franosch *et al.* (1997).

depend insensitively on ω. For $\omega t_{\text{mic}} \ll 1$, the fluctuation spectra are close to white noise ones: $\phi_q''(\omega) \approx c_q t_{\text{mic}}$. Here, c_q is a number of order unity, which depends smoothly on control parameters. However, the long-time tails of the critical correlations are not integrable. As a result, the correlation spectra at the critical state tend to infinity if the frequency tends to zero. For small frequencies, the fluctuation spectra exhibit a power-law divergency, which is specified by the exponent $(1-a)$. Using Eqs. (6.17b), (A.65b), (A.59d), one gets:

$$\lim_{\omega\to 0}\left[\phi_q^{c\prime\prime}(\omega)(\omega t_0)^{1-a}\right] = t_0 h_q\left[\sin(\pi a/2)\Gamma(1-a)\right]. \quad (6.21b)$$

In the limit of vanishing frequency, the critical fluctuation spectra exceed normal-liquid background spectra $c_q t_{\text{mic}}$ by an arbitrary large factor. If the state P approaches P^c, the spectra $\phi_q^{c\prime\prime}(\omega)$ develop a quasi-elastic peak.

The derivation of hydrodynamic-limit results and their generalizations are based on the understanding that fluctuating-force spectra are simpler than density-fluctuation spectra. As explained in Chapter 3, there is a time scale t^*, which separates low-frequency spectra from the rest. For $\omega t^* \ll 1$, a white-noise asymptote can be used to describe spectra for the kernels $m_q''(\omega)$. This white-noise approximation is adequate to describe the subtle structure of $\phi_q''(\omega)$ for small frequencies and small wave numbers. Such time-scale separation is not exhibited by the critical MCT dynamics. According to Eqs. (6.17b), (6.17c), the spectra for the density fluctuations are proportional to the ones for the fluctuating forces: $\phi_k^{c\prime\prime}(\omega)/h_k = m_p^{c\prime\prime}(\omega)/h_p^m$. One spectrum varies as strongly as the other. The power-law spectra do not define a time scale; a rescaling of the frequency by a factor s is equivalent to a rescaling of the amplitude by a factor $s^{(a-1)}$.

The above identified parts of the critical fluctuation spectra do not cause a low-frequency loss peak. Within the range of validity of the leading-order-asymptotics description, the critical spectra decrease with frequency:

$$\chi_q^{c\prime\prime}(\omega) \sim h_q\left[\sin(\pi a/2)\Gamma(1-a)\right](\omega t_0)^a, \quad \omega \to 0. \quad (6.21c)$$

For low-frequencies, there is power-law loss specified by the critical exponent a. For ω tending to zero, the $\log \chi_q^{c\prime\prime}$ versus $\log \omega$ curves tend to straight lines with slope a. For $\omega < 10^{-4}$, this behaviour is demonstrated in Fig. 4.13 for both spectra of a two-component model. A white-noise-induced loss spectrum exhibits a linear frequency variation: $\chi_q''(\omega) = c_q \omega t_{\text{mic}}$. Glassy MCT dynamics for the critical point manifests itself by the fact that the small loss spectra are enhanced compared to the even smaller white-noise-induced normal-liquid-dynamics spectra.

The two critical loss spectra for density fluctuations of a microscopic model for a hard-sphere system, which are shown in Fig. 4.15, are reproduced in the lower panel of Fig. 6.2. For the wave numbers q_1 and q_2, the power-law asymptote describes the results within a 10% accuracy for frequencies below about $10^{-4.5}$ and $10^{-2.8}$, respectively.

The following implications of the regularity properties of the MCT equations are explained in Secs. 4.4.1 and 4.4.2. For liquid states P near the critical point P^c, the correlators $\phi_q(t)$ cross the value f_q^c at some time τ^{pc}. This time is close to the time t_+, which is mentioned above. The decay to zero for $t > \tau^{pc}$ causes a low-frequency loss peak. The high-frequency wing of this peak approaches the critical spectrum $\chi_q^{c\prime\prime}(\omega)$ for frequencies of the order $1/\tau^{pc}$. If $1/\tau^{pc}$ is in the region of applicability of the leading-order asymptotic expansion, there is a loss minimum at some frequency $\omega_{\min} \approx 1/\tau^{pc} \approx 1/t_+$. With increasing $\omega/\omega_{\min}$, the spectrum approaches the asymptote (6.21c). This formula explains a strong enhancement of the loss intensity $\chi''_q(\omega_{\min})$ relative to what one would expect by superimposing the low-frequency loss peak on a white-noise-induced background spectrum. The described enhancement is demonstrated by the spectra in Figs. 4.13 and 4.15. This feature is discussed also in connection with the data shown in Fig. 1.3. Equation (6.21c) is the motivation for the high-frequency part of the fit formula (1.7a).

(ii) The leading asymptotic corrections

Within a given model, different critical correlators $\phi^c(t)$ can differ in their critical arrested parts f^c and in their critical amplitudes h. But, the long-time decay of the functions $[\phi^c(t)-f^c]/h$ is the same power law $(t_0/t)^a$ for all of them. Similarly, the critical loss spectra agree in the small frequency limit if they are normalized by the critical amplitude: $\chi^{c\prime\prime}(\omega)/h \sim \sin(\pi a/2)\Gamma(1-a)(\omega t_0)^a, \omega \to 0$. However, the range of times or of frequencies, respectively, where these functions are close to the common asymptote, can be quite different for different correlators. This fact is demonstrated in Fig. 6.2 for two correlators and two spectra evaluated for a microscopic model. It is also exhibited in Fig. 4.13 for two loss spectra of a schematic model. Awareness of this fact is of relevance for the discussion of comparisons of glassy-relaxation data of liquids with leading-asymptotic-formula results. In order to achieve some understanding of the specified feature of the dynamics, the asymptotic theory shall be extended by incorporation of the leading asymptotic corrections.

The first step beyond the leading asymptotic results for the correlators concerns the contribution $\tilde{g}_q(t)$ in Eq. (6.5a). It is obtained by substituting Eq. (6.16c) for the inhomogeneity into Eq. (6.5b). To write the result more transparently, Eq. (6.13) can be applied in order to convert the S-transform of $g_r^c(t)^2$ into a square of the S-transform: $\lambda^c ST[g_r^c(t)^2](z) \sim ST[g_r^c(t)](z)^2, z \to 0$. This result will be used repeatedly without emphasis in following calculations. Hence, one arrives at the pair of equivalent formulas

$$\tilde{g}_q(t) \sim [a_q K_q] g_r^c(t)^2, \quad t \to \infty, \tag{6.22a}$$

$$ST[\tilde{g}_q(t)](z) \sim [a_q K_q/\lambda^c] ST[g_r^c(t)](z)^2, \quad z \to 0. \tag{6.22b}$$

The contribution $\tilde{g}_q(t)$ is of second order in the expansion parameter η from Eq. (6.16b). The correction amplitude K_q is determined by the mode-coupling coefficients for the critical point P^c:

$$K_q = \sum_k R_{q,k} \Big\{ \Big[\sum_{k_1 k_2} A^{(2)c}_{k,k_1 k_2} a_{k_1} a_{k_2} \Big] - \lambda^c a_k^2 \Big\} / a_q. \tag{6.23}$$

It is the remaining task to extend the asymptotic result for the first term on the right-hand side of Eq. (6.5a). This extension shall be denoted by $\tilde{g}(t)$:

$$g^c(t) = g_r^c(t) + \tilde{g}(t). \tag{6.24a}$$

There holds the following proposition. The mode-coupling coefficients for the critical point P^c determine a function $\kappa(x)$ so that

$$\tilde{g}(t) \sim \kappa(a) g_r^c(t)^2, \quad t \to \infty. \tag{6.24b}$$

As explained above, this formula is equivalent to

$$ST[\tilde{g}(t)](z) \sim [\kappa(a)/\lambda^c] ST[g_r^c(t)](z)^2, \quad z \to 0. \tag{6.24c}$$

In the preceding subsection, it is shown that the ansatz $g_q^c(t) = a_q g_r^c(t)$ reduces the equations of motion to the request that the function $f(\eta) = \sum_q a_q^* J_q^c(z)$ vanishes identically. For small frequencies, the leading contributions to $f(\eta)$ are proportional to η^2. The leading asymptotic contribution results from the request of cancellation of the leading contributions to $f(\eta)$. The proposition means that Eq. (6.24b) ensures also the cancellation of the contributions to $f(\eta)$, which are proportional to η^3. A refinement of this statement will be formulated in the paragraph following Eq. (6.29).

It is not relevant for the proof of the proposition if such contributions $A_q(z)$ to $J_q^c(z)$ are dropped, which obey $\lim_{z\to 0} A_q(z)/\eta^3 = 0$. There are two types of such terms. Firstly, there are terms proportional to η^n and $\eta^n \ell n[-izt_0]$ for $n \geq 4$. Secondly, there are terms proportional to $z^n, n \geq 2$. The later can be dropped since $\lim_{z\to 0} z/\eta^2 = 0$. This implies in particular that the regular contribution in Eq. (6.4b) can be simplified to: $J_q^{\text{reg}}(z) = -iz\nu_q(1 - f_q^c)/\Omega_q^2$.

The proof starts by substituting the expression $g_q^c(t) = [g_r^c(t) + \tilde{g}(t)]a_q + \tilde{g}_q(t)$ into Eq. (6.4d) for the functions $J_q^{(n)c}(z), n \geq 2$. For $n = 2$, there are two types of relevant contributions:

$$J_q^{(2,1)}(z) = \Big[\sum_{k_1 k_2} A^{(2)c}_{q,k_1k_2} a_{k_1} a_{k_2}\Big] ST\big[g_r^c(t)^2 + 2g_r^c(t)\tilde{g}(t)\big](z)$$
$$- a_q^2 \Big\{ ST[g_r^c(t)](z)^2 + 2ST[g_r^c(t)](z) ST[\tilde{g}(t)](z)\Big\}, \tag{6.25a}$$
$$J_q^{(2,2)}(z) = 2\sum_{k_1 k_2} A^{(2)c}_{q,k_1k_2} a_{k_1} ST\big[g_r^c(t)\tilde{g}_{k_2}(t)\big](z)$$
$$- 2a_q ST[g_r^c(t)](z) ST[\tilde{g}_q(t)](z). \tag{6.25b}$$

There are further contributions to $J_q^{(2)c}(z)$. They consist of terms, which can be reduced to multiples of $ST[g_r^c(t)](z)^4$ or of $ST[g_r^c(t)^4](z)$. The former terms are proportional to η^4. The latter yield also such terms or terms proportional to $\eta^4 \ell n[-izt_0]$, as can be inferred from Eqs. (6.15a–d). All these contributions can be dropped. If $a > 1/4$, one concludes from Eqs. (6.15c,d) that there are polynomial additions. They can be simplified to the leading term, specified by the coefficients ν_n. The resulting function can be added to $J_q^{\text{reg}}(z)$, thereby renormalizing the white-noise constants ν_q. For $n = 3$, there appears the relevant term:

$$J_q^{(3,1)}(z) = \Big[\sum_{k_1k_2k_3} A^{(3)c}_{q,k_1k_2k_3} a_{k_1}a_{k_2}a_{k_3}\Big] ST[g_r^c(t)^3](z)$$
$$-a_q^3 ST[g_r^c(t)](z)^3. \qquad (6.25c)$$

As above, one shows that the not-specified parts of $J_q^{(3)c}(z)$ can be dropped or absorbed as renormalizations of the regular term. The contributions $J_q^{(n)c}(z)$ with $n \geq 4$ can contribute only to modifications of $J_q^{\text{reg}}(z)$. As a result, one arrives at

$$J_q^c(z) = J_q^{(2,1)}(z) + J_q^{(2,2)}(z) + J_q^{(3,1)}(z) - iz\tau_q' + O_q(\eta^4, \eta^4 \ell n[-izt_0], z^2). \qquad (6.25d)$$

The times τ_q' consist of $\nu_q(1 - f_q^c)/\Omega_q^2$ and possible additions. The latter decent from the contributions $izt_0\nu_n$ in the extended Tauberian theorems, which are discussed in Appendix A.7.

Substitution of the preceding result into the solubility condition (6.5c) and using the formula (6.13), the equation of motion reads

$$2\Big\{ST[g_r^c(t)](z)ST[\tilde{g}(t)](z) - \lambda^c ST[g_r^c(t)\tilde{g}(t)](z)\Big\} =$$
$$\sum_q a_q^*[J_q^{(2,2)}(z) + J_q^{(3,1)}(z)] - (iz\tau') + O(\eta^4, \eta^4\ell n[-izt_0], z^2). \quad (6.26)$$

Here, the coefficient for the regular contribution abbreviates: $\tau' = \sum_q a_q^*\tau_q' - \Gamma(1-a)^2 t_0 r^{1-2a}(2 - 2^{2a-1})$. There are four types of terms, which occur in the sum over the labels q. The η^3 contributions of two of them are identified by appeal to Eqs. (6.14b), (6.22b): $ST[g_r^c(t)](z)ST[\tilde{g}_q(t)](z) \sim \Gamma(1-a)^3\eta^3 a_q K_q/\lambda^c$ and $ST[g_r^c(t)](z)^3 \sim \Gamma(1-a)^3\eta^3$. Appeal to Eqs. (6.14a), (6.22a) and application of Eqs. (6.15a–c) yields the results for the other two terms. If $a < 1/3$, there holds: $ST[g_r^c(t)\tilde{g}_{k_2}(t)](z) \sim \Gamma(1-3a)a_{k_2}K_{k_2}\eta^3$ and $ST[g_r^c(t)^3](z) \sim \Gamma(1-3a)\eta^3$. If $a = 1/3$, the formulas are valid with $\Gamma(1-3a)$ replaced by $\ell n[i/(zt_0)]$. If $a > 1/3$, the formulas have to be complemented by a term proportional to (izt_0). These complements can be noted as a change of τ' to τ''. As a result, the equation of motion gets the form of the asymptotic identity:

$$ST[g_r^c(t)](z)ST[\tilde{g}(t)](z) - \lambda^c ST[g_r^c(t)\tilde{g}(t)](z) - iz\tau''$$
$$\sim (izt_0)^{3a}\{\xi\Gamma(1-3a) - \zeta\Gamma(1-a)^3\}, \quad a \neq 1/3, z \to 0. \qquad (6.27)$$

For $a = 1/3$, there holds the same formula with $\Gamma(1-3a)$ replaced by $\ell n[i/(zt_0)]$. There appear two parameters, which are determined by the mode-coupling coefficients for the critical point:

$$\xi = \sum_q a_q^* \left\{ \sum_{k_1 k_2} \left[A_{q,k_1 k_2}^{(2)c} K_{k_2} + \frac{1}{2} \sum_{k_3} A_{q,k_1 k_2 k_3}^{(3)c} a_{k_3} \right] a_{k_1} a_{k_2} \right\}, \tag{6.28a}$$

$$\zeta = \sum_q a_q^* \left\{ a_q^2 [K_q / \lambda^c] + \frac{1}{2} a_q^3 \right\}. \tag{6.28b}$$

The formula (6.24c) yields $ST[g_r^c(t)](z) ST[\tilde{g}(t)](z) \sim \Gamma(1-a)^3 [\kappa(a)/\lambda^c] \eta^3$. Using Eqs. (6.15a–c), one concludes the following. If $a < 1/3$, there holds $ST[g_r^c(t)\tilde{g}(t)](z) \sim \Gamma(1-3a)\kappa(a)\eta^3$. If $a = 1/3$, one has to replace the factor $\Gamma(1-3a)$ by $\ell n[i/zt_0)]$. If $a > 1/3$, one has to complement the formula by a term proportional to $(-izt_0)$. This complement shall be accounted for by replacing τ'' by some other constant to be denoted by τ_0. For $a \neq 1/3$, the cancellation of the contributions proportional to η^3 is achieved by using

$$\kappa(x) = \left[\xi \Gamma(1-3x) - \zeta \Gamma(1-x)^3 \right] / \left[\Gamma(1-x)\Gamma(1-2x) - \lambda^c \Gamma(1-3x) \right]. \tag{6.29}$$

For $a = 1/3$, the cancellation of the contributions proportional to $\eta^3 \ell n[i/(zt_0)]$ is achieved by choosing $\kappa(1/3) = -\xi/\lambda^c$. This result can be included in the preceding formula by requesting continuity: $\kappa(x \to 1/3) = \kappa(1/3)$. The expression for $\kappa(x)$ completes the proof of the proposition.

A comment shall be added with the aim to refine the discussion. For $a \neq 1/3$, the preceding manipulations have reduced the function $f(\eta)$ to $A\eta^2 + B\eta^3 + (izt_0)C$. Terms which are not written can be neglected either compared to η^3 or compared to z. Fixing the exponent a ensures $A = 0$; and fixing $\kappa(a)$ ensures $B = 0$. For $a < 1/3$, there holds $\lim_{z \to 0} z/\eta^3 = 0$. In this case, the two dominant contributions to $f(\eta)$ are eliminated. An analogous conclusion is valid for $a = 1/3$. Here, $B\eta^3$ has to be replaced by $B\eta^3 \ell n[i/(zt_0)]$. For $a > 1/3$, there holds $\lim_{z \to 0} \eta^3/z = 0$. The fixing of a and $\kappa(a)$ ensure the cancellation of the two dominant contributions to the low-frequency singularity of $f(\eta)$. This singularity is superimposed on a white-noise-induced holomorphic function $(izt_0)C$. An analytic evaluation of C is not possible. The discussion of Eqs. (3.20a,b) suggest the conjecture that t_0 is determined by the request of vanishing of C.

Combining the preceding results, one arrives at

$$\phi_q^c(t) - f_q^c - h_q (t_0/t)^a \sim h_q [K_q + \kappa(a)] (t_0/t)^{2a}, \quad t \to \infty. \tag{6.30a}$$

The formula means, that the relative difference between the functions on the two sides of the expression vanish in the long-time limit:

$$\lim_{t \to \infty} \left\{ \left[(\phi_q^c(t) - f_q^c)/h_q \right] - (t_0/t)^a \right\} (t/t_0)^{2a} = \left[K_q + \kappa(a) \right]. \tag{6.30b}$$

These formulas shall be referred to by the statement that

$$\phi_q^c(t) = f_q^c + h_q (t_0/t)^a \left\{ 1 + [K_q + \kappa(a)](t_0/t)^a \right\} \tag{6.30c}$$

presents the leading and next-to-leading asymptotic description of the critical density correlators. Analogous formulations shall be used for other functions and also for their Laplace transforms. For example, application of the Tauberian theorems (A.65a,b) yield the low-frequency critical loss spectra in leading and next-to-leading asymptotic description:

$$\chi_q^{c\prime\prime}(\omega) = h_q \sin(\pi a/2)\Gamma(1-a)(\omega t_0)^a \Big\{1 + [K_q + \kappa(a)]k_a(\omega t_0)^a\Big\}, \qquad (6.31\text{a})$$

$$k_a = 2\cos(\pi a/2)\Gamma(1-2a)/\Gamma(1-a). \qquad (6.31\text{b})$$

Within the microscopic MCT, the preceding results have their counterparts for the tagged-particle-density correlators $\phi_q^{sc}(t)$. To derive the formulas explicitly, one starts with Eqs. (6.6a), (6.7c) and concludes that the inhomogeneity $J_q^s(z)$ has to be evaluated correctly up to terms of order η^2. Three simplifications are possible for the expression (6.7a). The contributions of the last line can be dropped, since they are of order η^3 or since they are proportional to z and z^2. The correlators in the second line can be replaced by their leading asymptotic form: $g_q^c(t) \sim h_q(t_0/t)^a/(1-f_q^c), g_q^{sc}(t) \sim h_q^s(t_0/t)^a/[1-f_q^{sc}(P^s)]$. In the first contribution, one can substitute the result $g_k(t) - a_k(t_0/t)^a \sim a_k[K_k + \kappa(a)](t_0/t)^{2a}$. The desired formula for the leading and next-to-leading asymptotic representation of the correlators reads:

$$\phi_q^{sc}(t) = f_q^{sc} + h_q^s(t_0/t)^a\Big\{1 + [K_q^s + \kappa(a)](t_0/t)^a\Big\}. \qquad (6.32)$$

Here, Eq. (5.53) is used for the critical amplitude h_q^s; and Eq. (6.6d) is applied in order to obtain the correction amplitude:

$$K_q^s = [[1 - f_q^{sc}(P^s)]/h_q^s(P^s)] \sum_\ell R_{q,\ell}^s(P^s)\Big\{\Big[\sum_p B_{\ell,p}(P^s)a_pK_p\Big]$$
$$+ \Big[\sum_{kp}[1 - f_\ell^{sc}(P^s)]V_{\ell,kp}^s h_k h_p^s\Big] - \big[h_\ell^s/[1 - f_\ell^{sc}(P^s)]\big]^2\lambda^c\Big\}. \qquad (6.33)$$

Knowing the long-time expansions of the arrays $\boldsymbol{\phi}^c(t)$ and $\boldsymbol{\phi}^{sc}(t)$ up to terms of order $(t_0/t)^{2a}$, all mode-coupling functionals of these arrays can be expanded as well up to the specified order. For example, Eq. (5.28b) yields: $m_0^{sc}(t) = f_0^c + h_0(t_0/t)^a\{1 + [K_0 + \kappa(a)](t_0/t)^a\}$. The amplitudes are:

$$f_0^c = \int_0^\infty dq v_s(q) f_q^c f_q^{sc}, \qquad h_0 = \int_0^\infty dq v_s(q)\big[f_q^c h_q^s + h_q^s f_q^{sc}\big], \qquad (6.34\text{a})$$

$$K_0 = \int_0^\infty dq v_s(q)\big[f_q^c h_q^s K_q^s + h_q h_q^s + h_q K_q f_q^{sc}\big]/h_0. \qquad (6.34\text{b})$$

The low-frequency behaviour of the kernel's S-transform is described in leading and next-to-leading asymptotic expansion by:

$$ST[m_0^{sc}(t)](z) = f_0^c \tag{6.34c}$$
$$+h_0\Gamma(1-a)(-izt_0)^a\Big\{1 + [K_0 + \kappa(a)]\Gamma(1-a)(-izt_0)^a/\lambda^c\Big\}.$$

Corresponding expressions can be derived for the critical shear dynamics starting from Eqs. (5.6), (5.7).

The fluctuating-force correlator $m_0^s(z)$ determines the mean-squared displacement function $\Delta_s(z)$ and the velocity correlator $K_s(z)$. From Eqs. (3.32e), (5.29a), one gets: $ST[\Delta_s^c(t)](z) = d^2/\{ST[m_0^{sc}(t)](z) - (zd^2)[(z/(v_{th}^s)^2 + i/D_0^s]\}$. Since $\lim_{z\to 0} z/\eta^2 = 0$, the low-frequency asymptote of $\Delta_s^c(z)$ is determined in leading and next-to-leading order by

$$ST[\Delta_s^c(t)](z) = d^2/ST[m_0^{sc}(t)](z). \tag{6.35}$$

Restricting the discussion to non-degenerate A_2-bifurcation points, there holds $f_0^c > 0$. The preceding formula yields the low-frequency asymptotic expansion for $ST[\Delta_s^c(t)](z)$ up to terms proportional to η^2. This determines the long-time asymptotic expansion in leading and next-to-leading order:

$$\Delta_s^c(t) = r_{sc}^2 - h_{\text{MSD}}(t_0/t)^a\Big\{1 + \big[K_{\text{MSD}} + \kappa(a)\big](t_0/t)^{2a}\Big\}. \tag{6.36a}$$

Here Eqs. (5.57a,b) are used in order to introduce the critical localization length r_{sc} and the critical amplitude h_{MSD} for the mean-squared displacement: $(r_{sc}/d)^2 = 1/f_0^c, h_{\text{MSD}} = (r_{sc}^4/d^2)h_0$. The correction amplitude is given by

$$K_{\text{MSD}} = K_0 - \lambda^c h_0/f_0^c. \tag{6.36b}$$

Two paragraphs shall be added to the preceding derivations with the aim to emphasize general features which are shared by the asymptotic-expansion results. Let $\phi_X(t) = (X(t)|X)/(X|X)$ denote a normalized correlator for some dynamical variable X. The variable shall have definite parities for conjugations and time inversions. As explained repeated in chapter 3, $\phi_X(z)$ can be written as Zwanzig–Mori double-fraction in terms of a positive frequency Ω_X and a fluctuating-force kernel $M_X(z)$. The correlator and the kernel exhibit the standard symmetries; and they are Laplace transforms of continuous positive-definite functions. The kernel shall be split into a regular part $M_X^{\text{reg}}(z)$ and a remainder, which is denoted by $\Omega_X^2 m_X(z)$:

$$\phi_X(z) = -1\Big/\Big\{z - \Omega_X^2/\big[z + M_X^{\text{reg}}(z) + \Omega_X^2 m_X(z)\big]\Big\}. \tag{6.37a}$$

The first assumption to be made is that the dimensionless function $m_X(t)$ is given as mode-coupling polynomial. Consequently, for the relaxation kernel at the critical point, an asymptotic expression for the long-time behaviour can be formulated in leading and next-to-leading order:

$$m_X^c(t) = f_X^{mc} + h_X^m (t_0/t)^a \Big\{1 + \big[K_X^m + \kappa(a)\big](t_0/t)^a\Big\}. \qquad (6.37b)$$

Here, the positive numbers f_X^{mc} and h_X^m denote the critical arrested part and the critical amplitude, respectively, for the fluctuating-force correlator $m_X^c(t)$; and K_X^m denotes a correction amplitude for this function. A Tauberian theorem casts this formula into the expansion

$$ST[m_X^c(t)](z) = f_X^{mc} + h_X^m \Gamma(1-a)(-izt_0)^a \Big\{1 + \big[K_X^m + \kappa(a)\big]\Gamma(1-a)(-izt_0)^a/\lambda^c\Big\}. \qquad (6.37c)$$

This formula describes the low-frequency behaviour correctly up to order η^2. The second assumption to be made is that the regular kernel exhibits a finite zero-frequency limit: $M_X^{\rm reg}(z) \sim iM_X^{\rm reg\prime\prime}(\omega=0), z \to 0$. As a result, $M_X^{\rm reg}(z)$ can be dropped relative to the expression for $m_X^c(z)$. The formula (6.37a) simplifies to an expression, which is particularly transparent if written for the normalized dynamical susceptibility $\chi_X(z) = 1 + z\phi_X(z)$:

$$\chi_X^c = 1/\{1 + ST[m_X^c(t))](z)\}. \qquad (6.37d)$$

The S-transform of the kernel is the non-trivial part of the modulus for the response of variable X. This formula generalizes Eq. (5.137b) for the reorientational response of a dumbbell system.

If one substitutes the expansion (6.37c) into Eq. (6.37d), one can derive an asymptotic-expansion result for the correlator, which is correct up to terms of order η^2:

$$ST[\phi_X^c(t)](z) = f_X^c + h_X \Gamma(1-a)(-izt_0)^a \Big\{1 + \big[K_X + \kappa(a)\big]\Gamma(1-a)(-izt_0)^a/\lambda^c\Big\}.$$

Using a Tauberian theorem, the result leads to the equivalent asymptotic expression for the long-time decay including the leading correction:

$$\phi_X^c(t) = f_X^c + h_X (t_0/t)^a \Big\{1 + \big[K_X + \kappa(a)\big](t_0/t)^a\Big\}. \qquad (6.38a)$$

The amplitudes for the correlator are given by those for the kernel:

$$f_X^c = f_X^{mc}/(1 + f_X^{mc}), \qquad h_X = h_X^m/(1 + f_X^{mc})^2. \qquad (6.38b)$$

$$K_X = K_X^m - \lambda^c h_X^m/(1 + f_X^{mc}). \qquad (6.38c)$$

One can reverse the preceding derivation. Equation (6.37d) yields an expression for $m_X^c(z)$ in terms of $\phi_X^c(z)$. Assuming the asymptotic result (6.38b) for the correlator, the expression yields the result (3.37b) for the kernel. The amplitudes of the former determine those for the latter.

The two parameters ξ and ζ are quantities, which enter the formulas of all correlators for the description of the bifurcation dynamics at the A_2 singularity P^c. They determine the critical correlators via the number $\kappa(a)$. The correction amplitudes K_X are further general concepts, which quantify the distinction between different correlation functions.

Let us focus on the density-fluctuation correlators in order to discuss some implications of the asymptotic-expansion results. An onset time t_q^* for the leading-order result (6.21a) can be defined by the request that the relative error of this description reaches a margin δ^*. A similar δ^*-accuracy criterion can be used to define the onset frequency ω_q^* for the asymptotic description of the loss spectrum by Eq. (6.21c). Under the condition that higher than leading-order corrections can be neglected, one obtains:

$$t_q^*/t_0 = \left[|K_q + \kappa(a)|/\delta^*\right]^{1/a}, \qquad \omega_q^* = \left[1/t_q^*\right]/k_a^{1/a}. \tag{6.39}$$

The diamonds in Fig. 6.2 show onset parameters for the leading-asymptotic description of results for a hard-sphere system. They are defined with $\delta^* = 0.1$. The correction terms in Eqs. (6.30c), (6.31a) exhibit power law variation characterized by the critical exponent. Since $a < 1/2$, a moderate change of the factor $[K_q + \kappa(a)]$ can cause a large change of the onset parameters. This explains that the range of applicability of the asymptotic laws depends considerably on the correlator under discussion. The power-law variation does not define a time scale. As a result, $\omega_q^* t_q^*$ is not close to unity. For the hard-sphere system, the relevant factor $k_a^{1/a}$ is about 40. The onset value $\log[\omega_q^*]$ is separated much more from the natural-dynamics-scale value $\log[1/t_{\text{mic}}]$ than $\log[t_q^*]$ is separated from $\log[t_{\text{mic}}]$. Independent of the correlator chosen, the difference in the separation values for a discussion in the time domain and in the frequency domain is given by $\log[k_a^{1/a}]$. The difference is determined by the exponent parameter.

Figure 4.6 exhibits the correction amplitude $K_q + \kappa(a) \approx K_q$ for a hard-sphere-system model. It is explained in Sec. 4.3.5 that there are two wave-number regions of relevance for the arrest. Within the interval near but above the structure-factor-peak position, say $8 < qd < 14$, the critical arrested parts and the critical amplitude have intermediate-size values around 0.5. The correction amplitude has a small value of about -0.2. The leading-asymptotic formula explains the critical correlator for times exceeding about $10t_{\text{mic}}$. Inclusion of the leading correction provides a good description of $\phi_q^c(t)$ for $t \gtrsim 3t_{\text{mic}}$. This is demonstrated in Fig. 6.2 for the representative wave number q_2.

The second important wave-number interval is the structure-function-peak region, say $6.5 < qd < 7.5$. There is strong coupling of the density fluctuations with these wave numbers to the other fluctuations. This causes a pronounced peak of the f_q^c versus q curve. There are two trends, which prevent the increase of the critical correlator above unity. There appears a pronounced minimum of the h_q versus q curve and there is a large negative correction amplitude K_q. Figure 4.6 exhibits these two strong-coupling effects. Figure 6.2 demonstrates the implication of the increase of $|K_{q_1}|$ relative to $|K_{q_2}|$. The onset time $t^*_{q_1}$ exceeds $t^*_{q_2}$ by about two orders of magnitude. Inclusion of the leading corrections yields a description of $\phi^c_{q_1}(t)$ on a 10% accuracy level for all times exceeding about $8t_{\mathrm{mic}}$.

The mode-coupling coefficients for the tagged-particle-density dynamics diverge for vanishing wave number. Within the small-wave-number region, say $qd < 3$, the tagged-particle correlations deal with strong-coupling effects. Within this region, the same trends are exhibited, which are discussed in the preceding paragraph. Figure 5.5 demonstrates that there is a strong suppression of the critical amplitude h^s_q; and there is a large negative correction amplitude K^s_q.

Equations (4.32a,b), (4.33a,b) define the correlators $\phi(t)$ and $\phi_A(t)$ for a two-component model. The solutions for the second correlator have been used repeatedly for the quantitative description of experimental data for the glassy dynamics of liquids. This might justify the effort to understand in detail the results for this model. The manifold of glass-transition singularities is explained in connection with Figs. 4.2 and 4.4. Equations (4.67b) yield the critical arrested part f^c. For one-component models, Eq. (4.78b) implies for the critical amplitude $h = 1 - f^c$. The first correlator for transitions with critical exponents $a, 0 < a < 1/2$, is characterized by:

$$f^c = 1 - \lambda^c, \qquad h = \lambda^c, \qquad 1/2 < \lambda^c < 1. \tag{6.40a}$$

One infers from Eqs. (6.28a,b): $\xi = 0$, $\zeta = 1/2$. Hence, Eq. (6.29) specializes to

$$\kappa(x) = \frac{1}{2}\Gamma(1-x)^3/\big[\lambda^c\Gamma(1-3x) - \Gamma(1-x)\Gamma(1-2x)\big]. \tag{6.40b}$$

For one-component models, there is no correction amplitude K. The parameters for the second correlator are obtained by specializing the above derived formulas for $\phi^{sc}_q(t)$ to the case of $M = 1$. The positive coupling constant v_A plays the role of the control parameter point P^s. One gets for the critical arrest part of the second correlator and for its critical amplitude:

$$f^c_A = [1 - 1/(v_A f^c)], \qquad h_A = \lambda^c/(v_A f^{c2}), \qquad 1/f^c < v_A. \tag{6.40c}$$

The correction amplitude reads

$$K_A = (\lambda^c/f^c)[v_A\lambda^c f^c - 1]/[1 - v_A f^c]. \tag{6.40d}$$

These results exhibit the trends discussed above. If v_A increases to large values, f^c_A increases towards unity and h_A decreases towards zero. If v_A exceeds $1/[\lambda^c(1-\lambda^c)]$, the correction amplitude is negative. With increase of v_A to large values, K_A decreases towards its limit value $-\lambda^{c2}/(1-\lambda^c)$.

For $a = 1/3$, there holds $\kappa(a) = 0$. The corresponding exponent parameter is $\lambda^c = 0.684\ldots$. Consequently, for λ^c near 0.7, the asymptote (6.21a) describes

$\phi^c(t)$ very well for time decreasing to the end of the transient dynamics. This is demonstrated in Fig. 4.12 by the dashed-dotted line. The latter exhibits the leading-order asymptotic description for the exponent parameter $\lambda^c = 0.75$, which implies $a = 0.305$. The smallness of $\kappa(a)$ explains also the result shown in the upper panel of Fig. 4.13. The asymptote (6.21c) describes the critical loss for frequencies up to the edge of the band of normal-liquid excitations. For $v_A \geq 10$ the correction amplitude $K_A + \kappa(a)$ is close to its strong-coupling limit $-9/4$. Because of this large value, the onset frequency ω_A^* is decreased to small values. This explains the result exhibited in the lower panel of Fig. 4.13. The critical spectrum can be described by its asymptote (6.21c) only if $\omega < 10^{-5}$. There is the three-orders-of-magnitude interval of frequency variation between 10^{-4} and 10^{-1}, where the glassy relaxation spectrum $\chi_A^{c\prime\prime}(\omega)$ is strongly suppressed relative to the high-frequency extrapolation of the low-frequency asymptote.

Curves in Fig. 5.26 demonstrate the strong-coupling phenomenon in the time domain. For $v_A = 10$ and 20, the asymptote (6.21a) agrees with the critical correlators within the accuracy of the drawing for $t > 10^3$. But, for $t \leq 10^2$, the asymptote overestimates $\phi_A^c(t)$ considerably. Including the leading asymptotic corrections yields a description of the critical dynamics for all times outside the region of transient oscillations.

It is explained in Sec. 5.3.2 that $\phi_A(t)$ can serve as a model for the reorientational correlator $C_1(t)$ of a system of symmetric dumbbells. In this case, the regime $v_A \geq 10$ deals with the relaxation scenario for strong steric hindrance for reorientational motion. For smaller coupling constants v_A, one enters the region of weak steric hindrance. Here, the dynamics reflects precursor effects of a glass-transition singularity at the critical value $v_A^c = 1/f^c$. At this point, the correlator exhibits the critical decay towards zero, which is characteristic for the transition from glass 1 to glass 2. If v_A decreases towards $v_A^c = 4$, there appears a time interval of increasing length, where $\phi_A^c(t)$ is close to the critical decay for the glass 1-glass 2-transition point. The asymptotic-expansion theory for the discontinuous transition cannot describe the precursors of the nearby continuous glass transition. Such interval for the v_A^c-singularity precursor is exhibited in Fig. 5.26 by the curves for $v_A = 6$ and $10 < t < 200$.

(iii) The Cole–Cole susceptibility

The asymptotic law $\phi_X^c(t) = f_X^c + h_X(t_0/t)^a$ for the critical correlator of some dynamical variable X manifests itself in a decreasing upward-bent ϕ_X^c versus $\log t$ curve. Complementing the formula by a correction with a negative amplitude $[K_X + \kappa(a)]$ causes a maximum of the curve for small t. For times close to but above the maximum position, the curve is downward bent. Hence, there is an inflection point at some time t_{inf}. For times close to this point, the curve describes nearly-logarithmic decay: $\phi_X^c(t) - \phi_X^c(t_{\text{inf}}) \propto -\ell n[t/t_{\text{inf}}]$. Figure 6.2 exhibits this phenomenon for the correlator $\phi_{q_1}^c(t)$ within the interval of times between 3 and 300. In this subsection, a comprehensive description of the nearly-logarithmic decay of the critical correlators shall be discussed (Götze and Sjögren 1989a; Götze and Sperl 2004b; Sperl 2006).

The description of the dynamics in terms of the correlator is equivalent to that in terms of the dynamical susceptibility. Within the asymptotic-low-frequency expansion up to order η^2, Eq. (6.37d) expresses $\chi_X^c(z)$ in terms of the relaxation kernel. Equation (6.37c) represents the kernel as second-order polynomial in the parameter η. The polynomial is quantified by the three amplitudes f_X^{mc}, h_X^m, and K_X^m. One can rewrite the result as

$$\chi_X^c(z) = \chi_X^0/[1 + (-iz/\omega_X^c)^a + \hat{K}_X(-iz/\omega_X^c)^{2a}]. \tag{6.41}$$

The static susceptibility is given by

$$\chi_X^0 = (1 - f_X^c). \tag{6.42a}$$

The characteristic frequency for the critical dynamics of variable X reads:

$$\omega_X^c = t_0^{-1}\big[(1 - f_X^c)/(\Gamma(1-a)h_X)\big]^{1/a}. \tag{6.42b}$$

The correction amplitude is related to that for the kernel:

$$\hat{K}_X = \big[K_X^m + \kappa(a)\big]\big[(1 - f_X^c)/h_X\big]/\lambda^c. \tag{6.42c}$$

Formula (6.41) holds for Im $z > 0$; it represents also the analytic continuation on the half plane with Im $z < 0$. The standard symmetries $\chi_X^c(z) = \chi_X^c(-z) = \chi_X^c(z^*)^*$ can be used in order to get the susceptibility on the lower half plane of the frequencies. In particular, there holds $\chi_X^{c\prime\prime}(\omega) = -\chi_X^{c\prime\prime}(-\omega) = \text{Im}\,\chi_X^c(\omega + i0)$.

If one replaces $\hat{K}_X$ by zero, one gets the dynamical susceptibility, which is implied by the low-frequency asymptote of the modulus for the response:

$$\chi_X^{cc}(z) = \chi_X^0/[1 + (-iz/\omega_X^c)^a]. \tag{6.43a}$$

This function has been introduced in 1941 by K. S. Cole and R. H. Cole for the description of the dielectric response of glasses. The Cole–Cole loss spectrum

$$\chi_X^{cc\prime\prime}(\omega) = \big[\sin(\pi a/2)\chi_X^0\big]/\big[(\omega_X^c/\omega)^a + 2\cos(\pi a/2) + (\omega/\omega_X^c)^a\big] \tag{6.43b}$$

describes a hump with a maximum located at $\omega = \omega_X^c$. The function is invariant if (ω/ω_X^c) is changed to (ω_X^c/ω). The $\chi_X^{cc\prime\prime}$ versus $\log(\omega)$ curve is symmetric with respect to its maximum position. Since $a < 1/2$, it describes stretching. The stretching of the low-frequency wing of the peak is as pronounced as the one of the high-frequency wing. For a critical exponent a around 0.3, ω has to be varied by about five orders of magnitude in order to scan the interval with $\chi_X^{cc\prime\prime}(\omega)/\chi_X^{cc\prime\prime}(\omega_X^c) \geq 1/2$. In the original literature, such spectrum is referred to as a beta peak.

The normalized Cole–Cole susceptibility $\hat{\chi}_X^{cc}(z) = \chi_X^{cc}(z)/\chi_X^0$ shall be used to define $\hat{\phi}_X^{cc}(z) = [\hat{\chi}_X^{cc}(z) - 1]/z$. One checks that this function is positive analytic.

There holds $\lim_{z\to\infty}[-z\hat{\phi}_X^{cc}(z)] = 1$. From the results from Appendix A.3, one concludes that $\hat{\phi}_X^{cc}(z)$ is the Laplace transform of a continuous positive-definite function $\hat{\phi}_X^{cc}(t)$, which has initial value unity. It can be evaluated via its spectral representation (3.19b):

$$\hat{\phi}_X^{cc}(t) = \left(\frac{2}{\pi}\right)\int_0^\infty \cos(\omega t)\left[\chi_X^{cc\prime\prime}(\omega)/(\omega\chi_X^0)\right]d\omega. \tag{6.44a}$$

Since the spectrum is integrable and continuous, the long-time limit of the Fourier integral vanishes: $\hat{\phi}_X^{cc}(t\to\infty) = 0$. To evaluate this function for $t > 0$, one can also use the inversion formula (A.4). Performing the limit $\Omega\to\infty$ and $\epsilon\to 0$, one gets $\hat{\phi}_X^{cc}(t) = (-i/2\pi)\int\hat{\phi}_X^{cc}(\omega)\exp[-i\omega t]d\omega$. Formula (6.43a) provides an holomorphic continuation of $\hat{\phi}_X^{cc}(z)$ to all frequencies in the lower half plane, except for the points on the cut given by $-i\gamma, \gamma\geq 0$. Deforming the integration path from the real axis so that it encircles the cut, one arrives at

$$\hat{\phi}_X^{cc}(t) = \frac{1}{\pi}\int_0^\infty \exp[-\gamma\omega_X^c t]\Big\{\sin(\pi a)\gamma^{a-1}/ \\ \left[1+2\gamma^a\cos(\pi a)+\gamma^{2a}\right]\Big\}d\gamma. \tag{6.44b}$$

According to the discussion of Eqs. (1.2a,b), the function $\hat{\phi}_X^{cc}(t)$ is completely monotonic. Equation (6.43a) establishes a power-series $\hat{\phi}_X^{cc}(z) = \sum_n c_n\zeta^n, \zeta = [-iz/\omega_X^c]^{-a}$, which converges for $|\zeta| < 1$. From Eqs. (A.59a,b), one gets $\zeta^n = -zLT[(t\omega_X^c)^{na}](z)/\Gamma(1+na)$. Thereby, $\hat{\phi}_X^{cc}(z)$ is presented as Laplace transform of a power series in $(t\omega_X^c)^a$. Because of holomorphy, the presentation holds for all positive times: $\hat{\phi}_X^{cc}(t) = M_a[-(t\omega_X^c)^a]$. Here, $M_a(y) = \sum_n y^n/\Gamma(1+na)$ denotes the Mittag-Leffler function for index a.

Let us write $\phi_X^{cc}(z) = [\chi_X^{cc}(z) - 1]/z$ for the correlation function, which is equivalent to the dynamical susceptibility $\chi_X^{cc}(z)$. Combining results from the preceding paragraph, one gets $[-z\phi_X^{cc}(z)] = f_X^c + \chi_X^0[-z\hat{\phi}_X^{cc}(z)]$. Consequently, the Cole–Cole correlator in the time domain is given by

$$\phi_X^{cc}(t) = f_X^c + (1-f_X^c)\hat{\phi}_X^{cc}(t). \tag{6.45}$$

If $\hat{K}_X\geq 0$, Eq. (6.41) defines a function, which is holomorphic for Im $z > 0$. It can be used to define a positive analytic function $\hat{\phi}_X^c(z) = [(\chi_X^c(z)/\chi_X^0) - 1]/z$. The Eqs. (6.44a,b) remain valid in their proper generalization. These results define a Cole–Cole model with corrections. For large times and small frequencies, the model reproduces Eq. (6.38b) and Eq. (6.38a), respectively. If $\hat{K}_X < 0$, Eq. (6.41) does not define a positive analytic function.

The characteristic frequency ω_X^c is a scale, which separates regions of qualitatively different dynamical behaviour. For $|z/\omega_X^c| < 1$, one gets the convergent expansion $-z\phi_X^{cc}(z) = 1-\chi_X^{cc}(z) = 1-(1-f_X^c)[1-(-iz/\omega_X^c)^a+(-iz/\omega_X^c)^{2a}+\cdots]$.

This implies the low-frequency result for the loss spectrum

$$\chi_X^{cc\prime\prime}(\omega) = \left[\sin(\pi a/2)\chi_X^0\right](\omega/\omega_X^c)^a\left[1 - 2\cos(\pi a/2)(\omega/\omega_X^c)^a + \cdots\right]. \quad (6.46a)$$

Application of Eqs. (A.65a,b) yields the equivalent large-time expansion

$$\phi_X^{cc}(t) = f_X^c + h_X(t_0/t)^a\{1 - k_X(t_0/t)^a + \cdots\}, \quad (6.46b)$$

$$k_X = \lambda^c h_X^m/(1 + f_X^{mc}). \quad (6.46c)$$

The leading terms reproduce the asymptotic behaviour of the critical loss spectrum $\chi_X^{c\prime\prime}(\omega)$ and the critical correlator $\phi_X^c(t)$, respectively. For increasing ω, the correction term in Eq. (6.46a) describes a suppression of the loss spectrum below the power-law asymptote. This is a precursor for the approach of the spectrum towards the loss maximum. For decreasing t, the correction term in Eq. (6.46b) describes a suppression of the correlator below the small-time extrapolation of its long-time tail. It is a precursor for the monotonic approach of the correlator towards its initial value unity.

For $|\omega_X^c/z| < 1$, the Cole–Cole correlator can be represented by the convergent power series: $-z\phi_X^{cc}(z) = 1 - \chi_X^0(-iz/\omega_X^c)^{-a}[1 - (-iz/\omega_X^c)^{-a} + \cdots]$. The series implies a formula for the high-frequency tail of the loss spectrum. It is obtained from Eq. (6.46a) by the change of (ω/ω_X^c) to its inverse:

$$\chi_X^{cc\prime\prime}(\omega) = \left[\sin(\pi a/2)\chi_X^0\right](\omega_X^c/\omega)^a\left[1 - 2\cos(\pi a/2)(\omega_X^c/\omega)^a + \cdots\right]. \quad (6.47a)$$

Contrary to the low-frequency asymptote for the critical spectrum, this part of the loss spectrum increases with decreasing ω. For large frequencies, the variation is proportional to the power law ω^{-a}. The correction term suppresses this spectrum the more the smaller is (ω/ω_X^c). This is a precursor of the approach towards the maximum for the loss spectrum. Application of the formulas (A.64a,b) provides the expression for the short-time behaviour of the correlator:

$$\phi_X^{cc}(t) = 1 - h'_X(t/t_0)^a\{1 - k'_X(t/t_0)^a + \cdots\}, \quad (6.47b)$$

$$h'_X = (1 - f_X^c)^2/\left[h_X\Gamma(1+a)\Gamma(1-a)\right], \quad (6.47c)$$

$$k'_X = (1 - f_X^c)\Gamma(1+a)/\left[h_X\Gamma(1+2a)\Gamma(1-a)\right]. \quad (6.47d)$$

For increasing t, the correlator decreases below its initial value according to a power law. The latter is discussed in Eq. (1.4a) for von Schweidler's decay with $b = a$. Because of the negative correction term, the correlator increases above its pseudo-von Schweidler asymptote if t increases towards $1/\omega_X^c$. This increase is a precursor of the correlator's approach towards the arrest value f_X^c.

The dotted line in Fig. 6.3 exhibits the Cole–Cole correlator $\phi_A^{cc}(t)$ for the two-component model defined by Eqs. (4.32a,b), (4.33a,b). It is calculated for

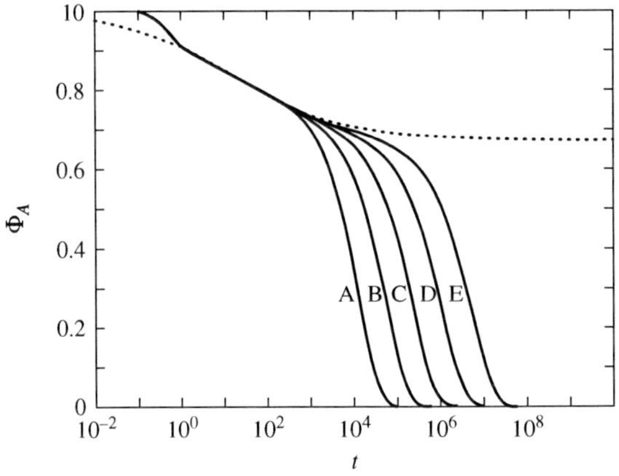

FIG. 6.3. Correlators $\phi_A(t)$ of the two-component model defined by Eqs. (4.32a,b), (4.33a,b) with $\Omega = \Omega_A = 1, \nu = \nu_A = 5, v_A = 10$. The states are located on a curve with $\lambda(P) = \lambda^c = 0.7$. The separation parameters for curve A to E are given by $-1000\sigma = 8, 4, 2, 1, 0.5$ from left to right. The critical arrested part reads $f_A^c = 2/3$. The dotted line presents the Cole–Cole correlator (6.45) for this model. Reproduced from Götze and Sjögren (1989a).

the critical point with exponent parameter $\lambda^c = 0.7$. This value implies a critical exponent $a = 0.327$. The function $\hat{\phi}_A^{cc}(t)$ decreases from 90% of its initial value to 10% within an interval, which deals with a time increase by about six orders of magnitude. For $t\omega_A^c \ll 1$, Eq. (6.47b) describes a strongly stretched downward-bent part of the $\phi_A^{cc}(t)$ versus $\log(t)$ curve. Equation (6.46b) describes a strongly stretched upward-bent part for $t\omega_A^c \gg 1$. Within the accuracy of the drawing, the curve cannot be distinguished from a straight line within the crossover interval $1 \leq t \leq 10^3$. This nearly logarithmic decay is reflected by nearly constant loss for the frequency interval $0.1 \leq \omega/\omega_A^c \leq 10$. Within this region, $\chi_A^{cc\prime\prime}(\omega)$ changes by about 15% only. The corresponding fluctuation spectrum exhibits nearly $(1/\omega)$ variation. Such spectrum is referred to as $(1/f)$ noise.

The Cole–Cole expression $\phi_X^{cc}(t)$ is based on the leading asymptotic expression for the fluctuating-force kernel $m_X^c(t) \sim f_X^{mc} + h_X^m(t_0/t)^a$. This expression provides a better description of the critical correlator than the asymptotic formula $\phi_X^c(t) \sim f_X^c + h_X(t_0/t)^a$ if and only if the onset time t_{mX}^* for the kernel is smaller than the onset time t_X^* for the correlator. This is the case if and only if $|K_X^m + \kappa(a)|$ is smaller than $|K_X + \kappa(a)|$. Equation (6.38d) yields $K_X^m = K_X + \lambda^c h_X^m/(1 + f_X^{mc})$. If K_X is large and negative, there is some cancellation of this amplitude by the positive second contribution to K_X^m, i.e., there is a trend for a decrease of t_{mX}^* relative to t_X^*. The Cole–Cole formulas (6.46b,c) provide the dominant negative contribution k_X to K_X. It is explained in the preceding subsection that strong-coupling effects cause a large negative amplitude K_X. Hence, one concludes the following. Strong coupling of the dynamical variable X to density fluctuations implies a trend that the Cole–Cole formulas provide a better description of the critical dynamics than the one provided by the leading asymptotic expansion formulas for the correlator.

The conclusion explains the results for the strong-coupling dynamics of the two-component model, which is discussed in connection with Eqs. (6.40a–d). There holds $\lim_{v_A\to\infty} K_A = -k_A$. For strong coupling, the correction amplitude $-k_A$ in the expression (6.46b) reproduces the amplitude K_A. For λ^c around 0.7, $\kappa(a)$ is very small. In this case, the asymptotic-limit description of the kernel $m_A^c(t)$ provides an equally perfect description of $\phi_A^c(t)$ as is provided by the asymptotic-limit description for $\phi^c(t)$. This is demonstrated in Fig. 5.26 by the curves, which refer to $v_A = 10$.

In order to understand further implications of the preceding results for the glassy-dynamics evolution of liquid states, let us recall the following. For small negative separation parameters σ, the correlators are close to the critical ones for times increasing up to the plateau-crossing value τ^{pc}. For σ tending to zero, τ^{pc} tends to infinity. The below-plateau decay for $t > \tau^{pc}$ causes a loss peak at some σ-sensitive position $\omega_{\max}$ obeying $\omega_{\max}\tau^{pc} \ll 1$. The discussion shall be restricted to: $10t_{\text{mic}} < t < \tau^{pc}, 1/\tau^{pc} < \omega < 0.1/t_{\text{mic}}$. The Cole–Cole correlators and loss spectra shall be used to characterize the critical dynamics. Depending on the ratio of the time scale $1/\omega_X^c$ for the Cole–Cole dynamics and the scale t_{mic} for the transient dynamics, four cases shall be distinguished. They can be demonstrated by the solutions of the mentioned schematic model. The parameter $\omega_A^c t_{\text{mic}}$ can be varied by changing the frequencies Ω_A and ν_A in the equation of motion (4.33a) relative to the frequencies Ω and ν in Eq. (4.32a).

The condition $1/\omega_X^c \ll t_{\text{mic}}$ specifies the first possibility. Within the whole time interval outside the transient, there holds $\phi_X^c(t) = \phi_X^{cc}(t) = f_X^c + h_X(t_0/t)^a$. For all frequencies below the normal-liquid-excitation band, the loss spectrum is described by $\chi_X^{cc''}(\omega) = h_X \sin(\frac{\pi}{2}a)\Gamma(1-a)(\omega t_0)^a$. The scenario deals with plateau crossings and loss minima, which are discussed at the end of subsection (i). The dashed line for the correlator $\phi(t)$ in Fig. 5.26 demonstrates an example. The dotted line in the upper panel of Fig. 4.13 demonstrates the case in the frequency domain. The specified scenarios for the plateau-crossings and loss minima occur for every model independent of the size of $\omega_X^c t_{\text{mic}}$. They occur for states so close to the arrest point that $1/\tau^{pc} \ll \omega_X^c$. Under this condition, the Cole–Cole correlators and loss spectra reproduce simple power-law variations. The first case deals with the possibility that there are no additional features for $t > t_{\text{mic}}$ or $\omega < 0.1/t_{\text{mic}}$, which complicate the scenario.

The second case deals with characteristic times $1/\omega_X^c$ within the centre of the transient regime, say $1/\omega_X^c \approx 3t_{\text{mic}}$. The frequency ω_X^c is located within the band of normal-liquid-dynamics excitations but near the low-frequency edge thereof. Consequently, there are strong deviations of the critical correlator from the power-law asymptote within the first decade of time increase outside the transient. These deviations are described by the dominant correction term of the Cole–Cole expression (6.46b). The correlator $\phi_A(t)$, which is shown in Fig. 5.26

for $v_A = 10$, demonstrates this possibility. Similarly, for frequencies below the edge of the microscopic-excitation band, the loss spectrum is suppressed below the extrapolated ω^a asymptote. This phenomenon is demonstrated in Fig. 6.10 below, and it is discussed there in another context.

A third possibility occurs if ω_X^c is about an order of magnitude smaller than the edge of the normal-liquid-excitation band. Adjacent to the band edge, there is a frequency interval of nearly constant loss. Equivalently, for times exceeding the terminal time for the transient dynamics, the critical correlator exhibits nearly logarithmic decay. The curves $A - E$ in Fig. 6.3 demonstrate the evolution of this decay upon decreasing the separation parameter. If $|\sigma|$ is not too small, the correlators deviate from $\phi_X^c(t)$ so that the $\phi_X(t)$ versus $\log t$ curves are downward bent. Curves A, B and C exhibit this property. Example C deals with glassy relaxation, which is stretched over six orders of magnitude time increase. The plateau crossing time τ^{pc} is about a factor 10^4 larger than the terminal time for the transient. Nevertheless, curve C does not exhibit a decay through the critical arrested part f_A^c, which manifests itself by an inflection point of the $\phi_X(t)$ versus $\log t$ diagram. This inflection point appears only if $|\sigma|$ is smaller than 10^{-3}, as is demonstrated by curve E.

The fourth possibility refers to frequencies ω_X^c, which are two or more orders of magnitude smaller than the microscopic-band-edge frequency. In this case, the critical loss spectrum exhibits a maximum at ω_X^c. The crossover of the high-frequency wing of the Cole–Cole peak to the microscopic excitation band leads to a loss minimum at some position ω'_{min}. The schematic-model solutions in Fig. 6.4 deal with this case for $\omega_A^c \approx 0.03$ and $\omega'_{\text{min}} \approx 1.5$. The spectra G and E refer to liquids with frequencies ω_{max} of the σ-sensitive low-frequency loss-peak maxima, which are more than five orders of magnitude smaller than ω_A^c. For frequencies above ω_{max} and below ω_A^c, the crossover from the high-frequency wing of the low-frequency loss peak to the low-frequency wing of the Cole–Cole peak causes a σ-sensitive minimum at some position ω_{min}. The upper part of the minimum converges to the critical spectrum, which is close to its ω^a asymptote. For such small separation parameters, the scenario for the evolution of the minimum is that of the first case.

Increasing the separation of the state from the critical one, the value $\chi''_X(\omega_{\text{min}})$ increases. There appears a characteristic negative separation parameter σ_c, which specifies a state with $\chi''_X(\omega_{\text{min}}) = \chi''_X(\omega_X^c)$. For σ near σ_c, there is a noteworthy frequency interval of nearly-constant loss, because the decrease of the loss-peak tail is nearly compensated by the increase of the critical spectrum. In Fig. 6.4, the phenomenon is demonstrated by curve C.

If $|\sigma|$ is larger than $|\sigma_c|$, the spectrum decreases monotonically for frequencies increasing from ω_{max} to ω'_{min}. The Cole–Cole peak is hidden under the low-frequency loss peak. The part of the spectrum for $\omega_{\text{max}} \ll \omega < 1/\tau^{pc}$ reflects the beginning of the below-plateau decay. The part for $1/\tau^{pc} \ll \omega \leq \omega'_{\text{min}}$ describes the approach of the spectrum towards the high-frequency wing of the Cole–Cole peak. If ω'_{min} would be much larger than ω_X^c, this part would approach

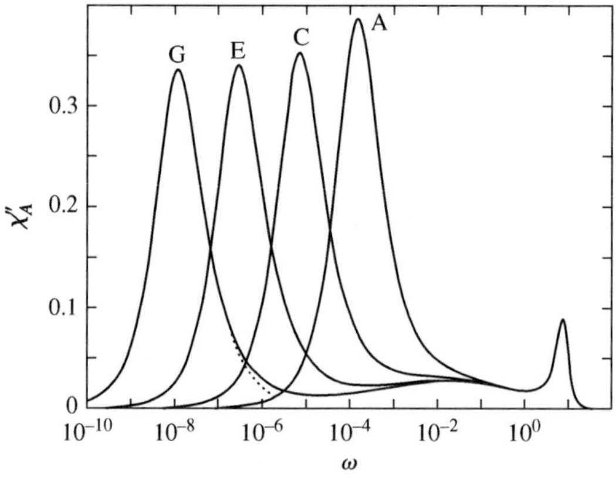

FIG. 6.4. Loss spectra $\chi''_A(\omega)$ of the two-component model defined by Eqs. (4.32a,b), (4.33b) with $\Omega = 1, \Omega_A = \nu_A = \nu = 5, v_A = 15$. The coupling coefficients are: $v_1 = v_1^c + \epsilon\lambda^c/[1+(1-\lambda^c)^2], v_2 = v_2^c + \epsilon\lambda^c(1-\lambda^c)/[1+(1-\lambda^c)^2]$. Here, v_1^c and v_2^c are the values for the critical point with the exponent parameter $\lambda^c = 0.7$. Curves A, C, E and G are evaluated for the distance parameter $\epsilon = -0.2/4^n$ with $n = 0, 1, 2$ and 3, respectively. The dotted line shows a Kohlrausch spectrum with stretching exponent $\beta = 0.85$ fitted to the low-frequency loss peak of spectrum G. Reproduced from Buchalla *et al.* (1988).

the pseudo-von Schweidler-law asymptote described by Eq. (6.47a). Curve A exhibits a crossover between the two parts for frequencies near $\omega = 0.005$. A double-logarithmic presentation of similar results demonstrate a crossover from a von Schweidler-law tail $\chi''_X(\omega\tau^{pc} < 1) \propto 1/\omega^b$ to a pseudo-von Schweidler-law behaviour $\chi''_X(\omega\tau^{pc} > 1) \propto 1/\omega^{b_{\text{eff}}}$. With increasing $|\sigma|$, the effective exponent b_{eff} increases from zero to a. The described scenario for the evolution of a loss peak with an excess wing is similar to that demonstrated by the dielectric-loss data in Fig. 1.14 (Cummins 2005). An example for this phenomenon is discussed in Fig. 6.9 below in another context.

Figure 6.5 exhibits results for the critical dynamics of a tagged particle of a simple system. They concern the critical mean-squared displacement $\delta r^2(t) = 6\Delta_s^c(t)$ for the hard-sphere system. The model and the units are defined in connection with Fig. 4.14; some relevant parameters are specified in the caption of Fig. 6.2. The amplitudes in Eq. (6.36a) have the values: $r_{sc}^2 = 0.00557, h_{\text{MSD}} = 0.0116, K_{\text{MSD}} = -1.23$ (Fuchs *et al.* 1998). Function $\Delta_s(t)$ quantifies the tagged-particle-density correlator for small wave numbers q. The small-q regime deals with strong coupling effects. As discussed in subsection (ii), this explains why K_{MSD} is negative and large. As a result, the asymptotic law $\Delta_s^c(t) - r_{sc}^2 \sim -h_{\text{MSD}}(t_0/t)^a$ can describe only the final 10% of the mean-squared displacement's increase towards its arrest value $6r_{sc}^2$. Inclusion of the leading asymptotic correction decreases the onset time for the analytic description of $\Delta_s^c(t)$ by about

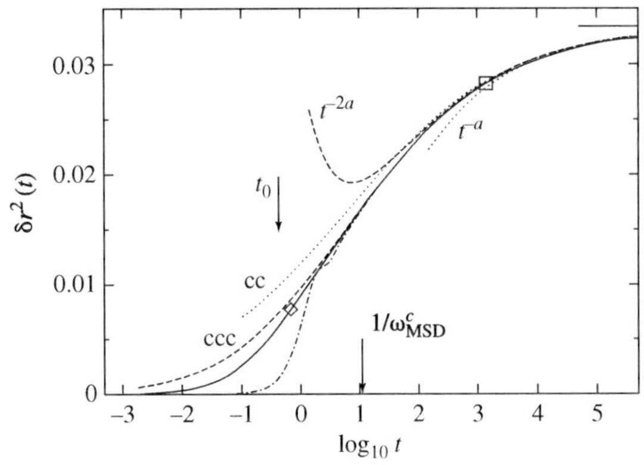

FIG. 6.5. The full line exhibits the mean-squared displacement at the transition point φ_c of a HSS with stochastic short-time dynamics. The model and the units are explained in connection with Figs. 4.6 and 4.14. The dashed-dotted line reproduces the corresponding result from Fig. 5.7 with the time rescaled so that it coincides with the full line for $t > 100$. The dotted line marked t^{-a} shows the function $6[r^2_{sc} - h_{\text{MSD}}(t_0/t)^a]$ and the dashed line marked t^{-2a} exhibits the result from Eq. (6.36a). The horizontal line marks the value $6r^2_{sc}$. The dashed line with label *ccc* is the result from Eqs. (6.48a–c) for the Cole–Cole law with correction. The dotted line marked *cc* shows the Cole–Cole law, which is obtained by using $\hat{K}_{\text{MSD}} = 0$. The diamond and the square mark the same position on the curve for the critical mean-squared displacement as shown in Fig. 5.9. Reproduced from Sperl (2005).

a factor 20. Equation (6.36a) describes the final third of the increase of $\delta r^2(t)$. The discussion of Fig. 4.24 suggests that structure relaxation occurs for times exceeding about $7t_0$. The dashed-dotted line in Fig. 6.5 supports this suggestion. It reproduces $\delta r^2(st)$ with $\Delta^c_s(t)$ taken from Fig. 5.7. The factor s is adjusted so that this function for a Newtonian transient dynamics and that for a Brownian short-time dynamics agree for large times. One concludes: the increase of $\delta r^2(t)$ from about 0.013 to about 0.022 deals with structural relaxation, which cannot be explained by Eq. (6.36a).

Equations (6.34c,6.35) provide the expressions:

$$ST[\Delta^c_s(t)](z) = r^2_{sc}/\big[1 + (-iz/\omega^c_{\text{MSD}})^a + \hat{K}_{\text{MSD}}(-iz/\omega_{\text{MSD}})^{2a}\big], \qquad (6.48a)$$

$$\omega^c_{\text{MSD}} = t_0^{-1}\big[f_0/(\Gamma(1-a)h_0\big]^{1/a}, \quad \hat{K}_{\text{MSD}} = \big[K_0 + \kappa(a)\big]f_0/(h_0\lambda^c). \qquad (6.48b)$$

These formulas are analogous to Eqs. (6.41), (6.42b,c) for the Cole–Cole susceptibility with correction. Specializing to $\hat{K}_{\text{MSD}} = 0$, Eq. (6.48a) is the analogue of the Cole–Cole susceptibility. The formulas can be obtained from the Cole–Cole results with and without correction, respectively, for the correlator $\phi^{sc}_q(z)$. The latter has to be specialized to the small-q limit (3.30). Therefore, all conclusions drawn above from Eq. (6.41) have their counterpart here. The strong-coupling induced large negative value $K_{\rm MSD}$ is partly compensated by $\lambda^c h_0/f_0$. Thereby, Eq. (6.36b) explains a reduction of the correction amplitude K_0, which leads to the small correction amplitude $\hat{K}_{\rm MSD} = 0.196$. Equation (6.36b) shows that the large negative contribution $(-\lambda^c h_0/f_0^c)$ is the punishment for simplifying the formula (6.35) to a power series in the small parameter $(-izt_0)^a$. Since $\hat{K}_{\rm MSD}$ is non-negative, Eq. (6.48a) defines a holomorphic function for Im $z > 0$. A spectral representation holds for the inversion of the S-transform:

$$\Delta_s^c(t) = r_{sc}^2 - \left(\frac{2}{\pi}\right) \int_0^\infty \cos(\omega t)\big[\mathrm{Im}\, ST[\Delta_s^c(t)](\omega)/\omega\big]\, d\omega. \tag{6.48c}$$

The curve *cc* in Fig. 6.5 exhibits the preceding results for $\hat{K}_{\rm MSD} = 0$. The curve describes $\Delta_s^c(t)$ as obtained from the leading asymptotic expression for $m_0^c(t)$. It reproduces the description of the mean-squared displacement by Eq. (6.36a). Contrary to the latter description, the former extrapolates reasonably to times below $1/\omega_{\rm MSD}^c$. The dashed line marked *ccc* exhibits the result of the preceding formulas. They describe the 70% increase of $\Delta_s^c(t)$, which occurs for $t \gtrsim 3t_0$. For the increase of t/t_0 between about 3 and about 300, $\delta r^2(t)$ increases from 0.010 to 0.023. Figure 5.9 demonstrates that this part is close to the one measured for a hard-sphere colloidal suspension with packing fraction $\varphi = 0.583$. This value of φ is very close to the glass-transition value φ_g.

Figure 6.5 exemplifies that the asymptotic-solution theory can provide an understanding of the glassy dynamics described by MCT. The figure is also an example for the relevance of the theory for the explanation of a subtle experimental result for the complex dynamics of a glass-forming liquid.

6.1.3 *Asymptotic description of the A_2-bifurcation dynamics*

In the following, the dynamics shall be discussed for the limit of states P approaching an A_2-glass-transition singularity P^c. A loosely-handled power-counting procedure will lead to simplified equations of motion. It is the goal of this section to discuss the dynamics described by these equations.

The considerations start by identifying a factorization property, which generalizes that discussed in Sec. 1.4. The dependence of the correlators on the time t and on the state P can be described by a function G, which is shared by all correlators. The dependence of G on P enters via that of the separation parameter $\sigma = \sigma(P)$. Function G has the form of a scaling law. This law is similar to that explained in Sec. 4.2.4, albeit with a more subtle shape function.

The plateau-crossing scenario for a generic liquid–glass transition is governed by two time scales t_σ and t'_σ. These scales exhibit a divergent power-law dependence of the separation parameter: $t_\sigma \propto 1/|\sigma|^\delta, t'_\sigma \propto 1/|\sigma|^\gamma, \gamma > \delta > 0$. The below-plateau decay of the correlation functions is dominated by von Schweidler's law. Like the critical exponent a and the exponents δ and γ for the time scales, the von Schweidler exponent b is determined by the exponent parameter λ^c.

(i) Reduction of the equations of motion

The equations of motion (6.4), (6.5) for the array $\boldsymbol{g}(t)$ are the basis for the following considerations. As explained in Sec. 6.1.2(i), the second contribution on the right-hand side of Eq. (6.5a) can be neglected compared to the first one. Hence, the leading contributions to the functions $g_q(t)$ is determined by the dominant part of $g(t)$, which shall be denoted by $G(t) = g_0^\sigma(t)$. The searched-for asymptotic-limit solution has the form

$$g_q(t) = a_q g_0^\sigma(t). \tag{6.49a}$$

As a result, Eqs. (6.4b–d) lead to the expression for the inhomogeneity

$$\begin{aligned} J_q(z) = {} & D_q^{(0)}(P) + \Big[\sum_k D_{q,k}^{(1)}(P)a_k\Big] ST\big[g_0^\sigma(t)\big](z) \\ & + \Big[\sum_{k_1 k_2} A_{q,k_1 k_2}^{(2)}(P)a_{k_1}a_{k_2}\Big] ST\big[g_0^\sigma(t)^2\big](z) - a_q^2 ST\big[g_0^\sigma(t)\big](z)^2 \\ & + J_q^{\rm reg}(z) + O_q(g^3). \end{aligned} \tag{6.49b}$$

If one substitutes this formula into the solubility condition (6.5c), one obtains a generalization of Eq. (6.10) to states P, which can be different from P^c:

$$\begin{aligned} & \Big[\sum_q a_q^* D_q^{(0)}(P)\Big] + \Big[\sum_{qk} a_q^* D_{q,k}^{(1)}(P)a_k\Big] ST\big[g_0^\sigma(t)\big](z) \\ & + \lambda(P) ST\big[g_0^\sigma(t)^2\big](z) - ST\big[g_0^\sigma(t)\big](z)^2 = -\sum_q a_q^*\big[J_q^{\rm reg}(z) + O_q(g^3)\big]. \end{aligned} \tag{6.49c}$$

Motivated by the results in Sec. 6.1.2, it shall be assumed that $J_q^{\rm reg}(z)$ can be neglected compared to term of order g^2. Consequently, for the intended leading-order discussion, the right-hand side of the preceding equation can be dropped. For a generic A_2 bifurcation, the first bracket in the equation is formula (4.86a) for the separation parameter. The anticipated classification of terms makes no sense, unless one assumes σ to be a term of order g^2. Then, the second term in Eq. (6.49c) is of order g^3 and can be dropped. The equation of motion reduces to the request of cancellation of three terms of order g^2:

$$\sigma(P) + \lambda(P) ST[g_0^\sigma(t)^2](z) - ST[g_0^\sigma(t)](z)^2 = 0. \tag{6.50a}$$

This is the simplified equation of motion in the frequency domain. It can be transformed into the time domain with the aid of Eqs. (6.2), (A.6), (A.7a):

$$\sigma(P) + \lambda(P) g_0^\sigma(t)^2 - \partial_t \int_0^t g_0^\sigma(t-t') g_0^\sigma(t')dt' = 0, \text{ generic } A_2. \tag{6.50b}$$

For a degenerate fold bifurcation, the first bracket in Eq. (6.49c) vanishes and the second one is the expression (4.96a) for the separation parameter. To give a meaning to the classification of the terms, one assumes that σ is a quantity of order g. The searched for simplified equation of motion results from the request that there is cancellation of the three remaining terms of order g^2:

$$\sigma(P)ST[g_0^\sigma(t)](z) + \lambda(P)ST[g_0^\sigma(t)^2](z) - ST[g_0^\sigma(t)](z)^2 = 0. \tag{6.51a}$$

This formula is equivalent to the equation in the time domain:

$$\sigma(P)g_0^\sigma(t) + \lambda(P)g_0^\sigma(t)^2 - \partial_t \int_0^t g_0^\sigma(t-t')g_0^\sigma(t')dt' = 0, \text{ degenerate } A_2. \tag{6.51b}$$

In Sec. 6.1.4, the solutions of the simplified equations shall be used as the basis for the definition of a limit procedure. Thereby, the formulas and the assumptions made for their derivation will be identified as parts of the first step of an asymptotic-expansion approach. Anticipating this result, the remaining part of this section deals with the implications of the preceding equations for $g_0^\sigma(t)$.

Equation (6.50b) or (6.51b) reflects the essence of MCT in a nutshell. Two terms formulate the interplay of nonlinearities with retardation phenomena. Their relative weight is specified by the exponent parameter $\lambda^c = \lambda(P^c)$. According to Eqs. (4.83a–d), $1 - \lambda^c$ characterizes the distance of the A_2 bifurcation singularity P^c from possible higher-order endpoint singularities of the bifurcation surface. The separation parameter $\sigma(P)$ is the relevant combination of the mode-coupling coefficients, which characterizes the distance between P and P^c.

Let us consider a generic path in parameter space, $\epsilon \to P^\epsilon$, as discussed in Eqs. (4.84)–(4.86). Because of Eq. (4.88a), $\lambda_\epsilon = \lambda(P^\epsilon)$ is a smooth function of the distance parameter ϵ. One can write

$$\lambda_\epsilon = \lambda^c + O(\epsilon). \tag{6.52}$$

For the intended leading-order calculations, one can drop the $O(\epsilon)$ contribution. Similarly, one can drop the $O(\epsilon^2)$ contribution in Eq. (4.86c). To ease the notations for the following formulas, $\sigma(P)$ and $\lambda(P)$ shall be noted as σ and λ,respectively. The letters can be interpreted as $\sigma = C\epsilon$ and $\lambda = \lambda^c$. However, one can imagine paths, where the truncation of σ_ϵ and λ_ϵ to their leading Taylor contributions implies an undesired restriction of the limit of applicability of the results to be derived. In this case, the abbreviations should be interpreted as $\sigma = \sigma_\epsilon$ and $\lambda = \lambda_\epsilon$. This means, for example, that the exponents, which will be derived from λ, exhibit a smooth ϵ-dependence.

Because of Eq. (6.5d), the formula (6.49a) is equivalent to the expressions for the density-fluctuation correlators:

$$\phi_q(t) - f_q^c = h_q g_0^\sigma(t), \quad q = 1, \ldots, M. \tag{6.53a}$$

This formula generalizes Eq. (6.17b) in the sense that $g_0^c(t) = (t_0/t)^a$ is replaced by the σ-dependent function $g_0^\sigma(t)$. Similarly, the derivation of Eq. (6.17c) can be generalized. The dependence on P of the functions $A_q^{(0)}(P)$ and $A_{q,k}^{(1)}(P)$ can be ignored; it would not contribute to the searched-for leading-order contributions. One obtains the expression for the fluctuating-force correlators

$$m_q(t) - f_q^{mc} = h_q^m g_0^\sigma(t), \quad q = 1, \ldots, M. \tag{6.53b}$$

The reasoning can be used also in order to derive the leading-order results for the other functions, which occur within the microscopic MCT for simple systems. The generalization of Eq. (6.18) is the formula for the shear correlator

$$K_T(t)/c_{is}^2 - f_T^c = h_T g_0^\sigma(t). \tag{6.54a}$$

The expression for the tagged-particle correlators generalizes Eq. (6.19):

$$\phi_q^s(t) - f_q^{sc} = h_q^s g_0^\sigma(t), \quad q = 1, \ldots, M. \tag{6.54b}$$

In particular, there holds for the mean-squared displacement function

$$\Delta_s(t) - r_{sc}^2 = h_{\text{MSD}} g_0^\sigma(t). \tag{6.54c}$$

The formulated results are examples for the factorization property. The deviations of the leading-order results for the functions $\phi_q(t), m_q(t), K_T(t)/c_{is}^2, \phi_q^s(t)$, and $\Delta_s(t)$ from their critical arrested parts are products of two factors. The first factor is the critical amplitude. It specifies the function under consideration, but it depends neither on the time nor on the separation of the state P from the critical point P^c. The second factor $g_0^\sigma(t) = G(t)$ is shared by all functions. It depends on the time and also on the state P, albeit via the separation parameter only. Positive separation parameters σ characterize glass states P, and negative values for σ refer to liquid states. The transition point is specified by $\sigma = 0$. The problem of solving the various MCT equations of motion is reduced to solving the single equation for $g_0^\sigma(t)$.

The factorization property is the basic general feature of the dynamics for states approaching a bifurcation singularity of type $A_\ell, \ell = 2, 3, \ldots$. Hence, the molecular-dynamics simulation data shown in Fig. 1.8 are a direct indication of the relevance of MCT for the understanding of the glassy dynamics of liquids.

(ii) The first scaling laws

The considerations of this subsection are restricted to generic A_2-bifurcation points P^c. It is the aim to derive scaling laws for the functions $g_0^\sigma(t)$.

Equation (6.50b) has solutions $g_\pm(t)$ for the special values $\sigma = \pm 1$ if and only if $\sqrt{|\sigma|}g_\pm(t)$ are solutions for general separation parameters $\sigma \gtrless 0$. Furthermore, the simplified equation of motion is scale invariant: $g_\pm(t)$ is a solution if and only if this is true for $g_\pm(t/\tau)$. Here, τ can be any positive number. Let us examine the ansatz $g_0^\sigma(t) = \sqrt{|\sigma|}g_\pm(t/\tau)$ with τ denoting a σ-dependent scale. The regularity properties of the MCT solutions require the following. For every fixed finite time interval, the correlator $\phi_q(t)$ for state P converges towards $\phi_q^c(t)$ if P tends to P^c.

Using the results for the leading asymptotic solution for $\phi_q^c(t)$ from Sec. 6.1.2, one has to request that $g_0^\sigma(t)$ approaches $g_0^c(t) = (t_0/t)^a$ if σ tends to zero. This holds if and only if $g_\pm(t/\tau)(t/\tau)^a$ tends to a positive constant C and if $(\tau/t_0)^a C\sqrt{|\sigma|}$ tends to unity. The value of C can be absorbed in a redefinition of τ. Deviations of the functions from their asymptotes can be ignored for the searched-for leading solutions.

The discussion shall be continued with the assumption that the equations

$$\pm 1 + \lambda g_\pm(\hat{t})^2 - \partial_{\hat{t}} \int_0^{\hat{t}} g_\pm(\hat{t} - \hat{t}') g_\pm(\hat{t}') d\hat{t}' = 0 \tag{6.55a}$$

have unique solutions, which obey the initial conditions

$$\lim_{\hat{t} \to 0} [g_\pm(\hat{t}) \hat{t}^a] = 1. \tag{6.55b}$$

A proof of the assumption is not available. One concludes:

$$g_0^\sigma(t) = c_\sigma g_\pm(\hat{t}), \qquad \hat{t} = t/t_\sigma, \qquad \sigma \gtrless 0, \tag{6.56a}$$

$$c_\sigma = \sqrt{|\sigma|}, \qquad t_\sigma = t_0/|\sigma|^\delta, \qquad \delta = 1/(2a). \tag{6.56b}$$

These equations formulate the first scaling laws of MCT. The laws have the same form as discussed in Eqs. (4.44b,c). The dependence of the function g_0^σ on the control parameters is quantified by the σ dependence of the correlation scale c_σ and of the time scale t_σ. The slowing-down of the dynamics upon approaching the critical point is described by the small-σ divergence of t_σ. It is specified by the exponent δ, which is determined by the critical exponent a. The shape functions $g_\pm(t)$ are determined by λ^c, i.e., by the mode-coupling coefficients at the critical point. Details of the transient dynamics enter only via the scale t_0 for the critical-correlation decay.

Using Eqs. (6.2), (A.5a), one obtains the scaling laws in the frequency domain:

$$ST[g_0^\sigma(t)](z) = c_\sigma ST[g_\pm(\hat{t})](\hat{z}), \qquad \hat{z} = z t_\sigma, \qquad \sigma \gtrless 0. \tag{6.57a}$$

Combined with the factorization result (6.53a), one gets the scaling law for the loss spectrum

$$\chi_q''(\omega)/h_q = c_\sigma \hat{\chi}_\pm(\hat{\omega}), \qquad \hat{\omega} = \omega t_\sigma, \qquad \sigma \gtrless 0. \tag{6.57b}$$

The shape function for the spectrum is determined by

$$\hat{\chi}_\pm(\hat{\omega}) = -\mathrm{Im} ST[g_\pm(\hat{t})](\hat{\omega} + i0). \tag{6.57c}$$

An S-transform casts the equation of motion (6.55a) into an equivalent one for the frequency domain. It is the specialization of Eq. (6.50a) to $\sigma = \pm 1$:

$$\pm 1 + \lambda ST[g_\pm(\hat{t})^2](\hat{z}) - ST[g_\pm(\hat{t})](\hat{z})^2 = 0. \tag{6.58a}$$

The Tauberian-theorem equations (A.64a,b) reformulate the initial condition (6.55b) to a condition for the large-frequency asymptote:

$$\lim_{\hat{z}\to\infty} ST[g_\pm(\hat{t})](\hat{z})/[\Gamma(1-a)(-i\hat{z})^a] = 1. \tag{6.58b}$$

Figure 6.6 exhibits shape functions for $\lambda^c = 0.735$, which are obtained as numerical solutions of Eq. (6.55a). Function $g_+(\hat{t})$ describes the crossover from the critical decay at short rescaled times, $g_+(\hat{t}\to 0) \sim \hat{t}^{-a}$, to arrest at large values of $\hat{t}$: $g_+(\hat{t}\to\infty) = ST[\hat{g}_+(\hat{t})](\hat{z}\to 0)$. Equation (6.58a) yields:

$$g_+(\hat{t}\to\infty) = 1/\sqrt{1-\lambda}. \tag{6.59}$$

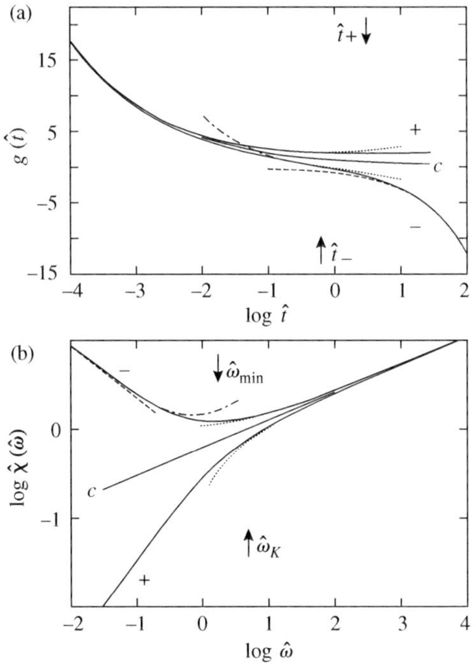

FIG. 6.6. Panel (a) shows the critical correlator $\hat{t}^{-a}$ (curve c) and the shape functions $g_\pm(\hat{t})$ of the first scaling law (curves $\pm$) for the exponent parameter $\lambda^c = 0.735$ of the hard-sphere-system model defined in connection with Fig. 4.6. The dotted lines present the small-$\hat{t}$ expansions (6.63a). The dashed line shows the von Schweidler asymptote (6.60a) and the result (6.63b) is presented as dashed-dotted curve. Panel (b) exhibits the shape functions $\hat{\chi}(\hat{\omega})$ for the loss spectra, which correspond to the specified curves in the upper panel. The parameters are: $A_1 = 1, 12, B = 0.836, B_1 = 0.431, a = 0.312, b = 0.583$. The crossover times $\hat{t}_+ = 3.33, \hat{t}_- = 0.704$ and the crossover frequencies $\hat{\omega}_K = 4.80, \hat{\omega}_{\min} = 1.56$ are explained in the text; they are marked by arrows. Reproduced from Franosch *et al.* (1997).

Combining this formula with Eqs. (6.53a), (6.56a), one obtains for the arrested parts of the density correlators $f_q = \phi_q(t \to \infty) = f_q^c + c_\sigma h_q/\sqrt{1-\lambda}$. Thereby, the small-$\sigma$ asymptote of Eq. (4.91a) is reproduced. The factorization theorem implies analogous formulas for the arrested parts of the other functions noted in Eqs. (6.53), (6.54). A rescaled crossover time $\hat{t}_+ = t_+/t_\sigma$ is marked in the figure. It is defined by the request that the shape function reaches its asymptote up to 0.1%: $g_+(\hat{t}_+) = 1.001/\sqrt{1-\lambda}$. With decreasing rescaled frequency, function $\hat{\chi}_+(\hat{\omega})$ describes a crossover from the critical loss spectrum at large rescaled frequencies, $\hat{\chi}_+(\hat{\omega} \to \infty) \sim \sin(\frac{\pi}{2}a)\Gamma(1-a)\hat{\omega}^a$, to a white-noise-induced spectrum at small $\hat{\omega}$: $\hat{\chi}_+(\hat{\omega} \to 0) \sim C_0\hat{\omega}, C_0 > 0$. As a result, the $\log \hat{\chi}_+$ versus $\log \hat{\omega}$ curve exhibits a knee. A characteristic rescaled crossover frequency $\hat{\omega}_K = \omega_K t_\sigma$ is marked in the figure. It denotes the maximum position of the function $\hat{\chi}_+(\hat{\omega})/\sqrt{\hat{\omega}}$.

For large times, the liquid correlators $\phi_q(t) = f_q^c + h_q g_0^\sigma(t)$ tend to zero, i.e., $g_0^\sigma(t)$ becomes negative. For times, where $g_0^\sigma(t)$ is below some fixed negative value, and for small values of $|\sigma|, -g_-(\hat{t}) = -g_0^\sigma(t)/c_\sigma$ becomes large. One can neglect the first term in Eq. (5.55a) compared to the second term. The equation reduces to one, which is equivalent to Eq. (6.11c). One concludes from the discussion of Fig. 6.1 that there is a solution $-Bt^b$. This holds under the conditions that B is positive and that the exponent b is a positive solution of the equation $\lambda(-b) = \lambda^c$. This condition fixes b uniquely. Assuming that the solution of the equation in the long-time limit determines the long-time limit of the solution, one arrives at the following result. The exponent parameter λ^c determines a positive constant B so that

$$g_-(\hat{t}) \sim -B\hat{t}^b, \quad \hat{t} \to \infty. \tag{6.60a}$$

The Tauberian-theorem equations (A.65a,b) reformulate this result as one for the asymptotic small frequency behaviour of the S-transform:

$$ST[g_-(\hat{t})](\hat{z}) \sim -B\Gamma(1+b)/(-i\hat{z})^b, \quad \hat{z} \to 0. \tag{6.60b}$$

The low-frequency loss demonstrates a von Schweidler-law variation (1.4b):

$$\hat{\chi}_-(\hat{\omega}) \sim B\sin(\pi b/2)\Gamma(1+b)/\hat{\omega}^b, \quad \hat{\omega} \to 0. \tag{6.60c}$$

One infers from Fig. 6.1 that exponent b increases monotonically from 0 to 1 if the exponent parameter λ^c decreases from 1 to 1/2. If λ^c would be smaller than 1/2, b would exceed 1. Using the second version of equations of motion for the transient dynamics, one gets completely monotonic solutions. In particular, there holds $0 < \partial_t^2\phi_q(t) = \partial_t^2 g_0^\sigma(t) \cdot h_q$. An exponent b larger than 1 would not be compatible with this property. One concludes that the von Schweidler exponent b is the unique solution of the equation:

$$\Gamma(1+b)^2/\Gamma(1+2b) = \lambda^c, \quad 0 < b \leq 1. \tag{6.61a}$$

This formula and Eq. (6.11d) show that exponent a is a monotonically increasing function of exponent b. The restriction for b implies one for a:

$$1/2 \leq \lambda^c; \quad 0 < a \leq 0.395\ldots. \tag{6.61b}$$

For small rescaled times, the shape function $g_-(\hat{t})$ is close to the critical asymptote $\hat{t}^{-a}$. The function is positive and the $g_-(\hat{t})$ versus $\log \hat{t}$ curve is upward bent. For large rescaled times $\hat{t}, g_-(\hat{t})$ is close to the von Schweidler asymptote. The function is negative and the $g_-(\hat{t})$ versus $\log \hat{t}$ curve is downward bent. Natural values for a rescaled crossover time are the inflection point $\hat{t}_{\text{inf}}$ or the zero $\hat{t}_-$ of the curve. The scaling law and the factorization theorem yield to the following result for the liquid correlators of variables X. All correlators cross their critical arrested part at the same time τ^{pc} and their semi-logarithmic representation exhibits an inflection point at the same time t_{inf}:

$$\phi_X(\tau^{pc}) = f_X^c, \qquad \tau^{pc} = \hat{t}_- t_\sigma \tag{6.62a}$$

$$\partial^2 \phi_X(t_{\text{inf}})/\partial(\log t)^2 = 0, \qquad t_{\text{inf}} = \hat{t}_{\text{inf}} t_\sigma. \tag{6.62b}$$

With decreasing rescaled frequencies $\hat{\omega}$, the function $\hat{\chi}_-(\hat{\omega})$ describes a crossover from the decreasing critical spectrum to the increasing von Schweidler-law spectrum. There occurs a minimum at some rescaled frequency $\hat{\omega}_{\text{min}}$. The combination of the scaling law with the factorization theorem yields a generalization of Eq. (6.57b). The loss minima of the liquid are located at the same frequency

$$\omega_{\text{min}} = \hat{\omega}_{\text{min}}/t_\sigma. \tag{6.62c}$$

The minimum intensity is given by

$$\chi_X''(\omega_{\text{min}}) = (c_\sigma h_X)\hat{\chi}(\hat{\omega}_{\text{min}}). \tag{6.62d}$$

The common control-parameter sensitive inflection point t_{inf} of the correlators and the common control-parameter sensitive loss-minimum position ω_{min} are connected by the control-parameter-independent product value $\hat{t}_{\text{inf}}\hat{\omega}_{\text{min}} = t_{\text{inf}}\omega_{\text{min}}$, which is determined by the exponent parameter λ^c. This theorem holds provided the separation parameter for the liquid states are so small that corrections to the leading-asymptotic laws can be ignored.

For small $\hat{t}$, Eq. (6.55a) can be solved by asymptotic series expansion $g_\pm(\hat{t}) - \hat{t}^{-a} \sim \pm A_1 \hat{t}^a + A_2 \hat{t}^{3a} \pm \cdots$ The coefficients can be determined recursively. The critical asymptote with first corrections reads

$$g_\pm(\hat{t}) = \hat{t}^{-a} \pm A_1 \hat{t}^a, \quad A_1 = \left\{2[\Gamma(1+a)\Gamma(1-a) - \lambda^c]\right\}^{-1}. \tag{6.63a}$$

Similarly, one can determine a leading correction to von Schweidler's law. Using the Tauberian-theorem formulas (A.65a,b) for the discussion of the large $\hat{t}$ asymptotes, one gets for $b < 1$:

$$g_-(\hat{t}) = -B\hat{t}^b + B_1/(B\hat{t}^b), \quad B_1 = \left\{2[\Gamma(1-b)\Gamma(1+b) - \lambda^c]\right\}^{-1}. \tag{6.63b}$$

For $0.2 < \hat{t} < 1$, Fig. 6.6 demonstrates that the approximation (6.63a) for $g_+(\hat{t})$ agrees with the limit value (6.59) within the accuracy of the drawing. Similarly, the asymptote (6.63a) for $g_-(\hat{t})$ cannot be distinguished from the approximation (6.63b). Splicing the two expansions of the $g_-(\hat{t})$ versus $\log t$ curves, one can determine the coefficient B on a 10%-accuracy level. In this sense, the solutions of Eqs. (6.55a,b) are understood quantitatively by analytical calculations. A similar statement holds for the discussions in the frequency domain. But, here one has to extend the expansion in Eq. (6.63a) up to terms $\pm A_3 \hat{t}^{5a}$. Tables are available, which present the various quantities $a, b, A_1, A_2, A_3, B, B_1, \hat{t}_-, \hat{\omega}_{\rm min}, \hat{\chi}(\hat{\omega}_{\rm min})$, and C_0 as function of λ^c (Götze 1990).

von Schweidler decay is the second general source for the stretching of the MCT dynamics for liquid states. The representative results in Fig. 6.6 show that the frequency has to be changed by about three orders of magnitude in order to scan the loss-minimum region where $\chi''_X(\omega_{\rm min}) < \chi''_X(\omega) < 2\chi''_X(\omega_{\rm min})$. Outside this region, the loss exhibits stretched variation nearly proportional to ω^{-b} and ω^a, respectively. The liquid correlator follows closely the t^b law for $t/\tau^{pc} > 10$, and it exhibits nearly the t^{-a} behaviour for $t/\tau^{pc} < 0.01$. The crossover from one asymptote to the other is stretched on an interval for a time increase by a factor 1000. Equations (1.7a,b) are reasonable interpolation formulas for the scaling-law description of the plateau-crossing dynamics.

In order to formulate von Schweidler's law more explicitly, Eq. (6.60a) can be combined with the scaling law and the factorization theorem. One obtains:

$$\phi_q(t) = f_q^c - h_q (t/t'_\sigma)^b, \tag{6.64a}$$

$$t'_\sigma = [t_0/B^{\frac{1}{b}}]/|\sigma|^\gamma, \qquad \gamma = (1/2a) + (1/2b). \tag{6.64b}$$

Three restrictions underly the derivation of these results. The times have to exceed the onset time (6.39) so that the critical correlator can be replaced by its asymptotic form. The rescaled time $\hat{t}$ has to be large so that the correction can be neglected in Eq. (6.63b). The function $|(\phi_q(t) - f_q^c)/h_q| = |g_0^\sigma(t)|$ has to be small so that the restriction to leading-order results is justified. The latter condition sets an upper limit for the time as is expected from condition (6.9c). The three conditions for the validity of the preceding formulas are

$$t_q^* < t_\sigma, \qquad t_\sigma \ll t, \qquad t \ll t'_\sigma. \tag{6.64c}$$

The first condition defines an upper bound for $|\sigma|$.

The evolution of the glassy dynamics for liquid states is governed by the two control-parameter-sensitive time scales t_σ and t'_σ. If P tends towards the arrest point P^c, both scales diverge:

$$\lim_{\sigma \to 0} t_\sigma / t_{\rm mic} = \infty, \qquad \lim_{\sigma \to 0} t'_\sigma / t_{\rm mic} = \infty. \tag{6.65a}$$

Since the exponent $\delta = 1/(2a)$ for the first critical time scale is smaller than the exponent γ for the second critical time scale, the latter scale diverges relative to the former:

$$\lim_{\sigma \to 0} t'_\sigma / t_\sigma = \infty. \tag{6.65b}$$

Viewing correlators as functions of $x = \log(t)$, von Schweidler decay is exhibited for x increasing from $x_- = \log(c_- t_\sigma)$ to $x_+ = \log(t'_\sigma / c_+)$. Here, $c_\pm$ are numbers large compared to unity. For vanishing separation parameters, the initial point x_- tends to infinity. But, the length $(x_+ - x_-)$ of the interval diverges as well. The description of a correlator $\phi_q(t)$ by Eq. (6.64a) is qualitatively incorrect for small times as well as for large ones. Nevertheless, the preceding formulas indicate the possibility for a subtle asymptotic regime of validity of von Schweidler's law. They suggest that there appears a time interval of finite length where this law holds within an arbitrarily chosen positive error margin. For increasing accuracy level, one has to shift the liquid state closer to the bifurcation singularity. Simultaneously, the initial point of the time interval has to be shifted to larger values. This is connected with an increase of the length of the interval. It will be the task of the following sections to substantiate these suggestions. Anticipating these results, one concludes that the scaling law of the generic A_2 bifurcation offers a derivation of von Schweidler's law as an essential feature of the bifurcation dynamics. The microscopic version of MCT provides a straightforward algorithm to evaluate von Schweidler's exponent b and the amplitudes f_q^c and h_q from the equilibrium-structure functions.

Combining the factorization-theorem result for the dynamics of some variable X with the scaling law, one arrives at the formulas for $\sigma \gtrless 0$:

$$\hat{\phi}_X(t) = [\phi_X(t) - f_X^c]/h_X = c_\sigma g_\pm(t/t_\sigma), \tag{6.66a}$$

$$\hat{\chi}_X(\omega) = \chi''_X(\omega)/h_X = c_\sigma \hat{\chi}_\pm(\omega t_\sigma). \tag{6.66b}$$

The full lines in Figs. 4.16 and 4.17 exhibit examples for functions $\hat{\phi}_q(t)$ and $\hat{\chi}_q(\omega)$, respectively. The variable $X = \rho_{\vec{q}}$ refers to density fluctuations with wave number q. The curves are evaluated for a microscopic model for the hard-sphere system. The dashed lines show the results for the right-hand sides of the preceding equations. They exhibit the shape functions from Fig. 6.6 after rescaling with the power-law expressions (6.56b) for the scales c_σ and t_σ. The diagrams demonstrate the evolution of intervals for the scaling-law description of the dynamics upon decreasing the separation-parameter modulus. For short times, the correlators are close to the critical ones. This explains why the range of validity of the leading-order results is ϵ-insensitive there. The q-dependence of the range of validity is caused by that of the onset time t_q^*; it is explained in the preceding section. Corresponding results hold for the range of validity of Eq. (6.66b) for large frequencies. The range of validity expands to large times proportional to t'_σ and to small frequencies proportional to $1/t'_\sigma$, respectively.

This follows from the above presented discussion of the von Schweidler law. In order to understand the q-dependence of the range of validity, one has to determine that of Eq. (6.64a).

The results shown in Figs. 4.16 and 4.17 for $n = 8$ refer to distance parameters $\epsilon = (\varphi - \varphi^c)/\varphi^c = \pm 0.0022$. Such small values are near the level of accuracy, which can be achieved by experimental determinations of the packing fraction φ for hard-sphere colloidal suspensions. For this ϵ, Fig. 4.14 demonstrates a time scale for the correlation decay of the liquid, which exceeds t_{mic} by a factor 10^6. Such large times are about the present-day limits, which can be reached by molecular-dynamics studies. The leading-order results describe $\phi_q(t)$ for time increases of about four orders of magnitude. The corresponding loss spectra $\chi''_q(\omega)$ are described for a frequency change of about a factor 100. This holds for wave numbers q_0, q_2, and q_3, but not for q_1. The strong-coupling effects for a wave number q near the structure-factor-peak position imply a large value for the onset time $t^*_{q_1}$. This is discussed above in connection with Fig. 6.2. As a result, the range of validity of Eqs. (6.66a,b) for q_1 is reduced compared to that for the other wave numbers. The $n = 8$ loss spectrum for q_1 barely exhibits a minimum at $\hat{\omega}_{\text{min}}/t_\sigma$.

The symbols in Fig. 6.7 exhibit crossover parameters Q for loss spectra as function of the external control parameter ξ. As discussed in Eqs. (4.84)–(4.86), a distance parameter is introduced as $\epsilon = \epsilon_0(\xi - \xi^c)$. It is proportional to the separation parameter: $\sigma = C\epsilon$. The parameters Q are deduced from the hard-sphere-model spectra shown in Fig. 4.15 for the labels $n = 6 - 10$. For this example, ξ abbreviates the packing fraction φ and ϵ_0 is the inverse of the critical value $\varphi^c = 0.516$. The full lines exhibit the power-law results $Q = Q_0|\sigma|^\beta$, which are obtained from the scaling laws. If Q denotes the position ω_{min} or ω_K, there holds $\beta = 1/(2a)$. The exponent $\beta = 1/2$ describes the correlation scale c_σ, i.e., the results for the spectral intensity for the minimum and for the knee. Comparing the presentation in panel (a) with that in panel (b) one concludes the following. The evolution of the power-law asymptotes upon decreasing $|\epsilon|$ is expressed more clearly by a $Q^{1/\beta}$ versus ξ diagram than by a Q versus ξ one. The former is called a rectification diagram, because the small $|\epsilon|$ limit appears as a straight line: $Q^{1/\beta} \sim [Q_0^{1/\beta} C\epsilon_0]|\xi - \xi^c|, \xi \to \xi^c$. The diagram eases the extrapolation of Q to zero, thereby identifying the critical value ξ^c. All rectified diagrams in Fig. 6.7 extrapolate to φ^c within an uncertainty of $\pm 0.1\%$.

The figure demonstrates that the scaling law describes the knee intensity $\chi''_q(\omega_K)$ adequately only if ϵ is smaller than 10^{-3}. For larger distance parameters, the spectra are smaller than the asymptotic value by a factor 3 or more. The loss-minimum position ω_{min} for the spectra for wave number q_2 are described well by the scaling law for the ϵ-interval exhibited in the figure. However, for wave-number q_1, the value ω_{min} at $\epsilon = -0.005$ is about 2.7 times larger than the scaling-law value $\hat{\omega}_{\text{min}}/t_\sigma$. This strong violation of a factorization-theorem implication is a consequence of the large value for the onset time $t^*_{q_1}$.

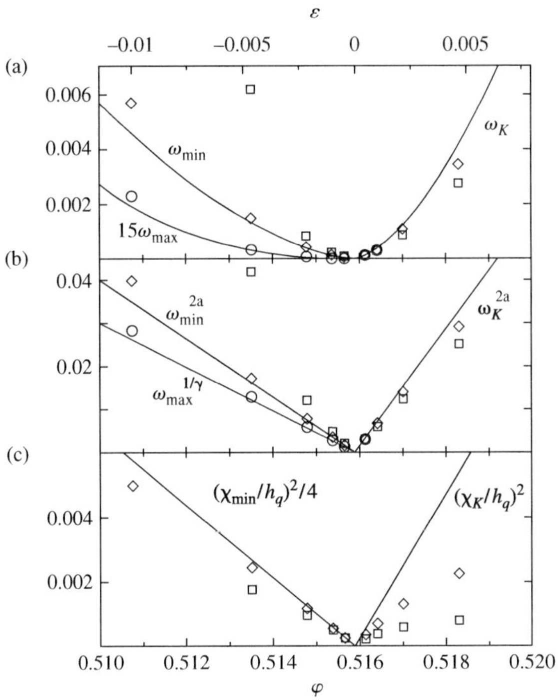

FIG. 6.7. The squares and diamonds exhibit crossover parameters (ω_{min}, $\chi_{\text{min}} = \chi''_q(\omega_{\text{min}})$) and ($\omega_K, \chi_K = \chi''_q(\omega_K)$) for loss spectra of density-fluctuations with wave numbers q_1 and q_2, respectively. The parameters are read off from the results shown in Fig. 4.15 for a hard-sphere-system model. The lines exhibit the leading-order results described by Eq. (6.66b) with amplitudes deduced from Fig. 4.6 and shape functions from Fig. 6.6. The circles in panels (a) and (b) exhibit the frequency ω_{max} and $\omega_{\text{max}}^{1/\gamma}$, respectively, for the low-frequency loss-peak maximum for the fluctuations with wave number q_2, and the lines represent the results for the asymptote $\omega_{\text{max}} \sim 3.74|\sigma|^{\gamma}, \gamma = (1/2a) + (1/2b) = 2.46$. Reproduced from Franosch *et al.* (1997).

Figures 5.30 and 5.31 exhibit mean-squared displacements $\Delta_s(t)$, which are calculated for microscopic MCT models for two molecular liquids. The dashed lines demonstrate the description of the plateau-crossing by the leading-order formulas (6.54c), (6.56a,b). The increase of $\Delta_s(t)$ towards the plateau is described only for a small interval of $\log t$. This reflects the strong-coupling-effect induced large value for the onset time, which is discussed in connection with Fig. 6.5.

(iii) First-scaling-law descriptions of some structure-relaxation data

The simplicity and generality of Eqs. (6.66a,b) obtrude the question of whether or not these formulas reflect features of the complex dynamics of glass-forming liquids. An adequate discussion of the published material on this question is beyond the scope of this book. But, a glimpse on the issue shall be made with the intention to relate the considerations of this section to the ones of Chapter 1.

van Megen (1995) presented comprehensive comparisons of glassy-dynamics data with the leading-order MCT formulas for ideal liquid-glass transitions. His discussions concern density correlators $\phi_q(t)$ measured for hard-sphere colloids. States for six packing fractions φ below the glass-transition value φ_g are examined and five states with $\varphi > \varphi_g$. Figure 1.15 reproduces examples of the original data. The first step of the cited analysis demonstrates that φ-independent numbers f_q^c and h_q can be fitted so that $\hat{\phi}_q(t) = [\phi_q(t) - f_q^c]/h_q$ is a q-independent function $G(\varphi, t)$. This holds for a time interval, where $\hat{\phi}_q(t)$ is not too large. The length of this interval expands upon decreasing $|\varphi - \varphi_g|$. Thereby, the evolution of the factorization property is demonstrated. The adjustment of the parameters f_q^c and h_q is one method to measure the critical arrested parts f_q^c and the critical amplitude h_q. The noteworthy q-dependence of the measured Debye–Waller factor f_q^c and of h_q reproduce well the MCT results shown in Fig. 4.6. As a second step, it is shown that scales c_φ and t_φ can be identified so that one can write $G(\varphi, t) = c_\varphi g_\pm(t/t_\varphi), \varphi \gtrless \varphi_g$. Here $g_\pm(\hat{t})$ denote the shape functions, which are evaluated for the exponent parameter of a microscopic model for the hard-sphere system. Hence, the measured evolution of the near-plateau relaxation is consistent with that described by the first scaling laws for the liquid and for the glass. The full lines in Fig. 1.15, which approximate the data for samples $H7$ and $H8$ outside the transient regime, exhibit fits by $f_q^c + h_q c_\varphi g_+(t/t_\varphi)$.

The measured slowing down of the dynamics due to an increase of the density of the liquid is described by an increase of t_φ by a factor of about 100. In order to describe the measured slowing down of the dynamics due to the decrease of the density of the glass, the time scale t_φ is increased by a factor of about 10. The final step of the cited analysis demonstrates that the fitted scales are in reasonable agreement with the power-law results $c_\varphi = |\sigma|^{1/2}, t_\sigma/t_0 = 1/|\sigma|^{1/2a}$. Here, the time t_0 is nearly φ-independent. For $\varphi < \varphi_g$ as well as for $\varphi > \varphi_g$, there holds $\sigma = C_0(\varphi - \varphi_g)/\varphi_g$. The fit value $C_0 = 1.2$ differs from the MCT value 1.5. As discussed in Chapter 4, the fit value $\varphi_g = 0.577 \pm 0.004$ differs from the MCT value $\varphi^c = 0.516$. These two discrepancies reflect genuine deficiencies of the asymptotic theory and possible problems with the measurements. The cited data analysis is extended by incorporating a description of the below-plateau decay of the liquid correlators. This is discussed below in Sec. 6.2.2(iv).

The full lines in Fig. 1.7 exhibit functions $\overline{f}_q^c + [\overline{h}_q c_\sigma] g_-(t/t_\sigma)$. They are constructed as fits to the anticipated plateau-crossing parts of correlators $\overline{\phi}_q^m(t)$, which are measured for liquid orthoterphenyl. The dynamical variable defining the correlator is a superposition of density fluctuations with wave-number q as explained for Eq. (1.13) in some other context. For every correlator measured at some temperature T for some q, the four numbers $\overline{f}_q^c, [\overline{h}_q c_\sigma], t_\sigma$ and λ are determined by optimization of the fit. The exponent parameter λ is needed in order to evaluate the shape function $g_-(\hat{t})$. The fit parameter λ does not exhibit a systematic dependence on T or on q. The possibility to describe all data with a common $\lambda = 0.77$ is consistent with the validity of the factorization property. Extrapolations of the rectification diagrams for the fitted scales $[\overline{h}_q c_\sigma]$ and t_σ suggest a critical temperature $T_c = 290\,\mathrm{K}$. The mentioned values for λ and T_c are consistent with conclusions of many other studies of the glassy OTP dynamics, which are discussed by Tölle (2001) in his review article.

Equation (6.66a) implies the following. For times tending to small values, correlators $\overline{\phi}_q^m(t)$ for different separation parameters converge to the critical asymptote $\overline{f}_q^c + \overline{h}_q(t_0/t)^a$ without pairwise crossings. The fits with crossing points in Fig. 1.7 anticipate that preasymptotic correction terms manifest themselves by a decrease of $\overline{f}_q^c$ with increasing temperature. The same phenomenon is anticipated by the Kohlrausch-law fits for the below-plateau decay, which are shown in the figure.

The full line in Fig. 1.4 exhibits a tagged-particle correlator $\phi_q^s(t)$. It is determined for a system of particles, which move according to a stochastic equation of motion. The dashed-dotted line exhibits the function $f_q^{sc} + [h_q^s c_\sigma] g_-(t/t_\sigma)$. This leading-order expression from Eq. (6.66a) describes the data for a time interval given by $0.6 \times 10^{-4} \le t/\tau \le 0.3$. Here, the shape function is that for the exponent parameter $\lambda = 0.71$, which is calculated for a microscopic MCT model for the binary Lennard-Jones mixture under consideration (Nauroth and Kob 1997). The fit value for the Lamb–Mössbauer factor f_q^{sc} is as close to the value calculated in the cited paper as is demonstrated in Fig. 5.13 for the Debye–Waller factor. Consequently, the fit in Fig. 1.4 is based on adjusting only two parameters, namely the scales $[h_q^s c_\sigma]$ and t_σ. One concludes that the shown stretching of the correlation decay does not exhibit any feature, which cannot be explained by the asymptotic results for the t^{-a}-to-t^b crossover.

The simulation results in Fig. 1.4 demonstrate the same phenomenon, as the MCT curves in Fig. 6.5. Replacing the stochastic equations of motion by Newtonian ones, the structure relaxation for $2 \times 10^{-5} < t/\tau < 6 \times 10^{-4}$ is masked by transient oscillations.

Equation (6.66b) can be reformulated as follows. The $\log[\chi''_X(\omega)/h_X]$ versus $\log \omega$ curve for the loss spectrum at state P of some variable X is connected with the $\log[\hat{\chi}(\hat{\omega})]$ versus $\log \hat{\omega}$ curve by a translation. The relative shift is $\log(t_\sigma)$ and $\log(c_\sigma)$ horizontally and vertically, respectively. The double-logarithmic presentation of loss spectra for different states of the liquid and different variables collapse if the curves are translated so that the position ω_{min} of the minima and their spectral strength $\chi_{\mathrm{min}} = \chi''_X(\omega_{\mathrm{min}})$ coincide. The upper panel of Fig. 6.8 exhibits the evolution of such a scaling-law regime. Upon decreasing the temperature T from $295\,\mathrm{K}$ to $180\,\mathrm{K}$, the scaling-law interval expands to one dealing with a frequency change of about three orders of magnitude. The shown data are obtained for liquid toluene by light-scattering spectroscopy as mentioned in connection with Fig. 1.1. The dashed line shows the approximation (1.7a) for the shape function $\hat{\chi}(\omega)$. The exponent parameter used is $\lambda^c = 0.69$, which implies the exponents $a = 0.331$ and $b = 0.657$. Increasing λ^c, both exponents a and b decrease, i.e., the minimum of the $\log[\hat{\chi}(\hat{\omega})]$ versus $\log(\hat{\omega})$ curve becomes flatter. Since the wings of the measured minimum are documented well, the fit value for λ^c is determined within an uncertainty not exceeding ± 0.02.

Figure 1.3 demonstrates translations of the shape function on the spectra of liquid CKN. In order to fit the evolution of the loss minimum, Li *et al.* (1992) have used Eq. (1.7a) with exponents $a = 0.27$ and $b = 0.46$. These exponents are derived from the exponent parameter $\lambda^c = 0.81$.

For the cited toluene spectra, a decrease of the temperature by 110 K is accompanied by a decrease of the loss intensity χ_{min} by about a factor 2.2 and by an increase of the time scale $1/\omega_{\text{min}}$ by about a factor 10. The rectification diagrams shown in Fig. 6.8 for the two scales extrapolate to a critical temperature $T_c = (153 \pm 3)$K. For $T > T_c + 15$ K, the measured evolution of the loss minimum can be described by the leading-order results for the MCT bifurcation. This conclusion can be corroborated by an analysis of the evolution of the low-frequency loss peak, as is discussed below in connection with Fig. 6.18. The cited description of the toluene spectra cannot cope with the data for $T \leq 160$ K.

Three paragraphs shall be added with the intention to specify a difficulty for a judgement of the relevance of a successful data description in terms of leading-order formulas for the bifurcation. The problem is demonstrated most clearly by Sciortino and Tartaglia (1999), who use the numerical solutions of the MCT equations for the hard-sphere system as 'data'. The analysed correlators are similar to those shown in Fig. 4.14 for labels $n \leq 8$. As emphasized above, these curves deal with times and separation parameters, which are studied by experiment or simulation. For systems with Newtonian dynamics, the interval $\log t < 2$ should be ignored, since transient oscillations mask structural relaxation. One should also imagine some noise of the data. Consequently, the data do not exhibit a clearly developed t^{-a} part. As a result, one cannot deduce the value for the exponent parameter directly. The authors showed that one can achieve data descriptions by Eq. (6.66a) for fit values λ, which differ strongly from the MCT value of about 0.74. This holds, provided one uses also a fit value for f_q^c, which differs strongly from the MCT value for the critical arrested part. The choice of a wrong value for λ does not worsen the quality of the rectification diagrams for the scales nor does it lead to unacceptable estimations of the critical packing fraction. The authors conclude: without additional information, it is not legitimate to conclude from the quality of the data description on the reliability of the fit values for λ^c and f_X^c.

Wuttke *et al.* (2000) demonstrate scaling-law descriptions of a large set of relaxation data measured for propylene carbonate. A fit of the experimental results obtained by depolarized-light-scattering measurements, by dielectric-loss spectroscopy, by solvation-dynamics studies, and by neutron-scattering work is possible using the common shape function for $\lambda^c = 0.72$. All rectification diagrams extrapolate to the critical temperature $T_c = 182$ K. Yet, contrary to what is implied by Eq. (6.62c), the frequencies ω_{min} for the spectral minimum differ

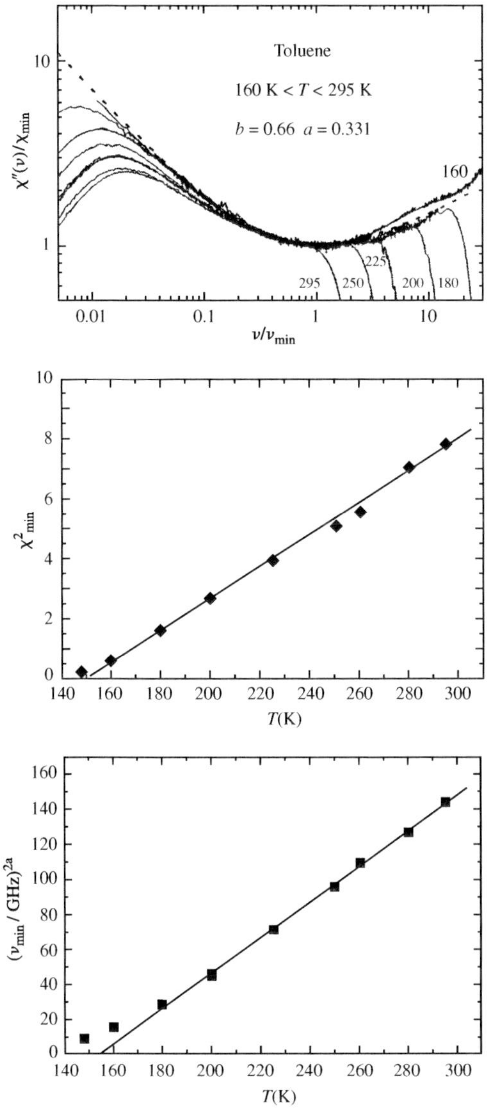

FIG. 6.8. The upper panel exhibits loss spectra $\chi''(\nu)$ measured by depolarized-light-scattering spectroscopy for liquid toluene. For every temperature T, the frequency ν is presented in units of $\nu_{\rm min}$ and the spectrum in units $\chi_{\rm min} = \chi''(\nu_{\rm min})$, where $\nu_{\rm min}$ denotes the position of the loss minimum. The dashed line exhibits the interpolation formula (1.7a) with exponents $a = 0.33$ and $b = 0.66$, which are implied by the exponent parameter $\lambda = 0.69$. The lower panels exhibit rectification diagrams for the scaling parameters $\chi_{\rm min}$ and $\nu_{\rm min}$, respectively. Reproduced from Wiedersich *et al.* (2000).

considerably for different probing variables X. The factorization theorem, which is the basis for deriving the scaling laws, is violated. Within the regime of separation parameters studied, the differences are as large as those demonstrated in Fig. 6.7 for the minima positions for the loss spectra for density fluctuations of the hard-sphere liquid with wave numbers q_1 and q_2.

It is possible to describe the above cited data for the propylene carbonate dynamics within the schematic model, which is discussed in connection with Fig. 4.29. The solutions can be compared with the leading-order results without introducing fit parameters. For the temperature range of interest, say $T > T_c + 20\,\mathrm{K}$, the pre-asymptotic corrections to the leading-order results are so large that a quantitative description of the solutions by Eqs. (6.66a,b) cannot be achieved (Götze and Voigtmann 2000). There is the question: why is it occasionally possible to describe solutions of the MCT equations by leading-order formulas for the bifurcation dynamics outside the regime of validity of the latter, provided one chooses the parameters in these formulas different from their correct values? An answer to this question is not available.

(iv) The extended Cole–Cole law

Strong mode-coupling of some dynamical variable X to density fluctuations causes large negative correction amplitudes for the critical loss spectrum $\chi_X^{c\prime\prime}(\omega)$. For high frequencies, this leads to a suppression of the critical spectrum below the asymptotic $(\omega t_0)^a$ variation. The phenomenon is discussed in connection with Fig. 6.2. The high-frequency part of the scaling-law description reproduces the $(\omega t_0)^a$ spectrum. Therefore, the strong-coupling effects are a source for the deviations of the loss spectra from the scaling-law results (6.66b). Figure 4.15 demonstrates such deviations. Equations (6.56b), (6.62c,d) imply: $\log[\chi_X''(\omega_{\min}] = a \log[\omega_{\min}] + C_X$. In a double-logarithmic presentation, loss minima for different states are located on a straight line of slope a. The line is parallel to the one for the critical asymptote: $\log[\chi_X^{c\prime\prime}(\omega)] = a \log[\omega] + C_X'$. The spectra $\chi_2''(\omega)$ exhibit this behaviour for $n \geq 7$, i.e., for $|\epsilon| \leq 0.005$. The strong-coupling spectra $\chi_1''(\omega)$ follow these leading-order law only for $n \geq 9$, i.e. for $|\epsilon| \leq 0.001$. The minima for the states $n = 8$ and $n = 7$, however, are shifted to larger frequencies and lower intensities compared to those of $\chi_2''(\omega)$. For the specified region of distance parameters ϵ, there occur strong deviations from the factorization property. For $n = 5$ or $n = 6$, the liquid spectrum $\chi_1''(\omega)$ does not exhibit a minimum at all; it exhibits an excess wing as shown by curve A in Fig. 6.4. In this subsection, a possibility will be considered for describing the mentioned deviations within the framework of the leading-order formulas.

According to the discussions in Sec. 6.1.2(iii), there is the following trend. The strong-coupling-induced correction amplitudes for the dynamic susceptibility $\chi_X(z)$ are reduced if they are transformed to those for the modulus for the response. As a result, the scaling-law description of the fluctuating-force correlator $m_X(t)$ has a larger range of applicability than that for the correlator $\phi_X(t)$. This holds for frequencies near and above the position of the scaling-law minimum $\omega_{\min} = \hat{\omega}_{\min}/t_\sigma$ and of the scaling-law knee $\omega_K = \hat{\omega}_K/t_\sigma$. The scaling-law result for the modulus follows from the combination of the factorization-property formula (6.53b) with Eq. (6.57a): $ST[m_X(t)](z) = f_X^{mc} + h_X^m C(z)$,

$$C(z) = c_\sigma ST[g_\pm(\hat{t})](\hat{z}), \qquad \hat{z} = zt_\sigma, \qquad \sigma \gtrless 0. \tag{6.67a}$$

Equations (6.38c) can be used to express the two amplitudes f_X^{mc} and h_X^m in terms of f_X^c and h_X. The assumptions made at the beginning of this section imply that Eq. (6.37d) remains valid also for states P different from but close to P^c. Introducing the characteristic frequency ω_X^c from Eq. (6.42b), one arrives at

$$\chi_X(z)/(1 - f_X^c) = 1/\{1 + C(z)/[\Gamma(1-a)(\omega_X^c t_0)^a]\}. \tag{6.67b}$$

This expression extends the Cole–Cole law for the critical susceptibility to one, which describes the dynamics for states P near some generic A_2 singularity P^c.

The relevance of the extended Cole–Cole law shall be demonstrated by two examples, which deal with the second and fourth of the four scenarios discussed in Sec. 6.1.2(iii). Figure 6.9 concerns the scenario exhibited in Fig. 6.4: the characteristic frequency ω_A^c for the response of variable A is located far below the edge of the microscopic-excitation band. Figure 6.10 exemplifies the possibility considered in Fig. 6.3: the characteristic frequency ω_A^c is located within the band of normal-liquid excitations; but, it is close enough to the edge to influence the structural relaxation. The MCT results shown for loss spectra $\chi_A''(\omega)$ and response functions $\chi_A(t)$ are obtained for the two-component model defined by Eqs. (4.32a,b), (4.33a,b). For the critical points considered, the time scale t_0 is determined by fitting the first critical correlator $\phi^c(t)$ to Eq. (6.17b). All other parameters, which quantify the leading-order results, are evaluated analytically.

The parameters for the model under discussion are chosen so that the correlators $\phi_A(t)$ describe accurately the evolution of the glassy dynamics of two van der Waals liquids. The experimental data are response functions $\chi_A(t) = -\partial_t \phi_A(t)$, which are obtained by Fayer and collaborators using optical-Kerr-effect spectroscopy. Figure 6.9 deals with results for benzophenon (BZP) (Cang *et al.* 2003). The data description by the schematic-model results (Götze and Sperl 2004b) is done as is demonstrated in Fig. 4.27 for liquid salol. Results for the latter system are presented in Fig. 6.10. The drifts of the mode-coupling coefficients with temperature changes, which are exhibited by the cited data analysis, can be extrapolated to identify the critical-point parameters. For BZP, it is found: $T_c = 235\,\text{K}$, $v_A^c = 30, \lambda^c = 0.70$. The latter value fixes the critical exponent and the von Schweidler exponent to: $a = 0.33, b = 0.64$. For salol, the parameters are: $T_c = 245\,\text{K}$, $v_A^c = 55, \lambda^c = 0.73, a = 0.31, b = 0.59$. The von Schweidler-law part of the 257 K-data for salol is exhibited in Fig. 1.2. The full line and the interpolating dotted line in the upper panel of Fig. 6.9 reproduce the experimental data and their schematic-model description, respectively, for BZP at 251 K. This is the lowest temperature analysed, which corresponds to a state with distance parameter $\epsilon = (T_c - T)/T_c = -0.068$. The lowest temperature analysed for salol is 247 K, which corresponds to the rather small distance parameter $\epsilon = -0.008$.

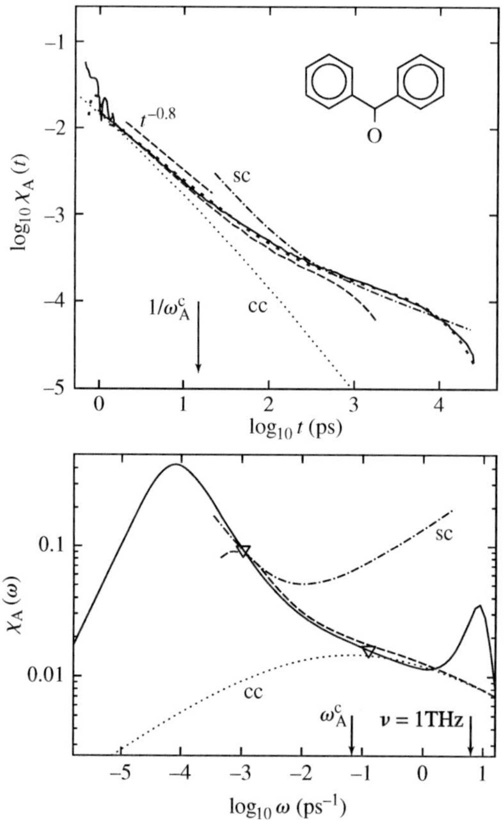

FIG. 6.9. The full line in the upper panel reproduces the optical-Kerr-effect response measured by Cang *et al.* (2003) for benzophenon at $T = 251\,\text{K}$. The dotted line is the description by the solution of the schematic model defined by Eqs. (4.32a,b), (4.33a,b). The parameters $\Omega = \Omega_A = 1.67\ \text{ps}^{-1}$, $\nu = \nu_A = 5\Omega$, and the chosen mode-coupling coefficients yield $t_0 = 0.378$ ps, $\lambda^c = 0.70$, $T_c = 235\,\text{K}$ (Götze and Sperl 2004b). The full line in the lower panel is the loss spectrum for the MCT response function. The dotted lines marked cc exhibit the Cole–Cole results for the critical dynamics; the characteristic frequency ω_A^c from Eq. (6.42b) is marked by arrows. The dashed-dotted lines marked sc exhibit the scaling-law results (6.66a,b) for $\chi_A(t) = -\partial_t\phi_A(t)$ and $\chi''_A(\omega)$, respectively. The dashed lines exhibit the results of the extended Cole–Cole law (6.67a,b). The spectrum of the latter deviates from $\chi''_A(\omega)$ by 10% at the points marked by triangles. The dashed straight line with label $t^{-0.8}$ exhibits the response for a pseudo-von Schweidler decay, which is characterized by an exponent $b_{\text{eff}} = 0.2 : \phi_A(t) -$ constant $\propto -t^{b_{\text{eff}}}$. Reproduced from Sperl (2006).

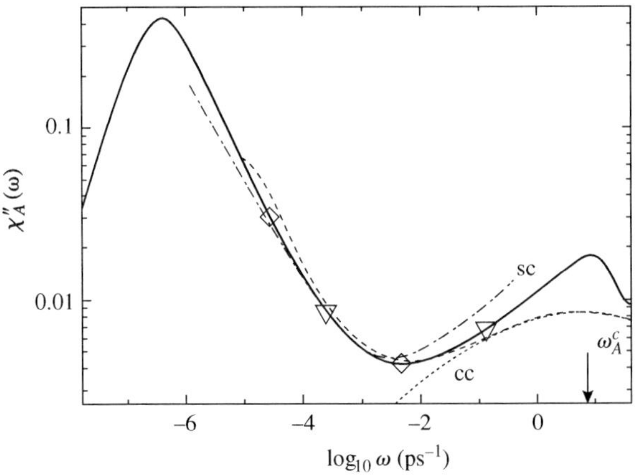

FIG. 6.10. The full line is the loss spectrum for the MCT function, which describes the glassy optical-Kerr-effect response measured by Hinze *et al.* (2000) for salol at $T = 247\,\mathrm{K}$. The data and their description are discussed in connection with Fig. 4.27. The chosen model parameters yield $t_0 = 0.00246$ ps, $\lambda^c = 0.73, T_c = 245\,\mathrm{K}$. The curves and the triangles have the same meaning as specified for Fig. 6.9. The diamonds mark the points of 10% deviations between the spectrum and its description by the scaling-law result (6.66b). Reproduced from Sperl (2006).

The Fourier-sine transformation of the above specified MCT results for $\chi_A(t)$ yields the loss spectrum $\chi''_A(\omega)$ of BZP shown in the lower panel of Fig. 6.9 and that of salol shown in Fig. 6.10. Since the MCT results reproduce perfectly the measured glassy-relaxation parts of the response functions for the two liquids, the shown loss spectra can be considered as a transformation of the experimental results from the domain of times to that of frequencies. This holds with two reservations. The response functions for $t > 10^{4.5}$ps have not been measured. Hence, the spectra for ω below about $1/(10^5\text{ps})$ reflect MCT extrapolations rather than the data. The MCT solutions do not reproduce the oscillatory transient dynamics for $t \leq 2$ ps. Therefore, the loss spectra, which are shown for the normal-liquid-excitation band, do not reflect properly the reality of the two molecular liquids under considerations. It is the intention to demonstrate that the above derived leading-order results for the MCT dynamics explain quantitatively the specified data for the plateau-crossing dynamics.

The large values v_A for the coupling of the reorientational motion of the molecules with the density-fluctuations leads to large negative correction amplitudes $K_A + \kappa(a)$. This is explained in Sec. 6.1.2. The values are -1.51 and -1.78 for BZP and salol, respectively. According to Eq. (6.39), this leads to large onset times t^*_A for the power-law asymptote for the critical decay, i.e., for the applicability of the scaling-law description. The upper panel of Fig. 6.9 shows that the scaling law accounts for the data for $t > 0.3$ ns. However, the large value of

$|\epsilon|$ causes a small offset. For salol, $|\epsilon|$ is so small for $T = 247\,\text{K}$ that the scaling law describes the response for $t > t_A^* \approx 30$ ps. The scaling law accounts also for the data for $T = 257\,\text{K}$, except for a small offset (Sperl 2006). There are large intervals between the end of the transient region and t_A^*. Within these intervals, the scaling-law expression (6.66a) for the correlator cannot explain the response.

Equation (6.38d) yields the values 0.12 and 0.18 for the correction amplitudes $K_A^m + \kappa(a)$ for the fluctuating force correlator for BZP and salol, respectively. From Eq. (6.42c), one gets for the correction amplitudes to the Cole–Cole law the values 0.08 and 0.09, respectively. These numbers quantify the trends explained in Sec. 6.1.2. The strong reduction of the correction amplitudes places the onset frequency ω_A^{m*} for the leading-order description of the critical fluctuating-force spectrum in the band of normal-liquid excitations. The Cole–Cole law describes the critical dynamics for $\omega < 0.2\ \text{ps}^{-1}$.

The results explained in the preceding paragraph have the following implications. For sufficiently small distance parameters $|\epsilon|$, the extended Cole–Cole law provides a quantitative description for the correlators for all times outside the transient region till the onset of the von Schweidler decay, say for 2 ps $< t < 10\hat{t}_{-}t_\sigma$. Similarly, the loss spectra are described well for $0.1[\hat{\omega}_{\text{min}}/t_\sigma] \leq \omega \leq 0.1$ ps^{-1}. Figure 6.10 demonstrates the good description of the salol-loss minimum at $T = 247\,\text{K}$. This holds for an interval of frequencies changes over about three orders of magnitude. Similarly, the formulas provide a description of the BZP response $\chi_A(t)$ and its loss spectrum $\chi''_A(\omega)$ at 251 K for changes of the time and the frequency, respectively, by a factor of about 300. This conclusion requires the reservation that the large distance parameter is connected with a small offset between data and analytic-formula results.

Within the von Schweidler-law regime, however, the description of the response and the loss of salol by the leading order formulas (6.66a,b) is superior to that by the extended Cole–Cole law. For this part of the dynamics, the corrections to the leading-order results for the fluctuating-force kernels $m_A(t)$ are larger than those for the correlator $\phi_A(t)$.

For BZP, the extended Cole–Cole law describes the approach of the correlator towards the pseudo-von Schweidler law (6.47b). Within the interval 2 ps $< t <$ 20 ps, this feature results in a decay $\chi_A(t) \propto t^x, x = b_{\text{eff}} - 1$. This is demonstrated for $b_{\text{eff}} = 0.2$ by the straight dashed line in the upper panel of Fig. 6.9. The lower panel exhibits this part of the dynamics as an excess wing. For 0.03 ps$^{-1} < \omega <$ 1 ps^{-1}, $\chi''_A(\omega)$ exhibits a decrease close to $C/\omega^{b_{\text{eff}}}$. The crossover from the von Schweidler-law asymptote to the excess wing occurs near the position $\hat{\omega}_{\text{min}}/t_\sigma$ of the minimum of the modulus spectrum.

Salol exemplifies the scenario where the relevant spectra are in the region with $\omega < \omega_A^c/10$. Therefore, the scaling-law results describe the qualitative features of the loss minimum. However, the deviations of the Cole–Cole spectrum from the $(\omega t_0)^a$ asymptote cause a shift of the loss minimum to frequencies above $\hat{\omega}_{\text{min}}/t_\sigma$ and values below $c_\sigma \hat{\chi}(\omega_{\text{min}})$. The high-frequency part of the loss minimum is suppressed relative to the one exhibited by the scaling-law result

$\hat{\chi}(\hat{\omega})$. The ad hoc modification of Eq. (1.7a) to $\chi''(\omega) = \chi_{\min}[b(\omega/\omega_{\min})^{a_{\text{eff}}} + a_{\text{eff}}(\omega_{\min}/\omega)^b]/(a_{\text{eff}}+b)$ with $a_{\text{eff}} = 0.24$ describes the minimum very well (Sperl 2006). Equivalently, the initial part of the structural relaxation can be fitted by a pseudo critical decay $\phi_A(t) = f'_A + h'_A(t_0/t)^{a'}$. Such behaviour is exhibited for $a' = 0.15$ as straight dashed line in Fig. 4.27.

(v) Scaling-law descriptions of degenerate A_2-bifurcations

For degenerate fold bifurcations, the formula (6.53a) simplifies to

$$\phi_q(t) = h_q g_0^\sigma(t), \quad q = 1, \dots, M. \tag{6.68}$$

There is no analogue for a below-plateau relaxation. Hence, restrictions (6.9c,d) do not occur. For small enough separation parameters σ, the leading order results are valid for $t \geq t_q^*$ and $\omega < \omega_q^*$. The onset parameter in Eq. (6.39) specify the regime of validity of the simplified equations of motion.

Let us express the function $g_0^\sigma(t)$ in terms of another function $\overline{g}^\sigma(t)$ according to: $g_0^\sigma(t) = \overline{g}^\sigma(t) + \sigma(P)/[2(1-\lambda(P))]$. Substitution of this expression into Eq. (6.51b) results in an equation for $\overline{g}^\sigma(t)$. This equation is obtained from Eq. (6.50b) if one replaces $g_0^\sigma(t)$ and $\sigma(P)$ by $\overline{g}^\sigma(t)$ and $\sigma_{\text{eff}}(P) = \sigma(P)^2/[4(1-\lambda(P))]$, respectively. Hence, the problem is reduced to the one discussed is subsection (ii) for the glass state. After substitution of $\sigma_{\text{eff}}(P)$ in place of σ, one gets the expressions for the correlation scale c_σ and time scale t_σ:

$$c_\sigma = |\sigma|/\big[2\sqrt{1-\lambda(P)}\big], \quad t_\sigma = t_0/\big[2\sqrt{1-\lambda(P)}/|\sigma|\big]^{1/a}. \tag{6.69a}$$

These scales describe the control-parameter dependence of the correlators. The shape function for $\overline{g}^\sigma(t)$ is the function $g_+(\hat{t})$. This yields

$$g_0^\sigma(t) = c_\sigma\Big\{g_+(t/t_\sigma) \pm \big[1/\sqrt{1-\lambda}\big]\Big\}, \quad \sigma \gtrless 0. \tag{6.69b}$$

For this leading-order result, the correlators for the liquid and the glass for the same $|\sigma|$ merely differ by an offset. Combining Eq. (6.59) with (6.68), one arrives at the expression for the arrested parts of the correlators of the glass states: $f_q = \phi_q(t \to \infty) = h_q[2c_\sigma/\sqrt{1-\lambda}]$. This result reproduces the leading-order contribution to Eq. (4.98). The shape function $g_+(\hat{t})$ is explained in connection with Fig. 6.6. The loss spectrum of the liquid agrees with that of the glass, provided the states agree in $|\sigma|$:

$$\chi''_X(\omega)/h_X = c_\sigma \hat{\chi}_+(\omega t_\sigma). \tag{6.69c}$$

The preceding equations generalise those discussed in Sec. 4.2.4 for an elementary model.

So far, scaling-law descriptions of degenerate fold bifurcations have not been applied for the explanation of data for the complex dynamics of liquids. Therefore, such transitions shall not be considered any further in this book.

6.1.4 *The scaling-limit description of the generic liquid–glass-transition dynamics*

In this section, the above discussed first scaling law shall be derived as the initial step of an asymptotic-solution approach for the equations of motion in the limit of states P tending towards a generic A_2-bifurcation point P^c. The leading correction for the scaling-law description shall be determined.

(i) A scaling limit

For all positive times and a positive regularization parameter r, a function shall be defined by:

$$g_r^\sigma(t) = c_\sigma C_r(t) g_\pm(\hat{t}), \quad \hat{t} = t/t_\sigma, \quad \sigma \gtrless 0. \tag{6.70a}$$

The notations for the correlation scale $c_\sigma = \sqrt{|\sigma|}$, the time scale $t_\sigma = t_0/c_\sigma^{1/a}$, and the shape functions $g_\pm(\hat{t})$ are used as introduced in Sec. 6.1.3.(ii). The cutoff function $C_r(t)$ from Eq. (6.12a) eliminates the small-time divergency of $g_\pm(\hat{t})$; all powers $g_r^\sigma(t)^n, n = 1, 2, \ldots$, are determining functions. The definition $g_r^\sigma(t) = C_r(t) g_0^\sigma(t)$ generalizes Eq. (6.12b) for the critical dynamics to one for states P, which are different from the glass-transition singularity. The regularized function $g_r^\sigma(t)$ shall be used to formulate an ansatz for the array $\boldsymbol{g}(t)$ in Eq. (6.1a):

$$g_q(t) = a_q g_r^\sigma(t). \tag{6.70b}$$

According to Eqs. (6.5a,d), the ansatz specifies a factorization expression for the correlator array $\boldsymbol{\phi}(t)$:

$$\phi_q(t) = f_q^c + h_q g_r^\sigma(t), \quad q = 1, \ldots, M. \tag{6.70c}$$

For times increasing outside the transient region, these correlators agree with the ones from Eq. (6.53a) up to exponentially decreasing deviations. Within the transient region, the correlators from Sec. 6.1.3 are modified by the cutoff factor.

In the following, the procedure from Sec. 4.3.4(i) for the solution of the fixed-point equation shall be generalized to one for the solution of the equations of motion. Again, the scale c_σ will be used as the small parameter. The limit of vanishing c_σ shall be analysed for fixed rescaled times $\hat{t}$ and for fixed rescaled frequencies $\hat{z} = z t_\sigma$, where $\hat{t} > 0$ and Im $\hat{z} > 0$. This first scaling limit implies that one studies solutions for times t tending to infinity and for frequencies z tending to zero. But, the three small quantities $c_\sigma, 1/t$, and z do not tend to zero independently. Rather, they are connected by $(1/t) = (1/\hat{t})/t_\sigma, z = \hat{z}/t_\sigma$. In Sec. 4.2.4, it is shown for an elementary example that scaling-limit results formulate the essence of the evolution of the correlators for states P tending to P^c. In this subsection, the same shall be demonstrated for the general discontinuous glass transition. This shall be done by proving that the ansatz from above solves the equations of motion in second order of c_σ. In this sense, the power-counting 'argumentation' from Sec. 6.1.3(i) shall be justified a posteriori.

To proceed, all functions occurring in the equations of motion have to be expressed in terms of the small parameter c_σ and of the rescaled variables $\hat{t}$ and $\hat{z}$. The cited definitions yield

$$\sigma = C\epsilon = \pm c_\sigma^2, \quad \sigma \gtrless 0, \tag{6.71a}$$

$$1/t = c_\sigma^\alpha/(t_0\hat{t}), \qquad z = c_\sigma^\alpha(\hat{z}/t_0), \qquad \alpha = 1/a. \tag{6.71b}$$

Since $a < 1/2$, there holds $\alpha > 2$. In the scaling limit, the inverse time and the frequency are small quantities of higher than second order in c_σ.

The exponential in the cutoff function reads: $\exp[-t/(rt_0)] = \exp[-\hat{t}/(rc_\sigma^\alpha)]$. In the scaling limit, the regularized function $g_r^\sigma(t)$ agrees with the scaling-law result $g_0^\sigma(t)$ within errors, which are smaller than any power of the small parameter c_σ:

$$\lim_{c_\sigma\to 0} \{g_r^\sigma(t) - g_0^\sigma(t)\}/c_\sigma^n = 0, \quad n = 1, 2, \ldots. \tag{6.72a}$$

Equations (A.3c), (A.5a) can be used to get: $ST[g_r^\sigma(t) - g_0^\sigma(t)](z)/c_\sigma = \hat{z}LT[\exp[-\hat{t}/(rc_\sigma^\alpha)]g_\pm(\hat{t})](\hat{z}) = -[\hat{z}/(\hat{z} + i/(rc_\sigma^\alpha))]ST[g_\pm(\hat{t})](\hat{z} + i/(rc_\sigma^\alpha))$. The leading result for small c_σ is obtained from Eq. (6.58b):

$$\lim_{c_\sigma\to 0} \{ST[g_r^\sigma(t)](z) - ST[g_0^\sigma(t)](z)\}/c_\sigma^\alpha = iK_1\hat{z}. \tag{6.72b}$$

In the same manner, one derives

$$\lim_{c_\sigma\to 0} \{ST[g_r^\sigma(t)^2](z) - ST[g_0^\sigma(t)^2](z)\}/c_\sigma^\alpha = iK_2\hat{z}. \tag{6.72c}$$

The numbers K_1 and K_2 are identical to those, which appear in Eqs. (6.12c,d) for the critical dynamics. Equivalently, there holds: $ST[g_r^\sigma(t)^n](z) - ST[g_0^\sigma(t)^n](z) \sim (izt_0)K_n, n = 1, 2$. In leading order for the scaling limit, the S-transforms of $g_r^\sigma(t)^n$ and $g_0^\sigma(t)^n$ agree up to a σ-independent linear function of z.

Function $g_0^\sigma(t)$ is a solution of Eq. (6.50a). Using the preceding equations in order to express $g_0^\sigma(t)$ in terms of $g_r^\sigma(t)$, one obtains:

$$\lim_{c_\sigma\to 0} \{\sigma + \lambda ST[g_r^\sigma(t)^2](z) - ST[g_r^\sigma(t)](z)^2\}/c_\sigma^\alpha = i\lambda K_2\hat{z}. \tag{6.73}$$

In the scaling limit, the regularized function $g_r^\sigma(t)$ obeys the simplified equation of motion up to corrections of order c_σ^α. Function $g_r^\sigma(t)$ solves Eq. (6.50a) up to a white noise-induced σ-independent background function. The latter is the same as the small-frequency limit occurring in Eq. (6.13) for the asymptotic discussion of the critical relaxation.

For the next-more-complicated expression $ST[g_r^\sigma(t)^3](z)/c_\sigma^3 = -\hat{z}LT[\{1 - \exp[-\hat{t}/(rc_\sigma^\alpha)]\}^3 g_\pm(\hat{t})^3](\hat{z})$, one has to distinguish three cases. If $g_\pm(\hat{t})^3$ is integrable for small times, the limit of vanishing c_σ can be commuted with the integral transform:

$$\lim_{c_\sigma\to 0} ST\big[g_r^\sigma(t)^3\big](z)/c_\sigma^3 = ST\big[g_\pm(\hat{t})^3\big](\hat{z}), \quad a < 1/3. \tag{6.74a}$$

If the critical exponent is not smaller than 1/3, two auxiliary functions $\delta f_\pm(\hat{t})$ and $f(\hat{t})$ can be introduced so that

$$g_\pm(\hat{t})^3 = \delta f_\pm(\hat{t}) + df(\hat{t})/d\hat{t}, \quad 1/3 \le a < 1/2. \tag{6.74b}$$

The first function is defined so that the non-integrable small time singularity of $g_\pm(\hat{t})^3$ is eliminated: $\delta f_\pm(\hat{t}) = g_\pm(\hat{t})^3 - (1/\hat{t}^a)^3$. Because of Eq. (6.63a), $\delta f_\pm(\hat{t})$ is integrable for small times. Also $f(\hat{t})$ is integrable for small times, since $f(\hat{t}) = -[(3a-1)\hat{t}^{3a-1}]^{-1}$ or $f(\hat{t}) = \ln[\hat{t}]$ for $a > 1/3$ and $a = 1/3$, respectively. Hence, both auxiliary functions have Laplace transforms. One gets: $ST[g_r^\sigma(t)^3](z)/c_\sigma^3 = ST[\{1-\exp[-\hat{t}/(rc_\sigma^\alpha)]\}^3\{\delta f_\pm(\hat{t}) - i\hat{z}f(\hat{t})\}](\hat{z}) + [3\hat{z}/(rc_\sigma^\alpha)]LT\big[\{\exp[-\hat{t}/(rc_\sigma^\alpha)] - 2\exp[-2\hat{t}/(rc_\sigma^\alpha)] + \exp[-3\hat{t}/(rc_\sigma^\alpha)]\}f(\hat{t})\big](\hat{z})$. Applying Eq. (A.3c), one can reduce the three Laplace transforms to the transform of $f(\hat{t})$. The latter is given by the elementary functions from Eq. (A.59b) or Eq. (A.69a), respectively.

$$\begin{aligned}\lim_{c_\sigma\to 0}&\Big\{ST[g_r^\sigma(t)^3](z) - \Gamma(1-3a)(3i\hat{z}c_\sigma^\alpha r^{1-3a})[1-2^{3a-1}+3^{3a-2}]\Big\}/c_\sigma^3\\ &= ST\big[\delta f_\pm(\hat{t}) - i\hat{z}f(\hat{t})\big](\hat{z}), \quad a > 1/3.\end{aligned} \tag{6.74c}$$

$$\lim_{c_\sigma\to 0}\Big\{ST[g_r^\sigma(t)^3](z) - i\hat{z}\ln[c_\sigma^\alpha]\Big\}/c_\sigma^3 = ST\big[\delta f_\pm(\hat{t}) - i\hat{z}f(\hat{t})\big](\hat{z}), \quad a = 1/3. \tag{6.74d}$$

The discussion of the preceding paragraph can be extended to one for the functions $F_n(z) = ST[g_r^\sigma(t)^n](z), n = 4, 5, \ldots$. If $na < 1$, the scaling limit yields the generalization of Eq. (6.74a): $F_n(z) = O(c_\sigma^n)$. If na is non-integer and larger than 1, there holds $F_n(z) + i\hat{z}c_\sigma^\alpha p(\hat{z}c_\sigma^\alpha) = O(c_\sigma^n)$. Here, $p(\zeta) = p_0 + \zeta p_1 + \cdots$ denotes some polynomial. If $na = 1, 2\ldots, p(c_\sigma^n)$ has to be replaced by $O(c_\sigma^n \ln[c_\sigma])$.

The formulas from above can be substituted into Eqs. (6.4b–d) in order to derive the scaling-limit expression for the inhomogeneity:

$$\begin{aligned}J_q(z) = D_q^{(0)}(P) + \Big[\sum_{k_1k_2} A_{q,k_1k_2}^{(2)}(P)a_{k_1}a_{k_2}\Big]ST\big[g_r^\sigma(t)^2\big](z) - a_q^2 ST\big[g_r^\sigma(t)\big](z)^2\\ + c_\sigma^2 O(c_\sigma, c_\sigma\ln[c_\sigma^\alpha], c_\sigma^{\alpha-2}).\end{aligned} \tag{6.75a}$$

The first three contributions to $J_q(z)$ yield finite terms for $\lim_{c_\sigma\to 0} J_q(z)/c_\sigma^2$. The remaining terms do not contribute to this limit. One concludes from Eq. (6.5b) that $\sum_k R_{q,k}J_k(z) = O(c_\sigma^2)$. Hence, the ansatz (6.70b) is consistent with the determination of the leading contribution for the solution of the equations of motion. The ansatz defines a solution in the scaling limit if and only if the condition (6.5c) is valid. This holds if and only if

$$\lim_{c_\sigma\to 0}\sum_q a_q^* J_q(z)/c_\sigma^2 = 0. \tag{6.75b}$$

Using the definitions (4.86a) and (4.88a) for the separation parameter σ and the exponent parameter λ, respectively, and observing the convention (4.74d), one concludes that the preceding formula is equivalent to

$$\lim_{c_\sigma \to 0} \{\sigma + \lambda ST[g_r^\sigma(t)^2](z) - ST[g_r^\sigma(t)](z)^2\}/c_\sigma^2 = 0. \tag{6.75c}$$

Since $\lim_{c_\sigma \to 0} c_\sigma^\alpha / c_\sigma^2 = 0$, the validity of Eq. (6.75c) is an implication of Eq. (6.73). This remark completes the derivation of the scaling-law results.

In Sec. 6.1.3(i), several examples for the factorization theorem are formulated. They are obtained as corollaries of the basic formula (6.53a) for the M components of the array $\boldsymbol{\phi}(t)$. The latter formula remains valid within the theory discussed here. But, the common factor $g_0^\sigma(t)$ is modified to $g_r^\sigma(t)$. Consequently, also within the scaling-limit approach, all previously specified factorization-property examples remain valid with the mentioned modification: $\phi_X(t) = f_X^c + h_X g_r^\sigma(t)$. Using Eq. (6.72a), one can write explicitly for $\sigma \gtrless 0$:

$$\lim_{c_\sigma \to 0} \left\{\phi_X(\hat{t}t_\sigma) - f_X^c\right\}/[h_X c_\sigma] = g_\pm(\hat{t}). \tag{6.76a}$$

Using Eq. (6.72b), one gets the formula in the frequency domain:

$$\lim_{c_\sigma \to 0} \left\{ST[\phi_X(t)](\hat{z}/t_\sigma) - f_X^c\right\}/[h_X c_\sigma] = ST[g_\pm(\hat{t})](\hat{z}). \tag{6.76b}$$

These limit results imply Eqs. (6.66a,b). Here and in the following, it is assumed that the explicitly constructed asymptotic solution $\boldsymbol{\phi}(t)$ agrees with the asymptote of the solution of the MCT equations of motion. A justification of this assumption is not available.

(ii) The leading asymptotic corrections

The factorization property and the scaling law are features of the correlators if these are evaluated correctly up to order c_σ. In this subsection, it will be shown that the asymptotic corrections to the mentioned result are proportional to c_σ^2. The leading scaling-limit results have been determined above from the request that there is cancellation of the leading terms in the equation of motion (6.5c). These leading terms are of second order in the small parameter c_σ. The corrections shall be determined from the request that there is additional cancellation of all terms proportional to c_σ^3. The calculations to be done are extensions of those explained for the critical dynamics in Sec. 6.1.2(ii).

The deviations of the functions $\phi_q(t) - f_q^c$ from the factorized form are described by the functions $\tilde{g}_q(t)$ in Eq. (6.5a). According to Eq. (6.5b), the leading contributions to these deviations are given by those of $J_q(z)$. The latter are of order c_σ^2; they are given by the first three terms on the right-hand side of Eq. (6.75a). The first term is due to the difference coefficient $D_q^{(0)}(P)$ defined in Eq. (4.71e). It varies smoothly with the distance parameter $\epsilon = \sigma(P)/C + O(\epsilon^2)$. One can introduce constants $d_q^{(0)}$ so that

$$D_q^{(0)}(P) = d_q^{(0)}\sigma + O(\sigma^2), \quad q = 1, \ldots, M. \tag{6.77}$$

Because of Eq. (4.86a), there holds $\sum_q a_q^* d_q^{(0)} = 1$. Within the intended leading-order calculations, one infers from Eq. (6.75c): $ST[g_r^\sigma(t)](z)^2 = \sigma + \lambda ST[g_r^\sigma(t)^2](z)$. This relation will be used here and throughout the following discussions in order to express $ST[g_r^\sigma(t)](z)^2$ in terms of $ST[g_r^\sigma(t)^2](z)$ and vice versa. As a result, there appears a second contribution to $J_q(z)$, which is proportional to σ. Using the expression (6.23) for the correction amplitudes K_q, one arrives at the leading-order results for the M components of $\tilde{\boldsymbol{g}}(t)$:

$$\tilde{g}_q(t) = a_q\{L_q^{(1)}\sigma + K_q g_r^\sigma(t)^2\} \tag{6.78a}$$

$$ST[\tilde{g}_q(t)](z) = a_q\{L_q^{(2)}\sigma + K_q ST[g_r^\sigma(t)](z)^2/\lambda\}. \tag{6.78b}$$

The factors of the linear-in-σ terms are given by:

$$L_q^{(1)} = \sum_k R_{q,k}\left[d_k^{(0)} - a_k^2\right]/a_q, \tag{6.79a}$$

$$L_q^{(2)} = \sum_k R_{q,k}\left[d_k^{(0)} - \sum_{k_1 k_2} A_{k,k_1k_2}^{(2)c} a_{k_1} a_{k_2}/\lambda\right]/a_q. \tag{6.79b}$$

There holds

$$L_q^{(1)} - L_q^{(2)} = K_q/\lambda, \quad q = 1, \ldots, M. \tag{6.79c}$$

The contribution $g(t)$ to Eq. (6.5a) shall be decomposed in the leading term $g_r^\sigma(t)$ and a remainder to be denoted by $\tilde{g}(t)$:

$$g(t) = g_r^\sigma(t) + \tilde{g}(t). \tag{6.80a}$$

The solution of the equation of motion will be constructed with the ansatz that $\tilde{g}(t)$ is a quantity of second order in the small parameter c_σ. For the solution to be determined, $\tilde{g}(t)/c_\sigma^2$ has a scaling limit for vanishing c_σ. Using the properties $g_r^\sigma(t) = O(c_\sigma), \tilde{g}_q(t) = O(c_\sigma^2)$, and $\tilde{g}(t) = O(c_\sigma^2)$, Eq. (6.5c) reads:

$$\sum_q a_q^*\left[J_q^\sigma(z) + J_q^{(2,1)}(z) + J_q^{(2,2)}(z) + J_q^{(3,1)}(z)\right] = (izt_0)p_0 + O(c_\sigma^4). \tag{6.80b}$$

The function $J_q^\sigma(z)$ is given by Eq. (6.4c). Here, one can use Eq. (6.77) in order to simplify $D_q^{(0)}(P)$ by $d_q^{(0)}\sigma$. The second difference term can be replaced by $\sigma \lim_{\sigma\to 0}\left[D_{q,k}^{(1)}(P)/\sigma\right] a_k ST[g_r^\sigma(t)](z)$. The other three contributions to the bracket in the preceding formula are given by Eqs. (6.25a–c) with $g_r^c(t)$ replaced by $g_r^\sigma(t)$. The further contributions to Eq. (6.5c) deal with functions of the type $F_n(z) = ST[g_r^\sigma(t)^n](z), n \geq 4$. They can be analysed as explained in the preceding subsection. To shorten the discussion, it shall be assumed $a \neq 1/3$. The case

$a = 1/3$ can be obtained by performing the limit $a \to 1/3$, as is demonstrated in Sec. 6.1.2(ii). Besides terms proportional to $c_\sigma^n, n \geq 4$, there occurs a polynomial contribution $izp(z) = i\hat{z}c_\sigma^\alpha p(\hat{z}c_\sigma^\alpha)$. The term due to $J_q^{\text{reg}}(z)$ in Eq. (6.4b) can be incorporated in the polynomial term. Terms of order $(c_\sigma^\alpha)^n, n \geq 2$, can be neglected on the right-hand side of Eq. (6.80b). Therefore, the polynomial can be reduced to its zero-frequency value $p_0 = p(z = 0)$.

The next step of the derivation deals with the identification of all contributions to the bracket on the left-hand side of the equation of motion (6.80b), which are of order c_σ^3 or larger. To proceed, one has to substitute Eqs. (6.78a,b), (6.80a) in the various expressions. As a result, one gets terms, which are proportional to σ, and other terms, which combine products of the functions $g_r^\sigma(t)$ and $\tilde{g}(t)$. These terms have the same form as discussed in Sec. 6.1.2(ii). Using the definitions (6.28a,b) for the parameters ξ and ζ, one arrives at:

$$\sum_q a_q^* \left[J_q^{(2,2)}(z) + J_q^{(3,1)}(z)\right]$$
$$= 2\xi ST[g_r^\sigma(t)^3](z) - 2\lambda\zeta ST[g_r^\sigma(t)](z) ST[g_r^\sigma(t)^2](z)$$
$$+ 2\sigma\left[C^{(1)} - C^{(2)} - \zeta\right] ST[g_r^\sigma(t)](z) + O(c_\sigma^4), \tag{6.81a}$$

$$\sum_q a_q^* \left[J_q^\sigma(z) + J_q^{(2,1)}(z)\right]$$
$$= 2\lambda ST[g_r^\sigma(t)\tilde{g}(t)](z) - 2ST[g_r^\sigma(t)](z) ST[\tilde{g}(t)](z)$$
$$+ \left\{\sigma + \lambda ST[g_r^\sigma(t)^2](z) - ST[g_r^\sigma(t)](z)^2\right\}$$
$$+ 2\sigma C^{(3)} ST[g_r^\sigma(t)](z) + O(c_\sigma^4). \tag{6.81b}$$

Three constants are defined by:

$$C^{(1)} = \sum_q a_q^* \sum_{k_1 k_2} A_{q,k_1 k_2}^{(2)c} a_{k_1} a_{k_2} L_{k_2}^{(1)}, \tag{6.82a}$$

$$C^{(2)} = \sum_q a_q^* a_q a_q L_q^{(2)}, \tag{6.82b}$$

$$C^{(3)} = \lim_{\sigma \to 0} \sum_{q,k} a_q^* D_{q,k}^{(1)}(P) a_k / (2\sigma). \tag{6.82c}$$

According to Eq. (6.73), the curly-bracket term in Eq. (6.81b) can be replaced by $\lambda K_2(i\hat{z}c_\sigma^\alpha)$. Substituted into Eq. (6.80b), the latter term alters the constant p_0 to some other value, which shall be denoted by $2p_1$. There appear terms in Eq. (6.80b), which are proportional to $\sigma ST[g_r^\sigma(t)](z)$. These can be eliminated by expressing the function $\tilde{g}(t)$ in terms of another function to be called $\tilde{H}_r^\sigma(t)$:

$$\tilde{g}(t) = \sigma\tilde{C} + \tilde{H}_r^\sigma(t), \tag{6.83a}$$

$$\tilde{C} = \left[C^{(1)} - C^{(2)} + C^{(3)} - \zeta\right]/(1-\lambda). \tag{6.83b}$$

The equation of motion is transformed to one for $\tilde{H}_r^\sigma(t)$:

$$\begin{aligned}&\lambda ST[g_r^\sigma(t)\tilde{H}_r^\sigma(t)](z) - ST[g_r^\sigma(t)](z)ST[\tilde{H}_r^\sigma(t)](z)\\ &= -\xi ST[g_r^\sigma(t)^3](z) + \lambda\zeta ST[g_r^\sigma(t)](z)ST[g_r^\sigma(t)^2](z)\\ &+ (i\hat{z}c_\sigma^\alpha)p_1 + O(c_\sigma^4).\end{aligned} \tag{6.83c}$$

Let us assume that there is a determining function $H_r^\sigma(t)$, which solves a simplified version of the preceding equation. The simplification is obtained by dropping $(i\hat{z}c_\sigma^\alpha)p_1 + O(c_\sigma^4)$. Using Eqs. (A.6a,b,) (A.7a) the simplified equation of motion is equivalent to

$$\begin{aligned}&\lambda g_r^\sigma(t)H_r^\sigma(t) - \partial_t\int_0^t g_r^\sigma(t-t')H_r^\sigma(t')dt'\\ &= -\xi g_r^\sigma(t)^3 + \lambda\zeta\partial_t\int_0^t g_r^\sigma(t-t')g_r^\sigma(t')^2dt'.\end{aligned} \tag{6.84}$$

According to Eq. (6.70a), there holds the scaling-limit result: $\lim_{c_\sigma\to 0} g_r^\sigma(t)/c_\sigma = g_\pm(\hat{t}), \sigma \gtrless 0$. Let us anticipate the existence of the analogous limit for the new function: $\lim_{c_\sigma\to 0} H_r^\sigma(t)/c_\sigma^2 = h_\pm(\hat{t}), \sigma \gtrless 0$. If one could commute the scaling limit with the integration and differentiation, the preceding linear integro-differential equation for $H_r^\sigma(t)$ would lead to one for the shape functions $h_\pm(\hat{t})$:

$$\begin{aligned}&\lambda g_\pm(\hat{t})h_\pm(\hat{t}) - \partial_{\hat{t}}\int_0^{\hat{t}} g_\pm(\hat{t}-\hat{t}')h_\pm(\hat{t}')d\hat{t}'\\ &= -\xi g_\pm(\hat{t})^3 + \lambda\zeta\partial_{\hat{t}}\int_0^{\hat{t}} g_\pm(t-\hat{t}')g_\pm(\hat{t}')^2d\hat{t}'.\end{aligned} \tag{6.85}$$

The motivation of this equation is the analogue of the loosely argued derivation of Eq. (6.55a). It is the goal to use the functions $h_\pm(\hat{t})$ for a construction of the searched for function $\tilde{H}_r^\sigma(t)$. It will be assumed that Eq. (6.85) has uniquely defined solutions $h_\pm(\hat{t})$. A justification of this assumption is not available.

Before proceeding towards the specified goal, five paragraphs shall be used for a discussion of the shape functions $h_\pm(\hat{t})$. Firstly, the short-time behaviour will be considered. The asymptotic series expansion for $g_\pm(\hat{t})$, which is mentioned in connection with Eq. (6.63a), yields a similar series for the right-hand side of Eq. (6.85): $\alpha_0\hat{t}^{-3a} \pm \alpha_1\hat{t}^{-a} + \alpha_2\hat{t}^a + \cdots$. This suggests the ansatz $h_\pm(\hat{t}) =$

$\kappa_0 \hat{t}^{-2a} \pm \kappa_1 + \kappa_2 \hat{t}^{2a} + \cdots$. A recursion relation can be derived for the coefficients. One gets in leading and next-to-leading order for vanishing $\hat{t}$:

$$h_\pm(\hat{t}) = \kappa(a)\hat{t}^{-2a} \mp \tilde{\kappa}(a). \tag{6.86a}$$

The function $\kappa(a)$ is defined in Eq. (6.29) and

$$\tilde{\kappa}(x) = \Big\{\kappa(x)\big[\Gamma(1+x)\Gamma(1-x) - \lambda^2\big]/\lambda - 3\xi + \zeta\big[2\lambda + \Gamma(1+x)\Gamma(1-x)\big]\Big\}/ \\ \Big\{2(1-\lambda)\big[\Gamma(1+x)\Gamma(1-x) - \lambda\big]\Big\}. \tag{6.86b}$$

If $a < 1/3$, all terms in Eq. (6.85) are integrable for small $\hat{t}$. In this case, one can get the equivalent equations in the frequency domain:

$$\lambda ST\big[g_\pm(\hat{t})h_\pm(\hat{t})\big](\hat{z}) - ST[g_\pm(\hat{t})](\hat{z})ST[h_\pm(\hat{t})](\hat{z}) \\ = -\xi ST[g_\pm(\hat{t})^3](\hat{z}) + \lambda\zeta ST[g_\pm(\hat{t})](\hat{z})ST[g_\pm(\hat{t})^2](\hat{z}). \tag{6.87}$$

In order to derive corresponding formulas, which are valid also for $1/3 < a < 1/2$, the shape functions shall be decomposed in the leading power-law contributions and remainders:

$$g_\pm(\hat{t}) = \hat{t}^{-a} + \hat{g}_\pm(\hat{t}), \qquad h_\pm(\hat{t}) = \kappa(a)\hat{t}^{-2a} + \hat{h}_\pm(\hat{t}). \tag{6.88a}$$

The asymptotes of the remainders are obtained from Eqs. (6.63a), (6.86a):

$$\hat{g}_\pm(\hat{t}) \sim \pm A_1 \hat{t}^a, \qquad \hat{h}_\pm(\hat{t}) \sim \mp\tilde{\kappa}(a), \quad \hat{t} \to 0. \tag{6.88b}$$

Formula (6.85) leads to an equivalent equation of motion for $\hat{h}_\pm(\hat{t})$:

$$\lambda g_\pm(\hat{t})\hat{h}_\pm(\hat{t}) - \partial_{\hat{t}} \int_0^t g_\pm(\hat{t} - \hat{t}')\hat{h}_\pm(t')dt' \\ = -\big(\xi + \lambda\kappa(a)\big)\hat{g}_\pm(\hat{t})\hat{t}^{-2a} - \xi g_\pm(\hat{t})\big(g_\pm(\hat{t})^2 - \hat{t}^{-2a}\big) \\ +\partial_{\hat{t}} \int_0^{\hat{t}} \Big\{(\lambda\zeta + \kappa(a))\hat{g}_\pm(\hat{t} - \hat{t}')\hat{t}'^{-2a} + \lambda\zeta g_\pm(\hat{t} - \hat{t}')\big[g_\pm(\hat{t}')^2 - \hat{t}'^{-2a}\big]\Big\}d\hat{t}'. \tag{6.88c}$$

In this formula, all functions of $\hat{t}$ are integrable for small $\hat{t}$. Hence, an S-transform provides an equivalent formula in the frequency domain.

$$\lambda ST\big[g_\pm(\hat{t})\hat{h}_\pm(\hat{t})\big](\hat{z}) - ST[g_\pm(\hat{t})](\hat{z})ST[\hat{h}_\pm(\hat{t})](\hat{z}) \\ = ST\big[-\big(\xi + \lambda\kappa(a)\big)\hat{g}_\pm(\hat{t})\hat{t}^{-2a} - \xi g_\pm(\hat{t})\big(g_\pm(\hat{t})^2 - \hat{t}^{-2a}\big)\big](\hat{z}) \\ +\big(\lambda\zeta + \kappa(a)\big)ST[\hat{g}_\pm(\hat{t})](\hat{z})ST[\hat{t}^{-2a}](\hat{z}) \\ +\lambda\zeta ST[g_\pm(\hat{t})](\hat{z})ST\big[g_\pm(\hat{t})^2 - \hat{t}^{-2a}\big](\hat{z}). \tag{6.88d}$$

Specializing the preceding formula to $z = 0$ and applying the identity $F(t \to \infty) = ST[F(t)](z \to 0)$, one gets an expression for the arrested part of $h_+(\hat{t})$ in terms of that of $g_+(\hat{t})$. Using Eq. (6.59), one obtains

$$h_+(\hat{t} \to \infty) = [\xi - \lambda\zeta]/(1-\lambda)^2. \tag{6.89}$$

Exploiting the result (6.63b) for the long-time behaviour of $g_-(\hat{t})$, one can apply Eqs. (A.65a,b) with $x = -b, b, 2b$ and $3b$ in order to work out the small-frequency asymptote of the right-hand side of Eq. (6.88d). In leading and next-to-leading order, one gets a sum of two terms, which are proportional to $(-i\hat{z})^{-3b}$ and $(-i\hat{z})^{-b}$, respectively. A similar result is obtained for the left-hand side if one substitutes the ansatz $h_-(\hat{t}) - \kappa_0'\hat{t}^{2b} \sim \kappa_1'$. Comparison of both sides fixes the coefficients κ_0' and κ_1'. As result, one finds the large-$\hat{t}$ asymptote for the shape function of the liquid states in leading and next-to-leading order:

$$h_-(\hat{t}) = \kappa(-b)(B\hat{t}^b)^2 + \tilde{\kappa}(-b). \tag{6.90}$$

Figure 6.11 exhibits the functions $h_\pm(\hat{t})$, which are obtained by solving Eq. (6.85) numerically for the parameters for the mode-coupling functional of the hard-sphere-system model defined in connection with Fig. 4.6. The shape functions $g_\pm(\hat{t})$ are shown in Fig. 6.6. For $\hat{t} < 0.3$, the functions $h_\pm(\hat{t})$ are very close

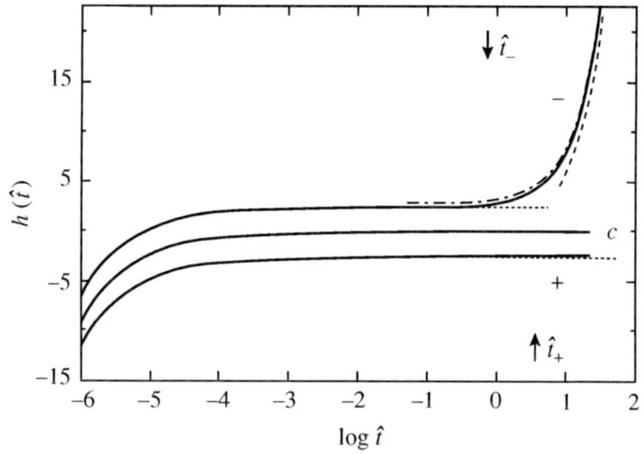

FIG. 6.11. The full lines with label $\pm$ show the shape functions $h_\pm(\hat{t})$ for the mode-coupling functional of the hard-sphere-system model defined in connection with Fig. 4.6. The line with label c exhibits the function $\kappa(a)/\hat{t}^{2a}$. The dotted lines present the expressions (6.86a) for the short-time expansion. The dashed dotted line shows the long-time expansion formula (6.90) for $h_-(\hat{t})$ and the dashed curve exhibits its truncation to the leading term $\kappa(-b)[B\hat{t}^b]^2$. There holds $\tilde{\kappa}(a) = 2.48, \tilde{\kappa}(-b) = 2.97, B = 0.836, B_1 = 0.431$ and the other parameters are noted in the caption of Fig. 6.2. The crossover times $\hat{t}_\pm$ are explained in connection with Fig. 6.6. Reproduced from Franosch *et al.* (1997).

to their short-time asymptotes. For $\hat{t} > 3$, the $h_\pm(t)$ are described well by their long-time asymptotes (6.89) and (6.90), respectively. Hence, on a 10%-accuracy level, the shape functions are explained by analytical calculations.

In order to resume the study of the leading-asymptotic corrections, the strategy of subsection (i) shall be followed. Regularized shape functions

$$H_r^\sigma(t) = c_\sigma^2 C_r(t) h_\pm(\hat{t}), \quad \sigma \gtrless 0, \tag{6.91a}$$

shall be defined. There holds the analogue of Eq. (6.72a): $\lim_{c_\sigma \to 0}\{H_r^\sigma(t) - H_0^\sigma(t)\}/c_\sigma^n = 0, n = 2, 3, \ldots$. Let us note this result explicitly for $n = 2$:

$$\lim_{c_\sigma \to 0} H_r^\sigma(t)/c_\sigma^2 = h_\pm(\hat{t}), \quad \sigma \gtrless 0. \tag{6.91b}$$

This formula shows that $H_r^\sigma(t)$ has a scaling limit, which is of second order in the small parameter c_σ. In analogy to Eq. (6.72b), there holds:

$$\lim_{c_\sigma \to 0} \Big\{ ST[H_r^\sigma(t)](z) - ST[H_0^\sigma(t)](z) \Big\} / c_\sigma^\alpha = \Gamma(1-2a)(i\hat{z}r^{1-2a})\kappa(a). \tag{6.91c}$$

Since $\alpha > 2$, one obtains

$$\lim_{c_\sigma \to 0} ST[H_r^\sigma(t)](z)/c_\sigma^2 = ST[h_\pm(\hat{t})](\hat{z}), \qquad \sigma \gtrless 0. \tag{6.91d}$$

In Eq. (6.91b), the scaling limit can be commuted with the S-transform.

The intended proof of

$$\begin{aligned}&\lambda ST\big[g_r^\sigma(t)H_r^\sigma(t)\big](z) + \xi ST[g_r^\sigma(t)^3](z)\\ &\quad -ST[g_r^\sigma(t)](z)ST[H_r^\sigma(t)](z) - \lambda\zeta ST[g_r^\sigma(t)](z)ST[g_r^\sigma(t)^2](z)\\ &\quad = (i\hat{z}c_\sigma^\alpha)p_2 + c_\sigma^3 O(c_\sigma^{\alpha-2})\end{aligned} \tag{6.92}$$

is the essential step towards the solution of Eq, (6.83c). Equations (6.72b,c), (6.91c) yield $\{-ST[g_0^\sigma(t)](z)ST[H_0^\sigma(t)](z) - \lambda\zeta ST[g_0^\sigma(t)](z)ST[g_0^\sigma(t)^2](z)\} + c_\sigma^3 O(c_\sigma^{\alpha-2})$ as a scaling-limit expression for the second-line contribution to Eq. (6.92). If $a < 1/3$, an analogous expression is obtained for the first-line contribution. In this case, the leading terms of the left-hand side of the preceding equation are of order c_σ^3. These leading terms cancel because of Eq. (6.87); and Eq. (6.92) is valid with $p_2 = 0$. The dominant small-time singularity of $g_r^\sigma(t)H_r^\sigma(t)$ or $g_r^\sigma(t)^3$is given by $c_\sigma^3\kappa(a)C_r(t)^2\hat{t}^{-3a}$ or $C_r(t)^3\hat{t}^{-3a}$, respectively. For $r = 0$ and $a > 1/3$, these singularities are not integrable. As explained in the preceding subsection, the S-transforms of the specified singularities can be evaluated conveniently via the detour of writing $\hat{t}^{-3a} = [d\hat{t}^{1-3a}/d\hat{t}]/(1-3a)$. Up to a white-noise-induced background term proportional to $(i\hat{z}c_\sigma^\alpha)$ and up to terms of order $(c_\sigma^3 c_\sigma^{\alpha-2})$, one gets for the S-transforms $c_\sigma^3\Gamma(1-3a)\kappa(a)(-i\hat{z})^{3a}$ or $c_\sigma^3\Gamma(1-3a)(-i\hat{z})^{3a}$, respectively. The two terms of the second line of Eq. (6.92) also yield two terms proportional to $c_\sigma^3(-i\hat{z})^{3a}$. The expression (6.29) for $\kappa(a)$

implies the cancellation of the four mentioned contributions. The remaining terms of Eq. (6.92) do not offer problems for the evaluation of the scaling limit. Up to terms of order $c_\sigma^3 c_\sigma^{\alpha-2}$, there appear terms of order c_σ^3. The latter terms cancel because of Eq. (6.88d). This remark finishes the proof.

Comparing Eq. (6.92) with Eq. (6.83c), one arrives at the following conclusion. The formula

$$\tilde{H}_r^\sigma(t) = H_r^\sigma(t) \tag{6.93}$$

reduces the equation of motion to the request of the vanishing of a sum of four kinds of terms. The first and second kind of terms are of order $c_\sigma^3 c_\sigma^{\alpha-1}$ and c_σ^4, respectively. The third term is a white-noise induced back-ground contribution $(i\hat{z}c_\sigma^\alpha)(p_1 - p_2) = iz(p_1 - p_2)$, which is quantified by the σ-independent numbers p_1 and p_2. Finally, there is a series of terms, which are proportional to c_σ^3. The constants of proportionality are functions of $\hat{z}$. Equation (6.93) implies that there is cancellation of all terms proportional to c_σ^3. Hence, the goal of this subsection is reached.

6.1.5 *Extended scaling-limit description of the generic A_2-bifurcation dynamics*

The results of the preceding section can be combined to a scaling-limit representation of the correlators, which reproduce the dynamics in a leading and next-to-leading order asymptotic description. For $q = 1, \ldots, M$ and $\sigma \gtrless 0$, there holds

$$\phi_q(t) - \left[f_q^c + \delta f_q^{(1)}\right] = h_q\left[c_\sigma g_\pm(\hat{t}) + c_\sigma^2 h_\pm(\hat{t})\right] \\ + h_q K_q \left[c_\sigma g_\pm(\hat{t})\right]^2, \quad \hat{t} = t/t_\sigma, \tag{6.94a}$$

$$ST[\phi_q(t)](z) - \left[f_q^c + \delta f_q^{(2)}\right] = h_q\left[c_\sigma ST[g_\pm(\hat{t})](\hat{z}) + c_\sigma^2 ST[h_\pm(\hat{t})](\hat{z})\right] \\ + h_q K_q \left[c_\sigma ST[g_\pm(\hat{t})](\hat{z})\right]^2/\lambda, \quad \hat{z} = z t_\sigma. \tag{6.94b}$$

The shape functions $g_\pm(\hat{t})$ for the leading asymptotic expression are determined by the exponent parameter λ. The shape functions $h_\pm(\hat{t})$ for the leading asymptotic corrections are determined by λ, ξ and ζ. These three parameters as well as the critical arrested parts f_q^c, the critical amplitudes h_q, and the corrections amplitudes K_q are determined by the mode-coupling functional at the critical point. There appear regular renormalizations of the arrested parts

$$\delta f_q^{(1,2)} = \sigma h_q[\tilde{C} + L_q^{(1,2)}]. \tag{6.94c}$$

The constant $\tilde{C}$, the amplitudes $L_q^{(1,2)}$ as well as the constant C in the expression (4.86c) for the separation parameter σ are given by first derivatives of the mode-coupling coefficients with respect to changes of the specified external control parameter. The derivatives have to be evaluated at the critical point P^c. The correlation scale c_σ and the scaling time t_σ are defined in Eq. (6.56b). There appears the single overall time scale t_0, which quantifies the effect of the transient dynamics on the results.

The fluctuating-force correlators $m_q(t)$ are determined as a mode-coupling polynomial of the density correlators. Within the microscopic MCT for simple systems, also the shear correlator is determined as mode-coupling functional. The tagged particle correlators $\phi_q^s(t)$ and the corresponding fluctuating-force correlators $m_q^s(t)$ can be discussed within an extension of the theory to one dealing with $2M$ correlators. Therefore, it is possible to derive expressions for the correlators $\phi_X(t)$ of the various mentioned dynamical variables X in analogy to the ones formulated above. One can follow the procedure, which is explained for the critical dynamics in Sec. 6.1.2. The amplitudes f_q^c, $\delta f_q^{(1,2)}$, h_q, and K_q have to be changed to other variable-specific numbers f_X^c, $\delta f_X^{(1,2)}$, h_X, and K_X, respectively. The quantities $\sigma, c_\sigma, t_\sigma, g_\pm(\hat{t})$ and $h_\pm(\hat{t})$ are shared by all functions. In this section, implications of the specified extended scaling-limit description shall be discussed.

It has been demonstrated in Figs. 6.6 and 6.11 that the shape functions $g_\pm(\hat{t})$ and $h_\pm(\hat{t})$ can be described well by the elementary formulas for their limit behaviour. These formulas can be used in order to identify the essential features of the bifurcation dynamics, which are expressed by the preceding equations. For small rescaled times $\hat{t}$, one can use the short-time expansions (6.63a) and (6.86a) in order to simplify Eq. (6.94a) to

$$\begin{aligned}\phi_q(t) - f_q^c = h_q(t_0/t)^a\Big\{1 + \big[K_q + \kappa(a)\big](t_0/t)^a\Big\} \\ + \delta f_q^{(a)} + h_q\big[\sigma A_1(t/t_0)^a\big],\end{aligned} \tag{6.95}$$

$\delta f_q^{(a)} = \delta f_q^{(1)} + \sigma h_q\big[2A_1K_q - \tilde{\kappa}(a)\big]$. The first-line contribution reproduces Eq. (6.30c). It describes the relaxation of the critical correlator $\phi_q^c(t)$ towards the plateau in a leading and next-to-leading asymptotic expansion in the small quantity $(t_0/t)^a$. The correction to this result are described by the two additional terms, which are proportional to σ. They demonstrate the regularity theorem for the MCT solutions: for a fixed finite time interval, $\phi_q(t) - \phi_q^c(t)$ is a smooth function of the coordinate differences for the states P and P^c. The correction term $\delta f_q^{(a)}$ describes a renormalization of the plateau. The correction term $h_q|\sigma|A_1(t/t_0)^a = h_q c_\sigma A_1(t/t_\sigma)^a$ increases with the time. For $\sigma \gtrless 0$, this correction to the critical decay describes the precursors for the arrest and the plateau crossing, respectively. This is demonstrated by the dotted lines in the upper panel of Fig. 6.6. Analogous simplified formulas can be derived for the loss spectra by performing the expansion of Eq. (6.94b) for large rescaled frequencies.

For large rescaled times, the shape functions of the glass can be replaced by their arrested part as noted in Eqs. (6.59), (6.89). The correlators exhibit arrest $\phi_q(t \to \infty) = f_q = f_q^c + h_q[c_\sigma/\sqrt{1-\lambda}][1 + c_\sigma k_q]$. The formulas (4.91a,b) are reproduced. The fluctuation part $\overline{K}_q$ and the averaged part κ of the correction amplitudes k_q are given by

$$\overline{K}_q = \big[L_q^{(1)}\sqrt{1-\lambda}\big] + \big[K_q/\sqrt{1-\lambda}\big], \tag{6.96a}$$

$$\kappa = \sqrt{1-\lambda}\Big\{\tilde{C} + [\xi - \lambda\zeta]/(1-\lambda)^2\Big\}. \tag{6.96b}$$

For large rescaled times $\hat{t}$ and $\sigma < 0$, one can use the expansions (6.63b) and (6.90) for the shape functions. For the liquid states, Eq. (6.94a) simplifies to

$$\begin{aligned}\phi_q(t) - f_q^c = &-h_q(t/t'_\sigma)^b\Big\{1 - \big[K_q + \kappa(-b)\big](t/t'_\sigma)^b\Big\} \\ &+ \delta f_q^{(b)} - \sigma h_q\big[B_1(t'_\sigma/t)^b\big],\end{aligned} \tag{6.97a}$$

$\delta f_q^{(b)} = \delta f_q^{(1)} + \sigma h_q\big[2B_1 K_q - \tilde{\kappa}(-b)\big]$. Here, the second critical time scale t'_σ is defined by Eq. (6.64b) and $\delta f_q^{(b)}$ denotes a smooth plateau modification. The dashed-dotted line in Fig. 6.6 demonstrates that the function $-(t/t'_\sigma)^b - \sigma[B_1(t'_\sigma/t)^b] = c_\sigma[-B\hat{t}^b + B_1/(B\hat{t})^b]$ is very close to $c_\sigma g_-(\hat{t})$, provided $\hat{t}$ exceeds about $0.3\hat{t}_-$. For $\hat{t} < \hat{t}_-$, the correction term $[K_q + \kappa(-b)](t/t'_\sigma)^{2b} = [K_q + \kappa(-b)]c_\sigma^2(B\hat{t})^{2b}$ is very small compared to $c_\sigma g_-(\hat{t})$. Hence, Eq. (6.97a) can be modified to

$$\phi_q(t) = \big[f_q^c + \delta f_q^{(b)}\big] + h_q c_\sigma g_-(t/t_\sigma) + h_q\big[K_q + \kappa(-b)\big](t/t'_\sigma)^{2b}. \tag{6.97b}$$

This hybrid formula describes an improvement of the scaling-law result (6.66a) for the plateau-crossing dynamics. Besides the plateau renormalization, it incorporates the time-dependent leading-asymptotic corrections for $t > 0.3\hat{t}_- t_\sigma$. The formula is inferior to Eq. (6.94a), because it ignores those time-dependent leading asymptotic corrections, which may be relevant for $t < 0.3\hat{t}_- t_\sigma$.

Multiplication of the hybrid formula with S_q yields $S_q\phi_q(t) = F_q + H'_q G(t) + H''_q(t/\tau)^{2b}$. The third term with amplitude $H''_q = S_q h_q[K_q + \kappa(-b)]$ and time scale $\tau = t'_\sigma$ denotes an extension of von Schweidler's law for the below-plateau relaxation. The second term formulates the factorization-theorem contribution. There is the correlator-specific amplitude $H'_q = S_q h_q$ and the scaling-law function $G(t) = c_\sigma g_-(t/t_\sigma)$. The formula for $S_q\phi_q(t)$ has the same form as formula (5.96b). The latter has been discussed above in connection with the construction of Figs. 1.4, 5.13 and 5.14. This observation requires the reservation that the first term $F_q = S_q[f_q^c + \delta f_q^{(b)}]$ is not the critical arrested part $F_q^c = S_q f_q^c$. Rather, it includes a plateau renormalization $S_q\delta f_q^{(b)}$, which varies linearly with σ. If a data description by Eq. (5.96b) could be done for different states P, one could determine F_q for different values of the separation parameter σ. In this case, a linear extrapolation would yield F_q^c and $\delta f_q^{(b)}$. Choosing P very close to P^c, one can write $F_q = F_q^c$. For the results of this and of the preceding paragraph there are analogous ones for the loss spectra.

The extended asymptotic description can be used in order to quantify the range of validity for the approximation of the correlators by the first-scaling-law formulas. As discussed in Sec. 6.1.2 for the critical correlators, the range of validity shall be characterized by an error margin δ^*. The admissible relative deviation of the correlators from the approximation must not exceed $2\delta^*$. The margin δ^* has to be chosen sufficiently small so that higher than second order contributions in the small parameter c_σ do not modify the formulas (6.94a–c). The numbers $n_1 = |\delta f_q^{(1)}/(\sigma f_q^c)|$, $n_{2,3} = |\delta f_q^{(a,b)}/(\sigma f_q^c)|$, and $n_4 = |k_q|$ quantify plateau renormalizations and deviations of the arrested parts from the scaling-law result, respectively. Restricting the separation parameter by the request

$$|\sigma| \leq \sigma_q^*, \tag{6.98a}$$

$\sigma_q^* = \delta^*/[\max(n_1, \ldots, n_4)]$, the mentioned linear-in-σ contributions to the correlators can be neglected. The request quantifies the condition that the state P is close to the critical point P^c.

For times smaller than the plateau-crossing value $\hat{t}_- t_\sigma$, the deviations of the correlators from the scaling-law result are given by the second term in the curly bracket of Eq. (6.95). Hence, the onset times t_q^* and onset frequencies ω_q^* for the scaling-law description are those identified in Eq. (6.39) for the onset parameters of the power-law description of the critical dynamics. The desired replacements of the symbolic conditions (6.9a,b) read:

$$t_q^* \leq t, \qquad \omega \leq \omega_q^*. \tag{6.98b}$$

For the glass states, the preceding two formulas provide the answer to the question under study. The diamonds on the short-time parts of the curves in Fig. 4.16 exhibit the σ-independence of the onset times t_q^* for results of a model for the hard-sphere system. The q-dependence of the onset parameters is discussed in Sec. 6.1.2(ii). The onset frequencies ω_q^* for the loss spectra for the mentioned correlators are marked by diamonds on the large-ω parts of the curves in Fig. 4.17. Comparison of the results for $n = 14$ and $n = 11$ with those for $n = 8$ demonstrates a σ dependence for ω_q^*. This holds in particular for the result $q = q_1$. For the state with $n = 8$ and wave-number q_1, the correction effects are so large that the formulas (6.94a,b) are not applicable.

For times exceeding the value for the plateau crossing, the deviations of the liquid correlators from their scaling-law description are determined by the second term in the curly bracket of Eq. (6.97a). This term quantifies corrections to von Schweidler's law for the below-plateau decay. A corresponding result can be obtained for the loss spectra for frequencies below the minimum position $\hat{\omega}_{\min}/t_\sigma$. The discussion of the correction effect follows that given in Sec. 6.1.2(ii) for the critical dynamics. One merely has to change the critical exponent a and the time t_0 to the negative von Schweidler exponent and to the second critical time scale, respectively. Let us define dimensionless onset times and frequencies by

$$\tilde{t}_q^* = \left[\delta^*/|K_q + \kappa(-b)|\right]^{1/b}, \qquad \tilde{\omega}_q^* = [1/\tilde{t}_q^*]k_b^{1/b}, \tag{6.98c}$$

with $k_b = 2\cos(\frac{\pi}{2}b)\Gamma(1+2b)/\Gamma(1+b)$. These numbers characterize the limit of validity of von Schweidler's law at large times and low frequencies, respectively. The searched-for replacements of the symbolic conditions (6.9c,d) are

$$t \leq \tilde{t}_q^* t'_\sigma, \qquad \tilde{\omega}_q^*/t'_\sigma \leq \omega. \tag{6.98d}$$

If the liquid state P approaches the bifurcation point P^c arbitrarily closely, t'_σ tends to infinity. In this case, the upper limit $\tilde{t}^* t'_\sigma$ of the time interval for the scaling-law description of the correlators $\phi_q(t)$ increases beyond any bound. Correspondingly, the lower limit $\tilde{\omega}^*/t'_\sigma$ of the frequency interval for the scaling-law description of the spectra $\chi''_q(\omega)$ tends to zero. This expansion of the scaling-law regions is demonstrated in Figs. 4.16 and 4.17.

In formulas (6.98a–d), the index q can be replaced by X in order to formulate the results for general dynamical variables.

A paragraph shall be used to formulate a corollary, which is of relevance for the judgement of attempts to interpret glassy-relaxation data of liquids in terms of the leading-order-asymptotics formulas of MCT. Let us compare the ranges of validity $t_X^* \leq t \leq \tilde{t}_X^* t'_\sigma$ and $\tilde{\omega}_X^*/t'_\sigma \leq \omega \leq \omega_X^*$. One gets

$$\left[\log(\tilde{t}_X^* t'_\sigma) - \log(t_X^*)\right] = \left[\log(\omega_X^*) - \log(\tilde{\omega}_X^*/t'_\sigma)\right] + \log(k_\lambda). \tag{6.99}$$

The constant of proportionality $k_\lambda = k_a^{1/a} \cdot k_b^{1/b}$ is determined by the exponent parameter. The factors $k_a^{1/a}$ and $k_b^{1/b}$ are caused by the fact that the corrections to the critical decay and to the von Schweidler decay, respectively, exhibit power-law variation. The first bracket denotes the length of the $\log(t)$ interval for the scaling-law description of the plateau-crossing dynamics of $\phi_X(t)$. The second bracket denotes the length of the $\log(\omega)$ interval for the scaling-law description of the minimum of $\chi''_X(\omega)$. The first length exceeds the second one by $\log(k_\lambda)$. For the cited hard-sphere-liquid model, there holds $k_a^{1/a} = 39.7$ and $k_b^{1/b} = 1.96$, and this implies $\log k_\lambda = 1.89$. In this case, the frequency interval of the validity of the scaling-law description of the loss spectra is about a factor 100 smaller than the corresponding time interval for that of the correlator decay.

According to Eq. (6.94a), the same factor K_q quantifies the violations of the factorization theorem for small as well as for large times. This and the following two paragraphs shall be used to identify general implications of this peculiarity. The discussion will be restricted to liquid states P, which are so close to the critical point P^c that the condition (6.98a) is obeyed. Then, within the above mentioned error level δ^*, all plateau-corrections can be ignored. Formula (6.94a) shall be rewritten as one for the function $\hat{\phi}_q(t) = [\phi_q(t) - f_q^c]/h_q$. And this formula shall be generalized to correlation functions $\phi_X(t)$ for the above specified more general dynamical variables X:

$$\hat{\phi}_X(t) = \left[c_\sigma g_-(\hat{t}) + c_\sigma^2 h_-(\hat{t})\right] + K_X[c_\sigma g_-(\hat{t})]^2. \tag{6.100a}$$

With increasing time, $g_-(\hat{t})$ decreases monotonically from large positive values to large negative ones. For times close to the plateau-crossing value $\tau^{pc} = t_- = \hat{t}_- t_\sigma$, there holds $g_-(\hat{t})^2 = [\partial g_-(\hat{t}_-)/\partial \log(\hat{t})]^2 \log[t/t_-]^2 + O(\log[t/t_-]^3)$. For all variables X, the $\hat{\phi}_X(t)$ versus $\log(t)$ curves meet at $t = t_-$ for the value $c_\sigma^2 h_-(\hat{t}_-)$. At this meeting point, all curves have the same slope. The difference of functions formed for two variables X and Y reads

$$\hat{\phi}_X(t) - \hat{\phi}_Y(t) = (K_X - K_Y)\big[c_\sigma g_-(\hat{t})\big]^2. \tag{6.100b}$$

For $K_X \neq K_Y$, the difference between the two function increases monotonically if t increases above t_- and also if t decreases below t_-.

Let us consider a finite sequence of correlators $\phi_1(t), \phi_2(t), \ldots$ formed for a set of variables $X_1, X_2, \ldots$. The largest value of the onset times $t_1^*, t_2^*, \ldots$ for the critical decay shall be denoted by t^*. The smallest value of the rescaled onset times $\tilde{t}_1^*, \tilde{t}_2^*, \ldots$ for the von Schweidler decay shall be abbreviated by $\tilde{t}^*$. Within the error margin δ^* and for times between t^* and $\tilde{t}^* t'_\sigma$, the correction terms can be dropped in Eq. (6.100a). The correlators can be represented by the scaling-law result (6.66a). The $\hat{\phi}_i(t)$ versus $\log(t)$ curves, $i = 1, 2, \ldots$, collapse on the function $c_\sigma g_-(\hat{t})$. The curves fan out if the times are outside the scaling-law interval. Let us choose two such times $t_{(a)}$ and $t_{(b)}$ so that $t_{(a)} < t^*$ and $\tilde{t}^* t'_\sigma < t_{(b)}$. Equation (6.100b) implies the rule: the sequence of short-time values of the functions exhibits the ordering $\hat{\phi}_1(t_{(a)}) < \hat{\phi}_2(t_{(a)}) < \cdots$ if and only if the sequence of long-time values exhibits the same ordering $\hat{\phi}_1(t_{(b)}) < \hat{\phi}_2(t_{(b)}) < \cdots$. This fanning-out rule is demonstrated in Fig. 4.16 for the sequence of density-fluctuation correlators with wave numbers q_1, q_2, q_0 and q_3. Equation (6.94b) can be used to derive two additional fanning-out rules for the renormalized loss spectra $\hat{\chi}_X(\omega) = \chi''_X(\omega)/h_X$. For frequencies around the loss-minimum position $\hat{\omega}_{\min}/t_\sigma$, the spectra collapse as expressed by the scaling-law-formula (6.66b). Choosing a frequency $\omega_{(a)}$ above the scaling-law interval, the spectra exhibit the same sequence as discussed for the correlators: $\hat{\chi}_1(\omega_{(a)}) < \hat{\chi}_2(\omega_{(a)}) < \cdots$. If a frequency $\omega_{(b)}$ is chosen below the scaling-law interval, the sequence of values is reversed $\hat{\chi}_1(\omega_{(b)}) > \hat{\chi}_2(\omega_{(b)}) > \cdots$. These rules are exhibited by Fig. 4.17.

A quantitative formulation of the fanning-out phenomenon (Aichele and Baschnagel 2001b) is based on the observation that the two correction terms in Eq. (6.100a) can be simplified by using the asymptotic expressions for the shape functions. For $t = t_{(a)}$, Eqs. (6.63a), (6.86a) imply $g_-(\hat{t}_{(a)}) = \hat{t}_{(a)}^{-a}$ and $h_-(\hat{t}_{(a)}) = \kappa(a)\hat{t}_{(a)}^{-2a}$. Consequently,

$$\hat{\phi}_X(t_{(a)})/c_\sigma = g_-(\hat{t}_{(a)}) + \big[\hat{t}_{(a)}^{-2a}/t_\sigma^a\big] D_X^{(a)}, \quad t_{(a)} \leq t^*, \tag{6.101a}$$

$D_X^{(a)} = [K_X + \kappa(a)]t_0^a$. For $t = t_{(b)}$, Eqs. (6.63b), (6.90) yield $g_-(\hat{t}_{(b)}) = -B\hat{t}^b$ and $h_-(\hat{t}_{(b)}) = \kappa(-b)B^2\hat{t}_{(b)}^{2b}$. One gets the result of the hybrid formula (6.97b) without plateau correction:

$$\hat{\phi}_X(t_{(b)})/c_\sigma = g_-(\hat{t}_{(b)}) + \big[\hat{t}_{(b)}^{2b}/t_\sigma^a\big] D_X^{(b)}, \quad \tilde{t}^* t'_\sigma < t_{(b)}, \tag{6.101b}$$

$D_X^{(b)} = B^2[K_X + \kappa(-b)]t_0^a$. Up to an offset $c = [\kappa(a) - \kappa(-b)]t_0^a$, the amplitude $D_X^{(a)}$ for the deviation from the scaling-law result at short times is proportional to the amplitude $D_X^{(b)}$ for the deviation at long times:

$$D_X^{(a)} = B^{-2} D_X^{(b)} + c. \tag{6.101c}$$

The constant of proportionality B^{-2} is determined by the exponent parameter λ.

Let us consider two figures, which deal with interpretations of molecular-dynamics-simulation data for glassy liquids in terms of MCT-bifurcation formulas. The first one refers to the binary Lennard-Jones mixture with stochastic short-time dynamics, which is explained in connection with Fig. 1.4. The evolution of the plateau-crossing dynamics upon cooling the liquid can be described with the first-scaling-law formulas (Gleim and Kob 2000) and this by using the exponent parameter $\lambda = 0.71$ calculated within a microscopic MCT-model for the system (Nauroth and Kob 1997). Figure 1.4 demonstrates an example for a comparison of a $\phi_X(t)$ versus $\log(t)$ curve with the result expected from Eq. (6.54b). The comparison is done for the lowest temperature analysed, which corresponds to a distance parameter $\epsilon = (T_c - T)/T_c \approx 0.03$. Data for the same temperature are exhibited in the upper inset of Fig. 6.12. The shown decay curves refer to fluctuations of the total densities and the tagged-particle densities for intermediate-size wave numbers for the two species. The corresponding results for the rescaled functions

$$R_X(t) = [\phi_X(t) - \phi_X(t')]/[\phi_X(t'') - \phi_X(t')] \tag{6.102a}$$

are shown in the main frame. The times t' and t'' are chosen close to the suspected ends t^* and $\tilde{t}^* t'_\sigma$ of the scaling interval. Hence, the functions $\phi_X(t')$ and $\phi_X(t'')$ can be evaluated from the scaling-law results (6.66a). Formula (6.100a) leads to

$$R_X(t) = \Big\{g_-(\hat{t}) - g_-(\hat{t}') + c_\sigma[h_-(\hat{t}) + K_X g_-(\hat{t})^2]\Big\}\Big/\Big\{g_-(\hat{t}'') - g_-(\hat{t}')\Big\}. \tag{6.102b}$$

The factorization property explains the collapse of the $R_X(t)$ versus $\log(t)$ curves on the function $\{g_-(\hat{t}) - g_-(\hat{t}')\}/\{[g_-(\hat{t}'') - g_-(\hat{t}')\}$. This demonstration is the same as explained in connection with Fig. 1.8, albeit with the role of the two factors interchanged. Figure 6.12 demonstrates this theorem within the noise level of the data for times between 10^2 and 3×10^4. For times outside the specified scaling-law interval, the $R_X(t)$ versus $\log(t)$ curves fan out as discussed above for the $\hat{\phi}_X(t)$ versus $\log(t)$ curves. In the cited paper, the 36 curves are labelled from top to bottom by i and j for $t = t_{(a)} = 3$ and $t = t_{(b)} = 10^5$, respectively. The lower inset exhibits the points (j, i). The fanning-out rule requires the relation $j = i$. The data exhibit this trend. It does not seem possible to judge whether or not the scatter of the points around the shown $j = i$ line is consistent with the noise of the data.

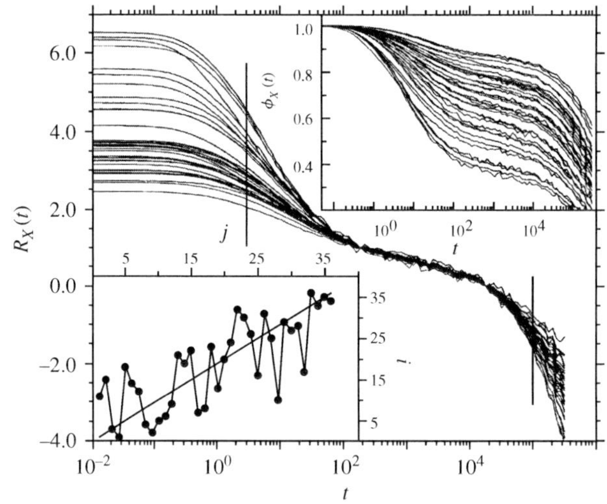

FIG. 6.12. The upper inset exhibits 36 normalized density-fluctuation correlators $\phi_X(t)$, which are determined by molecular-dynamics simulations for a binary Lennard-Jones mixture for the temperature $T = 0.446$. The system and the units for its description are defined in connection with Fig. 1.4. The main frame exhibits a rescaling of the correlators to functions $R_X(t)$, as defined in Eq. (6.102a); the reference times are $t' = 1.5 \times 10^4$ and $t'' = 2 \times 10^2$. The curves are labelled from top to bottom by i for the intersection with the left vertical line and by j for those with the right vertical one; $i, j = 1, \ldots, 36$. The lower inset shows the pairs (j, i) and the line $i = j$. Reproduced from Gleim and Kob (2000).

Let us contemplate the loss spectra for the above discussed correlators. Figure 1.4 demonstrates a 4.5-decade time interval for the applicability of the scaling-law results. But, this large range of validity of a leading-asymptotic formula is exceptional. According to Fig. 6.12, a typical range of validity is smaller by up to two decades. Equation (6.99) shows that there is a reduction of the scaling-law interval for the frequency domain by about two orders of magnitude relative to that in the time domain. Consequently, a typical interval for the description of the loss minimum by the scaling-law formula (6.66b) deals with a frequency variation by a factor 3 only. For the identified exceptional case of a tagged-particle-density-fluctuation correlator for a wave number near the structure-function-peak position, the range of validity deals with frequency variations of about 2.5 orders of magnitude. For the discussion of results obtained for simulations of liquids with Newtonian short-time dynamics, as opposed to those obtained from liquids with stochastic short-time motion, about another decade of frequency variations cannot be used for a comparison with scaling-law formulas. In this case, low-frequency tails of normal-liquid oscillation spectra cover considerable parts of the high-frequency structural-relaxation spectra.

This phenomenon is discussed in connection with Figs. 1.4, 4.24 and 6.5. One concludes the following. For the specified exceptional case, the scaling-law formula (6.66b) describes the low-frequency wing of the loss minimum for a frequency change of about a factor 30. For a more typical probing variables X, the scaling-law description is not applicable. Both conclusions could be proposed as explanation of the corresponding results for loss spectra, which are reported by Kob and Andersen (1995b).

The second example to be considered refers to the system of decamer chains, which is defined in connection with Fig. 1.5. Extensive glassy-dynamics studies for this system are reviewed by Baschnagel and Varnik (2005). The evolution of the plateau-crossing dynamics has been demonstrated for a series of variables X. It can be described by the first-scaling-law result (6.66a) if one chooses the exponent parameter $\lambda = 0.635$. This value implies the number $B = 0.475$. Hence, the constant of proportionality in Eq. (101c) reads $B^{-2} \approx 4.4$. For the lowest temperatures studied, $R_X(t)$ versus $\log(t)$ diagrams demonstrate a scaling-law interval of about 2.5 decades. The scaling-law fit provides the value for t_σ. It can be used to calculate $\hat{t}$ for every value for the time t and, thereby, the numbers $g_-(\hat{t}), \hat{t}^a$ and $\hat{t}^b$. Choosing times $t_{(a)}$ and $t_{(b)}$ as explained, the simulation data provide the numbers $\hat{\phi}_X(t)/c_\sigma - g_-(\hat{t})$ for $t = t_{(a)}$ and $t = t_{(b)}$, respectively. Using Eqs. (6.101a,b), one arrives at the parameters $D_X^{(a)}$ and $D_X^{(b)}$.

Figure 6.13 exhibits the pairs $D_X^{(a)}, D_X^{(b)}$ for three kinds of density fluctuations. The wave numbers vary between about 15% of the structure-function-peak position and about 2.7 times this value. The results for the incoherent function exhibit the linear relation (6.101a) and the expected coefficient B^{-2} for $-0.025 \leq D_X^{(b)} \leq 0.010$. The same is true for the chain correlators, except for the results for the small wave numbers $q = 1$ and $q = 2$. For larger values of the correction factor $D_X^{(b)}$, the data deviate from the linear law. Possibly, these results are influenced by higher-than-second-order terms for the asymptotic expansion. The results for the coherent function for intermediate wave numbers are close to the line defined by Eq. (6.101c). But, for small as well as for large wave numbers, there are strong deviations from the asymptotic result. The data for $D_X^{(a)}$ have to be considered with a reservation. The above mentioned slow-oscillation tails influence the values of the functions $\hat{\phi}_X(t)$ for times near $t_{(a)}$. This effect might mask the structure-relaxation parts.

Let us conclude the discussion of the first-scaling-law regime by a look at a quantitative comparison of numerical results for the solution of the equations of motion and the analytical results of the asymptotic-expansion approach. Figure 6.14 exhibits examples for two liquid states of the previously considered microscopic model for hard-sphere systems. The analysis of the critical dynamics is discussed in connection with Fig. 6.2. The dashed and dashed-dotted lines in Fig. 6.14 for the asymptotic results are fixed by the various above derived analytical expressions; there does not enter any fit parameter besides the value $t_0 = 0.425$, which had been determined from a fit of the critical dynamics.

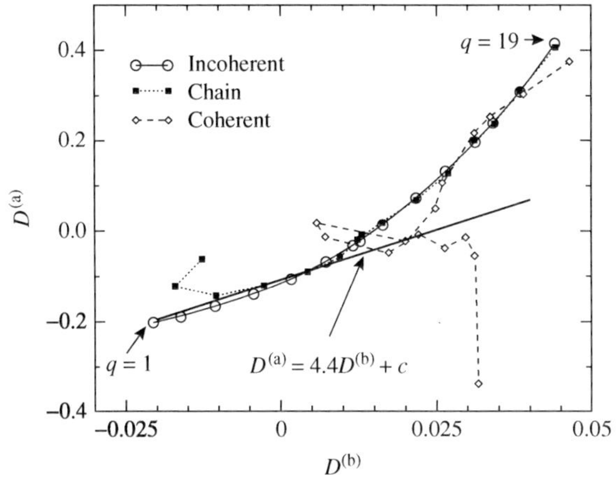

FIG. 6.13. The data points are extracted from correlators $\phi_X(t)$, which are obtained by molecular-dynamics simulations for a polymer-liquid model. The system and the units used for its description are defined in connection with Fig. 1.5. The amplitudes $D_X^{(a)}$ and $D_X^{(b)}$ specify the fanning out of the $\hat{\phi}_X(t)$ versus $\log(t)$ curves as defined in Eqs. (6.66a), (6.101a,b). The straight line exhibits the asymptotic-analysis result (6.101c) with the constant c adjusted. The correlators are formed with scalar-density fluctuations $A_{\vec{q}}$, which are combinations of the density fluctuations $\rho_{s\vec{q}}^{\alpha}$ of monomer number α of molecule number s. For the incoherent function, $A_{\vec{q}} = \rho_{s\vec{q}}^{\alpha}$, is used, for the chain function, there holds $A_{\vec{q}} = \sum_{\alpha} \rho_{s\vec{q}}^{\alpha}$, and the coherent function is formed with $A_{\vec{q}} = \sum_{\alpha s} \rho_{s\vec{q}}^{\alpha}$. The 15 values for the wave numbers cover the interval from 1 to 19. The structure-factor-peak position is near $q = 7$. Reproduced from Aichele and Baschnagel (2001b).

For the very small distance parameter $\epsilon = (\varphi - \varphi_c)/\varphi_c = -10^{-3}$, the figure demonstrates the following results for the density fluctuations with wave number $q_2 = 10.6/d$. The result (6.66a) describes the plateau crossing of $\phi_2(t)$ for a time increase by about five orders of magnitude. Similarly, the minimum of the spectrum $\chi_2''(\omega)$ is described on a 10% accuracy level for a frequency decrease by about a factor 10^3. The difference in size of the mentioned intervals of $\log(t)$ and $\log(\omega)$ is explained by Eq. (6.99). Including the leading-asymptotic corrections extends the range of validity of the description by about a factor 10 to smaller values of t and $1/\omega$ and also by about an order of magnitude to larger ones.

For wave number $q_1 = 7.0/d$ and $\epsilon = -0.001$, the leading-order results yield a qualitative explanation of the plateau crossing and for the loss minimum. But, there are pronounced shortcomings of the scaling-law descriptions. The inflection point of the $\phi_1(t)$ versus $\log(t)$ curve for the towards-plateau decay is not so pronounced as formulated by Eq. (6.66a). Similarly, the high-frequency part of the minimum of the $\log \chi_1''(\omega)$ versus $\log(\omega)$ curve is flatter than described by the

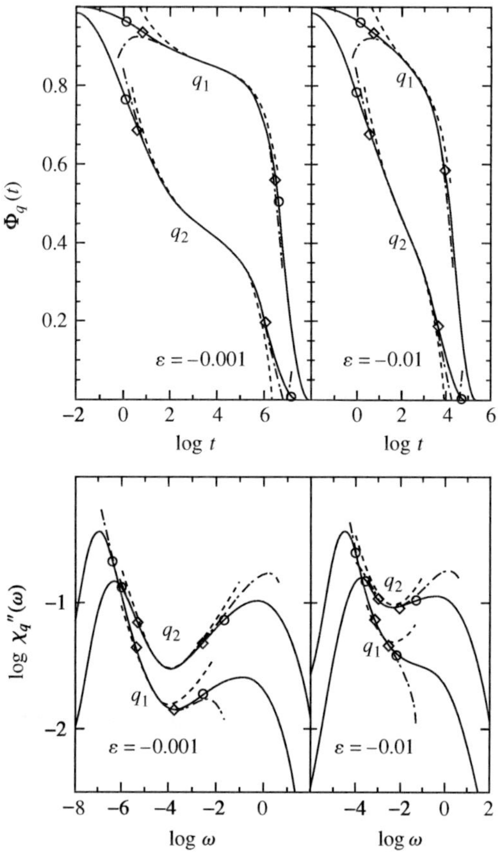

FIG. 6.14. The curves are calculated for the 100-component microscopic MCT model for hard-sphere liquids with stochastic short-time dynamics. Details are explained in the captions of Figs. 4.6, 4.14, 6.2, and 6.11. The full lines in the upper panels for the distance parameters $\epsilon = -0.001$ and $\epsilon = -0.01$ reproduce the correlators from Fig. 4.14 for label $n = 9$ and $n = 6$, respectively. The lower panels reproduce the corresponding loss spectra from Fig. 4.15. The dashed and dashed-dotted lines in the upper panels exhibit the results (6.53a) for the leading-order asymptotic description and for the formula (6.94a) with leading corrections included, respectively. The diamonds and the circles, respectively, denote the points of deviation by ± 0.05 between the cited asymptotic-expansion results and the correlators $\phi_q(t)$. The dashed and dashed-dotted lines in the lower panels denote the corresponding results for the loss spectra from Eq. (6.57b) and Eq. (6.94b), respectively. The diamonds and the circles, respectively, denote the points of 10% deviation between the asymptotic-expansion results and the loss spectra $\chi_q''(\omega)$. Reproduced from Franosch *et al.* (1997).

shape function $\hat{\chi}_-(\omega)$. These defects are explained in Sec. 6.1.2 as implications of the large modifications of the critical dynamics due to strong-coupling effects. Taking into account the leading asymptotic corrections, the range of validity for the analytic description expands considerably. The plateau crossing is reproduced for a time variation by five orders of magnitude. Similarly, the loss minimum is reproduced for a frequency interval of about three decades.

Increasing the separation of the state from the critical point, the range of applicability of the asymptotic-expansion formulas shrinks. This is demonstrated by a comparison of the two right panels in Fig. 6.14 with the corresponding left ones. For the small distance parameter $\epsilon = -0.01$, the plateau crossing of $\phi_2(t)$ is described for a time increase by about a factor 10^3. The minimum of $\chi_2''(\omega)$ is described up to some offset. Incorporating the leading corrections, the loss minimum is reproduced for an interval, which deals with a frequency change of about 3 orders of magnitude.

For $\epsilon = -0.01$, the decay of the correlations for wave number q_1 towards the plateau presents itself by a downward-bent $\phi_1(t)$ versus $\log(t)$ curve. This qualitative feature is not reproduced by the scaling-law result. The phenomenon is caused by strong-coupling effects for the critical dynamics. As explained in Sec. 6.1.2, the effect can be accounted for by incorporation of the leading asymptotic corrections. The results of Eq. (6.94a) reproduce the decay of $\phi_1(t)$ towards, through, and below the plateau for a time increase by more than a factor 10^3. It is discussed in Sec. 6.1.2 that the loss minimum may be eliminated due to the strong-coupling-induced modifications of the high-frequency parts of the critical spectrum. This phenomenon is demonstrated for $\chi_1''(\omega)$. Incorporation of the leading corrections reproduces this feature. The analytical formula describes the MCT result for frequency variations by almost two orders of magnitude.

The shown diagrams for the evolution of the stretched bifurcation dynamics for liquid states of a representative model suggest the following conclusion. The scaling laws and their extensions by incorporation of the leading-asymptotic corrections explain the essential features of the MCT scenario for freezing. The corresponding diagrams for positive distance parameters (Franosch *et al.* 1997) suggest the same conclusion for the melting scenario of the amorphous solid.

6.2 Dynamics of the second-scaling-law regime

In this section, the dynamics is discussed for liquid states P near some generic liquid-glass-transition point P^c. The times shall be restricted to be above the value $t_- = \hat{t}_- t_\sigma$, and the frequencies are requested to be located below $\omega_{\min} = \hat{\omega}_{\min}/t_\sigma$. Here t_- and $\omega_{\min}$ denote the plateau-crossing time and the loss-minimum frequency, respectively, which are explained within the first-scaling-law description of the transition scenario. It is the goal to derive asymptotic laws for the decay of the correlator arrays $\boldsymbol{\phi}(t) = (\phi_1(t), \ldots, \phi_M(t))$ from the plateau arrays $\boldsymbol{f}^c = (f_1^c, \ldots, f_M^c)$ to $\mathbf{0}$. It is the aim to identify general laws for the low-frequency peaks of the loss spectra $\chi_q''(\omega), q = 1, \ldots, M$.

In Sec. 6.2.1, a second scaling limit will be defined. The equations of motion for the shape functions and for the leading asymptotic corrections are derived. Properties of the shape functions are discussed in Sec. 6.2.2. Results for the corrections shall be used in Sec. 6.2.3 in order to specify the range of validity of von Schweidler's law.

The formulas to be derived apply also to states P with negative separation parameter σ, which are approaching a discontinuous glass–glass-transition point P^c. Such scenario is characterized by a non-vanishing array $\boldsymbol{f} = \boldsymbol{\phi}(t \to \infty)$ for $\sigma \to -0$. In this case, one can apply the covariance transformation (4.24a) with $\boldsymbol{f}^* = \boldsymbol{f}$ in order to reduce the problem to that to be discussed for $\boldsymbol{f} = \mathbf{0}$.

6.2.1 *Equations of motion for the second-scaling-law regime*

As done in all preceding discussions of the evolution of the glassy dynamics, the liquid states to be considered are assumed to be located on some smooth path P^ϵ in parameter space. According to the definitions discussed in connection with Eqs. (4.84)–(4.86), the states are specified uniquely by their negative separation parameters $\sigma = \sigma_\epsilon$. Let us assume that a time scale τ_σ is defined on the path, which increases smoothly with decreasing $|\sigma|$. The M correlation functions $\phi_q(t)$ of the model under discussion can be written as functions of σ and of the time rescaled with τ_σ:

$$\phi_q(t) = F_q(\sigma, \tilde{t}), \quad \tilde{t} = t/\tau_\sigma. \tag{6.103a}$$

Because of Eq. (A.5a), this definition in the time domain is equivalent to one in the frequency domain:

$$\phi_q(z) = \tau_\sigma F_q(\sigma, \tilde{z}), \quad \tilde{z} = z\tau_\sigma, \tag{6.103b}$$

$F_q(\sigma, \tilde{z}) = LT[F_q(\sigma, \tilde{t})](\tilde{z})$. The desired solution of the equations of motion shall be constructed with the ansatz of asymptotic expansions:

$$F_q(\sigma, \tilde{t}) = \tilde{\phi}_q(\tilde{t}) + \sigma(1 - f_q^c)\tilde{\psi}_q(\tilde{t}) + O(\sigma^2), \tag{6.103c}$$

$$F_q(\sigma, \tilde{z}) = \tilde{\phi}_q(\tilde{z}) + \sigma(1 - f_q^c)\tilde{\psi}_q(\tilde{z}) + O(\sigma^2), \quad q = 1, \ldots, M. \tag{6.103d}$$

The functions $\tilde{\phi}_q(\tilde{t})$ shall be continuous and positive definite and the functions $\tilde{\psi}_q(\tilde{t})$ shall be determining ones. These functions shall exhibit the standard symmetries so that the following considerations can be restricted to non-negative values for the rescaled times $\tilde{t}$. Similarly, it is sufficient to discuss the functions $\tilde{\phi}_q(\tilde{z}) = LT[\tilde{\phi}_q(\tilde{t})](\tilde{z})$ and $\tilde{\psi}_q(\tilde{z}) = LT[\tilde{\psi}_q(\tilde{t})](\tilde{z})$ for the upper half plane of rescaled frequencies $\tilde{z}$. The scale τ_σ will be determined below. For the discussion of this section, it is sufficient to request

$$\overline{\lim}_{\sigma \to 0} t_\sigma / \tau_\sigma < \infty. \tag{6.104}$$

If P approaches P^c, the time τ_σ diverges not less strongly than the scale t_σ, which enters the first-scaling-law description. The following calculations are done with the goal to justify the ansatz a posteriori by its success.

The preceding ansatz implies the existence of the two limits

$$\lim_{\sigma \to 0} \boldsymbol{\phi}(\tilde{t}\tau_\sigma) = \tilde{\boldsymbol{\phi}}(\tilde{t}), \tag{6.105a}$$

$$\lim_{\sigma \to 0} \boldsymbol{\phi}(\tilde{z}/\tau_\sigma)/\tau_\sigma = \tilde{\boldsymbol{\phi}}(\tilde{z}), \tag{6.105b}$$

and the statement that the limits can be interchanged with the Laplace transform and its inverse. These are scaling-limits similar to those considered in Sec. 6.1.4. They shall be referred to as the second scaling limits of MCT. The arrays $\tilde{\boldsymbol{\phi}}(\tilde{t})$ and $\tilde{\boldsymbol{\phi}}(\tilde{z})$ describe the correlators for times t tending to infinity, for frequencies z tending to zero and for liquid states P approaching the bifurcation point. But, the three quantities $\sigma, 1/t$, and z do not tend to zero independently; they are connected by the conditions $1/t = 1/(\tilde{t}\tau_\sigma)$ and $z = \tilde{z}/\tau_\sigma$. These conditions are quantified by fixed arbitrary pairs of finite non-vanishing numbers $\tilde{t}$ and $\tilde{z}$ for the rescaled time and rescaled frequency, respectively. Similarly, the ansatz formulates the limit properties

$$\lim_{\sigma \to 0} \{\phi_q(\tilde{t}\tau_\sigma) - \tilde{\phi}_q(\tilde{t})\}/\sigma = (1 - f_q^c)\tilde{\psi}_q(\tilde{t}), \tag{6.106a}$$

and the statement that the scaling limit can be interchanged with the integral transformation LT and its inverse:

$$\lim_{\sigma \to 0} \{\phi_q(\tilde{z}/\tau_\sigma) - \tilde{\phi}_q(\tilde{z})\}/\sigma = (1 - f_q^c)\tilde{\psi}_q(\tilde{z}), \quad q = 1, \ldots, M. \tag{6.106b}$$

The arrays of correction functions $\tilde{\boldsymbol{\psi}}(\tilde{t})$ and $\tilde{\boldsymbol{\psi}}(\tilde{z})$ quantify the evolution of the scaling limit upon decreasing $|\sigma|$.

Substitution of the formula (6.103c) into the Eq. (4.15a), yields the expansion for the mode-coupling polynomial

$$\begin{aligned} &\mathcal{F}_q\big[P, \boldsymbol{F}(\sigma, \tilde{t})\big] \\ &\quad = \tilde{m}_q(\tilde{t}) + \big[\sigma/(1 - f_q^c)\big]\big\{A_q(\tilde{t}) + \sum_k A_{q,k}(\tilde{t})\tilde{\psi}_k(\tilde{t})\big\} + O(\sigma^2). \end{aligned} \tag{6.107a}$$

In this formula,

$$\tilde{\boldsymbol{m}}(\tilde{t}) = \boldsymbol{\mathcal{F}}\big[P^c, \tilde{\boldsymbol{\phi}}(\tilde{t})\big] \tag{6.107b}$$

denotes the scaling limit of the array of fluctuating-force correlators: $\tilde{\boldsymbol{m}}(\tilde{t}) = \lim_{\sigma \to 0} \boldsymbol{m}(\tilde{t}\tau_\sigma)$. The amplitudes for the corrections are given by:

$$A_q(\tilde{t}) = (1 - f_q^c) \lim_{\sigma \to 0} \mathcal{F}_q \left[P - P^c, \tilde{\boldsymbol{\phi}}(\tilde{t})\right]/\sigma, \tag{6.107c}$$

$$A_{q,k}(\tilde{t}) = (1 - f_q^c)\big[\partial \mathcal{F}_q[P^c, \tilde{\boldsymbol{\phi}}(\tilde{t})]/\partial f_k\big](1 - f_k^c), \quad q, k = 1, \ldots, M. \tag{6.107d}$$

The M functions $A_q(\tilde{t})$ generalize the difference coefficients $D_q^{(0)}(P)$ from Eq. (4.71e) to time-dependent quantities. They reduce to these coefficients if $\tilde{\boldsymbol{\phi}}(\tilde{t})$ in Eq. (6.107c) is replaced by $\boldsymbol{f}^c$. Similarly, the M^2 functions $A_{q,k}(\tilde{t})$ generalize the elements of the critical stability matrix $A^c_{q,k}$, which is defined in Eq. (4.61c). Again, the former reduce to the latter if $\tilde{\boldsymbol{\phi}}(\tilde{t})$ is replaced by $\boldsymbol{f}^c$.

The equation of motion (4.29a) is equivalent to $F_q(\sigma,\tilde{z})/[1+\tilde{z}\tilde{F}_q(\sigma,\tilde{z})] = \tau_\sigma^{-1}[\tilde{z}\tau_\sigma^{-1}+i\nu_q]/\Omega_q^2 + LT[\mathcal{F}_q[P,\boldsymbol{F}(\sigma,\tilde{t})]](\tilde{z})$. Dropping $\tilde{z}\tau_\sigma^{-1}$, one gets a formula which is equivalent to Eq. (4.29b). Using the expression (6.56b) for t_σ, one can write: $t_0\tau_\sigma^{-1} = [t_\sigma/\tau_\sigma][|\sigma|^{(1-2a)/2a}]|\sigma|$. The request (6.104) ensures that the first-bracket term is bounded. Since the critical exponent a is smaller than 1/2, the second-bracket term vanishes with $|\sigma|$. One concludes the following. For a discussion of the scaling limit in a leading and next-to-leading asymptotic description, the contribution $\tau_\sigma^{-1}[\tilde{z}\tau_\sigma^{-1}+i\nu_q]/\Omega_q^2$ can be dropped relative to the contribution $LT[\mathcal{F}_q[P,\boldsymbol{F}(\sigma,\tilde{t})]](\tilde{z})$. Both versions of equations of motion reduce to

$$F_q(\sigma,\tilde{z})/\big[1+\tilde{z}F_q(\sigma,\tilde{z})\big] = LT\big[\mathcal{F}_q[P,\boldsymbol{F}(\sigma,\tilde{t})]\big](\tilde{z}), \quad q=1,\ldots,M. \qquad (6.108)$$

The scaling limit of the preceding formula leads an algebraic expression for the kernel $\tilde{m}_q(\tilde{z})$ in terms of the correlators $\tilde{\phi}_q(\tilde{z}) : \tilde{\phi}_q(\tilde{z})/[1+\tilde{z}\tilde{\phi}_q(\tilde{z})] = \tilde{m}_q(\tilde{z})$. Equivalently, there is an elementary expression of the scaling limit of the correlator in the frequency domain in terms of that for the fluctuating forces:

$$\tilde{\phi}_q(\tilde{z}) = \tilde{m}_q(\tilde{z})/[1-\tilde{z}\tilde{m}_q(\tilde{z})], \quad q=1,\ldots,M. \qquad (6.109)$$

These formulas together with the mode-coupling expression (6.107b) for the kernels $\tilde{m}_q(\tilde{t})$ constitute the searched for equations of motion for the second-scaling-limit description for the dynamics. Expanding Eq. (6.108) in linear order in σ, one gets the equations of motion for the correction functions:

$$\Big\{(1-f_q^c)/\big[1+\tilde{z}\tilde{\phi}_q(\tilde{z})\big]\Big\}^2\tilde{\psi}_q(\tilde{z}) - \sum_k LT\big[A_{q,k}(\tilde{t})\tilde{\psi}_k(\tilde{t})\big](\tilde{z}) = A_q(\tilde{z}). \qquad (6.110)$$

There are two trivial solutions of the equations of motion for $\tilde{\boldsymbol{\phi}}(\tilde{t})$. The first one reads $\tilde{\boldsymbol{\phi}}(\tilde{t}) = 0$. In this case, the classification of terms in leading and next-to-leading ones, which lead to the cited equations, is not justified. The second trivial solution is $\tilde{\boldsymbol{\phi}}(\tilde{t}) = \boldsymbol{f}^c$. In this case, Eq. (6.110) gets the form of Eq. (4.75a): $\sum_k[\delta_{q,k} - A^c_{q,k}]\tilde{\psi}_k(\tilde{z}) = I_q$, $I_q = -D_q^{(0)}/(\sigma\cdot z)$. The solubility condition (4.75b) cannot hold, since Eq. (4.86a) implies $\sum_q a_q^* I_q = -1/z$. Consequently, both trivial solutions have to be excluded from the discussion.

For every positive number x, one can use the array $\boldsymbol{F}(\sigma,\tilde{t})$ in order to define the array $\boldsymbol{F}^{(x)}(\sigma,\tilde{t}) = \boldsymbol{F}(\sigma,\tilde{t}/x)$. Equation (A.5a) implies the equivalent definition in the frequency domain $\boldsymbol{F}^{(x)}(\sigma,\tilde{z}) = x\boldsymbol{F}(\sigma,x\tilde{z})$. The mode-coupling functional transforms accordingly: $LT[\tilde{\mathcal{F}}_q[P,\boldsymbol{F}^{(x)}(\sigma,\tilde{t})]](\tilde{z}) = xLT[\mathcal{F}[P,\boldsymbol{F}(\sigma,\tilde{t})]](\tilde{z}x)$. Consequently, the equations of motion (6.108) are solved by the array $\boldsymbol{F}(\sigma,\tilde{t})$

if and only if they are solved by the array $\boldsymbol{F}^{(x)}(\sigma, \tilde{t})$. The equations of motion cannot determine the time scale for the dynamics. If $\tilde{\boldsymbol{\phi}}(\tilde{t})$ and $\tilde{\boldsymbol{\psi}}(\tilde{t})$ are solutions, the same is true for the pair of arrays

$$\tilde{\boldsymbol{\phi}}^{(x)}(\tilde{t}) = \tilde{\boldsymbol{\phi}}(\tilde{t}/x), \qquad \tilde{\boldsymbol{\psi}}^{(x)}(\tilde{t}) = \tilde{\boldsymbol{\psi}}(\tilde{t}/x). \tag{6.111a}$$

Since $\tilde{t} = t/\tau_\sigma$, the specified rescaling is equivalent to a renormalization of the time scale τ_σ to

$$\tau_\sigma^x = x\tau_\sigma. \tag{6.111b}$$

The discussion shall be continued with the following assumptions. The above formulated equations of motion have non-trivial solutions $\tilde{\boldsymbol{\phi}}(\tilde{t})$ and $\tilde{\boldsymbol{\psi}}(\tilde{t})$. They are defined uniquely up to a rescaling according to Eq. (6.111a). The components of the arrays are continuous, they have continuous derivatives for $\tilde{t} > 0$, and they are absolutely integrable. The functions $\tilde{\phi}_q(\tilde{t}), q = 1, \ldots, M$, are continuous also for $\tilde{t} = 0$ and they are positive definite. A proof of the legitimacy of these assumptions is not available. One concludes: for all scales τ_σ, which are consistent with condition (6.104), Eqs. (6.103a), (6.103c) provide a solution of the equations of motion in leading and next-to-leading order in the scaling limit.

The equations of motion (6.107b), (6.109) for the array $\tilde{\boldsymbol{\phi}}(\tilde{t})$ are quantified by the mode-coupling coefficients at P^c. Within the microscopic MCT models, $\tilde{\boldsymbol{\phi}}(\tilde{t})$ is determined by the equilibrium-structure functions at the glass-transition singularity. In particular, the solutions are the same for systems based on a Newtonian short-time dynamics and for ones based on stochastic transient motion. Differences of the details of the short-time motions can enter merely via the scale τ_σ. The same statement holds for the array of correction functions $\tilde{\boldsymbol{\psi}}(\tilde{t})$. However, the latter depends also on the leading-order changes of the mode-coupling coefficients due to changes of the external control parameters. These changes enter via the coefficients $A_q(\tilde{t})$ in Eq. (6.110).

The M kernels $\tilde{m}_q(\tilde{t})$ are polynomials of the correlators formed with positive coefficients. Hence, they are also positive definite continuous functions with continuous derivatives. The initial value of the correlator array $\boldsymbol{f}^* = \tilde{\boldsymbol{\phi}}(\tilde{t} \to 0)$ determines that for the kernel array: $\tilde{\boldsymbol{m}}(\tilde{t} \to 0) = \boldsymbol{\mathcal{F}}[P^c, \boldsymbol{f}^*]$. Equation (A.8c) yields $\lim_{\tilde{z}\to\infty}[-\tilde{z}\tilde{\boldsymbol{\phi}}(\tilde{z})] = \boldsymbol{f}^*$ and $\lim_{\tilde{z}\to\infty}[-\tilde{z}\tilde{\boldsymbol{m}}(\tilde{z})] = \boldsymbol{\mathcal{F}}[P^c, \boldsymbol{f}^*]$. Using the equations of motion (6.109), one gets $f_q^*/(1 - f_q^*) = \mathcal{F}_q[P^c, \boldsymbol{f}^*], q = 1, \ldots, M$. The M components of the initial-value array $\boldsymbol{f}^*$ form a fixed point for the critical state P^c. Because of Eq. (2.27b), the fixed point $\boldsymbol{f}^* = \boldsymbol{0}$ would lead to the first trivial solution, which has to be excluded. The array $\boldsymbol{f}^c$ of the critical arrested parts is a non-trivial fixed point for the state P^c. Generically, there is no other non-trivial fixed point. Consequently, there holds

$$\lim_{\tilde{t}\to 0} \tilde{\boldsymbol{\phi}}(\tilde{t}) = \lim_{\tilde{z}\to\infty} [-\tilde{z}\tilde{\boldsymbol{\phi}}(\tilde{z})] = \boldsymbol{f}^c. \tag{6.112a}$$

It has been requested that the correlations refer to liquid states. Thereby, the second trivial solution is excluded. One obtains

$$\lim_{\tilde{t}\to\infty} \tilde{\boldsymbol{\phi}}(\tilde{t}) = \lim_{\tilde{z}\to 0}[-\tilde{z}\tilde{\boldsymbol{\phi}}(\tilde{z})] = 0. \tag{6.112b}$$

The second-scaling-limit solution describes the decay of the correlations from the critical arrested parts to zero. Let us notice that the preceding limit values remain unchanged if a rescaling is performed.

The M formulas (6.109) can be rewritten as $\tilde{\phi}_q(\tilde{z}) = \tilde{m}_q(\tilde{z}) + \tilde{z}[\tilde{m}_q(\tilde{z})\tilde{\phi}_q(\tilde{z})]$. Equations (A.6a,b), (A.7a) can be used to derive an equivalent set of equations of motion in the time domain. For $q = 1, \ldots, M$, there holds

$$\tilde{\phi}_q(\tilde{t}) = \tilde{m}_q(\tilde{t}) - \partial_{\tilde{t}} \int_0^{\tilde{t}} \tilde{m}_q(\tilde{t}-\tilde{t}')\tilde{\phi}_q(\tilde{t}')d\tilde{t}', \tag{6.113a}$$

or, equivalently,

$$\int_0^{\tilde{t}} \tilde{\phi}_q(\tilde{t}')d\tilde{t}' = \int_0^{\tilde{t}} \tilde{m}_q(\tilde{t}')d\tilde{t}' - \int_0^{\tilde{t}} \tilde{m}_q(\tilde{t}-\tilde{t}')\tilde{\phi}_q(\tilde{t}')d\tilde{t}'. \tag{6.113b}$$

Here $\tilde{m}_q(\tilde{t})$ abbreviates the mode-coupling polynomial (6.107b) of the correlators. For every fixed positive value $\tilde{t}_+$, the preceding formula specifies a set of M coupled implicit functional equations for the set of M functions $\tilde{\phi}_q(\tilde{t}), q = 1, \ldots, M$, which are defined for the interval $0 \leq \tilde{t} \leq \tilde{t}_+$.

The structure of the equation of motion (6.109), $\tilde{\phi}_q(\tilde{z}) = -1/\{\tilde{z} - 1/[\tilde{m}_q(\tilde{z})]\}$, is similar to that of Eq. (4.29b). Therefore, it is possible to construct a sequence of approximants in analogy to that discussed in connection with Eq. (4.26b). To proceed, let $\tilde{\boldsymbol{\phi}}^{(r)}(\tilde{t})$ denote an array of M continuous positive definite functions. As explained in connection with Eq. (4.25b), an array of continuous positive definite functions is defined by

$$\tilde{\boldsymbol{m}}^{(r)}(\tilde{t}) = \boldsymbol{\mathcal{F}}\big[P^c, \tilde{\boldsymbol{\phi}}^{(r)}(\tilde{t})\big]. \tag{6.114a}$$

The results of Sec. 2.4.2 can be used in order to show that the functions

$$\tilde{\phi}_q^{(r+1)}(\tilde{z}) = -1/\big\{\tilde{z} - 1/\big[\tilde{m}_q^{(r)}(\tilde{z})\big]\big\}, \quad q = 1, \ldots, M, \tag{6.114b}$$

are Laplace transforms of continuous positive definite functions $\tilde{\phi}_q^{(r+1)}(\tilde{t})$. Choosing positive rates $\Gamma_q, q = 1, \ldots, M$, and writing $\tilde{\phi}_q^{(0)}(\tilde{t}) = f_q^c \exp[-\tilde{t}\Gamma_q]$, the preceding two equations define recursively the desired sequence of approximants with the properties specified for $\tilde{\phi}_q^{(r)}(\tilde{t})$. One can repeat the derivation of Eq. (6.112a) in order to show by induction that all approximants have the same initial value:

$$\tilde{\boldsymbol{\phi}}^{(r)}(\tilde{t} \to 0) = \boldsymbol{f}^c. \tag{6.114c}$$

Moreover, the discussion of Eqs. (D.14), (D.15) in appendix D imply that all approximants are completely monotonic. There holds

$$(-\partial_{\tilde{t}})^n \tilde{\phi}_q^{(r)}(\tilde{t}) > 0, \quad n = 0, 1, \ldots; \quad r = 0, 1, \ldots; \quad q = 1, \ldots, M. \tag{6.114d}$$

If one could prove the uniform convergence of the sequence of approximants for every finite time interval, one would have shown that $\lim_{r\to\infty} \tilde{\boldsymbol{\phi}}^{(r)}(\tilde{t}) = \tilde{\boldsymbol{\phi}}(t)$ is a solution of the equation of motion. The limit function would have all properties assumed above. In addition, the M components of the array would be completely monotonic.

6.2.2 *The second-scaling-law description of the liquid dynamics*

The second-scaling-limit results for the correlators and for the susceptibilities formulate the superposition principles of the MCT dynamics. In this section, von Schweidler's laws and their extensions will be identified as asymptotic expression for the shape functions for short rescaled times $\tilde{t} = t/\tau_\sigma$ and large rescaled frequencies $\tilde{z} = z\tau_\sigma$, respectively. This findings will be used to identify τ_σ as the second critical time scale t'_σ, which characterizes the long-time behaviour of the first-scaling-law correlators. Some examples for the interpretation of data for glassy liquids in terms of the second-scaling-law results shall be discussed also.

(i) Superposition principles

The leading-order results (6.105a,b) for the second-scaling-limit expansion yield

$$\boldsymbol{\phi}(t) = \tilde{\boldsymbol{\phi}}(\tilde{t}), \qquad \tilde{t} = t/\tau_\sigma, \tag{6.115a}$$

for the M-component arrays of correlators and

$$\boldsymbol{\chi}(z) = \tilde{\boldsymbol{\chi}}(\tilde{z}), \qquad \tilde{z} = z\tau_\sigma, \tag{6.115b}$$

for the arrays of dynamical susceptibilities $\chi_q(z) = 1 + z\phi_q(z) = 1 - ST[\phi_q(t)](z)$. Both formulas are equivalent, since $\tilde{\chi}_q(\tilde{z}) = 1 + \tilde{z}\tilde{\phi}_q(\tilde{z})$. For liquid states P, which approach the critical point P^c, the formulas describe the dynamics for fixed $\tilde{t}$ and fixed $\tilde{z}$, respectively, with errors, which vanish proportional to the separation parameter σ. The control-parameter dependence of the correlators in Eq. (6.115a) and susceptibilities in Eq. (6.115b) is described by that of the time scale τ_σ. The different correlators and susceptibilities differ by their control-parameter-independent shape functions $\tilde{\phi}_q(\tilde{t})$ and $\tilde{\chi}_q(\tilde{z})$, respectively. These second scaling laws of MCT are identical to the superposition principles, which have been discussed extensively in preceding chapters. Here, these principles are identified as laws for precursor phenomena for the arrest of correlations.

In Sec. 5.1 on the MCT for simple liquids, it is shown that the superposition principles for the density-fluctuation correlators lead to corresponding principles for other functions, which characterize the dynamics. There hold the superposition principle (5.16a) for the shear correlator, the second-scaling law (5.61a) for the tagged-particle density-fluctuation correlators, and the superposition principle (5.61b) for the mean-squared displacement. There is scale coupling: the same time τ_σ enters the different cited formulas.

So far, analytic expressions for the shape functions have been found only for a special class of one-component models. As discussed in Eqs. (4.32a–c), these models describe the dynamics by a single correlator $\phi(t)$. The speciality is the restriction of the mode-coupling polynomial to a mononomial. There holds $m(t) = v\phi(t)^n, n = 2, 3, \ldots$. The model exhibits a discontinuous glass transition for the critical control parameter $v^c = n^n/(n-1)^{n-1}$. The plateau value is $f^c = (n-1)/n$, and the exponent parameter has the marginal value $\lambda = 1/2$. Let us consider the sequence of approximants from Eqs. (6.114a–d). The simple relaxation law $\tilde{\phi}^{(0)}(\tilde{t}) = f^c \exp[-\Gamma\tilde{t}]$ yields $\tilde{m}^{(0)}(\tilde{z}) = -(n-1)/[\tilde{z} + in\Gamma]$. This implies $\tilde{\phi}^{(1)}(\tilde{z}) = \tilde{\phi}^{(0)}(\tilde{z})$. Hence, $\tilde{\phi}^{(r)}(\tilde{t}) = \tilde{\phi}^{(0)}(\tilde{t}), r = 1, 2, \ldots$; and $\tilde{\phi}(\tilde{t}) = \tilde{\phi}^{(0)}(\tilde{t})$ is the shape function. This speciality was noticed for the model with $n = 2$ by Leutheusser (1984) and Bengtzelius *et al.* (1984). The models describe a non-trivial stretched decay of the correlator towards f^c, which is characterized by the critical exponent $a = 0.395\ldots$. But, the low-frequency loss peak is the Lorentzian (1.1a); it does not exhibit stretching.

All shape functions to be discussed in this book are obtained by numerical solutions of Eqs. (6.107b), (6.113a). Let us consider some results, which have been obtained for schematic models (Fuchs *et al.* 1991). The shape functions $\tilde{\phi}(\tilde{t})$ and $\tilde{\chi}''(\tilde{\omega})$ for the one-component model, which is defined by Eqs. (4.32a,b), can be described very well by Kohlrausch functions. The model can be extended by a second correlator according to the definitions (4.33a,b). The corresponding shape functions can be fitted by Kohlrausch functions only less satisfactorily than the one cited for the first correlator. The two-component model, which is defined by Eqs. (4.34a–c), demonstrates liquid boundaries with corners. The latter are caused by the crossing of a smooth surface of transition points P^c with another such surface of transition points $P^{c\prime}$. The components of the critical arrested parts $\boldsymbol{f}^c$ for the former surface are larger than the ones of $\boldsymbol{f}^{c\prime}$ for the latter. This holds provided P^c and $P^{c\prime}$ are close to the corner. Fig. 4.5 demonstrates a corner for the cut with $v_3 = 45$. For such situation, the shape functions can exhibit a two-step decay. The first step describes the decay below $\boldsymbol{f}^c$. The second one describes the decay towards, through and below $\boldsymbol{f}^{c\prime}$. The loss spectra can exhibit a two-peak structure. In the cited paper, solutions for a three-component model are used to fit dielectric loss data. This model can exhibit loss spectra, which consist of three stretched peaks. The shape functions depend on the details of the mode-coupling functional. The cited examples make it evident that there is no generally valid parameterization of the shape functions for the superposition-principle description of the dynamics.

Figure 6.15 exhibits shape functions $\tilde{\phi}_q(\tilde{t})$ and $\tilde{\chi}''_q(\tilde{t})$ for a microscopic MCT model for the hard-sphere system. These functions are also shown as heavy lines in Fig. 4.21 and in the lower panels of Fig. 4.22. The corresponding reactive parts $\tilde{\chi}'_q(\omega)$ are reproduced in the upper panels of Fig. 4.22. The figures demonstrate

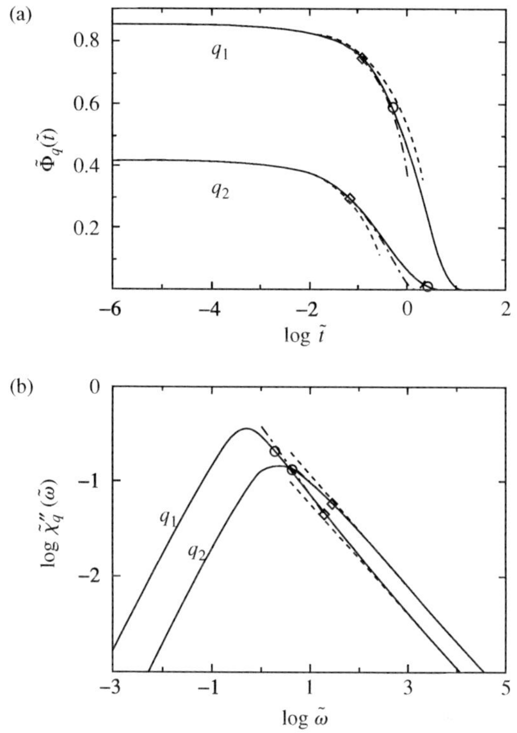

FIG. 6.15. The full lines are the second-scaling-law shape functions for the correlators (panel a) and the susceptibility spectra (panel b) for the wave numbers $q_1 = 7.0/d$ and $q_2 = 10.6/d$, which are evaluated for the hard-sphere-system model defined in the caption of Fig. 4.6. The dashed lines are the von Schweidler asymptotes according to Eqs. (6.120a,b); the von Schweidler exponent is $b = 0.583$. The diamonds mark the onset times $\tilde{t}_q^*$ and onset frequencies $\tilde{\omega}_q^*$, respectively; here, the dashed lines deviate from the full ones by 10%. The dashed-dotted lines exhibit the extended von Schweidler laws (6.122a,b). These lines deviate from the shape functions by 10% at the points marked by circles. Reproduced from Franosch *et al.* (1997).

the approach of the rescaled functions towards the shape function as formulated in Eqs. (6.105a,b) if the scale τ_σ is chosen as t'_σ.

Figure 6.16 shows a series of rescaled mean-squared displacements $\delta r_s^2(t) = 6\Delta_s(t)$ for decreasing distance parameters. The results are evaluated for the same hard-sphere-liquid model and for the same choice of the scale, which are mentioned in the preceding paragraph. The heavy line shows the shape function $6\tilde{\Delta}_s(\tilde{t})$. It describes the increase of $\delta r_s^2(t)$ from the critical arrested part $6(r_s^c)^2$ towards the diffusion limit. The evolution of the rescaled functions exhibits a scenario similar to that, which is demonstrated for glassy-liquid data in the lower panel of Fig. 1.9 and in the inset of Fig. 1.11.

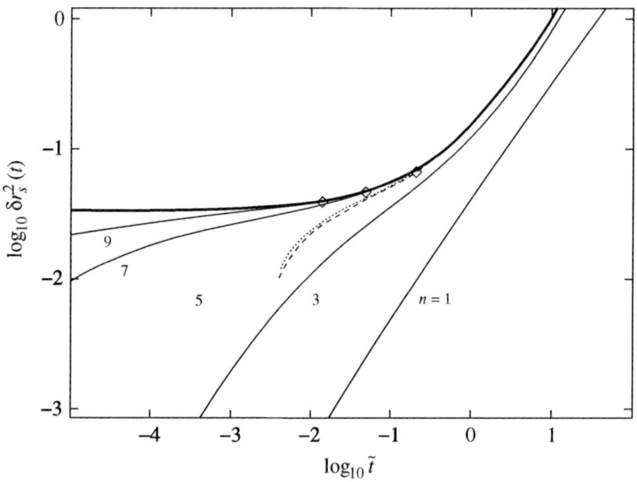

FIG. 6.16. Double-logarithmic representation of the mean-squared displacement $\delta r_s^2(t) = 6\Delta_s(t)$ as a function of the rescaled time $\tilde{t} = t/\tau$, evaluated for the hard-sphere-liquid model defined in the caption of Figs. 4.6 and 4.14. The scaling time is chosen according to Eq. (6.64b): $\tau = t'_\sigma = 0.578/|\sigma|^\gamma, \gamma = 2.46$. The state with label n refers to a relative deviation of the packing fraction φ from the critical value φ_c given by $\epsilon = (\varphi - \varphi_c)/\varphi_c = -10^{-n/3}$. The heavy uppermost line shows the shape function $6\tilde{\Delta}(\tilde{t})$ for the second-scaling law. The diamonds mark the points, where the rescaled curves with labels $n = 5$, 7 and 9 deviate from the shape function by 20%. The dashed line shows the shape function complemented by the correction for the $n = 5$ result according to Eq. (6.135). Dropping the $\tilde{t}$-independent term δr_{sc}^2 in this formula, one gets the result shown as dotted line. Reproduced from Fuchs *et al.* (1998).

(ii) von Schweidler's law and its extension

Equation (6.112a) ensures that $\tilde{\boldsymbol{\phi}}(\tilde{t})$ is close to $\boldsymbol{f}^c$ if the rescaled time $\tilde{t}$ is small. This motivates the attempt to solve the equations of motion for the shape functions asymptotically for the limit $\tilde{t} \to 0, \tilde{\boldsymbol{\phi}}(\tilde{t}) \to \boldsymbol{f}^c$. To proceed, let us express $\tilde{\boldsymbol{\phi}}(\tilde{t})$ in terms of an array $\boldsymbol{g}(\tilde{t})$, which is defined by

$$\tilde{\phi}_q(\tilde{t}) = f_q^c + (1 - f_q^c) g_q(\tilde{t}), \quad q = 1, \ldots, M. \tag{6.116a}$$

The limits (6.112a) can be rewritten as the pair of equivalent formulas

$$\lim_{\tilde{t} \to 0} \boldsymbol{g}(\tilde{t}) = \boldsymbol{0}, \quad \lim_{\tilde{z} \to \infty} ST[\boldsymbol{g}(\tilde{t})](\tilde{z}) = \boldsymbol{0}. \tag{6.116b}$$

Substituting Eq. (6.116a) into the formula (6.107b), the mode-coupling kernels are expressed as polynomials: $\tilde{m}_q(\tilde{t}) = \mathcal{F}_q[P^c, \boldsymbol{f}^c] + (1 - f_q^c)^{-1} \sum_{n=1}^{n_0} \sum_{k_1 \cdots k_n} A_{q,k_1 \cdots k_n}^{(n)c} g_{k_1}(\tilde{t}) \cdots g_{k_n}(\tilde{t})$. Equation (4.71c) defines the coefficients $A_{q,k_1 \cdots k_n}^{(n)c} =$

$A^{(n)}_{q,k_1\cdots k_n}(P^c)$. The fixed-point equation implies $\mathcal{F}_q[P^c, \boldsymbol{f}^c] = f^c_q/(1 - f^c_q)$. The equation of motion (6.109) can be written as $ST[m_q(\tilde{t})](\tilde{z}) = ST[\tilde{\phi}_q(\tilde{t})](\tilde{z})/[1 - ST[\tilde{\phi}_q(\tilde{t})](\tilde{z})]$. Substituting $ST[\tilde{\phi}_q(\tilde{t})](\tilde{z}) = f^c_q + (1 - f^c_q)ST[g_q(\tilde{t})](\tilde{z})$ and using the expression $A^c_{q,k} = A^{(1)c}_{q,k}$ for the critical stability matrix (4.61c), one arrives at an example of Eq. (4.75a): $\sum_{q,k}[\delta_{q,k} - A^c_{q,k}]ST[g_k(\tilde{t})](\tilde{z}) = I_q(\tilde{z}), q = 1, \ldots, M$. Here, the following abbreviations are used:

$$I_q(\tilde{z}) = \sum_{n=2}^{n_0+1} J^{(n)}_q(\tilde{z}), \tag{6.117a}$$

$$J^{(n)}_q(\tilde{z}) = \Big[\sum_{k_1\cdots k_n} A^{(n)c}_{q,k_1\cdots k_n} ST[g_{k_1}(\tilde{t})\cdots g_{k_n}(\tilde{t})](\tilde{z})\Big] - ST[g_q(\tilde{t})](\tilde{z})^n, \tag{6.117b}$$

for $2 \leq n \leq n_0$, and $J^{n_0+1}_q(\tilde{z}) = -ST[g_k(\tilde{t})](\tilde{z})^{n_0+1}/[1 - ST[g_k(\tilde{t})](\tilde{z})]$.

The formulas of the preceding paragraph can be reformulated as explained in connection with Eqs. (4.75a–e). The searched-for array splits uniquely in a part proportional to the distinguished eigenvector $\boldsymbol{a}$ and a remainder to be denoted by $\tilde{\boldsymbol{g}}(\tilde{t})$:

$$\boldsymbol{g}(\tilde{t}) = g(\tilde{t})\boldsymbol{a} + \tilde{\boldsymbol{g}}(\tilde{t}). \tag{6.118a}$$

The components of the array $\tilde{\boldsymbol{g}}(\tilde{t})$ are expressed by linear combinations of the inhomogeneity components. The coefficients are the elements $R_{q,k}$ of the distinguished resolvent of the critical stability matrix:

$$\tilde{g}_q(\tilde{t}) = \sum_k R_{q,k} I_k(\tilde{t}), \quad q = 1, \ldots, M. \tag{6.118b}$$

There holds the solubility condition

$$\sum_q a^*_q I_q(\tilde{t}) = 0. \tag{6.118c}$$

The manipulations leading to the preceding results are a simplification of those made in Sec. 6.1.1. Similarly, one can formulate the analogues of Eqs. (6.6), (6.7) for the shape functions $\tilde{\phi}^s_q(\tilde{t})$ for the tagged-particle correlators.

One can follow the procedure of Sec. 6.1.2 in order to construct an asymptotic solution for $\boldsymbol{g}(\tilde{t})$. Here, as opposed to in the discussion of the critical dynamics, integrals with short-time divergencies do not occur. One can appeal to the formulas (A.64a,b) in order to relate the short-time power-law asymptotes of functions $F(\tilde{t})$ to the high-frequency power-law behaviour of their S-transforms. Let us start by examining the ansatz

$$g_q(\tilde{t}) \sim -C_q \tilde{t}^{\beta}, \quad \tilde{t} \to 0. \tag{6.119a}$$

Because of Eq. (6.116b), one has to request that β is positive. The mentioned Tauberian theorem yields: $ST[g_q(\tilde{t})](\tilde{z}) \sim -C_q\Gamma(1+\beta)(-i\tilde{z})^{-\beta}, \tilde{z} \to \infty$. Using the theorem again, one gets $ST^{-1}[ST[g_q(\tilde{t})](\tilde{z})^n](\tilde{t}) \sim [-C_q\Gamma(1+\beta)]^n\Gamma(1+n\beta)^{-1}\tilde{t}^{n\beta}, \tilde{t} \to 0$. As a result, Eq. (6.117b) leads to the asymptotic law

$$J_q^{(n)}(\tilde{t}) \sim \Big\{\big[\sum_{k_1\cdots k_n} A^{(n)c}_{q,k_1\cdots k_n} C_{k_1}\cdots C_{k_n}\big] - \big[C_q\Gamma(1+\beta)\big]^n\Gamma(1+n\beta)^{-1}\Big\}(-\tilde{t}^{\beta})^n, \quad \tilde{t} \to 0. \quad (6.119b)$$

Equation (6.118b) shows that $\tilde{g}_q(\tilde{t})$ vanishes with $\tilde{t}$ proportional to $\tilde{t}^{2\beta}$. Hence, the second term on the right-hand side of Eq. (6.118a) does not contribute to the limit in Eq. (6.119a). There holds $g_q(\tilde{t}) \sim -Ca_q\tilde{t}^{\beta}, \tilde{t} \to 0$. Since the decay of $\tilde{\phi}(\tilde{t})$ below the plateau is to be studied, one has to request that C is positive. A rescaling according to Eq. (6.111a) with $x = C^{1/\beta}$, eliminates the factor C. Without loss of generality, one can choose $C = 1$, i.e.,

$$C_q = a_q, \quad q = 1, \ldots, M. \quad (6.119c)$$

Since $I_q(\tilde{t}) \sim J_q^{(2)}(\tilde{t})$ for $\tilde{t} \to 0$, the equations of motion reduce to the request of cancellation of those contributions to Eq. (6.118c), which are proportional to $\tilde{t}^{2\beta}$. Using the normalization conditions (4.74d) and the formula (4.88a) for the exponent parameter, the ansatz yields a solution if and only if there holds $\lambda^c = \Gamma(1+\beta)^2/\Gamma(1+2\beta)$. The discussion of Fig. 6.1 and Eq. (6.61a) shows that β agrees with von Schweidler's exponent:

$$\beta = b. \quad (6.119d)$$

The anticipated uniqueness of the equations of motion for the shape functions permits the summary: $\boldsymbol{g}(\tilde{t}) \sim -\boldsymbol{a}\tilde{t}^b, \tilde{t} \to 0$. Substituting this result into Eq. (6.116a) and using Eq. (4.78b) for the components of the array of critical amplitudes, von Schweidler's law is obtained as the asymptote for the correlator's shape function for short rescaled times:

$$\tilde{\boldsymbol{\phi}}(\tilde{t}) \sim \boldsymbol{f}^c - \boldsymbol{h}\tilde{t}^b, \quad \tilde{t} \to 0. \quad (6.120a)$$

The cited Tauberian theorem yields the formulation in the frequency domain:

$$\tilde{\chi}_q(\tilde{z}) \sim 1 - f_q^c + h_q\Gamma(1+b)/(-i\tilde{z})^b, \quad \tilde{z} \to \infty, \quad q = 1, \ldots, M. \quad (6.120b)$$

The formulas of the preceding paragraph can be combined with Eq. (6.118b) and the definition (6.23) of the correction amplitude K_q in order to get

$$\tilde{g}_q(\tilde{t}) \sim a_q K_q \tilde{t}^{2b}, \quad \tilde{t} \to 0, \quad q = 1, \ldots, M. \quad (6.120c)$$

Above, the ansatz (6.119a) is used to determine the small-$\tilde{t}$ asymptote for Eq. (6.118c). This asymptote is a combination of two terms proportional to

$\tilde{t}^{2b}$. The request of cancellation leads to the result (6.120a). Formula (6.120c) provides the motivation to extend the ansatz by using Eq. (6.118a) with $g(\tilde{t}) - \tilde{t}^b \sim \kappa \tilde{t}^{2b}, \tilde{t} \to 0$. One can follow the procedure of Sec. 6.1.2(ii) in order to cast Eq. (6.118c) in the form: $\sum_q a_q^* I_q(\tilde{t}) \sim K\tilde{t}^{3b}$. It is possible to choose κ so that the constant K vanishes. In this sense, the equations of motions are solved in a higher-asymptotic order than done above. The choice is $\kappa = \kappa(-b)$ with function $\kappa(x)$ defined in Eq. (6.29). The result can be noted as extension of Eq. (6.120a). For $q = 1, \ldots, M$, there holds

$$\tilde{\phi}_q(\tilde{t}) - f_q^c + h_q \tilde{t}^b \sim h_q \big[K_q + \kappa(-b)\big]\tilde{t}^{2b}, \quad \tilde{t} \to 0. \tag{6.121}$$

This formula is analogous to Eq. (6.30a). In particular, the same arrays $\boldsymbol{f}^c, \boldsymbol{h}$ and $\boldsymbol{K}$ determine the large-t asymptote of the critical-correlator array $\boldsymbol{\phi}^c(t)$ as they determine the small-$\tilde{t}$ expansion of the shape-function array $\tilde{\boldsymbol{\phi}}(\tilde{t})$. The procedure of Sec. 6.1.2 can be copied to generalize the result from above to correlation functions of other variables X. The result shall be referred to by the statement that the extended von Schweidler law

$$\tilde{\phi}_X(\tilde{t}) = f_X^c - h_X \tilde{t}^b \Big\{1 - \big[K_X + \kappa(-b)\big]\tilde{t}^b\Big\} \tag{6.122a}$$

describes the small-$\tilde{t}$ behaviour of the shape function in leading and next-to-leading order. Dropping the correction term $[K_X + \kappa(-b)]\tilde{t}^b$, one gets von Schweidler's law from Eq. (6.120a) as the leading description of the correlations of variable X. Application of the cited Tauberian theorem casts the result into a formula for $ST[\tilde{\phi}_X(\tilde{t})](\tilde{z})$. Specializing to real rescaled frequencies $\tilde{\omega}$, one gets the expression for the loss spectrum:

$$\tilde{\chi}_X''(\tilde{\omega}) = h_X \sin(\pi b/2)\Gamma(1+b)\tilde{\omega}^{-b}\Big\{1 - \big[K_X + \kappa(-b)\big]k_b \tilde{\omega}^{-b}\Big\}, \tag{6.122b}$$

$k_b = 2\cos(\frac{\pi}{2}b)\Gamma(1+2b)/\Gamma(1+b)$. The formula describes the large-$\tilde{\omega}$ behaviour of the shape functions for the loss spectra in leading and next-to-leading order in the small quantity $\tilde{\omega}^{-b}$. Dropping the second contribution in the curly bracket, one obtains the von Schweidler spectrum as leading-order approximation. Equation (6.122a) is a motivation for using formula (5.96a) for the description of glassy-dynamics data for liquids.

A special application of Eq. (6.122a) concerns the tagged-particle-density correlators $\phi_q^s(t)$. These functions together with the density correlators determine the mean-squared-displacement function $\Delta_s(t)$. Since the former functions exhibit a second-scaling limit, a corresponding result (5.61b) can be derived for $\Delta_s(t)$. The small-$\tilde{t}$ expansion of the density correlators implies a similar expansion for $\tilde{\Delta}_s(\tilde{t})$. Proper modifications of the derivation of Eq. (6.36a) for the expansion in powers of t^{-a} leads to the one in powers of $\tilde{t}^b$:

$$\tilde{\Delta}_s(\tilde{t}) = r_{sc}^2 + h_{\mathrm{MSD}}\tilde{t}^b \Big\{1 - \big[K_{\mathrm{MSD}} + \kappa(-b)\big]\tilde{t}^b\Big\}. \tag{6.123}$$

Let us consider some implications of the derived results for the description of the dynamics within the second-scaling-law regime. von Schweidler's law specifies a dimensionless scale $\tilde{\tau}_q$ for the below-plateau decay of the correlators. This scale quantifies the shape function within the region, where Eq. (6.121) can be simplified to its leading term:

$$\tilde{\phi}(\tilde{t})/f_q^c = 1 - (\tilde{t}/\tilde{\tau}_q)^b, \tag{6.124a}$$

$$\tilde{\tau}_q = (f_q^c/h_q)^{1/b}. \tag{6.124b}$$

The h_q versus q curve has a minimum near the pronounced maximum of the f_q^c versus q curve as is demonstrated in Fig. 4.6 for a hard-sphere system. This feature reflects a general trend in simple systems, because there is a trend to a suppression of the q-dependence of the ratio $(1 - f_q^c)/h_q$. This trend is explained in Sec. 4.3.5 in connection with the behaviour of the Debye–Waller factor and in Sec. 6.1.2 in connection with the critical decay. Consequently, the (f_q^c/h_q) versus q curve exhibits a maximum near the structure-factor-peak position, which is more pronounced than that for the f_q^c versus q curve. The peak of the $\tilde{\tau}_q$ versus q curve is even more enhanced, because the exponent $(1/b)$ in Eq. (6.124b) is larger than unity. These conclusions explain qualitatively the peak and the oscillations shown in the lower panel of Fig. 4.20 for the q-dependence of the scale for the shape functions $\tilde{\phi}_q(\tilde{t})$ of a hard-sphere-system model.

Mode-coupling implies the trend that all correlators reach their long-time limit zero at about the same value for $\tilde{t}$. This request is not consistent with an extrapolation to large times of the strongly q-dependent right-hand side of Eq. (6.124a). The trend can be accounted for by the correction amplitudes $[K_q + \kappa(-b)] = \tilde{k}_q$ if these are negative for the correlators with a large value for f_q^c and positive for the ones with intermediate and small values for the plateau. These two possibilities are demonstrated in Fig. 6.15 for the two wave numbers q_1 and q_2, respectively. In this case, $\kappa(-b) = 0.569$ and the results for K_q can be inferred from Fig. 4.6. The large enhancement of the scale $\tilde{\tau}_{q_1}$ relative to $\tilde{\tau}_{q_2}$ explains the following features of the shape functions. von Schweidler's law describes only about 10% of the decay of the correlator $\tilde{\phi}_q(\tilde{t})$ for the case with the large plateau value $f_{q_1}^c$; but, it accounts for about 30% of the decay for the case with the intermediate-size value $f_{q_2}^c$. An equivalent discrepancy is exhibited for the description of the large-$\tilde{\omega}$ wing of the loss peaks by the $\tilde{\omega}^{-b}$ tail. For q_1 and q_2, von Schweidler's law describes the tails of the spectra if these are below about 10% and 30%, respectively, of their values at the maximum.

The trends explained in the preceding paragraph are demonstrated also by the susceptibility spectra, which are shown in Fig. 4.13 for the two loss spectra of a two-component model. The plateau values for the first and second correlator are $f^c = 0.25$ and $f_A^c = 0.92$, respectively. For the state with label $n = 12$, the high-frequency wing of the peak of the $\log \chi_A''(\omega)$ versus $\log(\omega)$ curve exhibits the ω^{-b} behaviour only if the spectrum is below 10% of the peak-maximum value.

The corresponding part of $\chi''(\omega)$ is described by von Schweidler's law if the spectrum is below about 20% of the peak-maximum value (Fuchs *et al.* 1991). For the $n = 12$ state, this part of the $\tilde{\chi}''(\tilde{\omega})$ versus $\log \tilde{\omega}$ curve is covered by the large spectrum near the loss minimum.

The representative examples shown in Fig. 6.15 demonstrate the following. Extending von Schweidler's law by the leading correction, the range of validity of the description of $\tilde{\phi}_q(\tilde{t})$ by the analytic expressions of the asymptotic approach is expanded considerably. For times outside this interval, the correlators decay quickly to zero. Similarly, Eq. (6.122b) describes the high-frequency wing of the loss peak for rescaled frequencies $\tilde{\omega}$ decreasing close to the position of the loss maximum. The low-frequency wing of the loss peak exhibits a white-noise induced linear variation of the spectrum. One concludes that the results (6.122a,b) provide the complete explanation of the stretching of the below-plateau decay of the correlators and of the low-frequency loss peaks. This holds provided that the glass transition singularity P^c under discussion is not located near some other singularity $P^{c\prime}$.

If the correlators' shape functions could be described by the Kohlrausch function (1.3), von Schweidler's law would specify the three parameters: the strength factor f would agree with the plateau value f_q^c, the stretching exponent β would be the correlator-independent von Schweidler exponent b, and the scale τ would be identical with the parameter $\tilde{\tau}_q$ in Eq. (6.124b). The normalized shape function $\tilde{\phi}_q(\tilde{t})/f_q^c$ would agree with the Kohlransch function

$$\varphi_q^K(\tilde{t}) = \exp[-(\tilde{t}/\tilde{\tau}_q)^b]. \qquad (6.125)$$

This result would imply that the correction amplitude $[K_q + \kappa(-b)]$ for the asymptotic small-$\tilde{t}$ expansion is given by the positive number $[h_q/f_q^c]/2$. However, it has been explained in connection with the discussion of the hard-sphere-liquid results from Fig. 6.15 that the correction amplitudes are negative for the wave numbers near the structure-factor-peak position. Above this value for q, the amplitude has a zero. Consequently, for the intermediate-size-wave-number regime, the ad hoc representation of the shape function by Eq. (6.125) is inferior to that by the leading-asymptotic expression (6.124a). This conclusion is exemplified in Fig. 6.17 by the curves for $q = 4.4/a$ and $q = 7.8/a$.

The curves in Fig. 6.17 for $q = 20/a$ and $q = 29.9/a$ demonstrate a fascinating discovery by Fuchs (1994). For these large wave numbers, the normalized shape functions are close to the functions $\varphi_q^K(\tilde{t})$. The same statement holds also for the tagged-particle density correlators and for the correlators of a binary mixture of particles interacting with a regular potential. Closely related to the cited observation is the one that the functions $\tilde{\tau}_q^{-b} = h_q/f_q^c$ exhibit a linear increase with q for $q > 15/a$. In the cited paper subtle considerations are presented, which indicate that the microscopic MCT equations for the shape functions of simple liquids exhibit the asymptotic behaviour: $h_q/f_q^c \sim q/q_0, \tilde{\phi}_q(\tilde{t})/f_q^c \sim \varphi_q^K(\tilde{t}), q \to \infty$. It remains a challenge to expand these considerations to a proof.

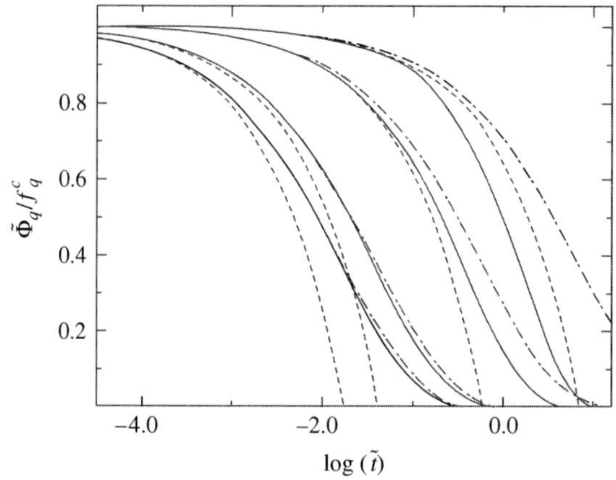

FIG. 6.17. The full lines represent normalized second-scaling-law shape functions for density-fluctuation correlators of the hard-sphere-liquid model, which is defined in the caption of Fig. 4.20. The dashed lines show the von Schweidler asymptotes described by Eqs. (6.124a,b). The dashed-dotted ones exhibit the Kohlrausch functions from Eq. (6.125). From right to left, the wave numbers are given by $qa = 4.4, 7.8, 20.0, 29.9$. Reproduced from Fuchs (1994).

(iii) The time scale for the superposition principle

Combining Eqs. (6.115a), (6.120a), one obtains the formulas $\phi_q(t) = f_q^c - h_q(t/\tau_\sigma)^b$, $q = 1, \ldots, M$. They describe the second-scaling-law asymptote for the below-plateau decay for small values of (t/τ_σ). For this process, the correlators $\phi_q(t)$ are close to the critical arrested parts f_q^c. The same near-plateau relaxation is described by Eq. (6.64a) for the first-scaling limit for large values of (t/t_σ). If one identifies τ_σ with the second critical time scale,

$$\tau_\sigma = t'_\sigma, \tag{6.126}$$

both formulas for the same process agree in the limit of vanishing separation parameter. Equation (6.65b) ensures the validity of the condition (6.104). The formula for τ_σ completes the construction of the second-scaling-law solution.

The first scaling-law regime deals with the decay of the correlators towards and through the plateaus. It deals with the dynamics on scale t_σ. Because of Eq. (6.63b), von Schweidler's law appears as the part of the below-plateau decay for times, which are large on scale t_σ:

$$\lim_{\hat{t}\to\infty} \lim_{\sigma\to -0} \left\{ \left[\phi_q(\hat{t}t_\sigma) - f_q^c\right] / \left[\sqrt{|\sigma|}B\hat{t}^b\right] \right\} = -h_q. \tag{6.127a}$$

The second-scaling-law regime deals with the decay of the liquid correlators from the plateau to zero. It deals with the dynamics on scale t'_σ. Because of

Eq. (6.121), von Schweidler's law appears as the part describing the dynamics for times, which are short on scale t'_σ:

$$\lim_{\tilde{t}\to 0}\lim_{\sigma\to -0}\left\{\left[\phi_q(\tilde{t}t'_\sigma)-f^c_q\right]/\left[\tilde{t}^b\right]\right\}=-h_q. \tag{6.127b}$$

The two limits in the preceding two formulas must not be interchanged. von Schweidler's relaxation process appears as the final part of the first-scaling-law process and as the initial part of the second-scaling-law process. The time interval for the first scaling-law description is bounded from above by a value which changes with σ proportional to t'_σ. This is implied by formula (6.98d). The time interval for the second-scaling-law description is bounded from below by the plateau-crossing time t_-, which changes with σ proportional to t_σ. Hence, the initial point of the overlap interval for the simultaneous description by both scaling laws tends to infinity if the liquid point P tends towards the arrest point P^c. But, because of Eq. (6.65b), the length of the overlap interval diverges as well. von Schweidler's law describes the dynamics within the overlap interval.

The demonstration of the evolution of the superposition principle for a hard-sphere-system model in Figs. 4.21, 4.22 and 6.16 is done for the scale $\tau_\sigma = t'_\sigma$. The exponent $\gamma = 2.46$ used for Eq. (6.64b) is calculated from the exponent parameter $\lambda^c = 0.735$ for this model. The rescaled functions start to follow the shape function for $|\epsilon| = (\varphi - \varphi_c)/\varphi_c \approx 0.05$.

The second scaling law is unvalid if the time is decreased so much that the correlators reach their plateau values. This explains qualitatively that the the deviations of the rescaled correlators $\phi_q(\tilde{t}t'_\sigma)$ increase with increasing value of $\tilde{\phi}_q(\tilde{t})$, i.e., with decreasing rescaled time $\tilde{t}$. Because of Eq. (6.65b) the rescaled plateau crossing time $\tilde{t}_- = t_-/t'_\sigma = \hat{t}_-(t_\sigma/t'_\sigma)$ decreases with $|\sigma|$. This explains that the interval of validity of the second scaling law expands from large rescaled times $\tilde{t}$ to small ones if the liquid state P approaches the arrest state P^c. The MCT evolution scenario is exhibited also by the data shown in Fig. 1.5. Equivalently, the interval of validity for the superposition principle for the loss spectra evolves from small rescaled frequencies $\tilde{\omega}$ to large ones if $|\sigma|$ decreases. This scenario is exhibited also by the data in Fig. 1.6.

The explanations given for the evolution scenario require a reservation. The glass-transition singularity P^c must not be located so close to another singularity $P^{c\prime}$ that changes of the separation parameter σ are causing control-parameter sensitive transition precursors for the approach towards $P^{c\prime}$.

(iv) Descriptions of some glassy-liquid data by asymptotic MCT laws

In this subsection, three studies shall be discussed with the intention to identify similarities between leading-order-asymptotic expansion results for the bifurcation scenario and features observed for glass-forming liquids. The first example concerns the evolution of the depolarized-light-scattering spectra of liquid toluene, which is indicated in Fig. 1.1. It is discussed in connection with Fig. 6.8 that the evolution of the loss minimum upon decreasing the temperature T from

295 K to 180 K can be described by the first-scaling-law formulas. This holds if one uses a separation parameter $\sigma \propto (T_c - T)/T_c$ with a fit value $T_c = (153 \pm 3)$ K for the critical temperature and if one chooses a fit value $\lambda = 0.69$ for the exponent parameter. The latter value implies the MCT exponents $a = 0.33, b = 0.66$ and $\gamma = 2.27$. The left upper panel of Fig. 6.18 reproduces the low-frequency parts of the loss spectra, which are represented as a function of a rescaled frequency $\nu \tau_{\text{LS}} \propto \tilde{\omega}$. For every temperature, the time scale τ_{LS} is adjusted so that the upper parts of the loss peaks collapse. The rescaled spectra coincide on

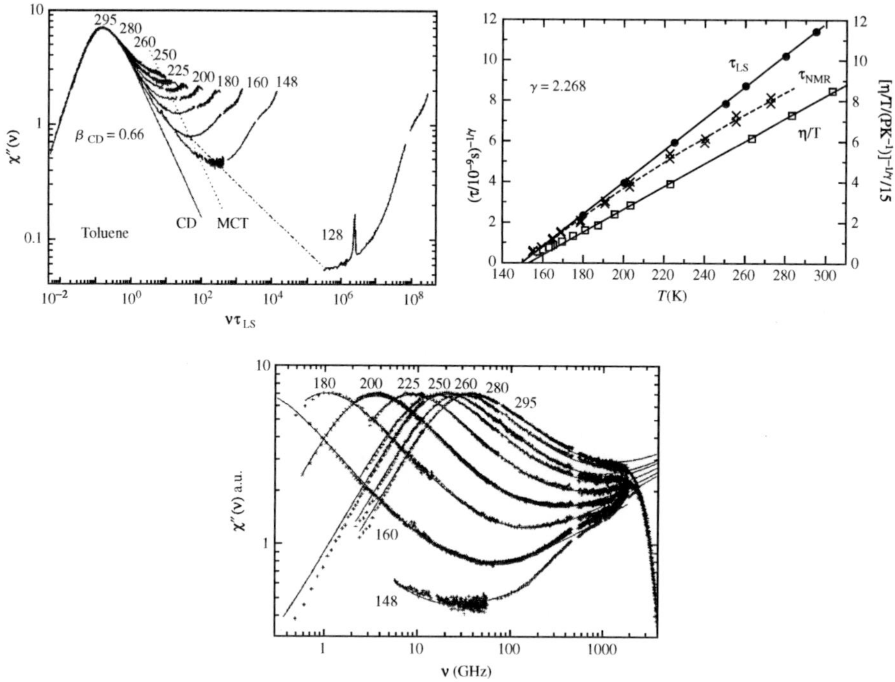

FIG. 6.18. The lower panel shows loss spectra of toluene, which are measured by depolarized light-scattering spectroscopy. They are discussed also in Figs. 1.1 and 6.8. The left upper panel exhibits the spectra as a function of the frequency, which is rescaled by a temperature-dependent time τ_{LS} so that the loss maxima coincide. The full line with label CD shows the spectrum of a Cole–Davidson function (1.5) with a stretching exponent $\beta_{\text{CD}} = b = 0.66$. The dotted straight line has slope $(-b)$ and connects the rescaled loss minima for $T > 160$ K. The straight dashed-dotted line indicates the positions of the loss minima for $T < 160$ K. The right upper panel exhibits rectification diagrams for scaling time τ_{LS}, relaxation times τ_{NMR} for nuclear-magnetic-resonance measurements, and relaxation times defined by the viscosity η. The full lines in the lower panel exhibit the interpolation spectrum defined by Eq. (6.128a). Reproduced from Wiedersich *et al.* (2000).

the low-frequency wings of the peaks, where a white-noise-induced spectrum is exhibited. Upon decreasing the temperature T from 295 K to 180 K, the high-frequency wings of the loss peak exhibit a scenario for the evolution of a shape function $\tilde{\chi}''(\tilde{\omega})$, which is similar to that explained in the preceding subsection. The full line demonstrates that the asymptotic spectrum can be fitted by that for the Cole–Davidson susceptibility: $\tilde{\chi}''(\omega) = C \text{ Im } [1 - i\tilde{\omega}\tau]^{-b}$. The constants C and τ are adjusted so that the height and the position of the spectral maximum are reproduced. The stretching exponent β_{CD} in Eq. (1.5) is chosen as the von Schweidler exponent cited above. The Cole–Davidson spectrum provides an adequate interpolation between the loss-peak wings for small and large rescaled frequencies. The wing asymptotes reproduce the asymptotic MCT results.

The measured time scale τ_{LS} for the loss-peak process is proportional to the scale t'_σ if and only if a $\tau_{\text{LS}}^{-(1/\gamma)}$ versus T diagram is a straight line, which intersects the horizontal axis at $T = T_{\text{c}}$. Scale coupling implies a corresponding result for the functions $\tau_{\text{NMR}}^{-(1/\gamma)}$ and $(\eta/T)^{-(1/\gamma)}$. Here, τ_{NMR} denotes a measured time scale for nuclear magnetic resonance spectroscopy and η abbreviates the measured viscosity of toluene. The rectification diagrams in the right upper panel are consistent with the general MCT laws.

The first scaling law implies that the loss-minimum intensity χ_{min} and the loss-minimum position ω_{min} decrease with temperature proportional to $|\sigma|^{1/2}$ and $|\sigma|^{1/(2a)}$, respectively. The second scale exhibits the behaviour $1/\tau_{\text{LS}} \propto |\sigma|^\gamma = |\sigma|^{1/(2a)} \cdot |\sigma|^{1/2b}$. Hence, there is some constant k so that $\log \chi_{\text{min}} = k - b \log(\nu\tau_{\text{LS}}), \nu = \omega/2\pi$. The MCT scenario for the interplay of the two scaling laws implies the following feature. The $\log \chi_{\text{min}}$ versus $\log(\nu\tau_{\text{LS}})$ curve is a straight line of slope $(-b)$. In the double-logarithmic presentation used in the left upper panel, this straight line is parallel to the von Schweidler-law tail of the loss peak. The dotted line indicates that the data are compatible with the specified asymptotic laws.

A more detailed result for the interplay of the two scaling laws can be obtained by combining the Cole–Davidson description of the second-scaling-law response with the interpolation formula (1.7a) for the first-scaling-law description of the loss minimum. von Schweidler's law appears as large-$\tilde{\omega}$ limit of the loss peak: $\chi''(\omega) \sim C \text{ Im } (-i\tilde{\omega}\tau)^{-b}, \omega t'_\sigma \to \infty$. It appears also as the small-$\hat{\omega}$ limit of the first-scaling-law description of the loss minimum: $\chi''(\omega) \sim \chi_{\text{min}}(\omega_{\text{min}}/\omega)^b a/(a+b), \omega t_\sigma \to 0$. Splicing the two spectra for $\omega_{\text{max}} \ll \omega \ll \omega_{\text{min}}$, one gets a description for the complete structural relaxation. For frequencies near ω_{max}, the small critical spectrum $\chi_{\text{min}}(\omega/\omega_{\text{min}})^a b/(a+b)$ can be neglected compared to that for the large loss peak. Hence, for frequencies near and above ω_{max}, the splicing is achieved by the expression

$$\chi''(\omega) = C\text{Im}\big[1 - i\tilde{\omega}\tau\big]^{-b} + \chi_{\text{min}}(\omega/\omega_{\text{min}})^a b/(a+b). \tag{6.128a}$$

This formula holds provided three conditions are obeyed. First, the separation parameter σ has to be so small that the two scaling laws are valid. Second, the shape function $\tilde{\chi}''(\tilde{\omega})$ has a form which can be described by the Cole–Davidson formula with $\beta_{\rm CD} = b$. Third, the frequencies have to be so large that the critical spectrum can be neglected compared to the white-noise-induced spectrum of the low-frequency wing of the loss peak, i.e.,

$$\chi_{\min}(\omega/\omega_{\min})^a b/(a+b) \ll C\omega\tau_{\rm LS}. \tag{6.128b}$$

The lower panel of Fig. 6.18 demonstrates that the splicing (6.128a) of the scaling-law formulas describe the evolution of the toluene loss spectra for temperatures decreasing to about $T_{\rm c} + 10\,{\rm K}$. This holds for the frequency region between 1 THz and 0.5 GHz. As emphasized in Sec, 5.4, MCT cannot describe the data for $T < T_{\rm c}$. The full line for $T = 250\,{\rm K}$ overestimates the corresponding spectrum measured for frequencies below 2 GHz. Here, condition (6.128b) is violated and Eq. (6.128a) does not describe a MCT result. It might be adequate to point out that, so far, no MCT model has been identified, which suggests the applicability of scaling-law descriptions of the spectra for distance parameters $|\epsilon| = (T - T_{\rm c})/T_{\rm c}$ large as those considered in Figs. 6.8 and 6.18.

The second study concerns the evolution of the glassy-dynamics parts of the density-fluctuation correlators $\phi_q(t)$, which have been measured by photon-correlation spectroscopy for hard-sphere colloids. In Sec. 6.1.3, it is reported that the experimental findings for the plateau crossing scenario can be described by the first-scaling-law expressions $\phi_q(t) = f_q^c + [h_q c_\sigma] g_-(t/t_\sigma)$. The shape functions used are calculated for the exponent parameter $\lambda = 0.77$, which implies a von Schweidler exponent $b = 0.55$ and an exponent $\gamma = 2.6$ for the second time scale t'_σ. The value for λ is not a result of curve fitting; it is evaluated for a microscopic MCT model for the hard-sphere system based on the Verlet–Weiss structure factor (Fuchs *et al.* 1992). The latter work provides also the normalized shape functions $\varphi_q(\tilde{t}) = \tilde{\phi}_q(\tilde{t})/f_q^c$ for the second scaling law for the wave numbers analyzed by van Megen and Underwood (1993). Figure 4.25 reproduces the measured evolution of the superposition principle for three representative wave numbers. For every measured packing fraction φ, the scaling time τ_σ is fitted so that the correlators coincide for long times with that for sample H6. The demonstrated scenario for the data is in qualitative agreement with that explained in the preceding subsection. With properly adjusted plateau values f_q^c, the MCT shape functions $f_q^c \varphi_q(\tilde{t})$ are added as lines. The figure demonstrates that the measured q-dependence of the shape functions is consistent with that calculated from the cited MCT model. In particular, the strong q-dependence of the stretching of the below-plateau decay is observed in agreement with that exhibited in Fig. 4.20. The data analysis is done for an increase of the packing fraction from 0.495 to 0.574. This change of the control parameter causes an increase of the scale τ_α by more than three orders of magnitude. The observed increase is consistent with the second-scaling-law formula $\tau_\alpha \propto 1/(\varphi_{\rm c} - \varphi)^\gamma$ (van Megen 1995).

Splicing the two cited scaling-law descriptions of the data, one gets a complete description of the measured relaxation curves for times outside the transient regime. The lines in Fig. 1.15 are examples for two wave numbers. Problems of the cited work on a quantitative data analysis in terms of asymptotic laws for the MCT bifurcation scenario can be summarized as follows. There is a 14% discrepancy between the calculated and the measured critical packing fraction. There are discrepancies between the calculated and the measured values for the critical Debye–Waller factor as exhibited by Fig. 1.16. There are fluctuating deviations of the fitted $(\varphi - \varphi_c)$ dependence of the scales $\sigma, c_\sigma, t_\sigma$ and t'_σ from their MCT-power-law forms, which are documented in the above cited paper by van Megen.

The third study to be discussed concerns the binary Lennard-Jones mixture, which is defined in connection with Eqs. (1.8a–c). The simulation data for the correlation function $\phi^s_q(t)$ for the density fluctuations of the tagged big particles exhibit the evolution of a second-scaling-law regime in qualitative agreement with what is explained in the preceding subsection. For intermediate-size wave numbers q and for the temperature $T = 0.466$, about 40% of the correlation decay can be described by the formula $\phi^s_q(t) = f^{sc}_q - h^s_q \tilde{t}^b$. This holds if the exponent b is chosen around 0.49. The von Schweidler-law fit deals with a time variation by more than three orders of magnitude (Kob and Andersen 1994). The small-q behaviour of $\phi^s_q(t)$ is determined by the mean-squared displacement $\delta r^2_s(t) = 6\Delta_s(t)$. As discussed in Sec. 6.1.2, these small-q fluctuations are coupled strongly to the density fluctuations with intermediate-size wave numbers. As a result, the correction amplitude $k_{\text{MSD}} = [K_{\text{MSD}} + \kappa(-b)]$ in Eq. (6.123) is big and negative. Within the interval of mean-squared-displacement increases, which is analysed in Fig. 1.9, one cannot discriminate between a $\tilde{t}^{2b}$ variation and a variation proportional to $\tilde{t}$. Within this interval, $\delta r^2_s(t)$ reaches the hydrodynamic-limit behaviour $6Dt$, which is quantified by the diffusion constant D. Assuming that higher-order corrections to the extended von Schweidler law (6.123) can be neglected, Eq. (6.123) is equivalent to $\tilde{\Delta}_s(\tilde{t}) = r^2_{sc} + h_{\text{MSD}}\tilde{t}^b + Dt$. Using the scale $\tau_\sigma = 1/D$, one obtains the shape function: $\delta \tilde{r}^2_s(t) = r^2_p + A\tilde{t}^b + 6\tilde{t}$. The lower panel of Fig. 1.9 shows that this expression describes the shape function very well.

For the hard-sphere-liquid model used to calculate the mean-squared displacements in Fig. 6.16, all parameters entering Eq. (6.123) have been evaluated (Fuchs *et al.* 1998). The results discussed in this work imply that the shape function $\tilde{\Delta}_s(\tilde{t})$ is described very well by the extended von Schweidler law. This holds even though the hard-sphere-system exponent $2b = 1.17$ differs more from 1 than that considered above for the binary mixture.

6.2.3 *Asymptotic corrections for the second scaling limit*

Within the second-scaling-limit approach towards the bifurcation dynamics, the evolution of the superposition principles is quantified by the array $\tilde{\boldsymbol{\psi}}(\tilde{t})$ of correction-functions. In this section, these functions shall be evaluated asymptotically for small rescaled times $\tilde{t}$. The results provide an extension of the superposition-principle formulas to such small times that also the plateau crossing of the correlators can be described.

The equation of motion for the correction functions requires as input the array $\tilde{\phi}(\tilde{t})$ of shape functions. Equation (6.116a) shall be used to express these functions in terms of the functions $g_q(\tilde{t})$. Equivalently, there hold the frequency-domain identities: $-\tilde{z}\tilde{\phi}_q(\tilde{z}) = ST[\tilde{\phi}_q(\tilde{t})](\tilde{z}) = f_q^c + (1-f_q^c)ST[g_q(\tilde{t})](\tilde{z}), q = 1, \ldots, M$. For the intended asymptotic expansions, the function $\tilde{t}^b$ serves as small parameter η. Equation (6.121) provides the input information correctly up to terms of order η^2:

$$g_q(\tilde{t}) = a_q\{-\tilde{t}^b + [K_q + \kappa(-b)]\tilde{t}^{2b}\}. \tag{6.129}$$

In the following calculations, the Tauberian-theorem equations (A.64a,b) shall be used. A function with the small-$\tilde{t}$ asymptote $\eta^n = \tilde{t}^{nb}, n = -1, 0, 1, \ldots$, is equivalent to its S-transform with the large-$\tilde{z}$ asymptote $\Gamma(1+nb)(-i\tilde{z})^{-nb}$. In this sense, the curly bracket in Eq. (6.110) can be denoted as an expansion up to order $\eta^2 : \{(1-f_q^c)/[1-ST[\tilde{\phi}_q(\tilde{t})](\tilde{z})]\}^2 = 1+2ST[g_q(\tilde{t})](\tilde{z})+3ST[g_q(\tilde{t})](\tilde{z})^2$. The expansion of the matrix with elements from Eq. (6.107d) can be expressed in terms of the critical mode-coupling coefficients defined in Eqs. (4.71b,c). Using the Eq. (4.61c) for the critical stability matrix, one gets the result up to order $\eta^2 : A_{q,k}(\tilde{t}) = A_{q,k}^c + 2\sum_p A_{q,kp}^{(2)c} g_p(\tilde{t}) + 3\sum_{p_1p_2} A_{q,kp_1p_2}^{(3)c} g_{p_1}(\tilde{t}) g_{p_2}(\tilde{t})$. In order to obtain the expansion for the functions in Eq. (6.107c), one uses Eqs. (4.71a-c) and the definition (4.71e) for the difference coefficients. There holds $A_q(\tilde{t}) = [\lim_{\sigma\to 0} D_q^{(0)}(P)/\sigma] + \sum_k [\lim_{\sigma\to 0} D_{q,k}^{(1)}(P)/\sigma] g_k(\tilde{t}) + O(\eta^2)$. Because of Eq. (6.77), the first bracket abbreviates the coefficients $d_q^{(0)}$.

Substituting the formulas of the preceding paragraph into the equation of motion (6.110) for the correction functions, one gets the structure of Eq. (4.75a): $\sum_k [\delta_{q,k} - A_{q,k}^c] ST[\tilde{\psi}_k(\tilde{t})](\tilde{z}) = I_q(\tilde{z}), q = 1, \ldots, M$. The inhomogeneity reads:

$$I_q(\tilde{z}) = I_q^{(0)}(\tilde{z}) + I_q^{(1)}(\tilde{z}) + \cdots, \tag{6.130a}$$

$$\begin{aligned} I_q^{(0)}(\tilde{z}) &= d_q^{(0)} - 2ST[g_q(\tilde{t})](\tilde{z}) ST[\tilde{\psi}_q(\tilde{t})](\tilde{z}) \\ &\quad + 2\sum_{kp} A_{q,kp}^{(2)c} ST[g_p(\tilde{t})\tilde{\psi}_k(\tilde{t})](\tilde{z}), \end{aligned} \tag{6.130b}$$

$$\begin{aligned} I_q^{(1)}(\tilde{z}) &= \sum_k [\lim_{\sigma\to 0} D_{q,k}^{(1)}(P)/\sigma] ST[g_k(\tilde{t})](\tilde{z}) - 3ST[g_q(\tilde{t})](\tilde{z})^2 \\ &\quad ST[\tilde{\psi}_q(\tilde{t})](\tilde{z}) + 3\sum_{kp_1p_2} A_{q,kp_1p_2}^{(3)c} ST[g_{p_1}(\tilde{t}) g_{p_2}(\tilde{t}) \tilde{\psi}_k(\tilde{t})](\tilde{z}). \end{aligned} \tag{6.130c}$$

In the limit of vanishing η, each of the three contributions to $I_q^{(1)}(\tilde{z})$ can be neglected compared to the corresponding contribution to $I_q^{(0)}(\tilde{z})$. The other contributions of $I_q^{(2)}(\tilde{z})$ can be neglected compared to those of $I_q^{(1)}(\tilde{z})$, etc.

Combining the results of the preceding paragraph with the discussion of Eqs. (4.75a–d), one arrives at the conclusion that the equation of motion for the array $\tilde{\boldsymbol{\psi}}(\tilde{t})$ is equivalent to the following three formulas. The array is a sum of one term proportional to the distinguished eigenvector $\boldsymbol{a}$ and a remainder:

$$\tilde{\boldsymbol{\psi}}(\tilde{t}) = X(\tilde{t})\boldsymbol{a} + \tilde{\boldsymbol{X}}(\tilde{t}). \tag{6.131a}$$

The components of the remainder are linear combinations of the inhomogeneities:

$$\tilde{X}_q(\tilde{z}) = \sum_k R_{q,k} I_k(\tilde{z}), \quad q = 1, \ldots, M. \tag{6.131b}$$

There holds the solubility condition

$$\sum_q a_q^* I_q(\tilde{z}) = 0. \tag{6.131c}$$

Because of Eq. (4.86a) for the separation parameter, the first term on the right-hand side of Eq. (6.130b) yields the contribution $1 = \sum_q a_q^* d_q^{(0)}$ to the solubility condition. Therefore, a small-$\tilde{t}$ asymptote of $\tilde{\boldsymbol{\psi}}(\tilde{t})$ proportional to $\tilde{t}^x$ yields a contradiction to Eq. (6.131c) unless $x = -b$. In this case, there are three leading contributions to the solubility condition, which are of order η^0. One can follow the reasoning in Sec. 6.2.2(ii) in order to show that $\tilde{\boldsymbol{X}}(\tilde{t})$ does not contribute to the $\tilde{t}^{-b}$ asymptote. Substituting the formulas $\tilde{\psi}_q(\tilde{t}) \sim \tilde{B}\tilde{t}^{-b} a_q, \tilde{t} \to 0, q = 1, \ldots, M$, into Eq. (6.131c) one concludes the following. For the limit of vanishing $\tilde{t}$, the solubility condition reads: $1 + 2\tilde{B}\left[\Gamma(1+b)\Gamma(1-b) - \lambda^c\right] = 0$. To get this result, the conventions (4.74d) and the formula (4.88a) for the exponent parameter have been used. The cancellation of the leading contributions to $\sum_q a_q^* I_q(\tilde{z})$ is achieved if $\tilde{B} = -B_1$. Here, B_1 is defined in Eq. (6.63b). With the repeatedly made assumption that the asymptotic solution of the equations of motion determines the asymptotic behaviour of the solution, one obtains:

$$X(\tilde{t}) \sim -B_1 \tilde{t}^{-b}, \quad \tilde{t} \to 0. \tag{6.132a}$$

One arrives at the asymptotic expression for the correction function:

$$\tilde{\boldsymbol{\psi}}(\tilde{t}) \sim -B_1 \tilde{t}^{-b} \boldsymbol{a}, \quad \tilde{t} \to 0. \tag{6.132b}$$

Substitution of this result together with the expression $\boldsymbol{g}(\tilde{t} \to 0) \sim -\tilde{t}^b \boldsymbol{a}$ into Eqs. (6.130a–c), one gets for the large-$\tilde{z}$ asymptote of the inhomogeneity: $I_q(\tilde{z} \to \infty) \sim I_q^{(0)}(\tilde{z} \to \infty) \sim d_q^{(0)} - 2B_1\{\Gamma(1+b)\Gamma(1-b)a_q^2 - \sum_{kp} A_{q,kp}^{(2)} a_k a_p\}$. Therefore, Eq. (6.131b) yields the small-$\tilde{t}$ limit for the components of the array $\tilde{\boldsymbol{X}}(\tilde{t})$:

$$\tilde{X}_q(\tilde{t}) \sim \left[L_q^{(1)} + 2B_1 K_q\right] a_q, \quad \tilde{t} \to 0, \quad q = 1, \ldots, M. \tag{6.133a}$$

Here, the amplitudes $L_q^{(1)}$ and K_q are defined in Eqs. (6.79a) and (6.23), respectively. This result motivates the ansatz for the extension of Eq. (6.132a) by a constant: $X(\tilde{t}) + B_1\tilde{t}^{-b} \sim X_0, \tilde{t} \to 0$. One obtains the small-$\tilde{t}$ expansion of the correction functions in leading and next-to-leading asymptotic description:

$$\tilde{\psi}_q(\tilde{t}) = a_q\Big\{-(B_1/\tilde{t}^b) + \big[\tilde{X}_q + X_0\big]\Big\}, \quad q = 1, \ldots, M. \tag{6.133b}$$

Substituting these formulas and Eqs. (6.129) into Eq. (6.130b), one finds an expansion of $I_q^{(0)}(\tilde{z})$ correctly up to order η. The expression for this contribution to the inhomogeneity is a linear function of X_0. In order to expand the expression (6.130c) for $I_q^{(1)}(\tilde{z})$, it is sufficient to replace the arrays $\boldsymbol{g}(\tilde{t})$ and $\tilde{\boldsymbol{\psi}}(\tilde{t})$ by their asymptotic limits. The other contributions to Eq. (6.130a) are of order η^2. The values of X_0 are determined by the request that $\sum a_q^* I_q(\tilde{z})$ does neither have contributions of order η^0 nor of order η^1.

It might be worthwhile to add a side remark to the preceding derivations. The function $X(\tilde{t})$ has been determined as solution of a linear equation. This solution is not unique. The discussion of Eq. (6.111c) implies that one can add multiples of the solution $\tilde{t}^b$ of the homogenous counterpart of the equation. This possibility of generalizing the formula for $X(\tilde{t})$ is irrelevant, since it is equivalent to a rescaling of the time $\tilde{t}$.

Combining Eq. (6.132b) with Eqs. (6.103a,c) and using formula (4.78b) for the critical amplitudes, one obtains the asymptotic result for the correlators

$$\phi_q(t) = \tilde{\phi}_q(\tilde{t}) + |\sigma| h_q(B_1/\tilde{t}^b). \tag{6.134a}$$

If one incorporates also the next-to-leading asymptotic contribution from Eq. (6.133b), one has to add a plateau-renormalization term. There holds

$$\phi_q(t) = \tilde{\phi}_q(\tilde{t}) + |\sigma| h_q(B_1/\tilde{t}^b) + \delta f_q^c, \quad q = 1, \ldots, M, \tag{6.134b}$$

with $\delta f_q^c = \sigma h_q[\tilde{X}_q + X_0]$. As explained before, these formulas can be generalized to ones referring to other correlators. An analogous formula holds also for the mean-squared-displacement function. Equation (5.61b) is extended to

$$\Delta_s(t) = \tilde{\Delta}_s(\tilde{t}) - |\sigma| h_{\text{MSD}}(B_1/\tilde{t}^b) + \delta r_{sc}^2. \tag{6.135}$$

The critical amplitude is defined in Eq. (5.57b) and δr_{sc}^2 denotes a plateau renormalization proportional to the separation parameter σ. The preceding results provide the desired extension of the superposition-principle description of the glassy-liquid dynamics for small rescaled times $\tilde{t} = t/t'_\sigma$.

The scaling-law result (6.120a) describes a monotonic increase of the correlators towards the plateau if the time tends to zero. Applying this second-scaling-limit formula for some fixed separation parameter, a small-time behaviour is described, which differs qualitatively from the behaviour of the MCT solutions. For decreasing time, the latter reach the critical arrested parts f_q^c at the plateau-crossing time; and for smaller times, they increase above the plateaus. This feature of the bifurcation dynamics is reproduced by inclusion of the correction as formulated in Eq. (6.134a). The term $|\sigma|h_q(B_1/\tilde{t}^b)$ causes an enhancement of $\phi_q(t)$ over the function $\tilde{\phi}_q(\tilde{t})$. This enhancement increases with increasing $|\sigma|$ and decreasing $\tilde{t}$. The formula describes the evolution scenario for the superposition principle, which is exhibited in Fig. 4.21 for a hard-sphere-liquid model. For distance parameter $|\epsilon|$ smaller than 0.10, the plateau crossing and the increase of the correlators somewhat above the plateaus is reproduced within the accuracy of the drawing. Similarly, Fig. 6.16 shows that Eq. (6.135) explains accurately the plateau crossing of the mean-squared displacement for the $n = 5$ state within the time interval: $-1.5 < \log \tilde{t} < -0.5$. For simple systems, the stretching of the control-parameter-sensitive loss peak comes about because the high-frequency part of the shape function $\tilde{\chi}_q''(\tilde{\omega})$ approaches the decay proportional to $\tilde{\omega}^{-b}$. The correction term in Eq. (6.134a) yields an enhancement of the wing proportional to $|\sigma|h_q B_1 \tilde{\omega}^b$. Figure 4.22 demonstrates that the increase of the enhancement with frequency and with $|\sigma|$ is a precursor of the approach towards the loss minimum.

The two scaling-limit approaches are defined quite differently. In particular, the respective leading-order results are different scaling laws. Nevertheless, both laws yield identical results for the von Schweidler-law part of the below-plateau relaxation. For fixed separation parameter and decreasing time, there appear increasing deviations of the correlators from the von Schweidler behaviour. Within the first-scaling-limit approach, these deviations are described by the correction term in Eq. (6.63b) for the shape function $g_-(\hat{t})$. The relative deviation reaches the margin δ^* for the rescaled time $\hat{t}^* = [B_1/(B^2\delta^*)]^{1/2b}$. Thereby, a first condition for the validity of von Schweidler's law is obtained:

$$\hat{t}^* t_\sigma \leq t. \tag{6.136a}$$

It is demonstrated in Eq. (6.97a) that the mentioned correction term can be written as $-\sigma h_q[B_1(t'_\sigma/t)^b]$. Hence, it is identical with the dominant correction term to the second-scaling law, which is noted in Eq. (6.134a).

For fixed $|\sigma|$ and increasing time, there also appear increasing deviations of the correlators from the von Schweidler behaviour. Within the second-scaling-limit approach, these deviations are described by the terms $h_q[K_q + \kappa(-b)]\tilde{t}^{2b}$ in Eq. (6.121). The relative deviation from the small-$\tilde{t}$ behaviour reaches the margin δ^* for the rescaled onset time $\tilde{t}_q^*$ from Eq. (6.98c). Thereby, a second condition for the validity of von Schweidler's law is specified:

$$t \leq \tilde{t}_q^* t'_\sigma. \tag{6.136b}$$

Within the first-scaling-limit approach, the same condition is inferred from the next-to-leading expansion term, which is noted as the second contribution in the curly bracket of Eq. (6.97a). The result marks the upper bound for the interval of validity of the first scaling law, which is identified by Eq. (6.98d).

For times between the bounds $\hat{t}^* t_\sigma$ and $\tilde{t}_q^* t'_\sigma$, both scaling-limit approaches yield identical results for the correlators in leading and next-to-leading asymptotic expansion. The lower and the upper bound diverge if the liquid state P approaches the glass-transition singularity P^c. But the length of the interval diverges as well. Because of Eq. (6.65b), the length diverges even if measured in units of the diverging lower bound. For fixed non-vanishing separation parameter, the lower bound is positive and the upper bound is finite, because von Schweidler's law differs qualitatively from the correlator $\phi_q(t)$ for very small and for very large times. But the identification of the two scaling laws permits the derivation of von Schweidler's law as limit results (6.127a,b) for the bifurcation dynamics. The derivation of the next-to-leading terms for the scaling-limit expansions permits the quantitative determination of the range of validity of von Schweidler's law. The preceding formulas for the interval hold under two restrictions. The error margin δ^* has to be so small that higher-than-next-to-leading-order correction terms can be neglected. Furthermore, the size of the separation parameter $|\sigma|$ has to be so small that plateau-renormalizations can be ignored.

von Schweidler's relaxation process appears as the essential part of the slowest relaxation process of the MCT glassy dynamics of the liquid state. The first-scaling-limit theory extends the analytic description of this process for times decreasing till the end of the transient regime. The second-scaling-limit theory continues the analytic description of the von Schweidler dynamics for times extending to such large values that all correlations have disappeared. The second-scaling-limit approach is self-contained with a single exception. Within this approach, one cannot explain the control-parameter dependence of the scale τ_σ. The result $\tau_\sigma \propto |\sigma|^{-\gamma}$ and the value $(1/2b) + (1/2a)$ for the exponent γ reflect the understanding of the stretched structural relaxation for times outside the transient and prior to the plateau crossing. This process is described by the first scaling-limit theory. It explains the existence of the second scale $\tau_\sigma = t'_\sigma$ as a corollary. On the basis of this insight, it seems adequate to consider the dynamics within the first-scaling-law regime as the primary part of the structural relaxation and the dynamics within the second-scaling-law regime as the secondary one. Within the interval specified by the preceding two inequalities, both parts are spliced.

6.3 Relaxation near higher-order singularities

This section presents some results for the evolution of the complex dynamics for states near higher-order glass-transition singularities. Section 6.3.1 deals with the asymptotic description of the arrested parts of the correlation functions for states near singularities of type A_3 and A_4. In Sec. 6.3.2, an intermediate-time interval is identified, where the equations of motion are solved by expansions of the correlators in powers of the logarithm of the time.

6.3.1 *Correlation arrest near higher-order singularities*

In every neighbourhood of a glass-transition singularity P^c of type $A_\ell, \ell = 3, 4, \ldots$, there are glass-transition singularities $P^{(k)c}$ of type $A_k, k = 2, \ldots, \ell - 1$. Understanding the topology of the manifold of these singularities near P^c is a prerequisite for a discussion of the complex dynamics described by the correlators for states P near P^c. It is the task to express the relevant parameters for a description of the manifold in terms of the mode-coupling coefficients. These relevant parameters appear as coefficients of a real polynomial of degree ℓ. For singularities of type A_3 and A_4, the coefficients shall be evaluated in this section. The manifold of coefficients, which characterize real roots of degeneracy k for the mentioned polynomial, specify the manifold of A_k-bifurcation points. The maximum property of the fixed-point equation will be used to identify those special roots of the polynomial, which determine the glass-transition singularities $P^{(k)c}$.

(i) Singularity parameters and separation parameters

Using the conventions of Sec. 4.2.1, the mode-coupling coefficients $V_\alpha, \alpha = 1, \ldots, N$, are considered as coordinates of the state P of the model to be analysed. The N coefficients V_α^c and the N differences $(V_\alpha - V_\alpha^c)$ specify the higher-order singularity under discussion and the separation of P from P^c, respectively. The differences shall be treated as small quantities of order ϵ. According to the discussion in Sec. 4.3.1, the singularity is connected with the array $\boldsymbol{f}^c$ of non-negative critical arrested parts $f_q^c, q = 1, \ldots, M$, of the correlators. Equation (4.61c) defines the critical stability matrix for the singularity. It is characterized by its M^2 non-negative elements $A_{q,k}^c, q, k = 1, \ldots, M$. This matrix has a non-degenerate maximum eigenvalue unity. The conventions (4.74c,d) are used to specify uniquely the arrays $\boldsymbol{a}^*$ and $\boldsymbol{a}$ for the distinguished eigenvectors. The distinguished resolvent for eigenvalue unity has matrix elements $R_{q,k}, q, k = 1, \ldots, M$, which are determined uniquely by the request (4.75e).

It is the goal to evaluate asymptotically the array $\boldsymbol{f}(P)$ of arrested parts of the correlators in regions of states, where $\lim_{P \to P^c} \boldsymbol{f}(P) = \boldsymbol{f}^c$. The derivations of this subsection are a continuation of those started in Sec. 4.3.3. Equations (4.78a,d), (4.80) imply a factorization property for the asymptotic behaviour of the arrested parts, namely

$$\boldsymbol{f} = \boldsymbol{f}^c + \boldsymbol{h} g(P). \tag{6.137}$$

Here, $\boldsymbol{h}$ abbreviates the array of the positive critical amplitudes defined in Eq. (4.78b). The problem is reduced to that of the determination of the asymptotic form of the single function of the states, $g(P)$, in regions where $\lim_{P \to P^c} g(P) = 0$. This function is a small quantity for the following discussions. The factorization theorem is a general result for all glass-transition singularities. The difference between different types of singularities is reflected by the difference of the singularities exhibited by function $g(P)$.

Within the microscopic versions of MCT for simple systems, f_q has the meaning of the Debye–Waller factor of the amorphous solid state P for wave number q.

It is shown in Sec. 5.1 that the factorization theorem can be generalized to describe other functions of interest. The simplest one is the shear correlator, which can be expressed as polynomial of the density correlators. Substituting Eq. (6.137) into Eq. (5.6b), one gets for its arrested part

$$f_T = f_T^c + h_T g(P). \tag{6.138a}$$

Equations (5.13b,c) provide the expressions for the critical arrested part f_T^c and the critical amplitude h_T. According to Eq. (5.12c), f_T is the normalized static shear modulus of the amorphous solid. The arrested parts $f_q^s(P)$ of the tagged-particle-density correlators are the Lamb–Mössbauer factors of the ideal glass state P for wave number q. Equation (5.48a) is generalized to

$$f_q^s(P) = f_q^{sc} + h_q^s g(P), \qquad q = 1, \ldots, M. \tag{6.138b}$$

In particular, there holds the generalization of Eq. (5.48b) for the square of the localization length:

$$r_s^2(P) = r_{sc}^2 - h_{\mathrm{MSD}} g(P). \tag{6.138c}$$

In Sec. 4.3.3, there has been demonstrated the possibility of a recursive construction of a sequence of arrays $\mathbf{c}^{(r)}(P) = (c_1^{(r)}(P), \ldots c_M^{(r)}(P)), r = 0, 1, \ldots,$ of smooth functions of P. Linear combination of the M components defines a sequence of smooth functions $c^{(r)}(P) = \sum_q a_q^* c_q^{(r)}(P)$. Let us introduce the abbreviations:

$$\mu_\ell = -c^{(\ell)}(P^c), \qquad \ell = 2, 3, \ldots. \tag{6.139a}$$

The formulas (4.83d) specify the type of the glass-transition singularity.

$$A_\ell : \mu_\ell \neq 0, \quad \mu_k = 0 \quad \text{for} \quad k < \ell. \tag{6.139b}$$

The number μ_ℓ shall be called the singularity parameter of P^c.

For the critical point $P = P^c$, Eq. (4.82a) simplifies, since the three kinds of difference coefficients vanish. Hence, the first two arrays vanish at the singularity: $\mathbf{c}^{(0)}(P^c) = \mathbf{c}^{(1)}(P^c) = \mathbf{0}$. There is only the contribution from the first line on the right-hand side of Eq. (4.82a), which determines the first non-trivial array: $c_q^{(2)}(P^c) = A_q^{(2,0)}(P^c)$. The singularity parameter for the A_2-glass-transition points is given by

$$\mu_2 = -\sum_q a_q^* A_q^{(2,0)c}. \tag{6.140a}$$

It is identical with μ_2^c, which fixes the exponent parameter λ^c via Eq. (4.89c).

Iteration of Eq. (4.82a) for $P = P^c$ yields the desired results for the higher-order singularities of interest (Götze and Sperl 2004a). There are two contributions to the third order array: $c_q^{(3)}(P^c) = A_q^{(3,0)}(P^c) + \sum_k A_{q,k}^{(1,1)}(P^c) A_k^{(2,0)}(P^c)$. Thus,

$$\mu_3 = -\sum_q a_q^* \Big[A_q^{(3,0)c} + \sum_k A_{q,k}^{(1,1)c} A_k^{(2,0)c} \Big]. \tag{6.140b}$$

One gets one contribution to $\boldsymbol{c}^{(4)}(P^c)$ from the first line on the right-hand side of Eq. (4.82a) and another one from the third line. The second line yields three terms. As a result, one obtains

$$\mu_4 = -\sum_q a_q^* \Big[A_q^{(4,0)c} + \sum_k \big(A_{q,k}^{(2,1)c} A_k^{(2,0)c} + A_{q,k}^{(1,1)c} A_k^{(3,0)c}\big) + \sum_{kp} \big(A_{q,k}^{(1,1)c} A_{k,p}^{(1,1)c} A_p^{(2,0)c} + A_{q,kp}^{(0,2)c} A_k^{(2,0)c} A_p^{(2,0)c}\big)\Big]. \qquad (6.140c)$$

For the determination of the array $\boldsymbol{c}^{(0)}(P)$, only the first terms of the brackets on the right-hand side of Eq. (4.82a) have to be considered. One gets: $c_q^{(0)}(P) = D_q^{(0)}(P) + \sum_k D_{q,k}^{(0,1)}(P) D_k^{(0)}(P) + \sum_{kp} A_{q,kp}^{(0,2)}(P) D_k^{(0)} D_p^{(0)} + \cdots$. The other terms are products of at least three difference coefficients. According to definition (4.71e), the difference coefficients are of order ϵ. For the intended leading-order calculation, only the first term matters: $c_q^{(0)}(P) = D_q^{(0)}(P) + O(\epsilon^2)$. One gets $c^{(0)}(P) = \sigma(P) + O(\epsilon^2)$ with Eq. (4.86a) for $\sigma(P)$. For a generic singularity, there holds $f_q^c > 0$ for $q = 1, \ldots, M$. Hence, $\sigma(P)$ has all the properties, which are explained in connection with Eqs. (4.84)–(4.86). Function $\sigma(P)$ plays the role of a first separation parameter characterizing the position of state P relative to the critical point P^c. It shall be denoted accordingly:

$$\text{generic } A_\ell : \sigma_1(P) = \sum_q a_q^* D_q^{(0)}(P). \qquad (6.141a)$$

A degenerate singularity is characterized by a trivial array of critical arrested parts: $\boldsymbol{f}^c = \boldsymbol{0}$. In this case, the difference coefficients $D_q^{(0)}(P)$ vanish. Consequently, there holds $\boldsymbol{c}^{(0)}(P) = \boldsymbol{0}$ and $c_q^{(1)}(P) = D_q^{(1,0)}(P) + O(\epsilon^2)$. The function $c^{(1)}(P)$ plays the role of a first separation parameter.

$$\text{degenerate} A_\ell : \sigma_1(P) = \sum_q a_q^* D_q^{(1,0)}(P); \qquad (6.141b)$$

it agrees with $\sigma(P)$, which is discussed in connection with Eqs. (4.94)–(4.96).

For a generic A_ℓ singularity, the result (4.83b) means that function $g(P)$ is a solution of the equation: $-\sigma_1(P) - c^{(1)}(P)g(P) \cdots - c^{(\ell-1)}(P)g(P)^{\ell-1} - c^{(\ell)}(P)g(P)^\ell + O(g(P)^{\ell+1}) = 0$. For the intended asymptotic calculation, one can replace $-c^{(\ell)}(P)g(P)^\ell + O(g(P)^{\ell+1})$ by $\mu_\ell g(P)^\ell$. For $r = 1, \ldots, \ell - 1$, the smooth functions $c^{(r)}(P)$ vanish for $P = P^c$; they are of order ϵ. It is sufficient to evaluate these functions correctly up to this order. Let us write

$$\text{generic } A_\ell : c^{(r)}(P) = \sigma_{r+1}(P) + O(\epsilon^2), \qquad r = 0, \ldots, \ell - 1. \qquad (6.142a)$$

The functions $\sigma_2(P), \ldots, \sigma_\ell(P)$ are further separation parameters for the description of the position of the state P. The asymptotic form of $g(P)$ is a solution of the equation $\mu_\ell g(P)^\ell - \sigma_\ell(P)g(P)^{\ell-1}\cdots - \sigma_1(P) = 0$. The function $\hat{g}(P) = g(P) - \sigma_\ell(P)/(\ell\mu_\ell)$ differs from $g(P)$ by a smooth function only. For the discussions of the asymptotic form of the singularities of the array $\boldsymbol{f}(P)$, it is irrelevant whether $g(P)$ or $\hat{g}(P)$ is substituted in Eq. (6.137). There holds the equation $\mu_\ell\hat{g}(P)^\ell - \hat{\sigma}_{\ell-1}\hat{g}(P)^{\ell-2}\cdots - \hat{\sigma}_1(P) = 0$. Here, $\hat{\sigma}_r(P) = \sigma_r(P) + O(\sigma_\ell(P)^2)$. For the intended asymptotic calculation, $\hat{\sigma}_r(P)$ can be replaced by $\sigma_r(P)$. One concludes that the searched-for function obeys the equation $\mu_\ell g(P)^\ell - \sigma_{\ell-1}g(P)^{\ell-2}\cdots - \sigma_1(P) = 0$. For a degenerate singularity, Eq. (4.83b) implies the equation $-\sigma_1(P)g(P) - c^{(2)}(P)g(P)^2\cdots - c^{(\ell-1)}(P)g(P)^{\ell-1} - c^{(\ell)}(P)g(P)^\ell + O(g(P)^{\ell+1}) = 0$. Again, $c^{(\ell)}(P)$ can be replaced by $(-\mu_\ell)$; and terms of order $g(P)^{\ell+1}$ can be neglected compared to $\mu_\ell g(P)^\ell$. The other coefficients of the polynomial are of order ϵ. For the asymptotic theory, it is sufficient to evaluate these coefficients correctly up to this order. As above, these coefficients specify separation parameters. The definition (6.141b) shall be extended.

$$\text{degenerate } A_\ell : c^{(r)}(P) = \sigma_r(P) + O(\epsilon^2), \qquad r = 1, \ldots, \ell - 1. \tag{6.142b}$$

The searched-for function $g(P)$ obeys the equation $\mu_\ell g(P)^\ell - \sigma_{\ell-1}(P)g(P)^{\ell-1}\cdots - \sigma_1(P)g(P) = 0$.

The preceding paragraph can be summarized as follows. The states P near an A_ℓ singularity P^c are characterized by $(\ell - 1)$ separation parameters $\sigma_1(P), \ldots, \sigma_{\ell-1}(P)$. These are smooth functions of order ϵ. They determine the coefficients x_r of a polynomial $P_\ell(u)$ of degree ℓ:

$$x_r = \sigma_r(P)/\mu_\ell, \qquad r = 1, \ldots, \ell - 1. \tag{6.143a}$$

The asymptotic form of function $g(P)$ is a real zero of the polynomial

$$P_\ell(g(P)) = 0. \tag{6.143b}$$

For a given state P, $g(P)$ is determined by the maximum of the roots of $P_\ell(u)$. There holds:

$$\text{generic } A_\ell : P_\ell(u) = u^\ell - x_{\ell-1}u^{\ell-2}\cdots - x_2 u - x_1, \tag{6.143c}$$

$$\text{degenerate } A_\ell : P_\ell(u) = \left[u^{\ell-1} - x_{\ell-1}u^{\ell-2}\cdots - x_2 u - x_1\right]u. \tag{6.143d}$$

A real root of degeneracy $k, k = 2, \ldots, \ell - 1$, specifies a bifurcation point of type A_k. Those bifurcation points, which exhibit the maximum property, determine the glass-transition singularities $P^{(k)c}$.

If P^c is a generic A_2 singularity, there holds $P_2(u) = u^2 - x_1$. The region of singular behaviour of the arrested parts is specified by $x_1 \geq 0$, and $g = \sqrt{\sigma_1(P)/\mu_2}$. If P^c is a degenerate A_2 singularity, there holds $P_2(u) = [u - x_1]u$. For $x_1 \leq 0$, one gets $g(P) = 0$. For $x_1 \geq 0$, one obtains $g(P) = \sigma_1(P)/\mu_2$. Substituting these formulas into Eq. (6.137), one reproduces the asymptotic limit of Eq. (4.91a) and of Eq. (4.98), respectively.

Generically, the equations $\sigma_r(P) = 0, r = 1, \ldots, \ell - 1$ define $(\ell - 1)$ smooth pieces of surfaces, which intersect transversally in P^c. For a neighbourhood of P^c, an invertible smooth mapping from the sets of $(V_1, \ldots, V_N)$ to sets of $(x_1, \ldots, x_N)$ can be performed so that there holds Eq. (6.143a) for the first $(\ell - 1)$ coordinates. The conditions $\sigma_r(P) = \mu_\ell x_r$ define the first $(\ell - 1)$ coordinate planes of the new system; and P^c has the coordinates $x_r^c = 0, r = 1, \ldots, (\ell - 1)$. The asymptotic form of $g(P)$ depends on the first $(\ell - 1)$ new coordinates only. The other $N - (\ell - 1)$ coordinates are irrelevant for the discussion; they can be replaced by those of P^c. In this sense, the singularity can be discussed in the $(\ell - 1)$-dimensional manifold of relevant control parameters $x_1, \ldots, x_{\ell-1}$.

In applications of MCT, the states are described by a set of external control parameters $\xi_1, \xi_2, \ldots$. An example is provided by the square-well system. It is discussed in Sec. 4.3.6 in connection with Fig. 4.9. Here, the packing fraction $\xi_1 = \varphi$, the attraction strength $\xi_2 = \Gamma$ and the attraction range $\xi_3 = \delta$ are used. Equivalently, one can use the distance parameters $\epsilon_1 = \xi_1 - \xi_1^c, \epsilon_2 = \xi_2 - \xi_2^c, \ldots$, where $\xi_1^c, \xi_2^c, \ldots$ are the external control parameters for the singularity P^c. To describe the A_ℓ singularity faithfully, one needs $(\ell - 1)$ external control parameters $\xi_1, \ldots, \xi_{\ell-1}$. There must exist a smooth invertible mapping of sets of small distance parameters on sets of small relevant control parameters

$$(\epsilon_1, \ldots, \epsilon_{\ell-1}) \leftrightarrow (x_1, \ldots, x_{\ell-1}). \tag{6.144a}$$

For the desired asymptotic description, the mapping can be linearized:

$$x_r = \sum_{s=1}^{\ell-1} C_{rs} \epsilon_s, \qquad r = 1, \ldots, \ell - 1. \tag{6.144b}$$

The $(\ell - 1)$-by-$(\ell - 1)$ matrix of coefficients C_{rs} has to be invertible. To proceed, one has to generalize Eqs. (6.141a,b) and determine the remaining relevant separation parameters. The result yields $\sigma_1(P), \ldots, \sigma_{\ell-1}(P)$ in terms of $\xi_1, \ldots, \xi_{\ell-1}$; and this establishes the mapping (6.144a). The coefficients $C_{rs} = [\partial \sigma_r(P^c)/\partial \xi_s]/\mu_\ell$ are obtained as corollary.

(ii) Generic A_3 singularities

Let us consider a generic glass-transition singularity P^c of type A_3. There are two relevant separation parameters. The first one is noted in Eq. (6.141a). According to Eq. (6.142a), the second one is given by $\mathbf{c}^{(1)}(P)$, which can be evaluated in leading order ϵ. There are two contributions descending from the first and second line, respectively, of the right-hand side of Eq. (4.82a): $c_q^{(1)}(P) = D_q^{(1,0)}(P) + \sum_k A_{q,k}^{(1,1)}(P^c) D_k^{(0)}(P) + O(\epsilon^2)$.

$$\text{generic } A_\ell : \sigma_2(P) = \sum_q a_q^* \Big[D_q^{(1,0)}(P) + \sum_k A_{q,k}^{(1,1)c} D_k^{(0)}(P) \Big]. \tag{6.145}$$

Thereby, the coordinates x_1 and x_2 of the plane of relevant control parameters are related to the mode-coupling coefficients.

The bifurcation manifold of the canonical cubic polynomial

$$P_3(u) = u^3 - x_2 u - x_1 \tag{6.146a}$$

shall be discussed by the inverse method. Let ζ denote a real degenerate root. Since the sum of the roots vanishes, -2ζ is the third root. There holds $P_3(u) = (u-\zeta)^2(u+2\zeta)$. The value ζ determines both coefficients

$$x_1 = -2\zeta^3, \qquad x_2 = 3\zeta^2. \tag{6.146b}$$

The value $\zeta = 0$ characterizes P^c as origin of the coordinate plane. For $\zeta \neq 0$, the preceding equation describes the manifold of A_2 bifurcations of $P_3(u)$. For $\zeta < 0$, Eq. (6.146b) defines a curve in parameter representation, which is located in the quadrant specified by $x_1 > 0$ and $x_2 > 0$. For $\zeta > 0$, a curve is defined, which is located in the quadrant with $x_1 < 0$ and $x_2 > 0$. The tangent vector with coordinates $t_1 = \partial x_1/\partial\zeta = -6\zeta^2$ and $t_2 = \partial x_2/\partial\zeta = 6\zeta$ does not vanish, i.e., the curves are smooth one-dimensional manifolds. They merge at the origin with a normalized tangent vector in the coordinate line specified by $x_1 = 0$. The two lines of A_2 bifurcations merge in the cusp singularity.

For the branch of bifurcation points with $\zeta < 0$, the second root -2ζ exceeds the degenerate one. These bifurcation points are not consistent with the maximum property; they do not describe glass-transition singularities. For the other branch, ζ exceeds -2ζ. These degenerate roots determine maximum fixed points $P^{(2)c}$ with $g(P^{(2)c}) = g^{(2)c} = \zeta$. Within the plane of relevant control parameters, a line of glass-transition points is identified.

$$\text{generic } A_2 : 0 < g^{(2)c} = \zeta, \quad x_1^{(2)c} = -2\zeta^3, \quad x_2^{(2)c} = 3\zeta^2. \tag{6.147}$$

Within the $x_1 - x_2$ plane, the curve of A_2 glass-transition singularities is determined by the formula $x_1^{(2)c} = -2(\sqrt{x_2^{(2)c}/3})^3, x_2^{(2)c} > 0$. The curvature of the transition line diverges if $P^{(2)c}$ approaches the endpoint.

Let us characterize a state P near $P^{(2)c}$ by the deviations $\hat{x}_{1,2}$ of the control parameters from their values at $P^{(2)c}$. Let us also introduce the deviation $\hat{g}(P)$ of the arrested part $g(P)$ from the value at $P^{(2)c}$.

$$x_{1,2} = x_{1,2}^{(2)c} + \hat{x}_{1,2}, \qquad g(P) = g^{(2)c} + \hat{g}(P). \tag{6.148}$$

For the original form of Eq. (4.83b), $\mu_3 P_3(g(P)) + O(g^4) = 0$, one gets $3\mu_3\zeta\hat{g}(P)^2 - \mu_3\hat{x}_2\hat{g}(P) - \mu_3(\hat{x}_1 + \zeta\hat{x}_2) + O(\hat{g}^3) = 0$. One can appeal to the discussion of the A_2 bifurcation done in the preceding subsection in order to arrive at the following conclusion. The singularity parameter of $P^{(2)c}$ is given by

$$\mu_2(P^{(2)c}) = 3\mu_3\zeta. \tag{6.149a}$$

The relevant separation parameter reads:

$$\hat{\sigma}_1(P) = \mu_3(\hat{x}_1 + \zeta\hat{x}_2). \tag{6.149b}$$

The asymptotic form of the arrested part is $\sqrt{\hat{\sigma}_1(P)/\mu_2}$. Hence,

$$g(P) = \zeta + \sqrt{\hat{\sigma}_1(P)/3\mu_3\zeta} + O(\epsilon), \qquad \hat{\sigma}_1(P)/\mu_3 \geq 0. \tag{6.149c}$$

For negative values of $\hat{\sigma}_1(P)$, the cubic equation exhibits a pair of non-real roots. The third root depends smoothly on ϵ. Consequently,

$$g(P) = -2\zeta + O(\epsilon), \qquad \hat{\sigma}_1(P)/\mu_3 < 0. \tag{6.149d}$$

The generic A_3 singularity P^c is the endpoint of a manifold of discontinuous glass-glass transition points $P^{(2)c}$. The discontinuity is given by 3ζ; it vanishes proportional to $g^{(2)c}$ if $P^{(2)c}$ tends to P^c. The approach of $P^{(2)c}$ towards P^c is connected with an approach of $\mu_2(P^{(2)c})$ to zero. The general inequality (4.89a) for μ_2 implies the inequality for the cusp parameter:

$$\mu_3 > 0. \tag{6.150}$$

Within the original parameter space, the surface of singularities $P^{(2)c}$ is located in a 'quadrant' specified by: $\sigma_1(P^{(2)c}) < 0, \sigma_2(P^{(2)c}) > 0$.

The upper panel of Fig. 4.3 demonstrates the above described singularity manifold for a one-component model. The dashed line exhibits the coordinate line $\epsilon_1(P) = \sigma_1(P) = 0$. The microscopic model for a square-well system exhibits generic A_3 singularities, provided the range parameter δ is sufficiently small. This is demonstrated in Fig. 4.9 for the attraction ranges of 3% and 5%.

(iii) Degenerate A_3 singularities

Two relevant separation parameters determine the position of the state P relative to a degenerate glass-transition singularity P^c of type A_3. The first one is noted in Eq. (6.141b). According to Eq. (6.142b), one has to evaluate $\mathbf{c}^{(2)}(P)$ in linear approximation in ϵ in order to get $\sigma_2(P) = \sum_q a_q^* c_q^{(2)}(P)$. Iteration of Eq. (4.82a) for vanishing $D_q^{(0)}(P)$ yields $c_q^{(2)}(P) = A_q^{(2,0)}(P) + \sum_k [D_{q,k}^{(0,1)}(P) A_k^{(2,0)}(P^c) + A_{q,k}^{(1,1)}(P^c) D_k^{(1,0)}(P)] + O(\epsilon^2)$. The first contribution on the right-hand side has to be evaluated correctly up to order ϵ. Introducing the difference coefficients for $n = 2$ from Eq. (4.71e), one gets $A_q^{(2,0)}(P) = A_q^{(2,0)}(P^c) + \sum_{kp} D_{q,kp}^{(2)}(P) a_k a_p$. There holds $\sum_q a_q^* A_q^{(2,0)}(P^c) = \mu_2 = 0$, i.e.:

$$\text{degenerate} A_\ell, \ell \geq 3 : \sigma_2(P) = \sum_q a_q^* \Big[\big(\sum_{kp} D_{q,kp}^{(2)}(P) a_k a_p \big) + \sum_k \big(D_{q,k}^{(0,1)}(P) A_k^{(2,0)c} + A_{q,k}^{(1,1)c} D_k^{(1,0)}(P) \big) \Big]. \tag{6.151}$$

The normalized polynomial for the discussion of the bifurcation points reads

$$P_3(u) = \big[u^2 - x_2 u - x_1 \big] u. \tag{6.152}$$

For all points $P, u = 0$ is a root. Because of the maximum property, glass-transition singularities cannot be connected with negative roots. Consequently, there are two possibilities for glass-transition singularities of type A_2. First, there are positive degenerate roots ζ of the polynomial in the bracket. These roots are generic A_2 singularities $P^{(2)c}$. Second, there are the points $(x_1, x_2) \neq (0,0)$, which yield $u = 0$ as root of the polynomial in the bracket. These bifurcations deal with degenerate glass-transition singularities $P^{(2)c\prime}$.

The first possibility implies $u^2 - x_2 u - x_1 = (u-\zeta)^2$, and this yields $x_1 = -\zeta^2$ and $x_2 = 2\zeta$. These formulas specify a parabola in the parameter plane. The branch for $\zeta < 0$ deals with A_2 bifurcations, which are no glass-transition states. The other branch characterizes the searched-for singularities $P^{(2)c}$.

$$\text{generic } A_2 : 0 < g^{(2)c} = \zeta, \quad x_1^{(2)c} = -\zeta^2, \quad x_2^{(2)c} = 2\zeta. \tag{6.153a}$$

The arrest for states P near $P^{(2)c}$ can be discussed by introducing new variables as defined in Eq. (6.148). One derives for the singularity parameter

$$\mu_2(P^{(2)c}) = \mu_3\zeta; \tag{6.153b}$$

and the relevant separation parameter reads $\hat{\sigma}_1(P) = \mu_3\zeta(\hat{x}_1 + \zeta\hat{x}_2)$. One concludes again that μ_3 is positive. The transition line is located in the 'quadrant' specified by $\sigma_1(P) < 0$ and $\sigma_2(P) > 0$. For $\hat{\sigma}_1(P) < 0$, there is the maximum fixed point specified by $g(P) = 0$. The identified manifold describes discontinuous liquid-glass transitions. States with $\sigma_1(P)/\mu_3 < x_1^{(2)c}$ or states with $\sigma_2(P)/\mu_3 < x_2^{(2)c}$ are liquids. If $P^{(2)c}$ approaches the singularity $P^c, \mu_2(P^{(2)c})/\mu_3 = g^{(2)c}$ tends to zero proportional to $x_2^{(2)c}$. The tangent vector of the transition line approaches that of the coordinate line $\sigma_1(P) = 0$.

The second possibility occurs if and only if $x_1 = 0$. The remaining zero is given by $u = x_2$. This specifies the second branch of transition points $P^{(2)c}$.

$$\text{degenerate } A_2 : x_1^{(2)c\prime} = 0, \qquad x_2^{(2)c\prime} < 0. \tag{6.154a}$$

The singularity parameter is given by

$$\mu_2(P^{(2)c\prime}) = -x_2^{(2)c\prime}\mu_3. \tag{6.154b}$$

It tends to zero if $P^{(2)c\prime}$ approaches P^c. There holds $\mu_3 > 0$. One obtains

$$g(P) = \sigma_1(P)/\mu_2(P^{(2)c\prime}) + O(\epsilon^2) \quad \text{if} \qquad \sigma_1(P) \geq 0, \tag{6.154c}$$

$$g(P) = 0 \quad \text{if} \qquad \sigma_1(P) \leq 0. \tag{6.154d}$$

The line of states with $\sigma_1(P) = 0$ deals with continuous liquid-glass transitions. The states $\sigma_1(P) < 0$ and $\sigma_2(P) < 0$ are liquids.

The degenerate singularity P^c is located on the boundary of the liquid states. Described in the plane of relevant control parameters, the boundary consists of two branches. One branch is the curve of discontinuous transitions. The other one is the coordinate line $x_1^{(2)c\prime} = 0$ of continuous transitions. These lines merge in P^c with continuous tangent but with a discontinuity of the curvature.

Figure 4.2 demonstrates the above described bifurcation manifold for a schematic model. For a microscopic model, the scenario has been identified by Krakoviack (2007) for a system of hard spheres. From representative equilibrium configurations, a subset of spheres with packing fraction φ_m is selected at random. These particles are treated as immobile. The ensemble of these particles is a porous-medium model. The other spheres of packing fraction φ_f represent a liquid confined by the medium. These particles move according the canonical equations of motion. Their phase-space coordinates are used to define dynamical variables like the density fluctuations. Canonical averaging and additional averaging over the selections of the ensembles of immobile particles is used to define correlation functions for a statistical description of the dynamics. The medium causes time-independent parts of the correlators formed with density fluctuations. These contributions have to be eliminated from the definitions of the correlators. The essential step in the motivation of closed equations is a factorization ansatz for the fluctuating-force kernel. The resulting mode-coupling functional consists of quadratic terms and of linear ones. The former are obtained as explained in Sec. 4.1.1. The latter are obtained by factorizing the forces due to interaction of the immobile particles of the medium with the mobile ones of the liquid as mentioned in connection with Eq. (4.13). The resulting mode-coupling constants are determined by the two control parameters φ_m and φ_f. The convolution approximation is used and the structure functions are determined in the Percus–Yevick approximation.

Curve I in Fig. 6.19 exhibits the diagram of glass-transition singularities for the model explained above. The degenerate A_3 singularity is located on the boundary of the liquid region for $\varphi_m^c \approx 0.16$ and $\varphi_f^c \approx 0.24$. The discontinuous transition points $P^{(2)c}$ are obtained for $\varphi_f > \varphi_f^c$. They denote transitions to a glass, whose arrest is dominated by the cage effect of the dense fluid. The medium particles are immobile parts of the cages and, therefore, φ_f^c decreases stronger with increasing φ_m^c than expected from the ad hoc estimation $\varphi_m^c + \varphi_f^c = \varphi_{\text{HSS}}^c \approx 0.52$. The continuous transition points $P^{(2)c\prime}$ are obtained for $\varphi_f < \varphi_f^c$. They describe a percolation transition for the liquid, which is modified by interactions. These transitions are connected with the appearance of a divergent localization length if $P^{(2)c\prime}$ is approached from the glass side. In Sec. 4.1.2, objections against the MCT treatment of the percolation transition are explained. At present, it is not known how important these objections are for a judgement of the relevance of the above cited work on the degenerate glass-transition singularity of type A_3.

Curve II exhibits the bifurcation manifold for a modified model for the medium. The centres of the immobile particles are chosen without any correlations.

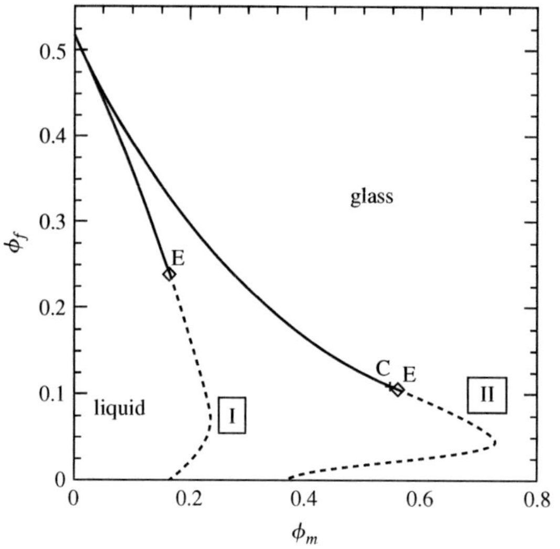

FIG. 6.19. Manifold of glass-transition singularities calculated for a microscopic model for a hard-sphere system of packing fraction φ_f, which is confined by a medium of immobile hard spheres of packing fraction φ_m. Curves I and II are obtained for different models for the distribution of the particles of the medium, as explained in the text. The dashed curves are lines of continuous liquid–glass transitions and the full ones show discontinuous transition points. For model I, the two lines terminate in a common endpoint E, which is a degenerate A_3 singularity. For model II, the dashed line terminates in a corner C, which it forms with the full line. The latter continues as glass–glass-transition line till the endpoint E, which is a generic A_3 singularity. Reproduced from Krakoviack (2007).

In this case, the boundary of the liquid region also consists of a branch of generic transitions $P^{(2)c}$ and a branch of degenerate points $P^{(2)c\prime}$. But, these branches form a corner. The branch of discontinuous liquid–glass transition points continues as line of generic glass–glass transition states till it terminates at a generic singularity of type A_3. The manifold is topologically equivalent to that, which is demonstrated in Fig. 4.3 for a schematic model.

(iv) Generic A_4 singularities

The determination of the arrest function $g(P)$ for states P near a generic glass-transition singularity P^c of type A_4 requires the expressions for three relevant separation parameters. The asymptotic formulas (6.141a) and (6.145) determine $\sigma_1(P)$ and $\sigma_2(P)$, respectively. According to Eqs. (4.83a), (6.142a), $\mathbf{c}^{(2)}(P)$ has to be determined in linear order in ϵ in order to relate $\sigma_3(P)$ to the mode-coupling coefficients. The iterative treatment of Eq. (4.82a) towards the formula for $\mathbf{c}^{(2)}(P)$ follows the one for the derivation of Eq. (6.151).

$$\text{generic } A_\ell, \ell \geq 3: \\ \sigma_3(P) = \sum_q a_q^* \Big[\big(\sum_{kp} D^{(2)}_{q,kp}(P) a_k a_p \big) \\ + \sum_k \big(D^{(0,1)}_{q,k}(P) A^{(2,0)c}_k + A^{(1,1)c}_{q,k} D^{(1,0)}_k(P) + A^{(2,1)c}_{q,k} D^{(0)}_k(P) \big) \\ + \sum_{kp} \big(2A^{(0,2)c}_{q,kp} A^{(2,0)c}_k + A^{(1,1)c}_{q,k} A^{(1,1)c}_{k,p} \big) D^{(0)}_p(P) \Big]. \tag{6.155}$$

For a state P with the relevant control parameters $x_r = \sigma_r(P)/\mu_4, r = 1, 2, 3$, the function $g(P)$ is the largest real root of the canonical quartic polynomial

$$P_4(u) = u^4 - x_3 u^2 - x_2 u - x_1. \tag{6.156a}$$

The coefficients, which lead to real roots of degeneracy $k, k = 2, 3$, specify the bifurcation points of type A_k for the polynomial. If such root has the maximum property, the coefficients $x_r^{(k)c}, r = 1, 2, 3$, determine a glass-transition singularity $P^{(k)c}$; and the value of the root is the corresponding critical arrested part $g^{(k)c} = g(P^{(k)c})$. Let ζ denote a degenerate real root. There holds $P_4(u) = (u-\zeta)^2 P_2(u)$ with $P_2(u)$ denoting a real polynomial of degree 2. It can be characterized by two real coefficients η and $\eta' : P_2(u) = (u-\zeta)^2 + \eta'(u-\zeta) + \eta$. Since the sum of the roots of $P_4(u)$ is zero, there holds $\eta' = 4\zeta$. One gets

$$P_4(u) = (u-\zeta)^2 \big[(u-\zeta)^2 + 4\zeta(u-\zeta) + \eta \big]. \tag{6.156b}$$

The numbers ζ and η parameterize the bifurcation manifold:

$$x_1 = \zeta^2(3\zeta^2 - \eta), \quad x_2 = 2\zeta(\eta - 4\zeta^2), \quad x_3 = 6\zeta^2 - \eta. \tag{6.156c}$$

P^c is an endpoint of the manifold, which is determined by $\zeta = \eta = 0$.

The parameter ζ characterizes a root of multiplicity larger than 2 if and only if $\eta = 0$. If this is the case, the remaining root is -3ζ. For $\eta = 0$ and $\zeta \neq 0$, Eq. (6.156c) describes smooth lines of generic A_3 bifurcation points. The two branches for $\zeta < 0$ and $\zeta > 0$ merge in P^c if ζ tends to zero. For the first branch, ζ is not a maximum root. For the second one, ζ denotes the arrested part of the $P^{(3)c}$.

$$\text{generic } A_3 : 0 < g^{(3)c} = \zeta, \quad x_1^{(3)c} = 3\zeta^4, \quad x_2^{(3)c} = -8\zeta^3, \quad x_3^{(3)c} = 6\zeta^2. \tag{6.157a}$$

Expanding the polynomial around $P^{(3)c}$, one finds for the singularity parameter

$$\mu_3(P^{(3)c}) = 4\mu_4\zeta. \tag{6.157b}$$

If $P^{(3)c}$ approaches P^c, this value tends to zero. Inequality (6.150) implies

$$\mu_4 > 0. \tag{6.158}$$

The line of singularities of type A_3 is located in the 'octant' specified by

$$\sigma_1(P^{(3)c}) > 0, \quad \sigma_2(P^{(3)c}) < 0, \quad \sigma_3(P^{(3)c}) > 0. \tag{6.159}$$

The surface of real roots of degeneracy 2 is determined by Eq. (6.156c) with the restriction $\eta \neq 0$. The above discussed points with $\eta = 0$ describe the end-points of this two-dimensional manifold. A complete overview can be obtained by considering the cuts of this manifold with the surfaces of constant third coordinates: $x_3 = x^*$. The other coordinates of the cut are given by

$$x_1 = \zeta^2(-3\zeta^2 + x^*), \qquad x_2 = 2\zeta(2\zeta^2 - x^*). \tag{6.160a}$$

These formulas describe a curve with tangent-vector coordinates $t_1 = \partial x_1/\partial\zeta = 2\zeta(-6\zeta^2 + x^*)$ and $t_2 = \partial x_2/\partial\zeta = 2\zeta(6\zeta^2 - x^*)$. The restriction to A_2 bifurcation points reads

$$6\zeta^2 - x^* \neq 0. \tag{6.160b}$$

In addition to the root ζ of multiplicity 2, there are the other roots $u_\pm = -\zeta \pm \sqrt{x^* - 2\zeta^2}$ of the polynomial. If $2\zeta^2$ exceeds x^*, ζ is a maximum root. The corresponding bifurcation points are glass-transition singularities.

$$P^{(2)c} : x^* < 2\zeta^2. \tag{6.161a}$$

If $x^* \geq 2\zeta^2$, ζ is a maximum root if and only if $2\zeta \geq \sqrt{x^* - 2\zeta^2}$. Observing condition (6.160b), one gets another possibility for glass-transition singularities.

$$P^{(2)c} : \zeta > 0, \qquad 6\zeta^2 > x^* \geq 2\zeta^2. \tag{6.161b}$$

Since $P_4(u) = (u - \zeta)^2(6\zeta^2 - x^*) + O((u - \zeta)^3)$, the singularity parameter is given by

$$\mu_2 = \mu_4(6\zeta^2 - x^*). \tag{6.161c}$$

Depending on the value x^*, there are three types of cuts through the bifurcation surface. These lead to three types of lines of glass-transition singularities $P^{(2)c}$. The first type deals with negative values for x^*. Let us introduce a positive number s in order to write $x^* = -s^2$. Condition (6.161a) is obeyed for all ζ. The tangent vector does not vanish. For ζ increasing from negative to positive values, a smooth line is specified by

$$x_1^{(2)c} = -\zeta^2(3\zeta^2 + s^2), \quad x_2^{(2)c} = 2\zeta(2\zeta^2 + s^2), \quad x_3^{(2)c} = -s^2, \tag{6.162a}$$

$$g^{(2)c} = \zeta, \qquad \mu_2 = \mu_4(6\zeta^2 + s^2). \tag{6.162b}$$

For negative ζ, the curve is located in the 'quadrant': $\sigma_1(P^{(2)c}) < 0, \sigma_2(P^{(2)c}) < 0$. For positive ζ it is located in the 'quadrant': $\sigma_1(P^{(2)c}) < 0, \sigma_2(P^{(2)c}) > 0$. For ζ tending to zero, the curve touches the coordinate line $x_1 = 0$:

$$x_1^{(2)c} \sim -\left[x_2^{(2)c}/(2s)\right]^2, \qquad x_2^{(2)c} \to 0. \tag{6.162c}$$

The curvature of the line diverges if s tends to zero. The singularity parameter has a minimum $\mu_4 s^2$ in the point of touch of the curve with the surface $\sigma_1(P) = 0$. If the state P is obtained from $P^{(2)c}$ by decreasing x_1 below $x_1^{(2)c}$, the degenerate root ζ is changed to a pair of non-real roots. Hence, the line separates liquid states with $x_1 < x_1^{(2)c}$ from glass states with $x_1 \geq x_1^{(2)c}$.

For a positive number ρ, a smooth invertible transformation of coordinates can be defined by

$$x_1 \to \tilde{x}_1 = \rho^4 x_1, \quad x_2 \to \tilde{x}_2 = \rho^3 x_2, \quad x_3 \to \tilde{x}_3 = \rho^2 x_3. \tag{6.163}$$

Transforming $u \to \tilde{u} = \rho u, \zeta \to \tilde{\zeta} = \rho u$ and $\eta \to \rho^2 \eta$, the preceding theory is covariant: all equations remain valid if the variables are changed to those with a tilde. The above defined scale transformation alters the curves for parameters s to ones for the parameter $\tilde{s} = \rho s$. In this sense, all curves for $x^* < 0$ are topologically equivalent.

The second type of cuts is obtained for $x^* = 0$. Condition (6.160b) excludes the point for $\zeta = 0$, which characterizes P^c. Condition (6.161a) is obeyed, i.e., all bifurcation points of the polynomial are glass-transition points $P^{(2)c}$.

$$x_1^{(2)c} = -3\zeta^4, \quad x_2^{(2)c} = 4\zeta^3, \quad x_3^{(2)c} = 0, \tag{6.164a}$$

$$g^{(2)c} = \zeta, \qquad \mu_2 = 6\mu_4\zeta^2. \tag{6.164b}$$

The manifold consists of two branches, specified by positive and negative values for $g^{(2)c}$.

$$P_+^{(2)c} : \zeta > 0; \qquad P_-^{(2)c} : \zeta < 0. \tag{6.164c}$$

Equation (6.137) shows that the arrested parts f_q of the correlators are larger than the critical values f_q^c on the branch of points $P_+^{(2)c}$, while they are smaller on the other branch, $q = 1, \ldots, M$. The two branches evolve continuously from the curves described by Eq. (6.162a) if s tends to zero. There holds

$$x_1^{(2)c} = -3\Big(\big[x_2^{(2)c}/4\big]^{1/3}\Big)^4, \quad \mu_2 = 6\mu_4\Big(\big[x^{(2)c}/4\big]^{1/3}\Big)^2, \quad g^{(2)c} = \big[x^{(2)c}/4\big]^{1/3}. \tag{6.164d}$$

If $P^{(2)c}$ approaches P^c, the normalized tangent of the curve approaches that of the surface $\sigma_1(P) = 0$, the curvature of the line diverges, and μ_2 tends to zero. The two branches and the point P^c form the manifold of discontinuous liquid-glass transition points. The liquid is located in the region $x_1 \le x_1^{(2)c}$. The manifold is invariant under the scale transformation (6.163).

The third type of cuts deals with positive x_3. Writing $x^* = s^2, s > 0$, the singularities $P^{(2)c}$ obey the equations

$$x_1^{(2)c} = -\zeta^2(3\zeta^2 - s^2), \quad x_2^{(2)c} = 2\zeta(2\zeta^2 - s^2), \quad x_3^{(2)c} = s^2, \tag{6.165a}$$

$$g^{(2)c} = \zeta, \qquad \mu_2 = \mu_4(6\zeta^2 - s^2). \tag{6.165b}$$

If $2\zeta^2$ exceeds s^2, condition (6.161a) is obeyed. For the limit case $\zeta = s/\sqrt{2}$, condition (6.161b) is valid. The bifurcation point $\zeta = -s/\sqrt{2}$ is not a glass-transition singularity. Hence, the preceding formulas define two branches.

$$P_+^{(2)c} : s/\sqrt{2} \le \zeta; \qquad P_-^{(2)c} : \zeta < -s/\sqrt{2}. \tag{6.166a}$$

Since the tangent vector does not vanish, the branches are smooth curves. For ζ increasing to $-s/\sqrt{2}$, $P_-^{(2)c}$ approaches the state $P_+^{(2)c}$ for $\zeta = s/\sqrt{2}$. The two branches form a corner. There is a

$$\text{crossing point} : x_1^{\text{cr}} = -(s^2/2)^2, \quad x_2^{\text{cr}} = 0, \quad x_3^{\text{cr}} = s^2. \tag{6.166b}$$

For s tending to zero, the crossing point tends to P^c and the two branches converge towards the ones defined in Eqs. (6.164a–c). As done above, one shows that the two branches form the boundary of a liquid. Liquid states are specified by the condition $x_1 < x_1^{(2)c}$. The states on the first branch have arrested parts of the correlators, which exceed those at the singularity P^c.

$$f_q(P_+^{(2)c}) \ge f_q^c + h_q(s/\sqrt{2}). \tag{6.166c}$$

For the arrested parts on the second parts, there holds

$$f_q(P_-^{(2)c}) < f_q^c - h_q(s/\sqrt{2}), \qquad q = 1, \ldots, M. \tag{6.166d}$$

If the transition points $P_\pm^{(2)c}$ approach the crossing point, the singularity parameter μ_2 decreases monotonically towards the minimal value $2\mu_4 s^2$.

If ζ increases from $-s/\sqrt{2}$ towards $-s/\sqrt{6}$, the line of bifurcation points of the polynomial continues smoothly. If $\zeta = -s/\sqrt{6}$, the tangent vector changes its sign. This point is an A_3 singularity. Increasing ζ further, the line continues smoothly. For ζ approaching $s/\sqrt{6}$, it approaches the singularity $P^{(3)c}$, which is specified by Eq. (6.157a). The bifurcation manifold described for $-s/\sqrt{2} \leq \zeta < s/\sqrt{6}$ does not deal with glass transition singularities. There remains a set of doubly degenerate roots, which obey the condition (6.161b).

$$P_{+}^{(2)c} : s/\sqrt{2} > \zeta > s/\sqrt{6}. \tag{6.167}$$

This line continues smoothly the one described by Eq. (6.166a). It deals with glass–glass transitions. The line ends at the singularity $P^{(3)c}$ for $x^{(3)c} = s^2$.

The manifold described for $x^* > 0$ exhibits the covariance property explained in connection with Eq. (6.163). Curves for different s are topologically equivalent since they are connected by a scaling transformation.

Figure 4.5 demonstrates glass-transition singularities, which are calculated for a two-component model. The mode-coupling polynomial is quantified by $N = 3$ coefficients v_1, v_2, and v_3. The shown cuts through the bifurcation manifold for $v_3 = 20, v_3 = v^*$, and $v_3 = 45$ are smooth mappings of those explained above for $x^* < 0, x^* = 0$, and $x^* > 0$, respectively.

The states of a microscopic model for the square-well system are specified by the external control parameters φ, Γ, and δ. Figure 4.9 exhibits cuts through the set of glass-transition singularities, which are defined by fixed values for the attraction-range parameter δ. The cuts for $\delta = 0.06$ and $\delta = \delta^*$ are smooth deformations of those explained above for $x^* < 0$ and $x^* = 0$, respectively. The cuts shown for $\delta = 0.03$ and $\delta = 0.02$ are smooth deformations of the manifolds for $x^* > 0$. Let us consider a state on the cut $\delta = \delta^*$. It can be specified by the distance parameter $\epsilon = (\Gamma - \Gamma^*)/\Gamma^*$, with Γ^* denoting the attraction strength parameter for the A_4 singularity. According to Eq. (6.164a), the dominant change of the relevant control parameters is due to $x_2^{(2)c}$. Hence, the asymptotic result for the arrested part is given by $g = g^* \epsilon^{1/3}$. Here, g^* denotes a positive constant. Substitution of this result into Eq. (6.138a) yields for the shear modulus: $f_T - f_T^c \sim h_T g^* \epsilon^{1/3}, \epsilon \to 0$. A similar result holds for the longitudinal modulus. These formulas quantify the continuous but singular change of the critical moduli, which are exhibited in Fig. 5.3 for Γ/Γ^* near unity. Equation (6.138c) yields $r_{sc} - r_s(P^c) \sim -[h_{\text{MSD}}/(2r_s(P^c))]g^* \epsilon^{1/3}, \epsilon \to 0$. This formula describes the singular asymptotic behaviour of the curves shown in Fig. 5.6 for Γ/Γ^* tending to unity.

6.3.2 *Logarithmic relaxation*

In this section, the solutions of the equations of motion shall be discussed for states from a special region near some generic glass-transition singularity P^c of type $A_\ell, \ell \geq 3$. For such states, there exists a time interval, where the correlation functions can be represented asymptotically as polynomials of the logarithm of the time. The size of this time interval expands beyond any bound if the state P approaches P^c. These correlators describe a scenario for the evolution of glassy dynamics (Götze and Sperl 2002; Sperl 2003, 2004b), which is qualitatively different from that explained in preceding sections.

(i) The logarithmic-relaxation regime

The following considerations shall start from the reformulation of the equations of motion, which is explained in Sec. 6.1.1. All quantities referring to the higher-order singularity P^c shall be denoted by a superscript c. The array $\boldsymbol{\phi}(t)$ of the M correlators shall be represented by the array $\boldsymbol{f}^c$ of critical arrested parts and by an array $\boldsymbol{g}(t)$ of functions defined in Eqs. (6.1a,b). The left eigenvector $\boldsymbol{a}^*$ of the critical stability matrix and the right one $\boldsymbol{a}$ shall be used with the conventions noted in Eqs. (4.74a–d). The array $\boldsymbol{h}$ of critical amplitudes is defined in Eq. (4.78b). The derivations are based on the assumption that there are intervals of the time t and regions of the complex frequency z, which can be characterized implicitly as indicated by the conditions (6.8a,b). Motivated by the two scaling-limit approaches, it shall be anticipated that the regular contribution $J_q^{\text{reg}}(z)$ to Eq. (6.4b) can be neglected. This assumption will be justified a posteriori in subsection *(v)*. As explained repeatedly before, the resulting simplified equations of motion are scale invariant. In order to ease the notations, this fact shall not be emphasized. In the formulation of the final results, the generality shall be restored by replacing the time t by the rescaled variable t/τ.

In order to specify the scope of the following approach, let us return to the results of Sec. 6.1.3(i). If one intends to use Eq. (6.50a) for the discussion of states near a higher-order singularity, one has to replace $\lambda = 1 - \mu_2$ by unity. Consequently, a solution $g_0^\sigma(t)$ of the equation $\sigma(P) + ST[g_0^\sigma(t)^2](z) - ST[g_0^\sigma(t)](z)^2 = 0$ defines a leading-order solution of the equations of motion. This holds provided such solution reproduces the correct general properties of correlation functions and if it is consistent with the conditions (6.8a,b). The formula (A.70a) shows that the cited equation of motion is solved by $g_0^\sigma(t) = -B_0 \ln(t)$ if and only if $\sigma(P) + (\pi^2 B_0^2/6) = 0$. This function $g_0^\sigma(t)$ is real and decreasing if and only if $B_0 > 0$. Hence, the approach is successful for all states P with negative values for $\sigma(P)$. The function $\sigma(P)$ is defined in Eq. (4.86a); it is identical with the first separation parameter $\sigma_1(P)$, which is introduced in Eq. (6.141a) for the general description of generic glass-transition singularities. For later convenience, it shall be renamed as $\epsilon_1(P)$. The general restriction for the states to be discussed is the inequality

$$0 > \epsilon_1(P) = \sum_q a_q^* D_q^{(0)}(P). \tag{6.168}$$

To be specific, one can consider an analytic path $\epsilon \to P^\epsilon$ in parameter space as discussed in Eqs. (4.84)–(4.86). The derivations of this section are restricted to states with negative distance parameters ϵ.

Combining the results of the preceding paragraph with those of Sec. 6.1.3(i), one concludes that a leading-order solution is defined by the formulas

$$g_0^\sigma(t) = -B_0 \ln(t), \tag{6.169a}$$

$$ST\big[g_0^\sigma(t)\big](z) = -B_0 \ln\big[i \exp(\Gamma_1)/z\big], \tag{6.169b}$$

$$B_0 = \sqrt{-6\epsilon_1(P)/\pi^2}. \tag{6.169c}$$

Here, Eq. (A.69a) is used to evaluate the S-transform of $\ln(t)$. The constant Γ_1 abbreviates the derivative of the gamma function at unity: $\Gamma_1 = d\Gamma(x=1)/dx$. The restrictions (6.8a,b) read

$$|B_0 \ln(t/\tau)| \ll 1, \tag{6.170a}$$

$$\big|B_0 \ln\big[\exp(\Gamma_1)/|z\tau|\big]\big| \ll 1. \tag{6.170b}$$

The formulas of Appendix A.8 guarantee that the power-counting argumentation of Sec. 6.1.3(i) is legitimate. The number B_0 is the small parameter for the discussion. The solubility condition (6.5c) consists of three terms for the leading contribution, which are of order B_0^2. The formulas (6.169a–c) ensure cancellation of these dominant terms. The condition (6.170a) suggests that the length of the time interval for the validity of Eq. (6.169a)) expands to infinity if $|\epsilon|$ tends to zero. Similarly, condition (6.170b) indicates that Eq. (6.169b) can be used for $|z|$ decreasing to zero if P tends to P^c.

Substituting $g_0^\sigma(t)$ into Eq. (6.5d), one obtains the asymptotic result for the logarithmic decay (Götze and Haussmann 1988):

$$\boldsymbol{\phi}(t) \sim \boldsymbol{f}^c - \boldsymbol{h} B_0 \ln(t/\tau), \qquad \epsilon \to -0. \tag{6.171a}$$

This equation is equivalent to the asymptotic expression for the constant loss:

$$\boldsymbol{\chi}''(\omega) \sim \boldsymbol{h} B_0 \pi/2, \qquad \epsilon \to -0. \tag{6.171b}$$

The scale τ denotes the time for the correlators' crossing of the critical arrested parts: $\boldsymbol{\phi}(t=\tau) = \boldsymbol{f}^c$. The results formulate a factorization property. As usual, they can be extended to describe other functions of interest like the fluctuating-force kernels. Since $B_0 \propto \sqrt{|\epsilon|}$, the slope of the above specified $\phi_q(t)$ versus $\log(t)$ curves depends singularly on the distance parameter ϵ. Similarly, the constant-loss intensity decreases singularly if P approaches P^c. This sensitive control-parameter dependence is an important difference of the above results and those for nearly logarithmic decay and nearly constant loss, which are discussed in Sec. 6.1.2(iii).

From the preceding discussion of the general behaviour of the manifolds of glass-transition singularities, one infers the following. The condition $\epsilon_1(P) = 0$ defines a smooth surface piece through P^c. This dividing surface specifies the weak-coupling neighbourhood of P^c by condition (6.168). The higher-order singularity is an endpoint of a surface of generic glass-transition singularities $P^{(2)c}$ of type A_2. This transition surface is located in the weak-coupling neighbourhood; it touches the dividing surface at P^c with infinite curvature. If $P^{(2)c}$ approaches P^c, the critical arrested parts $f_q^{(2)c} = f_q(P^{(2)c})$ vary strongly. According to Eqs. (6.147), (6.164d), the $f_q^{(2)c}$ exhibit a singular dependence on the coordinates of $P^{(2)c}$. As a result, the decay of $\phi_q(t)$ through f_q^c implies a decay towards, through and below a broad distribution of values $f_q^{(2)c}$. There is no distinguished critical arrested part, which could play the familiar role of a plateau. This explains why the relaxation scenario for states P^ϵ approaching P^c is qualitatively different from that, which is explained in Secs. 6.1 and 6.2.

Let us assume a given time interval and a given frequency region. Let P^{ϵ_0} be a state with such a small distance parameter ϵ_0 that the correlators and the spectra are described by Eq. (6.171a) and (6.171b), respectively. If this state is shifted transversally to the path $\epsilon \to P^\epsilon$, it may approach some state $P^{(2)c}$. Consequently, the familiar precursor effects of a generic glass transition will modify qualitatively the simple logarithmic decay law (6.171a) and the constant-loss spectrum (6.171b). Understanding of these modifications of the leading-order results is an essential part of understanding the relaxation scenario for states near P^c. The possibility to replace the symbolic conditions (6.170a,b) by inequalities, will appear as corollary of the intended extension of Eqs. (6.171a,b).

(ii) Extended logarithmic-decay formulas

In this subsection, the procedure of Sec. 6.1.3(i) shall be modified to an asymptotic-expansion approach for the correlators within the special region of states P specified by condition (6.168). The quantity $\sqrt{|\epsilon|}$ is used as the small expansion parameter. It is the goal to extend the leading-order result (6.171a) by incorporation of the next-to-leading expansion terms. The derivations are started by specializing Eq. (6.5a) to the ansatz

$$\boldsymbol{g}(t) = \boldsymbol{a}\big[g_0^\sigma(t) + \tilde{g}(t)\big] + \tilde{\boldsymbol{g}}(t) + O(|\epsilon|^{3/2}), \tag{6.172}$$

with $\tilde{g}(t)$ and $\tilde{g}_q(t), q = 1, \ldots, M$, being of order $|\epsilon|$. The function $g_0^\sigma(t)$ is of order $|\epsilon|^{1/2}$ and obeys the equation

$$\epsilon_1(P) + ST\big[g_0^\sigma(t)^2\big](z) - ST\big[g_0^\sigma(t)\big](z)^2 = 0. \tag{6.173}$$

It can be used to express S-transforms of $g_0^\sigma(t)^2$ in terms of $ST[g_0^\sigma(t)](z)^2$ and vice versa.

Formulas (6.4b–d) for the inhomogeneity components $J_q(z), q = 1, \ldots, M$, can be denoted in the form:

$$J_q(z) = \tilde{J}_q(z) + 2\big[J_q^{(3,1)}(z) + J_q^{(3,2)}(z)\big] + J_q^{(3,3)}(z) + O(\epsilon^2). \tag{6.174a}$$

The leading contributions are of order $|\epsilon|$:

$$\tilde{J}_q(z) = D_q^{(0)}(P) + \Big[\sum_{k_1 k_2} A_{q,k_1 k_2}^{(2)c} a_{k_1} a_{k_2}\Big] ST\big[g_0^\sigma(t)^2\big](z) - a_q^2 ST\big[g_0^\sigma(t)\big](z)^2. \tag{6.174b}$$

There are three types of contributions of order $|\epsilon|^{3/2}$. The first ones are linear expressions of the not yet determined function $\tilde{g}(t)$.

$$J_q^{(3,1)}(z) = \Big[\sum_{k_1 k_2} A_{q,k_1 k_2}^{(2)c} a_{k_1} a_{k_2}\Big] ST\big[g_0^\sigma(t)\tilde{g}(t)\big](z) - a_q^2 ST\big[g_0^\sigma(t)\big](z) ST\big[\tilde{g}(t)\big](z). \tag{6.174c}$$

The second ones are linear expressions of the not yet determined array $\tilde{\boldsymbol{g}}(t)$:

$$J_q^{(3,2)}(z) = \sum_{k_1 k_2} A_{q,k_1 k_2}^{(2)c} a_{k_1} ST\big[g_0^\sigma(t)\tilde{g}_{k_2}(t)\big](z) - a_q ST\big[g_0^\sigma(t)\big](z) ST\big[\tilde{g}_q(t)\big](z). \tag{6.174d}$$

The third type of terms are combinations of the leading-order functions:

$$J_q^{(3,3)}(z) = \Big[\sum_k D_{q,k}^{(1)}(P) a_k\Big] ST\big[g_0^\sigma(t)\big](z) \tag{6.174e}$$
$$+\Big[\sum_{k_1 k_2 k_3} A_{q,k_1 k_2 k_3}^{(3)c} a_{k_1} a_{k_2} a_{k_3}\Big] ST\big[g_0^\sigma(t)^3\big](z) - a_q^3 ST\big[g_0^\sigma(t)\big](z)^3.$$

In order to get $\tilde{\boldsymbol{g}}(t)$, one can use Eq. (6.5b) and replace the inhomogeneity by its leading contribution: $ST[\tilde{g}_q(t)](z) = \sum_k R_{q,k} \tilde{J}_k(z), q = 1, \ldots, M$:

$$\tilde{g}_q(t) = a_q\big\{K_q g_0^\sigma(t)^2 + L_q\big\}. \tag{6.175a}$$

The correction amplitudes are

$$K_q = \sum_k R_{q,k}\Big\{\Big[\sum_{k_1 k_2} A_{k,k_1 k_2}^{(2)c} a_{k_1} a_{k_2}\Big] - a_k^2\Big\}\Big/ a_q. \tag{6.175b}$$

They are specializations to $\lambda^c = 1$ of those defined in Eq. (6.23). The amplitudes $L_q = \sum_k R_{q,k}[D_k^{(0)}(P) - \epsilon_1(P)a_k^2]/a_q$ are functions of order ϵ. In this order, one can use Eqs. (6.77), (6.79a) and write

$$L_q = L_q^{(1)}\epsilon_1(P). \tag{6.175c}$$

An array $\boldsymbol{w}$ of positive normalized weights $w_q = a_q^* a_q, q = 1, \ldots, M$, is mentioned in connection with Eq. (4.78e). The identity (4.75e) for the distinguished-resolvent matrix can be written as $\sum_q w_q R_{q,k}/a_q = 0, k = 1, \ldots, M$. With respect to this weight distribution, the both amplitudes in Eq. (6.175a) describe fluctuations around a vanishing average.

$$\sum_q w_q K_q = 0, \qquad \sum_q w_q L_q = 0. \tag{6.175d}$$

The solubility condition reads $\sum_q a_q^* \tilde{J}_q(z) + 2\sum_q a_q^*[J_q^{(3,1)}(z) + J_q^{(3,2)}(z)] + \sum_q a_q^* J_q^{(3,3)}(z) + O(\epsilon^2) = 0$. The three terms of Eq. (6.174b) yield contributions of order $|\epsilon|$. It is explained in the preceding subsection that Eq. (6.173) ensures cancellation of these terms. The second and third sum collect all contributions to Eq. (6.5c), which are of order $|\epsilon|^{3/2}$. The request of cancellation of these contributions ensures that the equations of motion are solved in next-to-leading order. The request is equivalent to the validity of $(-1/B_0)\{ST[g_0^\sigma(t)\tilde{g}(t)](z) - ST[g_0^\sigma(t)](z)ST[\tilde{g}(t)](z)\} = f(z)$. Here, the abbreviation is introduced: $f(z) = \sum_q a_q^*[J_q^{(3,2)}(z) + J_q^{(3,3)}(z)]/B_0$. To arrive at this result, the identities $\sum_q a_q^* a_q^2 = 1$ and $\sum_q \sum_{k_1 k_2} a_q^* A_{q,k_1k_2}^{(2)c} a_{k_1} a_{k_2} = \lambda(P^c) = 1 - \mu_2(P^c) = 1$ are exploited.

In order to formulate the results of the preceding paragraph lucidly, a linear integral transformation $IT[F(t)](z)$ shall be defined for determining functions:

$$IT\big[F(t)\big](z) = ST\big[\ln(|t|)F(t)\big](z) - ST\big[\ln(|t|)\big](z)ST\big[F(t)\big](z). \tag{6.176}$$

The mapping will be applied to functions with standard symmetries $F(t) = F(t)^* = F(-t)$. Therefore, as done before, equations shall be noted for $t > 0$ and $\mathrm{Im}\ z > 0$ only. One arrives at the following conclusion. The ansatz (6.172) provides a next-to-leading-order solution of the equations of motion if Eqs. (6.175a–c) are used for the array $\tilde{\boldsymbol{g}}(t)$ and if function $\tilde{g}(t)$ is a solution with standard symmetries of the linear integral equation

$$IT\big[\tilde{g}(t)\big](z) = f(z). \tag{6.177a}$$

If one substitutes Eq. (6.175a) into Eq. (6.174d), one can express $J_q^{(3,2)}(z)$ in terms of $g_0^\sigma(t)$. The inhomogeneity $f(z)$ can be quantified by two numbers, which are determined by the mode-coupling coefficients at the critical point P^c. They shall be denoted by ζ and ξ. Furthermore, there appears $ST[g_0^\sigma(t)](z)$ multiplied by a smooth function of P, which vanishes at the singularity. This function shall be denoted by $\epsilon_2(P)$. For states on the above mentioned path, one can write $\epsilon_2(P^\epsilon) = C'\epsilon + O(\epsilon^2)$. The conventions are chosen so that

$$\begin{aligned} f(z) = \Big\{ & 2\zeta\Big(ST\big[g_0^\sigma(t)^3\big](z) - ST\big[g_0^\sigma(t)\big](z)^3\Big) \\ & - 2(\zeta - \xi)ST\big[g_0^\sigma(t)^3\big](z) + \epsilon_2(P)ST\big[g_0^\sigma(t)\big](z)\Big\}\Big/(2B_0). \end{aligned} \tag{6.177b}$$

For the number ξ, one finds Eq. (6.28a). The number ζ is given by Eq. (6.28b) with $\lambda^c = 1$. Disentangling the formula (6.140b), one finds

$$\mu_3 = 2(\zeta - \xi). \tag{6.178}$$

The function $\epsilon_2(P)$ is a second measure of the position of the state P relative to the singularity P^c.

$$\epsilon_2(P) = \sum_{q,k} a_q^* D_{q,k}^{(1)}(P) a_k + 2\epsilon_1(P) \sum_q a_q^* a_q^2 K_q$$
$$+ 2 \sum_q a_q^* \Big\{ \Big[\sum_{k_1 k_2} A_{q,k_1 k_2}^{(2)c} a_{k_1} a_{k_2} L_{k_2} \Big] - a_q^2 L_q \Big\}. \tag{6.179a}$$

Disentangling this formula and using Eqs. (6.141a), (6.145) for the first two relevant separation parameters, one can write

$$\epsilon_2(P) = \sigma_2(P) + c_{12}\sigma_1(P), \tag{6.179b}$$

$$c_{12} = 2 \sum_{qkk_1k_2} a_q^* \big[a_q R_{qk} A_{k,k_1 k_2}^{(2)c} a_{k_1} a_{k_2} - A_{q,k_1 k_2}^{(2)c} a_{k_1} R_{k_2 k} a_k^2 \big], \tag{6.179c}$$

Because of Eqs. (6.169a,c), the function $f(z)$ can be expressed in terms of S-transforms of powers of $\ln(t)$. According to Eqs. (A.69b), (A.70b), these S-transforms are polynomials of $\ln(i/z)$. One gets

$$f(z) = \sum_{n=0}^{3} a_n \ln(i/z)^n. \tag{6.180a}$$

The four coefficients are linear combinations of the functions $\epsilon_1(P)$ and $\epsilon_2(P)$:

$$a_0 = \big[(6\zeta/\pi^2)(\Gamma_3 - \Gamma_1^3) - (3\mu_3/\pi^2)\Gamma_3 \big] \epsilon_1(P) - (\Gamma_1/2)\epsilon_2(P), \tag{6.180b}$$

$$a_1 = \big[3\zeta - (9\mu_3/\pi^2)\Gamma_2 \big] \epsilon_1(P) - (1/2)\epsilon_2(P), \tag{6.180c}$$

$$a_2 = -(9\mu_3/\pi^2)\Gamma_1 \epsilon_1(P), \qquad a_3 = -(3\mu_3/\pi^2)\epsilon_1(P). \tag{6.180d}$$

The numbers Γ_k abbreviate the kth derivative of the gamma function at unity.

In Appendix A.8, it is shown that there are polynomials $p_n(x)$ of degree n with the following properties. There holds $p_n(x = 0) = 0$ and $IT[p_n(\ln(t))](z) = (n\pi^2/6)\ln(i/z)^{n-1}$. Consequently, $\tilde{g}(t) = \sum_{n=1}^{4} [6a_{n-1}/(n\pi^2)] p_n(\ln(t))$ is a solution of Eqs. (6.177a), (6.180a). The required four polynomials are listed in Eqs. (A.72a), (A.73a–c). One obtains

$$\tilde{g}(t) = \sum_{n=1}^{4} \tilde{B}_n \ln(t)^n \tag{6.181a}$$

with coefficients

$$\tilde{B}_1 = (0.44425\zeta - 0.065381\mu_3)\epsilon_1(P) - 0.22213\epsilon_2(P), \tag{6.181b}$$

$$\tilde{B}_2 = (0.91189\zeta + 0.068713\mu_3)\epsilon_1(P) - 0.15198\epsilon_2(P), \tag{6.181c}$$

$$\tilde{B}_3 = -0.13504\mu_3\epsilon_1(P), \qquad \tilde{B}_4 = -0.046197\mu_3\epsilon_1(P). \tag{6.181d}$$

Combining the preceding results for $\tilde{g}(t)$ and $\tilde{\boldsymbol{g}}(t)$ with Eq. (6.5d) and restoring the time scale τ, one obtains the desired formulas for the correlation functions for labels $q = 1, \ldots, M$ in leading and next-to-leading asymptotic description:

$$\phi_q(t) = \left[f_q^c + \hat{f}_q\right] + h_q\Big\{-B_0' \ln(t/\tau) + B_2(q)\ln(t/\tau)^2 + \tilde{B}_3 \ln(t/\tau)^3 + \tilde{B}_4 \ln(t/\tau)^4\Big\}. \tag{6.182a}$$

The numbers

$$\hat{f}_q = h_q L_q = h_q L_q^{(1)} \epsilon_1(P) \tag{6.182b}$$

specify corrections to the critical arrested parts f_q^c. The coefficient B_0' is a modification of the factor B_0 of the logarithmic decay law,

$$B_0' = B_0 - \tilde{B}_1. \tag{6.182c}$$

If P^c is a singularity of type A_ℓ for $\ell \geq 4$, the coefficients $\tilde{B}_3$ and $\tilde{B}_4$ are zero. For $\ell = 3$, the numbers $\tilde{B}_2$ and $\tilde{B}_3$ determine deformations of the leading order results by introducing an inflection point for the ϕ_q versus $\log(t)$ curve. For times t near τ, these deformations are small compared to the ones caused by the term proportional to $\ln(t/\tau)^2$. The latter contribution describes the qualitative change of the logarithmic-decay law. It is quantified by the coefficient

$$B_2(q) = \tilde{B}_2 + K_q B_0^2. \tag{6.182d}$$

The preceding equations formulate the desired extension of Eq. (6.171a). The error of the approximation (6.182a) has the form $|\epsilon|^{3/2}\tilde{g}_q'(t) + O_q(\epsilon^2)$. Here, $\tilde{g}_q'(t)$ denotes a polynomial of degree seven of the variable $x = \ln(t/\tau)$.

The results (6.175d) imply

$$\sum_q \left[w_q/h_q\right]\hat{f}_q = 0. \tag{6.183a}$$

Not all of the M corrections $\hat{f}_q$ can have the same sign, since $[w_q/h_q] > 0, q = 1, \ldots, M$. Similarly, the amplitudes K_q cannot have the same sign. The amplitudes $B_2(q)$ fluctuate around $\tilde{B}_2$:

$$\sum_q w_q B_2(q) = \tilde{B}_2. \tag{6.183b}$$

As explained repeatedly before, the result (6.182a) can be generalized to formulas for correlators of other variables X of interest and for the mean-squared displacement. The amplitudes $f_q^c, h_q, \hat{f}_q$ and K_q have to be altered to ones referring to X. But, the coefficients $B_0, \tilde{B}_1, \ldots, \tilde{B}_4$ are shared by all functions. Ignoring errors of order $\epsilon\mu_3 \ln(t/\tau)^3, \epsilon\mu_3 \ln(t/\tau)^4$ and $|\epsilon|^{3/2}$, there holds

$$\Big[\phi_X(t) - (f_X^c + \hat{f}_X)\Big]/h_X = -B_0' \ln(t/\tau) + B_2(X)\ln(t/\tau)^2, \tag{6.184a}$$

with $B_2(X) = \tilde{B}_2 + K_X B_0^2$. Using Eqs. (A.69a), (A.70a), one obtains the equivalent formula for the loss spectra:

$$\chi_X''(\omega)/h_X = (B_0'\pi/2) + (B_2(X)\pi)\ln\big[\omega\tau/\exp(\Gamma_1)\big]. \tag{6.184b}$$

Within the margin δ^* for the relative error, the logarithmic decay law is valid in a time interval specified by

$$|\ln(t/\tau)| \le \delta^* |B_0'/B_2(X)|. \tag{6.185a}$$

Modifications of the constant loss do not exceed the relative size δ^*, provided the frequencies are restricted by:

$$\Big|\ln\Big[\omega\tau/\exp(\Gamma_1)\Big]\Big| \le \frac{1}{2}\delta^* |B_0'/B_2(X)|. \tag{6.185b}$$

The inequalities are the searched-for substitute for the conditions (6.170a,b). If the state P tends to P^c, the size of the intervals of $\log t$ and of $\log\omega$ for the validity of the leading-order results expands proportional to $1/\sqrt{|\epsilon|}$. The size of the specified $\log(\omega)$ interval for the constant loss is half the size of the $\log(t)$ interval for the logarithmic correlation decay. This reduction of the range of validity for the asymptotic result in the frequency domain relative to that in the time domain is analogous the reduction formulated in Eq. (6.99) for the first-scaling-law asymptote. These formulas hold if the dominant corrections are determined by the $\ln(t/\tau)^2$ term. If variable X is chosen so that $B_2(X) = 0$, the range of validity of the leading-order formulas is determined by the corrections proportional to $|\epsilon|^{3/2}$, which are not evaluated here.

Within the error margin δ^* and within the specified time interval, the $[\phi_X(t) - (f_X^c + \hat{f}_X)]/h_X$ versus $\log(t)$ curves collapse on the variable-independent curve for the function $[-B_0' \ln(t/\tau)]$. Within the interval of frequencies characterized by the inequality (6.185b), the normalized susceptibility spectra $\chi_X''(\omega)/h_X$ agree with the constant $[B_0'\pi/2]$ up to a possible relative error not larger than δ^*. Outsides these intervals, the deviations of the curves for different X may exceed the specified error. There are the same fanning-out rules as explained in connection with Eq. (6.100b).

(iii) Logarithmic relaxation near an A_3 singularity

In this subsection, the evolution of the correlator $\phi(t)$ of a one-component model shall be considered in order to demonstrate some features of the glassy relaxation, which are explained by the above-derived logarithmic-expansion formulas. The second version of the equations of motion shall be used: $t_{\text{mic}}\partial_t\phi(t) + \phi(t) + \int_0^t m(t-t')\partial_{t'}\phi(t')dt' = 0$. The mode-coupling kernel is specified by two coupling coefficients v_1 and v_3 as noted in Eq. (4.32c). The manifold of glass-transition singularities is exhibited in Fig. 4.3. The parameters for the generic glass-transition singularity P^c of type A_3 are noted in Eq. (4.69a). Equation (6.28b) implies $\zeta = 1/2$ for all 1-component models. The deviations of the coupling constants from the critical-point values shall be used as external control parameters: $\xi_1 = \hat{v}_1 = v_1 - v_1^{cc}, \xi_2 = \hat{v}_3 = v_3 - v_3^{cc}$. There holds $\xi_1^c = \xi_2^c = 0$ and

$$\epsilon_1(P) = (2/81)\big[9\hat{v}_1 + \hat{v}_3\big], \qquad \epsilon_2(P) = (4/27)\big[3\hat{v}_1 + \hat{v}_3\big]. \tag{6.186a}$$

The dashed straight line shows the states obeying $\epsilon_1(P) = 0$. The weak-coupling neighbourhood of the critical point deals with the states below this line. There holds $\tilde{B}_2 = B_2$. The dashed-dotted line shows the states of vanishing dominant correction:

$$B_2 = 0 : \hat{v}_1 = 0.9298\epsilon, \qquad \hat{v}_3 = 3.3750\epsilon. \tag{6.186b}$$

Figure 6.20 exhibits the evolution of the dynamics for states P approaching P^c on the line of vanishing B_2. Within the error margin $\delta^* = 0.05$, the correlator with label $n = 2$ exhibits the logarithmic decay law (6.171a) for times between t_{mic} and 1000 t_{mic}. Decreasing ϵ by a factor 16, one gets the state with label $n = 4$. In this case, the logarithmic-decay law describes the correlator for a time increase of about six orders of magnitude. With decreasing $|\epsilon|$, there appears a time interval of increasing length between the end of the transient regime and the beginning of the logarithmic-decay regime. Here, the correlators are close to the one for the critical point. This behaviour is an implication of the regularity properties of the equations of motion. For the states with labels $n = 2, 3$, and 4, Eqs. (6.182a–d) describe $\phi(t)$ also within this interval of nearly critical decay. For large $|\epsilon|$, it may happen that the range of validity of the leading-order description of the correlator exceeds that for the description with inclusion of the leading-correction term. This is due to cancellation of higher-order correction terms. The figure demonstrates such possibility for the state with label $n = 1$. The correlator for this state is described within the accuracy of the drawing for times decreasing to $t = 1$ if the expansion (6.182a) is extended by inclusion of the seven terms, which are of order $|\epsilon|^{3/2}$ (Götze and Sperl 2002).

Let us consider the evolution of the relaxation due to shifts of the states transversely to the line specified by Eq. (6.186b). The upper panel of Fig. 4.3 exhibits such states by triangles. The states are labelled by n' and they are located on the straight line defined by $\epsilon_1(P) = -0.0182$. For seven such states, Fig. 6.21 exhibits the corresponding correlators. The $n' = 2$ state is identical

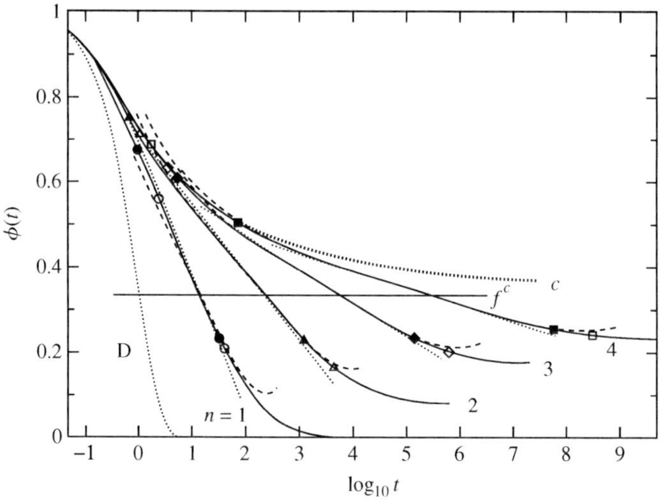

FIG. 6.20. The full lines show correlators $\phi(t)$ for a one-component model. The mode-coupling kernel is defined by Eq. (4.32c). The second version of the equation of motion is used with a time scale $t_{\text{mic}} = 1$. The states with label n are defined by Eq. (6.186b) with $\epsilon = -1/4^n$; they are marked by crosses in the upper panel of Fig. 4.3. The dotted line with label c shows the critical correlator for the A_3 singularity P^c. The horizontal line marks the critical arrested part $f^c = 1/3$. The dotted line with label D exhibits the correlator $\phi^{(0)}(t) = \exp[-(t/t_{\text{mic}})]$. The straight dotted lines show the leading-asymptotic results (6.171a) and the dashed lines present the results (6.182a) for the extended asymptotic description. The filled and open symbols, respectively, mark the points where the cited asymptotic-expansion results differ by 5% from the correlators. For every state, the time scale τ is fitted so that there holds $\phi(t = \tau) = f^c$. Reproduced from Götze and Sperl (2002).

with the one used above with label $n = 2$. The curves for $n' = 1$ and $n' = 3$ demonstrate deviations from the logarithmic decay, which are caused by the dominant correction term $B_2 \ln(t/\tau)^2$. The curves for the states for $n' = 1, 2, 3$ are described well by formulas (6.182a–c) for a time increase by about three orders of magnitude.

The condition $B_2(X) = 0$ defines a dividing surface of the weak-coupling neighbourhood of P^c into parts of qualitatively different $\phi(t)$ versus $\log(t)$ curves. For states with $B_2 > 0$ and times outside the transient regime, the $\phi(t)$ versus $\log(t)$ curves are upward bent. The results for states with $n' = 1$ and $n' = 2$ demonstrate that the glassy relaxation for states with $B_2 \geq 0$ is qualitatively different from that explained on preceding pages for states approaching a generic liquid–glass-transition point. There is control-parameter-sensitive

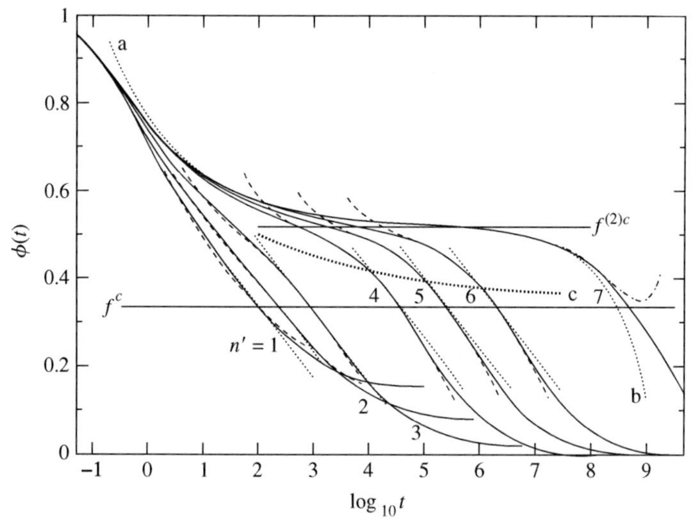

FIG. 6.21. The full lines show correlators calculated for the one-component model, which is defined in the caption of Fig. 6.20. The states labelled n' obey the condition $\epsilon_1(P) = -0.0182$. The states $n' = 1 - 4$ have coordinates (v_1, v_3) = (1.1169, 2.7141), (1.0669, 3.1641), (1.0169, 3.6141), and (0.9669, 4.0641), respectively, and they are marked by triangles in the upper panel of Fig. 4.3. The states with labels 5, 6 and 7 have the coordinates (0.9599, 4.1271), (0.9569, 4.1541), and (0.9549, 4.1721), respectively. The dotted line with label c shows the correlator at the A_3 singularity P^c. The line marked f^c shows the corresponding critical arrested part 1/3. The line labelled $f^{(2)c}$ exhibits the plateau for the A_2 singularity $P^{(2)c}$ explained in connection with Eq. (6.187). The asymptotic law (6.17b) for the state $P^{(2)c}$ is shown as dotted line with label a; the time scale t_0 is fitted to the numerical solution. The corresponding result (6.120a) for von Schweidler's law is shown as dotted line with label b, and its extension (6.121) is presented as dashed-dotted curve. The time scale for these second-scaling-law formulas is adjusted to the decay of the correlator for state $n' = 7$. The dotted straight lines and the dashed curves show the logarithmic decay law (6.171a) and the extended-asymptotic law (6.182a), respectively. The time scale τ is adjusted for each state with label $n' = 1, \ldots, 6$ so that $\phi(t = \tau) = f^c$. Reproduced from Götze and Sperl (2002).

stretched correlation decay without inflection point, which could be interpreted as precursor of a plateau. There is no possibility to identify meaningfully a part of the curves, which could be fitted by a Kohlrausch function.

A negative correction factor B_2 causes an inflection point of the $\phi(t)$ versus $\log(t)$ curves for some time smaller than τ. Figure 6.21 demonstrates this feature for states with labels $n' = 3$ and $n' = 4$. This feature is similar to that described by the first scaling law for the crossing of some plateau $f^{(2)c} > f^c$. This plateau value is the critical arrested part of the correlator at some generic glass-transition singularity $P^{(2)c}$. For times exceeding τ considerably, $\phi(t)$ approaches its long-time asymptote. Since the correlator is completely monotonic, the curve is upward bent for t/τ much larger than unity. Hence, there is a second inflection point for $t/\tau > 1$. The decay of the correlator for times beyond the first inflection-point position is similar to that exhibited by the below-plateau decay for states near $P^{(2)c}$. A Kohlrausch-law fit is possible for a considerable part of the correlators. For states with $B_2 < 0$, the relaxation pattern is similar to that explained in Secs. 6.1 and 6.2 for liquid states near some generic A_2 bifurcation.

Let us corroborate the preceding conclusion. The straight line through the states with label n' in Fig. 4.3 intersects the liquid-glass-transition line at a critical point $P^{(2)c}$ with coordinates $v^{(2)c} = 0.95466$ and $v^{(2)c} = 4.17407$. The corresponding values for the critical arrested part, the exponent parameter, the critical exponent, and the von Schweidler exponent, respectively, are:

$$f^{(2)c} = 0.520, \quad \lambda^c = 0.719, \quad a = 0.318, \quad b = 0.608. \tag{6.187}$$

It is demonstrated in Fig. 6.21 for the state with label $n' = 4$ that $\phi(t)$ exhibits the familiar decay towards the plateau $f^{(2)c}$. For times between $3t_{\text{mic}}$ and $30t_{\text{mic}}$, it is close to the critical correlator. This interval of critical decay expands if P approaches $P^{(2)c}$, as is demonstrated for the succession of states with labels 5, 6 and 7. Translating a replica of curve $n' = 6$ horizontally, one can check that the results exhibit the familiar evolution pattern for the superposition principle. Results for von Schweidler's law and its extension are added to the curve for $n' = 7$. Thereby, it is demonstrated that these formulas describe about 10% and 20%, respectively, of the below-$f^{(2)c}$ decay. The curves for states with $n' = 4, 5$ and 6 show that the extended-logarithmic-decay formula describes the about 60% of the below-plateau decay between 0.5 and 0.2. The asymptotic expansion result (6.182a) for states near the A_3 singularity P^c describes the below-plateau decay for states near $P^{(2)c}$ much better than does the asymptotic-expansion result (6.121) for states near A_2 singularities.

The relaxation of correlators for states with negative B_2 exhibits the interplay of features caused by the A_3 singularity P^c with those caused by the nearby A_2 singularities $P^{(2)c}$. The logarithmic decay laws appear as features of shape function $\tilde{\phi}(\tilde{t})$ for the superposition principle connected with the $P^{(2)c}$-induced glassy dynamics. The approach of $P^{(2)c}$ towards the nearby P^c is connected with an approach of $\mu_2 = 1 - \lambda^c$ to zero. The denominator of Eq. (6.86b) decreases and the correction term $\kappa(-b)$ in Eq. (6.121) increases. Approaching P^c, the range of applicability of the von Schweidler expansion (6.122a) shrinks.

The generic A_3 singularity P^c is an endpoint of a surface of discontinuous glass–glass transitions. Consequently, a liquid region near P^c is bounded by two surfaces of A_2 singularities, which form a corner. There is a surface of liquid–glass-transition states $P^{(2)c}$ with arrested parts $f^{(2)c}$, which are larger than f^c. The implications of such singularities are discussed in the preceding three paragraphs. In addition, there is a surface of liquid–glass-transition states $P^{(2)c\prime}$ with arrested parts $f^{(2)c\prime}$ smaller than f^c. There are liquid states in the weak-coupling neighbourhood of P^c, whose dynamics exhibits precursor effects of the transition at $P^{(2)c\prime}$. These effects influence the correlator for times where $\phi(t)$ is near and below $f^{(2)c\prime}$. The singularity diagram of Fig. 4.3 exhibits the second kind of surface as vertical line specified by $v_1 = 1$. For the model under discussion, the states $P^{(2)c\prime}$ are continuous liquid–glass transitions. As explained in Sec. 6.1.3 *(v)*, stretched relaxation is caused by the critical decay. Since $\lambda^c = 0$, the scaling law for the elementary model from Sec. 4.2.4 describes the evolution of a regime of $1/\sqrt{t}$ behaviour. The correlators for the states with label $n' \geq 3$ exhibit this tail (Götze and Haussmann 1988).

(iv) Correlation decay for a liquid with short-ranged attraction potentials

Within the microscopic MCT models, one can identify such implications of the above derived formulas, which reflect structure-specific details. In the following, this aspect shall be demonstrated for the wave-number dependence of the density-fluctuation correlators and for the time-dependence of the mean-squared displacement for states P near the glass-transition singularity P^c of type A_4 of the square-well system (SWS). The model and the units for its description are defined in connection with Figs. 4.9 and 4.18.

The considerations shall be restricted to states which have the attraction-range parameter δ^* of P^c. These states are specified by two external control parameters, namely the packing fraction φ and the attraction-strength parameter Γ. The values φ^* and Γ^* for P^c are discussed in connection with Eq. (4.106). The manifold of glass-transition singularities is shown in Fig. 4.9 as cut $\delta = \delta^*$, and it is reproduced in Fig. 6.22. It is obtained as a smooth mapping of the manifold described by Eqs. (6.164a–d). It is located within the region defined by the condition (6.168). The boundary of the liquid region consists of two branches of generic liquid–glass-transition points and of the A_4 singularity. For states $P^{(2)c}$ on the branch with $\Gamma > \Gamma^*$, the critical Debye–Waller factors $f_q^{(2)c}$ exceed f_q^c. For transition points $P^{(2)c\prime}$ on the branch with $\Gamma < \Gamma^*$, the critical arrested parts $f_q^{(2)c\prime}$ are smaller than the ones at P^c:

$$f_q^{(2)c} > f_q^c > f_q^{(2)c\prime}, \qquad q = 1, \ldots, M. \tag{6.188a}$$

The critical values for the Lamb–Mössbauer factors f_q^s demonstrate the same inequalities. For the values of the critical localization length, the inequalities are reversed

$$r_{sc}^{(2)} < r_{sc} < r_{sc}^{(2)\prime}. \tag{6.188b}$$

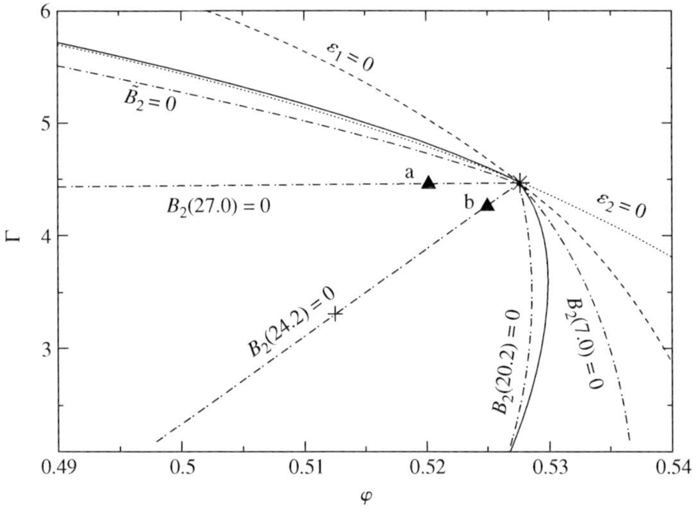

FIG. 6.22. Plane of states for a square-well system for an attraction-range parameter δ^* equal to that of the A_4 singularity P^c. The coordinates for the states P_a and P_b, which are shown by filled triangles, are noted in the caption of Fig. 4.9. The star shows the state P^c. The full lines exhibit the curves of generic A_2 glass-transition singularities, which form the liquid boundary. The dashed and dotted curve show the lines defined by $\epsilon_1(P) = 0$ and $\epsilon_2(P) = 0$, respectively. The dashed-dotted lines show the set of states of vanishing correction factor $B_2(q)$ and of vanishing $\tilde{B}_2$ as indicated. Reproduced from Sperl (2003).

The critical arrested parts f_q^c and the critical amplitudes h_q are exhibited in Fig. 4.19 as lines. For wave numbers above the structure-function-peak position, the quantities f_q^{sc} and h_q^s for the tagged-particle dynamics interpolate f_q^c and h_q, respectively. The gross features of the wave-number dependence of the mentioned quantities are the same as those demonstrated in Figs. 4.6, 5.4 and 5.5 for the hard-sphere system (HSS). However, the curves for the SWS at P^c are spread on a larger wave-number range. For example, the hump of the h_q versus q curve is located for values of q between 20 and 30 for the SWS. For the HSS, the corresponding interval extends from about 10 to 15. This is a manifestation of the fact that the critical localization length is shorter for the system with strong short-ranged attraction than for the system with pure repulsive forces. Figure 5.6 shows that r_{sc} at P^c is about half of the value at the critical point of the HSS.

Analogous statements hold for the correction amplitudes K_q and K_q^s. Because of (6.175d), these functions must have a zero $q^{(0)}$ and $q_s^{(0)}$, respectively, within the region of wave numbers relevant for the arrest. Figures 4.6 and 5.5 show that this distinguished wave number is located near 15 for the HSS. The amplitudes increase monotonically from -1 to 1 if q increases from about 5 to about 22. For the SWS, the zero $q^{(0)} \approx q_s^{(0)}$ is located near 30 and the amplitudes reach the value 1 at about 44 (Sperl 2003). The cited calculations yield the value $\xi = \zeta = 0.122$ and also the functions $\epsilon_1(P)$ and $\epsilon_2(P)$. Thereby, the coefficients $B_0, \tilde{B}_1$ and $\tilde{B}_2$ are determined. Equations (6.182a) for $\phi_q(t)$ and the corresponding ones for the tagged-particle functions can be used with $\tilde{B}_3 = \tilde{B}_4 = 0$.

For a neighbourhood of P^c, a new system of coordinates can be defined by $x = \epsilon_1(P)$ and $y = \epsilon_2(P)$. The weak-coupling part of the neighbourhood can be divided by a curve of states, which obey the condition $\tilde{B}_2 = 0$. The new system describes the curve by a straight line through P^c, which is located in the quadrant with $x < 0$ and $y < 0$. For the above identified distinguished wave number $q^{(0)}$, there holds

$$B_2(q^{(0)}) = \tilde{B}_2. \tag{6.189}$$

For states P on the dividing curve, the correction term on the right-hand side of Eq. (6.182a) vanishes. The extended-logarithmic description reduces to a logarithmic decay law: $\phi_{q^{(0)}}(t) = [f^c_{q^{(0)}} + \hat{f}_{q^{(0)}}] - B_0' \ln(t/\tau)$. If such state is shifted away from the dividing curve by increasing or decreasing its coordinate $x = \epsilon_1(P)$, function $\tilde{B}_2$ increases to positive values or decreases to negative ones, respectively. From the discussion of the preceding subsection one concludes also that a decreasing negative B_2 causes the evolution of the relaxation scenario of $\phi_{q^{(0)}}(t)$, which demonstrates the approach to a liquid-glass-transition point $P^{(2)c}$ on the branch with $\Gamma > \Gamma^*$. One concludes that the coordinate system is right handed. States with $\tilde{B}_2 < 0$ or $\tilde{B}_2 > 0$ are obtained by rotating the dividing curve clockwise or anti clockwise, respectively.

For a given wave number q, there is the set of distinguished states obeying the condition $B_2(q) = 0$. According to Eqs. (6.169c), (6.182d), these states are determined as solution of the equation

$$\tilde{B}_2 = K_q\big[6\epsilon_1(P)/\pi^2\big]. \tag{6.190a}$$

For these states, the extended logarithmic-decay formula (6.182a) simplifies to

$$\phi_q(t) = [f_q^c + \hat{f}_q] - B_0' h_q \ln(t/\tau). \tag{6.190b}$$

This logarithmic-decay law differs from the leading-order formulas (6.169a,c) by the correction $\hat{f}_q$ of the critical arrested parts and by the addition $(-\tilde{B}_1)$ to the factor B_0 in Eq. (6.182c). The range of validity of this formula is determined by corrections of order $\epsilon^{3/2}$. Together with Eq. (6.181c), the condition (6.190a) defines a linear relation between $\epsilon_1(P)$ and $\epsilon_2(P)$. Within the above introduced new coordinate system, the distinguished states form a straight line through P^c. For the region with $\epsilon_1(P) < 0$, one concludes from the preceding paragraph that the line rotates clockwise if K_q increases. If the wave number increases from a value near the structure-factor peak position, say $q = 7.5$, to $q^{(0)}$ and above, the line of states of vanishing $B_2(q)$ rotates clockwise. This bundle of lines is obtained as smooth mapping of a bundle of straight lines in the $x - y$ plane. Figure 6.22 exhibits five examples of such lines.

A bundle of lines with parameter k can be defined as solution of the equation

$$B_2(q) = k\epsilon_1(P). \tag{6.191}$$

These lines are obtained as smooth mapping of straight lines in the $x - y$ plane. With k increasing above zero, the lines are obtained by clockwise rotation of the line of distinguished states. On these lines, the correction coefficient $B_2(q)$ is negative. If k decreases below zero, lines with positive $B_2(q)$ are obtained by anticlockwise rotation of the line of distinguished states. In both cases, $|B_2(q)|$ increases monotonically with the angle of rotation.

Let us consider the state P_b on the line $B_2(24.2) = 0$, which is marked in Fig. 6.22 by a filled triangle with label b. Its coordinates differ by about 0.5% from the ones of P^c. Panel (b) of Fig. 4.18 shows a set of correlators for this state. On an accuracy level of 5%, Eq. (6.190b) describes the correlation decay of $\phi_{24.2}(t)$ between about 0.8 and about 0.2. Within the 5% error margin, the logarithmic decay is exhibited for a time increase of five orders of magnitude. The state marked in Fig. 6.22 by a plus has a packing fraction below the critical value φ^c of the HSS. If the time increases from 1 to 100, the correlator $\phi_{24.2}(t)$ for this state decreases also from 0.8 to 0.2. Also this decay can be described by Eq. (6.190b) within an error margin of 5% (Sperl 2003).

From the preceding discussion of Eq. (6.191), one concludes the following. If the wave number is decreases below 24.2, the correlator for state P_b evolves a decreasing negative coefficient for the correction term of the logarithmic decay; for example, $B_2(4.2) < B_2(20.2) < 0$. As explained in the preceding subsection, the correlators evolve the pattern for states near a liquid-glass transition point with critical arrested Debye–Waller factors $f_{4.2}^{(2)c} > f_{20.2}^{(2)c} > f_{24.2}^c$. The $\phi_q(t)$ versus $\log(t)$ for values near $f_q^{(2)c}$ differ from the corresponding ones shown in Fig. 6.21 for $n' = 3$ and $n' = 4$, however. The plateaus $f_q^{(2)c}$ are large and, therefore, the correction amplitudes K_q are large and negative. Consequently, the decay for state $P^{(2)c}$ exhibits the strong-coupling-induced modifications of the t^{-a} law, which are explained in Sec. 6.1.2(iii). The decay curves are similar to the ones exhibited in Fig. 5.26 for large coupling constant v_A. Thereby, a qualitative understanding is established for the evolution of the relaxation scenario, which is shown in panel (b) of Fig. 4.18 for the sequence of the three mentioned wave numbers. If the wave number increases above 24.2, the correlators for state P_b develop an increasing positive coefficient $B_2(q)$. As explained in connection with the result shown in Fig. 6.21 for the state $n' = 1$, the correlators get increasingly suppressed and the $\phi_q(t)$ versus $\log(t)$ curves are upward bent. This behaviour is demonstrated in panel (b) of Fig. 4.18 by the correlators for the three largest values for q. The analytical result (6.184a) describes quantitatively the correlation decay for times between $10t_{\text{mic}}$ and $10^5 t_{\text{mic}}$. For the region of wave numbers $q \geq 24.2$, these holds $B_2(q) \geq 0$, and the relaxation exhibits the unmistakable features caused by a glass-transition singularity of higher order.

The upper panel of 4.18 demonstrates the evolution scenario for the correlators for a state P_a, which is located on the line of distinguished states obeying the condition $B_2(27.0) = 0$. The considerations from above explain the scenario; and Eq. (6.184a) describes the functions for an increase of the time by three orders of magnitude.

The factorization property implies that the logarithmic-relaxation law for the mean-squared-displacement function reads

$$\Delta_s(t) \sim r_{sc}^2 + h_{\text{MSD}} B_0 \ln(t/\tau), \qquad \epsilon \to -0. \tag{6.192a}$$

Complementing this leading-order result by the leading corrections, one gets the extended-logarithmic description:

$$\Delta_s(t) = \left[r_{sc}^2 - \hat{r}_s^2\right] + h_{\text{MSD}}\Big\{B_0' \ln(t/\tau) - B_2(\text{MSD}) \ln(\text{t}/\tau)^2\Big\}. \tag{6.192b}$$

Here, $\hat{r}_s^2$ is a correction of the critical arrested part r_{sc}^2 and B_0' denotes the corrected amplitude of the logarithm according to Eq. (6.182c). The analogue of Eq. (6.182d) determines the coefficient of the correction:

$$B_2(\text{MSD}) = \tilde{\text{B}}_2 + \text{K}_{\text{MSD}}\text{B}_0^2. \tag{6.192c}$$

Using the conventional double-logarithmic presentation, Eq. (6.192a) determines a downward-bent curve. Examples are shown as dotted lines in Fig. 6.23. The curves in this figure are calculated for three states on a line, which is very close to the one shown in Fig. 6.21 for the states obeying $B_2(24.2) = 0$. The $\log \Delta_s$ versus $\log(t)$ curves differ qualitatively from their asymptotic results; they are almost straight for large time intervals. For the state with label $n = 2$, the glassy-relaxation part of $\Delta_s(t)$ is stretched over the interval $t_{\text{mic}} < t < 10^8 \times t_{\text{mic}}$. But the logarithmic asymptote describes the MCT solution for about a two-decade time interval only.

In connection with Fig. 6.5, the following features of the mean-squared-displacement function $\Delta_s(t)$ are explained for the HSS. Strong mode-coupling effects of the correlation decay of tagged-particle-density fluctuations with small wave number cause large negative correction amplitudes K_{MSD}. As a result, there are strong modifications of the leading-asymptotic result for the increase of $\Delta_s(t)$ towards the critical arrested part $[r_{sc}^{(2)}]^2$ for the liquid–glass-transition state $P^{(2)c}$. The cited reasoning applies also for the discussion of the SWS, where the result (6.188b) has to be observed. The line of states obeying $B_2(\text{MSD}) = 0$ is obtained from the $\tilde{B}_2 = 0$ line by anticlockwise rotation. It is located near the line, which is shown in Fig. 6.22 for states with $B_2(7.0) = 0$. Liquid states on

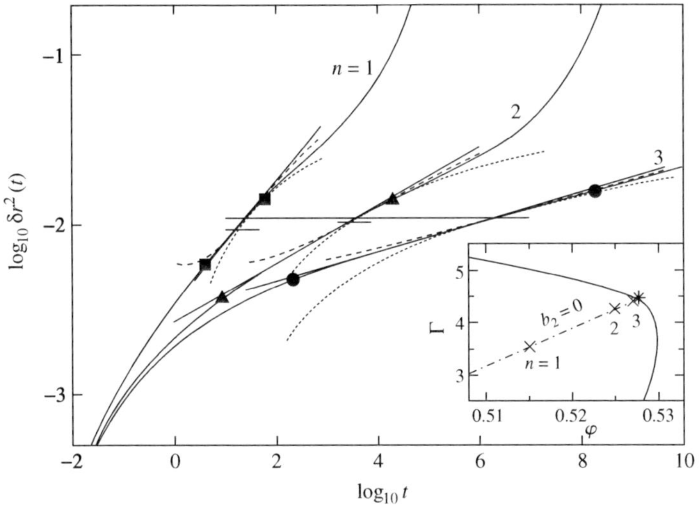

FIG. 6.23. The mean-squared displacement $\delta r^2(t) = 6\Delta_s(t)$ calculated for the SWS for states with attraction-range parameter $\delta^* = 0.04381$ of the A_4 singularity P^c. The inset exhibits the cut through the bifurcation manifold from Fig. 4.9. The states with label n are located on the line of vanishing coefficient b_2 from Eq. (6.194c). The long horizontal line exhibits $6r_{sc}^2 = 0.01086$ and the short ones show $6(r_{sc}^2 - \hat{r}_s^2)$. The dotted and dashed lines show the results of the asymptotic expansion theory Eq. (6.192a) and Eq. (6.192b), respectively. The amplitudes are $h_{\text{MSD}} = 0.004051$ and $k_{\text{MSD}} = -1.708$. The straight full lines have slopes $\alpha = 0.365$, 0.173, and 0.0878 for the states $n = 1$, 2, and 3, respectively. They show the leading-order result (6.193a,b) for the power-law relaxation. The filled symbols mark the points of 5% discrepancy between $6\Delta_s(t)$ and the power law. The time τ is fitted by the request that the analytical result and numerical solution of the equations of motion agree for the crossing of $6r_{sc}^2$. Reproduced from Sperl (2003).

this line are so close to P^c that the onset time for diffusion is beyond $10^{12}t_{\text{mic}}$. Consequently, for liquid states of possible interest for the interpretation of data, the correction factors $B_2(\text{MSD})$ are negative. The functions $\Delta_s(t)$ exhibit the precursors for the liquid–glass transitions on the branch with $\Gamma > \Gamma^*$. These precursor effects of the transition to the attraction-dominated glass cause a trend to compensate or to overcompensate the downward bending of the $\log \Delta_s(t)$ versus $\log(t)$ curve for the asymptote (6.192a).

The discussion of the preceding paragraph suggest to consider the asymptotic expansion for the function $F_s(t) = \ln[\Delta_s(t)/r_{sc}^2]$ rather than that for $\Delta_s(t)$. Equation (6.192a) yields $F_s(t) \sim \alpha \ln(t/\tau), \epsilon \to -0$. The leading-order result for the expansion approach yields a sublinear power-law increase of the mean-squared displacements, which is specified by an exponent α:

$$\Delta_s(t) \sim r_{sc}^2 (t/\tau)^{\alpha}, \qquad \epsilon \to -0, \tag{6.193a}$$

$$\alpha = h_{\mathrm{MSD}} B_0 / r_{sc}^2. \tag{6.193b}$$

The sensitive dependence of α on control parameters is a fascinating feature for the scenario discussed here. The leading corrections to formula (6.193a) are obtained from Eq. (6.192b). They yield the extended power-law description

$$\Delta_s(t) = \left[r_{sc}^2 - \hat{r}_s^2\right](t/\tau)^{\alpha'}\{1 + b_2 \ln(t/\tau)^2\}. \tag{6.194a}$$

The exponent α' is a renormalization of α, described by the coefficient (6.182c):

$$\alpha' = h_{\mathrm{MSD}} B_0' / r_{sc}^2. \tag{6.194b}$$

The qualitative change of the extended description relative to the asymptotic one is due to the $\ln(t/\tau)^2$ term in the curly bracket. The coefficient of this correction term is a smooth function of the coordinates of the state, given by

$$b_2 = \left[-h_{\mathrm{MSD}} B_2(\mathrm{MSD})/\mathrm{r}_{\mathrm{sc}}^2\right] - \left[\mathrm{h}_{\mathrm{MSD}} \mathrm{B}_0'/\mathrm{r}_{\mathrm{sc}}^2\right]^2/2. \tag{6.194c}$$

The coefficient b_2 exhibits the above discussed compensation effect. The first bracket presents the positive contribution due to strong-coupling effects. The second one presents the bending term due to the double-logarithmic presentation of Eq. (6.192a). The relative size of the correction term in the curly bracket of Eq. (6.192b) is proportional to the small quantity $\eta = \sqrt{|\epsilon|}\ln(t/\tau)$. The deviation from the power-law behaviour is described by the second term in the curly bracket of Eq. (6.194a), which is of order η^2. The important correction of Eqs. (6.193a,b) is the reduction of the exponent from α to α'.

The inset of Fig. 6.23 exhibits the lines of distinguished states, which obey the condition $b_2 = 0$. For the three states on this line with labels $n = 1, 2$ and 3, formula (6.193a) describes $\Delta_s(t)$ qualitatively correctly. For short times, the leading-order-power-law description is superior to the extended-logarithmic description. For large times, Eq. (6.192b) yields a better description of $\Delta_s(t)$ than does Eq. (6.193a). The latter formula overestimates the slope of the $\log \Delta_s$ versus $\log(t)$ line.

The state discussed above for label $n = 2$ is the same, which is considered in Fig. 6.24 with label $n' = 2$. Within the accuracy of the drawing, the lines defined by $B_2(24.2) = 0$ and $b_2 = 0$ cannot be distinguished. Within this accuracy, the states with label $n = n' = 2$ agree with the state P_b shown in Figs. 4.9 and 6.22. Also the straight full line with slope α and the filled triangles are reproduced from the preceding figure. The dashed straight line exhibits the result of Eq. (6.194a). On a 5% accuracy level, the extended power-law description accounts for $\Delta_s(t)$ for an increase of the time by about 5 orders of magnitude. The size of this $\log(t)$ interval is about the same as that, which is exhibited in the middle panel

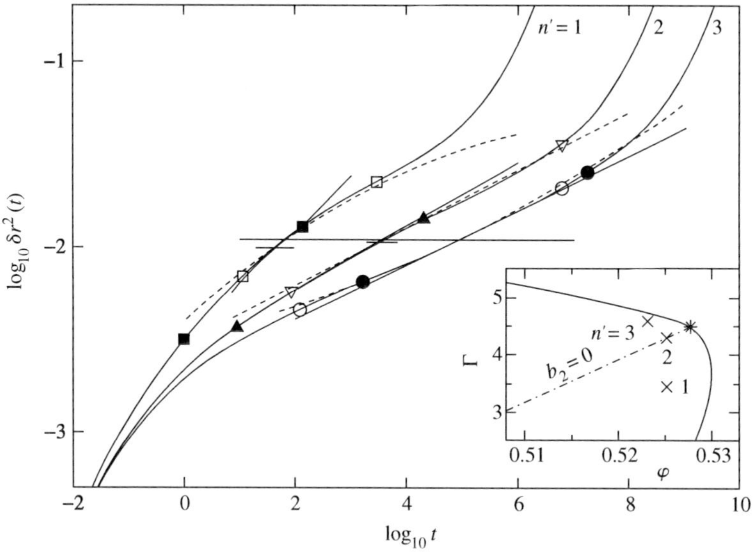

FIG. 6.24. Mean-squared displacements of the model for the SWS specified in the caption of the preceding figure. The states with labels $n' = 1, 2$ and 3 are shown by crosses in the inset. Within the accuracy of the drawing, the state with label $n' = 2$ agrees with that shown in Fig. 4.9 and Fig. 6.22 as triangle with label b. The straight full lines with slopes $\alpha = 0.285$ (0.173, 0.147) exhibit the result of Eqs. (6.193a,b) for the states $n' = 1$ (2, 3). The filled symbols mark the points of 5% from the solutions of the equations of motion. The dashed lines show the extended-power-law descriptions by Eqs. (6.194a–c). The open symbols mark the points of 5% deviations from $\delta r^2(t)$. The calculated parameters are $\alpha' = 0.214$ (0.155, 0.143) and $10^3 b_2 = -7.35$ (0.00, 3.63) for $n' = 1$ (2, 3). Reproduced from Sperl (2003).

of Fig. 4.18 for the logarithmic correlation decay of density fluctuations with wave number $q = 24.2$. But, the interval for the power-law increase of $\Delta_s(t)$ is shifted by a factor of about 100 to longer times relative to the interval for the logarithmic decrease of $\phi_{24.2}(t)$.

Figure 6.24 demonstrates the deformations of the power laws, which occur for states off the line determined by the condition $b_2 = 0$. The state with label $n' = 3$ is typical for one approaching a liquid–glass transition point $P^{(2)c}$, which is located on the branch of the liquid boundary with $\Gamma > \Gamma^*$. According to inequality (6.188b), $\Delta_s(t)$ rises above $[r_{sc}^{(2)}]^2$ before reaching $[r_{sc}]^2$. This causes precursors of the von Schweidler-law increase, which appears as upward bending of the $\log \Delta_s$ versus $\log(t)$ curve. Such bending is reproduced via the positive correction term b_2 in the curly bracket of Eq. (6.194a). The state with label $n' = 1$

demonstrates the deformation of the power law for states, which approach a transition point $P^{(2)c\prime}$ on the liquid boundary for $\Gamma < \Gamma^*$. After $\Delta_s(t)$ has increased above $[r_{sc}]^2$, it exhibits precursors of the critical increase towards $[r_{sc}^{(2)\prime}]^2$. This increase manifests itself by a downward-bent $\log \Delta_s$ versus $\log(t)$ curve. Such bending is reproduced by the negative correction term b_2.

The preceding considerations provide an understanding of the general features of the glassy dynamics, which are exhibited by the solution of the MCT equations of motion for simple liquids with short-ranged attraction forces for states near a higher-order glass-transition singularity P^c. In particular, the asymptotic expansion formulas explain the features, which are different from those characterizing the complex dynamics of van der Waals-like systems. The set of representative results in Fig. 4.18 can be described to a large extend quantitatively by the derived analytical expressions for the correlators and the mean-squared displacements.

The results in Fig. 4.18 are similar to those which are identified by Sciortino *et al.* (2003) for their molecular-dynamics-simulation results for a square-well system. In Fig. 1.22, power-law variation of the mean-squared displacement is demonstrated as can be described by Eqs. (6.194a–c). Their analysis of the density correlators in Fig. 1.21 is done with formula (1.14); and this formula has the same structure as Eq. (6.182a). The latter formulates the extended-logarithmic decay law, which is used to calculate the dashed lines in Fig. 4.18. The claim of similarity is corroborated by the data shown in Fig. 4.19. They demonstrate that the first two terms in Eq. (1.14) exemplify the factorization-property part of Eq. (6.184a). Namely, for both states S_1 and S_2, the critical arrested parts f_q^c are the same functions of q; and also the q-dependence of the critical amplitudes h_q is the same. Up to an overestimation of the critical localization length by about 20%, the MCT results for f_q^c and h_q are close to the ones obtained from the data analysis.

Colloidal suspensions with properly tuned depletion forces exhibit the subtle liquid-glass-transition diagram shown in Fig. 4.11. Molecular simulation studies for square-well systems with an attraction range below some critical value reproduce the essential peculiarities of this diagram as is shown in Fig. 5.8. Mode-coupling theory provides an explanation of the diagram in the high density region. The explanation implies the prediction that the high-density liquid states are close to a glass-transition singularity P^c of type A_4. The above derived formulas for density-fluctuation correlators and mean-squared displacements are a proposal for an understanding of the observed peculiarities of the complex dynamics of the class of glass-forming liquids specified.

(v) Some addenda

There are three closely related problems, which have not been analysed in the preceding discussions. First, the critical correlators $\phi_q^c(t), q = 1, \ldots, M$, have not been determined. Consequently, no understanding has been achieved for the control-parameter-insensitive correlation decay for times exceeding the scale

$t_{\rm mic}$ for the transient dynamics and preceding the onset of the logarithmic decay. Second, no result has been formulated for the control-parameter dependence of the scale τ for the logarithmic-relaxation laws. Third, the above derived formulas do not reflect the structure of the bifurcation manifold in the neighbourhood of P^c. Hence, Eq. (6.182a) does not reproduce the singular control-parameter dependence of the correlators for states P approaching an A_2-bifurcation state $P^{(2)c}$ near the higher-order singularity P^c. In particular, no results have been obtained for states, which violate the restriction (6.168). Some considerations about the solution of the specified problems shall be reported in this subsection.

Following the reasoning of Sec. 6.1.2, the evaluation of the long-time tails of the correlators can be started by simplifying Eq. (6.1a) with the ansatz $\boldsymbol{g}(t) = \boldsymbol{a}g(t)$. This implies the factorization-property formula

$$\boldsymbol{\phi}(t) = \boldsymbol{f}^c + \boldsymbol{h}G(t). \tag{6.195}$$

Thereby, the equations of motion can be reduced to an equation for the single function $G(t)$. The derivation of the extended-logarithmic decay formulas motivates the representation of $G(t)$ as a function of the logarithm of the time:

$$G(t) = g(x), \qquad x = \ln(t/t_0). \tag{6.196a}$$

The S-transform of $G(t)$ can be represented as conventional Laplace transform: $ST[G(t)](z) = \int ds\{\theta(s)\exp(-s)g(\ln s + \ln(i/zt_0))\}$. Assuming an asymptotic Taylor-series representation $g(\ln s{+}y) \sim \sum_k g^{(k)}(y)[\ln(s)]^k/k!$ and interchanging the sum with the integral transform, one gets a generalization of Eq. (A.69b):

$$ST[G(t)](z) \sim \sum_k g^{(k)}(y)\Gamma_k/k!, \qquad y = \ln(i/zt_0). \tag{6.196b}$$

These manipulations can be used to eliminate integral transforms of products $G_1(t)\cdots G_n(t)$. Thereby, the equation for $G(t)$ is reduced to the request of the vanishing of a combination of products of derivatives $g^{(k)}(y), k = 1, 2, \ldots$.

For $P = P^c$ and z tending to zero, the equation for $G(t)$ leads to the request that $g'(y)^2$ is proportional to $g(y)^3$ and $g(y)^4$ for the singularity of type A_3 and A_4, respectively. This yields the results for the long-time asymptotes of the critical correlators (Götze and Sjögren 1989b).

$$A_3 : G^c(t) \sim [2\pi^2/(3\mu_3)]/\ln(t/t_0)^2, \tag{6.197a}$$

$$A_4 : G^c(t) \sim [\pi^2/(6\mu_4)]^{1/2}/\ln(t/t_0). \tag{6.197b}$$

One infers from Eq. (6.5b) that the leading deviations from the factorization theorem are proportional to the amplitude K_q from Eq. (6.175b). The time dependence is given by $1/\ln(t/t_0)^4$ and $1/\ln(t/t_0)^2$ for bifurcation points of type A_3 and A_4, respectively. Up to this order, the critical correlators have been worked out and generalizations thereof have been discussed also (Götze and Sperl 2004a).

The scale τ can be estimated from the following requirement. For some time preceding τ, the $\phi_q(t)$ versus $\log(t)$ curves for the logarithmic-decay functions (6.171a) join with continuous derivatives the corresponding curves for the critical decay. One obtains for the limit of vanishing distance parameter ϵ:

$$A_3 : \ln(\tau/t_0) \sim c_3/[\mu_3^2 |\epsilon_1(P^\epsilon)|]^{1/6}, \tag{6.198a}$$

$$A_4 : \ln(\tau/t_0) \sim c_4/[\mu_4^2 |\epsilon_1(P^\epsilon)|]^{1/4}, \tag{6.198b}$$

with $c_{3,4}$ denoting numbers of order unity. The scales diverge exponentially if the states P^ϵ approach the critical point. (Sperl 2004c) has analysed the correlation decay for states near the higher-order singularities for the schematic model introduced in connection with Fig. 4.5. His diagrams exhibit the convergence of the $\log(\tau/t_0)$ versus ϵ curves towards the above specified asymptotes if $\log(\tau/t_{\mathrm{mic}})$ increases above 10^3.

The calculations of subsection (ii) are based on an iteratively constructed expansion of the inhomogeneity function (6.174a) in powers of the small parameter $B_0 \propto |\epsilon|^{1/2}$. The coefficients of the terms of order $B_0^n, n = 2, 3, \ldots$, are polynomials of $\ln(i/z\tau)$. The cited formula is obtained from the equations of motion with the ansatz that $J_q^{\mathrm{reg}}(z)$ can be dropped in Eq. (6.4b). There holds $J_q^{\mathrm{reg}}(z) = -[(1 - f_q^c)/\Omega_q^2]\{(z\tau)[(z\tau)\tau^{-1} + i\nu_q]\}\tau^{-1}$. One infers from Eqs. (6.198a,b) that $\lim_{\epsilon\to 0}[J_q^{\mathrm{reg}}(z)/B_0^n] = 0$ is valid for all fixed n and fixed $z\tau$. Thereby, the ansatz is justified a posteriori.

The procedure explained in connection with Eqs. (6.196a,b) has been used also to determine the small-frequency asymptote of $G(t)$ for states off an A_3 singularity P^c (Götze and Sjögren 1989b). The leading-order result is proportional to the Weierstrass function $p(\ln(t/t_0); g_2, g_3)$ (Gradshteyn and Ryzhik (2000), Sec. 8.16). The moduli g_2 and g_3 are proportional of the separation parameters discussed in Sec. 6.3.1(ii): $g_2 \propto \sigma_2(P), g_3 \propto \sigma_1(p)$. For the arrested parts $g(P) = G(t \to \infty)$, the formula $P_3(g(P)) = 0$ is obtained. Here P_3 denotes the canonical cubic polynomial (6.146a) with $x_1 = \sigma_1/\mu_3$ and $x_2 = \sigma_2/\mu_3$. Consequently, the solution reproduces the asymptotic form for the manifold of glass-transition states discussed above. For $\sigma_1 = \sigma_2 = 0$, the formula (6.197a) is reproduced. The results (6.171a) are reproduced together with the formula (6.198a) for the scale τ. For a state $P^{(2)c}$ on the line of generic A_2 singularities, the power-law asymptote $G \propto 1/t^a$ is obtained with the correct small-a solution of Eq. (6.11d). Analogous results hold for the generic A_4 singularity (Flach *et al.* 1992).

There remain two challenges. It has to be proven that the preceding considerations are legitimate parts of the first step of some asymptotic solution approach for the equations of motion. Furthermore, the asymptotic results for the corrections to the leading terms have to be evaluated in order to achieve a complete understanding of the complex dynamics for the states P near P^c.

APPENDIX A

MATHEMATICAL MISCELLANIES

A.1 Laplace transforms

In this section, some results for Laplace transforms will be listed, and the conventional formulations shall be translated to the ones used in this book.

The theory of Laplace transforms starts with a definition of the manifold of determining functions. These are complex-valued functions of the time $t, \varphi(t)$, defined for $t > 0$, which obey some regularity properties. For the applications intended in this book, it is sufficient to request that $\varphi(t)$ is continuous and has a continuous derivative $\dot{\varphi}(t)$. Furthermore, $|\varphi(t)|$ shall be integrable on the interval $0 \leq t \leq 1$. In addition, there should exist some exponent $\alpha > 0$ so that $|\varphi(t)/t^\alpha|$ is bounded for large t. Consider complex numbers s from the half plane Re $s \geq \eta$ for some $\eta > 0$. Obviously, the integrals over all positive times of $\exp[-st]t^n\varphi(t)$ converge absolutely and uniformly with respect to the parameter s for all $n = 0, 1, \ldots$. Therefore,

$$f(s) = \int_0^\infty \exp[-st]\varphi(t)dt \tag{A.1}$$

defines a function $f(s)$ that is holomorphic on the half plane Re $s > 0$. The differentiation with respect to s can be done under the integral. The function $f(s)$ is called the conventional Laplace transform of $\varphi(t)$.

The set of determining functions is a linear manifold. If φ_1 and φ_2 belong to the manifold, the same is true for any combination $\varphi = a_1\varphi_1 + a_2\varphi_2$ formed with complex coefficients a_1 and a_2. Similarly, if f_1 and f_2 are functions, which are holomorphic for Re $s > 0$, the same holds for $f = a_1 f_1 + a_2 f_2$. The Laplace transform is a linear transformation, i.e., if f_1 and f_2 are the transforms of φ_1 and φ_2, respectively, f is the transform of φ.

If $f(s)$ is holomorphic for Re $s > 0$, the following integral is well defined for every pair of positive numbers ϵ and Ω:

$$I_{\Omega,\epsilon}(t) = \left(\frac{1}{2\pi}\right) \int_{-\Omega}^{\Omega} f(\epsilon + ix) \exp[(\epsilon + ix)t]dx. \tag{A.2}$$

If $f(s)$ is the Laplace transform of $\varphi(t)$, there exists $\lim_{\Omega\to\infty} I_{\Omega,\epsilon}(t) = I(t)$. The limit is independent of ϵ and reads: $I(t) = 0$ for $t < 0$, and $I(t) = \varphi(t)$ for $t > 0$. If there exists the limit $\varphi(0+) = \lim_{t\to+0}\varphi(t)$, there holds $I(0) = \varphi(0+)/2$, (Widder (1946), Chapter II. 7). If f_1 and f_2 are the transforms of φ_1 and φ_2, respectively, φ_1 is identical with φ_2, if and only if f_1 is identical with f_2. The Laplace transform can be inverted, and $I(t)$ provides the inversion formula.

Let us introduce an extended manifold of determining functions $F(t)$. These functions are defined for all t and are requested to behave so that $\varphi_\pm(t) = \theta(t) F(\pm t)$ are determining functions in the restricted sense explained above. Here, $\theta(t)$ denotes the Heaviside function, which is defined in connection with Eq. (2.28). The functions $f_\pm(s)$ shall denote the Laplace transforms of $\varphi_\pm(t)$:

$$f_\pm(s) = \int_0^\infty \exp[-st]\varphi_\pm(t)dt. \qquad \text{(A.3a)}$$

Let us also rotate the complex half plane Re $s > 0$ into the upper or lower half plane of complex frequencies $z = \pm is$; Im $z \gtrless 0$. By

$$F(z) = \pm i f_\pm(\mp iz), \quad \operatorname{Im} z \gtrless 0, \qquad \text{(A.3b)}$$

a function is defined, which is holomorphic for all z off the real frequency axis. This function is the Laplace transform of $F(t)$ to be used in this book. It shall be denoted by $F(z) = LT[F(t)](z)$. The explicit transformation formula is noted in Sec. 2.2 as Eq. (2.28). Equations (A.3a,b) provide the translation formulas from the traditional notations to the ones used here. A multiplication of the determining function by an exponential specified by an positive parameter $\gamma, \varphi_\pm(t) \to \varphi_\pm^\gamma(t) = \exp[-\gamma t]\varphi_\pm(t)$, implies a translation of the Laplace transform: $f_\pm(s) \to f_\pm^\gamma(s) = f_\pm(s + \gamma)$. Hence, there holds

$$LT[\exp[-\gamma|t|]F(t)](z) = F(z \pm i\gamma), \quad \operatorname{Im} z \gtrless 0, \gamma \geq 0. \qquad \text{(A.3c)}$$

The Laplace transform provides an invertible linear mapping of the extended manifold of determining functions $F(t)$ into the manifold of functions $F(z)$, which are holomorphic for Im $z \neq 0$. The inversion formula explained in connection with Eq. (A.2) is translated to

$$F(t)\theta(\pm t) = [\mp i/2\pi] \lim_{\Omega\to\infty} \exp[\pm\epsilon t] \int_{-\Omega}^{\Omega} F(\omega \pm i\epsilon)\exp[-i\omega t]d\omega, \quad \epsilon > 0. \qquad \text{(A.4)}$$

Some formulas shall be noted, which will be used repeatedly. By variable substitution in Eq. (2.28), one derives the rules for the behaviour of the Laplace transform under rescaling,

$$F_x(t) = F(t/x), \quad F_x(z) = xF(zx), \quad x > 0, \qquad \text{(A.5a)}$$

under reflection,

$$F_r(t) = F(-t), \qquad F_r(z) = -F(-z), \qquad \text{(A.5b)}$$

and under conjugation,

$$F_c(t) = F(t)^*, \qquad F_c(z) = -F(-z^*)^*. \tag{A.5c}$$

If two functions, say $F_1(t)$ and $F_2(t)$, belong to the manifold of determining functions, the same holds for their convolution $F(t)$, which is defined by

$$F(t) = \int_0^t F_1(t-t')F_2(t')dt'. \tag{A.6a}$$

If $t > 0$, one can introduce the factor $\theta(t-t')\theta(t')$ to the integrand and extend the integral over all t'. This can be substituted into Eq. (2.28). Interchanging the sequence of integration, one arrives at the convolution formula for Im $z > 0$:

$$LT[F(t)](z) = -iF_1(z) \cdot F_2(z). \tag{A.6b}$$

Similar manipulations yield the same result for Im $z < 0$.

If the derivative $\partial_t F(t)$ belongs to the extended set of determining functions and if there exists $F(t=0) = \lim_{t\to\pm 0} F(t)$, one can write $\exp[\mp izt] = \partial_t \exp[\mp izt]/(\mp iz)$ and perform a partial integration in Eq. (2.28). This leads to the derivative formula

$$LT[i\partial_t F(t)](z) = F(t=0) + zF(z). \tag{A.7a}$$

The result can be iterated. If also the second derivative belongs to the extended set of determining functions and if there exists $\dot{F}(t=0) = \lim_{t\to\pm 0} \partial_t F(t)$, one gets

$$LT[-\partial_t^2 F(t)](z) = i\dot{F}(t=0) + zF(t=0) + z^2F(z). \tag{A.7b}$$

Since $|\exp[-st]| \leq \exp[-\eta t]$ for all s obeying Re $s \geq \eta > 0$, one concludes from Eq. (A.1): $f(s) \to 0$ if Re $s \to \infty$. To translate this result to the notation used here, a symbol $r(z)$ shall be introduced. It indicates a function that is holomorphic for Im $z \neq 0$ and obeys

$$\lim_{|y|\to\infty} r(x+iy) = 0. \tag{A.8a}$$

Using this notation, the asymptotic formula mentioned for $f(s)$ can be translated to

$$LT[F(t)](z) = r(z). \tag{A.8b}$$

If $\partial_t F(t)$ is a generalized determining function with finite $F(t=0)$, one derives from Eq. (A.7a) a more informative statement about the large-z asymptote of the Laplace transform:

$$F(z) = -\frac{1}{z}[F(t=0) + r(z)]. \tag{A.8c}$$

If $\partial_t^2 F(t)$ is a generalized determining function with finite $\dot{F}(t=0)$, one can iterate the result – or one can use Eq. (A.7b) – in order to get an even more informative formula for the behaviour at large frequencies:

$$F(z) = -\frac{1}{z}F(t=0) - \frac{1}{z^2}[i\dot{F}(t=0) + r(z)]. \tag{A.8d}$$

A.2 Fourier transforms

In this section, some basic formulas for Fourier transformations will be listed and their connection with Laplace transformations shall be specified.

Fourier transformations shall be defined as integral transform on the linear space $\mathbb{V}$ of complex-valued functions defined for all times t, which are stepwise continuous, and which are absolutely integrable. The latter property means that $\int_{-\infty}^{\infty} |F(t)|dt = C$ is a finite number. For every real frequency $\omega, \exp[i\omega t]F(t)$ is in $\mathbb{V}$. The integral over t is the Fourier transform of $F(t)$,

$$FT[F(t)](\omega) = \int \exp[i\omega t]F(t)dt. \tag{A.9}$$

As is done always in this book, an integral over time t with unspecified limits is meant to extend over the whole range $-\infty < t < \infty$. If it is convenient, also the notation $\hat{F}(\omega) = FT[F(t)](\omega)$ will be used.

Since $|F(t)|$ is an ω-independent majorant of the integrand in Eq. (A9), $\hat{F}(\omega)$ is a continuous function of ω and $|\hat{F}(\omega)| \leq C$. The Fourier transformation defines a linear mapping of $\mathbb{V}$ in the linear manyfold of continuous bounded functions. Let $F_0(t)$ be defined by: $F_0(t) = 1$ for $|t - t_0| < T$ and $F_0(t) = 0$ for $|t - t_0| \geq T$. One gets as Fourier transform $\hat{F}_0(\omega) = 2\exp[i\omega t_0]\sin[\omega T]/\omega$. Thus, $\hat{F}_0(\omega)$ tends to zero for ω tending to infinity; the oscillating parts of the integrand in Eq. (A.9) cancel the better, the larger is $|\omega|$. Indicating by $r(\omega)$ a function of the real frequency ω that vanishes for $|\omega| \to \infty$, one can write $\hat{F}_0(\omega) = r(\omega)$. This result can be generalized: according to the Riemann–Lebesque theorem,

$$\hat{F}(\omega) = r(\omega) \tag{A.10}$$

holds for all functions of $\mathbb{V}$.

With $F(t)$ from $\mathbb{V}$, the same is true for the rescaled function $F_x(t)$, the reflected function $F_r(t)$, and the conjugate function $F_c(t)$, which are defined in Eqs. (A.5a–c). One obtains the transformation rules

$$\hat{F}_x(\omega) = x\hat{F}(\omega x), \tag{A.11a}$$

$$\hat{F}_r(\omega) = \hat{F}(-\omega), \tag{A.11b}$$

$$\hat{F}_c(\omega) = \hat{F}(-\omega)^*. \tag{A.11c}$$

With $F_1(t)$ and $F_2(t)$ from $\mathbb{V}$, also the convolution $F(t) = \int F_1(t-t')F_2(t')dt'$ is in $\mathbb{V}$. The evaluation of $\hat{F}(\omega)$ requires the integration of $F_1(t-t')\exp[i\omega(t-t')]$ $F_2(t')\exp(i\omega t')$ over t' and t. Interchanging the sequence of integrations, one gets the convolution theorem

$$FT\left[\int F_1(t-t')F_2(t')dt'\right](\omega) = \hat{F}_1(\omega)\hat{F}_2(\omega). \tag{A.12}$$

If $t^n F(t)$ is absolutely integrable for some $n = 1, 2, \ldots$, one can differentiate Eq. (A.9) m times, $m = 1, 2, \ldots, n$, under the integral. Hence, $\hat{F}(\omega)$ has continuous derivatives up to order n, and

$$[-i\partial/\partial\omega]^m \hat{F}(\omega) = FT[t^m F(t)](\omega), \qquad m = 1, \ldots, n. \tag{A.13}$$

If $F(t)$ has n stepwise continuous derivatives that are absolutely integrable, $\partial_t^n F(t)$ is in $\mathbb{V}$. Partial integration in Eq. (A.9) yields

$$[-i\omega]^n \hat{F}(\omega) = FT[\partial_t^n F(t)](\omega). \tag{A.14a}$$

Using Eq. (A.10), one gets for the large-ω-asymptote,

$$\hat{F}(\omega) = r(\omega)/\omega^n. \tag{A.14b}$$

A central task of the theory of Fourier transforms is the derivation of inversion formulas. Here, only one result shall be cited. If $\hat{F}(\omega)$ is absolutely integrable, it is a member of $\mathbb{V}$. The function $\hat{F}(\omega)$ can be Fourier transformed, and the result is $2\pi F(-t)$, i.e., there holds the inversion formula

$$F(t) = \frac{1}{2\pi}\int \exp[-i\omega t]\hat{F}(\omega)d\omega. \tag{A.15}$$

Here as well as elsewhere in this book, an integral over the frequency ω without specified integration limits is meant to extend over the whole axis $-\infty < \omega < \infty$. Closely related to this formula is Parseval's theorem. It deals with a pair of functions $F_{1,2}(t)$ in $\mathbb{V}$, whose Fourier transforms $\hat{F}_{1,2}(\omega)$ are also in $\mathbb{V}$. $\hat{F}_1(\omega)^*\hat{F}_2(\omega)$ is absolutely integrable. Expressing in $\int \hat{F}_1(\omega)^*\hat{F}_2(\omega)d\omega$ the function $\hat{F}_2(\omega)$ by $F_2(t)$ according to Eq. (A.9) and interchanging the sequence of integrations over

ω and t, one gets

$$\int F_1(t)^* F_2(t) dt = \frac{1}{2\pi} \int \hat{F}_1(\omega)^* \hat{F}_2(\omega) d\omega. \tag{A.16}$$

Let us consider a so-called test function $\xi(t)$. The test functions are complex-valued continuous functions, which are defined for all t. They have continuous derivatives of all orders, and the functions $|t^m \partial_t^n \xi(t)|$ have finite bounds for all $n, m = 0, 1, 2, \ldots$. From the discussion of Eqs. (A.13, A.14) one concludes that the Fourier transform $\hat{\xi}(\omega)$ of a test function is a test function as well. Equations (A.9 and (A.15) describe a one-to-one mapping of the linear space of test functions on itself. If $F(t)$ is in $\mathbb{V}$, the Fourier transform of the convolution $F_2(t) = \int F(t-t')$ $\xi(t')dt'$ is in $\mathbb{V}$. This is because Eq. (A.12) yields $\hat{F}_2(\omega) = \hat{F}(\omega)\hat{\xi}(\omega)$, and the strong decrease of $\hat{\xi}(\omega)$ for large $|\omega|$ ensures that $\hat{F}_2(\omega)$ is absolutely integrable. Therefore Eqs. (A.15, A.16) can be applied with $F_1(t) = \xi(t)$ in order to obtain

$$\iint \xi(t)^* F(t-t')\xi(t')dt'dt = \frac{1}{2\pi} \int |\hat{\xi}(\omega)|^2 \hat{F}(\omega) d\omega. \tag{A.17}$$

Let us assume that $F(t)$ is a function in $\mathbb{V}$. Writing $z = \omega \pm i\epsilon, \epsilon > 0$, one notices that $|F(t)|$ is an ϵ-independent absolutely integrable majorant of $\theta(\pm t)\exp[izt]F(t)$ in Eq. (2.28). Hence, the limit ϵ to zero can be performed under the integral in this formula. One identifies $F(\omega \pm i0) = \lim_{\epsilon \to 0} F(\omega \pm i\epsilon)$ as Fourier transform

$$F(\omega \pm i0) = \pm i FT[\theta(\pm t)F(t)](\omega). \tag{A.18}$$

This formula implies that the Laplace transforms approach finite limit values if the complex frequency $z = \omega \pm i\epsilon$ approaches the real value ω from the upper or lower half plane, respectively. The limit values $F(\omega \pm i0)$ are continuous functions of ω, which are bounded. From Eq. (A.18) one gets an expression of the Fourier transform of $F(t)$ in terms of the Laplace transform

$$\hat{F}(\omega) = (-i)[F(\omega + i0) - F(\omega - i0)]. \tag{A.19}$$

If $\hat{F}(\omega)$ is absolutely integrable, Eq. (A.15) can be used and substituted into Eq. (2.28) for the Laplace transform. The sequence of integrations can be interchanged. Since $LT[\exp[-i\omega t]](z) = 1/[\omega - z]$, one arrives at an expression for the Laplace transform in terms of the Fourier transform

$$F(z) = \frac{1}{2\pi} \int [\omega - z]^{-1} \hat{F}(\omega) d\omega. \tag{A.20}$$

A.3 Positive-definite and positive-analytic functions

For an understanding of correlations functions it is necessary to understand the connection between positive definite and positive analytic functions. In this section, this connection will be derived under restrictive conditions leading to continuous spectra. The general result will be explained also.

Let $\phi(t)$ be a continuous and absolutely integrable complex-valued function defined for all times t. Hence, its Fourier-transform $\hat{\phi}(\omega)$ exists and is a continuous function of frequency ω. It shall be requested in addition that also $\hat{\phi}(\omega)$ is absolutely integrable. There is a one-to-one relation between these two functions provided by Eqs. (A.9), (A.15). Let $\eta(t)$ and $\hat{\eta}(\omega)$ be a pair of test functions, as defined above in connection with Eq. (A.17), which are also related Fourier transformations. The normalization $\int|\hat{\eta}(\omega)|^2 d\omega = 2\pi$ shall be requested. For every real frequency Ω and every $\epsilon > 0$, another pair of test functions $\xi_{\Omega,\epsilon}(t)$ and $\hat{\xi}_{\Omega,\epsilon}(\omega)$, which are related by Fourier transformations, shall be introduced by

$$\hat{\xi}_{\Omega,\epsilon}(\omega) = \hat{\eta}((\omega - \Omega)/\epsilon)/\sqrt{\epsilon}. \tag{A.21a}$$

For ϵ tending to zero, $|\hat{\xi}_{\Omega,\epsilon}(\omega)|^2$ exhibits a peak for ω near Ω. For decreasing ϵ, the width of the peak decreases to zero proportional to ϵ and the height increases proportional to $1/\epsilon$. But the area under the peak is ϵ-independent: $\int|\hat{\xi}_{\Omega,\epsilon}(\omega)|^2 d\omega = \int|\hat{\eta}(\omega)|^2 d\omega = 2\pi$. The integral

$$I_R = \frac{1}{2\pi}\int|\hat{\xi}_{\Omega,\epsilon}|^2 \hat{\phi}(\omega)d\omega \tag{A.21b}$$

can be written as $I_R = \int|\hat{\eta}(\omega)|^2 \hat{\phi}(\Omega + \epsilon\cdot\omega)d\omega/2\pi$. Since the Fourier transform $\hat{\phi}(\omega)$ is bounded, say $|\hat{\phi}(\omega)| \le C$, the integrand has an ϵ-independent absolutely integrable majorant $C|\hat{\eta}(\omega)|^2$. Thus, the limit ϵ tending to zero can be performed under the integral, and this leads to

$$\lim_{\epsilon\to 0} I_R = \hat{\phi}(\Omega). \tag{A.21c}$$

The quantity I_R is an example for the right-hand side of Eq. (A.17). The left-hand side shall be denoted by I_L. It is a two dimensional integral over an absolutely integrable integrand. Hence, it can be written as limit of Riemann sums: $I_L = \lim_{N\to\infty}\sum_{j,k=1}^{N}\xi_j^*\phi(t_j - t_k)\xi_k$. Here $t_1 < t_2 \cdots < t_N$ are the centers of N adjacent time intervals of length $\Delta_1, \Delta_2, \ldots \Delta_N$, respectively, and $\xi_k = \xi_{\Omega,\epsilon}(t_k)\Delta_k$. The limit is to be done so that the Δ_i tend to zero and that t_1 and t_N tend to $-\infty$ and $+\infty$, respectively. The sum is the same as entering the definition of a positive-definite function in Eq. (2.27). Thus, one concludes from Eqs. (A.17), (A.21b), (A.21c) that $\hat{\phi}(\Omega) \ge 0$ if $\phi(t)$ is positive definite. If $\phi(t)$ has a non-negative Fourier transform $\hat{\phi}(\omega)$, one gets from Eq. (A.15): $\sum_{j,k} \xi_j^* \phi(t_j - t_k)\xi_k = \int |\sum_j \xi_j \exp[i\omega t_j]|^2 \hat{\phi}(\omega) d\omega/2\pi \geq 0$. This holds for all choices of $t_1 < t_2 < \cdots < t_N$ and all sets of complex numbers $\xi_1, \xi_2, \ldots, \xi_N$. This means that $\phi(t)$ is positive definite.

The preceding two paragraphs provide the conclusion: the continuous function $\phi(t)$ is positive definite if and only if it can be written as Fourier integral of a positive function $\hat{\phi}(\omega)$:

$$\phi(t) = \frac{1}{2\pi} \int \exp[-i\omega t] \hat{\phi}(\omega) d\omega, \tag{A.22a}$$

$$\hat{\phi}(\omega) \geq 0. \tag{A.22b}$$

This is Bochner's theorem, albeit formulated under the restrictive assumptions on $\phi(t)$ and $\hat{\phi}(\omega)$ specified above.

Let $\phi(z)$ be a function of the complex frequency z, which is holomorphic for $\text{Im}\, z \neq 0$. In addition, two requests shall be obeyed. There exists a finite positive number C so that there holds: $|y\phi(x + iy)| < C$ for $|y| > 0$. The limits $\phi(\omega \pm i0)$ exist as continuous functions of ω. For given z from the area of holomorphy, two numbers ϵ and R shall be chosen such that $0 < \epsilon < |\text{Im} z|, |z| < R$. Cauchy's theorem allows to write the function $\phi(z)$ as contour integral, $\phi(z) = \int [\zeta - z]^{-1} \phi(\zeta) d\zeta/(2\pi i)$. Here, the contour consists of a closed path in the upper half plane and a closed path in the lower one, both oriented anti-clock wise. The paths consist of straight lines parallel to the real axis, $\zeta = \pm i\epsilon + \omega, |\omega| < \sqrt{R^2 - \epsilon^2}$, which are closed by semicircles of radius R. The limit $\epsilon \to 0$ can be performed to get $\phi(z) = I_1(z) + I_2(z)$. Here, $I_1 = \int \theta(R^2 - \omega^2)[\omega - z]^{-1} \phi''(\omega) d\omega/\pi$ is due to the straight-line pieces with $\phi''(\omega)$ denoting

$$\phi''(\omega) = [\phi(\omega + i0) - \phi(\omega - i0)]/(2i). \tag{A.23a}$$

The second contribution is due to the circle of radius R, described by $\zeta = R \exp(i\alpha), -\pi \leq \alpha \leq \pi$. This contribution vanishes in the limit $R \to \infty$, i.e.,

$$\phi(z) = \lim_{R \to \infty} \int \theta(R^2 - \omega^2)[\omega - z]^{-1} \phi''(\omega) d\omega/\pi. \tag{A.23b}$$

If $\phi''(\omega)$ is absolutely integrable, $|\phi''(\omega)|/|\text{Im} z|$ is an R independent majorant. One can perform the limit $R \to \infty$ under the integral and gets a representation of $\phi(z)$ as Cauchy integral

$$\phi(z) = \frac{1}{\pi} \int [\omega - z]^{-1} \phi''(\omega) d\omega. \tag{A.23c}$$

Remembering the discussion of the concept of a positive-analytic function presented in connection with Eq. (2.48), one can summarize Eqs. (A.23a,c) as follows. Let $\phi(z)$ denote a function, which is holomorphic for $\text{Im}\, z \neq 0$ and obeys the additional requests formulated above. This function is positive analytic if and only if it allows for a representation in form of Eq. (A.23c) with $\phi''(\omega) \geq 0$.

Combining the preceding findings with those of Sec. A.2, one arrives at the theorem: A continuous function $\phi(t)$ of the time t is positive definite if and only if its Laplace transform $\phi(z) = LT[\phi(t)](z)$ is positive analytic and obeys $|y\phi(x+iy)| \leq C$ for $|y| > 0$.

The theorem was proven under restrictive additional assumptions. These lead to the spectral representations for $\phi(t)$ and $\phi(z)$ connected by $\phi''(\omega) = \hat{\phi}(\omega)/2$. To proceed to the general case, let us introduce the so-called integrated spectral density

$$\sigma(\omega) = \frac{1}{\pi}\int \theta(\omega - \Omega)\phi''(\Omega)d\Omega. \tag{A.24}$$

This is a monotonically increasing function that is bounded by $\sigma(\omega \to -\infty) = 0$ and $\sigma(\omega \to \infty) = C^{(0)} = \phi(t=0)$. A variable transformation from ω to σ can be performed in Eqs. (A.22a), (A.23c), understanding ω as function of σ. For all bounded continuous functions $C(\omega)$ of frequency ω, the integral $I(C;\sigma) = \int C(\omega)d\sigma$ shall be introduced. This expression yields $\phi(t)$ and $\phi(z)$ if one chooses $C(\omega) = \exp[-i\omega t]$ and $1/[\omega - z]$, respectively. These integrals can be evaluated as limits of Riemann sums

$$I(C;\sigma) = \lim_{N\to\infty}\sum_{k=1}^{N} C(\omega_k)\Delta\sigma_k. \tag{A.25}$$

Here $-\infty = \omega_0 < \omega_1 < \omega_2 < \cdots < \omega_N$ is a sequence of points defining N adjacent intervals: $-\infty < \omega < \omega_1, \omega_1 < \omega < \omega_2, \ldots, \omega_{N-1} < \omega < \omega_N$. The $\Delta\sigma_k = \sigma(\omega_k) - \sigma(\omega_{k-1})$ denote the increase of $\sigma(\omega)$ within the kth interval. The limit is to be done so that ω_1 tends to $-\infty$, ω_N to $+\infty$, and the maximum of the interval lengths tends to zero.

The limit in Eq. (A.25) exists for all bounded continuous functions $C(\omega)$ even if $\sigma(\omega)$ is replaced by a general function $\alpha(\omega)$ of bounded variation. A function $\alpha(\omega)$ is called of bounded variation if its real and imaginary part can be written as difference of two monotonically increasing bounded functions: $\sigma_1(\omega), \sigma_2(\omega)$ and $\sigma_3(\omega), \sigma_4(\omega)$, respectively: $\alpha(\omega) = \alpha_R(\omega) + i\alpha_I(\omega)$, $\alpha_R(\omega) = \sigma_1(\omega) - \sigma_2(\omega)$, $\alpha_I(\omega) = \sigma_3(\omega) - \sigma_4(\omega)$. The corresponding limit is called the Stieltjes integral of function $C(\omega)$ with the weight function $\alpha(\omega)$. It is denoted by

$$I(C;\alpha) = \int C(\omega)d\alpha(\omega). \tag{A.26}$$

The Stieltjes integral is a linear functional in the linear manifold of bounded continuous functions

$$I(\gamma_1 C_1 + \gamma_2 C_2;\alpha) = \gamma_1 I(C_1;\alpha) + \gamma_2 I(C_2;\alpha). \tag{A.27a}$$

It is also a linear functional in the linear manifold of functions of bounded variation

$$I(C;\gamma_1\alpha_1+\gamma_2\alpha_2)=I(C;\alpha_1)\gamma_1+I(C;\alpha_2)\gamma_2. \tag{A.27b}$$

Here γ_1,γ_2 are arbitrary complex numbers.

If $\alpha(\omega)$ has a continuous derivative, the Stieltjes integral reduces to a Riemann integral, as explained above in connection with Eq. (A.24)

$$I(C;\alpha)=\int C(\omega)[d\alpha(\omega)/d\omega]d\omega. \tag{A.28}$$

If $\alpha(\omega)$ exhibits a discontinuity, such reduction is impossible. For the Heaviside function $\sigma(\omega)=\theta(\omega-\Omega)$, one derives from Eq. (A.25) the result $I(C;\sigma)=C(\Omega)$. For a step function $\alpha(\omega)$ with discontinuities α_m at $\omega=\Omega_m, m=1,\ldots,d$, and $\alpha(\omega)=0$ for $\omega<\Omega_1$, $\alpha(\omega)=\sum_{m=1}^{d}\alpha_m$ for $\Omega>\omega_d$, one gets

$$I(C;\alpha)=\sum_{m=1}^{d}\alpha_m C(\Omega_m). \tag{A.29}$$

In this case, the Stieltjes integral reduces to a sum. An introduction to the theory of Stieltjes integrals can be found in Chapter I of Widder (1946).

Bochner's theorem reads: a continuous function $\phi(t)$ is positive definite if and only if there is a bounded monotonically increasing function $\sigma(\omega)$ so that $\phi(t)$ can be expressed as Fourier–Stieltjes transform of $\sigma(\omega)$:

$$\phi(t)=\int \exp[-i\omega t]d\sigma(\omega). \tag{A.30}$$

The weight function increases by $\phi_0=\phi(t=0)=\sigma(\infty)-\sigma(-\infty)$. It is uniquely determined by $\phi(t)$ if one normalizes so that

$$\sigma(\omega\to-\infty)=0, \tag{A.31a}$$

and if one imposes the convention for the behaviour at the discontinuities Ω:

$$\sigma(\Omega)=[\sigma(\Omega+0)+\sigma(\Omega-0)]/2. \tag{A.31b}$$

A theorem that rests on a theorem for holomorphic functions proven by Riesz and Herglotz reads: a function $\phi(z)$ is positive analytic and obeys $|y\phi(iy)|\le C<\infty$ for all $|y|>0$ if and only if there is a bounded monotonically increasing function $\sigma(\omega)$ so that

$$\phi(z)=\int [\omega-z]^{-1}d\sigma(\omega). \tag{A.32}$$

Imposing the conventions from Eqs. (A.31a,b), the function $\sigma(\omega)$ is uniquely determined by $\phi(z)$. The Laplace transform of Eq. (A.30) yields Eq. (A.32), and this proves the theorem formulated above, but now for the general case. Formulas expressing $\sigma(\omega)$ in terms of $\phi(t)$ or in terms of $\phi(z)$ are implied by the proofs of Eqs. (A.30, A.32) (Akhiezer and Glazman (1993), Sects. 59 and 60), but they will not be needed in this book.

Let us introduce a number F by

$$F = \sigma(0+) - \sigma(0-). \tag{A.33a}$$

The increasing function $\sigma(\omega)$ is continuous for $\omega = 0$ if and only if $F = 0$. By construction,

$$\phi(t = 0) \geq F \geq 0. \tag{A.33b}$$

The weight function $\sigma(\omega)$ can be decomposed as

$$\sigma(\omega) = F\theta(\omega) + \tilde{\sigma}(\omega). \tag{A.33c}$$

The function $\tilde{\sigma}(\omega)$ increases between $\tilde{\sigma}(-\infty) = 0$ and $\tilde{\sigma}(\infty) = \phi(t = 0) - F$. Like $\sigma(\omega)$ and $\theta(\omega), \tilde{\sigma}(\omega)$ obeys the convention formulated in Eq. (A.31b). The function $\tilde{\sigma}(\omega)$ is continuous at $\omega = 0$, i.e.,

$$\lim_{\alpha \to 0} [\tilde{\sigma}(\alpha) - \tilde{\sigma}(-\alpha)] = 0. \tag{A.33d}$$

Substituting the decomposition of $\sigma(\omega)$ into the spectral representations of Eqs. (A.30) and (A.32), one gets corresponding decompositions of the positive definite function $\phi(t)$ and the positive analytic function $\phi(z)$:

$$\phi(t) = F + \tilde{\phi}(t), \tag{A.34a}$$

$$\phi(z) = [-F/z] + \tilde{\phi}(z). \tag{A.34b}$$

Here $\tilde{\phi}(t)$ and $\tilde{\phi}(z)$ can be written as spectral integrals with weight function $\tilde{\sigma}(\omega)$. In particular, $\tilde{\phi}(z)$ is the Laplace transform of $\tilde{\phi}(t)$. If $F = \phi(t = 0)$, one gets the trivial functions $\tilde{\phi}(t) = 0, \tilde{\phi}(z) = 0$. If this is not the case, $\tilde{\phi}(z)$ is a positive analytic function and $\tilde{\phi}(t)$ is a continuous non-vanishing positive-definite function.

Bochner's theorem implies that F is determined uniquely by $\phi(t)$. This connection between $\phi(t)$ and F is provided by the theorem

$$\lim_{\epsilon \to 0} \epsilon \int_0^{1/\epsilon} \phi(t) dt = F. \tag{A.35}$$

Similarly, F is determined by $\phi(z)$ according to the theorem

$$\lim_{\epsilon\to 0+}(\mp i\epsilon)\phi(\pm i\epsilon) = F. \tag{A.36}$$

Because of Eqs. (A.30, A.34a), the derivation of Eq. (A.35) requires the proof that $I(\tau) = (\frac{1}{\tau})\int_0^\tau \tilde{\phi}(t)dt = \int f(\omega\tau)d\tilde{\sigma}(\omega)$ vanishes in the limit of $\tau = 1/\epsilon$ tending to infinity. Here $f(2x) = \exp[-ix]\sin(x)/x$. Because of Eq. (A.32, A.34b), the derivation of Eq. (A.36) requires the same proof, but $f(x)$ has to be replaced by $g(x) = \mp i/[x \mp i]$. To proceed, one chooses some $\alpha > 0$ and splits the integral in one term $I_1(\tau)$ due to the integration interval $-\alpha \leq \omega \leq \alpha$ and another term $I_2(\tau)$ due to the remaining integrations for $\alpha \leq \omega$ and $\omega \leq -\alpha$. Within the first interval, one estimates $|f(x)| \leq 1$ and $|g(x)| \leq 1$ in order to find $|I_1(\tau)| \leq [\tilde{\sigma}(\alpha) - \tilde{\sigma}(-\alpha)]$. For the other intervals, one estimates $|f(x)| \leq 2/|x|$ and $|g(x)| \leq 2/|x|$ in order to find $|I_2(\tau)| \leq 2 \cdot \phi(t = 0)/(\alpha\tau)$. For every $\eta > 0$, one can choose α so small that $[\tilde{\sigma}(\alpha) - \tilde{\sigma}(-\alpha)] \leq \eta/2$, because of Eq. (A.33e). Then, one can choose a τ obeying $\tau \geq 4 \cdot \phi(t = 0)/(\alpha \cdot \eta)$. As a result, one gets $|I(\tau)| \leq \eta$. Since η is arbitrary, one gets $\lim_{\tau\to\infty} I(\tau) = 0$, and this finishes the proof of Eqs. (A.35), (A.36).

A.4 Harmonic-oscillator correlators

In this section, the properties of an elementary function of the complex frequency z are listed, which is defined by the double-fraction representation

$$\phi(z) = -1/\big[z - \Omega^2/(z \pm i\Gamma)\big], \quad \operatorname{Im} z \gtrless 0. \tag{A.37a}$$

Here, Ω and Γ denote two positive frequencies.

Function $X(z) = -1/\phi(z) = z - \Omega^2/(z \pm i\Gamma)$ is holomorphic for Im $z \neq 0$ and Im $X(z) \gtrless 0$ for Im $z \gtrless 0$. Thus, $\phi(z) = -1/X(z)$ is holomorphic and Im $\phi(z) = -\mathrm{ImX}^*(\mathrm{z})/|\mathrm{X}(\mathrm{z})|^2 \gtrless 0$. Hence, $\phi(z)$ is a positive-analytic function. The large-frequency asymptote reads

$$\phi(z) = -(1/z)\Big\{1 + (\Omega/z)^2\big[1 \mp i(\Gamma/z)\big] + O(1/z^4)\Big\}. \tag{A.37b}$$

From Appendix A.3 one concludes that $\phi(z)$ is the Laplace transform of a continuous positive-definite function $\phi(t)$ of the time t. This function can be written as Fourier transform of its spectrum $\phi''(\omega)$. The latter is obtained from the boundary values $\phi(\omega \pm i0) = \phi'(\omega) \pm i\phi''(\omega) = [-X'(\omega) \pm iX''(\omega)]/[X'(\omega)^2 + X''(\omega)^2]$. It is an analytic function of ω given by

$$\phi''(\omega) = \Gamma\Omega^2/\big[\omega^4 + (\Gamma^2 - 2\Omega^2)\omega^2 + \Omega^4\big]. \tag{A.37c}$$

Since the spectrum is a real and even function of ω, $\phi(t)$ is a real and even function of t. Functions $\phi(z)$ and $\phi(t)$ have the properties of correlation functions in the frequency domain and time domain, respectively. They provide a model for a dynamics whose qualitative features are determined by the frequency ratio

$$r = \Gamma/\Omega. \tag{A.37d}$$

The functions $\omega^\ell \phi''(\omega)$ are absolutely integrable for $\ell = 0, 1$, and 2. One concludes from Eq. (A.15) that $\phi(t)$ as well as its first and second derivative are continuous functions of t. Comparing Eq. (A.8d) with Eq. (A.37b), one gets

$$\phi(t=0) = 1, \quad \partial_t \phi(t=0) = 0. \tag{A.38a}$$

One can rewrite Eq. (A.37a) as $[z^2\phi(z) + z\phi(t=0) + i\partial_t\phi(t=0)] \pm i\Gamma[z\phi(z) + \phi(t=0)] - \Omega^2\phi(z) = 0$. Using Eqs. (A.7a,b), one gets for the Laplace back transformation for $t \geq 0$:

$$\partial_t^2 \phi(t) + \Gamma \partial_t \phi(t) + \Omega^2 \phi(t) = 0. \tag{A.38b}$$

This equation together with the initial conditions (A.38a) is equivalent to Eq. (A.37a). The correlator describes the dynamics of a harmonic oscillator, which is specified by the oscillator frequency Ω and a friction coefficient Γ. The correlator for negative times is obtained from the symmetry property $\phi(-t) = \phi(t)$. Asymptotic solution of the differential equation for small times yields the counterpart of Eq. (A.37b):

$$\phi(t) = 1 - \tfrac{1}{2}(\Omega t)^2 \left[1 - \tfrac{1}{3}\Gamma|t| + O(t^2)\right]. \tag{A.38c}$$

If the friction constant is so large that $\Gamma^2 \geq 2\Omega^2$, the denominator in Eq. (A.37c) increases monotonically with ω^2. Hence, the spectrum describes a central peak. The $\phi''(\omega)$ versus ω curve is bell shaped. If the damping constant is so small that $\Gamma^2 < 2\Omega^2$, the denominator has a minimum at $\omega^2 = \omega_{\text{peak}}^2$, where

$$\omega_{\text{peak}} = \sqrt{\Omega^2 - \Gamma^2/2}, \quad r < \sqrt{2}. \tag{A.39}$$

The spectrum describes a doublet. The $\phi''(\omega)$ versus ω curve has maxima at $\omega = \pm\omega_{\text{peak}}$, which are separated by a minimum at $\omega = 0$.

For Im $z > 0$, there holds $\phi(z) = -(z + i\Gamma)/[z(z + i\Gamma) - \Omega^2]$. This formula also defines a meromorphic function for all frequencies and provides the analytic continuation of the correlator from the upper half plane onto the complete frequency plane. If $\Gamma \neq 2\Omega$, the singularities of $\phi(z)$ are two simple poles, say $z_\pm$, located in the lower half plane:

$$z_\pm = -i(\Gamma/2) \pm \sqrt{\Omega^2 - (\Gamma/2)^2}, \tag{A.40a}$$

For $r \neq 2$, there is the partial-fraction representation

$$\phi(z) = -[A_+/(z - z_+)] - [A_-/(z - z_-)], \tag{A.40b}$$

with

$$A_\pm = \tfrac{1}{2}\left[1 \pm i(\Gamma/2)/\sqrt{\Omega^2 - (\Gamma/2)^2}\right]. \tag{A.40c}$$

For Im $z > 0$, the Laplace transform of $\exp[-iz_\pm|t|]$ is $-1/(z - z_\pm)$. Since the inverse Laplace transformation is uniquely determined, Eq. (A.40b) is equivalent to

$$\phi(t) = A_+ \exp[-iz_+|t|] + A_- \exp[-iz_-|t|]. \tag{A.40d}$$

The limit $\Omega^2 - (\Gamma/2)^2 \to 0$ yields the result of the correlator for $r = 2$.

For positive argument of the root in Eq. (A.40a), the poles have a non vanishing real part. Introducing the parameters

$$\gamma = \Gamma/2, \qquad \omega_r = \sqrt{\Omega^2 - \gamma^2}, \tag{A.41a}$$

one can write:

$$z_\pm = -i\gamma \pm \omega_r, \qquad A_\pm = \frac{1}{2}[1 \pm i(\gamma/\omega_r)]. \tag{A.41b}$$

Since $A_\pm = (\Omega/2\omega_r)\exp[\pm i\delta]$ with the phase δ given by $tg\delta = \gamma/\omega_r$, one gets

$$\phi(t) = (\Omega/\omega_r)\exp[-\gamma|t|]\cos(\omega_r t - \delta), \quad r < 2. \tag{A.42}$$

There are oscillations with frequency ω_r, which are damped with a decay rate γ. Equation (A.40b) suggests a splitting of the spectrum $\phi''(\omega) = \text{Im}\ \phi(\omega + i0)$ as sum of two contribution $\phi^\pm(\omega) = -\ \text{Im}\ [A_\pm/(\omega - z_\pm)]$:

$$\phi''(\omega) = \phi^+(\omega) + \phi^-(\omega), \tag{A.43a}$$

$$\phi^\pm(\omega) = \frac{1}{2}\Big\{1 + [(\omega_r \mp \omega)/\omega_r]\Big\}\Big\{\gamma/[(\omega_r \mp \omega)^2 + \gamma^2]\Big\}. \tag{A.43b}$$

Each contribution consists of two factors. One is a Lorentzian with a peak position $\pm\omega_r$ and a half width γ at half height. The other factor is a linear function of ω, which shifts the peak positions of $\phi^\pm(\omega)$ away from ω_r.

For negative arguments of the square root in Eq. (A.40a), both poles are located on the negative imaginary axis of the frequency plane. Introducing the parameters

$$\gamma' = \sqrt{\gamma^2 - \Omega^2}, \qquad \gamma_\pm = \gamma \pm \gamma', \tag{A.44a}$$

one can write $z_{\pm} = -i\gamma_{\mp}$ and $A_{\pm} = \pm\gamma_{\pm}/(2\gamma')$. The correlator in the time domain is the difference of two exponentials

$$\phi(t) = (\gamma_+/2\gamma')\exp[-\gamma_-|t|] - (\gamma_-/2\gamma')\exp[-\gamma_+|t|], \quad r > 2. \tag{A.44b}$$

Since $\gamma_+ > \gamma_- > 0$, function $\phi(t)$ decreases with increasing $|t|$ monotonically from unity to zero.

A.5 Matrix correlators

In this section, some notation is introduced and some formulas are noted, which are useful for the discussions of matrices of correlation functions. Bold letters like $\boldsymbol{M}$ and $\boldsymbol{N}$ are used to abbreviate d-by-d matrices with elements $M_{k,\ell}$ and $N_{k,\ell}$, respectively; $k, \ell = 1, \ldots, d$.

According to the discussion of Eq. (2.11), a matrix $\boldsymbol{M}$ is called positive semidefinite if there holds

$$\sum_{k,\ell} a_k^* M_{k,\ell} a_\ell \geq 0 \tag{A.45a}$$

for every array of complex numbers $a_k, k = 1, \ldots, d$. This property shall be abbreviated by writing

$$\boldsymbol{M} \geq \boldsymbol{0}, \tag{A.45b}$$

where $\boldsymbol{0}$ denotes the null matrix. The condition implies that $\boldsymbol{M}$ is hermitian. Matrix $\boldsymbol{M}$ is called positive definite if $\sum_{k,\ell} a_k^* M_{k,\ell} a_\ell = 0$ implies $a_k = 0$ for all k. In this case, it shall be written $\boldsymbol{M} > \boldsymbol{0}$. One shows that $\boldsymbol{M} \geq \boldsymbol{0}$ or $\boldsymbol{M} > \boldsymbol{0}$ holds if

$$\boldsymbol{P}^\dagger \boldsymbol{M} \boldsymbol{P} \geq \boldsymbol{0} \tag{A.45c}$$

or $\boldsymbol{P}^\dagger \boldsymbol{M} \boldsymbol{P} > \boldsymbol{0}$, respectively, is valid for every invertible matrix $\boldsymbol{P}$. The principal-axis theorem states that there is a unitary matrix $\boldsymbol{U}$, which transforms $\boldsymbol{M}$ on diagonal shape

$$(\boldsymbol{U}^{-1}\boldsymbol{M}\boldsymbol{U})_{k,\ell} = \delta_{k,\ell} M_k. \tag{A.45d}$$

Consequently, $\boldsymbol{M} \geq \boldsymbol{0}$ holds if and only if $M_k \geq 0$, $k = 1, \ldots, d$. Matrix $\boldsymbol{M}$ is positive definite if and only if all d eigenvalues are positive. In this case, $\boldsymbol{M}^{-1}$ exists and its eigenvalues are $1/M_k, k = 1, \ldots, d$. The statements $\boldsymbol{M} > 0$ and $\boldsymbol{M}^{-1} > 0$ are equivalent.

Within the set of matrices $\boldsymbol{L}, \boldsymbol{M}, \boldsymbol{N}, \ldots$, a semi-ordering

$$\boldsymbol{N} \geq \boldsymbol{M} \tag{A.46a}$$

can be introduced by the request that $\boldsymbol{N} - \boldsymbol{M}$ is positive semidefinite:

$$\boldsymbol{N} - \boldsymbol{M} \geq \boldsymbol{0}. \tag{A.46b}$$

The notations are considered as equivalent to $\boldsymbol{0} \leq \boldsymbol{N} - \boldsymbol{M}$ and $\boldsymbol{M} \leq \boldsymbol{N}$. There holds the fundamental condition: if $\boldsymbol{L} \leq \boldsymbol{M}$ and $\boldsymbol{M} \leq \boldsymbol{N}$, then $\boldsymbol{L} \leq \boldsymbol{N}$. If $\boldsymbol{N} \geq \boldsymbol{0}$, there holds for every matrix $\boldsymbol{M} : \boldsymbol{M} \leq \boldsymbol{M} + \boldsymbol{N}$. If $\alpha > 0$ and $\boldsymbol{M} \geq 0$, there holds $\alpha\boldsymbol{M} \geq 0$ and $(-\alpha)\boldsymbol{M} \leq \boldsymbol{0}$.

Let $\boldsymbol{\phi}(t)$ denote a continuous matrix function, which is defined for all times t. It shall be called a positive-definite matrix function if $\phi_a(t) = \sum_{m,n} a_m^* \phi_{m,n}(t) a_n$ is a positive-definite function for all arrays of complex numbers $a_1, \ldots, a_d$. Using Eq. (2.27a) and the notations introduced there, one can write the condition explicitly as

$$\sum_{j,k=1}^{N} \sum_{m,n} (\xi_j a_m)^* \phi_{m,n}(t_j - t_k)(\xi_k a_n) \geq 0. \tag{A.47}$$

The conditions (2.27b) imply that $0 \leq \phi_a(0)$; and $\phi_a(0) = 0$ can hold only if $\phi_a(t)$ vanishes for all times. Since the coefficients a_k are arbitrary, there holds

$$\boldsymbol{\phi}(t = 0) > \boldsymbol{0}, \tag{A.48a}$$

unless $\boldsymbol{\phi}(t) = \boldsymbol{0}$ for all t. The general symmetry of positive-definite functions $\phi_a(t)^* = \phi_a(-t)$ implies the symmetry for the matrix function

$$\boldsymbol{\phi}(t)^\dagger = \boldsymbol{\phi}(-t). \tag{A.48b}$$

According to Bochner's theorem there is the spectral representation $\phi_a(t) = \int \exp[-i\omega t]\phi_a''(\omega) d\omega/\pi$, with $\phi_a''(\omega) \geq 0$. The inversion formula (2.35a) identifies $\phi_a''(\omega)$ as hermitian quadratic form of the array $a_1, \ldots, a_d$. Hence, there is a hermitian matrix $\boldsymbol{\phi}''(\omega)$ with elements $\phi_{k,\ell}''(\omega)$ so that $\phi_a''(\omega) = \sum_{k,\ell} a_k^* \phi_{k,\ell}''(\omega) a_\ell$. This form is positive semidefinite. Consequently, there exists a spectral matrix $\boldsymbol{\phi}''(\omega)$ obeying

$$\boldsymbol{0} \leq \boldsymbol{\phi}''(\omega), \tag{A.49a}$$

so that the matrix identity

$$\boldsymbol{\phi}(t) = \frac{1}{\pi} \int \exp[-i\omega t] \boldsymbol{\phi}''(\omega) d\omega \tag{A.49b}$$

can be formulated as abbreviation for each of the d^2 matrix elements. A less loosely formulated statement is obtained if $\boldsymbol{\phi}''(\omega) d\omega/\pi$ is replaced by a Stieltjes measure $d\boldsymbol{\sigma}(\omega)$. Here, $\boldsymbol{\sigma}(\omega)$ is an increasing matrix function: $\boldsymbol{\sigma}(\omega_1) \leq \boldsymbol{\sigma}(\omega_2)$ for $\omega_1 \leq \omega_2$.

The Laplace transformation $\phi_{k,\ell}(z) = LT[\phi_{k,\ell}(t)](z)$ maps the elements of $\boldsymbol{\phi}(t)$ on the ones of a matrix $\boldsymbol{\phi}(z)$, which are holomorphic for Im $z \neq 0$. From Eq, (A.49b), one gets the spectral representation for the matrix function in the frequency domain

$$\boldsymbol{\phi}(z) = \frac{1}{\pi} \int [\omega - z]^{-1} \boldsymbol{\phi}''(\omega) d\omega. \tag{A.50a}$$

Again, the integral abbreviates d^2 expressions for the matrix elements. Since $\boldsymbol{\phi}''(\omega)$ is hermitian, there holds the symmetry

$$\boldsymbol{\phi}(z)^\dagger = \boldsymbol{\phi}(z^*). \tag{A.50b}$$

Function $\boldsymbol{\phi}''(\omega)$ can vanish for all ω only if $\boldsymbol{\phi}(t)$ is identical to zero. This would imply also that $\boldsymbol{\phi}(z)$ vanishes for all z. Excluding this trivial case, one gets

$$\mathrm{Im}\, \boldsymbol{\phi}(z) \gtrless \mathbf{0}, \qquad \mathrm{Im}\, z \gtrless 0. \tag{A.50c}$$

Here, the imaginary part of some matrix $\boldsymbol{X}$ is defined by

$$\mathrm{Im}\, \boldsymbol{X} = (\boldsymbol{X} - \boldsymbol{X}^\dagger)/(2i). \tag{A.51}$$

A matrix function $\boldsymbol{\phi}(z)$, which is defined for all non-real frequencies z, is called positive analytic, if $\phi_a(z) = \sum_{k,\ell} a_k^* \phi_{k,\ell}(z) a_\ell$ is positive analytic for all arrays $a_1, \ldots, a_d$. Reasoning in analogy to that for $\boldsymbol{\phi}(t)$ shows that $\boldsymbol{\phi}(z)$ is positive analytic if it is holomorphic for Im $z \neq 0$ and if it obeys the conditions (A.50b,c). A function, which exhibits the spectral representation (A.50a) with a positive semidefinite spectral matrix $\boldsymbol{\phi}''(\omega)$ is positive analytic. In addition, there exists a finite positive number C so that there holds for all $k, \ell = 1, \ldots, d$:

$$|\phi_{k,\ell}(z)| \leq C/\mathrm{Im}\, |z|. \tag{A.52}$$

The theorems of Appendix A.3 imply the reversed conclusion. If a positive-analytic function $\boldsymbol{\phi}(z)$ obeys the preceding inequality, it is the Laplace transform of a positive-definite continuous function $\boldsymbol{\phi}(t)$. The Eqs. (A.49b), (A.50a) provide another one-to-one mapping of the functions in the time domain $\boldsymbol{\phi}(t)$ on functions in the frequency domain $\boldsymbol{\phi}(z)$.

Let $\boldsymbol{\phi}(t)$ be defined for $t \geq 0$. This matrix function is called completely monotonic if $\phi_a(t)$ is completely monotonic for all a. This means that all derivatives exist for $t > 0$ and $(-\partial_t)^\ell \phi_a(t) > 0$. Consequently, all derivatives of the matrix correlators exist and they obey

$$(-\partial_t)^\ell \boldsymbol{\phi}(t) > \mathbf{0}, \qquad \ell = 0, 1, \ldots. \tag{A.53a}$$

According to Eqs. (1.2a,b), there is a one-to-one correspondence between $\phi_a(t)$ and a non-negative weight function $\rho_a(\gamma)$, defined for $\gamma \geq 0$. Since $\rho_a(\gamma)$ is a quadratic form of the array $a_k, k = 1, \ldots, d$, Bernstein's theorem is generalized as follows. The matrix function $\boldsymbol{\phi}(t)$ is completely monotonic if an only if d^2 elements can be written as Laplace integrals of the d^2 elements $\rho_{k,\ell}(\gamma)$ of a matrix $\boldsymbol{\rho}(\gamma)$:

$$\boldsymbol{\phi}(t) = \int_0^\infty \exp[-\gamma t]\boldsymbol{\rho}(\gamma)d\gamma. \tag{A.53b}$$

Here, the weight is a positive semidefinite function:

$$\boldsymbol{\rho}(\gamma) \geq \mathbf{0}. \tag{A.53c}$$

More precisely, $\boldsymbol{\rho}(\gamma)d\gamma$ has to be read as Stieltjes weight $d\boldsymbol{\sigma}(\gamma)$ with $\boldsymbol{\sigma}(\gamma)$ denoting a monotonically increasing function: $\boldsymbol{\sigma}(\gamma_1) \leq \boldsymbol{\sigma}(\gamma_2)$ for $\gamma_1 \leq \gamma_2$.

A.6 Product correlators

This section will start with the demonstration that the product

$$\phi(t) = \phi_1(t)\phi_2(t) \tag{A.54}$$

of two positive-definite functions $\phi_1(t)$ and $\phi_2(t)$ is a positive-definite function as well. In order to proceed, one can choose a sequence of times $t_1 < t_2 < \cdots < t_N$ of arbitrary length N. This sequence can be used to define an N-by-N matrix $\boldsymbol{g}$ by its elements $g^{j,k} = \phi(t_j - t_k), j, k = 1, \ldots, N$. According to the definition (2.27a), one has to show that $\boldsymbol{g}$ is positive semidefinite. This is equivalent to the property that for every sequence of complex numbers $\xi_1, \ldots, \xi_N$, the sum $g = \sum_{j,k} \xi_j^* g^{j,k} \xi_k$ is not negative. Let us define two N-by-N matrices $\boldsymbol{g}_1$ and $\boldsymbol{g}_2$ by their elements $g_q^{j,k} = \phi_q(t_j - t_k), q = 1, 2$. According to the discussion of Eqs. (2.11), (2.27a), these matrices are hermitian and positive semidefinite. Because of Eq. (A.45d), there are unitary matrices $\boldsymbol{U}_q$ and non-negative eigenvalues $E_q^\ell, \ell = 1, \ldots, N, q = 1, 2$ so that $g_q^{j,k} = \sum_\ell U_q^{j,\ell} E_q^\ell U_q^{k,\ell *}$. One finds $g^{j,k} = \sum_{\ell,n} b_{j,\ell n} E_1^\ell E_2^n b_{k,\ell n}^*$, where $b_{j,\ell n} = U_1^{j,\ell} U_2^{j,n}$. Since $g = \sum_{\ell,n} |a_{\ell n}|^2 E_1^\ell E_2^n$ with $a_{\ell n} = \sum_j \xi_j^* b_{j,\ell n}$, the desired result $g \geq 0$ is demonstrated.

Completely monotonic functions $\phi_q(t)$, which obey $\phi_q(t) = \phi_q(-t)$ and $\phi_q(0) < \infty$, are special positive definite functions. They can be written as superposition of elementary relaxation functions, Eq. (1.2a). They have continuous derivatives of all orders. Because of Bernstein's theorem, they can be defined by Eq. (1.2b). These inequalities hold also for $\phi_1(t)\phi_2(t)$. One concludes that also the product function $\phi(t)$ is completely monotonic; it obeys $\phi(t) = \phi(-t)$ and $\phi(0) < \infty$.

One can iterate the preceding discussions so that there are more than two factors on the right-hand side of Eq. (A.54). One can also introduce a positive factor there. As a result, one concludes the following. If $\phi_q(t), q = 1, 2, \ldots, M$, are continuous positive-definite functions and if $V_{k_1 \cdots k_n}$ denote non-negative coefficients,

$$\phi(t) = \sum_{k_1,\ldots,k_n} V_{k_1\cdots k_n}\phi_{k_1}(t)\cdots\phi_{k_n}(t) \tag{A.55}$$

is a positive-definite continuous function. The analogous statement holds if the $\phi_q(t)$ are even functions of the time, which are completely monotonic with finite values for $t = 0$.

The preceding results are crucial for understanding the basis of MCT. In this context, the class of functions can be restricted to elements of a linear function space to be called $\mathbb{V}'$. For this restricted class, an alternative proof shall be presented. The $\mathbb{V}'$ denotes the set of complex-valued functions $F(t)$, which are defined for all real t, which are continuous and absolutely integrable, and which have an absolutely-integrable Fourier transform $\hat{F}(\omega) = FT[F(t)](\omega)$. According to Appendix A.2, $\hat{F}(\omega)$ is continuous, i.e., $\hat{F}(\omega)$ is also a function in $\mathbb{V}'$. Equations (A.9, A.15) show that the Fourier transform defines a linear invertible mapping of $\mathbb{V}'$ onto itself. If $\hat{c}_F = \int |F(t)|dt$ and $c_F = \int |\hat{F}(\omega)|d\omega/(2\pi)$ there holds $|\hat{F}(\omega)| < \hat{c}_F$ and $|F(t)| \leq c_F$. Equation (A.12) shows that the convolution of two functions from the space $\mathbb{V}'$ is an element of $\mathbb{V}'$. Applying the convolution theorem with the role of $F(t)$ and $\hat{F}(\omega)$ interchanged, one concludes the following. If $F_1(t)$ and $F_2(t)$ are in $\mathbb{V}'$, the same is true for the product function

$$F(t) = F_1(t)F_2(t). \tag{A.56a}$$

This holds, because this definition of $F(t)$ is equivalent to the formula for the Fourier transforms:

$$\hat{F}(\omega) = \int \hat{F}_1(\omega - \omega')\hat{F}_2(\omega')d\omega'/2\pi. \tag{A.56b}$$

Let $\phi_1(t)$ and $\phi_2(t)$ denote two positive-definite functions in $\mathbb{V}'$. These functions have the general properties of autocorrelation functions with the additional request of having absolutely integrable Fourier transforms. One can apply the restricted form of Bochner's theorem (A.22a,b). There hold the elementary forms of the spectral representation from Sec. 2.3, and the spectra $\hat{\phi}_q(\omega)/2$ are non-negative:

$$\phi_q''(\omega) \geq 0; \quad q = 1, 2. \tag{A.57a}$$

Equation (A.56b) yields the spectrum for the product function:

$$\phi''(\omega) = \int \phi_1''(\omega - \omega')\phi_2''(\omega')d\omega'/\pi. \tag{A.57b}$$

This implies

$$\phi''(\omega) \geq 0. \tag{A.57c}$$

Using Bochner's theorem again, one concludes that $\phi(t)$ is positive definite. Thereby, the alternative proof is completed.

Using the findings from Appendix A.5, the preceding results shall be generalized for matrix correlators. Bold letters shall denote d-by-d matrices. Let $\boldsymbol{\phi}_q(t), q = 1, 2$, denote two positive-definite matrix functions, with elements $\phi_q^{\alpha\beta}(t)$ from $\mathbb{V}'$; $\alpha, \beta = 1, \ldots, d$. This means that the spectral matrices are positive semidefinite:

$$\boldsymbol{\phi}_q''(\omega) \geq 0; \quad q = 1, 2. \tag{A.58a}$$

A product function $\boldsymbol{\phi}(t)$ shall be defined by its matrix elements:

$$\phi^{\alpha\beta}(t) = \sum_{\alpha_1\alpha_2\beta_1\beta_2} v_{\alpha,\alpha_1\alpha_2}\phi_1^{\alpha_1\beta_1}(t)\phi_2^{\alpha_2\beta_2}(t)v^*_{\beta,\beta_1\beta_2}. \tag{A.58b}$$

The coefficients $v_{\alpha,\beta\gamma}$ can be arbitrary complex numbers. Every term in the preceding sum is an element of $\mathbb{V}'$ and Eq. (A.56b) yields for the spectra: $\phi^{\alpha\beta\prime\prime}(\omega) = \sum_{\alpha_1\alpha_2\beta_1\beta_2} v_{\alpha,\alpha_1\alpha_2}\{\int \phi_1^{\alpha_1\beta_1\prime\prime}(\omega - \omega')\phi_2^{\alpha_2\beta_2\prime\prime}(\omega')d\omega'\}v^*_{\beta,\beta_1\beta_2}/\pi$. According to Eq. (A.45d), there are unitary matrices $\boldsymbol{U}_q(\omega)$ and sets of non-negative eigenvalues $E_q^\gamma(\omega), q = 1, 2, \gamma = 1, \ldots, d$ so that the spectral matrices can be written in the form: $\phi_q^{\alpha\beta\prime\prime}(\omega) = \sum_\gamma U_q^{\alpha\gamma}(\omega)E_q^\gamma(\omega)U_q^{\beta\gamma}(\omega)^*$. Substitution of this representation in the preceding equation yields $\phi^{\alpha\beta\prime\prime}(\omega) = \sum_{\gamma_1\gamma_2}\int b_{\alpha,\gamma_1\gamma_2}(\omega,\omega')$ $E_1^{\gamma_1}(\omega - \omega')E_2^{\gamma_2}(\omega')b_{\beta,\gamma_1\gamma_2}(\omega,\omega')^* d\omega'/\pi$. Here, the abbreviation is used: $b_{\alpha,\gamma_1\gamma_2}$ $(\omega,\omega') = \sum_{\alpha_1\alpha_2} v_{\alpha,\alpha_1\alpha_2}U_1^{\alpha_1\gamma_1}(\omega - \omega')U_2^{\alpha_2\gamma_2}\omega')$. For every set of complex numbers $a_\alpha, \alpha = 1, \ldots, d$, one finds for the number $g(\omega) = \sum_{\alpha\beta} a^*_\alpha \phi^{\alpha\beta\prime\prime}(\omega)a_\beta$ the result: $g(\omega) = \sum_{\gamma_1\gamma_2}\int|\sum_\alpha a^*_\alpha b_{\alpha,\gamma_1\gamma_2}(\omega,\omega')|^2 E_1^{\gamma_1}(\omega - \omega')E_2^{\gamma_2}(\omega')d\omega'/\pi$. Hence, $g(\omega) \geq 0$. Thereby, it is shown that the spectral matrix for the product function is positive semidefinite:

$$\boldsymbol{\phi}''(\omega) \geq 0. \tag{A.58c}$$

This result is the matrix generalization of Eq. (A.57c), which implies that $\boldsymbol{\phi}(t)$ is a positive-definite matrix function.

A.7 Power-law variations

In this section, some formulas shall be compiled which relate power-law variations of functions $F(t)$ for large and small times t with the power-law variations of their Laplace transforms $F(z) = LT[F(t)](z)$ for small and large frequencies z, respectively. Standard symmetries $F(t) = F(-t) = F^*(t)$ are assumed, which are equivalent to $F(z) = -F(-z) = F(z^*)^*$. Therefore, in the following text, the restrictions $t > 0$ and Im $z > 0$ can be imposed.

A power-law function shall be specified by some positive time scale t_0 and by a real exponent x:

$$F_x(t) = (t_0/t)^x, \quad x < 1. \tag{A.59a}$$

Since this function is integrable for $0 < t/t_0 < 1$, it can be added to the set of determining functions. For real positive s, the conventional Laplace transform (A.1) of $F_x(t)$ can be expressed in terms of Euler's second integral for the gamma function: $sf_x(s)/(st_0)^x = \int_0^\infty \exp[-\xi]\xi^{-x}d\xi = \Gamma(1-x)$. This formula defines also the analytic continuation to all complex s, except for $s = -\sigma, \sigma \geq 0$. Here and in the following, the power of a complex number $\zeta = |\zeta|\exp[i\varphi], -\pi < \varphi \leq \pi$, is defined by $\zeta^x = |\zeta|^x \exp[i\varphi x]$. Using Eq. (A.3b) one gets

$$-zF_x(z) = \Gamma(1-x)(-izt_0)^x. \tag{A.59b}$$

This equation defines also the analytic continuation from the upper half plane for the frequencies to the complete plane, except for the points $z = -i\sigma, \sigma \geq 0$. For real positive frequencies ω, one obtains $F_x(\omega + i0) = F_x'(\omega) + iF_x''(\omega)$ with

$$F_x'(\omega) = -\cos(\pi x/2)\Gamma(1-x)(\omega t_0)^x/\omega, \tag{A.59c}$$
$$F_x''(\omega) = \sin(\pi x/2)\Gamma(1-x)(\omega t_0)^x/\omega. \tag{A.59d}$$

Let us introduce a positive parameter r in order to modify the power law to

$$F_{x,r}(t) = \exp[-t/(rt_0)]F_x(t). \tag{A.60a}$$

According to Eq. (A.3c), the modification of the Laplace transform reads

$$F_{x,r}(z) = \Gamma(1-x)(it_0r^{1-x})/[1-izt_0r]^{1-x}. \tag{A.60b}$$

For $x > 0$, function $F_x(t)$ exhibits a power-law tail for long times and a power-law divergency for small ones. Introducing the exponential cutoff factor in Eq. (A.60a), the tail is eliminated, but the small-time divergency is not changed. Function $F_x(z)$ exhibits power-law variation for small as well as for large frequencies. The cutoff factor implies that $F_{x,r}(z)$ is holomorphic in a neighbourhood of the frequency origin, which is specified by $|zt_0r| < 1$. The large-frequency asymptote is not changed: $\lim_{z\to\infty} F_{x,r}(z)/F_x(z) = 1$.

In the remainder of this appendix, the preceding results will be generalized. To proceed, this and the following paragraph are used to define concepts of asymptotic equality. Let $F(t)$ and $G(t)$ denote two continuous functions, which are defined for all positive times t. They are called asymptotically equal for large or small times, respectively, if

$$\lim_{t\to\infty \text{ or } 0} F(t)/G(t) = 1. \tag{A.61a}$$

The formula means that the relative deviations between these functions vanishes if the times tend to infinity or to zero, respectively: $\lim_{t\to\infty \text{ or } 0}[F(t) - G(t)]/G(t) = 0$. This limit property shall be denoted by the symbol $\sim$:

$$F(t) \sim G(t), \quad t \to \infty \quad \text{or} \quad t \to 0. \tag{A.61b}$$

Let $\hat{F}(z)$ and $\hat{G}(z)$ denote holomorphic functions, which are defined for Im $z > 0$. These functions are called asymptotically equal for small or large frequencies, respectively, if

$$\lim_{z\to 0 \text{ or } \infty} \hat{F}(z)/\hat{G}(z) = 1. \tag{A.61c}$$

The formula means that the relative difference of these functions vanishes for frequencies z tending to zero or infinity, respectively. This property shall be denoted as:

$$\hat{F}(z) \sim \hat{G}(z), \quad z \to 0 \quad \text{or} \quad z \to \infty. \tag{A.61d}$$

Let $L(t)$ denote a continuous function, which is defined for all positive times. This function is said to exhibit slow variation for large or small times, respectively, if there holds $L(\rho t) \sim L(t)$ for $t \to \infty$ or for $t \to 0$, respectively. Here ρ can be any positive factor. Explicitly, the function has the property

$$\lim_{t\to\infty \text{ or } 0} L(\rho t)/L(t) = 1. \tag{A.62a}$$

Let $\hat{L}(z)$ be a holomorphic function, which is defined for all frequencies obeying Im $z > 0$. This function is said to exhibit slow variation for small or large frequencies, respectively, if $\hat{L}(\rho z) \sim \hat{L}(z)$ for $z \to 0$ or for $z \to \infty$, respectively. Equivalently, for all positive ρ, there holds

$$\lim_{z\to 0 \text{ or } \infty} \hat{L}(\rho z)/\hat{L}(z) = 1. \tag{A.62b}$$

Function $L(t) = \ell n[t/t_0]$ exhibits slow variation for large as well as for small times. Similarly, $\hat{L}(z) = \ell n[i/(zt_0)]$ exhibits slow variation for small and for large frequencies.

Tauberian theorems relate asymptotic properties of determining functions $F(t)$ with those of their Laplace transform $F(z)$. Two examples shall be cited (Feller (1971), Sec. XIII. 5). The asymptotics shall be specified by some real exponent x and by some continuous function $L(t)$, which is defined for all positive times. Function $L(t)$ shall have an analytic continuation to all complex t, which obey $\mathrm{Re}\, t > 0$. Consequently, $L(i/z)$ defines a function, which is holomorphic for $\mathrm{Im}\, z > 0$. A positive time t_{reg} shall exist so that intervals for regularity properties can be specified. For the first example the following is requested for $t > t_{\mathrm{reg}}$. The functions $F(t), \partial_t F(t)$, and $\partial_t^2 F(t)$ are continuous and they do not change sign. Furthermore, $L(t)$ shall be slowly varying for large times. Under these conditions,

$$F(t) \sim (t_0/t)^x L(t/t_0), \quad x < 1, \quad t \to \infty \tag{A.63a}$$

is valid if and only if there holds

$$-zF(z) \sim \Gamma(1-x)(-izt_0)^x L(i/zt_0), \quad z \to 0. \tag{A.63b}$$

For the second example, the regularity is obeyed for $t < t_{\mathrm{reg}}$, and $L(t)$ is slowly varying for small times. Then, the two equations

$$F(t) \sim (t_0/t)^x L(t/t_0), \quad x < 1, \quad t \to 0 \tag{A.64a}$$
$$-zF(z) \sim \Gamma(1-x)(-izt_0)^x L(i/zt_0), \quad z \to \infty \tag{A.64b}$$

are equivalent. If $F(t)$ is completely monotonic, the regularity properties and the possibility to continue analytically $F(t)$ are insured. Above, one can choose two arbitrary positive numbers ρ_1 and ρ_2 and replace $L(t/t_0)$ by $L(t/\rho_1 t_0)$ and $L(i/zt_0)$ by $L(i/z\rho_2 t_0)$.

Let $G(t)$ denote a determining function, which exhibits the above specified regularity properties for large times. Equations (A.63a,b) imply the equivalence of its non-integrable long-time power-law part

$$G(t) \sim (t_0/t)^x, \quad x < 1, \quad t \to \infty \tag{A.65a}$$

and the small-frequency power-law part of its Laplace transform $G(z)$:

$$-zG(z) \sim \Gamma(1-x)(-izt_0)^x, \quad z \to 0. \tag{A.65b}$$

Two corollaries formulate generalizations of the preceding Tauberian theorems. The first one deals with integrable long-time tails, which are specified by non-integral exponents larger than unity. The behaviour

$$G(t) \sim (t_0/t)^x, \quad n < x < n+1, \qquad n = 1, 2, \ldots, \quad t \to \infty, \tag{A.66a}$$

and regularity for $t > t_{\mathrm{reg}}$ implies

$$-zG(z) + (izt_0)P_n(-izt_0) \sim \Gamma(1-x)(-izt_0)^x, \quad z \to 0. \tag{A.66b}$$

Here, a polynomial is abbreviated by $P_n(\zeta) = A_0 + A_1\zeta + \cdots + A_{n-1}\zeta^{n-1}$ and

$$A_0 = \int_0^\infty G(t)dt/t_0. \tag{A.66c}$$

For small z, the Laplace transform $G(z)$ exhibits the algebraic function (A.65b). However, this function is added to a polynomial contribution $it_0P_n(-izt_0)$. For small frequencies, the Laplace transform does not diverge. The coefficient A_0 quantifies the zero-frequency spectrum: $G''(\omega \to 0) = t_0A_0$.

The proof of the preceding equations starts with introducing the function $\overline{G}(t) = \int \theta(t'-t)G(t')dt'/t_0$. It is a determining function with determining derivative and initial condition A_0. Equation (A.7a) yields: $G(z) - it_0A_0 = (izt_0)$ $LT[\overline{G}(t)](z)$. For large times, $\overline{G}(t)$ exhibits the regularity properties and $\overline{G}(t) \sim (t_0/t)^y/y$ with $y = x - 1$. If $x < 2$, Eqs. (A.65a,b) can be used to get: $LT[\overline{G}(t)]$ $(z) \sim \Gamma(1-y)(-izt_0)^y/y$. This yields the desired result for $n = 1$. If $x > 2$, the procedure can be iterated $(n-2)$ times in order to get Eq. (A.66b).

The second corollary deals with integer exponents. Let $H(t)$ denote a determining function, which exhibits the regularity properties for $t > t_{\text{reg}}$. The function $\overline{H}(t) = \int \theta(t-t')H(t')dt'/t_0$ exhibits the same properties. Using Eq. (A.7a), one gets $H(z) = (-izt_0)LT[\overline{H}(t)](z)$. If $H(t) \sim (t_0/t)$, $t \to \infty$, one gets $\overline{H}(t) \sim \ell n[t/t_0]$. Application of Eqs. (A.63a,b) with $x = 0$ and $L(t) = \ell n[t/t_0]$ leads to the proposition:

$$H(t) \sim (t_0/t), \quad t \to \infty \tag{A.67a}$$

implies

$$-zH(z) \sim (-izt_0)\ell n[i/(zt_0)], \quad z \to 0. \tag{A.67b}$$

A $(1/t)$-long-time tail causes a logarithmic small-frequency divergence of the Laplace transform $H(z)$. If the tail is integrable,

$$H(t) \sim (t_0/t)^n, \quad n = 2, 3, \ldots, \quad t \to \infty, \tag{A.68a}$$

the Laplace transform is finite for all frequencies. There holds:

$$-zH(z) + (izt_0)Q_n(-izt_0) \sim -(izt_0)^n \ell n[i/(zt_0)]/(n-1)! \quad z \to 0. \tag{A.68b}$$

A polynomial appears as regular part $Q_n(\zeta) = B_0 + B_1\zeta + \cdots + B_{n-2}\zeta^{n-2}$ with

$$B_0 = \int_0^\infty H(t')dt'/t_0. \tag{A.68c}$$

The proof follows the pattern explained above. Function $\overline{H}(t) = \int \theta(t'-t)H(t')dt'$ is related to $H(t)$ by: $[-zH(z)] + (izt_0)\overline{H}(t=0) = (izt_0)[-z\overline{H}(z)]$. It exhibits the long-time asymptote: $\overline{H}(t) \sim (t_0/t)^{n-1}/(n-1)$. For $n = 2$, the asymptote of $\overline{H}(z)$ is determined by Eq. (A.67b). If $n > 2$, one can iterate the procedure $(n-2)$ times in order to arrive at Eq. (A.68b).

A.8 Logarithmic variations

In this section, some formulas shall be listed for the Laplace transforms of the powers of logarithms. It will be convenient to note the results in terms of S-transforms, which are defined by $ST[F(t)](z) = -zLT[F(t)](z)$. Standard symmetries $F(t) = F(t)^* = F(-t)$ shall be assumed for all functions to be considered. The equations shall be noted only for $t > 0$ and Im $z > 0$.

For $y > 0$, Eqs. (A.59a,b) yield $ST[(t/t_0)^y](z) = \Gamma(1+y)(i/zt_0)^y$. Differentiation with respect to y can be commuted with the integral transform. For $y = 0$, one gets

$$ST\big[\ell n(t/t_0)\big](z) = \Gamma_1 + \ell n(i/zt_0). \tag{A.69a}$$

Here, $\Gamma_1 = \Gamma'(1) = -0.57721\ldots$ denotes the negative of Euler's constant. The nth derivative, $n = 1, 2, \ldots$, for $y = 0$ generalizes the preceding formula to

$$ST\big[\ell n(t/t_0)^n\big](z) = \sum_{k=0}^{n} \big[n!/(n-k)!k!\big]\Gamma_k \ell n(i/zt_0)^{n-k}, \tag{A.69b}$$

with $\Gamma_k = d^k\Gamma(y=1)/dy^k, k = 0, 1, \ldots$. There holds $\Gamma'(y) = \Gamma(y)\psi(y)$ with $\psi(y)$ denoting the digamma function. Iterating this formula, one can express Γ_k in terms of the first $(k-1)$ derivatives of $\psi(y)$ for $y = 1$. These numbers are given in terms of the tabulated zeta function. For example: $\Gamma_2 - \Gamma_1^2 = \zeta(2) = \pi^2/6$.

The mentioned formulas imply

$$ST\big[\ell n(t/t_0)^2\big](z) - ST\big[\ell n(t/t_0)\big](z)^2 = \pi^2/6. \tag{A.70a}$$

More generally, there holds for $n = 2, 3, \ldots$

$$\begin{aligned} &ST\big[\ell n(t/t_0)^n\big](z) - ST\big[\ell n(t/t_0)\big](z)^n \\ &\quad = (\pi^2/12)n(n-1)\ell n(izt_0)^{n-2} + \sum_{k=3}^{n} \big[n!/(n-k)!k!\big](\Gamma_k - \Gamma_1^k)\ell n(i/zt_0)^{n-k}. \end{aligned} \tag{A.70b}$$

If one defines a linear integral transform of some determining function $F(t)$ by

$$IT\big[F(t)\big](z) = ST\big[\ell n(t/t_0)F(t)\big](z) - ST\big[\ell n(t/t_0)\big](z)ST\big[F(t)\big](z), \tag{A.71a}$$

one gets for $n = 1, 2, \ldots$

$$IT\left[\ell n(t/t_0)^n\right](z) = \zeta(2)\left\{n\ell n(i/zt_0)^{n-1} + \sum_{k=2}^{n}(n-k+1)\Gamma_{n,k}\ell n(i/zt_0)^{n-k}\right\}. \tag{A.71b}$$

Here, the abbreviations are introduced $\Gamma_{n,k} = \left[n!/(n-k)!k!\right]\left[\Gamma_{k+1} - \Gamma_k\Gamma_1\right]/\left[\zeta(2)(n-k+1)\right]$

A sequence of polynomials $p_n(x)$ of degree $n, n = 1, 2, \ldots$, can be defined recursively by the equations:

$$p_1(x) = x, \tag{A.72a}$$

$$p_n(x) = x^n - \sum_{k=2}^{n}\Gamma_{n,k}p_{n+1-k}(x), \quad n \geq 2. \tag{A.72b}$$

One shows by induction that

$$IT\left[p_n(\ell n(t/t_0))\right](z) = \zeta(2)n\ell n(i/zt_0)^{n-1}. \tag{A.72c}$$

The simplest examples are (Götze and Sperl 2002):

$$p_2(x) = 2.6160x + x^2, \tag{A.73a}$$

$$p_3(x) = -2.1482x + 3.9239x^2 + x^3, \tag{A.73b}$$

$$p_4(x) = -12.813x - 4.2964x^2 + 5.2319x^3 + x^4. \tag{A.73c}$$

APPENDIX B

SYMMETRIES OF FLUCTUATION CORRELATORS

In this section, the implications of symmetry relations shall be derived, which hold for correlation functions formed with scalar and vector fluctuations. As is discussed in Sec. 3.1.1, the symmetry relations are invariance properties, which are explained with the aid of certain transformations $\mathcal{S}$. The latter are unitary operators defined on the space of dynamical variables so that

$$([\mathcal{S}A](t)|\mathcal{S}B) = (A(t)|B). \tag{B.1}$$

Requesting the preceding equation for all translations $\mathcal{S}_{tr}(\vec{d})$ is equivalent to requesting homogeneity. This request gets a transparent form if applied to correlators formed with field fluctuations, say, $A = X_{\vec{k}}$ and $B = Y_{\vec{p}}$. In Sec. 3.1.2, a field fluctuation $Z_{\vec{q}}$ is defined as an array of dynamical variables explained for every wave vector $\vec{q}$ so that there holds the transformation law

$$\mathcal{S}_{tr}(\vec{d})Z_{\vec{q}} = \exp[-i\vec{q}\vec{d}]Z_{\vec{q}}. \tag{B.2a}$$

Equation (B.1) is valid if and only if there holds $(X_{\vec{k}}(t)|Y_{\vec{p}})\exp[i\vec{q}\vec{d}] = (X_{\vec{k}}(t)|Y_{\vec{p}})$, $\vec{q} = \vec{k} - \vec{p}$, for all translation vectors $\vec{d}$. Differentiating with respect to $\vec{d}$, one gets for $\vec{d} = 0 : (X_{\vec{k}}(t)|Y_{\vec{p}})\vec{q} = 0$. Hence

$$(X_{\vec{k}}(t)|Y_{\vec{p}}) = 0, \quad \text{if } \vec{k} \neq \vec{p}. \tag{B.2b}$$

Conversely, this property implies Eq. (B.1) for all $\mathcal{S} = \mathcal{S}_{tr}(\vec{d})$.

The isotropy condition implies more subtle restrictions for the correlations of two field fluctuations. A general discussion of these restrictions requires the application of the representation theory for the rotation group. But, for the cases to be used in this book, the restrictions can be derived elementarily. To proceed, some auxiliary concepts shall be introduced. For every wave vector $\vec{q}$, a wave vector $\vec{q}^*$ shall be defined by:

$$q^{*1} = q^{*2} = 0, \quad q^{*3} = q, \quad q = |\vec{q}|. \tag{B.3a}$$

The vector $\vec{q}^*$ has the same modulus as $\vec{q}$, but it points in the third coordinate direction. A rotation matrix D_* shall be defined as function of $\vec{q}$, which rotates the third coordinate direction in the direction of $\vec{q}$:

$$D_*\vec{q}^* = \vec{q}. \tag{B.3b}$$

Furthermore, a rotation $D(\varphi)$ shall be introduced, which has the elements: $D^{31} = D^{32} = D^{13} = D^{23} = 0$, $D^{33} = 1$, $D^{11}(\varphi) = D^{22}(\varphi) = \cos(\varphi)$. $D^{12}(\varphi) = -D^{21}(\varphi) = \sin(\varphi)$. These matrices specify rotations by an angle φ around the third coordinate axis. There holds

$$D(\varphi)\vec{q}^{\,*} = \vec{q}^{\,*}. \tag{B.3c}$$

The simplest case deals with the correlations of two scalar fields, say $A_{\vec{q}}$ and $B_{\vec{q}}$. A field fluctuation $Z_{\vec{q}}$ is called a scalar if there holds the transformation law

$$\mathcal{S}_{ro}(D)Z_{\vec{q}} = Z_{D\vec{q}} \tag{B.4a}$$

for all rotation matrices D. Application of Eq. (B.1) with $\mathcal{S} = \mathcal{S}_{ro}(D)$ yields $(A_{\vec{q}}(t)|B_{\vec{q}}) = (A_{D\vec{q}}(t)|B_{D\vec{q}})$. Specializing to $D = D_*^{-1}$, Eq. (B.3b) yields $D\vec{q} = \vec{q}^{\,*}$, and one arrives at

$$(A_{\vec{q}}(t)|B_{\vec{q}}) = F_q^{A,B}(t). \tag{B.4b}$$

Here, the right-hand side is a special correlator, which is uniquely defined by the wave number q:

$$F_q^{A,B}(t) = (A_{\vec{q}*}(t)|B_{\vec{q}*}). \tag{B.4c}$$

The reverse statement holds also: If Eq. (B.4b) is valid, there holds the invariance relation (B.1) for the pair of variables $A_{\vec{q}}$ and $B_{\vec{q}}$ and for all $\mathcal{S} = \mathcal{S}_{ro}(D)$.

The next case concerns the correlations formed with a scalar field, say $A_{\vec{q}}$, and a vector field, say $\vec{V}_{\vec{q}}$. The latter is defined as a triple of field fluctuations $V_{\vec{q}}^m$, $m = 1, 2, 3$, which obeys the transformation law

$$\mathcal{S}_{ro}(D)V_{\vec{q}}^m = \sum_k V_{D\vec{q}}^k D^{km} \tag{B.5a}$$

for all rotation matrices D. For isotropic systems, Eq. (B.1) yields

$$(A_{\vec{q}}(t)|V_{\vec{q}}^m) = \sum_k (A_{D\vec{q}}(t)|V_{D\vec{q}}^k)D^{km}. \tag{B.5b}$$

Choosing $D = D_*^{-1}$, the formula specializes to $(A_{\vec{q}}(t)|V_{\vec{q}}^m) = \sum_k g^k D_*^{mk}$. Here, the abbreviation $g^k = (A_{\vec{q}*}(t)|V_{\vec{q}*}^k)$ is introduced.

Specializing the preceding equation to $\vec{q} = \vec{q}^{\,*}$ and $D = D(\varphi)$, one gets $g^m = \sum_k g^k D(\varphi)^{km}$. The triple of g^m, $m = 1, 2, 3$, is the common eigenvector of all rotations $D(\varphi)$. This means that $g^1 = g^2 = 0$, i.e., $g^k = gq^{*k}$. For later reference, it shall be mentioned that the correctness of this result can be seen by calculating the derivative of the eigenvector equation for $\varphi = 0 : 0 = \sum_k g^k \epsilon^{km3}$. Here ϵ^{km3} denotes the Levi-Civita tensor: it is zero if two indices are equal, and otherwise it is $+1$ or -1 depending on whether (k, m, ℓ) is an even or odd permutation of (1, 2, 3), respectively. The special eigenvector of the 3 by 3 matrix ϵ^{km3} is given by $(g^1, g^2, g^3) = (0, 0, g)$.

Since $\sum_k D_*^{mk} q^{*k} = q^m$, the preceding two paragraphs imply the formula

$$(A_{\vec{q}}(t)|V_{\vec{q}}^m) = G_q^{AV}(t)(q^m/q). \tag{B.6a}$$

The function on the right-hand side is the special correlator defined by

$$G_q^{AV}(t) = (A_{\vec{q}*}(t)|V_{\vec{q}*}^3). \tag{B.6b}$$

It depends on the wave vector via its modulus q only. The analogous result holds for the correlator of a vector and a scalar

$$(V_{\vec{q}}^m(t)|A_{\vec{q}}) = G_q^{VA}(t)(q^m/q), \tag{B.6c}$$

$$G_q^{VA}(t) = (V_{\vec{q}*}^3(t)|A_{\vec{q}*}). \tag{B.6d}$$

One checks that also the reverse statement is true: if the scalar–vector correlator can be represented in the form of Eq. (B.6a), the condition (B.5b) is valid. Hence, the pair of Eqs. (B.6a,b) is equivalent to the symmetry condition for scalar–vector correlators. The analogous statement holds for vector–scalar correlators.

The third case deals with the correlations of two vector fields, say $\vec{V}_{\vec{q}}$ and $\vec{W}_{\vec{q}}$. Using Eq. (B.5a) the isotropy condition (B.1) leads to

$$(V_{\vec{q}}^m(t)|W_{\vec{q}}^n) = \sum_{k\ell}(V_{D\vec{q}}^k(t)|V_{D\vec{q}}^\ell)D^{km}D^{\ell n}. \tag{B.7a}$$

This result has to hold for all rotation matrices D. The result can be specialized by using $D = D_*^{-1}$ in order to get $(V_{\vec{q}}^m(t)|W_{\vec{q}}^n) = \sum_{k\ell} g^{k\ell} D_*^{mk} D_*^{n\ell}$. Here $g^{k\ell} = (V_{\vec{q}*}^k(t)|W_{\vec{q}*}^\ell)$ denote the correlators for the distinguished wave vector $\vec{q}^*$. This wave vector is invariant for all rotations $D(\varphi)$. Hence, Eq. (B.7a) yields $g^{mn} = \sum_{k\ell} g^{k\ell} D(\varphi)^{km} D(\varphi)^{\ell n}$ for arbitrary φ. The derivative with respect to φ for $\varphi = 0$ leads to nine homogeneous linear equations for the nine quantities g^{mn}, $m, n = 1, 2, 3 : 0 = \sum_{k\ell} g^{k\ell}(\epsilon^{km3}\delta^{\ell n} + \delta^{km}\epsilon^{\ell n3})$. One derives that this system has a three-dimensional space of solutions. A solution is given by: $g_{(1)}^{33} = g_1$ and $g_{(1)}^{mn} = 0$ for $(m, n) \neq (3, 3)$. Another solution is specified by: $g_{(2)}^{\prime nn} = g_2^\prime$ for $n = 1, 2, 3$ and $g_{(2)}^{\prime mn} = 0$ for $n \neq m$. A third solution is determined by: $g_{(3)}^{mn} = g_3\epsilon^{mn3}$. It is more convenient to use as second independent solution the matrix with elements $g_{(2)}^{mn} = g_{(2)}^{\prime m,n} - g_{(1)}^{m,n}$. Hence, one arrives at $g^{mn} = g_1[q^{*m}q^{*n}/q^2] + g_2[\delta^{mn} - q^{*m}q^{*n}/q^2] + g_3\sum_\ell \epsilon^{mn\ell}[q^{*\ell}/q]$. Using Eq. (B.3b), one gets

$$\begin{aligned}(V_{\vec{q}}^m(t)|W_{\vec{q}}^n) &= H_{1,q}^{VW}(t)\big[(q^m/q)(q^n/q)\big] \\ &+ H_{2,q}^{VW}(t)\big[\delta^{mn} - (q^m/q)(q^n/q)\big] \\ &+ H_{3,q}^{VW}(t)\sum_\ell \epsilon^{mn\ell}(q^\ell/q).\end{aligned} \tag{B.7b}$$

The three time-dependent factors depend on the wave vector only via the modulus. They are special correlators given by

$$H_{1,q}^{VW}(t) = (V_{\vec{q}*}^3(t)|W_{\vec{q}*}^3), \tag{B.7c}$$
$$H_{2,q}^{VW}(t) = (V_{\vec{q}*}^1(t)|W_{\vec{q}*}^1) = (V_{\vec{q}*}^2(t)|W_{\vec{q}*}^2), \tag{B.7d}$$
$$H_{3,q}^{VW}(t) = (V_{\vec{q}*}^1(t)|W_{\vec{q}*}^2) = -(V_{\vec{q}*}^2(t)|W_{\vec{q}*}^1). \tag{B.7e}$$

As above, one can show that a representation in form of Eq. (B.7b) ensures the validity of the transformation law (B.7a).

Every vector fluctuation, say $\vec{V}_{\vec{q}}$, can be split uniquely in its projection parallel to $\vec{q}$, called the longitudinal part $\vec{V}_{L,\vec{q}}$, and the projection perpendicular to $\vec{q}$, called the transverse part $\vec{V}_{T,\vec{q}}$:

$$\vec{V}_{\vec{q}} = \vec{V}_{L,\vec{q}} + \vec{V}_{T,\vec{q}}. \tag{B.8a}$$

There holds $\vec{V}_{L,\vec{q}}\vec{V}_{T,\vec{q}} = 0$. The longitudinal part is characterized uniquely by the condition

$$\vec{q} \times \vec{V}_{L,\vec{q}} = 0, \tag{B.8b}$$

and the transverse one by the equation

$$\vec{q}\,\vec{V}_{T,\vec{q}} = 0. \tag{B.8c}$$

For the special wave vector $\vec{q}^*$, the longitudinal part $\vec{V}_{L,\vec{q}*}$ is given by the triple of coordinates $(0, 0, V_{\vec{q}*}^3)$, and the transverse part has the coordinates $(V_{\vec{q}*}^1, V_{\vec{q}*}^2, 0)$. Let us split $\vec{W}_{\vec{q}}$ in the same manner. The expressions in Eqs. (B.7d,e) vanish for the correlators of the longitudinal parts. Thus,

$$(V_{L,\vec{q}}^m(t)|W_{L,\vec{q}}^n) = H_{1,q}^{VW}(t)\left[q^m q^n/q^2\right]. \tag{B.9a}$$

For the correlators of the transverse parts, there holds $H_{1,q}^{VW}(t) = 0$, i.e.,

$$(V_{T,\vec{q}}^m(t)|W_{T,\vec{q}}^n) = H_{2,q}^{VW}(t)\left[\delta^{mn} - (q^m/q)(q^n/q)\right] + H_{3,q}^{VW}(t)\sum_{\ell}\epsilon^{mn\ell}(q^\ell/q). \tag{B.9b}$$

All three function in Eqs. (B.7c–e) vanish if a longitudinal part is correlated with a transverse one. There are no correlations between transverse and longitudinal fields in a homogeneous isotropic system

$$(V_{T,\vec{q}}^m(t)|W_{L,\vec{q}}^n) = (V_{L,\vec{q}}^m(t)|W_{T,\vec{q}}^n) = 0. \tag{B.9c}$$

Similarly, one finds that the right-hand sides of Eqs. (B.6b,d) vanish if $\vec{V}_{\vec{q}}$ is replaced by $\vec{V}_{T,\vec{q}}$. There are no correlations between scalar fields and transverse-vector fields,

$$(A_{\vec{q}}(t)|V^m_{T,\vec{q}}) = (V^m_{T,\vec{q}}(t)|A_{\vec{q}}) = 0. \tag{B.10}$$

Suppose that the two vector fields $\vec{V}_{\vec{q}}$ and $\vec{W}_{\vec{q}}$ have the same space-inversion parity. The absence-of-chirality condition (3.15b) holds for the correlator of the two fields if and only if the last term in Eq. (B.7b) is absent:

$$H^{VW}_{3,q}(t) = 0 \quad \text{if } \epsilon^V_P \epsilon^W_P = 1. \tag{B.11}$$

APPENDIX C

SMOOTHENED CORRELATORS

The replacement of some fluctuating-force spectrum by a white-noise one is an essential step in all derivations of hydrodynamic-limit descriptions of correlation functions. In Sec. 3.2.2, this is demonstrated for the simplest example, namely for the diffusion law. Also Langevin's extension of the hydrodynamic-limit theory for the tagged-particle motion is built on such replacement, as is discussed in Sec. 3.2.3. These theories motivate the replacement of some correlator $\phi(t)$ with a spectrum $\phi''(\omega)$ by a model correlator $\overline{\phi}(t)$ with a spectrum $\overline{\phi}''(\omega)$ by arguing that $\overline{\phi}''(\omega)$ approximates $\phi''(\omega)$ for small frequencies ω. The mode-coupling-theory models for correlation functions are more involved than the above-mentioned models. But, also within this approach, the appeal to a low-frequency restriction is an essential point for the motivation of the equations of motion for the model correlators. This section shall be used to explain in which sense the model correlator $\overline{\phi}(t)$ in the time domain can be considered as an approximation of the correlator $\phi(t)$.

In this paragraph, properties of a pair of real auxiliary functions $\hat{c}(\hat{\omega})$ and $c(\hat{t})$ shall be specified, which are connected by Fourier transformation. The functions are defined for all values of $\hat{\omega}$ and $\hat{t}$, respectively. The first function depends smoothly on $\hat{\omega}$ and it obeys: $0 \leq \hat{c}(\hat{\omega}) = \hat{c}(-\hat{\omega}) \leq c_{\max} < \infty$. It shall exhibit the normalization $\hat{c}(\hat{\omega} = 0) = 1$. There shall be a constant c_0 and some exponent $n_0 \geq 2$ so that

$$\hat{c}(\hat{\omega}) \leq c_0/|\hat{\omega}|^{n_0}. \tag{C.1}$$

This condition ensures the integrability. The Fourier transform exists as even continuous function, as discussed in Appendix A.2. It defines the second function: $c(\hat{t}) = \int_0^\infty \cos[\hat{\omega}\hat{t}]\hat{c}(\hat{\omega})d\hat{\omega}/\pi$. According to Bochner's theorem, $c(\hat{t})$ is positive definite. In particular, there holds $|c(\hat{t})| \leq c(\hat{t} = 0)$. It is requested that $c(\hat{t})$ is absolutely integrable. Hence, $\hat{c}(\hat{\omega}) = 2\int_0^\infty \cos[\hat{\omega}\hat{t}]c(\hat{t})d\hat{t}$.

For every positive time t_{cg}, a function of frequency ω can be defined by

$$\hat{c}_{cg}(\omega) = \hat{c}(\omega t_{cg}). \tag{C.2a}$$

The normalization and smoothness condition imply

$$\hat{c}_{cg}(\omega) = 1 + O((\omega t_{cg})^2). \tag{C.2b}$$

This function $\hat{c}_{cg}(\omega)$ specifies a filter on a frequency scale $1/t_{cg}$. It is unity for $\omega \ll 1/t_{cg}$ and, because of Eq. (C.1), it is small for $\omega \gg 1/t_{cg}$. It is the Fourier transform of

$$c_{cg}(t) = c(t/t_{cg})/t_{cg}. \tag{C.2c}$$

This is a function with its maximum value at time zero: $|c_{cg}(t)| \leq c_{cg}(t=0) = c(\hat{t}=0)/t_{cg}$. The area under the c_{cg} versus t curve is normalized:

$$\int c_{cg}(t)dt = 1. \tag{C.2d}$$

Let $F(t)$ denote a continuous function, which is explained for all times t so that $|F(t)| \leq F_{\max} < \infty$. The convolution integral

$$F_{cg}(t) = \int c_{cg}(t-t')F(t')dt' \tag{C.3a}$$

is defined for all t_{cg}. There holds $F_{cg}(t) = \int F(t - t_{cg}x)c(x)dx$. The function $F_{\max}|c(x)|$ is an integrable majorant for the integrand. Consequently, $F_{cg}(t)$ is continuous and $\lim_{t_{cg}\to 0} F_{cg}(t) = F(t)$. The value $F_{cg}(t)$ is obtained by averaging $F(t')$ with weight $c_{cg}(t-t')$. The main contributions stem from the interval $|t-t'| < t_{cg}$. In this sense, $F_{cg}(t)$ is obtained by smoothening $F(t)$ on a time scale t_{cg}. The smoothening is also referred to as coarse graining.

Suppose that $F(t)$ is a measurable quantity. Every experimental set up is specified by some resolution function and some characteristic resolution scale. Equation (C.3a) is a mathematical description of such measuring result, where $c(\hat{t})$ models the resolution function of the spectrometer and t_{cg} quantifies the resolution scale. The smoothening can be defined also for unbounded functions $F(t)$, provided one specifies proper conditions for the large-$\hat{t}$ behaviour of the resolution function $c(\hat{t})$.

If $F(t)$ is absolutely integrable and has an absolutely integrable Fourier transform $\hat{F}(\omega)$, there is the one-to-one correspondence between these functions, which is given by Eqs. (A.9), (A.15). The convolution theorem (A.12) casts Eq. (C.3a) into the equivalent relation

$$\hat{F}_{cg}(\omega) = \hat{F}(\omega)\hat{c}_{cg}(\omega). \tag{C.3b}$$

Hence, coarse graining of $F(t)$ on a time scale t_{cg} is equivalent to eliminating the Fourier components of $\hat{F}(\omega)$ for $|\omega| \gg 1/t_{cg}$.

Let $\phi(t)$ and $\overline{\phi}(t)$ denote two continuous positive-definite functions of the time. In order to relate them by elementary formulas to their spectra $\phi''(\omega)$ and $\overline{\phi}''(\omega)$, respectively, all four functions are requested to be absolutely integrable. According to the discussions of Sec. 2.3, the spectra are continuous non-negative functions of the frequency. To avoid too many technicalities, it shall be requested in addition that there is a constant $c_1 > 0$ so that

$$\phi''(\omega) \leq \pi c_1/|\omega|, \qquad \overline{\phi}''(\omega) \leq \pi c_1/|\omega|. \tag{C.4}$$

Motivated by the introductory remarks above, $\phi(t)$ shall be considered as some correlator and $\overline{\phi}(t)$ as some model for this quantity. For convenience, the initial value of the two functions shall be assumed to be the same: $\phi(t=0) = \overline{\phi}(t=0) = g$. There holds $|\phi(t)| \leq g$ and $|\overline{\phi}(t)| \leq g$. The number g defines the natural scale for the values of the two functions considered.

The essential request for the relation of the two spectra $\phi''(\omega)$ and $\overline{\phi}''(\omega)$ is the following. For every positive error margin ϵ, there is a positive frequency ω_ϵ so that

$$|\phi''(\omega) - \overline{\phi}''(\omega)| \leq \epsilon \cdot \overline{\phi}''(\omega)/(2c_{\max}) \quad \text{if } |\omega| \leq \omega_\epsilon. \tag{C.5a}$$

In this sense, the model spectrum $\overline{\phi}''(\omega)$ approximates the spectrum $\phi''(\omega)$ for small frequencies with a relative error $\epsilon/(2c_{\max})$. One obtains $|\phi_{cg}(t) - \overline{\phi}_{cg}(t)| \leq \int |\phi''(\omega) - \overline{\phi}''(\omega)|\hat{c}_{cg}(\omega)d\omega/\pi$. For the contribution I_1 to the integral, which is due to the small-frequency interval $-\omega_\epsilon \leq \omega \leq \omega_\epsilon$, one can estimate: $0 \leq I_1 \leq c_{\max} \int_{-\omega_\epsilon}^{\omega_\epsilon} |\phi''(\omega) - \overline{\phi}''(\omega)|d\omega/\pi \leq (\epsilon/2) \int_{-\omega_\epsilon}^{\omega_\epsilon} \overline{\phi}''(\omega)d\omega/\pi \leq \epsilon g/2$. For the remaining contribution I_2, which is due to the large-frequency intervals $\omega \leq -\omega_\epsilon$ and $\omega \geq \omega_\epsilon$, one can use Eqs. (C.1), (C.4) in order to estimate: $0 \leq I_2 \leq [4c_1 c_0/t_{cg}^{n_0}] \int_{-\omega_\epsilon}^{\infty} [1/\omega^{n_0+1}]d\omega$. Consequently,

$$|\phi_{cg}(t) - \overline{\phi}_{cg}(t)|/g \leq \epsilon \quad \text{if } c_\epsilon/\omega_\epsilon \leq t_{cg}, \tag{C.5b}$$

$$c_\epsilon = [8c_0c_1/(gn_0\epsilon)]^{1/n_0}. \tag{C.5c}$$

Measured on the naturale scale g, the smoothened model correlator $\overline{\phi}_{cg}(t)$ approximates uniformly the smoothened correlator $\phi_{cg}(t)$ with an error bound ϵ. This holds if the smoothening scale t_{cg} is chosen coarser than $c_\epsilon/\omega_\epsilon$.

An ideal description of $\phi(t)$ by $\overline{\phi}(t)$ would require the performance of the limit t_{cg} tending to zero. Naturally, this would require also that the description of the spectrum of $\phi''(\omega)$ by $\overline{\phi}''(\omega)$ would be valid for all frequencies. The smoothening can be done on a smaller scale t_{cg} if the high-frequency contributions of the spectra decrease and if the accuracy of the approximation is decreased. Similarly, t_{cg} can decrease if the filter $\hat{c}(\hat{\omega})$ suppresses more efficiently the high-frequency spectra. This explains why c_ϵ decreases with $c_1, 1/\epsilon$, and c_0 as well as with increasing n_0. If n_0 tends to infinity, c_ϵ tends to unity.

The result (C.5b) deals with an absolute and not with a relative deviation of the two smoothened functions. For times, where both functions are small compared to g, $\phi_{cg}(t)$ may differ qualitatively from $\overline{\phi}_{cg}(t)$. For example, a good approximation of the low-frequency part of $\phi''(\omega)$ by $\overline{\phi}''(\omega)$ in the sense of Eq. (C.5a) does not exclude that $\overline{\phi}(t)$ decays exponentially for large times, while $\phi(t)$ exhibits a power-law long-time tail.

APPENDIX D

THEOREMS ON MCT EQUATIONS

D.1 Convergence of the approximant sequences

A sequence of arrays of approximants $\boldsymbol{\phi}^{(r)}(t), r = 0, 1, \ldots$, is constructed in connection with Eqs. (4.25)–(4.27). In the present section, the uniform convergence of this sequence will be demonstrated. The proofs are due to Haussmann (1990) for the first version of the equations of motion and due to Götze and Sjögren (1995) for the second one. For the discussions, which go beyond the ones presented in Sec. 4.2.2, the label q for the components of the arrays merely occurs as a fixed parameter. There is no additional reasoning necessary for the extension of the proofs from the case dealing with a single component, $M = 1$, to the general case with a larger value of M. Therefore, only the one-component case shall be considered.

It is the goal to prove the uniform convergence of the sum $S = \sum_{r=0}^{\infty} X_r(t)$, which is formed with the positive functions

$$X_r(t) = |\phi^{(r+1)}(t) - \phi^{(r)}(t)|/2. \tag{D.1}$$

This result ensures the absolute and uniform convergence of the sum $\sum_{r=0}^{\infty} [\phi^{(r+1)}(t) - \phi^{(r)}(t)]$ and, consequently, the existence of the desired uniform limit formulated in Eq. (4.25a). The goal shall be reached by constructing a sequence of positive numbers b_r, which define a convergent sum $B = \sum_{r=0}^{\infty} b_r$. The numbers b_r depend neither on the time t nor on the coefficients $V_\alpha, \alpha = 1, \ldots, N$, nor on the frequencies Ω and ν or the time τ, which specify the equations of motion. This holds provided the coefficients are restricted to the closed finite intervals specified at the beginning of Sec. 4.2.2 and provided $0 \leq t \leq t_{\max}$. Here, $t_{\max}$ is some arbitrary positive finite upper limit for the time interval to be considered. The numbers b_r shall serve as majorants:

$$X_r(t) \leq b_r, \quad r = 0, 1, \ldots \tag{D.2}$$

The approximant $\phi^{(0)}(t)$ reads $\exp[-\Gamma t], \Gamma > 0$. For the other approximants, the recursion relations (4.27a,b) shall be written in the form:

$$\phi^{(r+1)}(t) = 1 + \nu t - \nu \int_0^t \phi^{(r+1)}(t')dt'$$
$$+ \Omega^2 \int_0^t \Big\{ \int_0^{t'} K\big[\phi^{(r)}(t''), \phi^{(r)}(t' - t''), \phi^{(r+1)}(t'')\big] dt'' \Big\} dt', \tag{D.3a}$$

$$\phi^{(r+1)}(t) = 1 + (1/\tau) \int_0^t K\big[\phi^{(r)}(t'), \phi^{(r)}(t - t'), \phi^{(r+1)}(t')\big] dt', \tag{D.3b}$$

respectively. The kernel abbreviates the function of three variables ξ, η and ζ:

$$K[\xi,\eta,\zeta] = \mathcal{F}[P,\xi] - \zeta\{1 + \mathcal{F}[P,\eta]\}. \tag{D.4a}$$

One infers from Eq. (4.25c) that the three variables are located in finite closed intervals:

$$-1 \le \xi, \eta, \zeta \le 1. \tag{D.4b}$$

K is a polynomial and, thus, it obeys a Lipschitz condition. Since the variables as well as the coefficients of the mononomials of $\mathcal{F}[P,\xi]$ are restricted to finite closed intervals, a Lipschitz constant $L_{\max}$ can be chosen so that there holds

$$|K[\xi_1,\eta_1,\zeta_1] - K[\xi_2,\eta_2,\zeta_2]| \le L_{\max}(|\xi_1 - \xi_2| + |\eta_1 - \eta_2| + |\zeta_1 - \zeta_2|) \tag{D.4c}$$

for all variables occurring and all coefficients admitted. Using this estimation, one obtains from Eq. (D.3a) for the first version of equations of motion:

$$X_r(t) \le \nu \int_0^t X_r(t')dt' + (L_{\max}\Omega^2) \int_0^t \Big\{ \int_0^{t'} \big[2X_{r-1}(t'') + X_r(t'')\big]dt'' \Big\} dt'. \tag{D.5a}$$

From Eq. (D.3b), one gets for the second version:

$$X_r(t) \le (L_{\max}/\tau) \int_0^t \big[2X_{r-1}(t') + X_r(t')\big]dt'. \tag{D.5b}$$

Let us define recursively two versions of sequences of functions $a_r(t), r = 0, 1, \ldots$ One writes $a_0(t) = 1$. For the first version, there holds for $r \ge 1$:

$$a_r(t) = \nu \int_0^t a_{r-1}(t')dt' + 3\mu \int_0^t \Big\{ \int_0^{t'} a_{r-1}(t'')dt'' \Big\} dt' \tag{D.6a}$$

with $\mu = L_{\max}\Omega^2$. For the second one, there holds for $r \ge 1$:

$$a_r(t) = 3\kappa \int_0^t a_{r-1}(t')dt' \tag{D.6b}$$

with $\kappa = L_{\max}/\tau$. One shows by induction that the functions $a_r(t)$ are non-negative and non-decreasing functions of t. The important inequality (4.25c) implies

$$X_r(t) \le a_0(t). \tag{D.7}$$

Let us prove by induction that the preceding inequality can be extended to: $X_r(t) \le a_1(t), a_2(t), \ldots, a_r(t), r \ge 1$. As first step, one substitutes the inequality $X_1(t) \le a_0$ into the Eq. (D.5a) in order to get: $X_1(t) \le \nu \int_0^t a_0(t')dt' + \mu \int_0^t \{\int_0^{t'} [2X_0(t'') + a_0(t'')]dt''\}dt' = \nu \int_0^t a_0(t')dt' + 3\mu \int_0^t \{\int_0^{t'} a_0(t'')dt''\}dt' = a_1(t)$.

Using Eq. (D.5b), one obtains $X_1(t) \leq 3\kappa \int_0^t a_0(t')dt' = a_1(t)$. Repetition of this step yields $X_2(t) \leq a_1(t)$. This can be substituted into Eq. (D.5a) in order to get: $X_2(t) \leq \nu \int_0^t a_1(t')dt' + \mu \int_0^t \{\int_0^{t'} [2X_1(t'') + a_1(t'')]dt''\}dt'$. Using the preceding result, $X_1(t'') \leq a_1(t'')$, and formula (D.6a), one gets $X_2(t) \leq a_2(t)$. The second version of inequalities (D.5b,D.6b) yields the same conclusion. This reasoning can be continued. In step number r, one uses Eq. (D.7) to derive $X_r(t) \leq a_1(t)$. From this, there follows $X_r(t) \leq a_2(t)$. This leads to $X_r(t) \leq a_3(t)$, etc. The resulting inequality

$$X_r(t) \leq a_r(t) \tag{D.8}$$

establishes the sequence $a_r(t), r = 0, 1, \ldots$, as majorant of the sequence of functions $X_r(t)$.

Since $0 \leq a_r(t'') \leq a_r(t')$ for $0 \leq t'' \leq t' \leq t_{\max}$, one gets the estimation: $\int_0^{t'} a_{r-1}(t'')dt'' \leq t_{\max} a_{r-1}(t')$. Defining a further sequence of functions $b_r(t)$, $r = 0, 1, \ldots$ recursively by $b_0(t) = 1$ and $b_r(t) = (\nu + 3\mu t_{\max}) \int_0^t b(t')dt'$, one derives from Eq. (D.6a): $a_r(t) \leq b_r(t)$. One concludes for both versions of equations of motion

$$X_r(t) \leq (At)^r/r! \tag{D.9}$$

Here $A = (\nu + 3\mu t_{\max})$ and $A = 3\kappa$, respectively. The desired result (D.2) is established. One can choose $b_r = B^r/r!$ with $B = (\nu_{\max} + 3L_{\max}\Omega^2_{\max} t_{\max})t_{\max}$ and $B = 3L_{\max}\gamma_{\max}t_{\max}$, respectively.

D.2 Completely monotonic approximants

In this section, the sequence of approximants $\phi^{(r)}(t)$ shall be discussed, which is constructed in Sec. 4.2.2 for the second version of equations of motion; $r = 0, 1, \ldots$ It will be demonstrated that all functions $\phi_q^{(r)}(t), q = 1, \ldots, M$, are finite sums of elementary relaxation correlators.

Let us use this paragraph to note some formulas with the aim to simplify the following discussions. All functions $F(t)$ to be considered shall be continuous and exhibit the standard symmetries: $F(t) = F(t)^* = F(-t)$. Therefore, all times can be restricted to $t \geq 0$. Function $F(t)$ is called completely monotonic if there is a bounded monotonically increasing weight function $\sigma(\gamma)$ so that

$$F(t) = \int_0^\infty \exp[-\gamma t] d\sigma(\gamma). \tag{D.10a}$$

This is equivalent to the formula for the Laplace transform

$$F(z) = \int_0^\infty \{-1/[z + i\gamma]\} d\sigma(\gamma), \quad \mathrm{Im}\, z > 0. \tag{D.10b}$$

The expression describes also the analytical continuation on the whole plane of complex frequencies z, except for the values $z = -i\gamma, \gamma \geq 0$. From Eq. (2.46b), one gets for the spectrum

$$F''(\omega) = \int_0^\infty \{\gamma/[\omega^2 + \gamma^2]\} d\sigma(\gamma). \tag{D.10c}$$

It is an analytic function of the frequency for all $\omega \neq 0$. The Laplace transform in the conventional notation is defined by Eq. (A.1) for the half plane of values s, which obey Re $s > 0$: $\hat{F}(s) = \int_0^\infty \exp[-st]F(t)$. From Eq. (D.10a), one gets

$$\hat{F}(s) = \int_0^\infty \{1/[s + \gamma]\} d\sigma(\gamma), \quad \mathrm{Re}\, s > 0. \tag{D.11a}$$

This expression defines the analytical continuation on the whole complex plane except, possibly, for the values $s = -\gamma, \gamma \geq 0$. In agreement with the general formula (A.3b), there holds the relation

$$F(z) = i\hat{F}(s), \quad s = -iz. \tag{D.11b}$$

It is demonstrated in Sec. 4.2.2 that the recursion relation (4.26b) is equivalent to the double-fraction representation: $\phi_q^{(r+1)}(z) = -1/\{z - 1/[i\tau_q + m_q^{(r)}(z)]\}$, Im $z > 0$. The index q shall be dropped in the following in order to simplify the formulas. According to Eq. (A.3b), the recursion relation can be written in the conventional notation for Re $s > 0$ as

$$\hat{\phi}^{(r+1)}(s) = [\tau + \hat{m}^{(r)}(s)]/\{1 + s[\tau + \hat{m}^{(r)}(s)]\}. \tag{D.12}$$

Let us assume that the kernel $m^{(r)}(t)$ is a sum of n elementary relaxation functions, $n = 1, 2, \ldots$ This means that there are $n + 1$ positive amplitudes μ_k and positive rates $\gamma_k, k = 1, \ldots, n$ so that

$$m^{(r)}(t) = \sum_{k=1}^{n} \mu_k \exp[-\gamma_k t]. \tag{D.13a}$$

The rates shall be ordered:

$$0 < \gamma_1 < \gamma_2 < \cdots < \gamma_n. \tag{D.13b}$$

It is the goal to derive the proposition: there are $n + 1$ positive amplitudes f_k and positive rates $\gamma'_k, k = 0, 1, \ldots, n$ so that

$$\phi^{(r+1)}(t) = \sum_{k=0}^{n} f_k \exp[-\gamma'_k t]. \tag{D.13c}$$

The sequence of rates for $\phi^{(r+1)}(t)$ is separated by the one for $m^{(r)}(t)$:

$$0 < \gamma'_0 < \gamma_1 < \gamma'_1 < \cdots \gamma_n < \gamma'_n. \tag{D.13d}$$

Suppose that the proposition is correct. Then, one can show by induction the statement made in the first paragraph of this section. For $r = 0$, the statement is correct since the iteration in Sec. 4.2.2 is started with $\phi_q^{(0)}(t) = \exp[-\Gamma_q t]$, $\Gamma_q > 0, q = 1, \ldots, M$. If the statement holds for $\phi_q^{(r)}(t)$, one can write this function as a finite sum $\sum_j f_{q,j}^{(r)} \exp[-\gamma_{q,j}^{(r)\prime}]$, with all amplitudes and rates being positive. Equation (4.25b) implies $m_q^{(r)}(t) = \sum_{k=1}^{n} \mu_{q,k}^{(r)} \exp[-\gamma_{q,k}^{(r)} t]$. Formula (4.15a) shows that the rates $\gamma_{q,k}^{(r)}$ are positive, since they are sums of positive terms of the kind $\gamma_{k_1,j_1}^{(r)\prime} + \gamma_{k_2,j_2}^{(r)\prime} + \cdots$ The amplitudes are sums of products of the kind $V_{q,k_1\cdots k_n}^{(n)} f_{k_1,j_1}^{(r)} \cdots f_{k_n,j_n}^{(r)}$ and, because of Eq. (4.15b), they are not negative. The proposition yields the desired formula (4.28) for $\phi_q^{(r+1)}(t)$.

The proof of the proposition starts by rewriting Eq. (D.13a) in the equivalent form

$$\hat{m}^{(r)}(s) = \sum_{k=1}^{n} \mu_k/(s + \gamma_k). \tag{D.13e}$$

One can present this function as ratio of a denominator polynomial of degree n, $D(s) = (s+\gamma_1)(s+\gamma_2)\cdots(s+\gamma_n) = s^n + O(s^{n-1})$, and a numerator polynomial of degree $n-1$, $N(s) = (\sum_{k=1}^{n} \mu_k)s^{n-1} + O(s^{n-2}) : \hat{m}^{(r)}(s) = N(s)/D(s)$. Because of Eq. (D.12), the approximant can be presented as ratio of polynomials of degree n and $(n+1)$: $\hat{\phi}^{(r+1)}(s) = [\tau D(s) + N(s)]/[\tau s D(s) + D(s) + sN(s)]$. Function $\hat{\phi}^{(r+1)}(s)$ is meromorphic, and it can have at most $(n+1)$ poles. Since $D(s = 0) = \gamma_1 \cdots \gamma_n \neq 0$, the value $s = 0$ cannot be a pole. One concludes that the poles are the zeros of the function $\varphi(s) = \tau + (1/s) + \hat{m}^{(r)}(s)$. Restricted to a function of the real variable x, $\varphi(x)$ is a real function. It decreases strictly: $\partial\varphi(x)/\partial x < 0$. It has $(n+1)$ simple poles for the positions $-\gamma_n < -\gamma_{n-1} < \cdots < -\gamma_1 < -\gamma_0 = 0$. If x increases from $-\gamma_{\ell+1}$ to $-\gamma_\ell, \varphi(x)$ decreases from ∞ to $-\infty, \ell = 0, \ldots, n-1$. Hence, there are n zeroes γ'_ℓ, obeying the conditions (D.13d). If x decreases from $-\gamma_n$ to $-\infty, \varphi(x)$ increases from $-\infty$ to τ. Consequently, there is a zero $-\gamma'_n$, obeying $-\gamma'_n < -\gamma_n$. Thereby, $(n+1)$ simple poles of the approximant are identified, and one can write the partial-fraction representation:

$$\hat{\phi}^{(r+1)}(s) = \sum_{k=0}^{n} f_k/(s + \gamma'_k). \tag{D.13f}$$

Since $\partial\hat{\phi}^{(r+1)}(x)/\partial x < 0$, there holds $f_k > 0, k = 0, \ldots, n$. Since the representation is equivalent to Eq. (D.13c), the proof is completed.

The discussion of Eq. (D.12) can be modified to one for the recursion defined in Sec. 6.2.1 for the shape functions. Equation (6.114b) can be noted as

$$\hat{\phi}^{(r+1)}(s) = \hat{m}^{(r)}(s)/\big[1 + s\hat{m}^{(r)}(s)\big]. \tag{D.14}$$

This expression is obtained from Eq. (D.12) by specializing to $\tau = 0$. The kernel is quantified by n pairs of numbers $(\mu_k, \gamma_k), k = 1, \ldots, n$, as is explained in connection with Eqs. (D.13a,b). The following proposition shall be proven. There are n pairs of positive numbers $(f_k, \gamma'_k), k = 0, \ldots, n-1$ so that

$$\hat{\phi}^{(r+1)}(t) = \sum_{k=0}^{n-1} f_k \exp[-\gamma'_k t]. \tag{D.15a}$$

The sequence of rates for function $\hat{\phi}^{(r+1)}(t)$ separates that for the function $m^{(r)}(t)$:

$$0 < \gamma'_0 < \gamma_1 < \gamma'_1 < \cdots < \gamma'_{n-1} < \gamma_n. \tag{D.15b}$$

Contrary to what is discussed above for Eq. (D.12), the number of relaxators contributing to $\hat{\phi}^{(r+1)}(t)$ is the same as that of the relaxators contributing to $m^{(r)}(t)$.

The induction proof presented in the paragraph preceding Eq. (D.13e) remains valid. Hence, one can proceed as above and write the kernel as ratio of a polynomial $N(s)$ of degree $(n-1)$ and a polynomial $D(s)$ of degree $n : \hat{m}^{(r)}(s) = N(s)/D(s)$. Equation (D.14) yields $\hat{\phi}^{(r+1)}(s) = N(s)/[D(s) + sN(s)]$. This function is meromorphic. Different to what is deduced above from Eq. (D.12), $\hat{\phi}^{(r+1)}(s)$ cannot have more than n poles. The poles are the zeros of the function $\varphi(s) = (1/s) + \hat{m}^{(r)}(s)$. One continues as above. For real values $x = s$, the function $\varphi(x)$ is strictly decreasing. It has $(n+1)$ simple poles at $-\gamma_n < -\gamma_{n-1} < \cdots < \gamma_1 < 0 = \gamma_0$. Consequently, there are n poles at $\gamma'_0, \ldots, \gamma'_{n-1}$, which obey the condition (D.15b). As above, one shows that the residues f_k are positive. Hence, there holds the partial fraction representation

$$\hat{\phi}^{(r+1)}(s) = \sum_{k=0}^{n-1} f_k/(s + \gamma'_k). \tag{D.15c}$$

Since this formula is equivalent to Eq. (D.15a), the proof is completed.

D.3 The maximum-eigenvalue inequality

In this section, the inequality $E(P) \leq 1$ for the maximum eigenvalue $E(P) = E[P, \boldsymbol{f}(P)]$ for the maximum fixed point $\boldsymbol{f}(P)$ shall be derived. The proof (Götze and Sjögren 1995) is done indirectly. It will be assumed that there is some positive δ so that $E(P) = 1 + \delta$. The assumption shall be used to construct a fixed point $\boldsymbol{g}(P)$ with a component of some label $q_0, 1 \leq q_0 \leq M$, obeying $g_{q_0}(P) > f_{q_0}(P)$. This result contradicts the maximum property (4.52e). Consequently, the assumption is illegitimate; and this conclusion implies the desired result (4.60c).

In order to formulate the assumption more explicitly, let $\boldsymbol{r}$ denote an eigenvector of the stability matrix (4.60a) for state P with eigenvalue $E(P)$, i.e.,

$$\sum_p A_{q,p}(P) r_p = (1+\delta) r_q, \quad \delta > 0, q = 1, \ldots, M. \tag{D.16a}$$

There is a non-zero component of $\boldsymbol{r}$, say, $r_{q_0} \neq 0$. One of the Frobenius–Perron theorems formulates that one can choose the eigenvector so that none of its components is negative (Gantmacher 1974). Hence, one can request

$$0 \leq r_q, \quad q = 1, \ldots, M; \quad 0 < r_{q_0}. \tag{D.16b}$$

The transformation (4.22a) shall be applied for the maximum fixed point $\boldsymbol{f}^*(P) = \boldsymbol{f}(P)$:

$$x_q = f_q(P) + [1 - f_q(P)]\hat{x}_q, \quad q = 1, \ldots, M. \tag{D.17a}$$

The formula (4.22c) for the transformed coefficients $\hat{V}^{(1)}_{q,p}$ agrees with the corresponding one for the elements of the stability matrix in Eq. (4.60a). Equation (4.22b) can be written as

$$\hat{\mathcal{F}}_q[P, \boldsymbol{x}] = \sum_p A_{q,p} x_p + B_q[P, x]. \tag{D.17b}$$

The last term combines the contribution due to the coefficients $\hat{V}^{(n)}_{q,k_1 \cdots k_n}$ with $n \geq 2$. Since these coefficients are non-negative, there holds

$$B_q(P, \boldsymbol{x}) \geq 0, \quad q = 1, \ldots, M, \quad \text{if} \quad x_k \geq 0, \ k = 1, \ldots, M. \tag{D.17c}$$

Let us choose some positive ϵ, which is smaller than the inverse of the largest component of $\boldsymbol{r}$. There holds $0 \leq \xi r_q < 1$ for all q and all ξ obeying $0 < \xi \leq \epsilon$. The polynomials $\hat{\mathcal{F}}_q[P, \boldsymbol{x}]$ vanish for $\boldsymbol{x} = \boldsymbol{0}$. Hence, there is some positive Lipschitz constant C so that: $0 \leq \hat{\mathcal{F}}_q[P, \xi \boldsymbol{r}] \leq C\xi, q = 1, \ldots, M$. Introducing the positive number $\xi_0 = \min(\epsilon, \delta/(2C))$, an array shall be defined by

$$\hat{\boldsymbol{f}}^{(0)} = \xi_0 \boldsymbol{r}. \tag{D.18a}$$

There holds

$$0 \leq \hat{f}^{(0)}_q < 1, \quad q = 1, \ldots, M, \quad 0 < \hat{f}^{(0)}_{q_0}. \tag{D.18b}$$

Array $\hat{\boldsymbol{f}}^{(0)}$ is used as starting element of a sequence of arrays $\hat{\boldsymbol{f}}^{(n)}, n = 0, 1, \ldots,$ which shall be defined recursively by

$$\hat{\boldsymbol{\mathcal{T}}}[P, \hat{\boldsymbol{f}}^{(n)}] = \hat{\boldsymbol{f}}^{(n+1)}, \quad n = 0, 1, \ldots \tag{D.18c}$$

The covariance theorem implies that the mapping $\hat{\boldsymbol{\mathcal{T}}}$ has the same general properties as $\boldsymbol{\mathcal{T}}$. Therefore, Eq. (4.20b) ensures the restrictions

$$0 \leq \hat{f}_q^{(n)} < 1, \quad q = 1, \ldots, M, \quad n = 0, 1 \ldots . \tag{D.18d}$$

The choice of ξ_0 yields the estimation: $1 + \hat{\mathcal{F}}_q[P, \hat{\boldsymbol{f}}^{(0)}] \leq 1 + C\xi_0 \leq (1 + \delta/2)$. From Eqs. (D.17b,c), one gets $\hat{\mathcal{F}}_q[P, \hat{\boldsymbol{f}}^{(0)}] \geq \sum_p A_{q,p} \hat{f}_p^{(0)} = (1+\delta)\hat{f}_q^{(0)}$. Combining both inequalities, one derives from Eq. (4.23a): $\hat{\mathcal{T}}_q[P, \hat{\boldsymbol{f}}^{(0)}] \geq [(1 + \delta)/(1+\delta/2)]\hat{f}_q^{(0)} \geq \hat{f}_q^{(0)}$. Consequently, there holds

$$\hat{f}_q^{(0)} \leq \hat{f}_q^{(1)}, \quad q = 1, \ldots, M. \tag{D.19a}$$

Using Eq. (4.20c) for the mapping $\hat{\mathcal{T}}$, one shows by induction that the preceding result can be extended to:

$$\hat{f}_q^{(n)} \leq \hat{f}_q^{(n+1)}, \quad q = 1, \ldots, M, \quad n = 0, 1, \ldots . \tag{D.19b}$$

The increasing sequence of numbers $f_q^{(n)}, n = 0, 1, \ldots$, which is bounded from above, converges towards some non-negative number, say $\hat{g}_q$. These numbers form the components of $\hat{\boldsymbol{g}}$:

$$\lim_{n \to \infty} \hat{\boldsymbol{f}}^{(n)} = \hat{\boldsymbol{g}}. \tag{D.20a}$$

Since $\hat{\mathcal{T}}[P, \boldsymbol{x}]$ is a continuous function of $\boldsymbol{x}$, Eq. (D.18c) implies that $\hat{\boldsymbol{g}}$ is a fixed point:

$$\hat{\boldsymbol{\mathcal{T}}}[P, \hat{\boldsymbol{g}}] = \hat{\boldsymbol{g}}. \tag{D.20b}$$

Equation (4.20b) yields $0 \leq \hat{g}_q < 1, q = 1, \ldots, M$. By construction, there holds $\hat{f}_q^{(0)} \leq \hat{g}_q$ for all q; and one gets from Eq. (D.18b):

$$0 < \hat{g}_{q_0}. \tag{D.20c}$$

The covariance theorem (4.23b) implies that $\boldsymbol{g}$ is a fixed point: $\boldsymbol{\mathcal{T}}[P, \boldsymbol{g}] = \boldsymbol{g}$. From Eq. (D.17a), one derives the desired inequality: $g_{q_0} = f_{q_0}(P) + (1 - f_{q_0}(P))\hat{g}_{q_0} > f_{q_0}(P)$.

D.4 Further properties of stability matrices

The M-component eigenvectors $\boldsymbol{a}^*$ and $\boldsymbol{a}$ and the M-by-M matrix R are defined in Eqs. (4.74), (4.75). These quantities are determined by the stability matrix A^c at the bifurcation point P^c. The quantities enter all formulas for the asymptotic expansions of solutions of MCT equations for states P near P^c. For the most relevant case of a primitive irreducible critical stability matrix, the quantities

$a_q^*, a_p, R_{q,p}$ for $q, p = 1, \ldots, M$ can be expressed as limits of powers of the M^2 elements $A_{q,p}^c$ of the critical stability matrix A^c. The relevant formulas shall be derived in this section. Furthermore, the asymptotic behaviour of the maximum eigenvalue $E(P)$ will be determined for P approaching P^c.

According to Eq. (4.76b), the maximum η of the moduli of the first $(M-1)$ eigenvalues of A^c is smaller than unity.

$$|e_k| \leq \eta, \quad 0 \leq \eta < 1, \quad k = 1, \ldots, M-1. \tag{D.21}$$

The Jordan form of A^c consists of a 1-by-1 block for the maximum eigenvalue $e_M = 1$. The other blocks correspond to the other eigenvalues. The nth power of A^c leaves fixed the block for e_M, and reduces the diagonal elements of the other blocks to e_k^n. Consequently, $(A^{cn})_{q,k} = a_q a_k^* + O(n^\ell \eta^n), 0 \leq \ell \leq M-1$. With increasing exponent n, the matrix elements of A^{cn} converge exponentially towards the product of the distinguished eigenvectors:

$$\lim_{n\to\infty} (A^{cn})_{q,k} = a_q a_k^*, \quad q, k = 1, \ldots, M. \tag{D.22}$$

Imposing the conventions (4.74c,d), the M^2 products on the right-hand side determine the eigenvectors $\boldsymbol{a}$ and $\boldsymbol{a}^*$ uniquely.

The eigenvectors shall be used to define an M-by-M matrix A' by its elements

$$A'_{q,k} = A^c_{q,k} - a_q a_k^*. \tag{D.23a}$$

The Jordan form of A' agrees with that of A^c, except that the distinguished 1-by-1 block is replaced by zero. Hence, the spectral radius of A' does not exceed $\eta : (A'^n)_{q,k} = O(n^\ell \eta^n)$. The Neumann series for $R' = (1 - A')^{-1}$ converges exponentially and determines the matrix elements

$$R'_{q,k} = \delta_{q,k} + \sum_{n=1}^{\infty} (A'^n)_{q,k}. \tag{D.23b}$$

There holds

$$\sum_k [\delta_{q,k} - A'_{q,k}] R'_{k,p} = \delta_{q,p}. \tag{D.23c}$$

Finally, an M-by-M matrix R shall be defined by

$$R_{q,k} = R'_{q,k} - a_q a_k^*. \tag{D.24a}$$

Using Eqs. (4.74b,d), one gets $\sum_q a_q^* A'_{q,k} = 0$. From Eq. (D.23b) one concludes: $\sum_q a_q^* R'_{q,k} = a_k^*$. Similarly, one derives $\sum_k R'_{qk} a_k = a_q^*$. As a result, one obtains:

$$\sum_q a_q^* R_{q,k} = 0, \quad \sum_q R_{k,q} a_q = 0; \quad k = 1, \ldots, M. \tag{D.24b}$$

Consequently, R is the distinguished resolvent.

Let $\boldsymbol{I}$ denote some M-component array. It shall be used to define another array $\boldsymbol{Y}$ with the components

$$Y_q = \sum_k R_{q,k} I_k, \quad q = 1, \ldots, M. \tag{D.25a}$$

If an array $\boldsymbol{F}$ is introduced by $F_q = \sum_k [\delta_{q,k} - A^c_{q,k}] Y_k$, one gets $F_q = I_q - a_q \sum_p a^*_p I_p$. One concludes: if and only if

$$\sum_q a^*_q I_q = 0, \tag{D.25b}$$

there holds $\boldsymbol{F} = \boldsymbol{I}$, i.e.,

$$\sum_k [\delta_{q,k} - A^c_{q,k}] Y_k = I_k, \quad q = 1, \ldots, M. \tag{D.25c}$$

If the solubility condition (D.25b) is obeyed, $\boldsymbol{Y}$ is a special solution of the preceding M linear equations. This result constitutes the non-trivial part of the discussions of Eqs. (4.75a–e).

The results of the preceding paragraph are used in Sec. 4.3.4 in order to derive the $\sqrt{\epsilon}$-law for the change of the form factor of the glass for states near a generic liquid–glass-transition point. The states P are evolving along a path P^ϵ as specified in Eqs. (4.84)–(4.86). The result (4.91a) can be denoted as

$$f_q(P^\epsilon) = f^c_q + (1 - f^c_q) a_q \sqrt{C\epsilon/\mu^c_2} + O(\epsilon), \quad \epsilon > 0. \tag{D.26a}$$

For sufficiently small ϵ, the maximum eigenvalue $E(P^\epsilon)$ remains non-degenerate. Hence, it is a continuous function of ϵ. It can be characterized by a positive function δE_ϵ, which vanishes for $\epsilon = 0$:

$$E(P^\epsilon) = 1 - \delta E_\epsilon. \tag{D.26b}$$

There hold the equations: $\sum_p [E(P^\epsilon)\delta_{q,p} - A_{q,p}(P^\epsilon)] b_p = 0, q = 1, \ldots, M$. The eigenvector $\boldsymbol{b}$ for the maximum eigenvalue can be chosen as continuous function of ϵ, which agrees with $\boldsymbol{a}$ for $\epsilon = 0$. It shall be written as $\boldsymbol{b} = \boldsymbol{a} + \boldsymbol{\delta b}$; the continuous function $\boldsymbol{\delta b}$ vanishes for ϵ tending to zero. Writing for the stability matrix $A_{q,p}(P^\epsilon) = A^c_{q,p} + \delta A_{q,p}$, the equation for the eigenvalue gets the form of Eq. (D.25c): $\sum_p [\delta_{q,p} - A^c_{q,p}] \delta b_p = I_q$. Here, the inhomogeneity reads $I_q = \delta E b_q + \sum_p \delta A_{q,p} b_p$. The condition (D.25b) is equivalent to $\delta E = -\sum_{q,p} a^*_q \delta A_{q,p} b_p / \sum_q a^*_q b_q$. The leading-order result is obtained from $\delta A_{q,p}$; the eigenvector $\boldsymbol{b}$ can be replaced by $\boldsymbol{a}$. According to Eq. (4.60a), the leading contribution to $\delta A_{q,p}$ is of order $\sqrt{\epsilon}$; and this is due to the $\sqrt{\epsilon}$-contribution to $f_q(P^\epsilon)$. Using the abbreviations from Eq. (4.71c), one gets $\delta A_{q,p} = \sqrt{C\epsilon/\mu^c_2}[-a_q A^c_{q,p} - A^c_{q,p} a_p + 2\sum_k A^{(2)c}_{q,kp} a_k] + O(\epsilon)$. Equation (D.25a) is used in order to justify the assumption $\boldsymbol{\delta b} = O(\sqrt{\epsilon})$. Hence, $\delta E = -\sum_{q,p} a^*_q \delta A_{q,p} a_p + O(\epsilon)$. With the aid of the abbreviations (4.88a,b), one arrives at a square-root law:

$$\delta E = 2\sqrt{C\epsilon\mu^c_2} + O(\epsilon). \tag{D.26c}$$

BIBLIOGRAPHY

S. Adichtchev, T. Blochowicz, C. Tschirwitz, V. N. Novikov, and E. A. Rössler. Reexamination of the evolution of the dynamic susceptibility of the glass former glycerol. *Phys. Rev. E*, 68:011504, 2003.

M. Aichele and J. Baschnagel. Glassy dynamics of simulated polymer melts: Coherent scattering and van Hove correlation functions. Part II: Dynamics in the α-relaxation regime. *Eur. Phys. J. E*, 5:245, 2001a.

M. Aichele and J. Baschnagel. Glassy dynamics of simulated polymer melts: Coherent scattering and van Hove correlation functions. Part I: Dynamics in the β-relaxation regime. *Eur. Phys. J. E*, 5:229, 2001b.

N. I. Akhiezer and I. M. Glazman. *Theory of Linear Operators in Hilbert Space.* Dover, New York, 1993.

C. Alba-Simionesco and M. Krauzman. Low frequency Raman spectroscopy of supercooled fragile liquids analyzed with schematic mode coupling models. *J. Chem. Phys.*, 102:6574, 1995.

B. J. Alder, D. M. Gass, and T. E. Wainwright. Studies in molecular dynamics. VIII. The transport coefficients for a hard-sphere fluid. *J. Chem. Phys.*, 53: 3813, 1970.

V. I. Arnol'd. Critical points of smooth functions and their normal forms. *Russ. Math. Survey*, 30:1, 1975.

V. I. Arnold. *Catastrophe Theory.* Springer, Berlin, 3rd edition, 1992.

U. Balucani and M. Zoppi. *Dynamics of the Liquid State.* Clarendon Press, Oxford, 1994.

J.-L. Barrat and A. Latz. Mode coupling theory for the glass transition in a simple binary mixture. *J. Phys.: Condens. Matter*, 2:4289, 1990.

J. Baschnagel. private communication, 2003.

J. Baschnagel, C. Bennemann, W. Paul, and K. Binder. Dynamics of a supercooled polymer melt above the mode-coupling critical temperature: cage versus polymer-specific effects. *J. Phys.: Condens. Matter*, 12:6365, 2000.

J. Baschnagel and F. Varnik. Computer simulations of supercooled polymer melts in the bulk and in confined geometry. *J. Phys.: Condens. Matter*, 17: R851, 2005.

R. J. Baxter. Percus–Yevick equation for hard spheres with surface adhesion. *J. Chem. Phys.*, 49:2770, 1968.

R. J. Baxter. Ornstein–Zernike relation and Percus–Yevick approximation for fluid mixtures. *J. Chem. Phys.*, 52:4559, 1970.

U. Bengtzelius. Dynamics of a Lennard-Jones system close to the glass transition. *Phys. Rev. A*, 34:5059, 1986a.

U. Bengtzelius. Theoretical calculations on liquid-glass transitions in Lennard-Jones systems. *Phys. Rev. A*, 33:3433, 1986b.

U. Bengtzelius, W. Götze, and A. Sjölander. Dynamics of supercooled liquids and the glass transition. *J. Phys. C*, 17:5915, 1984.

J. Bergenholtz and M. Fuchs. Nonergodicity transitions in colloidal suspensions with attractive interactions. *Phys. Rev. E*, 59:5706, 1999.

J. Bergenholtz, W. C. K. Poon, and M. Fuchs. Gelation in model colloid–polymer mixtures. *Langmuir*, 19:4493, 2003.

B. J. Berne, J. P. Boon, and S. A. Rice. On the calculation of autocorrelation functions of dynamical variables. *J. Chem. Phys.*, 45:1086, 1966.

B. J. Berne and R. Pecora. *Dynamic Light Scattering.* John Wiley, New York, 1976.

B. Bernu, J. P. Hansen, Y. Hiwatari, and G. Pastore. Soft-sphere model for the glass transition in binary alloys: Pair structure and self-diffusion. *Phys. Rev. A*, 36:4891, 1987.

G. Biroli and J.-Ph. Bouchaud. Diverging length scale and upper critical dimension in the mode-coupling theory of the glass transition. *Europhys. Lett.*, 67: 21, 2004.

J. P. Boon and S. Yip. *Molecular Hydrodynamics.* McGraw-Hill, New York, 1980.

J. Bosse, W. Götze, and M. Lücke. Current fluctuation spectra of liquid argon near its triple point. *Phys. Rev. A*, 17:447, 1978a.

J. Bosse, W. Götze, and M. Lücke. Mode-coupling theory of simple classical liquids. *Phys. Rev. A*, 17:434, 1978b.

J. Bosse and U. Krieger. Relaxation of a simple molten salt near the liquid–glass transition. *J. Phys. C: Solid State Phys.*, 19:L609, 1987.

J. Bosse and J. S. Thakur. Delocalization of small particles in a glassy matrix. *Phys. Rev. Lett.*, 59:998, 1987.

A. C. Branka and D. M. Heyes. Time correlation functions of hard sphere and soft sphere fluids. *Phys. Rev. E*, 69:021202, 2004.

G. Buchalla, U. Dersch, W. Götze, and L. Sjögren. α- and β-relaxation for single-particle motion near the glass transition. *J. Phys. C: Solid State Phys.*, 21: 4239, 1988.

H. Cang, V. N. Novikov, and M. D. Fayer. Logarithmic decay of the orientational correlation function in supercooled liquids on the Ps to Ns time scale. *J. Chem. Phys.*, 118:2800, 2003.

S.-H. Chong. private communication, 2005.

S.-H. Chong. Role of structural relaxations and vibrational excitations in the high-frequency dynamics of liquids and glasses. *Phys. Rev. E*, 74:031205, 2006.

S.-H. Chong and M. Fuchs. Mode-coupling theory for structural and conformational dynamics of polymer melts. *Phys. Rev. Lett.*, 88:185702, 2002.

S.-H. Chong and W. Götze. Idealized glass transitions for a system of dumbbell molecules. *Phys. Rev. E*, 65:041503, 2002a.

S.-H. Chong and W. Götze. Structural relaxation in a system of dumbbell molecules. *Phys. Rev. E*, 65:051201, 2002b.

S.-H. Chong, W. Götze, and M. R. Mayr. Mean-squared displacement of a molecule moving in a glassy system. *Phys. Rev. E*, 64:011503, 2001a.

S.-H. Chong, W. Götze, and A. P. Singh. Mode-coupling theory for the glassy dynamics of a diatomic probe molecule immersed in a simple liquid. *Phys. Rev. E*, 63:011206, 2001b.

S.-H. Chong and F. Hirata. Mode-coupling theory for molecular liquids based on the interaction-site model. *Phys. Rev. E*, 58:6188, 1998.

S.-H. Chong, A. J. Moreno, F. Sciortino, and W. Kob. Evidence for the weak steric hindrance scenario in the supercooled-state reorientational dynamics. *Phys. Rev. Lett.*, 94:215701, 2005.

S.-H. Chong and F. Sciortino. Structural relaxation in a supercooled molecular liquid. *Europhys. Lett.*, 64:197, 2003.

S.-H. Chong and F. Sciortino. Structural relaxation in supercooled orthoterphenyl. *Phys. Rev. E*, 69:051202, 2004.

B. Cichocki and B. U. Felderhof. Linear viscoelasticity of semidilute hard sphere suspensions. *Phys. Rev. A*, 43:5405, 1991.

S. Ciuchi and A. Crisanti. Different scenarios for critical glassy dynamics. *Europhys. Lett.*, 49:754, 2000.

H. Z. Cummins. Dynamics of supercooled liquids: excess wings, β peaks, and rotation-translation coupling. *J. Phys.: Condens. Matter*, 17:1457, 2005.

H. Z. Cummins, G. Li, W. Du, Y. H. Hwang, and G. Q. Shen. Light scattering spectroscopy of orthoterphenyl. *Prog. Theor. Phys. Suppl.*, 126:21, 1997.

P. S. Damle and A. D. Tillu. Memory function of the velocity auto-correlation in liquid argon. *Indian J. Pure Appl. Phys.*, 7:539, 1969.

S. P. Das. Mode-coupling theory and the glass transtion in supercooled liquids. *Rev. Mod. Phys.*, 76:785, 2004.

S. P. Das and G. F. Mazenko. Fluctuating nonlinear hydrodynamics and the liquid–glass transition. *Phys. Rev. A*, 34:2265, 1986.

K. Dawson, G. Foffi, M. Fuchs, W. Götze, F. Sciortino, M. Sperl, P. Tartaglia, Th. Voigtmann, and E. Zaccarelli. Higher-order glass-transition singularities in colloidal systems with attractive interactions. *Phys. Rev. E*, 63:011401, 2001.

P. G. de Gennes. Liquid dynamics and inelastic scattering of neutrons. *Physica (Utrecht)*, 25:825, 1959.

H. De Raedt and W. Götze. Scaling properties of correlation functions at the liquid-glass transition. *J. Phys. C: Solid State Phys.*, 19:2607, 1986.

I. M. de Schepper and E. G. D. Cohen. Very-short-wavelength collective modes in fluids. *J. Stat. Phys.*, 27:223, 1982.

P. K. Dixon, L. Wu, S. R. Nagel, B. D. Williams, and J. P. Carini. Scaling in the relaxation of supercooled liquids. *Phys. Rev. Lett.*, 65:1108, 1990.

M. Doi and S. F. Edwards. *The Theory of Polymer Dynamics.* Clarendon Press, Oxford, 1986.

W. M. Du, G. Li, H. Z. Cummins, M. Fuchs, J. Toulouse, and L. A. Knauss. Light-scattering study of the liquid-glass transition in propylene carbonate. *Phys. Rev. E*, 49:2192, 1994.

T. Eckert and E. Bartsch. Re-entrant glass transition in a colloid–polymer mixture with depletion attractions. *Phys. Rev. Lett.*, 89:125701, 2002.

S. F. Edwards and P. W. Anderson. Theory of spin glasses. *J. Phys. F: Metal Phys.*, 5:965, 1975.

L. Fabbian, W. Götze, F. Sciortino, P. Tartaglia, and F. Thiery. Ideal glass-glass transitions and logarithmic decay of correlations in a simple system. *Phys. Rev. E*, 59:R1347, 1999.

W. Feller. *An Introduction to Probability Theory and Its Applications*, volume II. Wiley & Sons, New York, 2nd edition, 1971.

M. Fixman. Viscosity of critical mixtures. *J. Chem. Phys.*, 36:310, 1962.

S. Flach, W. Götze, and L. Sjögren. The *A*4 glass transition singularity. *Z. Phys. B – Condensed Matter*, 87:29, 1992.

G. Foffi, K. A. Dawson, S. V. Buldyrev, F. Sciortino, E. Zaccarelli, and P. Tartaglia. Evidence for unusual dynamical-arrest scenario in short-ranged colloidal systems. *Phys. Rev. E*, 65:050802, 2002.

G. Foffi, W. Götze, F. Sciortino, P. Tartaglia, and Th. Voigtmann. Mixing effects for the structural relaxation in binary hard-sphere liquids. *Phys. Rev. Lett.*, 91:085701, 2003.

G. Foffi, W. Götze, F. Sciortino, P. Tartaglia, and Th. Voigtmann. α-relaxation processes in binary hard-sphere mixtures. *Phys. Rev. E*, 69:011505, 2004.

D. Forster. *Hydrodynamic Fluctuations, Broken Symmetry, and Correlation Functions*. Benjamin, Reading, Massachusetts, 1975.

T. Franosch, M. Fuchs, W. Götze, M. R. Mayr, and A. P. Singh. Asymptotic laws and preasymptotic correction formulas for the relaxation near glass-transition singularities. *Phys. Rev. E*, 55:7153, 1997.

T. Franosch and W. Götze. A theory for a certain crossover in relaxation phenomena in glasses. *J. Phys.: Condens. Matter*, 6:4807, 1994.

T. Franosch, W. Götze, M. R. Mayr, and A. P. Singh. Structure and structure relaxation. *J. Non-Cryst. Solids*, 235–237:71, 1998.

T. Franosch and R. M. Pick. Transient grating experiments on supercooled molecular liquids II: microscopic derivation of phenomenological equations. *Eur. Phys. J. B*, 47:341, 2005.

T. Franosch and Th. Voigtmann. Completely monotone solutions of the mode-coupling theory for mixtures. *J. Stat. Phys.*, 109:237, 2002.

B. Frick, D. Richter, W. Petry, and U. Buchenau. Study of the glass transition order parameter in amorphous polybutadiene by incoherent neutron scattering. *Z. Phys. B*, 70:73, 1988.

M. Fuchs. The Kohlrausch law as a limit solution to mode coupling equations. *J. Non-Cryst. Solids*, 172–174:241, 1994.

M. Fuchs and M. E. Cates. Theory of nonlinear rheology and yielding of dense colloidal suspensions. *Phys. Rev. Lett.*, 89:248304, 2002.

M. Fuchs, W. Götze, I. Hofacker, and A. Latz. Comments on the α-peak shapes for relaxation in supercooled liquids. *J. Phys.: Condens. Matter*, 3: 5047, 1991.

M. Fuchs, W. Götze, and M. R. Mayr. Asymptotic laws for tagged-particle motion in glassy systems. *Phys. Rev. E*, 58:3384, 1998.

M. Fuchs, I. Hofacker, and A. Latz. Primary relaxation in a hard-sphere system. *Phys. Rev. A*, 45:898, 1992.

F. R. Gantmacher. *The Theory of Matrices*, volume II. Chelsea Publishing, New York, 1974.

T. Geszti. Pre-vitrification by viscosity feedback. *J. Phys. C: Solid State Phys.*, 16:5805, 1983.

T. Gleim and W. Kob. The β-relaxation dynamics of a simple liquid. *Eur. Phys. J. B*, 13:83, 2000.

T. Gleim, W. Kob, and K. Binder. How does the relaxation of a supercooled liquid depend on its microscopic dynamics? *Phys. Rev. Lett.*, 81:4404, 1998.

M. Goldstein. Viscous liquids and the glass transition: A potential energy barrier picture. *J. Chem. Phys.*, 51:3728, 1969.

W. Götze. An elementary approach towards the Anderson transition. *Solid State Commun.*, 27:1393, 1978.

W. Götze. The mobility of a quantum particle in a three-dimensional random potential. *Philos. Mag. B*, 43:219, 1981.

W. Götze. Some aspects of phase transitions described by the self consistent current relaxation theory. *Z. Phys. B – Condensed Matter*, 56:139, 1984.

W. Götze. Properties of the glass instability treated within a mode coupling theory. *Z. Phys. B – Condensed Matter*, 60:195, 1985.

W. Götze. Comments on the mode coupling theory of the liquid glass transition. In E. Lüscher, G. Fritsch, and G. Jacucci, editors, *Amorphous and Liquid Materials*, volume 118 of *NATO ASI Series E*, page 34, Dordrecht, 1987. Nijhoff Publishers.

W. Götze. The scaling functions for the β-relaxation process of supercooled liquids and glasses. *J. Phys.: Condens. Matter*, 2:8485, 1990.

W. Götze. Aspects of structural glass transitions. In J. P. Hansen, D. Levesque, and J. Zinn-Justin, editors, *Liquids, Freezing and Glass Transition*, volume Session LI (1989) of *Les Houches Summer Schools of Theoretical Physics*, 287, Amsterdam, 1991. North Holland.

W. Götze and R. Haussmann. Further phase transition scenarios described by the self consistent current relaxation theory. *Z. Phys. B – Condensed Matter*, 72:403, 1988.

W. Götze, E. Leutheusser, and S. Yip. Correlation functions of the hard-sphere Lorentz model. *Phys. Rev. A*, 24:1008, 1981a.

W. Götze, E. Leutheusser, and S. Yip. Dynamical theory of diffusion and localization in a random, static field. *Phys. Rev. A*, 23:2634, 1981b.

W. Götze and M. Lücke. Dynamical current correlation functions of simple classical liquids for intermediate wave numbers. *Phys. Rev. A*, 11:2173, 1975.

W. Götze and M. Lücke. Dynamical structure factor $S(q,\omega)$ of liquid helium II at zero temperature. *Phys. Rev. B*, 13:3825, 1976.

W. Götze and M. R. Mayr. Evolution of vibrational excitations in glassy systems. *Phys. Rev. E*, 61:587, 2000.

W. Götze and L. Sjögren. A dynamical treatment of the spin glass transition. *J. Phys. C: Solid State Phys.*, 17:5759, 1984.

W. Götze and L. Sjögren. α-relaxation near the liquid-glass transition. *J. Phys. C: Solid State Phys.*, 20:879, 1987a.

W. Götze and L. Sjögren. The glass transition singularity. *Z. Phys. B – Condensed Matter*, 65:415, 1987b.

W. Götze and L. Sjögren. Scaling properties in supercooled liquids near the glass transition. *J. Phys. C: Solid State Phys.*, 21:3407, 1988.

W. Götze and L. Sjögren. β relaxation near glass transition singularities. *J. Phys.: Condens. Matter*, 1:4183, 1989a.

W. Götze and L. Sjögren. Logarithmic decay laws in glassy systems. *J. Phys.: Condens. Matter*, 1:4203, 1989b.

W. Götze and L. Sjögren. Relaxation processes in supercooled liquids. *Rep. Prog. Phys.*, 55:241, 1992.

W. Götze and L. Sjögren. General properties of certain non-linear integro-differential equations. *J. Math. Analysis and Appl.*, 195:230, 1995.

W. Götze and M. Sperl. Logarithmic relaxation in glass-forming systems. *Phys. Rev. E*, 66:011405, 2002.

W. Götze and M. Sperl. Higher-order glass-transition singularities in systems with short-ranged attractive potentials. *J. Phys.: Condens. Matter*, 15:S869, 2003.

W. Götze and M. Sperl. Critical decay at higher-order glass-transition singularities. *J. Phys.: Condens. Matter*, 16:S4807, 2004a.

W. Götze and M. Sperl. Nearly logarithmic decay of correlations in glass-forming liquids. *Phys. Rev. Lett.*, 92:105701, 2004b.

W. Götze and Th. Voigtmann. Universal and nonuniversal features of glassy relaxation in propylene carbonate. *Phys. Rev. E*, 61:4133, 2000.

W. Götze and Th. Voigtmann. Effect of composition changes on the structural relaxation of a binary mixture. *Phys. Rev. E*, 67:021502, 2003.

I. S. Gradshteyn and I. M. Ryzhik. *Table of Integrals, Series, and Products.* Academic Press, San Diego, 2000.

G. Gripenberg, S. O. Londen, and O. Staffans. *Volterra Integral and Functional Equations*, volume 34 of *Encyclopedia of Mathematics and Its Applications.* Cambridge University Press, Cambridge, 1990.

J. P. Hansen and I. R. McDonald. *Theory of Simple Liquids.* Academic Press, London, 2nd edition, 1986.

R. Haussmann. Some properties of mode coupling equations. *Z. Phys. B – Condensed Matter*, 79:143, 1990.

S. I. Henderson, T. C. Mortensen, S. M. Underwood, and W. van Megen. Effect of particle size distribution on crystallisation and the glass transition of hard sphere colloids. *Physica (Amsterdam) A*, 233:102, 1996.

G. Hinze, D. D. Brace, S. D. Gottke, and M. D. Fayer. A detailed test of mode-coupling theory on all time scales: Time domain studies of structural relaxation in a supercooled liquid. *J. Chem. Phys.*, 113:3723, 2000.

F. Höfling, Th. Franosch, and E. Frey. Localization transition of the three-dimensional Lorentz model and continuum percolation. *Phys. Rev. Lett.*, 96: 165901, 2006.

J. Horbach and W. Kob. Static and dynamic properties of a viscous silica melt. *Phys. Rev. B*, 60:3169, 1999a.

J. Horbach and W. Kob. Structure and dynamics of sodium disilicate. *Philos. Mag. B*, 79:1981, 1999b.

J. Horbach and W. Kob. The relaxation dynamics of a viscous silica melt: The intermediate scattering functions. *Phys. Rev. E*, 64:041503, 2001.

J. Horbach and W. Kob. The structural relaxation of molten sodium disilicate. *J. Phys.: Condens. Matter*, 14:9237, 2002.

J. Horbach, W. Kob, and K. Binder. Dynamics of sodium in sodium disilicate: Channel relaxation and sodium diffusion. *Phys. Rev. Lett.*, 88:125502, 2002.

H. W. Jackson and E. Feenberg. Energy spectrum of elementary excitations in helium II. *Rev. Mod. Phys.*, 34:686, 1962.

L. P. Kadanoff and P. C. Martin. Hydrodynamic equations and correlation functions. *Ann. Phys. (NY)*, 24:419, 1963.

L. P. Kadanoff and J. Swift. Transport coefficients near the liquid–gas critical point. *Phys. Rev.*, 166:89, 1968.

K. Kawasaki. Correlation-function approach to the transport coefficients near the critical point. I. *Phys. Rev.*, 150:291, 1966.

K. Kawasaki. Kinetic Equations and time correlation functions of critical fluctuations. *Ann. Phys. (NY)*, 61:1, 1970.

T. R. Kirkpatrick. Large long-time tails and shear waves in dense classical liquids. *Phys. Rev. Lett.*, 53:1735, 1984.

T. R. Kirkpatrick and D. Thirumalai. p-spin-interaction spin-glass models: Connections with the structural glass problem. *Phys. Rev. B*, 36:5388, 1987.

T. R. Kirkpatrick and P. G. Wolynes. Connections between some kinetic and equilibrium theories of the glass transition. *Phys. Rev. A*, 35:3072, 1987.

W. Knaak, F. Mezei, and B. Farago. Observation of scaling behaviour of dynamic correlations near liquid–glass transition. *Europhys. Lett.*, 7:529, 1988.

W. Kob. Supercooled liquids, the glass transition, and computer simulations. In J.-L. Barrat, M. Feigelman, J. Kurchan, and J. Dalibard, editors, *Slow Relaxations and Nonequilibrium Dynamics in Condensed Matter*, volume Session LXXVII (2002) of *Les Houches Summer Schools of Theoretical Physics*, 199, Berlin, 2003. Springer.

W. Kob and H. C. Andersen. Scaling behavior in the β-relaxation regime of a supercooled Lennard-Jones mixture. *Phys. Rev. Lett.*, 73:1376, 1994.

W. Kob and H. C. Andersen. Testing mode-coupling theory for a supercooled binary Lennard-Jones mixture: The van Hove correlation function. *Phys. Rev. E*, 51:4626, 1995a.

W. Kob and H. C. Andersen. Testing mode-coupling theory for a supercooled binary Lennard-Jones mixture. II. Intermediate scattering function and dynamic susceptibility. *Phys. Rev. E*, 52:4134, 1995b.

V. Krakoviack. Mode-coupling theory for the slow collective dynamics of fluids adsorbed in disordered porous media. *Phys. Rev. E*, 75:031503, 2007.

V. Krakoviack, C. Alba-Simionesco, and M. Krauzman. Study of the depolarized light scattering spectra of supercooled liquids by a simple mode-coupling model. *J. Chem. Phys.*, 107:3417, 1997.

G. J. Kramer, A. J. M. de Man, and R. A. van Santen. Zeolites versus alumino silicate clusters: The validity of a local description. *J. Am. Chem. Soc.*, 113: 6435, 1991.

R. Kubo. Statistical-mechanical theory of irreversible processes. I. General theory and simple applications to magnetic and conduction problems. *J. Phys. Soc. Japan*, 12:570, 1957.

J. L. Lebowitz and J. S. Rowlinson. Thermodynamic properties of mixtures of hard spheres. *J. Chem. Phys.*, 41:133, 1964.

M. Letz, R. Schilling, and A. Latz. Ideal glass transitions for hard ellipsoids. *Phys. Rev. E*, 62:5173, 2000.

E. Leutheusser. Diffusion blocking in a frozen rigid-sphere fluid. *Phys. Rev. A*, 28:2510, 1983a.

E. Leutheusser. Self-consistent kinetic theory for the Lorentz gas. *Phys. Rev. A*, 28:1762, 1983b.

E. Leutheusser. Dynamical model of the liquid-glass transition. *Phys. Rev. A*, 29:2765, 1984.

D. Levesque, L. Verlet, and J. Kürkijarvi. Computer 'experiments' on classical fluids. IV. Transport properties and time-correlation functions of the Lennard-Jones liquid near its triple point. *Phys. Rev. A*, 7:1690, 1973.

L. J. Lewis and G. Wahnström. Molecular-dynamics study of supercooled *ortho*terphenyl. *Phys. Rev. E*, 50:3865, 1994.

G. Li, W. M. Du, X. K. Chen, H. Z. Cummins, and N. J. Tao. Testing mode-coupling predictions for α and β relaxation in Ca0.4K0.6(NO3)1.4 near the liquid-glass transition by light scattering. *Phys. Rev. A*, 45:3867, 1992.

G. Li, H. E. King Jr., W. F. Oliver, C. A. Herbst, and H. Z. Cummins. Pressure and temperature dependence of glass–transition dynamics in a 'Fragile' glass former. *Phys. Rev. Lett.*, 74:2280, 1995.

S. W. Lovesey. *Theory of Neutron Scattering from Condensed Matter.* Vol. 1. Clarendon Press, Oxford, 1984.

F. Mallamace, P. Gambadauro, N. Micali, P. Tartaglia, C. Liao, and S.-H. Chen. Kinetic glass transition in a micellar system with short-range attractive interaction. *Phys. Rev. Lett.*, 84:5431, 2000.

F. Mallamace, P. Tartaglia, W. R. Chen, A. Faraone, and S. H. Chen. A mode coupling theory analysis of viscoelasticity near the kinetic glass transition of a copolymer micellar system. *J. Phys.: Condens. Matter*, 16:S4975, 2004.

T. G. Mason and D. A. Weitz. Linear viscoelasticity of colloidal hard sphere suspensions near the glass transition. *Phys. Rev. Lett.*, 75:2770, 1995.

A. Meyer, J. Horbach, W. Kob, F. Kargl, and H. Schober. Channel formation and intermediate range order in sodium silicate melts and glasses. *Phys. Rev. Lett.*, 93:027801, 2004.

A. Meyer, H. Schober, and D. B. Dingwell. Structure, structural relaxation and ion diffusion in sodium disilicate melts. *Europhys. Lett.*, 59:708, 2002.

F. Mezei, W. Knaak, and B. Farago. Neutron spin-echo study of dynamic correlations near the liquid–glass transition. *Phys. Rev. Lett.*, 58:571, 1987.

A. J. Moreno, S.-H. Chong, W. Kob, and F. Sciortino. Dynamic arrest in a liquid of symmetric dumbbells: Reorientational hopping for small molecular elongations. *J. Chem. Phys.*, 123:204505, 2005.

H. Mori. A continued-fraction representation of the time-correlation functions. *Prog. Theor. Phys.*, 34:399, 1965a.

H. Mori. Transport, Collective motion, and Brownian motion. *Prog. Theor. Phys.*, 33:423, 1965b.

S. Mossa, E. La Nave, E. H. Stanley, C. Donati, F. Sciortino, and P. Tartaglia. Dynamics and configurational entropy in the Lewis–Wahnström model for supercooled orthoterphenyl. *Phys. Rev. E*, 65:041205, 2002.

R. D. Mountain. Thermal relaxation and Brillouin scattering in liquids. *J. Res. Natl. Bur. Stand.*, 70A:207, 1966.

T. Munakata and A. Igarashi. Dynamical correlation functions of classical liquids. II. A Self-consistent approach. *Prog. Theor. Phys.*, 60:45, 1978.

G. Nägele. On the dynamics and structure of charge-stabilized suspensions. *Phys. Rep.*, 272:215, 1996.

G. Nägele, J. Bergenholtz, and J. K. G. Dhont. Cooperative diffusion in colloidal mixtures. *J. Chem. Phys.*, 110:7037, 1999.

M. Nauroth and W. Kob. Quantitative test of the mode-coupling theory of the ideal glass transition for a binary Lennard-Jones system. *Phys. Rev. E*, 55: 657, 1997.

D. Oxtoby. Crystallization of liquids: a density functional approach. In J. P. Hansen, D. Levesque, and J. Zinn-Justin, editors, *Liquids, Freezing and Glass Transition*, volume Session LI (1989) of *Les Houches Summer Schools of Theoretical Physics*, page 145, Amsterdam, 1991. North Holland.

W. Petry, E. Bartsch, F. Fujara, M. Kiebel, H. Sillescu, and B. Farago. Dynamic anomaly in the glass transition region of orthoterphenyl. A neutron scattering study. *Z. Phys. B*, 83:175, 1991.

K. N. Pham, S. U. Egelhaaf, P. N. Pusey, and W. C. K. Poon. Glasses in hard spheres with short-range attraction. *Phys. Rev. E*, 69:011503, 2004.

R. M. Pick, C. Dreyfus, A. Azzimani, R. Gupta, R. Torre, A. Taschin, and T. Franosch. Heterodyne detected transient gratings in supercooled molecular liquids: A phenomenological theory. *Eur. Phys. J. B*, 39:169, 2004.

A. Pimenov, P. Lunkenheimer, H. Rall, R. Kohlhaas, A. Loidl, and R. Böhmer. Ion transport in the fragile glass former 3KNO3−2Ca(NO3)2. *Phys. Rev. E*, 54:676, 1996.

W. C. K. Poon, J. S. Selfe, M. B. Robertson, S. M. Ilett, A. D. Pirie, and P. N. Pusey. An experimental study of a model colloid-polymer mixture. *J. Phys. II (Paris)*, 3:1075, 1993.

A. M. Puertas, M. Fuchs, and M. E. Cates. Comparative simulation study of colloidal gels and glasses. *Phys. Rev. Lett.*, 88:098301, 2002.

A. M. Puertas, E. Zaccarelli, and F. Sciortino. Viscoelastic properties of attractive and repulsive colloidal glasses. *J. Phys.: Condens. Matter*, 17:L271, 2005.

P. N. Pusey. Colloidal Suspensions. In J. P. Hansen, D. Levesque, and J. Zinn-Justin, editors, *Liquids, Freezing and Glass Transition*, volume Session LI (1989) of *Les Houches Summer Schools of Theoretical Physics,* 763, Amsterdam, 1991. North Holland.

A. Rahman. Correlations in the motion of atoms in liquid argon. *Phys. Rev.*, 136:A405, 1964.

A. Rinaldi, F. Sciortino, and P. Tartaglia. Dynamics in a supercooled molecular liquid: Theory and simulations. *Phys. Rev. E*, 63:061210, 2001.

J. N. Roux, J. L. Barrat, and J.-P. Hansen. Dynamical diagnostics for the glass transition in soft-sphere alloys. *J. Phys.: Condens. Matter*, 1:7171, 1989.

B. Rufflé, C. Ecolivet, and B. Toudic. Dynamics in the supercooled liquid Na0.5Li0.5PO3: A schematic mode coupling model analysis. *Europhys. Lett.*, 45:591, 1999.

R. Schilling. Reference-point-independent dynamics of molecular liquids and glasses in the tensorial formalism. *Phys. Rev. E*, 65:051206, 2002.

R. Schilling and T. Scheidsteger. Mode coupling approach to the ideal glass transition of molecular liquids: Linear molecules. *Phys. Rev. E*, 56:2932, 1997.

W. Schirmacher and H. Sinn. Collective dynamics of simple liquids: A mode-coupling description. *Conds. Matts. Phys.*, 11:127, 2008.

U. Schneider, P. Lunkenheimer, R. Brand, and A. Loidl. Dielectric and far-infrared spectroscopy of glycerol. *J. Non-Cryst. Solids*, 235–237:173, 1998.

U. Schneider, P. Lunkenheimer, R. Brand, and A. Loidl. Broadband dielectric spectroscopy on glass-forming propylene carbonate. *Phys. Rev. E*, 59:6924, 1999.

F. Sciortino, P. Gallo, P. Tartaglia, and S.-H. Chen. Supercooled water and the kinetic glass transition. *Phys. Rev. E*, 54:6331, 1996.

F. Sciortino and W. Kob. Debye–Waller factor of liquid silica: Theory and simulation. *Phys. Rev. Lett.*, 86:648, 2001.

F. Sciortino and P. Tartaglia. On the mode-coupling-theory β-correlator. *J. Phys.: Condens. Matter*, 11:A261, 1999.

F. Sciortino, P. Tartaglia, and E. Zaccarelli. Evidence of a higher-order singularity in dense short-ranged attractive colloids. *Phys. Rev. Lett.*, 91: 268301, 2003.

D. Sherrington and S. Kirkpatrick. Solvable model of a spin-glass. *Phys. Rev. Lett.*, 35:1792, 1975.

G. F. Signorini, J.-L. Barrat, and M. L. Klein. Structural relaxation and dynamical correlations in a molten state near the liquid–glass transition: A molecular dynamics study. *J. Chem. Phys.*, 92:1294, 1990.

N. B. Simeonova and W. K. Kegel. Gravity-induced aging in glasses of colloidal spheres. *Phys. Rev. Lett.*, 93:035701, 2004.

A. P. Singh, G. Li, W. Götze, M. Fuchs, T. Franosch, and H. Z. Cummins. Structural relaxation in orthoterphenyl: a schematic mode-coupling-theory model analysis. *J. Non-Cryst. Solids*, 235–237:66, 1998.

L. Sjögren. Kinetic theory of current fluctuations in simple classical liquids. *Phys. Rev. A*, 22:2866, 1980a.

L. Sjögren. Numerical results on the density fluctuations in liquid rubidium. *Phys. Rev. A*, 22:2883, 1980b.

L. Sjögren. Diffusion of impurities in a dense fluid near the glass transition. *Phys. Rev. A*, 33:1254, 1986.

L. Sjögren. Mode coupling theories of the glass transition. In J. Colmenero and A. Alegría, editors, *Basic Features of the Glassy State*, page 137, Singapore, 1990. World Scientific.

A. Sjölander and L. A. Turski. On the properties of supercooled classical liquids. *J. Phys. C: Solid State Phys.*, 11:1973, 1978.

M. Sperl. Logarithmic relaxation in a colloidal system. *Phys. Rev. E*, 68:031405, 2003.

M. Sperl. private communication, 2004a.

M. Sperl. Dynamics in colloidal liquids near a crossing of glass- and gel-transition lines. *Phys. Rev. E*, 69:011401, 2004b.

M. Sperl. Logarithmic decay in a two-component model. In M. Tokuyama and I. Oppenheim, editors, *Slow Dynamics in Complex Systems*, volume 708 of *AIP Conference Proceedings*, 559, New York, 2004c. AIP.

M. Sperl. Nearly logarithmic decay in the colloidal hard-sphere system. *Phys. Rev. E*, 71:060401, 2005.

M. Sperl. Cole–Cole law for critical dynamics in glass-forming liquids. *Phys. Rev. E*, 74:011503, 2006.

G. Szamel and H. Löwen. Mode-coupling theory of the glass transition in colloidal systems. *Phys. Rev. A*, 44:8215, 1991.

P. Taborek, R. N. Kleinman, and D. J. Bishop. Power-law behavior in the viscosity of supercooled liquids. *Phys. Rev. B*, 34:1835, 1986.

A. Tölle. Neutron scattering studies of the model glass former orthoterphenyl. *Rep. Prog. Phys.*, 64:1473, 2001.

A. Tölle, H. Schober, J. Wuttke, and F. Fujara. Coherent dynamic structure factor of orthoterphenyl around the mode coupling crossover temperature T_c. *Phys. Rev. E*, 56:809, 1997.

A. Tölle, H. Schober, J. Wuttke, O. G. Randl, and F. Fujara. Fast relaxation in a fragile liquid under pressure. *Phys. Rev. Lett.*, 80:2374, 1998.

R. Torre, P. Bartolini, and R. M. Pick. Time-resolved optical Kerr effect in a fragile glass-forming liquid, salol. *Phys. Rev. E*, 57:1912, 1998.

B. W. H. van Beest, G. J. Kramer, and R. A. van Santen. Force fields for silicas and aluminophosphates based on *ab initio* calculations. *Phys. Rev. Lett.*, 64: 1955, 1990.

J. C. van der Werff, C. B. de Kruif, C. Blom, and J. Mellema. Linear viscoelastic behavior of dense hard-sphere dispersions. *Phys. Rev. A*, 39:795, 1989.

W. van Megen. Crystallisation and the glass transition in suspensions of hard colloidal spheres. *Transp. Theory Stat. Phys.*, 24:1017, 1995.

W. van Megen, T. C. Mortensen, S. R. Williams, and J. Müller. Measurement of the self-intermediate scattering function of suspensions of hard spherical particles near the glass transition. *Phys. Rev. E*, 58:6073, 1998.

W. van Megen and S. M. Underwood. Glass transition in colloidal hard spheres: Mode-coupling theory analysis. *Phys. Rev. Lett.*, 70:2766, 1993.

W. van Megen and S. M. Underwood. Glass transition in colloidal hard spheres: Measurement and mode-coupling-theory analysis of the coherent intermediate scattering function. *Phys. Rev. E*, 49:4206, 1994.

W. van Megen, S. M. Underwood, and P. N. Pusey. Nonergodicity parameters of colloidal glasses. *Phys. Rev. Lett.*, 67:1586, 1991.

F. Varnik and O. Henrich. Yield stress discontinuity in a simple glass. *Phys. Rev. B*, 73:174209, 2006.

R. Verberg, I. M. de Schepper, and E. G. D. Cohen. Viscosity of colloidal suspensions. *Phys. Rev. E*, 55:3143, 1997.

L. Verlet. Computer 'experiments' on classical fluids. II. Equilibrium correlation functions. *Phys. Rev.*, 165:201, 1968.

Th. Voigtmann. private communication, 2003a.

Th. Voigtmann. Dynamics of colloidal glass-forming mixtures. *Phys. Rev. E*, 68:051401, 2003b.

Th. Voigtmann. *Mode coupling theory of the glass transition in binary mixtures (dissertation.de, 2003), URL http://tumb1.biblio.tu-muenchen.de/publ/diss/ph/2003/voigtmann.html.* PhD thesis, TU München, 2003c.

Th. Voigtmann and J. Horbach. Slow dynamics in ion-conducting sodium silicate melts: Simulation and mode-coupling theory. *Europhys. Lett.*, 74:459, 2006.

Th. Voigtmann, A. M. Puertas, and M. Fuchs. Tagged-particle dynamics in a hard-sphere system: Mode-coupling theory analysis. *Phys. Rev. E*, 70:061506, 2004.

F. Wegner. On the Heisenberg model in the paramagnetic region and at the critical point. *Z. Phys.*, 216:433, 1968.

D. V. Widder. *The Laplace Transform.* Princeton University Press, Princeton, 1946.

S. Wiebel and J. Wuttke. Structural relaxation and mode coupling in a non-glassforming liquid: depolarized light scattering in benzene. *New J. Phys.*, 4: 56, 2002.

J. Wiedersich. private communication, 2003.

J. Wiedersich, N. V. Surovtsev, and E. Rössler. A comprehensive light scattering study of the glass former toluene. *J. Chem. Phys.*, 113:1143, 2000.

S. R. Williams and W. van Megen. Motions in binary mixtures of hard colloidal spheres: Melting of the glass. *Phys. Rev. E*, 64:041502, 2001.

J. Wuttke, M. Ohl, M. Goldammer, S. Roth, U. Schneider, P. Lunkenheimer, R. Kahn, B. Rufflé, R. Lechner, and M. A. Berg. Propylene carbonate reexamined: Mode-coupling β scaling without factorization? *Phys. Rev. E*, 61:2730, 2000.

Y. Yang and K. A. Nelson. Impulsive stimulated thermal scattering study of α relaxation dynamics and the Debye–Waller factor anomaly in Ca0.4K0.6(NO3)1.4. *J. Chem. Phys.*, 104:5429, 1996.

J. L. Yarnell, M. J. Katz, R. G. Wenzel, and S. H. Koenig. Structure factor and radial distribution function for argon at 85 K. *Phys. Rev. A*, 7:2130, 1973.

E. Zaccarelli, G. Foffi, K. A. Dawson, S. V. Buldyrev, F. Sciortino, and P. Tartaglia. Confirmation of anomalous dynamical arrest in attractive colloids: A molecular dynamics study. *Phys. Rev. E*, 66:041402, 2002.

R. Zwanzig. Memory effects in irreversible thermodynamics. *Phys. Rev.*, 124: 983, 1961a.

R. W. Zwanzig. Statistical mechanics of irreversibility. In W. E. Brittin, B. W. Downs, and J. Downs, editors, *Lectures in Theoretical Physics Volume III*, 106, New York, 1961b. Interscience.

INDEX